# Soilborne Microbial Plant Pathogens and Disease Management, Volume One

## Nature and Biology

T0199241

# Soilborne Microbial Plant Pathogens and Disease Management, Volume One

## Nature and Biology

Prof. P. Narayanasamy

**CRC Press**
Taylor & Francis Group
Boca Raton  London  New York

CRC Press is an imprint of the
Taylor & Francis Group, an **informa** business

CRC Press
Taylor & Francis Group
6000 Broken Sound Parkway NW, Suite 300
Boca Raton, FL 33487-2742

First issued in paperback 2021

© 2020 by Taylor & Francis Group, LLC
CRC Press is an imprint of Taylor & Francis Group, an Informa business

No claim to original U.S. Government works

Printed on acid-free paper

ISBN-13: 978-0-367-17875-8 (hbk)
ISBN-13: 978-1-03-208731-3 (pbk)

**Visit the Taylor & Francis Web site at**
**http://www.taylorandfrancis.com**

**and the CRC Press Web site at**
**http://www.crcpress.com**

*Dedicated*

*To*

*the memory of my parents*

*for their*

*Love and affection*

# Contents

# Preface

The earth functions as a habitat that supports the existence of countless organisms ranging from simple subcellular forms to complex multicellular forms, including human beings. The soil, occupying about 70% of the surface of the earth, has three phases – solid, water and air. Solid phase (minerals and organic matter) generally makes up about 50% of soil volume, whereas soil solution (liquid phase) and air (gaseous phase) constitute the other half of soil volume. Microorganisms present in the soil, water and air, are either beneficial or harmful to the development of animals and plants that have been growing wild or domesticated to serve the needs of humans. Soilborne microbial plant pathogens – oomycetes, fungi, bacteria and viruses – are capable of inducing various kinds of diseases in economically important crops, accounting for quantitative and/qualitative losses which may be enormous, when epidemics occur. Microbial plant pathogens exist in the form of several races, varieties or pathotypes differing in their virulence, whereas the crop cultivars and genotypes vary in their levels of susceptibility/resistance to different isolates of the pathogens, necessitating precise identification of the microbial pathogen involved in the disease(s). Techniques based on the morphological, biochemical, immunological and genetic characteristics of soilborne microbial pathogens have been developed to detect, differentiate and quantify them in soil, plants, seeds and water, even when they are present in low populations. Advancements in nanotechnology are likely to improve the sensitivity of pathogen detection and identification in the future. Comprehensive investigations on the biology of microbial plant pathogens to determine the pathogenic potential, host range and the influence of environments on the pathogen population buildup, are essentially required to identify effective strategies for reducing incidence and spread of the diseases caused by them.

Soil environments provide both opportunities and challenges to microbial plant pathogens for their survival and perpetuation. They are capable of producing variants or overwintering propagules to overcome the adverse conditions existing in the soils in different agroecosystems. Potato late blight pathogen *Phytophthora infestans*, with multiple dispersal modes, remains as an elusive problem, because of its ability to cause epidemics under favourable conditions. This pathogen ruined the livelihood of millions of Irish people, who had to migrate to different countries in the 18th century. Management of diseases caused by soilborne microbial plant pathogens is based on the principles of exclusion, eradication and immunization. Threats to crop cultivation posed by soilborne microbial plant pathogens have to be dealt with firmly by intensification of research on determination of the pathogenic potential of different races/varieties of the pathogens and the susceptibility/resistant levels of crop cultivars/genotypes for formulation of effective strategies for management of the diseases caused by them. Studies on precise identification of microbial plant pathogens and assessment of variability in their virulence have laid the foundation for development of crop disease management strategies. Furthermore, responses of the pathogens to changes in the environment and chemicals, selection of cultivars/genotypes with built-in genetic resistance and deploying biological control agents have received greater attention resulting in reduction in disease incidence/ severity and enhancement of returns for the grower and preservation of the environment. Investigations on the development of integrated disease management systems by combining two or more compatible strategies are focused on for achieving greater levels of pathogen suppression and consequent reduction in the adverse effects of diseases caused by soilborne microbial plant pathogens.

This book is designed to provide information gathered through extensive literature review with a view to providing comprehensive information to the researchers, teachers and graduate level students in the Departments of Plant Pathology, Microbiology, Plant Breeding and Genetics, Agriculture, and Horticulture as well as the certification and quarantine personnel, culminating in the provision of safe food and a safe environment to all.

**P. Narayanasamy**

# Acknowledgment

The author wishes to place on record his humble salutation with reverence to his *alma mater* that continues to provide inspiration, knowledge, wisdom and competitive ability to establish his identity and to share his concepts with fellow plant pathologists. The author recognizes the immense benefits derived from the interactions with the staff and students of the Department of Plant Pathology, Tamil Nadu Agricultural University, Coimbatore, India. The author considers himself to be very fortunate to have Mrs. N. Rajakumari as his life partner and it would not have been possible to devote his time exclusively for the preparation of the manuscript of this book without her enormous kindness, patience, love and affection. Further, the author heartily thanks his family members Mr. N. Kumar Perumal, Mrs. Nirmala Suresh, Mr. T. R. Suresh, and Mr. S. Varun Karthik, who are the fulcrum by which he can smoothly achieve his life goals.

Above all, the author sincerely thanks the various researchers, editors and publishers for the free access to required information included in this publication. Credits have been given for the materials included in the book at appropriate places.

**P. Narayanasamy**

# Author

**Prof P. Narayanasamy** was awarded a BSc (Ag) in 1958, an MSc (Ag) in 1960, and a PhD in 1963 by the University of Madras. Later, with a Rockefeller Foundation Fellowship, he pursued postdoctoral research on rice virus diseases at the International Rice Research Institute, the Philippines, during 1966–1967. He served as the virus pathologist at the Indian Agricultural Research Institute, New Delhi, during 1969–1970. He returned to his *alma mater*, which was upgraded to the Tamil Nadu Agricultural University (TNAU), Coimbatore, in 1970. He was appointed associate professor and promoted as professor and Head of the Department of Plant Pathology. He was elected fellow of the Indian Phytopathological Society, New Delhi. He functioned as the editor of the *Madras Agricultural Journal*, published from the Tamil Nadu Agricultural University Campus, and as a member of editorial committees of plant pathology journals published in India. He was invited to participate as the lead speaker and chairman of sessions in national seminars held in India.

As a researcher in plant pathology, he was the leader of the research projects on diseases of rice, legumes and oilseeds. He organized the national seminar for the management of diseases of oilseed crops. His research on antiviral principles yielded practical solutions for the management of virus diseases affecting various crops. He has published over 200 research papers in national and international journals. He was the principal investigator of several research projects funded by the Department of Science and Technology, Government of India, New Delhi and the International Crops Research Institute for Semi-Arid Tropics, Hyderabad. He continues to share his experience and knowledge with the staff and graduate students of the Department of Plant Pathology, TNAU, Coimbatore. As a teacher, he taught courses on plant virology, molecular biology, physiopathology and crop disease management for masters and doctoral programs. Under his guidance, 25 graduate students and 15 research scholars earned masters and doctoral degrees, respectively. With many years of experience and in-depth knowledge on various aspects of microbial plant pathogens and crop disease management, he has authored 18 books published by leading publishers including Marcel Dekker, John Wiley & Sons, Science Publishers, The Haworth Press and Springer Science. These publications cover various aspects of plant pathology and serve as valuable sources of information and have been well received by the intended audience.

He is deeply involved in social welfare activities to help the orphaned, old and infirm, as well as children through Udavum Karangal (Coimbatore), HelpAge (New Delhi), Global Cancer Concern (New Delhi), Children Rights and You (CRY, Bangalore) and The Hindu Mission Hospital (Chennai) to lessen the sufferings of needy persons. As one interested in literature and spirituality, he has composed poems in Tamil and English and published two collections of poems to encourage self-confidence in young people and spiritual exploration for mature people.

# 1 Introduction

Microbial plant pathogens – oomycetes, fungi, bacteria, phytoplasmas, viruses and viroids – infect crops, weeds and wild plants in all agroecosystems. Microbial plant pathogens induce various types of symptoms, the severity of which may depend on the levels of susceptibility/resistance of the cultivar, virulence of the pathogen species/strains/varieties/isolates, and prevailing environmental conditions. Microbial plant pathogens may be disseminated predominantly via seeds/propagules, soil, irrigation water and wind. Soilborne microbial plant pathogens commonly cause damping-off, stem rot, crown rot and root rot in planta and soft rot in tubers and other storage organs. Sheath blight, foliar or flower blight symptoms may be induced, when infected by the spores released from infected plant debris left in the soil after harvest. Transmission of some soilborne pathogens may also occur through seeds/propagules finding their way rapidly even to distant locations.

## 1.1 ECONOMIC IMPORTANCE AND DISTRIBUTION OF DISEASES CAUSED BY SOILBORNE MICROBIAL PLANT PATHOGENS

Soilborne microbial plant pathogens generally infect hypogeal (belowground) plant organs – crowns and roots, stems, storage organs like tubers and pods in some pathosystems. When the root system is damaged, transport of water and nutrients to the epigeal (aboveground) plant organs is reduced to a different extent, depending on the growth stage of the plants at infection, levels of susceptibility/resistance of the plants, and virulence of the pathogen species, and prevailing environmental conditions. The formation of characteristic lesions on the affected parts and necrosis of the vascular tissues and the ultimate death of the infected plants are commonly observed. Thus, the extent of losses depends on several major factors such as host responses to pathogens, and presence of other pathogens acting synergistically. The term 'pathometry', as introduced by Large (1966, refers to the measurement of disease, based on reliable methods of assessment of crop losses, since the magnitude of losses is directly related to disease severity. Disease severity represents the extent of pathogen invasion in the host plants and the amount of host tissues destroyed or made nonfunctional/less functional by the pathogen, resulting in reduction in crop productivity to varying degrees. When soilborne pathogens such as *Fusarium oxysporum*, *Verticillium* spp. and *Rhizoctonia solani* infect the roots, discoloration of leaves (foliage) and loss of flaccidity are observed, indicating the adverse effect on photosynthesis and absorption of nutrition and water and translocation of photosynthates to storage organs. Injury to the crop may be defined as the visual or measurable symptoms induced by

pathogens (Gaunt 1995). Disease assessment provides quantitative information required for determining the effectiveness of disease management strategies, surveys for losses and breeding programs to develop disease resistant cultivars.

Assessment of losses caused by rice sheath blight (induced by *R. solani*) and sugar beet rhizomania (caused by *Beet necrotic yellow vein virus*) was taken up in the earlier years (Cu et al. 1996; Henry 1996). Assessment of losses due to diseases caused by soilborne microbial plant pathogens is generally made by determining the percentages of plants infected out of total number of plants examined. Methods have been developed to assess the disease incidence and severity quantitatively in certain pathosystems. The severity of root damage due to *Phytophthora fragariae* var. *fragariae* in strawberry plants was assessed based on the percentage of invaded root tissue (PIRT) showing rotting symptoms, using the light microscope. Based on disease severity assessments, the levels of resistance of strawberry cultivars and genotypes were also determined (Van de Weg et al. 1997). Among the soilborne bacterial pathogens, *Ralstoinia solanacearum* is more widely distributed and capable of infecting more than 200 plant species. *R. solanacearum* causes enormous losses in several crops including banana, potato, tomato, groundnut (peanut) and tobacco (Alvarez et al. 2010). Incidence of banana Xanthomonas wilt caused by *Xanthomonas campestris* pv. *musacearum* was observed in 60% of the sites, where banana was grown in 35 districts in Uganda between 2001 and 2007. Production loss was estimated to be about 53% over a ten-year period amounting to between two and eight billion dollars (Tripathi et al. 2009). Different techniques such as image analysis, remote sensing technology and molecular methods have been employed to quantify the extent of tissue damage by soilborne microbial pathogens (Narayanasamy 2002, 2011). The effects of infection by soilborne pathogens have to be assessed using specialized procedures, since the interactions are very complex. It has been found to be difficult to determine the contributions of different factors to disease development. Invasive methods are commonly followed to assess the intensity of root lesions and crown infections by taking samples at different intervals after challenge inoculation. Yield losses have been estimated in various pathosystems to help the taking of appropriate preventive/curative activities to lessen the losses to the extent possible (see Table 1.1).

## 1.2 CONCEPTS AND IMPLICATIONS OF INFECTION BY SOILBORNE MICROBIAL PLANT PATHOGENS

Soilborne fungal plant pathogens infecting belowground plant organs are classified as root inhabitants or soil inhabitants. Root inhabitants are considered as ecologically obligate

1

**TABLE 1.1**

**Assessment of Yield Losses Due to Diseases Caused by Soilborne Microbial Plant Pathogens in Different Ecosystems**

| Pathogen | Crop | Disease | Yield loss | References |
|---|---|---|---|---|
| **Fungal pathogens** | | | | |
| *Fusarium oxysporum* f.sp. *ciceris* | chickpea | Fusarium wilt | up to 100% | Navas-Cortés et al. 2000 |
| *Rhizoctonia solani* | rice | sheath blight | 50% reduction in grain quality | Groth 2005 |
| *Sclerotinia sclerotiorum* | tomato | Sclerotinia rot | 100% depending on age of plants | Jnr and Silva, 2000 |
| | Canola | Sclerotinia stem rot | 0.5% reduction per unit % infection | del Rio et al. 2007 |
| *Sclerotium rolsfii* | peanut | southernblight and other soilborne diseases | 50% yield | Melouk and Backman 1995 |
| **Bacterial pathogen** | | | | |
| *Ralsotonia solanacearum* | tomato | bacterial wilt | very high | Alvarez et al. 2010 |
| **Viral pathogen** | | | | |
| *Soilborne cereal* Mosaic virus | wheat | soilborne cereal mosaic disease | 50% of grainmosaic disease | Rubies-Autonell et al. 2006 |

parasites. On the other hand, soil inhabitants can exist as saprophytes in the absence of susceptible host plants. The nature of invasive force of the pathogen was redefined and the term 'inoculum potential' was suggested as the energy of growth of a parasite available for infection of a host, at the surface of the host organ to be infected. The energy of growth implies the infective hyphae of one sort or other are actually growing out from the inoculum. Root-infecting fungi may be of two types. Unspecialized root-infecting fungi are usually characterized by wide host range, which confers the advantage of abundance and wide distribution upon the parasite. In contrast, specialized parasites have evolved in such a way as to cause minimum possible damage to the functioning of host plant. They may be host-specific and they have discarded the ability to continue as competitive saprophytes. The growth rate of specialized parasites seems to have been much reduced in the absence of susceptible plants. Damping-off and root rot diseases are caused by *Pythium* spp. and *R. solani*, which can exist as saprophytes for longer periods on organic matter and become parasitic, when susceptible crops are planted. Vascular wilt fungi *F. oxysporum* and *Verticillium* spp. are specialized soilborne pathogens. Saprophytic survival of these pathogens is more due to the formation of sclerotia or chlamydospores remaining viable in soil for long periods (Garrett 1970). The soilborne microbial pathogens are able to infect annuals, as well as perennials. Fungal pathogens infecting annuals may have different modes of survival in the absence of crop hosts: (i) pathogens that survive as sclerotia have a wide host range; (ii) pathogens surviving as resting spores have restricted host range; and (iii) pathogens surviving as mycelium in crop debris may have a somewhat restricted host range. Among the fungal pathogens, *Phytophthora* spp., *Pythium* spp., *F. oxysporum*, *Verticillium* spp., *R. solani* and *Sclerotinia sclerotiorum* have worldwide distribution and are responsible for total crop loss, when susceptible cultivars are grown in large areas and favorable environmental conditions prevail. Bacterial pathogens causing soilborne diseases are less numerous and they include *Ralstonia solanacearum*, *Clavibacter michiganensis* subsp.

*sepedonicum*, *Pectobacterium* spp. and *Streptomyces scabies*. Viruses, except for a few, cannot remain infective in a free state in the soil environment. They have to be protected in the body of the vectors. These viruses, transmitted by soilborne fungal or nematode vectors, have been found to be responsible for considerable losses in cereals, vegetable and fruit crops.

As the soils are opaque, examination of the pathogens *in situ* is not possible and hence, special biological and molecular techniques have been applied for detection of their presence and quantification of their populations. Bait tests, isolation-based methods are useful to detecting and quantifying viable populations of the pathogens, but they require a long time to yield results. On the other hand, molecular methods are more sensitive and rapid in providing results. Distribution of the soilborne microbial pathogens is not uniform and variations in population may be significantly influenced by soil depth, pH, soil moisture, temperature, and organic matter composition and content. The soil is a heterogenous medium consisting of many microhabitats that vary in size, microbial activity, nutrient availability and presence of toxicants at different concentrations. Soil microorganisms mask the pathogen population due to their ability to survive saprophytically and proliferate more rapidly. In addition, following initiation of infection, the pathogen multiplies to the required level in the host plant and expression of disease symptoms occurs after completion of the required incubation period. Symptoms on the roots and the crown of infected plants can be observed directly only after destructive sampling. At the time of symptom expression, it may be nearly impossible to save infected plants, as the pathogen would have already well established itself in the internal plant tissues. However, constant and concerted research efforts taken up in different countries yielded useful results that are applicable under different agroecosystems. The pathogens may reach the soil via infected seeds/propagules, irrigation water and farm equipments. Once the pathogen reaches the soil, it is very difficult to disinfest the soils, making the field unsuitable for the cultivation of susceptible crop(s). These factors form formidable limitations and make the investigations more difficult, compared with those meant for pathogens infecting aerial plant organs.

## 1.3 NATURE OF SOILBORNE MICROBIAL PLANT PATHOGENS CAUSING CROP DISEASES

### 1.3.1 PATHOGEN BIOLOGY

It is essential to have a basic knowledge of soilborne microbial plant pathogens – oomycetes, fungi, bacteria and viruses. Different techniques may be employed to detect, identify, differentiate and quantify the populations of the pathogens in plants, seeds/propagules, soil and waterbodies, based on their morphological, biochemical, immunological and genetic characteristics. The usefulness of various techniques that can be applied to soilborne fungal pathogens is discussed in Chapter 2. The biological properties such as production of various kinds of spores formed during different stages of the life cycle of fungal pathogens, survival structures produced in the plant tissues or soil for overwintering during the absence of crop (primary host) plants, host range and process of infection and disease development (pathogenesis), variability in virulence and sensitivity due to chemicals and changes in the soil environment are described in Chapter 3. Various biological and molecular methods available for detection, identification and quantification of bacterial pathogens are presented in Chapter 4 for selecting suitable techniques for application in the investigations on these cellular pathogens. Bacterial pathogens show variations in the pathogenic potential (virulence) and phenomenon of pathogenesis, host range, ability to survive under different adverse environments prevailing in the soil, as discussed critically to facilitate development of disease management systems for containing bacterial diseases caused by soilborne bacterial pathogens in Chapter 5. Viral pathogens, ultramicroscopic and smallest of the soilborne microbial pathogens, are pathogenic to a variety of crop plants grown under different environmental conditions. Most of the viruses cannot exist in a free state in the soil, and they need fungal or nematode vectors for transmission from infected to healthy plants. As they cause frequently asymptomatic infections in plants, especially in asexually propagated plant materials, highly sensitive methods have to be applied for their detection, differentiation and quantification in plants, soil and irrigation water. They may have either restricted or wide host range including crop plants, weeds and wild plants that can serve as sources of inoculum for crops, when they are planted in the next season. Variations in their virulence and transmission specificity by vectors are highlighted in Chapter 6.

### 1.3.2 ECOLOGICAL AND EPIDEMIOLOGICAL PERSPECTIVES

Soil, formed from different parent materials, provides minerals and organic materials needed for plant growth and the development of all organisms, including soilborne microbial pathogens. The activities of microbial pathogens are greatly influenced by the host plant and other biotic and abiotic agents. Soil microbial communities interact with plants and also compete among themselves for the available nutrients and niches for establishment in the rhizosphere/rhizoplane. Root exudates of plants significantly influence the structure and composition of microbial community in the rhizosphere. Soils may be either conducive or suppressive to specific soilborne microbial plant pathogens. Soils suppressive to different soilborne fungal pathogens have been identified in different agroecosystems. The cultural practices such as tillage, irrigation, application of amendments and nutrients, and cropping systems practiced in the location, significantly influence the rhizosphere microflora, resulting in either increase or decrease in disease incidence/severity. Epidemiological investigations are confronted by challenges posed by complex effects of several factors operating simultaneously or sequentially in the soil environments. Early detection and identification of soilborne microbial plant pathogens, under conditions existing in the soil, are yet to reach the desired level. However, available techniques have been applied to gather useful information to analyze epidemiological factors influencing disease incidence and spread. The development of epidemics may be influenced by host, pathogen and environmental factors that favor disease development. Forecasting systems have been formulated to predict incidence of some soilborne diseases in certain particular geographical locations. The effectiveness of predictive models, in facilitating decision making by the growers, is discussed in Chapter 7.

## 1.4 PRINCIPLES OF MANAGEMENT OF CROP DISEASES CAUSED BY SOILBORNE MICROBIAL PATHOGENS

Crop disease management strategies revolve around the basic principles of exclusion, eradication and immunization. Disease management practices may be directed toward either the host plant or the pathogen and sometimes toward both. Disease management strategies may be of two kinds: host management strategies and pathogen management strategies. Host management practices aim at prevention of pathogen introduction through seeds and/ propagules, and reduction of inoculum production by the pathogen, whereas pathogen management practices tend to reduce the development of and elimination of pathogen(s).

### 1.4.1 PREVENTION OF PATHOGEN INTRODUCTION AND INOCULUM PRODUCTION

The production and use of disease-free certified seed and propagules is the basic step in preventing the introduction of native and exotic strains of soilborne microbial pathogens. Regulatory methods have been imposed by several countries by applying certification standards and establishing domestic and international plant quarantines. Maintenance of soil health is important to attaining targeted production levels. The selection of a suitable farming system will be useful to enhancing soil fertility and also to preventing accumulation of inoculum of pathogens in soils. Soil solarization using polyethylene sheets has been shown to be effective in reducing the incidence of diseases caused by soilborne pathogens by raising the soil temperature. This nonchemical soil disinfestation tactic is effective through differentially suppressing many soilborne microbial plant pathogens and favoring development of microorganisms with biocontrol potential against the pathogens (see Chapter 8). Among the various strategies that are available, the development of cultivars with built-in resistance is the most preferred strategy for limiting infection and loss inflicted by soilborne microbial plant pathogens. However, nonavailability of reliable sources of resistance for improving resistance levels in cultivars, linkage of disease resistance gene(s) with undesirable characteristics and interspecific sterility have been important impediments preventing the wider applicability of this approach. Mutation breeding has been useful to modifying genetic constitution of the cultivars, leading to an increase in resistance level against the target pathogen. The marker-assisted selection (MAS) procedure has been effectively applied to select genotypes with resistance to disease(s) rapidly. Biotechnological methods have widened the scope of transferring resistance gene(s) from diverse sources, which is not possible through conventional breeding methods. Several countries have planted large areas with transgenic crops, although reservations on the cultivation of transgenic crops have been stumbling blocks in many countries (see Chapter 9).

### 1.4.2 INHIBITION OF PATHOGEN DEVELOPMENT

Plant pathogen management strategies – application of biotic and abiotic biocontrol agents and chemicals – are employed to reduce the rate of increase in pathogen populations present in different substrates such as seeds/propagules, plants, soil and irrigation water sources. Biotic biological control agents constitute fungi, bacteria and viruses, which can act on the pathogens via different mechanisms such as antibiosis, competition for nutrition and niches in plant roots/rhizosphere, prevention of colonization of host root tissues and induction of resistance in treated plants against the target pathogen(s). Abiotic biological control agents include organic compounds of plant and animal origin and inorganic compounds that have either direct inhibitory activity against soilborne microbial plant pathogens or act on the host plant to enhance the resistance of treated plants against the pathogens. Use of biocontrol agents (BCAs) is preferred, because of their ecofriendly nature and ability to activate host gene(s) through stimulation of mechanisms as in genetically resistant cultivars (see Chapter 10). Chemicals belonging to different classes have been applied on various crops to suppress the development of pathogens infecting them and consequently to reduce disease incidence and severity. The effectiveness of chemicals against target pathogen(s) under *in vitro*/growth room, greenhouse and field conditions is determined, based on the level of inhibition of growth, spore germination and reduction in pathogen population in soil, and disease incidence/severity. The effective chemicals are applied as seed treatment, foliar spray or soil application. The extent of disease suppression by chemicals may depend on the levels of susceptibility/resistance of the cultivars, virulence of the pathogen species/strains and

prevailing environmental conditions that influence the disease pressure level significantly. Under high disease pressure conditions, increase in frequency of fungicide application may be warranted. The need for restricting chemical use for the management of diseases caused by soilborne microbial pathogens to avoid environmental pollution and accumulation of chemical residues in grains, vegetables and fruits is emphasized (see Chapter 11).

## 1.5 DEVELOPMENT OF INTEGRATED DISEASE MANAGEMENT SYSTEMS

Interactions between microbial plant pathogens and crop plant species may lead to different levels of disease incidence and rapidity of disease spread resulting in epidemics, if appropriate disease management strategies are not applied at the right time. Several investigations have demonstrated the effectiveness of disease management strategies applied individually in some locations under certain conditions. Hence, it is essential to examine the possibility of enhancing the efficiency of disease suppression, by integrating two or more strategies that are compatible with each other. Unfortunately, such attempts have been fewer than how many are necessary. Nevertheless, the efforts to integrate host resistance with biological control agents (BCAs) or chemicals (applied at lower concentrations) have provided encouraging results. Application of the cultural practices to reduce pathogen inoculum levels, is the foundation for the integrated disease management (IDM) system and when combined with other strategies, it may be employed to raise the levels of disease control to sustain profitable crop production in different ecosystems (see Chapter 12).

This book aims to provide comprehensive information derived from an extensive literature search on various aspects of soilborne microbial plant pathogens and the diseases induced by them on various crops grown in different agroecosystems. The need for development of IDM systems for effective suppression of disease development is emphasized to achieve production levels that can be profitable to the grower. Further, the IDM system is capable of providing healthy and chemical residue-free food, in addition to preservation of environment. The information presented in this book is expected to be useful for the graduate students, researchers and teachers in the Departments of Plant Pathology, Microbiology, Plant Breeding and Genetics, Agriculture and Horticulture and to personnel of certification and quarantine programs.

### REFERENCES

Alvarez B, Biosca E, Lopez MM (2010) On the life of *Ralstonia solanacearum*, a destructive bacterial plant pathogen. In: Mendez-Vilas A (ed.), *Current Research Technology and Education Topics in Applied Microbiology and Microbial Technology*. FORMATEX 2010© pp. 267–269.
Cu RM, Mew TW, Cassman KG, Teng PS (1996) Effect of sheath blight on yield in tropical intensive rice production system. *Plant Dis* 80: 1103–1108.
del Rio LE, Bradley CA, Hanson RA et al. (2007) Impact of Sclerotinia stem rot on yield of canola. *Plant Dis* 91: 191–194.

Garrett SD (1970) *Pathogenic Root-infecting Fungi*. Cambridge University Press, Cambridge, UK.

Gaunt RE (1995) New technology in disease measurement and yield loss appraisal. *Canad J Plant Pathol* 17: 185–189.

Groth DE (2005) Azoxystrobin rate and timing effects on rice sheath blight incidence and severity and rice grain and milling yields. *Plant Dis* 89: 1171–1174.

Henry C (1996) Rhizomania – its effect on sugar beet yield in UK. *British Sugar Beet Rev* 64: 24–26.

Jnr L, Silva L (2000) Sclerotinia rot losses in processing tomatoes grown under centre pivot irrigation in central Brazil. *Plant Pathol* 49: 51–56.

Large EC (1966) Measuring plant disease. *Annu Rev Phytopathol* 4: 9–28.

Melouk HK, Backman PA (1995) Management of Soilborne Fungal pathogens. In: Melouk HA, Shokes FM (eds.), *Peanut Health Management*, The American Phytopathological Society Press, St. Paul, MN, USA.

Narayanasamy P (2002) *Microbial Plant Pathogens and Crop Disease Management*. Science Publishers, Enfield, USA.

Narayanasamy P (2011) *Microbial Plant Pathogens – Detection and Disease Diagnosis*, Volume 1, Springer Science, Heidelberg, Germany.

Navas-Cortés JA, Hau B, Jimenez-Diaz RM (2000) Yield loss in chickpea in relation to development of Fusarium wilt epidemics. *Phytopathology* 90: 1269–1278.

Rubies-Autonell C, Ratti C, Pisi A, Vallega V (2006) Behaviour of durum and common wheat cultivars in regard to *Soilborne cereal mosaic virus* (SBCMV). Results of fifteen trials. *Proc 12th Conf Mediterr Phytopathol Union*, Rhodes, Greece, pp. 100–102.

Tripathi L, Mwangi M, Aritua V, Tushemerierwe WK, Abele S, Bandyopadhyay R (2009) Xanthomonas wilt – A threat to banana production in east and central Africa. *Plant Dis* 93: 440–451.

Van de Weg WE, Henken B, Giezen S (1997) Assessment of the resistance to *Phytophthora fragariae* var. *fragariae* of the USA and Canadian differential series of strawberry genotypes. *J Phytopathol* 145: 1–6.

# 2 Detection and Identification of Soilborne Fungal Plant Pathogens

Soils present in various agroecosystems differ in their composition of abiotic and biotic components which exert remarkable influence on the development of plants from seed/propagule germination, to growth and maturity of grains and other plant produce. Soil provides anchorage and nutrients for the emerging seedlings and transplants till they attain harvest stage. The availability of nutrients from soil to plants depends on the fertility of the soil derived from the organic matter and minerals (fertilizers) that are added naturally or artificially. In addition, the microorganisms present in the soil may be beneficial to the development of plants by fixing atmospheric nitrogen and decomposing complex organic matter into easily available forms. In contrast, some of them may be harmful by infecting roots/stems or aerial parts of plants at different stages of plant growth, causing a range of adverse effects (diseases), resulting in the debilitation and ultimately the death of plants. As these microorganisms cause several economically important diseases, it is essential to establish their identity rapidly and reliably by employing appropriate technique(s), depending on their morphological, biological, biochemical, immunological and genetic properties.

Soilborne microbial plant pathogens include oomycetes, fungi, bacteria and actinomycetes which may lead either saprophytic or parasitic life during the entire or part of the life cycle. In contrast, plant viruses are obligate parasites and some of them are transmitted by fungi or nematodes that are soilborne. The fungal/nematode species that are involved in virus transmission can also cause root diseases of crops. Among the soilborne microbial plant pathogens, fungi cause large number of diseases affecting almost all crops grown in different ecosystems. These pathogens cause damping-off, root rots, collar rots, stem rots and wilt diseases in plants and rot diseases of tubers and grains of various agricultural and horticultural crops, as well as in forest trees. The soilborne fungal pathogens may be placed in three groups: (i) soil-inhabiting pathogens that can survive as saprophytes independently in the absence of a susceptible host plant species; (ii) soil-invading or root-inhabiting pathogens that may have a saprophytic or a parasitic phase depending on the availability of susceptible plant species and (iii) obligate parasites that require the presence of a living susceptible plant species during their entire parasitic life. The microbial plant pathogens may reach the soil environment through infected seed/propagules, transplants, infected plant residues (debris) left on the soil after harvest, infected volunteer plants and weeds, irrigation/rain water moving from field to field and organic amendments incorporated into the soil.

## 2.1 DETECTION OF FUNGAL PATHOGENS IN SOIL ENVIRONMENTS

The soil-invading fungal pathogens may move from infected seeds/propagules (planting materials) which may be symptomatic or asymptomatic (as in many cases) to emerging young seedlings. As the plants grow, the fungal pathogens proliferate, reaching high populations, inducing symptoms characteristic of the disease. During this phase of disease development (pathogenesis), the fungal pathogens may reach the top soil layer and colonize the organic matter/plant debris left after harvest. Alternatively, stubbles and foliages, if infected, may add significant amounts of resting spores such as chlamydospores, sclerotia and other asexual spore-forming structures and sexual spores that are resistant to adverse environmental conditions, existing in the soil. The soilborne diseases pose a formidable obstacle in recognizing infected plants before visible symptoms appear on the plants. The symptoms of infection on aerial plant parts can be observed in most cases, only after the fungal pathogen has established well inside infected plants. Root infection by these pathogens can be seen only by destructive sampling (pulling out the suspected plants). At this stage, it may be difficult to save the infected plants, although further spread of the disease to neighboring plants may be restricted by applying appropriate management practices. Hence, the soilborne fungal pathogens have to be detected, differentiated and quantified in the infected plants, field soils and water, including river, irrigation water and water bodies.

### 2.1.1 ISOLATION-BASED METHODS

Fungal pathogens, other than obligate pathogens, are conventionally isolated from the infected plant parts such as seeds, roots, stems, leaves or inflorescences, using appropriate media that favor their growth and sporulation. In the case of fungal pathogens which are slow-growing, semiselective/selective media that support differential growth of the target pathogen(s), are employed. The infected plant tissues, after surface sterilization, are plated on the suitable medium for development of the fungal pathogen. The culture, after allowing time required for production of asexual/sexual spores, is examined and the cultural characteristics are recorded and compared with those of known pathogens. It may be difficult to identify the fungal pathogens based only on morphological characteristics. Additional tests have to be performed for precise identification of the pathogen to be investigated. Isolation of *Rhizoctonia solani* anastomosis group (AG)-8, causing root rot and bare patch disease of wheat and barley was difficult even in a semiselective medium containing water agar amended with benomyl and chloramphenicol, because of its slow-growing nature. On the other hand, *R. oryzae* could be easily isolated from symptomatic wheat roots (Paulitz et al. 2003). Isolation of *Phytophthora* sp. causing decline of young almond trees (*Prunus dulcis*) from cankers, roots and soil around affected trees was achieved and its pathogenicity was proved by artificial inoculation (Perez-Saierra et al. 2010). *Fusarium oxysporum* and *F. avenaceum* were isolated from field pea roots with rotting symptoms. *F. avenaceum* was virulent causing root rot among the nine *Fusarium* spp. isolated from pea (Chitten et al. 2015).

A detached leaf assay was developed to determine the pathogenicity of *Pythium* isolates to cut-flower chrysanthemum roots. Leaves of young plants were excised and inoculated by insertion of a plug of mycelium into the slit cut in the excised petiole. The presence and extent of necrosis were assessed after the incubation of leaves. Well-defined necrosis was induced in detached leaves of chrysanthemum cv. Splendid Reagan by pathogenic isolates of *Pythium* spp., causing root rot and the necrosis rapidly spread across entire leaves within a few days. All isolates of *P. sylvaticum* and *P. ultimum* were more virulent than *P. irregulare*, *P. oligandrum* and *P. aphanidermatum*. In contrast, the majority of nonpathogenic isolates did not induce necrosis in detached leaf assays. The pathogenicity of *Pythium* isolates obtained from the roots of chrysanthemum could be determined by this simple detached leaf assay efficiently (Pettitt et al. 2011). Soybean seedlings were infected by a large number (29) of *Pythium* spp. The three most abundantly recovered isolates from infected soybean roots belonged to *P. ultimum*, *P. heterothallicum* and *Pythium* sp. The majority of *Pythium* spp. caused pre-emergence damping-off of soybean seedlings (Zitnick-Anderson and Nelson Jr 2015). The population density of *Phomopsis sclerotioides*, causal agent of cucumber black root rot, in soil was determined to assess its effect on disease incidence and development. Cucumber plants were grown in soil containing 1, 10, 100 and 1,000 CFU/g of soil. Incidence of disease at three to seven weeks after transplanting on infested soil was strongly correlated with *P. sclerotioides* density in soil (P > 0.05). Root rot of squash rootstock occurred in soil with very low inoculum densities (0.1 CFU/g) and was strongly related to *P. sclerotioides* density at eight weeks after transplanting. Cucumber plants exhibited wilt symptoms in soil containing 1 CFU/g. Wilt symptoms in cucumber plants occurred four to seven days earlier in soil at 22°C than in soil at 27°C or 17°C. Root rot development could be predicted based on the density of *P. slcerotioides* in soil and the prevailing soil temperature. The determination of threshold of pathogen in soil might be useful to assess the effects of disease management strategies (Shishido et al. 2016).

*Verticillium dahliae*, causal agent of Verticillium wilt of potato, is soilborne and can be found in soil carried on the potato seed tubers. In order to quantify *V. dahliae* populations in soil associated with certified seed tubers and to determine the possibilities of disease development in the field, *V. dahliae*

was isolated from tuber surfaces of seed lots from *V. dahliae*-infested soil. Over 82% of seed lots contained *V. dahliae* in loose seed lot soil obtained from bags and trucks used to transport seed tubers. Most samples had < 50 CFU/g, while some contained >500 CFU/g. Most isolates (93%) belonged to vegetative compatibility group (VG)-4A. The results indicated that seed tubers contaminated with *V. dahliae*-infested soils might introduce the pathogen into fields not previously cropped to potato or recontaminate the fields that received preplant management practices to reduce *V. dahliae* populations in soil. Further, effective long-term management of *V. dahliae* for reducing pathogen populations associated with seed lots has to be considered (Dung et al. 2013). The potential role of *V. dahliae* inoculum, introduced into the soil via infested spinach seed, in the incidence of Verticillium wilt in subsequent lettuce crops was studied under microplot conditions. Pathogen isolates were identified, using *Verticillium* species-specific polymerase chain reaction (PCR) assay. Soil, collected before the first spinach crop from all microplots, was free of *V. dahliae*. Microsclerotial colonies of *V. dahliae* were observed on NP-10 agar containing soil samples collected from microplots planted with infested seed at 67, 33 or 15% infestation levels in treatments 1, 2, and 3. At this time point, the amount of *V. dahliae* inoculum recovered was the highest in microplots sown with about 67% infested spinach seed, followed by 33% and 15% (see Figure 2.1). During the first lettuce crop that followed three spinach crops, Verticillium wilt symptoms were visible in a few plants. Plating of infected lettuce tissues and PCR assay revealed the infection of lettuce plants by *V. dahliae* inoculum present in the soil. Following the first lettuce planting, the incidence of vascular discoloration in lettuce tap roots in treatments with approximately 10, 15, 33 and 67% infested spinach seed was 4.44, 5.26, 9.02 and 9.29, respectively. In the second lettuce crop, the disease incidence significantly increased to levels ranging from 21.43 to 31.82. The green fluorescent protein (GFP)-tagged strain of *V. dahliae* VdS0925-316 colonized lettuce root surface at fives weeks after planting lettuce in soil, following two prior plantings of infested spinach seeds and incorporation of spinach residue (Short et al. 2015).

### 2.1.1.1 Baiting Assays

The presence and location of soilborne fungal pathogens cannot be recognized in most cases, because of the low pathogen populations in the field soils. Hence, highly susceptible plant species that develop visible characteristic symptoms, are used as baits or indicators. *Phytophthora capsici*, infecting pepper and other vegetable crops, was detected by employing the pepper leaf disk assay. Pepper leaf disks were floated on saturation water held on soil samples placed in plastic cups for 24 h. The leaf disks were disinfested with 0.5% sodium hypochlorite solution for 1 min, rinsed in sterile water and plated on Masago's Phytophthora selective medium (Masago et al. 1977) and incubated for 72 h. Pathogen colonies developing on leaf disks were counted and the percent infection of leaf disks was considered to reflect pathogen population (Larkin et al. 1995). Isolation of nonculturable obligate

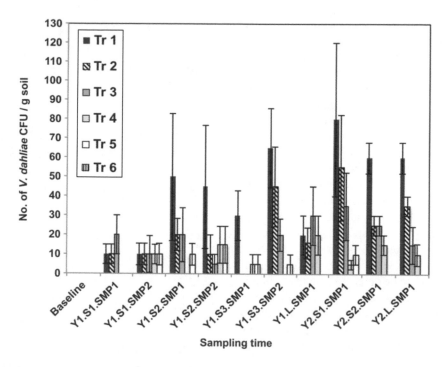

**FIGURE 2.1** Recovery of *V. dahliae* from microplots planted with spinach seeds with different levels of pathogen infestation at different times of sampling. Treatments: Tr1–Tr6 treatments with varying seed infestation ranging from 0 to 67% and no spinach seed planted; Y1- first year (2011), S1- first spinach crop, SMP1- first sampling, SMP2- second sampling, S2- second spinach crop, S3- third spinach crop, L- lettuce crop and Y2- second year (2012).

**[Courtesy of Short et al. (2015) and with kind permission of the American Phytopathological Society, MN, United States]**

pathogens like *Plasmodiophora brassicae* from soil samples demands special methods. Sensitive indicator (bait) plant species is grown in the test soil samples from field considered to have been infested by the target pathogen. The development of characteristic symptoms in the indicator plants shows the presence of the pathogen in the field soil. For the detection of *Phytophthora fragariae* var. *fragariae*, infected root tissues were incorporated in soilless planting mix and then highly susceptible *Fragaria semperflorans* var. *alpine* (Baron Solemacher) plants were planted as baits. After a period of three to six weeks, the infected plants collapsed and stele tissue of roots showed red coloration as well as the presence of oospores of the pathogen. Infection by *P. fragariae* var. *rubi* of raspberry plants was also detected by using bait plants (Duncan 1980, 1990).

A quantitative baiting assay was developed for recovering *Phytophthora cinnamomi* from soil samples, using blue lupin (*Lupinus angustifolium*) radicles as baits. The optimum temperatures for baiting and bait incubation were ~ 25°C and 20–20°C, respectively. The assay showed greater baiting efficiency, when the subsample size was decreased. The peak of zoospore release from naturally infested soil occurred on the first day of incubation and infection on radicles decreased as a function of distance from radical tip (Eden et al. 2000). The efficacy of baiting in sudden oak death-affected regions of California was assessed. Rhododendron leaves and pears were employed as baits for the detection of soilborne pathogen populations. Natural inoculum associated with bay laurel (*Umbellularia californica*), tanoak (*Lithorcarpus densiflorus*)

and red wood (*Sesquoia sempervirens*) was determined by monthly baiting. Rhododendron leaf baits were more efficient than pear baits for the detection of sporangia, but neither of the baits could detect chlamydospores. *P. ramorum* survived through chlamydospores in soil over summer (Fichtner et al. 2007). The baiting method was found to be effective in recovering *P. cinnamomi* and *P. heveae* from soil of Southern Appalachian Mountain forest (Meadows et al. 2011).

The inoculum density of *Phytophthora sojae*, causal agent of soybean Phytophthora blight, in naturally infested soils was estimated. Two bioassays applied to quantify zoospores and another bioassay for estimating oospores of *P. sojae* in soil. A known density of zoospores or oospores of enhanced green fluorescent protein (EGFP)-labeled *P. sojae* Eps597-3 (pathotype La3c7) was spiked into the soil. The zoospores were quantified by using leaf disks as baits and by counting them at the margins of leaf disks, using a light microscope. Likewise, the oospore concentrations in soil were determined. Linear regression equations for calculating the zoospores and oospores contents in the soil samples spiked with different concentrations were formulated. The equations provided highly reproducible and accurate estimations of zoospores/oospores in different soil types (Suo et al. 2015). A combination of baiting and plating procedure was followed for the detection of oomycetes, *Pythium* spp. and *Phytophthora* spp. Seeds of susceptible cultivar, after sterilization, were placed for baiting the pathogen in water extract of soil samples and incubated for about 12 h. The seeds were then transferred to plates containing suitable medium. The developing fungal colonies were examined and identified based on morphological characteristics. The identity of the pathogens was confirmed by performing enzyme-linked immunosorbent assay (ELISA) (Yuen et al. 1998). Several crops are grown using hydroponic culture and ebb-and-flow irrigation systems. Infection by *Pythium helicoides*, *P. aphanidermatum* and *P. myriotylum* in these crops was observed frequently. These pathogens could tolerate high temperatures ( > 40ºC). A baiting technique, involving the use of seeds of cucumber, tomato and radish and leaves of bent grass and rose, was developed. These baits were found to be very efficient in recovering zoospores produced by these pathogens. Bent grass leaf traps (BLTs) could detect three *Pythiu*m spp. after a short exposure (one day) to suspensions containing 40 zoospores/l of water and the frequency of detection increased with increase in zoospore density and baiting period. By employing this technique, the pathogen could be detected in nutrient solution as early as 23 days before disease spread to noninoculated roses. The baiting technique was shown to be an effective approach for monitoring the development of thermotolerant *Pythium* spp. in recirculating hydroponic systems (Watanabe et al. 2008).

*Rhizoctonia soalni* AG-8 could not be easily isolated from root systems of wheat and barley and quantified in soil samples, due to its slow-growing nature and low population densities in soils. A quantitative assay for estimating active hyphae present in soil was developed, using wooden toothpicks as baits inserted into soil samples. After a two-day incubation period, the toothpicks were placed on a selective medium consisting of water agar amended with benomyl (1 μg/ml) and chloramphenicol (100 μg/l). Using a dissecting microscope, the developing pathogen colonies were counted at two days after inoculation. *R. solani* and *R. oryzae* (causing root rot of wheat) could be distinguished from other fungi, based on hyphal morphology (see Figure 2.2 and Figure 2.3). This toothpick assay was successfully employed to assess the activity of *R. solani* inside and outside of bare patches in wheat fields. Further, this technique was also useful to compare the activity of *R. solani* over time in direct-seeded plots versus

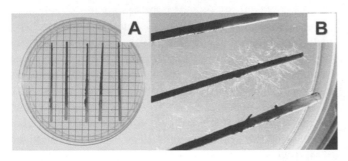

**FIGURE 2.2** Isolation of *R. solani* from soil by toothpick assay. A – Petriplate with Rhizoctonia selective medium on which the toothpicks (5) removed from test soil sample are placed; B – Development of *R. solani* colony growing out of toothpick after 24 hours.

[Courtesy of Paulitz and Schroeder (2005) and with kind permission of the American Phytopathological Society, MN, United States]

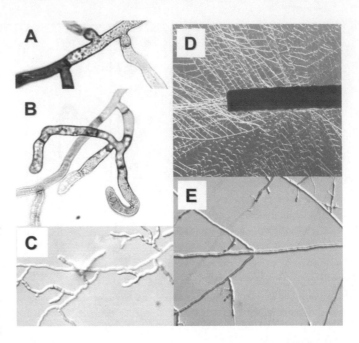

**FIGURE 2.3** Differentiation of *Rhizoctonia* spp. based on hyphal morphologic characteristics observed under light microscope. *R. solani* with typical right-angle branching of main hyphae with constriction at the point of branching (A), curling growth (B) and growth in different directions (C) of secondary hyphae; D and E- *R. oryzae* with unidirectional growth of main hyphae and angular branching (30–50 degrees) of secondary hyphae.

[Courtesy of Paulitz and Schroeder (2005) and with kind permission of the American Phytopathological Society, MN, United States]

conventionally tilled plots under field conditions. The hyphal activity of *R. solani* AG-8 and *R. oryzae* from soil cores taken from various positions in and around Rhizoctonia bare patches at two locations was determined. Activity of *R. solani* was maximum at the center and inside the edge of the patch. This method might provide an accurate estimate of inoculum potential (IP) of a soil at a given time, since the hyphae initiate the infection process. The toothpick method is simple and inexpensive with potential for application to the detection and diagnosis of *Rhizoctonia* infection in cultivator's fields and ecological and epidemiological investigations (Paulitz and Schroeder 2005). In order to quantitatively determine populations of *R. solani* in soil from soybean fields, toothpick-baiting method, the multiple-pellet soil sampler and pour-plate procedure were evaluated for their efficacy. The radius of isolation with toothpick-baiting technique was approximately 1 cm. The toothpick method was found to be the most reliable procedure for assaying samples with most isolates across space and greater recovery of *R. solani* AG-IA, *R. solani* AG11 and *R. oryzae*. The recovery of three AG groups of *Rhizoctonia* did not differ significantly by toothpick-baiting method. The toothpick-baiting method was preferred, because it was inexpensive, nondestructive and rapid. Further, use of the toothpick-baiting method allowed for determination of the depth of inoculum of isolated fungi from intact soil cores. The mean depth of activities of *R. solani* AG1-IA, *R. solani* AG11 and

*R. oryzae* was 1.15, 1.55 and 1.47 cm, respectively (Spurlock et al. 2015).

### 2.1.1.2 Direct Plating of Soil Samples

The soilborne fungal pathogens present in the field soil under natural conditions have to be studied both qualitatively and quantitatively, in order to assess the influence of epidemiological factors on disease incidence and spread. Soil-dilution procedure was applied for the isolation of *Verticillium* spp. from naturally infested fresh soils (Jordan 1971). By adopting a modified Anderson Sampler technique, *V. dahliae* populations persisting in air-dried soil were estimated. Populations of *V. dahliae* increased significantly in soils cropped to cotton. This technique was more precise and sensitive by detecting 2.8 times greater a number of microsclerotia/g of soil (Butterfield and De Vay 1977). It may be difficult to isolate slow-growing fungal pathogens from soil using a direct plating method, since fast-growing fungal pathogens/saprophytes may overgrow the slow-growing pathogens, making it difficult to obtain their axenic cultures. Hence, semiselective/selective media that differentially favor the development of the slow-growing pathogens have to be used for isolating them from soils. Standard soil-dilution plating on selective medium, saturation water and pepper leaf disk assay were compared, after 'sample saturation and resaturation of soil with water' for detection and quantification of specific propagule types of *P. capsici* in soil. None of the techniques tested could detect and quantify all propagules types viz., zoospores, oospores or mycelial fragments of the pathogen. All techniques detected the sporangia at a concentration of one propagule per gram (PPG) of soil. However, the pepper leaf disk assay could detect different propagule types, but failed to quantify the propagules with required accuracy (Larkin et al. 1995). In the case of soilborne oomycetes, the soil-dilution plate method was found to be unsuitable, because of failure of germination of oospores on any agar medium. Hence, methods that stimulate oospore germination had to be employed to improve detection efficiency of the pathogens present in the soil. Baiting methods were found to be more efficient than direct plating procedure (Goodwin et al. 1990).

The usefulness of selective media for isolating pathogens from soil has been reported in many pathosystems. Modified Nash and Snyder's medium (MNSM) was employed for the isolation of *Fusarium solani* f. sp. *glycines (Fsg)*, incitant of sudden death syndrome ) in soybean. The isolates developing on MNSM medium were pathogenic and induced —sudden death-like symptoms in soybeans in the greenhouse. The results showed that soil populations of *Fsg* could be accurately enumerated using MNSM medium (Cho et al. 2001). In a later investigation, potato dextrose agar (PDA) amended with hymexazoll, ampicillin and rifampicin (HAR) was employed to quantify pathogen propagules in tap roots of symptomatic soybean plants. The fungal colonies grew slowly in the medium and produced masses of blue macroconidia, characteristic of *Fsg*. The pathogenicity of the isolates was proved by using sorghum seeds infested with the isolates for inoculating the soil on which soybean plants were grown in the greenhouse. The identity of the pathogenic isolates was established by performing real-time quantitative polymerase chain reaction (qPCR) assay. Wide distribution of sudden death syndrome (SDS) in soybean in Nebraska was confirmed by the results of this investigation (Ziems et al. 2006). *Fusarium oxysporum* f.sp. *vasinfectum*, causative agent of cotton wilt disease, is soilborne and seedborne. Long distance dispersal may occur through infected seeds, because of movement of seed for commerce and germplasm exchange. A Fusarium-selective liquid medium that favored differentially the development of *F. oxysporum* f. sp. *vasinfectum* was used for the isolation of the pathogen. The identity of the fungus isolated was confirmed by employing race-specific primers in PCR assay, after extracting the DNA from isolates growing on the selective medium (Bennett et al. 2008).

A semiselective medium was employed for quantifying populations *of Verticillium dahliae* in soil samples from 87 and 51 fields intended for potato and mint production respectively. The pathogen could be isolated in 89% and 80% of potato and mint fields respectively. Population densities ranged from 0 to 169 and 0 to 75 propagules per g of air-dried soils from potato and mint fields respectively, indicating the efficiency of recovering the pathogen propagules from soils using the semiselective medium (Omer et al. 2008). The population density of *V. dahliae* in soil was estimated by using Sorensen's NP-10 medium (NP-10) which contained polygalacturonic acid (PGA). Of the different types of PGA salts tested, the NP-10 with P-3889 and 0.025N NaOH yielded consistently higher numbers of microsclerotia MS) of *V. dahliae* from soil samples and supported the growth and production of MS to a level similar to that of NP-10 with P-1879, allowing the replacement of the original NP-10 medium developed previously (Kabir et al. 2004). The ingredients used to amend media previously were inhibitors of MS germination and subsequent growth of *V. dahliae*. Hence, an improved medium designated Ethanol Potassium Amoxicillin Agar (EPAA) was developed. The EPAA medium recovered 98% of the MS added to the soil samples with high accuracy (SE = 1.37%). This medium was highly specific, allowing only minor growth of other contaminating microorganisms. Distinct identifiable colonies of *V. dahliae* were recovered within seven days of incubation at 20°C (Mansori 2011).

*Pythium* spp. cause damping-off or root rot diseases in many vegetable nurseries. In order to isolate these pathogens from soil, the standard Pythium selective medium (PARP) [pimaricin + ampicillin + rifampicin + PCNB + agar] was modified by replacing PCNB and pimaricin with other antifungal agents by screening them. The NARM medium containing nystatin + ampicillin + miconazole + agar was in equivalence to PARP medium in inhibiting non-pythiaceous microorganisms, except *F. oxysporum* and it was significantly better than the PARP medium in encouraging mycelial growth of five of eight isolates of *Pythium* obtained from field soils (Morita and Tojo 2007). *Thielaviopsis basicola*, infecting pansy and other ornamental plants was isolated from the soil by directly plating on the selective medium TB-CEN medium. One ml of soil suspension was dispensed to each plate containing TB-CEN medium and incubated in darkness for one to two weeks for the development of pathogen colonies which were compact and dark gray to black in color (Avanzato and Rothrock 2010). The prevalence of damping-off disease affecting conifer seedlings in the Pacific Northwest (PNW) region of the United States was investigated by taking field surveys in three forest nurseries. *Pythium* spp. were isolated by plating soil samples onto semiselective medium or by baiting with rhododendron leaf disks and Douglas fir needle segments. Of the 300 isolates obtained, *P. irregulare* was predominantly isolated from the nursery in Washington (65%), whereas *P. vipa* and *P. dissotocum* were more frequently isolated in the two nurseries in Oregon (53 and 47%, respectively) (Weiland 2011). Shatangju mandarin seedlings were infected by stem rot disease in China. A *Pythium*-like oomycete was isolated from symptomatic seedlings in the nurseries. Morphological and cultural characteristics and combined phylogenetic analysis of three DNA sequences, including internal transcribed spacer (ITS), mitochondrial *CoxI* and *CoxII* genes of 21 isolates were determined. The isolates causing stem rot of mandarin seedlings were identified as *Phytopythium helicoides*. The isolates were pathogenic to the scions, but not to the rootstocks of Shantangju mandarin seedlings producing symptoms similar to that which were observed in the field. Different steps of Koch's postulates were applied to prove the pathogenicity of *P. helicoides* on mandarin seedlings in the greenhouse (Chen et al. 2016).

A semiselective medium was employed to isolate *Ggt* from soil. The diagnostic medium (R-PDA) containing dilute PDA amended with rifampicin (100 µg/ml) and tolclofosmethyl (10 µg/ml) facilitated clear differentiation of *Ggt* from other soilborne fungi. The presence of *Ggt* could be recognized by its ability to change the color of rifampicin in R-PDA medium from orange to purple in about 24 h. This semiselective medium was more effective in isolating *Ggt*, compared with SM-GGT3 medium used earlier. *Rhizoctonia* spp. commonly present in soil was inhibited by the combination of rifampicin and tolclofosmethyl, differentially supporting the development of *Ggt* (Duffy and Weller 1994). A marker, tannic acid (300 ppm) was incorporated into the Czapek's agar/liquid medium for isolation, identification and quantification of *R. solani*. The color of the medium turned yellow to dark brown as the biomass of the pathogen increased. Absorption values at 363 nm were positively correlated with mycelia biomass (Hsieh et al. 1996). The selective medium FSM was reported to be useful for the isolation and identification of *Fusarium solani* f. sp. *cucurbitae* infecting cucurbits (Mehl and Epstein 2007). Wilt disease caused by *Nalanthamala psidii* is an important disease of guava (*Psidium guajava*) in Taiwan impacting its production. The modified sucrose-glycine semiselective medium (mSGSSM) was developed for detecting *N. psidii* in soil and infected tissues of guava, using the modified Czapek-Dox medium containing sucrose (3%), glycine (0.3%), iprodione (5 µg/ml), azoxystrobin (1 µg/ml), streptomycin (200 µg/ml) and neomycin (200 µg/ml). *N. psidii* was detected more efficiently on mSGSSM than on other media tested. The mSGSSM was able to detect the pathogen

efficiently in various parts of guava trees as well as in the soil samples (Hong et al. 2013).

A slow-growing *Phytophthora* sp. was isolated from the roots of young rose cuttings showing root rot symptoms in the Netherlands, by plating root pieces on CMA-P$_{10}$ARP selective medium containing pimaricin, ampicillin, rifampicin and pentachloronitrobenzene (PCNB) with or without hymexazol or CMP-P$_{10}$VP selective medium containing pimaricin, vancomycin and PCNB with or without hymexazol. The pathogen was identified as a new species designated *P. bisheria* sp. nov. (Abad et al. 2008). A homothallic *Fusarium* sp. was isolated from greenhouse tomato plants with brown lesions at graft points and pruning sites and these plants wilted and died within a few weeks. PDA amended with chloramphenicol (80 mg/l) and streptomycin sulfate (80 mg/l) was employed to isolate the pathogen. After incubation for four to five weeks at 28°C, the colonies were exposed to near UV light to induce sporulation. The fungus isolated was pathogenic to tomato and reproduced disease symptoms similar to those observed on tomato plants from which the fungus was isolated. The pathogen was identified as *Fusarium striatum* and found to be a new record on tomato grown in greenhouses in Canada (Moine et al. 2014).

### 2.1.1.3 Plating of Water Samples

Water is recycled or reclaimed surface water is used for irrigation purposes to conserve water and preserve the surrounding ecosystem and ground water from pesticides and nutrient runoff. A major deterent of recycling water is the potential for spread of waterborne plant pathogens. Irrigation water contaminated with microbial pathogens can be a potential source of inoculum transporting the pathogen propagules from one field to other fields. Quality of irrigation water is important to prevent infection of different crops by *Phytophthora* spp. and *Pythium* spp. Plant pathogens may be spread through recycled effluent. *Phytophthora* spp. and *Pythium* spp. present in a water-recycling irrigation system at a container nursery were characterized, using filtering and baiting techniques with two selective media. *Pythium* spp. were recovered more frequently and in greater number than *Phytophthora* spp. Members of *Phytophthora* were identified up to species level, whereas *Pythium* spp. could be identified up to genus level only. *P. capsici*, *P. citricola*, *P. citrophthora*, *P. cryptogea*, *P. drechsleri* and *P. nicotianae* were recovered by filtering assays. *P. cryptogea* and *P. drechsleri* only were recovered from baits placed on the surface of irrigation reservoir, while *P. cactorum*, *P. capsici*, *P. citricola*, *P. citrophthora*, *P. cryptogea* and *P. drechsleri* were recoverable at depths (1.0 to 1.4 m). These techniques may be used for detecting plant pathogens in irrigation water (Bush et al. 2003).

*P. capsici*, infects cucurbitaceous and solanaceous crops. In free water, sporangia of *P. capsici* release zoospores which may be dispersed by moving surface water. In Michigan counties with a history of *P. capsici* susceptible crop production, surface irrigation sources such as river system, ponds and ditches were monitored for the presence of the pathogen during four growing seasons (2000–2005). Pear and cucumber baits were suspended in water at monitoring sites at three to seven day intervals. The baits, after washing, were plated on water agar amended with rifampicin and ampicillin. *P. capsici* was detected at monitoring sites in two or more years, even when nonhost crops were planted nearby. *P. capsici* isolates resistant to mefenoxam were common in water sources. Most monitoring sites yielded isolates of a 1:1 ratio of A1:A2 mating types (MTs). Frequent detection of *P. capsici* in surface water used for irrigation in vegetable growing areas suggested that irrigation water could be an important mode of dispersal (Gevens et al. 2007). Contaminated irrigation water has been demonstrated, using an *in situ* baiting bioassay, to be an important source of inoculum for *Phytophthora* spp., capable of infecting several economically important crops. The effects of bait depth and duration and growth media used for detection of *Phytophthora* spp. in irrigation run-off containment basin on the efficiency of bait species viz., *Camellia japonica*, *Ilex crenta* and *Rhododendron catawbiense* and bait type (whole leaf/leaf disk) were investigated. Rhododendron leaf baiting for seven days was more efficient in recovering the greatest diversity and populations of *Phytophthora* spp. with minimum interference due to the preference of another oomycetes *Pythium* spp. The flexible (two-rope) bait deployment system was more effective than the fixed (one-rope) system, minimizing the risk of bait loss and allowing the baits to remain at designated depths from the surface under inclement weather. In addition, the flexible bait-deployment system might also be employed to investigate the horizontal and vertical distribution of *Phytophthora* spp. in containment basins (Ghimire et al. 2009). Different *Phytophthora* species have been recovered from plants, soil and waterways in nursery production systems. Species of *Phytophthora* recovered from samples vary with location, sample source and season. Some of them have been shown to be pathogens of crop plants, whereas other species were found to be nonpathogenic to plant species tested. Hence, the survey for *Phytophthora* species in the containment basins at one nursery each in Alabama and Mississippi showed that eight species and one taxon of *Phytophthora* were present in the water samples. *Phytophthora gonapodyides* was dominant in cooler months and *P. hydropathica* in warmer months accounting for 39.6% and 46.6% overall recovery, respectively. Three new species, *P. macilentosa*, *P. mississippiae* and *P. stricta* were recovered from a small lake (M4) at the Mississippi nursery. None of the new species and other *Phytophthora* spp. was able to infect any of the plant species tested. However, the possibility of any one of the *Phytophthora* spp. Becoming a pathogen of crop plants in future cannot be ruled out (Copes et al. 2015).

*Fusarium* species were isolated from the Andarax River and coastal sea water of the Mediterranean provinces of Spain. Forty-one isolates representing nine *Fusarium* spp. isolated from sea and river waters were found to be pathogenic to barley, kohlrabi, melon and tomato. *F. acuminatum*, *F. chlamydosporum*, *F. culmorum*, *F. equiseti*, *F. verticillioides*, *F. oxysporum*, *F. proliferatum*, *F. solani* and *F. sambucinum* caused pre- and post-emergence damping-off diseases. Pathogenicity of *Fusarium* isolates did not appear to be related to the origin of the isolates (sea water or fresh water). However, the presence of pathogenic species of *Fusarium* in river water flowing to the sea might indicate long-distance dissemination of the pathogens in natural water environments (Palmero et al. 2009). Susceptible cultivars of *Rhododendron* and *Pieris japonica* were used as trap plants at two commercial nurseries irrigated with water naturally infested with *Phytophthora* spp. during 2011 and 2012 growing seasons to assess the frequency of disease incidence. *Phytophthora* spp. were consistently recovered from water samples at every collection time, but detected on only two of 384 trap plants during the two growing seasons. *P. hydropathica* and *Phytophthora* taxon Pg Chlamydo recovered from water were foliar pathogens and neither species was able to cause root rot on these hosts. On the other hand, *Rhododendron* plants inoculated with *P. cinnamomi* (positive control) developed root rot symptoms. Although none of the *Phytophthora* taxa tested caused root infection on *Rhododendron* and *Pieris* spp., three of the four taxa were detected in the root zone at six weeks postinoculation by the *Rhododendron* leaf bait method. *P. pini* isolate was recovered from 100% of the root balls of asymptomatic *Rhododendron* and *Pieris* plants baited with *Rhododendron* leaves. The results indicated that *Phytophthora* spp.-infested irrigation water did not act as a primary source of inoculum for *Rhododendron* and *Pieris* root infection leading to root symptoms (Loyd et al. 2014).

### 2.1.2 PHYSICAL AND CHEMICAL METHODS

Techniques based on the physical and biochemical properties of fungal pathogens have been evaluated for their usefulness for the detection and identification of soilborne fungal pathogens. Recognizing the infection of plant roots by soilborne microbial pathogens is very difficult, as infection of roots/collar regions is initiated much earlier than the appearance of symptoms on aerial plant parts. Visible symptoms may be observed, only after the pathogen has spread well within the infected plants, making the possibility of saving such plants remote. Remote sensing is an alternative method for nondestructively assessing plant diseases rapidly, repeatedly and over a large area without physical contact with the sample. This technique involves measurement of reflectance of electromagnetic radiation from the vegetation (plant population) usually in the invisible (390–770 nm), near infrared (NIR, 770–1,300 nm) or middle infrared (1,300–2,500 nm) ranges (Nilsson 1995). The potential for detecting rice sheath blight disease using multispectral remote sensing was demonstrated by Qin and Zhang (2000). The comparative efficacy of visual and infrared application to assess the extent of root colonization of apple trees by *Phymatotrichopsis omnivora* was determined. The canopy temperature and air temperature differed significantly ($P < 0.01$) and this parameter was utilized as a basis for predicting infection in asymptomatic apple trees. Extensive tap root decay and infection of lateral roots were observed in these asymptomatic plants revealing the potential of infrared technique for detection of infection of roots early, facilitating the time of application of disease management strategies to contain further spread of root diseases in apple orchards (Watson et al. 2000). Rhizoctonia crown and root rot (RCRR) disease caused by *R. solani* AG2-2 became a serious limiting factor for sugar beet production. Remote sensing technique was evaluated for its efficacy in assessing RCRR disease of sugar beet. The sugar beet field plots were inoculated with *R. solani* AG2-2 IIIB at different inoculum densities at the ten-leaf stage of plant growth. Green, red and near-infrared reflectance and several calculated narrow band and wide band vegetation indices (VIs) were correlated with visual RCRR ratings, resulting in strong nonlinear regressions. Values of VIs were constant until at least 26 to 50% of root surface had rotted and then decreased significantly as RCRR ratings (0 to 7 scale) increased and plants began dying. The results showed that the remote sensing technique might be useful to detecting infection by RCRR, but not before initial appearance of foliar symptoms (Reynolds et al. 2012).

A gas chromatography technique was applied to characterize the fatty acid methyl esters (FAMEs) of isolates of *R. solani* AG-4 and AG-7 and analyzed with Microbial Identification System software. Palmitic, stearic and oleic acids were common in all isolates of both AGs AG-4 and AG-7 and accounted for 95% of C14 and C18 fatty acids present. Oleic acid present in both AG-4 and AG-7 isolates accounted for the highest percentages of total FAMEs and Key FAME ratios were analyzed with cluster analysis and principal components were plotted. Isolates of AG-7 from the United States appeared to be more closely related to each other than to AG-7 isolates from Japan and Mexico. These differences in FAMEs were sufficiently distinct, facilitating determination of geographical variability of the isolate. The results, however, indicated that variability of the FAMEs would not be suitable as the main diagnostic tool for distinguishing individual isolates of *R. solani* AG-4 from AG-7 (Baird and Mullinix 2000).

Fungal taxa may be distinguishable by the fatty acid profiles, if the growth substrate, incubation temperature and incubation time are kept constant. Fatty acid profiling is a relativey rapid, cost-effective and efficient method for characterization and differentiation of isolates of *Phytophthora infestans*, incitant of late blight diseases of potato and tomato. Two libraries of FAME profiles, ie., one representing average genotype characteristics and another representing individual isolate characteristics, were established from at least eight replicate samples of each of 25 different isolates of *P. infestans*, including representative isolates of US-1, US-6, US-7, US-8, US-11, US-14 and US-17 genotypes. All isolates of *P. infestans* produced eight primary fatty acids which accounted for 97 to 99% of total fatty acid contents. In addition, four additional fatty acids were also detected in a majority of isolates, but as minor components (< 1% of total content). Further, other oomycetes, *P. capsici*, *P. erythroseptica*, and *Pythium ultimum* also contained these predominant fatty acids, but the relative percentages showed variations. *P. ultimum* contained minor components (< 1% content) fatty acids that were not present in any *Phytophthora* isolates. Principal component analysis of the FAME profiles revealed distinct differences among *Phytophthora* spp., as well as considerable variability

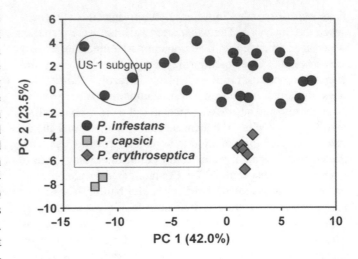

**FIGURE 2.4** Principal component (PC) analyses of fatty acid methylesterase (FAME) profiles of *Phytophthora* spp. Relationship of FAME profiles of *P. infestans*, *P. capsici* and *P. erythroseptica*, as represented by an ordination plot of PC1 and PC2 from PC analysis; values in parenthesis represent percentage of total variability accounted for by each PC.

**[Courtesy of Larkin and Groves (2003)) and with kind permission of the American Phytopathological Society, MN, United States]**

among *P. infestans* isolates (see Figure 2.4). Profiles of individual isolates of *P. infestans* maintained consistency and reproducibility. FAME profiling could be very useful for the differentiation of oomycetes pathogens. However, FAME profiling may be applied as a supplemental and complementary tool and cannot be considered as an alternative or replacement for more established means of cultural and genetic characterization (Larkin and Groves 2003).

Composition of cellular fatty acid analysis has been shown to be an useful basis to characterize, differentiate and identify other fungal pathogens also. Fatty acid profiles have been successfully used to characterize and differentiate *R. solani* AG 4. Fatty acid profiles of 70 isolates of *R. solani* AG-4 clustered into three groups corresponding to the homogeneous group (HG)-I, HG-II and HG-III. Isolates from Georgia peanuts exhibiting limb rot were characterized as gas chromatography (GC) subgroup 1 (GC-1) and contained HG-I isolates. Isolates from diseased soybean hypocotyls grown in North Dakota and sugar beet seedlings tap roots and tare soil in Minnesota and North Dakota were characterized as GC subgroup 2 (GC-2) and contained predominantly HG-II isolates, but included three distinct isolates based on FAME analysis and morphological characteristics. Fatty acid profiles of AG-4 isolates were compared with FAME library profiles of AG-1, AG-2 type 2 and AG-3. They were sufficiently different that they could be used to support speciation of this group from *R. solani*. The results suggested that the isolates of AG-4 identified as HG-II might be included in *R. practicola* (Stevens Johnk and Jones 2001). The efficacy of in planta quantification of *P. brassicae* present in plant or soil samples by whole-cell fatty acid (WCFA) estimation and real-time chain reaction (PCR). Arachidionic acid (ARA, 20:4) was the most abundant fatty acid in resting spores and was only present in infected roots, indicating the potential of ARA as a biomarker for *P. brassicae*. The concentration of ARA determined by WCFA analysis and the DNA assessed by real-time PCR in infected plants were well correlated. These assessments also correlated with soil spore content and the assessed disease incidence and disease severity scores. The results indicated the potential of WCFA analysis for the detection and differentiation of fungal pathogens (Sundelin et al. 2010).

Volatile organic compounds (VOCs) produced by healthy and infected plants may vary. The VOCs profiles of six *Phytophthora* spp. and one *Pythium* spp. were characterized by optimized head space solid-phase microextraction (HS-SPME)-GC method. The combination of the VOCs created a unique pattern for each pathogen. The chromatograms of different isolates of *P. cinnamomi* were the same and the specific VOC pattern of *P. cinnamomi* remained consistently independent of the growth medium used. The chromatograms and morphological studies showed that *P. cinnamomi* released specific VOCs at different stages of colony development. Using the optimized HS-SPME-GC procedure, identification of *P. cinnamomi* from 15 *in vivo* infested soil samples was as high as 100%, indicating the potential of this method for physiological investigations on *P. cinnamomi* (Qiu et al. 2014a). The HS-SPME technique was applied to collect VOCs from *P. cinnamomi*-infected lupin seedling. The HS-SPME samples taken at different periods after inoculation with the pathogen were analyzed by gas chromatography-flame ionization detector (GC-FID) procedure to determine the difference in the volatile compounds between healthy and infected plants. Three specific peaks were identified after 52 to 68 h with infected plants at this sampling time, and no visible symptom on aerial plant organs could be observed. Of the 87 VOCs identified, the five most abundant compounds viz., 4-ethyl-2-methoxyphenol, 4-ethylphenol butyrolactone, phenylethylalocohol and 3-hydroxy-2-butanone were unique to all five inoculated substrates. These metabolites could be used as markers for the identification of *P. cinnamomi* in different growth environments. The results indicated that volatile organic profiles produced by the pathogen might be used as an early warning system to predict the presence of the pathogen in contaminated field samples (Qiu et al. 2014b).

Acid electrolyzed water (AEW) is a germicidal product of electrolysis of a dilute solution of sodium chloride (0.4%, v/v) and it contains a low concentration of hypochlorous acid which is a highly effective sanitizer at low pH. AEW containing 0.4% sodium chloride could be used to disinfest wheat seed or soil samples of *Tilletia indica*, causal agent of Karnal bunt disease of wheat. AEW and sodium hypochlorite (NaOCl), when tested individually, eliminated more than 99.6% of all contaminating fungi and bacteria from extracts of four different soils on antibiotic water agar (AWA). PDA containing ampicillin sodium salt and streptomycin sulfate (PDAA) medium was used to score total viable counts of fungi and in nutrient broth yeast agar containing cycloheximide (NYBC) medium to determine total viable counts of bacteria. The results showed that 98.47 to 99.05% of fungi and

99.64 to 99.94% of bacteria were eliminated by AEW treatment. NaOCl (0.4%) treatement eliminated 97.12 to 99.10% of fungi and 79.94 to 98.23% of bacteria. A comparison of numbers of fungal and bacterial colonies growing on PDAA dishes seeded with wheat washes, treated with pH 2.5-buffered water, and untreated controls indicated that acidic conditions in combination with antibiotics eliminated 46 to 87% of fungi and more than 99% of bacteria, respectively, whereas both AEW and 0.4% NaOCl treatments eliminated 100% of both fungi and bacteria. The procedure developed in this investigation was useful to detect and enumerate viable teliospores of *T. indica* in wheat seed or soil extracts. The germicidal effects of AEW revealed its potential as an alternative to seed disinfectants used previously (Bonde et al. 2003).

Isozymes are multiple molecular forms of a single enzyme and these forms have similar enzymatic activities, but have slightly different amino acid sequences. Isozyme analysis has been shown to be a powerful biochemical approach effective for the detection, identification and differentiation of morphologically similar or closely related species, varieties or *formae speciales*. Protein profiles of a fungal pathogen may be obtained by extracting proteins from fungal cells and subjecting the extract to sodium dodecylsulfate (SDS)-polyacrylamide gel electrophoresis (PAGE) analyses. The separated proteins are stained and the patterns of protein bands are compared visually or by computer-based analysis for determining the differences between isolates of a fungal pathogen. Fingerprints obtained by SDS-PAGE technique may indicate the presence of multiple molecular forms (isozymes) of certain enzyme(s) produced by the fungal pathogen concerned. This technique is relatively less expensive and faster than immunoassays or restriction fragment length polymorphisms (RFLP) technique (Schulze and Bahnweg 1998). Downy mildew diseases infect cereals, reducing yield levels drastically. It is difficult to identify three species of *Peronosclerospora*, since wide variations in the morphological characteristics exist within a species. Isozymes from ten fungal cultures of *P. sorghi*, *P. sacchari*, and *P. sacchariphilippinensis* were compared electrophoretically on starch gels for species identification. No influence could be seen due to host plant species on which the pathogen was maintained, the time of year of conidial collection or growth conditions (greenhouse or growth chamber) on 11 enzymes investigated. Isozyme analysis was effective for differentiating all three *Peronosclerospora* spp. complex. Isozyme analysis has to be used as an adjunct to morphological characteristics that constitute the basis of classical systematics of fungal pathogens (Bonde et al. 1984). In a later investigation, isozyme patterns were found to be useful to distinguishing teliospores of *T. indica* and *T. barclayana* present in stored grains (Bonde et al. 2003). These two pathogens could not be differentiated based on morphological characteristics resulting in rejection of wheat consignment, since many countries had banned import of wheat grains, because of the presence of *T. indica* which is a quarantinable pathogen.

Isolates (162) of three species of *Phytophthora*, *P. cambivora*, *P. cinnamomi* and *P. cactorum* from a wide range of host plants and geographical locations were investigated to determine isozyme variation within and among them. These isolates were compared, using 18 isozyme loci separated by starch-gel electrophoresis. *Phytophthora* spp. could be separated clearly based on isozyme analysis and each species was further subdivided into electrophoretic types (ETs). In *P. cinnamomi* and *P. cactorum*, eight and two ETs were recognized, while *P. cambivora* separated into eight ETs. By employing cellulose acetate electrophoresis (CAE) technique, three enzymes viz., phosphoglucose isomerase, malate dehydrogenase and lactose dehydrogenase were shown to possess diagnostic value, permitting clear discrimination of these three *Phytophthora* spp. Isolates of *P. cactorum* showed the lowest level of diversity compared to other two *Phytophthora* spp. Two MTs of A1 and A2 of *P. cinnamomi* exhibited distinct isozyme phenotypes, based on the migration pattern of isocitrate dehydrogenase (Oudemans and Coffey 1991a). Intraspecific relationships could be established among 12 papillate species of *Phytophthora* spp. based on the ability to cluster together, indicating close genetic relatedness, whereas *P. kasturae* and *P. heveae* formed single cluster (Oudemans and Coffey 1991b). Isozyme analysis may be useful to resolve the differences in the pathogenic potential (virulence) of some fungal pathogens. The majority of the isolates (68) of *Leptosphaeria maculans*, infecting canola (oilseed rape) had a single isozyme of glucose phosphate isomerase (GPI) which migrated faster in starch gel (70 mm) than the isozyme present in the rest of the isolates (24) that could move only 65 mm in 11 h. The isolates of *L. maculans* were grouped into ETs, ET1 and ET2 clusters, indicating their fast and slow movement in the electrophoretic gels. Highly virulent strains contained fast moving GPI (ET1). On the other hand, weakly virulent strains were able to move only slowly forming the ET2 cluster (Sippel and Hall 1995). In a later investigation, a reliable procedure to identify *L. maculans* strains, based on GPI electrophoresis on starch gels was developed. The extracts of lesions induced by *L. maculans* were directly subjected to electrophoresis. Four different ET patterns, ET1, ET2, ET3 and ET4 of allozymes were recognized and these patterns correlated with virulence of the isolates of the pathogen. Highly virulent and weakly virulent isolates were placed in group A and B, respectively, while isolates inducing a few typical and atypical leaf lesions produced ET3 pattern. *Pseudocercosporella capsellae* isolates infecting canola produced ET4 pattern which was distinct from those produced by isolates of *L. maculans*. The allozyme of isolates belonging to ET4 pattern was the fastest, permitting the differentiation of *L. maculans* from *P. capsellae*, infecting canola plants (Braun et al. 1997).

Incidence of white root rot (WRR) disease of avocado, posing serious production threat was investigated, using a nondestrtuctive procedure. Progression of WRR disease was monitored at the leaf and root levels by a combination of chlorophyll fluorescence measurements and conforcal laser-scanning microscopy on avocado genotypes susceptible to *Rosellinia necatrix*. Leaf photochemistry was affected at early stages of disease development prior to the appearance of above ground symptoms. Significant decreases in the trapping efficiency of phytosystem-II ($Fv'/Fm'$) and in the steady-state of chlorophyll fluorescence yield (Fs) were revealed. Decreases in $Fv'/Fm'$) were associated with different degrees of fungal penetration, primarily in the lateral roots, but not in areas next to the main root collar. Above ground symptoms were observed only when the fungus reached the root collar. Leaf physiology was also tracked in a tolerant genotype where no changes were observed during disease progression, despite the presence of the pathogen in the root system. The results revealed the usefulness of this method for early detection of *R. necatrix* infection of avocado plants, facilitating rapid eradication of highly susceptible genotypes in root- stock avocado breeding programs (Martinez-Ferri et al. 2016). The ability of *Colletotrichum coccodes* and *Helminthosporium solani*, infecting potato tubers to modify the pH in around the infected site via alkalization or acidification was investigated. Direct visualization and estimation of pH changes near the inoculum site was accomplished by using pH indicators and image analysis. The pH of the area infected by either fungal pathogen increased from potato native pH of approximately 6.0–7.4 to 8.0. Using the image analysis, the growth curves of the fungal pathogens and estimates of the log phase of their radial growth could be obtained: ten days for *C. coccodes* and 17 days for *H. solani*. Further, a distinctive halo (an edge area with increased pH) was observed only during the lag phase of *H. solani* infection. During pathogen-host interaction, pH modulation is a major factor and the changes in pH could be monitored by the methods employed in this investigation (Tardi-Ovadia et al. 2017).

### 2.1.3 IMMUNOASSAYS

Isolation of fungal pathogens by soil dilution plating often provided inconsistent and variable results, based on the methods and media used for isolation. In order to obtain more reliable and reproducible results, immunoassays, depending on the reaction between pathogen-specific antibodies and antigenic proteins of the fungal pathogen species, were developed. Both polyclonal antibodies (PAbs) and monoclonal antibodies (MAbs) have been prepared for specific detection of various fungal pathogens present in plant tissues and/ soil on which infected plants have been grown in the greenhouse/field. The principles of immunological reactions, their applications, advantages and limitations, different immunoassays available for detection and identification of fungal pathogens have been discussed in detail in previous publications (Narayanasamy 2005, 2011). The salient features of immunoassays that have been applied for detection and differentiation of soilborne fungal pathogens, are highlighted.

Antigens are fairly large molecules or particles containing proteins or polysaccharides of pathogen origin and they are capable of inducing formation of specific antibodies (immunoglobulins) in the blood stream of animals immunized with the antigens. The ability of the antigen to induce antibodies is known as immunogenicity, whereas the property of the antigen to react with the antibody is termed as antigenicity. The antigenicity is due to the existence of specific sites in the antigen known as epitopes or antigenic determinants. The antibodies are produced by lymphoid (B) cells or ß-lymphocytes and they are circulated in the bloodstream of immunized animals. Antibodies belonging to a group of proteins designated immunoglobulins (IgG) have similar basic structure constituted by one pair of identical light (L) chain (MW 25,000) and heavy (H) chain (MW 50,000). These two chains are linked together by noncovalent forces and disulfide bonds. Each light and heavy chain has a variable ($V_L$) and constant regions. The VL region shows extensive variation. There are different classes of antibodies (IgG, IgM, IGA, IgD and IgE) defined based on the constant region located in the C-terminus of the antibody molecules. The sensitivity of the immunoassays may be substantially enhanced by using purified immunoglobulins instead of whole antiserum (Van Regenmortel 1982). PAbs are prepared by immunizing different animals such as rabbits, mice, fowl and horses, depending on the quantity of antiserum required. Preservatives such as glycerol, phenol or sodium azide may be added to preserve the serological activities of antisera. MAbs have been prepared by employing hybridoma technology introduced by Kohler and Milstein (1975). The hybridoma, obtained by the fusion of β-lymphocytes (antibody- producing cells) and myeloma cells (capable of multiplying indefinitely), secretes and ensures continuous supply of specific MAbs. Each hybridoma clone produces identical antibodies that are specific for a single epitope present in the antigen. Phage-displayed recombinant antibodies are produced by employing M13 phage infecting *Escherichia coli*. Components of M13 such as phage DNA, gene 8 coat proteins, gene 3 attachment proteins and other proteins that may be fused with the phage components, can be continuously produced in the bacterial cells without undergoing lysis. By genetically linking the DNA from antibody-producing β-lymphocytes or hybridomas with phage gene 3 DNA, the recombinant antibodies can be produced. Any antibody DNA linked to phage DNA and antibody protein fused to phage proteins may be assembled and secreted just like phage proteins. Antibodies displayed on phage can be produced more rapidly and less expensively compared to MAbs. These recombinant antibodies are as specific as MAbs for identification of microbial plant pathogens in different substrates.

### 2.1.3.1 Enzyme-Linked Immunosorbent Assay

ELISA and its variants have been applied extensively for the detection of soilborne fungal pathogens present in plants, soil and water (see Table 2.1). MAbs have been useful in discriminating biomass components or metabolic products like mycotoxins by raising the antibodies against constitutively expressed antigens that were secreted during hyphal development. The specificity of the antibodies primarily determines the sensitivity of pathogen detection. As the immunoassays fail to differentiate viable and dead pathogen cells, they have to be employed as a secondary confirmatory step, in addition to baiting and plating assays. Species-specific and subspecies-specific MAbs were employed for the detection of *P. cinnamomi* and inclusion of glutaraldehyde in fixative improved the specificity of the reaction, permitting *in situ* detection

of the pathogen in plant tissues (Hardham et al. 1986, 1991). Comparative efficacy of two commercial ELISA kits and the culture plate method for detection of *P. cinnamomi* in root tissues of inoculated azaleas was assessed. Both multiwell kits and rapid assay F kit detected *P. cinnamomi* on azalea roots, beginning one week after inoculation. The results of ELISA kits were in agreement with that of culture plates consistently from three to five weeks after inoculation. Root symptoms of PRR were visible at this stage, but not foliar symptoms of infection. A positive correlation was observed between root rot severity and ELISA absorbance values in the greenhouse trials. The multiwelll kit detected *P. cinnamomi* in root samples containing 10% infected root tissue. The rapid assay kit was faster and easier to use, providing results in a short time (Benson 1991).

The comparative efficacy of multiwell ELISA Kit (Agri-Screen) and 'flow through' immunoassay in detecting other *Phytophthora* spp. has been assessed. Multiwell ELISA Kit was effective in detecting *P. capsici* in pepper tissues only, but had higher absorbance values for healthy tissues. On the other hand, 'flow through' immunoassay could be performed easily and the results could be obtained rapidly in about ten min for the detection of *P. capsici* in cucurbit crops. The results of ELISA tests were in agreement with those of plating assay, using a semiselective medium (Miller et al. 1994). For the detection of *Phytophthora* spp. in commercial nurseries equipped with water recirculation system, double antibody sandwich (DAS)-ELISA format was useful. However, *Rhododendron* leaf test was found to trap the widest range of *Phytophthora* spp. and it was more efficient than DAS-ELISA format (Themann et al. 2002). A polyclonal antiserum was raised against water soluble proteins from *P. cinnamomi* mycelium. When the antiserum at a dilution of 1:10,000 was used in the ELISA test, clear-cut discrimination among congenerous species was not observed, although the antiserum showed higher activity against *P. cinnamomi*. The antiserum gave positive reactions in Western blot analyses against mycelial proteins from nine species of *Phytophthora* and *Pythium*, but not with *R. solani*, *V. dahliae*, *F. oxysporum* and *Cryphonectria parasitica*. All *Phytophthora* species showed common epitopes on proteins of molecular masses 77-, 66-, 51- and 48-kDa. However, the species-specific protein of 55-kDa was immunodecorated only in *P. cinnamomi*, providing reliable identification of *P. cinnamomi*. When tested against total proteins from the fungi grown on water, the antibody revealed diagnostic bands of 55- and 51-kDa only in *P. cinnamomi*. The results showed that the antiserum might be employed for specific identification of *P. cinnamomi* emerging in distilled water from infected tissues of chestnut, blueberry and azalea (Ferraris et al. 2004). Antiserum produced with β-D-galactosidase-labeled antirabbit IgG was employed as secondary antibody for the detection of *F. oxysporum* f. sp. *cucumerinum* (FOC) causing wilt disease in cucumber. Cell fragments of the pathogen attached to balls (Amino Dylark) were used as solid-phase antigen in ELISA tests. This test provided highly specific and sensitive detection of FOC (Kitagawa et al. 1989). The PAbs reacting with purified exopolygalacturonase (exoPG) produced by

**TABLE 2.1**

**Detection of Fungal Pathogens in Different Substrates by Immunoassays**

| Immunoassay | Pathogen detected | Substrate | References |
|---|---|---|---|
| ELISA formats | | | |
| DAS-ELISA | *Aphanomyces euteiches* | pea | Kraft and Boge 1994 |
| | *Fusarium oxysporum* f.sp. *cucumerinum* | culture | Platiño-Álvarez et al. 1999 |
| | *Phytophthora* spp. | water circulating system | Themann et al. 2002 |
| | *P. cinnamomi* | roots of azalea | Benson 1991 |
| | | water | Ferraris et al. 2004 |
| | *P. citrophthora* | citrus roots, soil | Timmer et al. 1993 |
| | *P. fragariae* var. *rubi* | raspberry roots | Olsson and Heiberg 1997, Pekarova et al. 2001 |
| | *P. ramorum* | culture | Kox et al. 2007 |
| | *Pythium ultimum* | sugar beet roots | Yuen et al. 1993, 1998 |
| | *P. violae* | carrot roots | Lyons and White 1992 |
| | *Macrophomina phaseolina* | cowpea roots, seeds | Afouda et al. 2009 |
| | *Rhizoctonia solani* | culture | Thornton et al. 1993, 1999 |
| | *Spongospora subterrranea* f.sp. *subterranea* | potato tubers, soil | Merz et al. 2005 |
| | *Verticillium dahliae* | potato, cotton, soil | Sundaram et al. 1991 |
| Indirect ELISA | *Pythium sulcatum* | carrot roots | Kageyama 2002 |
| | *Plasmodiophora brassicae* | cabbage roots | Oriharra and Yamamoto 1998 |
| Direct antigen coating (DAC)-ELISA | *Polymyxa graminis* | sorghum roots | Delfosse et al. 2000 |
| ELISA Kits | *Phytophthora* spp. *Pythium* spp. | recycled water | Ali-Shtayeh et al. 1991 |
| Fungal capture sandwich ELISA | *Thielaviopsis basicola* | tobacco roots | Holtz et al. 1994 |
| Dot immunobinding | *Plasmodiophora brassicae* | cabbage roots | Orihara and Yamamoto 1998 |
| | *Phytophthora nicotianae* | tobacco roots | Robold and Hardham 1998 |
| | *Fusarium oxysporum* f.sp. *lycopersici* | tomato | Arie et al. 1998 |
| SssAgriStrip | *Spongospora subterranea* f.sp. *subterranea* | potato tubers, soil | Bouchek-Mechiche et al. 2011 |
| Immunofluorescence assay (IFA) | *Thielaviopsis basicola* | tobacco roots | Holtz et al. 1994 |
| | *Phytophthora cinnamomi* | chestnut, blueberry azalea | Gabor et al. 1993 |
| | *Polymyxa graminis* | cysts and resting spores | Delfosse et. al. 2000 |
| Dipstick immunoassay | *Phytophthora cinnamomi* | soil | Cahill and Hardham 1994 |
| Tissue blot immunoassay | *F. oxysporum* f.sp. *lycopersici* | Tomato | Arie et al. 1998 |
| | *P. brassicae* | turnip roots, stems | Yamamoto et al. 1998 |
| Zoospore trapping immunoassay (ZTI) | *Phytophthora* spp. | water | Pettitt et al. 2002 |
| | *Pythium* spp. | water | |
| Lateral flow devices (LDFs) | *Rhizoconia solani* | soil | Thornton et al. 2004 |
| | *Phytophthora ramorum* | plants | Lane et al. 2007 |
| | *P. kernoviae* | plants | |

*F. oxysporum* f.sp. *radicis-lycopersisci* (FORL) were effective in detecting the pathogen infecting tomato plants (Platiño-Álvarez et al. 1999). Commercial ELISA kits were used to determine propagule densities of *Phytophthora citrophthora* infecting citrus and the pathogen populations were quantified in citrus roots and soil samples (Timmer et al. 1993). For the detection of *P. fragariae* var. *rubi* in raspberry root tissues, a commercial multiwell ELISA assay was employed. The pathogen was detected at four days after inoculation (DAI) with a detection limit of about 0.25% of simulated infection level (percentage of infected tissue/healthy tissue, w/w) (Olsson and Heiberg 1997). Both PAbs and MAbs raised against specific proteins of *P. fragariae* var. *rubi* were used for the efficient detection of the pathogen in strawberry roots (Pekarova et al. 2001). The comparative diagnostic efficiency of ELISA and TaqMan PCR assays was assessed for the detection of

*Phytophthora* at genus level and *P. ramorum* at species level. Both assays had higher sensitivities for genus level detection than species-specific detection, indicating their usefulness for prescreening of isolates of *Phytophthora* spp. (Kox et al. 2007).

ELISA formats have been applied for the successful detection of *Pythium* spp., causing damping-off diseases in the nurseries of several crops. *Pythium violae*, a soilborne oomycete, was detected in field-grown carrots using PAbs (Lyons and White 1992). *P. ultimum*, infecting sugar beet seedlings could be detected by employing a MAb highly specific to 21 isolates of *P. ultimum*. The MAb did not react with any of the 16 *Pythium* species tested by ELISA. Presence of *P. ultimum* was efficiently detected in the roots of sugar beet seedlings with more than two infections/10 cm of root (Yuen et al. 1993). The PAb generated against cell walls of *P. ultimum* as the capture antibody and an MAb specific for recognition were employed in indirect DAS-ELISA format for the detection and quantification of the pathogen biomass. Culture filtrates of seven isolates of *P. ultimum* showed positive reactions in the tests. The pathogen could be detected in the roots of sugar beet, beans and cabbage seedlings grown in infested soils. Roots with one infection/100 cm of root tissues could be recognized using the ELISA test (Yuen et al. 1998). For the detection of *Pythium sulcatum*, infecting carrots, MAbs were generated against surface antigens from the pathogen. The MAbs showed high specificity to seven isolates of *P. sulcatum* among 26 species of soilborne fungi. Weak cross-reactions with *P. aristosporum*, *P. myriotylum* and *P. zingiberum* were observed in indirect ELISA format. However, no positive reactions with these *Pythium* spp. was indicated in Western blot analysis. *P. sulcatum* was detected in naturally infected carrot tissues and soil using indirect competition ELISA in which the glycoproteins present in cell wall of the pathogen were recognized by the MAbs, resulting in positive results (Kageyama et al. 2002).

*R. solani* isolates are grouped into AGs, based on their ability or inability to anastomose with known standard isolates. The possibility of distinguishing the AGs of *R. solani*, using PAbs in immuodiffusion tests was indicated by Adams and Butler (1979). In a later investigation, MAbs raised against total secreted proteins were generated. These MAbs showed higher levels of specificity for AG-8 isolates and cross-reacted with only few isolates of other AGs. An IgM MAb recognized a 40-kDa protein specific to AG-8 isolates, while the IgG MAb was more specific by reacting with 38-, 40- and 55-kDa proteins from AG-8 isolates (Matthew and Brooker 1991). ELISA test was applied for the detection of *R. solani* infection in poinsettia (Benson 1992). Specific recognition of the antigen from *R. solani* by MAbs was utilized for the detection of the pathogen present in the field soil. In a further investigation, the effectiveness of combining baiting and double MAb-ELISA test was demonstrated. The protocol developed, allowed recovery of *R. solani* isolates from colonized baits for determination of their AG affiliation and pathogenicity. Isolates of *R. solani* pathogenic to lettuce were identified as AG-4 group (Thornton et al. 1993, 1999). *Macrophomina phaseolina*, causing charcoal rot disease of cowpea is soil- and seedborne and has a wide host range of over 300 plant species, including legumes and cereals. A double DAS-ELISA was developed for specific detection and quantification of the pathogen biomass in plants. Both PAbs and MAbs generated against immunogens from mycelium and culture filtrates of *M. phaseolina* detected the pathogen in mycelial and plant extracts. Two antisera raised against mycelial antigens showed highest titer in DAS-ELISA for detection of the pathogen. The detection limit of antiserum against mycelial extracts was 15–30 ng protein of antigen/ml. DAS-ELISA test detected *M. phaseolina* in cowpea seed lots. The sensitivity of ELISA was greater than the agar plating assay, since ELISA detected the pathogen in a higher percentage of seeds. Symptomless infection of cowpea plants could be recognized by ELISA. Detection of *M. phaseolina* by ELISA

was possible in samples in which the β-1,3-glucanase test failed. Using ELISA test, *M. phaeolina* was found to be essentially localized in roots and hypocotyls of infected plants. In symptomatic cowpea plants, the pathogen was quantified in leaves, epicotyl, hypocotyls and roots by DAS-ELISA at one month after seed or soil inoculation. The protocol for ELISA test was particularly suitable for the detection of the pathogen in asymptomatic plants that may serve as source of inoculum for further spread of the disease (Afouda et al. 2009).

Purified mycelial proteins of *V. dahliae* were used as immunogens to generate PAbs. The antiserum reacted positively with 11 of 12 isolates from potato, cotton and soil samples. But the PAbs did not react with the tomato isolate of *V. dahliae*. However, the antiserum could be employed for the detection of *V. dahliae* or *V. albo-atrum* in infected roots and stems of potato (Sundaram et al. 1991). The effectiveness of ELISA test for the detection of *Aphanomyces euteiches*, causing common root rot of peas was demonstrated. The polyclonal antiserum strongly reacted with mycelial antigens and zoospores of *A. euteiches*. The specificity of PAbs was revealed by the absence of reaction with other species of *Phytophthora*, *Fusarium* and *Pythium* tested. The levels of resistance of pea germplasm lines were assessed by determining pathogen biomass in pea lines with differing levels of resistance/susceptibility to *A. euteiches*. Lines PI 180693 and 90–2131 showed lower absorbance values at 405 nm, indicating the presence of lower amounts of biomass, compared to susceptible pea lines. The ELISA readings were correlated positively to amounts of antigen (Kraft and Boge 1994). A fungal capture sandwich ELISA involving the use of a polyclonal antiserum prepared against soluble protein extracts of chlamydospores and mycelium of *T. basicola*, causative agent of cotton black rot disease, was applied. The purified IgG fraction from the antiserum was biotin-labeled and employed in this assay. Both brown and gray cultural types of *T. basicola* could be detected by the ELISA format. There was negligible cross-reaction with other soilborne fungi present in cotton field soil. The detection limit of this assay was between 1 and 20 ng of pathogen protein. The pathogen could be detected in the root tissues at 2 DAI of roots. The result of DAS-ELISA and immunofluorescence assay corroborated each other (Holtz et al. 1994). The ELISA tests were effective in detecting the resting spores of *P. brassicae brassicae* in plant tissues. Using the purified resting spores as immunogen, antiserum was prepared by immunizing rabbits. The ELISA format using the antiserum detected the pathogen in samples containing a concentration of $1 \times 10^2$ to $1 \times 10^3$ resting spores in the homogenates of club root-infected roots of cabbage. Further, the concentration of resting spores of *P. brassicae* in soil was quantified, using the seedling bait assay which was time-consuming and inconsistent in providing reliable results. Hence, an indirect ELISA format was developed for specific detection of *P. brassicae* in artificially infested soil samples by the antiserum prepared against purified resting spores of *P. brassicae*. The detection limit of the ELISA format was $1 \times 10^4$/ml of the soil suspension. The resting spore density in soil that could cause economical loss was estimated as $1 \times 10^3$

to $1 \times 10^4$ spores/g of dry soil. The comparative efficacy of dot immunobinding assay (DIBA) was also assessed. DIBA test could detect *P. brassicae* in soil samples as efficiently as ELISA format with the same detection limit of $1 \times 10^4$ spores /ml in artificially infested soil samples (Orihara and Yamamoto 1998).

*Spongospora subterranea* f.sp. *subterranea (Sss)* causes the powdery scab disease of potato tubers and it is difficult to control the disease, as it is soil- and tuberborne. Early detection and precise identification of the pathogen in soil and tubers will be helpful to initiate disease management measures early. A MAb was generated, using cystosori of *Sss* as immunogen. ELISA test detected less than one *Sss* cytosorus and recognized *Sss* infected material from different countries. No cross-reaction was observed with other members of Plasmodiophoromycetes, including *P. brassicae*, *Polymyxa betae*, *Polymyxa graminis* and different *Streptomyces* species, causal agents of common and netted scab of potatoes. A novel method of taking tuber samples was developed, using a kitchen peeling machine. This procedure detected infection in two tubers mixed with 18 uninfected tubers. When soil samples spiked with cystosori were tested with MAb, different *Sss* infestation levels could be differentiated (Merz et al. 2005). Visual assessment of potato tuber infection by *S. subterranea* f.sp. *subterranea* (Sss) might lead to misidentification, due to the difficulty of distinguishing lesions induced by Sss/*Streptomyces* spp. DAS-ELISA and '*Sss*AgriStrip', a rapid and laboratory-independent assay based on lateral flow immunoassay were compared for their efficacy in detecting *Sss* in tubers, showing typical and atypical (suspicious) symptoms. DAS-ELISA and *Sss*AgriStrip assays were equally sensitive with a detection limit between 1 and 10 cystosori per ml buffer. Both methods failed to assign the tubers with doubtful symptoms to either powdery or common scab. Isolation and molecular methods (PCR and real-time PCR assays) confirmed that the atypical lesions were mostly caused by *Streptomyces* spp. (Bouchek-Mechiche et al. 2011).

*P. graminis* and *P. betae* are obligate root pathogens of cereals and sugar beet, respectively. In addition, they can function as vectors of many plant viruses attracting the attention of plant pathologists as pathogens and vectors to develop strategies to reduce their populations in soil. The resting spores of *P. graminis* were extracted from sorghum root tissues and used as immunogen to raise PAbs. In direct antigen coating (DAC)-ELISA format, the antiserum could detect one sporosorus per well of the ELISA plate. Likewise, DAC-ELISA test detected one sporosorus/mg of dried sorghum roots spiked with pathogen propagules. Isolates of *P. graminis* from Europe, North America and India reacted strongly with the antiserum. The antiserum cross-reacted with *P. betae* to an extent of 40%. However, there was no reaction with other root-infecting fungi or with the obligate parasite *Olpidium brassicae*. But the assay showed similar absorbance values, when two isolates of *Spongospora subterranea* were tested. DAC-ELISA format was useful to detect various stages in the life cycle of *P. graminis* and to detect infection under *in vivo* and *in vitro* conditions (Delfosse et al. 2000). The presence of

*P. betae* was detected in infected beet roots and quantified the pathogen population using ELISA format. The MAb and PAb raised to a recombinantly expressed glutathione-S-transferase (GST) from *P. betae* were used in the test. A close correlation was observed between the number of *P. betae* zoospore in serially diluted suspension and absorbance values in the assay. The resistance levels of accessions of wild sea beet (*Beta vulgaris* ssp. *maritima*) could be assessed, using the ELISA format (Kingsnorth et al. 2003).

The fungal pathogens may spread through irrigation water within a field and also from one field to another field, if the fields are irrigated with water from canal, river or ponds/lakes. The oomycetes require free water for zoospore release and movement to reach plant roots. Recycled water contaminated with fungal pathogens can be an important source of inoculum for nurseries. Plating and baiting assays for detection of fungal pathogens in water are time-consuming and often they do not provide repeatable results. Hence, commercially available ELISA tests, Pythium C Kits and Phytophthora E Kits were employed to detect pathogen propagules in water samples. The propagules were concentrated by filtering through 0.45 μm filters. Heating filter residues to 100°C for 6 min, yielded extracts that were equal or superior in reactivity to those obtained by liquid nitrogen disruption. Phytophthora E Kit gave reactions significantly (P < 0.05) above background, when extracts contained 30 or more zoospores. Absorbance values increased nearly linearly as zoospore numbers increased from 150 to 1,200 and began to level off at zoospore number > 1,200. A water sample collected in late winter from a recycling pond in a northern California nursery did not contain any viable pythiaceous fungi or detectable antigen. On the other hand, water samples collected in early spring from a southern California contained 442 viable propagules per liter of water. Using agar media *Phytophthora parasitica*, *P. citrophthora*, *P. cryptogea*, and an unidentified *Phytophthora* sp., in addition to *Pythium coloratum*, *P. rostratum*, *P. middletonii*, *P. ultimum* var. *sproangiferum* and *Pythium* L groups were recovered from water samples. All *Phytophthora* species recovered, reacted positively with Pythium Kits. All *Pythium* spp. recovered also reacted positively with Phytophthora Kits, but not with Pythium Kits. It appeared that *Pythium* spp. lacked detectable amounts of the antigen recognized by Pythium C Kit antibody, while they shared the antigen(s) detected by Phytophthora E Kit. An isolate of *P. coloratum* gave reactions very similar to those of *Phyophthora citrophthora*. The results indicated that appropriate detection kit should be used to detect the pythiaceous pathogens in water (Ali-Shtayeh et al. 1991).

### 2.1.3.2  Dipstick Immunoassay

Although ELISA formats have been found to be useful and effective for the detection and quantification of several soilborne fungal pathogens, these tests cannot be performed in the field. Hence, simple procedures that produce visible reactions under field conditions, were developed. In place of the 96-well microplates used in ELISA tests, dipstick immunoassay was found to be a suitable alternative for on-site detection of fungal pathogens. The dipstick immunoassay is based on the phenomena of chemotaxis and electrotaxis to attract the zoospores of fungus-like oomycete pathogens to a membrane on which they encyst and they are subsequently detected by the immunoassay. The chemotactic properties of zoospores are influenced by several chemicals such as amino acids, alcohols, phenols, pectin and abscissic acid (phytohormone). The nylon membranes with positive charge attract the zoospores released from the sporangia. Dipstick immunoassay had a detection limit of about 40 zoospores/ml and provided the results within 45 minutes. Immunolabeled cysts, attached to the membrane, could be visualized under the low power lens of light microscope, after silver enhancement of the gold-labeled secondary probe. *P. cinnamomi* was detected rapidly by employing specific MAbs (Cahill and Hardham 1994).

The efficiency of detection of *P. ultimum* var. *sporangiferum*, *Phytophthora cactorum* and *P. cryptogea* by two immunodiagnostic assays was compared with two conventional assays in dilution series in sterile water. The most sensitive assay for all four pathogen species was the zoospore trapping immunoassay (ZTI). Conventional membrane filtration dilution plating gave similar results, as in ZTI with two *Phytophthora* spp., but was less sensitive in *Pythium* detection. Immunodiagnostic dipstick assays and conventional bait test showed similar results in the dilution series and were generally about two orders of magnitude less sensitive than ZTI. For detection of the pathogens in water samples obtained from horticultural nurseries and *in situ* tests of infected root zones of tomato and chrysanthemum, ZTI was the most sensitive test for water samples. Dipstick and baiting assays were more effective for *in situ* testing. The antibody-based dipstick assay provided results similar to the baiting assays, using plant tissues for the detection of both *Pythium* spp. and *Phytophthora* spp. The dipstick immunoassays are simple, rapid and effective for *in situ* detection of soilborne pathogens under field conditions and have the potential for application in epidemiological investigations (Pettitt et al. 2002).

### 2.1.3.3 Lateral Flow Devices

Antibody-based diagnostic tests rely on enzyme reactions for the final visualization of results. To remove unbound enzyme, washing steps are necessary, thereby making the tests more time-consuming. Enzymes can be replaced by binding antibodies to particulate labels such as latex, colloidal gold or silica. Using these labels, the target antigen is visualized, when sufficient numbers of particles accumulate in one area, as in slide agglutination tests. Alternatively, particles can be captured by antibodies immobilized to membranes in lines or dots. Specific antigens are 'sandwiched' between the immobilized antibodies and the antibody-sensitized particles and immune complexes. As the amount of captured antigen increases, the reaction site becomes visible to the naked eye, once a sufficiently high density is reached. This principle underlies all lateral flow devices (LFD), rapid immunofilter paper assays (RIPAs) or immunochromatographic assays that employ particulate labels.

Colloidal gold was used as the reporter molecule, due to the restriction on the use of latex as a label in immunochromatographic LFD tests. A murine hybridoma cell line GD2 secreting surface antigens from an AG 4 isolate of *R. solani* was employed to develop a rapid immunochromatographic LFD for the detection of antigens from *R. solani* and other *Rhizoctonia* spp. The antigens from representative isolates of *R. solani* AGs 1, 2-1, 2-t, 3, 4, 5, 6, 7, 8, 9, 10, 11 and BI reacted positively in LFD tests. The antigens from related fungi *R. fragariae* and *Thanetophorus orchidicola* and *T. practicola* gave a positive reaction. However, some related fungi *R. cerealis*, *R. crocorum*, *R. oryzae* and *R. zeae* and other unrelated fungi failed to react with the MAb tested. The usefulness of the LFD to detect *R. solani* in soils naturally infested with *R. solani* AG 3 was demonstrated. The results of LFD tests correlated with that of conventional plate enrichment tests employing selective medium. The specificity of detection by LFD tests was confirmed by PCR assays, using specific primer pairs. LFD tests could be used to quantify *R. solani* AG 4 in artificially infested soil samples. Estimates of CFU obtained in LFD tests and those derived from plate-trapped antigen (PTA)-ELISA format incorporating MAbG2 were identical, indicating accuracy of results of LFD tests (Thornton et al. 2004). On-site equipment required expensive equipment and considerable expertise and they were not suitable for plant health inspection for screening a large number of plant samples at the time of inspection. Hence, the LFD was evaluated for detection of *Phytophthora* species in 634 samples. The assay was simple to use, provided results in a few minutes and a control line reacted positively, confirming the validity of LFD results. The diagnostic sensitivity of the LFD (87.6%), compared favorably with the standard laboratory methods, although the diagnostic specificity was not as stringent (82.9%). Overall efficiency of the LFD test was 95.6%, as compared with visual assessment of symptoms of between 20–30% for *Phytophthora ramorum* and *P. kernoviae*. The results revealed the usefulness of the LFD test for diagnosis of *Phytophthora* species in a large number of samples at the time of inspection and for employing it as a primary screen for selecting samples for laboratory testing to determine the identity up to species level (Lane et al. 2007).

### 2.1.3.4 Dot Immunobinding Assay

Nitrocellulose or nylon membrane is used in DIBA, in place of microtiter plates used in ELISA formats. The free protein-binding sites in the membrane have to be blocked with bovine serum albumin (BSA) or nonfat dry milk powder or gelatin. The unconjugated pathogen-specific antibody is allowed to react with immobilized antigen in the membrane. The trapped antibody is then probed with appropriate enzyme and in suitable substrate which permits visual detection of a colored product. The presence of resting spores of *P. brassicae*, incitant of club root disease of cabbage and other cruciferous crops was detected in infected root tissues by DIBA procedure (Orihara and Yamamoto 1998). The monoclonal antibody generated against zoospores of *Phytophthora nicotianae* recognized a 40-kDa glycoprotein present on the surface of zoospores provided specific detection of the pathogen infecting tobacco (Robold and Hardham 1998). The metabolite of pathogen origin may be detected in infected tissues. DIBA technique was employed to monitor development of tomato wilt disease caused by *F. oxysporum* f.sp. *lycopersici* race 2 by detecting endopolygalacturonase (endoPG) produced by the pathogen involved in pathogenesis (Arie et al. 1998). DIBA technique offers certain advantages over ELISA tests: (i) the membranes can be stored easily and transported as required during surveys; (ii) membranes can be more easily processed than the microtiter plates; (iii) DIBA can be completed in a shorter period compared to ELISA test; (iv) direct tissue blotting does not need any sample preparation and (v) antigen can be quantified, using a reflectance densitometer.

### 2.1.3.5   Tissue Blot Immunoassay

The antigen from freshly cut tissue surface from infected plant is transferred to nitrocellulose membrane to perform tissue blot immunoassay (TBIA). A tissue imprint of the cut tissue surface is made on the membrane by gently pressing the cut end. The free protein-binding sites present on the membrane are blocked as in DIBA test. Specific antibodies are allowed to react with the immobilized antigen from infected plant tissues. A direct tissue blot immunoassay (DTBIA) procedure was developed to detect *Fusarium* spp., causing wilt diseases, in the transverse sections from stems or crown of tomato and cucumber plants by employing the MAb (AP19-2) and FITC-conjugated antimouse IgM-sheep IgG. The results may be observed within 4 hours, facilitating early application of measures to check disease spread (Arie et al. 1995). Involvement of endopolygalacturonase (endoPG) produced by *F. oxysporum* f. sp. *lycopersici* race 2 during pathogenesis was indicated by applying TBIA procedure. The purified endoPG enzyme of the pathogen was used to generate the PAb APG1. Presence of the enzyme endoPG produced by *F. oxysporum* f.sp. *lycopersici* race 2 was detected by TBIA in the stem tissue of inoculated tomato plants (Arie et al. 1998). For the detection of *P. brassicae* by TBIA, roots including hypocotyls from healthy and club root-infected turnip plants were cut transversely starting 1 cm downward from the basal portion of the stems. The cut surfaces of root disks from healthy and infected plants were stamped into nitrocellulose membrane sheet and probed with pathogen-specific antibodies. The membranes blotted with infected tissue disks exhibited strong positive reactions throughout the blotting marks, indicating the presence of resting spores of *P. brassicae* in infected root tissues. The possibility of distinguishing the live from the dead fungal spores is a distinct advantage over ELISA tests (Orihara and Yamamoto 1998).

### 2.1.3.6   Western Blot Analysis

Fungal culture, plant or soil extracts are subjected to gel electrophoresis for separation of proteins of pathogen origin and the proteins of interest are transferred from gel to nitrocellulose membrane. They are subsequently probed by specific antibodies generated against the pathogen protein (antigen). Water soluble mycelial proteins from *P. cinnamomi* were employed as antigens to prepare a polyclonal antiserum (A379). This antiserum showed positive reactivity with mycelial proteins of *Phytophthora* spp. and *Pythium* spp. However, two protein bands of 55- and 51-kDa with diagnostic value were present only in mycelial extracts of *P. cinnamomi*. It was possible to detect and identify *P. cinnamomi*, emerging in distilled water from infected tissues of chestnut, blueberry and azalea (Ferraris et al. 2004).

### 2.1.3.7   Immunofluorescence Assay

Aldehyde-fixed zoospores and cysts of *P. cinnamomi* were employed to generate 24 MAb clones. Using immunofluorescence assay (IFA), 11 MAbs were found to be species-specific, reacting specifically with zoospores and cysts of *P. cinnamomi* only, but not with other fungi such as *Phytophthora* spp., *Pythium* spp., *Fusarium* spp., *Verticillium* spp., and *Rhizoctonia* spp. tested. One MAb was genus-specific, reacting with other *Phytophthora* spp. in both IFA and ELISA techniques (Gabor et al. 1993). The presence of *T. basicola* in cotton roots was also detected by IFA procedure (Holtz et al. 1994). Fluorescence antibody technique (FAT), involving the use of fluorescein-5-isothiocyanate (FITC)-labeled antibodies was evaluated for its efficacy in detecting resting spores of *P. graminis*, an obligate root pathogen of sorghum. The sporosori of *P. graminis* fluoresced with typical apple green color following staining with FITC-labeled specific antibodies. The majority of specific staining was restricted to the outer layers of the resting spores and the inner parts of the spores were orange-brown in color. FAT assay was employed for the detection of different isolates of *Polymyxa*, including isolates that produced weak reaction in ELISA test. A clear distinction between those that reacted strongly and those that reacted weakly was revealed by the results of FAT assay. Isolates showing weak reaction in ELISA test, showed specific staining in FAT test (Delfosse et al. 2000).

### 2.1.3.8   Immunosorbent Electron Microscopy

Immunosorbent electron microscopy (ISEM), previously known as serologically specific electron microscopy, involves the use of antisera produced against fungal pathogen-specific antigen and linkage of the antibodies to protein A-gold complexes to visualize the pathogen in infected plant tissues or tissue extracts. The MAbs were generated against the species-specific epitopes on the surface of zoospores and cysts of *P. cinnamomi*. The MAbs contained a spectrum of specificities. The MAbs were tested against six isolates of *P. cinnamomi* and six species of *Pythium*. Reactions of MAbs with antigens in the pathogen isolates indicated that spatially restricted antigens were present on the surface of zoospores and the MAbs might be useful for investigations on the biology and taxonomic relationships of *P. cinnamomi* (Hardham et al. 1986).

### 2.1.4   Nucleic Acid-Based Techniques

All living organisms contain RNA and/DNA as their genomes that carry all necessary information for the synthesis of all compounds required for their growth and reproduction, in addition to determining their characteristics and expression of genes at different stages of development. It is generally considered that closely related organisms share a greater nucleotide sequence similarity than those that are distantly related. A highly specific nucleotide sequence present in an isolate/strain of a microbial pathogen may be useful for the detection and identification of the isolate/strain concerned. The nature of the nucleic acid present in a microbial pathogen remains constant at all stages of its life cycle. On the other hand, the nature of antigenic determinants of fungi or bacteria, which are complex antigens, may change depending on the stage of their development. Hence, the antisera produced against one type of spore or mycelium at one stage may not react positively with spores or mycelium produced at different stages in the life cycle. In addition, the presence or absence of spore-bearing structures and slow-growing nature of certain fungal pathogens will not affect their detection by nucleic acid-based techniques, since all fungal cells (mycelial, asexual/sexual spores) contain the same genomic DNA. This fact gives a distinct advantage for using nucleic acid-based technique(s) for detection, identification, differentiation and quantification of fungal plant pathogens in plants and soil environments. Using appropriate probes containing pathogen-specific sequences, the fungal pathogens can be identified reliably and efficiently. Nucleic acid-based techniques are effective in detecting soilborne pathogens belonging to genera such as *Phytophthora*, *Pythium*, *Verticillium*, *Gaeumannomyces*, *Sclerotinia*, *Rhizoctonia* and *Leptosphaeria* in plant, soil and water samples. Nucleic acid-based techniques are based either on hybridization of probes with similar nucleotide sequences present in pathogen DNA or amplification of specific sequences/ fragments of DNA present in the fungal DNA by the primers prepared from specific sequences of the pathogen to be detected and identified. Generally, nucleic acid-based techniques possess higher levels of sensitivity, specificity, reliability and rapidity, compared with conventional isolation methods and immunoassays (Narayanasamy 2001, 2011).

### 2.1.4.1 Fluorescence Microscopy

The presence of microbial plant pathogens can be visualized in infected tissues, using fluorescence microscopy. Patch disease of turf grass is caused by four *Rhizoctonia* spp. The number of nuclei present in each cell of *Rhizoctonia* sp. a primary characteristic, was used as the basis of differentiating these pathogens. Fluorescence microscopy, involving the use of DNA-binding dye, 4,6-diamidino-2-phenylindole (DAPI) was employed for the detection and identification of *Rhizoctonia* spp. The stained preparations were examined under fluorescence microscope (Zeiss Epiillumination System), using ultraviolet excitation. Based on the number of nuclei present in the cells, *R. solani*, *R. cerealis*, *R. zeae* and *R. oryzae* could be differentiated. As these *Rhizoctonia* spp. could be tentatively identified within 24 h after the receipt of the sample, the technique might be employed for application as preliminary diagnostic tool (Martin 1987).

### 2.1.4.2 Hybridization Methods

Detection of soilborne microbial plant pathogens by nucleic acid hybridization methods is based on the formation of double-stranded (ds) target nucleic acid sequence (denatured DNA or RNA) and complementary single-stranded (ss) target nucleic acid probe. Sequences of either RNA or DNA have been used as probes. Probes of different types such as cloned and uncloned nucleic acid molecules, oligonucleotides, *in vitro* RNA transcripts, radioactive and nonradioactive probes have been evaluated for their efficacy in detecting the soilborne microbial pathogens in plants, soil and water samples. A slot blot hybridization procedure was employed for the detection of *Ggt*, causal agent of wheat take-all disease by using a plasmid DNA (pG158) probe. Strong hybridization of this probe pG158 specifically to pathogenic isolates of *Ggt* was observed, whereas the probe showed moderate hybridization signal with *G. graminis* var. *avenae*, infecting barley. In contrast, no hybridization of the probe occurred to non-pathogenic isolates of *Ggt* and other soil fungi associated with wheat rhizosphere. The probe pG158 could detect the pathogenic isolates of *Ggt* both in the soil and wheat root samples. Differentiation of pathogenic isolates of *Ggt* from morphologically similar nonpathogenic isolates has epidemiological significance, because it may be possible to relate soil populations of pathogenic *Ggt* isolates with the incidence of wheat take-all disease. In addition, the probe pG158 might be useful for intraspecific classification of *G. graminis* isolates (Harvey and Ophel-Keller 1996). *R. solani* AG-8 causes root rot and damping-off diseases in several economically important crops. The DNA probe pRAG12 was specific and efficiently detected the pathogen in soil, because of the high copy number of AG-8 probe (Whisson et al. 1995). *R. solani* AG-2-IV strain was detected using a plasmid DNA fragment PE-42 as the probe which hybridized to DNA of all isolates of *R. solani* AG2-2-IV, but not to the DNA of other pathogens of Zoysia grass, indicating the specificity of the hybridization assay with PE-42 plasmid DNA. The results of southern hybridization employing PE-42 DNA fragment as a probe, indicated the possibility of using this fragment as a marker for differentiating *R. solani* AG2-2-IV from other intraspecific groups of *R. solani* (Takamatsu et al. 1998).

The effectiveness of detection and quantification of *P. cinnamomi* in avocado roots was demonstrated by employing a probe selected from a library of genomic DNA. This specific probe detected as little as 5 pg of *P. cinnamomi* DNA in dot blot and slot blot assays. The extent of colonization of avocado roots by *P. cinnamomi* was assessed by determining relative amounts of pathogen and host DNAs, over a period of time (Judelson and Messenger-Routh 1996). A reverse dot blot procedure for the identification of oomycetes was based on assays, involving the use of oligonucleotides labeled with nonradioactive digoxigenin (DIG). This method showed for fewer cross-hybridization than the one based on entire amplified ITS fragments. Different species of oomycetes could be identified by recording the positive reactions between the DNA labeled directly from the sample and the specific oligonucleotides immobilized on nylon membrane. Using this assay, *Pythium aphanidermatum*, *P. ultimum*, *P. acanthium* and *P. cinnamomi* could be identified precisely (Lévesque et al. 1998). *Pythium* spp. infect seedlings in the nurseries of vegetable crops, causing damping-off diseases or rotting of rhizomes. Universal primers for amplifying sequences of specific regions like ITS of rDNA by PCR have been employed for the detection and differentiation of *Pythium* spp. that are difficult to differentiate based on morphological characteristics. The restriction fragment probes from ITS1 showed high degree of species specificity to *P. ultimum*, when tested by dot blot hybridization against 24 different *Pythium* spp. No distinct differences could be recognized among 13 isolates of *P. ultimum* var. *ultimum* and var. *sporangiferum* from eight countries and two isolates of *Pythium* group G, later classified as *P. ultimum* (Lévesque et al 1994). Probes specific to *P. ultimum* var. *ultimum* and *P. ultimum* var. *sporangiferum* were species- or variety-specific. The results indicated that 5S rRNA gene spacers might be useful in defining species boundaries in the genus *Pythium*, as these sequences diverged rapidly after speciation (Klassen et al. 1996).

*Sporisorium reiliana*, causing maize head smut disease, invades susceptible maize (corn) during emergence or at seedling stage through soilborne teliospores and grows systemically along with meristem. A highly specific and sensitive DNA-based procedure was developed for the detection of *S. reiliana* and its differentiation from *Ustilago maydis*, inducing similar symptoms as *S. reiliana*. Clones containing strongly hybridizing species-specific DNA were selected by screening libraries with their own labeled genomic DNA, followed by cross-hybridization with genomic DNA of maize and other maize-pathogenic fungi. Using DIG-dUTP labeled inserts, detection of less than 0.16 ng of fungal DNA was possible by dot blot hybridization. The primer pairs SR1 and SR3 were specific to *S. reiliana*, while UM11 was specific to *U. maydis* and the PCR assay detected fungal DNA of less than 1.6 pg of *S. reiliana* and 8.0 pg of *U. maydis*. Both dot blot hybridization and PCR assay were highly sensitive and reliable for the detection and differentiation of *S. reiliana* and *U. maydis* (Xu et al. 1999). *F. oxysporum* f.sp. *basilici* induces wilt disease of basil. Amplified DNA fragments (69) generated from the pathogen were evaluated for the specificity in hybridizing with pathogen DNA using dot blot hybridization procedure. A fragment of 1,038-bp specifically hybridized to the DNA of all *F. oxysporum* f.sp. *basilici* isolates, but not to DNA of isolates nonpathogenic to basil or representatives of other *formae speciales* of *F. oxysporum* or other *Fusarium* spp., *R. solani*, *Sclertoinia sclerotiorum*, *S. minor* and *Pythium ultmum* isolated from basil. Absence of hybridization to the DNA of other pathogenic and nonpathogenic isolates tested, indicated the reliability and specificity of the probes employed in the dot blot assay for the detection of *F. oxysporum* f.sp. *basilici* (Chiocchetti et al. 2001). *Synchytrium endobioticum*, causing wart disease of potatoes has worldwide distribution with the capacity for catastrophic impact on agricultural economy in endemic regions. A method for colorimetric detection of *S. endobioticum* from potato wart samples, using a peptide nucleic acid (PNA) probe and a cyanine dye was developed. A segment of the 18S ribosomal RNA gene was sequenced for several pathotypes of *S. endobioticum* and employed to design a species-specific PNA probe. This probe was used in

a rapid colorimetric hybridization assay that detected RNA from infected wart tissue down to $10^{-17}$ mol within 15 min. This technique has the potential for rapid identification of potato wart pathogen in non-laboratory settings with minimally trained staff (Duy et al. 2015).

Fluorescent *in situ* hybridization (FISH) assay has the potential for species-specific identification of microbial plant pathogens. The FISH technique was applied for the visualization of *P. cinnamomi*, causing root rot disease in many crops, in infected root tissues, using a light microscope. A species-specific fluorescently labeled DNA probe was developed for use without damage to plant or pathogen cell integrity or the need for isolation of the pathogen in pure culture. The probe was specific to *P. cinnamomi* and there was no hybridization with the DNA of other *Phytophthora* spp., *Pythium* spp. or bacterial species tested. *Phytophthora cambivora* and *P. niederhauserii*, closely related to *P. cinnamomi*, also did not show positive hybridization with the probe, indicating the specificity of the probe. The positive reaction was inferred by the nuclei fluorescence occurring only in the cells of *P. cinnamomi*. Nuclei fluorescence with chlamydospores and hyphae of *P. cinnamomi* was clearly distinct from the dark background. The probe hybridized with the mycelium of *P. cinnamomi* in inoculated *L. angustifolius* cv. Mandalup seedlings, but not with uninoculated plant tissues. Likewise, naturally infected field plant samples, when assayed by FISH procedure, showed blue fluorescence revealing the presence of *P. cinnamomi*, differentiating it from other fungal or oomycete pathogens present in plants. Fluorescence of *P. cinnamomi* nuclei could also be observed in woody root material of *Stylidium diuroides* suspected to be infected with *P. cinnamomi*, when plant cells and hyphal structures could not be seen under light microscope. This study indicates the possibility of detecting *P. cinnamomi* at different stages of life cycle within infected plant materials (Li et al. 2014).

### 2.1.4.3 Polymerase Chain Reaction-Based Assays

PCR is a sensitive, reliable and versatile technique that can yield results rapidly. The assay essentially involves heat denaturation of the target ds-DNA and hybridization of a pair of synthetic oligonucleotide primers to both strands of the target DNA, one to the 5′ end of the antisense strand by an annealing step. Then by using a thermostable *Taq* DNA of polymerase enzyme from *Thermus aquaticus* (*Taq*), new DNA strand is synthesized on templates to produce twice the number of target DNA strands. Newly synthesized DNA strands are used as targets for subsequent DNA synthesis and the steps mentioned above are repeated up to 50 times. There is an exponential increase in the number of target DNA molecules. In this procedure, DNA sequences between the primers are produced with high fidelity and the protocol can be automated. *Taq* DNA polymerase is heat stable and can be employed at temperatures between 60°C and 85°C (Weier and Gray 1988). The PCR-amplified products may be useful as a target for hybridization, for direct sequencing of the DNA to assess strain variations and as a specific probe. Hundreds of different primers may be synthesized at a cost comparable to that required for the development of only a few MAbs for the detection of *Gaeumannomyces graminis* in plants grown in infested soil by immunoassay (Henson and French 1993). PCR-based assays have been applied more frequently than other diagnostic methods for the detection of microbial plant pathogens due to several advantages offered by these assays. There is no need for isolation of the pathogen(s) to be detected and identified, from plants, as only the pathogen DNA or RNA is required for PCR amplification. It is possible to detect by PCR-based assays from plants infected by single or multiple pathogens simultaneously, as in the case of immunoassays which are more expensive and time-consuming. In addition, high levels of sensitivity and specificity and simplicity of PCR-based assays have been the factors that made these procedures of choice for researchers. However, the performance of the PCR-based assays is hampered by the inhibitors of PCR amplification that are coextracted with pathogen DNA from plants/soil. Further, the PCR-based assays cannot differentiate the living from dead pathogen propagules present in the infected plants/soil/water (Keer and Birceh 2003). BIO-PCR procedure involving enrichment culturing prior to PCR amplification may alleviate the problem to some extent (Schaad et al. 2003).

### 2.1.4.3.1 Conventional PCR Assay

Soilborne microbial pathogens cause several crop diseases, symptoms of which can be recognized only after the pathogens have well established in the root and vascular tissues of infected plants. In addition, at the time of infection, the pathogen presence in the soil cannot be recognized and the IPin the soil may be insufficient just to initiate root infection in susceptible host plants. Hence, detection techniques have to be very efficient and sensitive to detect the soilborne pathogens at low population levels in plants which may remain asymptomatic and also in the soils. PCR-based assays have been shown to be sensitive enough for the detection of soilborne fungal pathogens in plants and soils effectively. PCR assays have been found to be effective in detecting *P. parasitica* in infected tomato roots and soil (Goodwin et al. 1990), *Ggt* in wheat (Henson et al. 1993; Fouly and Wilkinson 2000; Rachawong et al. 2002), *Verticillium* spp. in potato (Moukhamedov et al. 1994), *P. fragariae* var. *fragariae* in roots of strawberry (Hughes et al. 1998; Ioos et al. 2006; Suga et al. 2013), *P. cinnamomi* in ornamentals and soilless media (Kong et al. 2003), *A. euteiches* in alfalfa (Vandemark et al. 2002), *Fusarium solani* f.sp. *cucurbitae* (Mehl and Epstein 2007), *Phytophthora melonis* (Wang et al. 2007) in cucurbits, *P. capsici* in pepper (Lan et al. 2013) and *F. oxysporum* f.sp. *vasinfectum* in cotton (Crutcher et al. 2016).

Sequences of ITS regions of the fungal pathogens have been used as targets for their detection in plants and soils, since they appear to be conserved within species, but may differ significantly among several plant pathogenic species, allowing the construction of unique primer sequences. Highly specific primer pairs could be designed for the detection and identification of *P. violae* and *P. myriotylum*, based on the amplification of a fragment from the ITS region by PCR assay (Wang and White 1996; Wang et al. 2003a). Five species-specific primer pairs were selected from the sequences of ITS region of rDNA of 34 *Pythium* spp., after testing the specificity of these forward primers paired with IST2 or IST4 and reverse universal primers. The specific amplification, using these species-specific primers, allowed identification and differentiation of nine *Pythium* spp. *P. aphanidermatum* in infected carrot tissues and *P. dimorphum* in infested soil were detected rapidly and reliably (Wang et al. 2003a, b).

Primers for PCR amplification have been designed based on the specific sequences or gene sequences of the target pathogen DNA. Sequences of ITS regions have been frequently employed for the construction of primers. *Pythium intermedium*, *P. sulcatum*, *P. sylvaticum*, *P. violae* and *P. vipa*, causing carrot cavity spot disease were detected in soils using primers designed based on ITS sequences of rDNA. In soils known to have incidence of cavity spot disease on carrots, relatively strong signals were obtained and in several of the soils more than one of the five *Pythium* spp. could be detected. In a field where infection level was 79.7%, all five *Pythium* spp. could be detected in soil samples by the PCR procedure developed in this investigation, indicating its usefulness for wide application to reduce the loss due to cavity spot disease in carrots (Klemsdal et al. 2008).

The ITS regions within *Pythium* spp. are conserved, but variable between species, prompting the frequent use of ITS sequences of rDNA for detection and identification of *Pythium* spp. They, however, were not useful for quantitative assessment of pathogen population in plants and soils. Hence, variable regions of ITS regions of rDNA were used for designing species-specific primers for the detection and quantification of nine species of *Pythium* in soils in eastern Washington. These primers could be employed in PCR assays from artificially and naturally infested soils using a soil DNA extraction kit and real-time PCR assay (Schroeder et al. 2006). A sensitive and specific PCR-based method was developed for the detection of *Pythium myriotylum*, causal agent of ginger soft rot disease, in soil samples. Oospores of *P. myriotylum* were separated from large soil particles by flotation in sucrose solution. The thick-walled oospores were disrupted by vortexing with sea sand. Pathogen DNA was extracted by CTAB method. PCR amplification of a 150-bp target sequence of *P. myriotylum* revealed its presence in field soil samples with a detection limit of 10 oospores/g of soil (Wang and Chang 2003). DNA from *P. capsici* infecting pepper and weed plant species was extracted, using a modified phenol-chloroform extraction procedure (French-Monar et al. 2006). Commercial kits for extracting DNAs of fungal pathogens are available. PUREGENE DNA isolation kit (Gentra Systems Inc., United States) for *Phytophthora bisheria* (Abad et al. 2008) and EZNA Plant DNA Kit (Omega Biotek) for *F. striatum* (Moine et al. 2014) were employed for efficient extraction of DNAs from the fungal pathogens. Irrigation water contaminated with oomycetes propagules may be the primary source of inoculum for disease incidence and further spread. Irrigation ponds, providing water for vegetable production areas in Georgia, United States were examined for the presence of

*Pythium litorale*, causal agent of seedling damping-off and fruit rot of yellow squash. Camellia and rhododendron leaves were used as baits for recovery of oomycetes which were identified based on morphological characteristics and analysis of ITS regions. *P. litorale* was frequently isolated from all irrigation ponds sampled. The pathogen had optimum and maximum temperatures of 30°C and 40°C for growth, respectively. The results showed that *P. litorale* isolates were pathogenic to yellow squash and the irrigation water could be an important source of inoculum for infection of yellow squash and outbreaks of the disease in Georgia (Parkunan and Ji 2013).

Primers based on the sequences of ITS2 region of rDNA of *P. infestans* (causing potato late blight) and *P. erythroseptica* (causing potato pink rot), primers were employed for the detection of these pathogens by PCR assay in potato tubers as early as 72 h after inoculation, well ahead of the appearance of visible symptoms (Tooley et al. 1998). The primers P-FRAGINT and the universal ITS4 were effective in detecting *P. fragariae* var. *fragariae* and *P. fragariae* var. *rubi* in the roots of strawberry and raspberry, respectively (Hughes et al. 1998). Primers designed based on the sequences of ITS region of ribosomal gene repeat (rDNA) were successfully employed for the detection of *P. fragariae* and the PCR detection was more efficient than ELISA test (Bonants et al. 2000). The comparative efficacy of PCR assay and bait test was assessed for the detection of *P. fragariae* var. *rubi*. Primers designed on DNA sequences of various parts of the rDNA were employed for the detection of the pathogen in raspberry roots. PCR assay was rapid, specific and sensitive and the results correlated well with that of bait tests. However, bait test detected the pathogen in a higher number of samples (12) than PCR assay which detected the pathogen only in ten samples, although bait test required a long time (six weeks), when compared to PCR, requiring only three days (Schlenzig et al. 2005).

For the detection of *F. oxysporum* f.sp. *vasinfectum*, causing cotton wilt disease, the primers Fov1 and Fov2, were designed on the differences in the ITS sequences between 18S, 5.8S and 28S rDNA, amplified a PCR product of 500-bp DNA fragment from all isolates of the pathogen, indicating the reliability and potential of the primers for disease diagnosis and monitoring pathogen movement (Morrica et al. 1998). *F. oxysporum* f.sp. *niveum* (FON) and *Mycosphaerella melonis* (MM) cause the destructive diseases, Fusarium wilt and gummy stem blight, respectively in watermelon. In order to detect and differentiate these pathogens rapidly and precisely in infected plants and infested soils, PCR assays were developed. Two pairs of species-specific primers Fn-1/Fn-2 and Mm-1/Mm-2 were designed based on the differences in the ITS sequences of FON and MM. The primers specific to FON amplified a single PCR product of ~ 320-bp only from 24 FON isolates. On the other hand, MM-specific primers produced an amplicon of ~ 420-bp only from 22 MM isolates. No product was amplified, when 72 isolates of other Ascomycotina, Basidiomycotina and Deuteormycotina were tested, indicating the specificity and reliability of the PCR assays. The sensitivity of detection was one fg of genomic

DNA of both pathogens, revealing the usefulness of the PCR assays for detecting and differentiating the important fungal pathogens infecting melons (Zhang et al. 2005).

*Phytophthora megasperma* BHR-type isolates, infecting olive trees were detected effectively by using primers designed, based on ITS regions of rDNA of the pathogen. Pathogenicity tests showed that these isolates were highly virulent on roots of olive trees (Sánchez-Hernández et al. 2001). PCR primers specific to *P. melonis* (Pm1 and Pm2) were designed, based on sequences of ITS of nuclear ribosomal DNA. PCR amplification with these primers produced a product of ~ 545-bp exclusively from DNAs of isolates of *P. melonis* only, when 115 isolates representing 26 *Phytophthora* spp. and 29 other pathogenic species were tested, indicating the specificity of the primers. The detection sensitivity of Pm1 and Pm2 primers was 100 fg of genomic DNA, revealing the usefulness of the PCR assay for the specific and sensitive detection of *P. melonis*, infecting cucumber, causing Phytophthora blight disease (Wang et al. 2007). Isolates of *P. capsici*, infecting pepper and other vegetable crops as well as several weed species were identified by PCR assay using PCAP primer, in combination with universal primer ITS1. A 172-bp product was obtained from the DNAs of all the isolates after amplification, establishing their identity as *P. capsici* (French-Monar et al. 2006). For the detection in soil of *P. sojae*, causing root and stem rot of soybean, a PCR assay was developed, using primers designed based on ITS regions of eight *P. sojae* isolates. The PCR products were sequenced and compared with published sequences of 50 other *Phytophthora* species and a region specific to *P. sojae* was used to design specific primers, PS1 and PS2. The primers PS1 and PS2 were specific in amplifying a product of approximately 330-bp exclusively from the isolates of *P. sojae*, but not from the DNAs of more than 245 isolates representing 25 *Phytophthora* spp. and other pathogens tested. The sensitivity of PCR detection by PS primers was approximately 1 fg. The PCR assay combined with a single soil screening method developed in this investigation, enabled testing of the pathogen within 6 h, with a detection sensitivity of 200 spores in 20-g soil sample. The PCR procedure could be used for the detection of *P. sojae* in diseased soybean tissues and residues in soil, indicating the usefulness of the technique for large-scale application (Wang et al. 2006).

For the specific detection of *P. brassicae*, an obligate endoparasite, causing club root disease in cabbage and other cruciferous crops by PCR, the ITS region of pathogen DNA was targeted to identify a unique DNA sequence. Two sets of primers designed from ribosomal repeat and ITS regions were employed. The pathogen could be detected in soil and water samples with high levels of sensitivity because of the high copy number of rDNA sequences in the pathogen genome. The specificity of the PCR assay was tested against more than 40 common soil organisms, host plants and spore suspension contaminants, as well as *P. brassicae* isolates from around Australia and the other countries in the world. Amplification of pathogen sequence occurred only in samples containing *P. brassicae*. The sensitivity of the PCR assay was 0.1 fg (fg = $10^{-15}$ g) for pure template and as low as 1,000 spores/g of potting mix. *P. brassicae* was detected in all soils, where IP was sufficient to induce club root symptoms (Faggian et al. 1999). The resting spores of *P. brassicae* present in plant tissues and soil may remain viable for over seven years. The resting spores released from decayed galls may facilitate inoculum buildup in areas where crucifers are repeatedly cultivated. The PCR could be used for the detection of *P. brassicae* in infected plant tissues. In a later investigation, an oligonucleotide primer set designed based on sequences of the small subunit (18S-like) and ITS region of rDNA was employed for specific detection of *P. brassicae* in plant tissues. This primer set amplified a 1,000-bp fragment in the PCR assay. The high copy number of rDNA gene sequences from which the primer set was designed, allowed detection of small quantities of target pathogen DNA in the total DNA extracts of infected Chinese cabbage plants. The DNAs of test plant and other soilborne fungi and bacteria were not amplified, indicating the high level of specificity of the PCR assay (Kim and Lee 2001). *S. endobioticum*, causal agent of potato wart disease, produces two kinds of sori in the galls. Summer sori have thin cell wall and form sporangia and resting sori (sporangia) with thick walls enriched with a chitin layer. They can survive in the soil for more than 30 years. In order to detect the resting sori in the soil, PCR procedure was developed by generating primers from the sequences of pathogen ITS primer # 4. *S. endobioticum*-specific primer Kbr1, generated an amplification product of 543-bp from all four pathotypes. The zoospores emerging from resting sori were used for DNA extraction. PCR assay was also useful to differentiate weakly resistant and moderately susceptible responses of potato cultivars, based on the pathogen population (Niepold and Stachewicz 2004).

A PCR assay, based on an amplification of 5.8S rDNA gene and ITS4 and ITS5 primers was employed for the rapid detection and identification of economically important pathogens *P. capsici*, *P. cinnamomi*, *P. citricola*, *P. citrophthora*, *P. erythroseptica* and *P. fragariae*. *P. medicagines* was detected, using a pair of oligonucleotide primers (PPED04 and PPED05) for amplifying a specific fragment in the intergenic spacer (IGS)-2 region of pathogen DNA (Liew et al. 1998). *P. nicotianae*, causative agent of tobacco black shank (TBS) disease, was detected by employing two specific primers designed from ITS regions ITS1 and ITS2. A product of 737-bp was amplified by the PCR from the DNA of all isolates of P. *nicotianae*, but not from any other *Phytophthora* spp. tested, indicating the specificity of the PCR assay (Grote et al. 2000). *P. sojae*, incitant of stem rot and root rot of soybean, could be detected rapidly in plant and soil samples by employing primers based on ITS region sequences DC6 and ITS4 in the PCR assay. By aligning sequences of PCR products, a region specific to *P. sojae* was identified and specific primers amplified a product of a 330-bp exclusively from the isolates of *P. sojae* among 245 isolates of 25 different *Phytophthora* spp. The pathogen could be detected in infected soybean tissues and residues in the soils. The detection limit of the assay was 1 fg of purified DNA of *P. sojae* (Wang et al. 2006). Precise detection and identification methods for

*P. ramorum* and *P. fragariae* which are of quarantine importance, are required for monitoring their movement through plant materials. Introns are generally highly polymorphic regions and considered to be useful for the discrimination of phylogenetically close *Phytophthora* species. A series of PCR primers were designed from introns located in different copy genes (GPA1, RAS-like and TRP1) of *P. ramorum* and *P. fragariae*. The specificity of the primers were determined using a wide collection of *Phytophthora* isolates and an efficient protocol was formulated for detecting both pathogens directly in infected plant tissues (Ioos et al. 2006). *Phytophthora erythroseptica*, causing potato pink rot disease, can survive in the soil for many years as oospores which are dispersed from diseased potato tubers. Since the soils are one of the most challenging environmental matrices to extract microbial DNA, the method combining baiting method with PCR assay was applied to detect *P. erythroseptica* DNA in soil samples. Hairy nightshade (*Solanum sarrachoides*) and bitter nightshade (*Solanum dulcamara*) two-leaf stage seedlings and cotyledonary leaves efficiently baited *P. erythroseptica* from zoospores suspensions, artificially inoculated soils and naturally infested soils. *P. erythroseptica* was detected in the bait tissue, using PCR assays which increased the sensitivity of detection tests. *P. erythroseptica* could be detected in some bait plants at two days after incubation. However, an incubation period of ten days was required to have consistent results across the replicates with hairy and bitter nightshade cotyledonary leaves and seedlings. The time-lag needed for the detection *P. erythroseptica* makes the procedure less preferable (Nanayakkara et al. 2010).

*G. graminis* varieties cause take-all diseases of wheat and oats. NS5 and NS6 universal primers amplified the middle region of 18S rDNA of *Gaeumannomyces* species and varieties. Primers GGT_RP and GGA_RP were developed by sequence analysis of cloned NS5-NS6 fragments. The primer pair NS5:GGT_RP amplified a single 410-bp fragment from isolates of *G. graminis* var. *tritici (Ggt)* and a single 300-bp fragment from isolates of *G. graminis* var. *avenae (Gga)*. No amplification of products occurred from the DNA of *G. graminis* var. *graminis (Ggg)*, infecting turf grass or other species of *Gaeumannomyces*. Two sets of primer pairs, (NSG:GGT_ RP and NS5:GGA_RP) were employed for the detection and identification of *Ggt* and *Gga* colonizing wheat or oats or in culture. There was no amplification from DNA extracted from roots infected with eight other soilborne fungal pathogens or from uninoculated plants, indicating the specificity of the PCR assays (Fouly and Wilkinson 2000). A single-tube PCR procedure was developed for the detection and differentiation of take-all pathogens *G. graminis* varieties *Gga*), *Ggg* and *Ggt*. Nucleotide base sequence analyses of avinacinase-like genes from *Gga*, *Ggg* and *Ggt* isolates provided the basis for designing variety-specific primers. Three 5′ primers specific for *Gga*, *Ggg* and *Ggt* and a single 3′ common primer allowed amplification of variety-specific fragments of 617-, 870- and 1,086-bp, respectively. Each 5′ primer was specific in mixed populations of primers and templates. No amplification of PCR product was observed, when tested using DNAs of other related fungi, indicating the specificity and reliability of the one-tube PCR format for the detection and identification of varieties of *G. graminis*, infecting wheat, oats and grasses (Rachdawong et al. 2002).

*P. nicotianae* causes several economically important diseases in tobacco, tomato and citrus. A PCR assay, using species-specific primers was employed for rapid and precise detection of *P. nicotianae* in irrigation water, a primary source of inoculum and an efficient mode of propagule dissemination. The PCR assay consists of a pair of species-specific primers (PN) and customization of a commercial soil DNA extraction kit for purification of DNA from propagules in irrigation water. The PN primers were specific for *P. nicotianae*, allowing evaluation of 131 isolates of *P. nicotianae*, 102 isolates of other *Phytophthora* species and 64 isolates of other oomycetes, fungi and bacteria. The diagnostic PCR product was amplified only from the DNA of *P. nicotianae* isolates. The sensitivity of the PCR assay was between 80 and 800 fg DNA/μl. The PCR assay protocol developed in this investigation, detected *P. nicotianae* in naturally infested water samples from Virginia and South Carolina nurseries more rapidly compared with conventional isolation methods (Kong et al. 2003a). Presence of *P. capsici*, causative agent of Phyophthora blight of pepper, in diseased plant tissues, soil and artificially infested irrigation water was detected by employing species-specific PCR assay. One pair of species-specific primers PC-1/PC-2 was designed based on the differences in the ITS sequences of *Phytophthora* spp. and other oomycetes. The primers PC-1/PC-2 amplified a single product of ca. 560-bp only from the DNA of *P. capsici*, but not from the DNAs of 77 isolates from other true fungi tested. The detection sensitivity of the assay was 1pg of genomic DNA (equivalent to half the genomic DNA of a single zoospore) per 25 μl PCR reaction volume. The conventional PCR assay could detect *P. capsici* in naturally infected plant tissues, infested field soil and artificially infested irrigation water. Using ITS1/ITS4 as the first-round primers and PC-1/PC-2 in the second round, the nested PCR format was shown to have higher level of sensitivity (1 fg/25 μl reaction volume) in detecting *P. capsici*. The PCR assays have the potential for accurate detection of *P. capsici*, pathogen monitoring and making effective disease management decisions (Zhang et al. 2006).

A species-specific PCR assay was developed for rapid and precise detection of *P. capsici* in pepper plants, soil and water. The PCR primers were designed based on Ras-related protein (*Ypt1*) gene. The specificity of the primers was tested on 236 isolates representing 12 *Phytophthora* spp. and 26 fungal plant pathogens. Amplification of 364-bp product occurred only with all isolates of *P. capsici*, but not with any other fungal pathogens tested. The detection sensitivity of the species-specific primer pair Pc1F/Pc1R was 10 pg of genomic DNA. The sensitivity of the test could be enhanced by 1,000-fold by employing nested PCR format, involving the uses of PC1F/Pc1R, as the first-round primers combined with Pc1F/Pc2R as the second round primers. The PCR assay effectively detected *P. capsici* in naturally infected plant tissues, soil and water samples. This protocol has the potential for use in

disease diagnosis and pathogen monitoring for effective management of diseases caused by *P. capsici* (Lan et al. 2013). *P. ramorum*, causing sudden oak death (SOD) disease, has multiple modes of dispersal. Propagules of *P. ramorum* could be recovered from infested stream and nursery irrigation runoff, using baiting and infiltration methods. Five methods of detection, including pear and rhododendron leaf baits, Bottle O'Bait, filtration and quantitative PCR assay performed on zoospores trapped on a filter, were compared simultaneously in laboratory assays, using laboratory or creek water spiked with known quantities of *P. ramorum* zoospores. The detection threshold for each method was determined. Filtration and qPCR procedure were the most sensitive in detecting low levels of zoospores, followed by wounded rhododendron baits. Filtration, qPCR and leaf disk methods quantified *P. ramorum* zoospores ranging from two to 451 direct-plate CFU/liter, while wounded leaves and pear bits appeared to be better at detection rather than quanfication. The methods of detection and quantification of *P. ramorum* propagules in water would be useful for assessing the risk of spreading the pathogen in nurseries where untreated infested water might be used for irrigation (Rollins et al. 2016).

*Fusarium solani* f.sp. *cucurbitae* (*Fsc*) includes isolates of two biologically and phylogenetically distinct species FSC race 1 [Fsc-1= *Nectria haematococca*, mating population I (MP I)] and FSC race 2 [Fsc-2= *N. haematococca* mating population V (MP V)]. Both species infect the fruit (causing fruit rot), but Fsc-1 also infects the crown and stem (causing stem rot). Pathogenic isolates (156) were isolated from affected plants and soil under diseased fruits and they were identified morphologically as members of the *F. solani* species complex. Based on the pathogenicity tests, 81 isolates were considered as Fsc-1. The remaining 75 isolates were either nonpathogenic or only weakly pathogenic on fruit. The PCR assay showed that all of the Fsc-1 isolates were MAT-1-2, suggesting that the pathogen might be strictly clonal in affected fields. The results conclusively showed that the isolates of Fsc-1 were responsible for Fusarium foot rot and fruit rot in pumpkin (Elmer et al. 2007). As isolates of race 1 (*Fsc* 1) and race 2 (*Fsc* 2) are not easily distinguishable by morphological characteristics, a PCR assay was developed for differentiating these two races, involving the use of taxon-specific primers designed from translation elongation factor 1-α sequences. Isolates obtained from lesions induced were identified as *Fsc* 1 based on the specific amplicons generated (Mehl and Epstein 2007). A new disease, causing reminiscent of infection by *F. oxysporum* f.sp. *radicis-lycopersici*, was observed in the greenhouse tomato plants in Canada and the United States. A homothallic *Fusarium* spp. was isolated from infected tomato plants. A PCR assay, using pathogen-specific primes ITS1F and ITS4, amplified a product of 605-bp from group 1 isolates which were 99% identical to *F. striatum*, while a fragment from group 2 isolates showed 100% homology with *F. oxysporum*. A second PCR analysis was performed using primers EF1/EF2 based on fungi-specific translation elongation factor (*tef*) 1-α gene sequences. A 733-bp amplicon was generated from seven strains. Sequences of group 1 showed 99% homology with *tef* sequences of *F. striatum* and hence, the pathogen was identified as *F. striatum* (Moine et al. 2014).

Early and reliable detection and identification of *F. oxysporum* f.sp. *cubense* (FOC), causative agent of Panama wilt of banana was found to be essential, since the pathogen was widespread and negatively impacted the banana production all over the world. FOC exists in the form of four races, varying in their virulence to banana cultivars. A primer set Foc-1/Foc-2 was designed from the sequence of a random primer OP-AO2-amplified fragment. A 242-bp (Foc$_{242}$) DNA fragment was specifically amplified from FOC race 4 present in tropical countries. The primer set Foc-1/Foc-2 amplified the marker fragment Foc$_{242}$ at a concentration as low as 10 pg of pathogen DNA, indicating the high level of sensitivity of the assay. The primer set specific to Foc race 4 isolates could detect the pathogen in field-collected naturally infected banana pseudostem tissues. The race-specific primer set detected the marker fragment in symptomless banana leaves also, although the bands in the gels were very faint. In combination with Southern hybridization technique, the sensitivity of the PCR could be enhanced 100- fold. In addition, the PCR assay using race-specific primer set could effectively differentiate race 4 from other races of *F. oxysporum* f.sp. *cubense* (Lin et al. 2009). *F. oxysporum* f.sp. *fragariae* (*Fof*), causing Fusarium wilt disease of strawberry, is present along with nonpathogenic *F. oxysporum*. A PCR assay using primers that could discriminate *Fof* from nonpathogenic isolates, was developed. Transposable elements *Fot3*, *Han Hop*, *Hornet1* and *Skippy* were characterized and used for designing a specific set of PCR primers. Portions of these transposable elements were detected in all 33 strains of *Fof* tested by PCR assay. In addition, the transposable element *Foxy* was detected in 32 strains and Impala sequences were present in 30 strains. The PCR primers developed in this investigation could discriminate *Fof* strains from nonpathogenic *F oxysporum* strains and five other *formae specials*. Conidia of *Fof* were detected in brown lowland type soil by the PCR assay (Suga et al. 2013).

For the detection and identification of *F. oxysporum* f.sp. *vasinfectum (Fov)*, causing cotton wilt disease, a PCR assay was developed. PCR amplification of ribosomal IGS regions combined with digestion of three restriction enzymes *Alu*I, *Hae*III and *Rsa*I resulted in three unique restriction profiles (IGS-RFLP haplotypes) for Australian *Fov* isolates and restriction profiles could differentiate them from other *formae speciaels* of *F. oxysporum*. In addition, two specific real-time PCR-based assays were used for absolute quantification of genomic DNA of isolates from soil substrates and cotton root tissues. The limit of detection of *Fov* DNA was 5 pg/µl of genomic DNA. The detection sensitivity of inoculum spiked into sterile soil was lower than $10^4$ conidia/g of soil. In the plant tissues, the detection limit was 30 pg to 1 ng/100 ng of total plant genome DNA for quantification of *Fov* DNA (Zambounis et al. 2007). *Fov* race 4 is seedborne and persists in infested soils and has become a potential threat to all cotton producing regions of the United States. *Fov* 4 has spread among soils planted to cotton in the San Joaquin Valley of

California, necessitating the development of methods to rapidly identify race 4 in the field. Race 4-specific PCR assay and a test kit were employed to identify race 4 in the field. Both PCR and Kit tests were evaluated on a panel of 36 *Fov* isolates representing all known races and many genotypes from several geographic regions. Race 4 isolates were positively detected, in addition to race 7 isolates. Except one race 7 isolate, other isolates were vegetatively compatible with race 4 isolates. The diagnostic tests clearly distinguished race 4 from races 1, 2, 6 and 8, but not race 3 and race 7 isolates. Race 3 and race 7 isolates induced symptoms similar to that of race 4. Hence, the need for development of a method to differentiate race 3 and race 7 was indicated by this investigation (83Crutcher et al. 2016).

Different methods for extraction of DNA from soil samples for DNA analyses have been developed. The method of extracting DNA from soil, using polyvinylpyrrolidone to remove soil organic matter from cell preparations and repetitive cesium chloride density gradient centrifugation to purify DNA was found to be cumbersome and labor intensive (Holben et al. 1983). A direct method for extraction of DNA from soil samples was developed, using the rDNA sequences of *V. dahliae*, causing Verticillium wilt diseases in cotton, potato and other crops, as the target for PCR-mediated diagnostics, without the need for further DNA purification or culturing the pathogen in artificial media. The fungal pathogen was disrupted by grinding in liquid nitrogen with the natural abrasives in soils. Losses due to degradation and absorption were largely eliminated by the addition of skim milk powder. The DNA from disrupted cells was extracted with sodium dodecylsulfate (SDS)-phenol and collected by ethanol precipitation. After diluting the DNA solution to the required level, the extracted DNA solution could be used for PCR amplification employing appropriate *V. dahliae*-specific oligonucleotide primers. This procedure was found to be rapid, cost-effective and when combined with suitable internal controls, it could be applied for the detection and quantification of specific soilborne pathogens for wide application (Volossiouk et al. 1995). Based on the differences in the sequences of the rRNA genes of *Verticillium* spp., appropriate primers were designed. These primers efficiently amplified the specific PCR products from *Verticillium albo-atrum*, *V. dahliae* and *V. tricorpus*. A diagnostic set of primers was employed for monitoring the pathogen development in potato-*Verticillium* pathosystems (Moukhamedov et al. 1994). Two arbitrarily primed oligonucleotide primers (15–16-mer) were used in the arbitrarily primed patterns from total DNA extracted from *V. dahliae*. A 350-bp fragment (named as MGC), unique for the recognition of *Verticillium* spp. proved specific for *V. dahliae* in Southern blot. This fragment had diagnostic value for the detection of *V. dahliae* (Cipriani et al. 2000). Three species of *Verticillium*, *V. albo-atrum*, *V. dahliae* and *V. tricorpus* were detected in field soils by PCR assay. Relative efficacies of PCR assay and plating methods were assessed. Plating method required a long time (four to five weeks) to provide results and it could not detect *V. tricorpus*, whereas PCR assay was rapid in yielding results within one to two days and detected all three *Verticillium* spp. However, quantification of pathogen propagules in soil samples was possible, by using plating method. On the other hand, PCR assay failed to quantify pathogen populations in soil samples. The principal advantage of PCR assay is the possibility of detecting the pathogen rapidly in the soils, in the absence of disease symptoms in plants growing in the infested soils (Platt and Mahuku 2001). The PCR was shown to be useful for the detection and identification of pathotypes of *V. dahliae* from hop rapidly, using the primers PG-1 and PG-2 (Radišek et al. 2004).

As the disease symptoms induced by *R. solani* appear only well after the establishment of infection, early detection and precise identification of the pathogen strains become essential to reduce the loss due to the disease. *R. solani* has many AGs within the morphologic species. *R. solani* AG-IA, causing rice sheath blight disease could be detected and identified rapidly by employing primers designed, based on the sequences of ITS region of rDNA. The pathogen was detected in infected rice plant tissues and also in rice field soils by PCR assay (Matsumoto and Matsuyama 1998). For the detection of *R. solani* AG2, infecting radish, the primers designed based on sequences of 5.8S rDNA and part of ITS region in combination with general fungal primers ITS IF and ITS 4B were successfully employed for the detection of the pathogen in infected radish plants and also in axenic cultures of the pathogen (Salazar et al. 2000). A universally primed (UP)-PCR cross hybridization assay was applied for the detection and identification of 16 isolates of *Rhizoctonia* in tissues of sugar beet and potatoes. The results were corroborated by RFLP analysis of the ITS1-5.8S ITS region of nuclear ribosomal DNA (Lübeck and Poulsen 2001).

A simple, one-step PCR procedure was developed for efficient detection of *P. brassicae*, causal agent of club root disease of cruciferous crops, in infected plant tissues and soil. The primers, based on rRNA gene sequence of *P. brassicae*, amplified a 548-bp product in the optimized PCR. A second pair of primers TC2F and TC2R amplified a fragment of the 18S and ITS region of the rDNA repeat, with amplicon of 519-bp. The specificity of the primers was indicated by the absence of the marker products, when noninfected plant hosts, noninfested soil or common soil fungi and bacteria were tested by this protocol. The sensitivity of the PCR was 100 fg of total *P. brassicae* DNA or $1 \times 10^3$ resting spores/g of soil and this might be due to the high copy numbers of rDNA sequences. The PCR procedure was effective in detecting the pathogen in asymptomatic root tissues at 3 DAI with *P. brassicae*. The PCR assay yielded the same or nearly the same size amplicon for all four *P. brassicaae* populations tested, suggesting that the fragment amplified in the rRNA repeat was conserved among the populations of *P. brassicae*. This PCR assay procedure showed potential for use in routine detection of *P. brassicae* in plant and soil samples (Cao et al. 2007). *S. endobioticum*, incitant of potato wart disease, is a A2 quarantine pathogen which is soil- and tuber-borne. As extraction of winter sporangia of *S. endobioticum* was found to be difficult, a novel technique, using the zonal centrifugation procedure known as Hendrickx centrifuge method, was developed.

The efficiency of extraction of winter sporangia from soil by Hendrickx centrifuge method was compared with a procedure used by the Dutch Plant Protection Service. Naturally and artificially contaminated soil samples were used for extraction of the pathogen. The Hendrickx centrifuge method provided better extraction recovery (60% higher), lower measurement error (50% lower) and lower detection level (down to 0.02 sporangia per g of soil). In addition, the Hendrickx centrifuge method was much less labor intensive than the procedure used by the Dutch Plant Protection Service. Another advantage of the Hendrickx centrifuge method is the possibility of using the supernatant for extraction of DNA of *S. endobioticum* required for PCR assays (Wander et al. 2007). Specific primers and a TaqMan probe designed from the ITS region of the multicopy rDNA gene of the pathogen were tested, using the extracts from artificially and naturally infested soils. Coamplification of target DNA along with an internal competitor DNA fragment resulted in a more reliable diagnosis of *S. endobioticum*, avoiding possible false negative results (van Gent-Pelzer et a. 2010).

Potato tubers are infected by *H. solani* and *C. coccodes*, causing respectively, silver scurf and black dot diseases that adversely affect processing and fresh market trade. Tuber incubation and PCR assays were performed with respective specific primers to detect these pathogens in asymptomatic tubers from organic farms. Infected tubers could be recognized after incubation, based on the symptoms of the diseases. Tuber incubation and PCR assays were in slight to fair agreement for detecting both pathogens. Most asymptomatic tubers tested positive for one or both assays for *H. solani* (75%) or *C. cocodes* (94%) (Mattupalli et al. 2013). *P. graminis* infects roots of cereals and also peanuts. In addition, *Polymyxa* spp. function as vectors of plant viruses. *P. graminis* was detected in roots of barley plants in Queensland by the presence of sporosori in the roots stained with trypanblue. *P. graminis* f.sp. *tepida* was detected in the roots of wheat, oats and barley by PCR assays, using pathogen-specific primers. *P. graminis* is known to be the vector of several destructive virus diseases caused by the members of the genera *Furovirus*, *Bymovirus* and *Pecluvirus*. No virus particles could be detected in the extracts of leaves of infected plants (Thompson et al. 2011).

Potato smut caused by *Thecaphora solani* is one of the most destructive diseases in the Andean region of South America, accounting for about 85% yield losses frequently. The development of galls specifically located on lower stems, stolons and tuber is the principal symptom induced by *T. solani*. The production of oval or irregular locular sori of variable sizes within the galls and reddish dark, granular-powdery mass of teliospores in the sori are distinctive characteristics of the pathogen. DNA fingerprinting and partial sequencing of the large subunit (LSU) rDNA region procedures were applied to establish the identities of teliospores and sponge-like mycelia mass produced by them under *in vitro* conditions. DNA was extracted from both structures and used to generate PCR DNA profiles. DNAs extracted from different teliospores collected throughout Chile were also analyzed. All primers used, generated similar DNA amplification products. The profiles obtained were the same for all teliospore collections and mycelia analyzed, but different from those obtained from potato (*Solanum tuberosum*) plant DNA (see Figure 2.5), indicating the close genetic relationship among all isolates of *T. solani* tested. Further, LSU sequencing also showed that the same 658-bp sequence was generated from the teliospores isolated from galls of different geographical origin and the sponge-like mycelia mass produced in cultures. The identity of the pathogen causing smut disease in potato could be confirmed by these nucleic-acid-based techniques (Andrade et al. 2004).

### 2.1.4.3.2 Nested PCR Assay

The sensitivity of PCR may be significantly enhanced by employing two different primer pairs in the process of amplification of specific sequences of the pathogen DNA. A single-tube nested PCR (STN-PCR) was employed for the detection of *P. brassiccae*. The outer primer PBTZS-2 was employed for amplifying a 1,457-bp fragment from *P. brassicae* DNA, whereas nested primers PBTZS-3 and PBTZS-4 could amplify a 398-bp fragment internal of the 1,457-bp fragment.

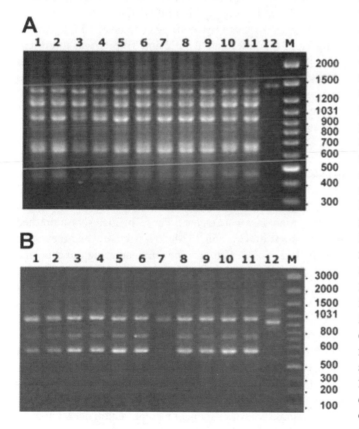

**FIGURE 2.5** DNA profiles of teliospores and mycelium of *Thielaviopsis solani* generated by PCR assay using sequence repeated primers DNA fragments amplified from DNA of teliospores and mycelium of isolates of TS-Q01 (lanes 1 and 8), TS-15 isolates (lanes 2 and 9), TS-16 isolates (lanes 3 and 10), and TS-21 (lanes 4 and 11), and teliospores of isolates TS-4, TS-5 and TS-25 (lanes 5, 6 and 7, respectively along with DNA profile of potato plant. Standard DNA markers are placed in lane M.

**[Courtesy of Andrade et al. (2004) and with kind permission of the American Phytopathological Society, MN, United States]**

The assay could detect even a single resting spore present in 1 g of soil. The sensitivity of detection could be further increased by subjecting the STN-PCR product to second PCR amplification (double PCR), using nested primer PBTZS-3 and PBTZS-4 (176Ito et al. 1999). *P. nicotianae*, infecting tobacco and other solanaceous crops, is difficult to tackle, because of its soilborne and waterborne nature and ability to infect several crops. Conventional and single closed-tube nested PCR formats were applied for the specific and sensitive detection of *P. nicotianae*. Two new specific primers designed, based on the sequences of ITS1 and ITS2 regions, internal to the nucleotide sequence flanked by universal primers ITS4 and ITS6 were employed. A positive reaction was indicated by the presence of amplification product of 737-bp from all *P. nicotianae* isolates and *P. nioctianae/cactorum* hybrids. In contrast, no amplification occurred of any fragment from the DNAs of other *Phytophthora* spp. tested. Nested PCR format was 1,000 times more sensitive than conventional PCR assay. The protocol was validated by testing samples from different sites, origins and crops, samples from nutrient solution, water and rockwool used in hydroponic cultures (Grote et al. 2002). In another investigation, the PCR assay was developed for the detection of *P. nicotianae* and *P. citrophthora* in citrus roots and soils, using primers based on the nucleotide sequences of ITS1 and ITS2 regions of 16 different *Phytophthora* spp. Two pairs of primers, Pn5B-Pn6 and Pc2B-Pc7 were designed specifically to amplify DNA from *P. nicotianae* and *P. citrophthora*, respectively. Another primer pair, Ph2-ITS4 was designed to amplify DNA from many *Phytophthora* spp. The specificity of the primers was assessed against 118 isolates of *Phytophthora* and 82 isolates of other common soil fungi. In conventional PCR format, the detection limit for all pairs of primers was found to be 1 pg/μl DNA. In the nested PCR, with primers Ph2-ITS4 in the first round, the detection limit was 1 fg/μl for both primer sets, Pn5B-Pn6 and Pc2B-Pc7. Further, simple, inexpensive and rapid procedures for DNA extraction from soil and roots were also developed enabling the completion of PCR assay within 2–3 h. With nested PCR format, utilizing primers Ph2-ITS4 in the first round and primer pairs Pn5B-Pn6 and Pc2B-Pc7 in the second round, *P. nicotianae* and *P. citrophthora* were effectively detected, revealing the suitability of molecular methods for rapid and reliable detection of *P. nicotianae* and *P. citrophthora* in plants and soil samples (Ippolito et al. 2002). A species-specific PCR assay was developed for the detection of *P. nicotianae* in infected plants and soil. A pair of primers based on the sequences of a Ras-related protein gene *Ypt1* (a single-copy gene) was employed, because of high levels of sensitivity and specificity provided by these primers. Single round amplification by PCR using the primer pair Ypt1-F/Tpt1-R resulted in efficient detection of *P. nicotianae* with a detection limit of 1ng pure DNA of *P. nicotianae* per 25 μl PCR reaction mixture. For enhancing detection sensitivity for *P. nicotianae* in soil and irrigation water samples with low quantities of pathogen DNA, a pair of *Phytophthora* universal primer pair Ypt1-F/Ypt1-R was used for the first round amplification in the nested PCR format. The amplicons were then amplified with species-specific primer pair Pn1/Pn2. The detection limit of the nested PCR was 10 pg DNA of *P. nicotianae*. This protocol is very sensitive for detecting *P. nicotianae* in soil samples and has the potential for monitoring movement of *P. nicotianae* in soil and irrigation water (Meng and Wang 2010).

*P. fragariae* var. *fragariae* is a strawberry pathogen of quarantine importance, requiring rapid and reliable detection methods to prevent the introduction and spread of the disease to other locations or countries. Nested PCR procedure was developed for specific detection of the pathogen with a sensitivity of 100 ag (atogram = $10^{-16}$ g) of pure pathogen DNA which is equivalent to 1/60 part of the nucleus. The high level of sensitivity was because of rDNA being a multicopy gene. However, in practice the nested PCR assay could detect between five and ten zoospores of *P. fragariae* var. *fragariae* consistently (Bonants et al. 2004). For the detection of *P. cactorum* infecting strawberry plants by nested PCR assay, necrotic strawberry tissues were soaked in 5% alconox solution for >12 h before DNA extraction for removing PCR inhibitors. The extracted DNA was embedded in an agarose gel chamber and subjected to electrophoresis. Nested PCR assay was employed to detect a portion of rRNA gene of *P. cactorum* in samples. In the first round of PCR amplification, primers ITS1 and ITS4 amplified fragments of varying sizes from total genomic DNA from diseased strawberry plants. In the second round PCR, a 1:25 dilution of the first round PCR product was used as template with two *P. cactorum*-specific primer pairs, PhycacL87FRG and PhycacR87RRG which amplified a 340-bp fragment and a 480-bp fragment from rRNA gene. The primer pair BPhycacL87FRG and BPhycacR176RRG amplified a 431-bp fragment. The results indicated that the primer sets employed in the nested PCR assay were specific to *P. cactorum* and reliably amplified the pathogen-specific product from a portion of rRNA gene of *P. cactorum* obtained from necrotic root, crown and petiole tissues of strawberry (Bhat and Browne 2010).

The detection and identification of *P. capsici*, causal agent of Phytophthora blight disease of pepper, were possible by performing a PCR-based method using three primers, CAPRW, CAPRV1 and CAPRV2 specific for *P. capsici*. These primers were designed based on the sequences of ITS regions. The primer set CAPW/CAPV1 amplified a PCR product of 452-bp, whereas the primer set CAPFW/CAPRV2 generated an amplicon of 595-bp fragment from the genomic DNA of *P. capsici*. Neither set amplified any fragment from the DNAs of host plants or several fungi pathogenic to pepper, indicating the specificity of the primers. Conventional (single round) PCR format had a detection limit of 5 pg DNA for both primer sets. On the other hand, nested PCR assay had a detection limit of 0.5 fg for both primer sets. The pathogen could be detected as early as 8 h postinoculation in stem samples, when the plants remained asymptomatic. Further, *P. capsici* could also be detected in soil samples from infested fields (Silvar et al. 2005b). *P. melonis*, causal agent of Phytophthora blight of cucumber, was detected by a PCR-based assay. The primers specific to *P. melonis*, Pm1 and Pm2, designed based on ITS

of rDNA sequences amplified a product of ca. 545-bp exclusively from the isolates of *P. melonis*, but not from the isolates representing 26 *Phytophthora* spp. and 29 other pathogenic species. The sensitivity of detection of *P. melonis* was 100 fg of genomic DNA. A nested PCR protocol with DC6 and ITS4, as first-round primers, followed by Pm1 and Pm2 primers, as second round primers showed a higher level of sensitivity (100 ag, 100-fold) than the conventional PCR format. Using nested PCR, ten zoospores in 0.5 g of artificially inoculated soils could be detected. The PCR with primers Pm1 and Pm2 was also able to detect *P. melonis* in naturally infected cucumber tissues, soil and irrigation water. The usefulness of PCR-based assays has been demonstrated for the detection of *P. melonis* in all possible sources of inoculum and pathogen monitoring for guiding disease management activities (Wang et al. 2007). *P. sojae*, causing the destructive root rot disease in soybean, has a transposon-like element, A3aPro. The primer pairs, TrapF1/TrapR1 were designed based on the unique sequences derived from transposon-like element. This primer pair amplified a 267-bp fragment only from *P. sojae* in conventional PCR format, but not from 72 isolates of other *Phytophthora* spp. and other fungi tested. The PCR had a detection sensitivity of 10 pg. The nested PCR format, using TrapF1/TrapR1 and TrapF2 and TrapR2 was more sensitive than the conventional PCR format, and it could detect 10 fg of pathogen DNA. A single oospore or zoospore could be detected by the nested PCR assay, indicating that this PCR format might be effectively employed for the detection of *P. sojae* in soybean tissues, residues and soils (Dai et al. 2013).

A PCR-based assay was found to be effective in detecting *Fusarium solani* f.sp. *glycines*, incitant of soybean sudden death disease, by using two primers, Fsg1/Fsg2 designed based on the sequences of mitochondrial subunit ribosomal RNA gene and Fsg EF1/Fsg EF2, based on the sequences of elongation factor 1-α gene. These primer sets amplified PCR products of 438- and 237-bp, respectively and they were specific to *F. solani* f.sp. *glycines* in generating the expected amplicons and they did not react with DNAs of 55 isolates of *F. solani* non-SDS isolates, 43 isolates of 17 soybean fungal pathogens and the oomycete *P. sojae* and soybean. The sensitivity of the primer sets Fsg1/Fsg2 and Fsg EF1/Fsg EF2 were 10 pg and 1 ng, respectively, when total genomic DNA or 1,000 macroconidia/g soil was used for testing. Nested PCR format improved the sensitivity of detection by 1,000-fold. The DNA of *F. solani* f.sp. *glycines* was detected in the roots of field-grown soybean plants and soil by PCR assay, using either a single pair of primers or a combination of two pairs of primers. Nested PCR assay was employed for the detection of the pathogen in 47 soil samples collected from soybean fields in 70 counties of Illinois and also in soil samples from all five Illinois Agricultural Districts, revealing the wide distribution of *F. solani* f.sp. *glycines* in Illinois State of the United States (Li and Harman 2003). *F. oxysporum* f.sp. *lactucae* race 1, incitant of lettuce Fusarium wilt disease, has worldwide distribution, causing significant yield losses. A molecular diagnostic assay for detecting and identifying race 1 in vegetables was developed. Inter-retroposon amplified polymorphism

PCR format based on the amplification of genomic regions between long terminal repeats was effective in grouping race 1 isolates of *F. oxysporum* f.sp. *lactucae*. A specific set of PCR primers generated from inter-retroposon-sequence characterized amplified regions (IR-SCAR) was employed for the differentiation of *F. oxyspoum* f.sp. *lactuace* from other *F. oxysporum* isolates. The primers designed in this investigation amplified unique fungal genomic DNA fragment from race 1 isolates collected from Italy, Portugal, the United States, Japan and Taiwan. In addition, the presence of the pathogen could be detected in artificially inoculated seeds and naturally and artificially infected lettuce plants (Pasquali et al. 2007).

Isolates of *V. dahliae,* infecting olive plants are grouped as defoliating (D) or nondefoliating (ND) pathotypes, depending on the type of symptoms induced in infected plants. The duplex, nested PCR format, was developed for simultaneous amplification of both ND- and new D-specific markers. It was possible to rapidly detect the D and ND pathotypes of *V. dahliae* with certainty in both artificially inoculated, own-rooted olive plants and naturally infected adult olive trees of different cultivars, age and growing conditions. In addition, the duplex, nested PCR assay could detect D and ND pathotypes simultaneously in adult olive trees naturally infected by both pathotypes and in young olive plants that were double-inoculated with D and ND pathotypes under controlled conditions (Mercado-Blanco et al. 2003). In a later investigation, the comparative sensitivity of single and nested PCR assays along with standard plating method was assessed for the detection of *V. dahliae* infection, especially in asymptomatic olive trees at different tree height (at 10–130 cm from the trunk base). Oligonucleotide primers derived from the ITS regions of nuclear ribosomal RNA (rRNA) genes of *V. dahliae*. The standard (single) PCR assay was highly sensitive in detecting *V. dahliae* at most tree heights, compared with the plating method and this format detected *V. dahliae* in a higher number of symptomatic olive trees than in asymptomatic trees and the intensity of PCR product in the gel decreased with increasing distance from the trunk base. The nested PCR assay most specifically detected infection of *V. dahliae* in a greater number of asymptomatic trees at all heights tested, compared with standard PCR format. The results indicated that both standard and nested PCR formats could be employed for reliable and specific detection of *V. dahliae* infection in both symptomatic and asymptomatic olive trees to prevent further spread of the disease in newly established orchards (Karajeh and Masoud 2006).

A nested PCR assay, employing general fungal primers, ITS1/ITS4 in the primary reaction, was developed to detect *Phaeoacremonium* spp. in the grapevine wood. The pathogen was identified by digesting PCR amplicons with restriction enzymes *Bss*KI, *Eco*1091 and *Hha*I. Different species of *Phaeoacremonium, P. australiensis, P. krajdenii, P. scolytii* and *P. subulatum* were identified by RFLP patterns. A species-specific PCR amplification of partial β-tubulin gene using the primer pair was required to differentiate *P. angustius* and *P. viticola* (Aroca and Raposo 2007). *Phaeomoniella chlamydospora*, causing Petri disease of grapevine, is well distributed in grape-growing regions all over the world. Nested PCR assay was employed to detect the pathogen at different stages of propagation process. *P. chlamydospora*-specific primers used in the nested PCR assays were specific for the amplification of pathogen DNA and did not amplify the DNA from *Phaeoacremonium aleophilum* and *R. solani*. The ITS PCR produced a 600-bp from all water samples, indicating the presence of the pathogen DNA. The nested PCR assay was the most sensitive method of detection of *P. chlamydospora*, followed by quantitative PCR and conventional PCR formats. Nested PCR increased the yield of the target template or reduced the effects of inhibition associated with the DNA extraction by using the cDNA of the first PCR template for the second PCR. The risk of carryover contamination, resulting in false positive results was greater with nested PCR than with conventional PCR format. The modified protocol minimized this risk by conducting both rounds in a single tube (Edwards et al. 2007).

### 2.1.4.3.3 Competitive PCR Assay

Competitive PCR assay, as a means of quantifying propagules of soilborne pathogens, employs the highly sensitive PCR and bypasses quantification problems caused by differences in the exponential PCR amplification of DNA by using an internal standard. The internal standard is a competitive DNA template that shares two primers and thus is coamplified in competition with specific DNA sequences in the sample. As the sizes of the fragments vary, amplification products from the internal standard and sample DNAs may be clearly separated on a gel. Use of competitive PCR for analysis of soil samples may be limited, since soil is heterogeneous and consists of large amounts of inhibitory compounds, resulting in difficulty with obtaining reliable results.

A competitive PCR assay was developed for the detection and quantification of *Verticillium tricorpus* directly from infested soil samples. Coamplification of *V. tricorpus* DNA with a competitor DNA provided accurate quantification in the range of $10^2$–$10^6$ spores and 1–500 MS in the inoculated soil. A strong correlation (r = 0.944) was found between the number of spores added to pathogen-free soil under controlled conditions and the number of spores estimated by competitive PCR format. The number of propagules determined at harvest time was not correlated with the initial amount of inoculum spiked at planting time, as determined by the competitive PCR assay (Heinz and Platt 2000). The competitive PCR assay was developed to quantify *V. dahliae* propagules directly in soil samples collected from fields where potato crops were grown, along with the soil dilution (SD) method used previously. PCR assay consistently detected a higher number of sclerotia (4.9 to 18.9) than did the classical SD method (0.06 to 0.51) in all the years (1995–1997). A strong correlation (r = 0.97) was observed between pathogen propagules estimated by competitive PCR and SD methods. Coamplification of *V. dahliae* DNA with competitor DNA provided accurate quantification in the range of $10^2$ to $10^7$ spores and 1 to 100 MS/g of soil. The differences in the values between replications were not significant, suggesting that the competitive PCR was reliable and reproducible. This procedure was fast and with no dependence on the subjectiveness of SD method, offering improvement in speed and precision over the methods employed previously (Mahuku and Platt 2002). *Sss* was detected and quantified by using a PCR-based assay employing SsF and SsR primers designed from ITS regions of the pathogen DNA. The PCR assay was modified with improved soil DNA extraction methods to detect *Sss* in soil samples. The DNA was extracted directly from aqueous extracts of field soil samples and a noninfested field soil sample, using a Bead-beating/CTAB method in addition to the UltracleanTM Soil DNA Kit. After amplification with specific primers, the pathogen was detected in infested soil only. A competitive PCR was employed to quantify *Sss* in naturally infested field soil samples. The ratios of *Sss* product of 434-bp and competitor product of 541-bp were determined for each reaction. The amounts of amplicon (434-bp) of pathogen origin increased with increasing numbers of spore balls of *Sss* in the reaction mixture. The levels of spore ball-counts in soil samples determined by competitive PCR showed a positive correlation with the extent of powdery scab disease incidence. The PCR formats may be employed for the routine detection and quantification of *S. subterranea* f.sp.*subterranea* in infested soils and also in plant tissues (Qu et al. 2006). The occurrence of powdery scab in the minituber production facility necessitated to find sources of inoculum, as the mintubers were from certified laboratory. By using PCR assay, the presence of *Sss* sporosori in the tunnel in the absence of potatoes could be detected. Further, the sporosori that remained in the run-off channels, water troughs and drain pipes were likely to be primary sources of inoculum, resulting in recurrence of disease (Wright et al. 2012).

### 2.1.4.3.4 Multiplex PCR Assay

The multiplex PCR assay enables detection of soilborne pathogens directly from plants and soil, without the need for culturing them in artificial media. The process of detection and identification of fungal pathogens may be accelerated by using multiplex PCR format. The sequences of a repetitive satellite DNA fragment of *P. infestans*, causative agent of the dreadful late blight disease of potato, were utilized for designing specific primers. A one-tube PCR multiplex protocol was formulated. All known A1 MTs of *P. infestans* races 1, 3, 4 and 7–11, occurring in Germany, and A2 MTs could be detected by multiplex PCR format (Niepold and Schober-Butin 1995). In another investigation, specific primers using sequences of ITS2 region of DNA amplified DNA fragments specific to *P. infestans* and *P. erythroseptica* (causing pink rot of potato tubers), facilitating the detection of these pathogens even prior to the development of disease symptoms (Tooley et al. 1998). A multiplex PCR format was developed for detecting two or more fungal pathogens infecting hops (*Humulus lupulus*). PCR primers R1 and R2 amplified a fragment of 305-bp nuclear DNA (mainly ITS1 region) from hop plants, a fragment of 297-bp nuclear rDNA from *Pseudoperonospora humuli* (downy mildew pathogen), a fragment of 248-bp

nuclear rDNA from *Sphaerotheca humuli* (powdery mildew pathogen), a fragment of 204-bp from nuclear rDNA of *V. albo-atrum* (Verticillium wilt pathogen) and a fragment of 222-bp from nuclear rDNA of *Fusarium sambucinum*. Likewise, R3 and R4 primers amplified 397, 598-, 312-, 331- and 317-bp fragments respectively from *S. humuli*, *P. humuli*, *V. albo-atrum* and *F. sambucinum*. By employing primer combinations, R1 + R2 + R3 and R4, all four pathogens could be detected by multiplex PCR format, thus saving considerable time required for diagnostic tests for individual pathogens (Patzak 2003). In a later investigation, the multiplex PCR format was applied for the detection and differentiation of two pathotypes of *V. albo-atrum*, infecting hop. SCAR primer pairs were employed in multiplex PCR assay. The pairs of SCAR primers, 9-21-For/9-21-Rev and 11-For/11-Rev amplified the fragments specific to hop PG1 and PG2 pathotypes. The amplified PCR products corresponded to the SCAR markers, revealing the specificity of the primers which remained unchanged after multiplex reaction. The sensitivity and specificity of diagnosis of pathotypes was improved by simultaneous amplification of two specific loci for PG2 and one locus for PG1, making the pathotypes screening by multiplex PCR format more reliable (Radišek et al. 2004).

*P. ramorum*, causal agent of sudden oak death (SOD) disease, has stream water as an important mode of dispersal. Leaf baits were used for monitoring pathogen movement. Leaves of rhododendron (*Rhododendron macrophyllum*) and tanoak (*Lithocarpus densiflorus*) were placed in mesh bags for trapping the propagules of *P. ramorum*. Leaves were assayed by isolation on selective medium and by using the multiplex PCR format with primers based on sequences of ITS of rDNA of the pathogen. The assay methods provided comparable results, but multiplex PCR assay was more sensitive. *P. ramorum* was recovered regularly at all seasons of the year from streams draining infested sites even after five years after eradication treatment. In streams with lower inoculum densities, recovery was much higher in summer than in winter. *P. ramorum* was isolated from streams in 23 watersheds. *P. ramorum* was detected, during intensive ground surveys that located tanoaks or other host plants at an average 306 meters upstream from the bait stations. *P. ramorum* was isolated from stream baits up to 1,091 meters from the probable inoculum source (Sutton et al. 2009).

*P. nicotianae* and *P. cactorum* infect strawberry crops, reducing the yield levels significantly. Simultaneous detection of these two pathogens by applying the multiplex PCR assay could be very useful for the survey to assess their occurrence in the main strawberry production areas in Japan. New species-specific primers were designed based on the sequences of ITS regions of rDNA and the ras-related protein gene *Ypt1*, respectively for *P. nicotianae* and *P. cactorum*. Multiplex PCR assay showed the required level of specificity, as indicated by tests against 68 isolates of other *Phytophthora* spp., *Pythium* spp. and other soilborne pathogens. The multiplex PCR format differentiated *P. nicotianae* and *P. cactorum* in DNA mixtures of mycelia of these two species. In addition, *P. nicotianae* and *P. cactorum* were detected in artificially and naturally infested soils, indicating the effectiveness of the multiplex PCR assay and the potential of the identified markers for use in diagnosis of the diseases caused by these pathogens. In order to determine the distribution of these pathogens, soil samples from 89 strawberry fields were analyzed by the multiplex PCR assay. The assay was successfully applied to survey the incidence of diseases caused by *P. nicotianae* and *P. cactorum* in strawberry plantings in eight prefectures of Japan (Li et al. 2011). Stoechas lavender (*Lavandula stoechas*) plants and other ornamentals are affected by root and collar rot diseases caused by *Phytophthora × pelgrandis*, a natural hybrid of *P. nicotianae x P. cactorum* in Italy. Polyphasic tests including morphological and cultural characteristics, sequencing the ITS rDNA region, the *Pheca* and the mitochondrial *coxI* genes, multiplex PCR with primers specific to *P. nicotianae* or *P. cactorum*, as well as random amplified polymorphic DNA (RAPD)-PCR assays were performed. *Phytophthora* isolates with atypical morphological and biological characteristics infecting Stoechas lavender were identified as *P. × pelgrandis*. This pathogen was already known to occur in European countries, the Americas and Taiwan (Faedda et al. 2013).

In order to detect a highly virulent race 4 (Cal race 4) of *F. oxysporum* f.sp. *vasinfectum (Fov)*, occurring in California cotton fields, the multiplex PCR (mPCR) assay was developed, based on a unique *Tfo1* transposon insertion in the PHO gene present in Cal race 4 isolates. A panel of *Fov* isolates representing different vegetative compatibility groups (VCGs) and DNA sequence types was assembled to test the specificity of the mPCR format. Of the 17 Cal race 4 isolates, 16 isolates produced a 583-bp amplicon, while the other isolates produced a 396-bp amplicon, indicating the absence of the Tfo1 insertion. This isolate was moderately virulent among the Cal race 4 isolates. In addition, 80 other *F. oxysporum* isolates associated with cotton and 11 other *formae speciales* of *F. oxysporum* produced only the 396-bp amplicon. The mPCR format could distinguish Cal race 4 isolates from other countries such as India race 4 isolates and China race 7 isolates, which did not contain the unique *Tfo1* insertion in their genome. Except for the *Tfo1* insertion, other isolates tested, had identical DNA sequences and all of them belonged to VCG 0114. Using the mPCR format, the pathogen could be detected directly from infected stem tissues even before expression of visible symptoms, providing an efficent technique for the precise identification of infested fields and seed lots resulting in reduction in dissemination of Cal race 4 isolates of *F. oxysporum* f.sp. *vasinfectum* in the cotton belt of the United States (Ortiz et al. 2017).

Contamination of food materials with aflatoxin produced by *Aspergillus flavus* is a potential threat to human and animal health. All strains of *A. flavus* do not produce aflatoxin. A multiplex PCR format was, therefore, developed to discriminate aflatoxin producers and nonproducers among the strains of *A. flavus*. Five genes of the aflatoxin gene cluster of *A. flavus*, two regulatory (*aflR* and *aflS*) and three structural (*aflD*, *aflO* and *aflQ*) were targeted with specific primers to highlight their expression in mycelia cultivated under inducing conditions for aflatoxin production. Expression of aflatoxin genes analyzed by multiplex PCR assay was well correlated with aflatoxin production. This investigation paved the way for screening *A. flavus* strains for their ability to produce aflatoxins (Degola et al. 2007).

### 2.1.4.3.5 Polymerase Chain Reaction (PCR)-ELISA

Combination of PCR and ELISA was applied in microplates for the detection of *Didymella bryoniae*, causing gummy stem blight disease along with *Phoma* spp. in cucurbits. Primers modified by addition of a fluorescein and a biotin label to the 5′ ends of the forward and reverse primers, respectively were employed. PCR amplification products were detected, employing ELISA test, using horseradish peroxidase-conjugated antifulorescein antibody and three substrates that yielded colored product. PCR-ELISA format reliably detected *D. bryoniae* in 45 of 46 samples and *Phoma* isolates (13) tested. The results obtained, were similar to those with the gel electrophoretic technique, when blind fungal samples were tested. PCR-ELISA format detected and differentiated all seven isolates of *D. bryoniae* and *Phoma* sp. PCR-ELISA format was found to be simple, specific, fast and convenient, in addition to saving time required for performing two different assays (Somai et al. 2002).

### 2.1.4.3.6 Real-Time Polymerase Chain Reaction

The conventional PCR format requires considerable time for post-PCR manipulation and processing of the reaction with slab gel and electrophoresis or hybridization to immobilized oligonucleotides, making the assay cumbersome. Further, quantification of PCR amplicons by image analysis requires additional time. To overcome these limitations, real-time PCR assay is preferred for obtaining results rapidly without compromising detection sensitivity. The real-time PCR involves the use of the fluorogenic 5′-nuclease (TaqMan) and a spectrofluorometric thermocycler. TaqMan, representing a homogeneous PCR, employs a fluorescence resonance energy transfer (FRET) probe typically consisting of a green fluorescent 'reporter' dye (6-carboxy-fluorescein, FSM) at the 5′ end and an orange 'quencher' dye (6-carboxy tetramethylrhodomine, TAMARA) at 3′ end. In addition to quencher, the presence of a phosphate group prevents PCR-derived elongation. During PCR amplification, the probe anneals to complementary strand of an amplified product, whereas *Taq* polymerase cleaves the probe during extension of one of the primers and the dye molecules are displaced and separated. As the electronically excited reporter is no longer suppressed by the quencher dye, variations in green emission can be monitored by a fluorescence detector. The measurements of green fluorescence intensity reflect the concentrations of the PCR amplicons in the reaction. The synthesis of amplicons can be monitored continuously during thermocycling and no post-PCR handling is required for product quantification. A specific 'Ct' value is worked out for each sample. The Ct value is the number of cycle at which a statistically significant increase in the reporter fluorescence can be first detected. The Ct values are then used to calculate the starting copy number of the target for each sample. This calculation can be done automatically by comparing Ct values of standards with known amounts of target copies. The chief advantage of real-time PCR is that amplification products can be monitored, as they are accumulated in the long-linear phase of amplification. In addition, the rapidity and greater accuracy of real-time PCR are the other advantages over conventional PCR format (Bohm et al. 1999; Schaad et al. 2003; Vandemark and Barker 2003). The real-time PCR assay applied for detection of microbial plant pathogens may be either amplicon sequence nonspecific (SYBR) or sequence-specific (TaqMan, Molecular Beacon or Scorpion PCR) (Mumford et al. 2006; Cooke et al. 2007). SYBR Green is a nonspecific dye that fluoresces, when intercalated into ds-DNA. On the other hand, amplicon sequence-specific methods are based on labeling of primers or probes with fluorogenic molecules that permit the detection of a specific amplified target fragment. Real-time PCR assays performed by either of the specific or nonspecific methods have been reported to show greater levels of sensitivity and specificity and provide results more rapidly than conventional PCR formats. Further, the risk of false positives is at lower levels in real-time PCR assay, in addition to being amenable for multiplex detection and quantitative assessments of pathogen populations in plants, soil and water samples (Schena et al. 2006).

Specific primers based on ITS region of rDNA of *P. capsici* were employed for pathogen detection by PCR in artificially inoculated and naturally infected pepper plants. A real-time PCR assay was developed for reliable and quantitative detection of *P. capsici* in pepper tissues. DNA levels of highly virulent and a less virulent isolates were determined in different pepper genotypes with varying degrees of resistance. Using SYBR Green and specific primers for *P. capsici*, the minimal amount of pathogen DNA quantified was 10 pg. The presence of pathogen DNA could be detected as early as 8 h postinoculation after which period, the increase in pathogen DNA in susceptible cultivar was rapid and slower in resistant cultivars. The pathogen DNA concentration in each pepper genotype was negatively correlated with resistance to Phytophthora root rot (PRR) disease. Similarly, the virulence of *P. capsici* isolates was positively correlated to the degree of host tissue colonization. Differences in the amount of pathogen biomass among pepper tissues were observed, with stems containing maximum amount of biomass of *P. capsici*. The real-time PCR format developed in this investigation was sensitive and robust enough to assess both pathogen development and levels of resistance of pepper genotypes to *P. capsici* (Silvar et al. 2005a). Oospores of *P. capsici* present in the soil were detected and quantified by combining sieving-centrifugation and a real-time quantitative PCR (QPCR) format. Soil samples (5) representing three different soil textures infested with oospores of *P. capsici* were suspended in water and then passed through 100-, 60- and 38-μm metal sieves and finally through a 20 μm filter. The material retained in the filter was resuspended in water and centrifuged. The pellet was subjected to a sucrose density gradient process and the concentration of oospores was determined using a hemacytometer. Primers based on the rDNA sequence were employed in QPCR assay for quantifying oospores recovered from the soil samples. A regression equation was formed based on the positive relationship between the number of oospores and the amount of DNA extracted from oospores recovered from the soil samples. The pathogen population represented by oospores could be determined, using the QPCR protocol developed in this investigation (Pavon et al. 2008).

*Phytophthora cryptogea* causes root rot disease of gerbera which affects cutflower production seriously in Europe. Conventional PCR and SYBR Green-based real-time PCR assays were applied for the detection and quantification of pathogen population in infected plants. The primer pair Cryp1 and Cryp2 was designed based on the sequences of *Ypt1* gene of *P. cryptogea*. The highly polymorphic nature of *Ypt1* gene sequences obtained from different *Phytophthora* spp. facilitated the differentiation of closely related species that have identical ITS regions. The primer pair Cryp1/Cryp2 amplified a product of 369-bp from the DNA of 17 isolates of *P. cryptogea*, but not from DNA of 34 other *Phytophthora* spp., water molds, true fungi and bacteria tested. This primer pair Cryp1/Cryp2 was adapted to real-time PCR format and comparative efficiency of conventional and real-time PCR formats was assessed. The PCR assay detected *P. cryptogea* in naturally infected gerbera plants at 21 DAI, the detection limit being 5 × $10^3$ zoospores or 16 fg DNA of the pathogen. On the other hand, real-time PCR format was more sensitive (100-fold) with a detection limit of 50 zoospores or 160 ag of the pathogen DNA. The pathogen presence could be detected in symptomless roots by real-time PCR format, indicating its effectiveness to provide sensitive detection and precise quantification of *P. cryptogea* DNA in symptomatic and asymptomatic plant tissues (Minerdi et al. 2008). *P. erythroseptica*, causal agent of pink rot of potatoes, infects all underground potato tissues, such as roots, stolons, tubers and basal stems. Conventional PCR and real-time PCR formats were applied to detect *P. erythroseptica* in different potato tissues. *P. erythroseptica* was detected in stem pieces, leaf tissue, aerial tubers and roots of Shepody plantlets grown in potting mixtures amended with pathogen zoospores (23 × $10^4$/ml) by PCR and real-time PCR formats. About 95% of the samples showed positive reaction with real-time PCR, while the pathogen was detected by conventional PCR format in 45% of the inoculated plantlets only, indicating that real-time PCR assay was more sensitive in detecting *P. erythroseptica*. The pathogen was detected in progeny tubers and stolons produced by infected potato plants. The presence of *P. erythroseptica* could be detected in a few samples of debris taken from naturally senesced above ground potato tissue after harvest. Pathogen propagules in leaves and stems were viable as indicated by successful isolation of *P. erythroseptica* from these tissues. The results indicated the potential danger of using asymptomatic tubers for planting in the next season (Nanayakkara et al. 2009).

The appearance of new genotypes of *P. infestans*, causal agent of late blight disease of potato and tomato, with different MTs and sensitivity to metalaxyl fungicide, had altered the genetic composition of the pathogen population during the past few years. Hence, development of methods employing genetic markers, enabling rapid assessment of genotypes of *P. infestans* from small quantities of biological material would be useful for the early detection and control of the pathogen throughout Canada. Mining of the *P. infestans* genome revealed the presence of several regions containing single nucleotide polymorphisms (SNPs) within both nuclear genes and flanking sequences of microsatellite loci. Allele- specific oligonucleotide polymerase chain reaction (ASO-PCR) assays were developed from 14 of the 50 SNPs found by sequencing. Nine optimized ASO-PCR assays were validated using a blind test comprising *P. infestans* and other *Phytophthora* spp. The assays revealed diagnostic profiles unique to each of the five dominant genotypes of *P. infestans* present in Canada. The markers could be employed with environmental samples such as infected leaves and had the potential to contribute to the toolbox available to assess the genetic diversity of *P. infestans* at the intraspecific level. Further, early warnings about the genotypes of *P. infestans* present in potato and tomato fields based on the results of the protocol developed in this investigation, would help growers to use appropriate fungicide application schedule (Gagon et al. 2016).

*Helminthosporiun solani*, causative agent of silver scurf disease, induces blemishes on potato tubers, affecting the market quality of tubers seriously. For the detection of *H. solani* in plant tissues and soil, two sets of PCR primers, Hs1F1/Hs2R1 (outer) and Hs1NF1/Hs2NR1 (nested) were designed, based on the unique sequences of the nuclear ribosomal ITS1 and ITS2 regions of *H. solani*. Nested PCR assay was employed to enhance the levels of sensitivity and specificity of single round PCR. Each primer set amplified a single product of 447-bp and 371-bp, respectively with DNA from 71 European and North American isolates of *H. solani* and the specificity of primers was demonstrated by testing against DNA from other fungal and bacterial plant pathogens. A simple and rapid procedure for direct extraction of DNA from soils and potato tubers was developed. The sensitivity of PCR for the specific detection of *H. solani* in seeded soils was 1.5 spores/g of soil. The pathogen was detected in naturally infested soil and from tuber peel and peel extract from infected and apparently healthy tubers. Pathogen-specific primers and a TaqMan™ fluorogenic probe were designed using original primary sequences to perform real-time quantitative (TaqMan™) PCR. The same levels of sensitivity for specific detection of *H. solani* in soil and tubers were obtained during first round TaqMan-based PCR as with conventional nested PCR and gel electrophoresis. The quantitative PCR format was rapid, enabling precise estimation of tuber and soil contamination with *H. solani* (Cullen et al. 2001).

*P. nicotianae* and *P. citrophthora* infect roots of citrus and survive in nursery soil. Two primers, specific for *P. nicotianae* (Pnb) and *P. citrophthora* (Pc2B) were modified to obtain Scorpion primers for the detection and identification of the pathogen by real-time PCR assay. Multiplex PCR assay with dual labeled fluorogenic probes allowed concurrent identification of both species of *Phytophthora* among 150 isolates including 14 *Phytophthora* spp. Using *P. nicotianae*-specific primers, a delayed and lower fluorescence increase was obtained also from *P. cactorum* DNA. However, in separate real-time amplification, a specific increase in fluorescence from *P. cactorum* was avoided by increasing the annealing temperature. In multiplex PCR, with a series of ten-fold DNA dilutions, the detection limit was 10 pg/μl for *P. nicotianae* and 100 pg for *P. citrophthora*, whereas in separate reaction, DNA up to 1 pg/μl was detected for both pathogens. Simple and rapid procedures for direct DNA extraction from soil and roots were utilized to yield DNA whose purity and quality was suitable for PCR assays. By combining these protocols with a double amplification (nested Scorpion-PCR) using primers Ph2-ITS4 amplifying DNA from the main *Phytophthora* species (first round) and PnB5-Pn6 Scorpion and Pc2BScorpion-Pc7 (second round), it was possible to achieve real-time detection of *P. nicotianae* and *P. citrophthora* from roots and soil. The sensitivity level was similar to that of traditional detection methods based on selective media. A high and significant correlation between the pathogen population (propagules) and the real-time PCR cycle threshold was observed for analysis of artificially and naturally infested soil samples (Ippolito et al. 2004). In a later investigation, a rapid, specific and sensitive real-time PCR detection procedure was developed for *P. nicotianae*, employing primers targeting ITS regions of rDNA genes of *Phytophthora* spp. Based on the nucleotide sequences of ITS2 of 15 different species of *Phytophhora*, the primers and probe were designed specifically to amplify DNA from *P. nicotianae*. Using a series of ten-fold dilutions of DNA extracted from *P. nicotianae*, the detection limits of conventional PCR and SYBR Green I PCR and TaqMan PCR formats were found to be, respectively 10 pg/μl, 0.12 fg/μl and 1.2 fg/μl. The real-time PCR formats were more sensitive ($10^3$-$10^4$-fold) than conventional PCR assay. The simple and rapid protocols developed in this investigation maximized the yield and quality of recovered DNA from infested soil and enabled processing of many samples in a short time. The sensitivity of detection of *P. nicotianae* from soil, by real-time PCR format, was 1.0 pg/μl. The detection protocol was tested for surveying soil samples from tobacco fields in China and the procedure showed practical utility providing reliable results for naturally infested soil samples (Huang et al. 2010).

*P. ramorum* causes the destructive sudden death in oak and several tree species *Lithocarpus densiflora* and *Quercus* spp. and foliar infections on shrubs such as *Vaccinum* spp. and *Rhodendron* spp. commonly. European Union emergency restrictions have been applied since 2002, to prevent the introduction and dispersion of *P. ramorum* in Europe. The need for developing sensitive and reliable detection methods was emphasized. Real-time and nested PCR formats were developed for detecting and quantifying the pathogen DNA in infected plants as an environmental screen through California. The real-time PCR assay had a detection limit of >12 fg of pathogen DNA and was specific to *P. ramorum*, when tested against 21 *Phytophthora* spp. Hundreds of symptomatic samples from 33 sites in 14 California counties were assayed, resulting in the recognition of ten new hosts species and 23 infested areas, including four new counties, indicating a wide distribution of *P. ramorum* in California State

(Hayden et al. 2004). In a later investigation, a real-time PCR assay using 5′ fluorogenic exonuclease (TaqMan) chemistry was applied for the detection and quantification of *P. ramorum* in diseased tissues. A universal primer and probe set was included as an internal control and this format was sensitive with a detection limit of 15 fg of target DNA, when used in nested design or 50 fg, when used in single round of PCR. The specificity of the assay was indicated by the absence of amplification from DNA of 17 other *Phytophthora* spp. tested. The nested methods were shown to be more sensitive than non-nested procedure. The host substrate had significant influence on detection sensitivity. Field testing revealed that the nested TaqMan protocol could detect *P. ramorum* in 255 of 874 plants in California Woodlands. On the other hand, single round TaqMan procedure was able to detect the pathogen, in significantly lower number of plants only (Hayden et al. 2006).

A real-time fluorescent PCR assay was developed for the detection of *P. ramorum* based on mitochondrial DNA sequence with an ABI Prism 7700 (TaqMan) Sequence Detection System. Primers and probes for detecting *P. pseudosyringe* (causing symptoms similar to *P. ramorum* on some host plant species) were also designed. The species-specific primer probe systems were combined in a multiplex assay with a plant primer-probe system to allow plant DNA present in extracted samples to serve as positive control in each reaction. The lower limit of detection of *P. ramorum* DNA was 1 fg of genomic DNA which was lower than many of the PCR assays employed earlier for *Phytophthora* spp. The assay could be applied in a three-way multiplex format for simultaneous detection of *P. ramorum, P. pseudosyringae* and plant DNA in a single tube. *P. ramorum* was detected at a dilution of $10^{-5}$ of extracted tissues of artificially inoculated rhododendron and the amount of pathogen DNA from infected tissue was estimated using a standard curve. Field samples of several hosts showed the presence of *P. ramorum* tested by the protocol developed in this investigation which was highly sensitive and specific for confirmation of infection by *P. ramorum* (Tooley et al. 2006). Field samples from 41 wild plant species were tested for the presence of *P. ramorum*, employing five detection methods viz., two culture-based assays, DAS-ELISA (using PAbs), nested PCR assay and TaqMan real-time assay. Diagnostic values including sensitivity, specificity, positive predictive value and negative predictive values were considered for determining the comparative efficacy of the five diagnostic methods applied. Significant effects of season, host species and laboratory were evident. A combination of either culture-based and molecular diagnosis or of two molecular assays was preferable to reliably detect and identify *P. ramorum*. Diagnosis of infection by *P. ramorum* should be taken up during wet and warm periods favorable for pathogen development. Length of time interval between sample collection and processing has to be reduced to the bare minimum to obtain reliable results, especially for culture-based methods (Vettraino et al. 2010).

A single-round real-time PCR assay based on TaqMan chemistry, designed within the ITS region of nuclear ribosomal (nr) RNA gene was developed for the detection of *P. ramorum*, causing bleeding cankers, die-back and leaf blight on trees and shrubs in Europe and North America. This protocol did not require any post-amplification steps or multiple rounds of PCR. This one-step real-time PCR format had a detection limit of 10 pg of *P. ramorum* DNA and could detect the pathogen in plant materials containing 1% infected material by weight within 36 cycles of PCR. The assay was specific to *P. ramorum*, as indicated by the absence of positive reaction with the DNA of 28 other *Phytophthora* spp. tested. Further, a rapid and simple method for extraction of DNA directly from host plants was developed and the extracted DNA was amplified using the multiplex protocol. The results showed that the amplified DNA was extracted from 84.4% of samples, as demonstrated by amplification of host plant DNA. The real-time protocol was used to test 320 samples from 19 different plant species from which the DNA was successfully extracted (Hughes et al. 2006). An intra-laboratory procedure was developed for detection of *P. ramorum*. The real-time procedure was applied for rapid screening host plant species, *Rhododendron* spp. *Viburnum* spp. and *Pieris* spp. for infection by *P. ramorum*. The limit of detection was 50 fg. The procedure was effective for the detection of the pathogen in asymptomatic plant tissues also (Chandelier et al. 2006).

Molecular methods available earlier failed frequently to reliably distinguish *P. ramorum* from closely related species. In order to circumvent this limitation and to increase the confidence of tests, ITS, β-tubulin and elicitin gene regions were sequenced and searched for polymorphism in a collection of *Phytophthora* spp. Molecular beacons, TaqMan and SYBR Green approaches were compared for their efficacies. The assays differentiated *P. ramorum* from 65 species of *Phytophthora* tested. DNA extracts from 48 infected and noninfected plant samples that were used for testing. All three real-time PCR methods detected *P. ramorum* in all environmental samples. Sequence analysis of the *coxI* and *coxII* spacer regions confirmed the identity of the pathogen in most samples. The assays based on detection of ITS and elicitin regions using TaqMan tended to have lower cycle threshold values than those using β-tubulin and seemed to be more sensitive (Bilodeau et al. 2007). In the later investigation, employing multiple gene regions of *P. ramorum* was shown to increase the reliability of detection of this pathogen that had affected tens of thousands of oak trees over large areas as well as more than 100 other plant species, indicating the urgency of finding an effective solution to this problem. In order to improve the sensitivity and reliability of detection of *P. ramorum*, three different TaqMan assays were multiplexed with a fourth TaqMan specific to *Phytophthora* genus in a single reaction. A second multiplex TaqMan PCR assay to detect oomycetes and give a positive PCR amplification in the presence of plant DNA was also designed and tested in conjunction with the *P. ramorum* ITS and *Phytophthora* genus TaqMan assays. These assays were tested on different *Phytophthora* spp. and were verified on two different sets of field samples previously assayed by other laboratories. These were obtained from multiple field hosts infected by various *Phytophthora* species and the DNA from one set was extracted from ELISA lysates. All known *P. ramorum* samples from pure cultures or field samples were detected using the multiplex real-time PCR assays. The detection sensitivity of TaqMan multiplex assays was invariably at a lower level, compared to single separate reactions. However, reduction in test costs and increased throughput appear to be factors in favor of the multiplex formats for adoption (Bilodeau et al. 2009).

A semiquantitative PCR (QPCR) assay was developed for efficient determination of in planta DNA of the necrotrophic pathogen *P. cinnamomi*, causative agent of root rot diseases in several crops. This protocol effectively avoided problems caused by variation in the DNA extraction efficiency and degradation of host DNA during host tissue necrosis. Normalization of pathogen DNA to sample fresh weight or host DNA in samples with varying degrees of necrosis resulted in overestimation of pathogen biomass. Purified plasmid DNA, containing the pScFvB1 mouse genes was added during DNA extraction and pathogen biomass was normalized on plasmid DNA rather than on host DNA or sample fresh weight. The QPCR was found to be robust and improved the accuracy of pathogen quantification in both resistant (non-host *Arabidopsis thaliana*) and susceptible (*Lupinus angustifolius*) interactions even in the presence of substantial host cell necrosis induced by *P. cinnamomi* (Eshraghi et al. 2011). *P. cinnamomi*, causes PRR which is responsible for severe economic losses to the avocado industry all over the world. A low cost and precise assay was developed for the reliable detection and quantification of *P. cinnamomi* in avocado cultivars/genotypes to determine the levels of their tolerance to PRR disease. A nested real-time PCR assay was developed for sensitive detection of pathogen DNA in avocado plant tissues. Root samples of a highly tolerant (Dusa) and less tolerant (RO.12) rootstocks were collected at 0, 3, 7, 14 and 21 DAI with *P. cinnamomi* and assayed for pathogen DNA contents. Nested primers developed in this investigation were specific and sensitive in detecting *P. cinnamomi* in root tissues. The pathogen DNA contents of less tolerant RO.12 plants were significantly higher than that of tolerant Dusa plants at all sampling times. The assay protocol might be useful for breeding programs for identification of rootstocks tolerant/resistant to PRR disease of avocado (Engelbrecht et al. 2013).

*Pythium* spp., causing damping-off diseases in nurseries of vegetable crops are widespread in all countries. Species-specific primers based on variable regions of ITS of rDNA were employed for the detection and quantification of nine *Pythium* spp. in eastern Washington. Identification and quantification of *Pythium* spp. in soil using conventional methods of isolation in selective media and dilution plating are time-consuming and enumeration and counts may not be representative of the pathogenic populations of *Pythium* spp. A real-time PCR format using SYBR Green I fluorescent dye was developed for the detection and species-specific primers designed based on variable regions of the ITS of rDNA. Standard curves were generated for each *Pythium* spp. Isolates of *Pythium* (77) were screened and the specificity of each primer set was confirmed. The populations of *P. irregulare*

group I, *P. irregulare* group IV and *P. ultimum* were correlated with the DNA contents of soil samples. Other *Pythium* spp. identified in soil samples by the real-time PCR assay were *P. abappressorium*, *P. attrantheridium*, *P. heterothallicum*, *P. paroecandrum*, *P. rostratifingens* and *P. sylvaticum*. The real-time PCR format developed in this investigation was rapid and precise in identifying and quantifying *Pythium* spp. in soils (Schroeder et al. 2006). Replant and decline diseases affecting grapevines inflict heavy qualitative and quantitative losses as well as extra costs for replanting. The role of *Pythium* spp. in decline syndrome in South Africa was investigated in the nurseries and established vine gardens. Quantitative real-time PCR (qPCR) assays were applied for the detection of most prevalent oomycete pathogens. Of the 26 *Pythium* spp. and 3 *Phytophthora* spp. isolated from grapevines, the most infections in the nurseries were caused by *P. vexans* (16.7%), followed by *P. ultimum* var. *ultimum* (15%) and *P. irregulare* (11.7%). In established vineyards, *P. irregulare* (18%), and *P. vexans* (6.2%) were more prevalent along with *P. heterothallicum* (7.3%). *P. cinnamomi* (5.1%) was predominant, followed by *P. niederhauserii* (1.1%). In established vineyards, a higher incidence and more diverse species composition were recorded in spring and winter than in summer. *Ph. cinnamomi* and *P. irregulare* were aggressive and *Ph. niederhauserii* and *P. vexans*, newly recorded pathogens, were also equally aggressive. The qPCR assays were effective in detecting and quantifying the pathogens involved in replant and decline diseases of grapevine caused by *Pythium* spp. and *Phytophthora* spp. (Spies et al. 2011).

Crops grown under hydroponic culture systems are seriously damaged by high temperature-tolerant *Pythium* spp. in Japan and diseases induced by them are difficult to contain, because the zoosporic pathogens spread rapidly. A real-time PCR format was, therefore, developed for monitoring the spread of zoospores of *P. aphanidermatum*, *P. helicoides* and *P. myriotylum*. Specific primers and probes were designed based on the sequences of ITS regions of rDNA of these pathogens. The assay was specific to target *Pythium* spp. individually and also failed to generate a positive reaction when closely related nontarget species were tested. The detection limit for each pathogen was 10 fg DNA which corresponded to 4, 3, and 4 zoospores of *P. aphanidermatum*, *P. myriotylum* and *P. helicoides*, respectively. The real-time PCR format was successfully employed to evaluate and monitor zoospore populations in the nutrient solutions of ebb-and-flow irrigation systems for potted flower production and closed hydroponic culture systems for tomato production. The results clearly revealed the presence of these pathogens in the hydroponic culture systems all the time in a year and their spread before appearance of disease symptoms on infected plants (Li et al. 2014).

*P. brassicae*, causal agent of club root disease of cabbage and other cruciferous crops is soilborne and the resting spores survive in the soil for several years. A multiplex TaqMan qPCR assay, including a competitive internal positive control (CIPC) was developed to detect and quantify the resting spores of *P. brassicae* in the soil. The CIPC amplicon was developed by modifying a sequence coding for GFP so that it could be amplified along with *P. brassicae*-specific primers. Addition of CIPC at 5 fg/μl to the singleplex qPCR format designed to quantify *P. brassicae* genomic DNA, did not reduce the sensitivity, specificity or reproducibility of the assay. Amplification of CIPC did not occur with soil samples either artificially or naturally infested with *P. brassicae* resting spores. When the samples were diluted and reassessed, the quantification cycle of CIPC relative to the control (water only) was delayed in each sample. The magnitude of delay was used to adjust the estimate of resting spore concentration. The correlated concentration estimates were significantly higher than unadjusted estimate which indicated the presence of DNA inhibitors in samples even after dilution. The assay was optimized for use on a range of soil types. The protocol developed in this investigation showed an improvement over the methods available previously (Deora et al. 2015). Genetic marker(s) useful as a diagnostic tool were identified to differentiate and predict certain pathotypes of *P. brassicae*, causing club root disease. The PCR assay employing gene-specific primers was used to identify 117 non-housekeeping *P. brassicae* genes for their presence/absence in isolates representing pathotypes 2, 3, 5, 6 or 8 (Williams' system) and in new strains of the pathogen from Alberta, Canada which were highly virulent on club root resistant canola (*Brassica napus*) cultivars. The gene *Cr811* was detected in isolates of pathotype 5 and also in new strains of *P. brassicae* which caused severe club root on resistant canola. Three additional newly identified virulent populations, similar to pathotype 3, also carried the *Cr811* gene. In contrast, the *Cr811* gene was absent in all isolates representing the original pathotypes 2, 3, 6 and 8. Quantitative real-time PCR analysis indicated the presence of greater copy numbers of *Cr811* in the newly virulent pathotype 3-like and pathotype 5-like populations versus the original pathotype 5. The results revealed the potential of the method to identify new pathotypes of *P. brassicae*, before the breakdown of resistance in a canola crop. The association of *Cr811* with infection by *P. brassicae* of previously resistant canola cultivars, as revealed by qPCR analysis, suggested that *Cr811* might play an important role in the interaction between host resistance and *P. brassicae* pathogenicity (Feng et al. 2016). Quantitative PCR assay used for quantifying long-lived resting spores of *P. brassicae*, causal agent of clubroot disease of *Brassica* crops was unable to differentiate viable and nonviable resting spores. Propidium monoazide (PMA) can inhibit amplification of DNA from nonviable microbes. Spore suspensions from mature and immature clubs were heat-treated followed by application of PMA-PCR assay. Bioassays were also performed to assess viability of resting spores. Prior to heat treatment, assessments by PMA-PCR and qPCR assays were compared and most of the mature spores were found to be viable. However, only a small percentage of immature spores (< 26%) were amplified in PMA-PCR assay. Bioassays showed that clubroot severity was much higher in plants inoculated with mature spores than in plants inoculated with immature spores. Heat treatment could be considered to affect significantly the estimates of mature spores from qPCR assay,

but spore estimates from PMA-PCR and clubroot severity in bioassays were both substantially reduced. Estimates of spore concentration with PMA-PCR were less consistent for immature spores. A protocol for extracting resting spores from soil was developed for using PMA-PCR on infested soil to estimate the viable spore counts of *P. brassicae* more precisely (Al-Daoud et al. 2017).

*S. endobioticum*, causing potato wart disease, is soilborne and PCR-based methods have been applied for the detection and quantification of the pathogen in plants and soils. Conventional and real-time PCR formats, using primers targeting ITS region of the multicopy gene rDNA were evaluated for their specificity, sensitivity and reproducibility. The primers amplified a product of 472-bp from *S. endobioticum* DNA, but not DNA from other potato pathogens and related species, indicating the specificity of the primer. Standard cell disruption and DNA extraction and purification methods were optimized for amplification of pathogen DNA from resting sporangia. Positive amplification was achieved from a single resting sporangium and equivalent preparations from soil extracts. A real-time PCR assay developed for soil-based extracts using primers and probes based on the rDNA gene sequences, involved coamplification of target DNA along with an internal DNA fragment. Both conventional and real-time PCR methods had a threshold sensitivity of ten sporangia per PCR assay. Among the three soil extraction procedures tested, Hendrickx centrifugation (HC) method could process 100-g samples efficiently in one single PCR assay, showing its high capacity for use in routine soil analysis for disease risk assessments (van den Boogert et al. 2005). In another investigation, specific primers and a TaqMan probe, designed from the ITS region of the multicopy rDNA gene were employed for the detection of *S. endobioticum* in extracts from artificially and naturally infested soil. Coamplification of target DNA along with an internal competitor DNA fragment ensured more reliable diagnosis by guarding against false negative results. The TaqMan assay was also applied for the detection of *S. endobioticum* in infected potato stolons along with cytochrome oxidase gene as a potato endogenous control. The sensitivity of TaqMan assay was improved (>100-fold), providing reliable and precise diagnosis of potato wart disease in infected plant tissues (van Gent-Pelzer et al. 2010).

*S. subterranea*, the incitant of potato powdery scab and root galling, functions also as the vector of *Potato mop-top virus* (PMTV). A specific primer pair and a fluorogenic TaqMan® probe were designed for quantification of *S. subterranea* in soil, water and plant tissue samples. The DNA from cystosori, zoospores, plasmodia and zoosporangia of *S. subterranea* was used as template. The DNA was extracted directly from cystosori, suspended in water and clay soil spiked with varying numbers of cystosori. The DNA obtained from zoospores released into nutrient solution by cystosori in the presence of tomato bait plants, was also tested, as was DNA from plasmodia and zoosporangia in infected tomato roots. The pathogen DNA was detected even at low inoculum levels. This protocol has the potential for application in investigations on biology of *S. subterranea* (van de Graaf et al. 2003). In a later study, the effects of levels of soil inoculum and three environmental factors such as soil type, soil moisture regime and temperature on the incidence and severity of potato powdery scab disease were assessed under controlled conditions. Real-time PCR assay using primers and a TaqMan® probe specific to *S. subterranea* was applied to quantify the pathogen populations in soils. Soil inoculum levels of the pathogen did not significantly affect the incidence and severity of either tuber infection or powdery scab symptoms at maturity. Detection of *S. subterranea* DNA in asymptomatic potato tubers did not increase the soil inoculum levels. This suggested that most latent infections were due to the presence of the pathogen in tuber tissues rather than by cystosori on tuber surface. The high frequency of latent tuber infections have to be considered as a potential danger for taking appropriate management measures to contain the spread of powdery scab disease in potatoes (van de Graaf et al. 2005). A fast automated procedure was combined with a real-time PCR assay using TaqMan® chemistry to improve the effectiveness of diagnosis of powdery scab on potato tubers. An internal control consisting of cytochrome oxidase gene was included to guard against false negative results. Scab tuber samples (37) taken from different potato cultivars collected from various locations in the United Kingdom were tested for validation of the real-time assay protocol. Presence of *S. subterranea* was detected in all tuber samples with asymptomatic and symptomatic infections. The TaqMan format was more sensitive (100-fold) than conventional PCR and ELISA techniques for the detection of *S. subterranea* in potato tubers (Ward et al. 2004). In a later investigation, a real-time PCR assay for the detection and quantification of *S. subterranea* in soil was developed. The DNA from spore balls, zoospores and plasmodia/zoosporangia of powdery scab pathogen was detected and quantified using the protocol developed in this investigation. The viability of spore balls in soil could be determined by this real-time PCR format. Further, by combining real-time PCR with a tomato plant bait test, zoospore release and infection levels in potato tubers under different environmental conditions could be studied (Lees et al. 2008).

*N. haematococca* (anamorph- *Fusarium solani* f.sp. *pisi*) causes foot rot disease of pea impacting the production levels adversely in most countries. A PCR-based assay was developed using primers based on the sequences of the pathogenicity gene *PEP3* present exclusively in highly pathogenic forms of *N. haematococca* for the detection and quantification of the pathogen in field soils. The quantitative PCR format was applied to estimate the pathogen population in soil samples and to establish the relationship between *PEP3* gene numbers and incidence of foot rot disease. The quantitative PCR format was efficient and specific in quantification of pathogen population with an amplification efficiency of 92%. The gene copy numbers differed significantly between fields and were positively correlated to the number of spores of pathogenic *N. haematococca* to pea causing foot rot disease. *PEP3* numbers up to 100/g of soil constituted the threshold number of infection, potentially capable of causing foot rot disease. The density of virulent *N. haematococca* in

field soil capable of causing foot rot disease could be determined with a high degree of accuracy with this assay (Etebu and Osborn 2010).

*Fusarium virguliforme (Fv,* syn. *Fusarium solani* f.sp. *glycines)*, primary incitant of soybean sudden death syndrome (SDS) is difficult to isolate and identify, because of its slow-growing nature. Hence, with the aim of developing a fast and reliable method of detection of *Fv*, protocols were optimized for the extraction of pathogen DNA from cultures or infected dry roots. A procedure to detect the presence of PCR inhibitors in DNA extracts was also developed. Real-time quantitative PCR (QPCR) assays were applied for both absolute and relative quantification of *Fv*. The pathogen population was quantified based on the detection of the mitochondrial small subunit rRNA gene and the host plant based on detection of the *cyclophilin* gene of the soybean plants, both with and without SDS foliar symptoms with contents as low as $9 \times 10^{-5}$ ng in the absolute QPCR assays. The relative QPCR assay results are reliable, if care is taken to avoid PCR inhibition (Gao et al. 2004). *F. virguliforme*, causes variable symptoms of soybean sudden death syndrome (SDS) in North America, making disease diagnosis difficult. As the pathogen has plastic morphology, identification of the cause of SDS became problematic. Despite the availability of several real-time PCR formats, accurate quantification of *F. virguliforme* in soil and plant tissues was not possible. A TaqMan qPCR assay was developed based on the ribosomal DNA (rDNA) IGS region of *F. virguliforme*. The detection limit of the assay was 100 fg of pure *F. virguliforme* DNA or 100 macroconidia in 0.5 g of soil. The specificity of the assay was demonstrated by testing with genomic DNA of closely related *Fusarium* spp. and other soilborne fungal pathogens. An exogenous control was multiplexed with the assay to evaluate the possible inhibition of PCR amplification. In addition, a conventional PCR format was also shown to be useful for the detection of *F. virguliforme*, under conditions where qPCR format could not be applied (Wang et al. 2015a). Although several quantitative real-time PCR (qPCR) assays have been shown to be effective for the detection and quantification of *F. virguliforme*, their comparative sensitivity and specificity has not been assessed. Six qPCR assays were, therefore, compared in five independent laboratories, using the same set of DNA samples from fungi, plants and soil. Multicopy gene based assays targeting the ribosomal DNA, IGS or the mitochondrial small subunit (mtSSU) showed relatively high sensitivity [limit of detection (LOD = 0.05 to 5 pg)], compared with single-copy gene (FvTox1)-based assay (LOD = 5 to 50 pg). Specificity varied greatly among assays, with the FvTox1 assay ranking the highest (100%) and two IGS-based assays being slightly less specific (95 to 96%). Another IGS-based assay targeting four SDS-causing fusaria showed lower specificity (70%), while the two mtSSU-based assays were lowest (41 and 47%). An IGS-based assay showed consistently highest sensitivity (LOD = 0.05 pg) and specificity and inclusivity above 94%. This format was suggested as the most useful qPCR format for diagnosis of *F. virguliforme* infection and its quantification in plants and soil samples (Kandel et al. 2015).

A reliable and rapid method of detection and quantification of *F. oxysporum* f.sp. *ciceris (Foc)*, causing chickpea wilt disease, in soil and chickpea tissues, was developed. The real-time quantitative PCR (qPCR) format allowed quantification of *Foc* DNA down to 1 pg in soil as well as in roots and stem of chickpea. The protocol could quantify as low as 45 CFU of *Foc*/g of dry soil from naturally infested field soil. It was possible to differentiate races of *Foc*, with differing virulence, based on the reactions of chickpea cultivars that contained variable concentrations of pathogen DNA, as determined by the qPCR assay. Further, the assay detected early asymptomatic root infection and distinguished differences in the level of resistance of 12 chickpea cultivars that were grown in the field infested with several races of *Foc*. The possibility of recognizing early asymptomatic root infection of chickpea by applying this protocol will be very useful for investigations on the process of infection, epidemiology and resistance breeding in chickpea (179Jiménez-Fernandez et al. 2011). *F. oxysporum* f.sp. *phaseoli (Fop)*, infecting common bean (*Phaseolus vulgaris*), may be introduced into the field through infected seeds. A real-time PCR (qPCR) assay was developed for the detection and quantification of *Fop* in common bean seeds. Seed lots of seven cultivars with infection incidence ranging from 0.25 to 20% were prepared by mixing infected and healthy seeds in required proportions. The efficacy of SYBR Green and TaqMan PCR formats were compared using primers based on the *Fop* virulence factor *ftf1*. The primers amplified a product of 63-bp for highly virulent strains of *Fop*, but did not generate this amplicon from nonpathogenic or weakly virulent isolates of *F. oxysporum* from *P. vulgaris* or other hosts. Under optimized conditions, both qPCR formats detected *Fop* infection at the lowest level of infection (0.25%) tested. The results suggested the superiority of TaqMan format to be more reliable at quantification than the SYBR Green format (de Souza et al. 2015).

Fusarium wilt disease affecting cucurbit crops caused by *F. oxysporum* f.sp. *cucumerinum* is of significant economic importance. The pathogen is soilborne and it is capable of infecting all cucurbits grown in infested fields. A real-time quantitative PCR assay and an ultra-sensitive diagnostic pseudo-nested PCR formats were developed for specific and reliable detection and quantification of the pathogen in environmental samples. Standard curves were prepared, using serial dilutions of copy numbers in a large complex environmental DNA samples. The presence of nontarget background DNA did not affect the amplification efficiency, sensitivity and reproducibility of qPCR assays. Quantitative real-time PCR format could reliably detect as few as 100 copies of target DNA and quantify the pathogen DNA content in various samples. This simple and pseudo-nested PCR procedure detected as little as 10 pg and 10 fg, respectively. The qPCR protocol might be useful for investigations on plant-pathogen interactions, epidemiology and development of disease management systems (Scarlett et al. 2013). *F. oxysporum* f.sp. *melonis (Fom)*, infecting melons can persist in soil over extended periods on crop residues and nonhost crops and by forming chlamydospores that are resistant to adverse conditions

existing in the soils. A reliable and species-specific real-time quantitative PCR (qPCR) assay was developed for the detection and quantification of the complex soilborne anamorphic fungus *F. oxysporum*. A new primer pair designed based on the sequences of the translation elongation factor 1-α gene was utilized in the qPCR assay and an amplicon of 142-bp was generated by this primer pair. This protocol was applied to grafted melon plants for the detection and quantification of *Fom* in melon tissues. Grafting technique is widely applied in melon to confer resistance against new virulent races of *Fom*, while maintaining desirable characteristics of commercial varieties. Pathogen development was monitored using qPCR assay in Charentis-T (susceptible) and Nad-1 (resistant) melon varieties, both used either as rootstock and scion inoculated with race 1 and race 1,2 of *Fom*. The pathogen could be detected in early asymptomatic plants with low DNA content of *Fom*, the quantification limit being 1 pg of fungal DNA. The qPCR assay showed that pathogen development was significantly affected by the nature of host-pathogen interaction (compatible/incompatible) and the interval after inoculation. The amount of DNA of race 1,2 was significantly higher compared with that of race 1 in the incompatible interaction at 18 days postinoculation. Both races of *Fom* could be detected in both rootstock and scion of grafted plants in either the compatible or incompatible interaction. The qPCR protocol was faster than the isolation-based method and could detect *Fom* in melon stems where no pathogen could be isolated, indicating higher sensitivity of this assay (Haegi et al. 2013).

Fusarium wilt disease of banana, caused by *F. oxysporum* f.sp. *cubense (Foc)* race 4, is a devastating disease impacting adversely banana production the world over. Disease monitoring depends largely on rapid, reliable and sensitive detection of *Foc* race 4 in field-infected banana plants from which suckers (planting materials) are selected for planting in the next season. The PCR parameters were optimized using the SCAR primers, FocSc-1/FocSc-2 and a real-time PCR format. The assay procedure showed high reproducibility and sensitivity in detecting extremely low concentrations of pathogen genomic DNA (gDNA). Quantitative estimation of pathogen genomic DNA showed that *Foc* gDNA in severely infected banana pseudostems and leaves was very high (6945-fold and 26.69-fold, respectively) compared to plants showing mild symptoms (Lin et al. 2013). Fusarium wilt pathogen *F. oxysporum* f.sp. *spinaciae (Fos)* is soilborne and occurs in several countries growing spinach. The pathogen was found to be persistent in acid soils of maritime Western Oregon and Washington where commercial seed production is primarily done. TaqMan real-time PCR format was developed based on the sequencing of IGS region of rDNA of isolates of *Fos*. A guanidine single nucleotide polymorphism (GSNP) was detected in the IGS sequences of 36 geographically diverse isolates of *Fos*, but not in the sequences of 64 isolates of other *formae speciales* and 33 isolates representing other fungal species or genera. The SNP was used to develop a probe for a real-time PCR assay. The pathogen could be detected by the real-time PCR assay in the soil samples with a concentration of 3–14,056 CFU/g of soil samples collected over three years from naturally infested spinach seed production sites. However, the reliable detection limit of the assay was 11 CFU/g of soil. There was a significant correlation between the enumeration of *Fos* population on Komada's agar medium and quantification of *Fos* by TaqMan assay. Pathogen DNA levels, Fusarium wilt disease ratings and spinach biomass were significantly correlated for one set of naturally infested soils. However, pathogen DNA levels, wilt incidence ratings and spinach biomass did not show any relationship among them. The results suggested that wilt disease incidence was not dependent only on soilborne pathogen populations. The presence of GSNP in other *F. oxysporum formae speciales*, however, limited the application of this protocol, although all isolates of *F. oxsporum* f.sp *spinaciae* tested positive with the protocol developed in this investigation (Okubara et al. 2013). *F. oxysporum* f.sp. *cepae (Foc)*, incitant of onion basal rot disease, has a *secreted in xylem (SIX) 3* homolog (*FocSIX3*) with 91.4% identity to *SIX3* gene (*FolSIX3*) of *F. oxysporum* f.sp. *lycopersici (Fol)*. For the detection and quantification of *Foc*, a primer pair (P1) was designed based on the differences in the nucleotide sequences between *FocSIX3* and *FolSIX3*. The primer pair P1 amplified a 106-bp DNA fragment exclusively from the isolates of *Foc* pathogenic to onions, but no amplification of any PCR product occurred from *Foc* isolates from either Welsh onion or *F. oxysporum* strains belonging to *formae speciales* other than *cepae*. Real-time quantitative PCR format was employed to quantify pathogen DNA contents in root and basal plate of onion, using P1 primer (Sasaki et al. 2015).

Pathogenicity and VCG tests, although reliable, are laborious for identification of *F. oxysporum* f.sp. *lycopersici* (FOL). Hence, a rapid, sensitive and quantitative real-time PCR assay was developed for detecting and quantifying FOL in soil. An inexpensive and relatively simple method for soil DNA extraction and purification was developed, based on bead-beating and a silica-based DNA-binding method. A TaqMan probe and PCR primers were designed, using the DNA sequence of the species-specific virulence gene *SIX1* which was present only in the isolates of FOL, but not in isolates of other *formae speciales* or nonpathogenic isolates of *F. oxysporum*. The real-time PCR assay efficiently amplified DNA of isolates of three races of FOL used and quantified FOL DNA in soils. The assay had a detection limit of 0.44 pg of genomic DNA of FOL in 20 µl of the real-time PCR reaction mixture. A spiking test performed by adding different concentrations of conidia to soil, showed significant linear relationship between the amount of genomic DNA of FOL detected by the real-time PCR assay and the concentration of conidia added to the soil samples. The DNA extraction protocol and real-time PCR assay showed potential for determining FOL populations in soil, developing threshold models to predict Fusarium wilt severity, identifying high-risk fields and assessing the impact of cultural practices on FOL populations in infested soil (Huang et al. 2016).

*F. oxysporum* f.sp. *basilici*, causes wilt and crown rot disease of basil, reducing the production levels considerably. A real-time PCR assay using a TaqMan® probe was developed for the rapid detection of the pathogen in roots and seeds of basil, as the nested PCR assay required a long time to provide results. The sensitivity of the real-time PCR format was at the same level as the nested PCR assay. But the reduction of testing time by half, ability to detect both external and internal seed infection with a sensitivity of 1 pg of pathogenic genomic DNA or 24 CFU per 100 seeds were the advantages of employing the real-time PCR for the detection of *F. oxysporum* f.sp. *basilici* in basil seeds and root tissues (Pasquali et al. 2006). *Fusarium commune*, incitant of Fusarium wilt disease of Chinese water chestnut (*Eleocharis dulcis*) is considered as a major limiting factor for its production in China. A SYBR Green I real-time PCR (qPCR) assay was developed based on the mitochondrial small subunit rDNA of *F. commune*. A single PCR amplicon of 178-bp was generated only from the genomic DNA of the pathogen, but not from any of the 41 other fungal isolates tested, indicating the specificity of the primer pair FO1/FO2. The limits of detection of the assay were 1fg/µl of pure genomic DNA of *F. commune*, 1 pg/µl of *F. commune* DNA (0.5ng/µl) or 1,000 conidia/g of soil artificially infested. The amount of pathogen DNA quantified by qPCR assay was significantly correlated with the disease severity (DS) ratings. But no significant relationship between spore densities in soil and Fusarium wilt DS ratings. However, when tested on 76 soil samples from commercial fields of *E. dulcis*, the spore density of the pathogen showed positive correlations with disease index in one growing season (2012), but not in the previous year (2011). The results indicated the effectiveness and specificity of detection of *F. commune* in plant and soil samples, facilitating the monitoring of pathogen movement within/outside the geographic location (Zhu et al. 2016). A primer set for PCR and a primer-probe set for real-time PCR assay, based on its endopolygalacturonase gene sequence were used for the specific detection of *F. oxysporum* f.sp. *conglutinans (Foc)*, incitatnt of cabbage yellows disease, in soil and differentiation of *Foc* from other pathogenic *formae speciales* and nonpathogenic *F. oxysporum*. The real-time PCR format could detect *Foc* in DNA extracted from soil samples. Sensitivity of real-time PCR assay was improved by ten to 10,000 times by adding BSA or by performing pre-PCR assay. The real-time PCR assay could detect *F. oxysporum* f.sp. *conglutinans* in soil at a population density that caused only slight yellows symptoms in plants (Kashiwa et al. 2016).

*V. dahliae* causes a seasonal yield-limiting potato early dying (PED) disease in most potato growing countries. For the detection and quantification of *V. dahliae* in potato tissues, a real-time quantitative PCR assay was developed, using primer pair VertBt-F/VertBt-R derived from the sequence of the β-tubulin gene. The detection efficiency of this primer pair was greater (95%) than in monoplex QPCR and duplex formats with primers PotAct-F/PotAct-R based on the actin gene sequence designed for potato. QPCR format could quantify 148 fg of *V. dahliae* DNA which was equivalent to five nuclei. This assay detected *V. dahliae* in naturally infected air-dried potato stems and fresh stems of inoculated plants. The results of QPCR assay were similar to those of plating assays which had a higher level of variability. Further, QPCR assay detected *V. dahliae* in 10% of stem samples from which the pathogen could not be isolated by plating assay. The results showed that QPCR format was more sensitive and had the potential for assessing the response of breeding lines to colonization by *V. dahliae* that will be useful for breeders (Atallah et al. 2007). The comparative efficacy of the real-time quantitative PCR (QPCR) assay developed in this investigation was compared with other PCR formats for sensitive detection and quantification of *V. dahliae* in potato stem tissues. In this new method, the trypsin gene was targeted, whereas the earlier QPCR assay amplified a region of the β-tubulin gene of *V. dahliae*. The new QPCR procedure was highly sensitive, capable of detecting 0.25 pg of pathogen DNA. Application of the duplex real-time PCR assay utilizing the potato actin gene to normalize quantification resulted in clear differentiation of levels of resistance among eight russet-skinned potato cultivars inoculated in the greenhouse tests, when compared with traditional plating assays using a semiselective medium. However, relative levels of resistance among cultivars were similar between traditional plating and QPCR procedures with correlation coefficiency greater than 0.93. The QPCR assay developed in this investigation was rapid, efficient and precise in quantifying the DNA of *V. dahliae* and it could be useful for breeding programs meant for screening large number of potato clones and selections for resistance to Verticillium wilt disease (Pasche et al. 2013).

*V. dahliae* infects *Capsicum annuum* cultivars Luesia, Padron, SCM331 and PI201234 and *Capsicum chinense* cultivars C118 with varying susceptibility and symptom expression. Real-time PCR assay was developed to detect and quantify *V. dahliae* DNA in the roots at 23 and 34 days postinoculation (dpi). The pathogen DNA contents were higher in cv. C118 than in other cultivars, followed by SCM 331, Padron and PI201234. Cv. Luesia had the lowest contents of pathogen DNA. In the hypocotyls, the highest concentration of the pathogen was detected in SCM 331, whereas Luesia, Padron and PI201234 had much lower amounts and C118 had intermediate levels. The effects of compatible and incompatible interactions were assessed by inoculating near-isogenic tomato lines LA3030 (susceptible) and LA3038 (resistant) with *V. dahliae*. The pathogen/host plant DNA ratio was lower in LA3038 than in LA3030 and it decreased with time in LA3038. *V. dahliae* DNA contents in the roots of LA 3030 remained constant between 23 and 34 dpi, but increased tenfold in collars. In addition, the real-time PCR assay was also able to detect *V. dahliae* in pepper plants, soil and water collected from farms in northwest Spain. The pathogen could be detected in asymptomatic pepper plants, indicating the usefulness of real-time PCR assay in disease monitoring and risk assessments (Gayso et al. 2007). *V. dahliae*, a soilborne pathogen with a wide host range, induces Verticillium wilt disease in several crops such as potato and cotton grown on infested soils. Soil plating assay used traditionally for estimating the inoculum density of *V. dahliae* required six to eight weeks for providing the results. In order to hasten the process of estimation of inoculum density in soil, a multiplexed TaqMan real-time PCR format was developed based on ribosomal DNA (rDNA) IGS region of the pathogen and an internal control

was also included to estimate the effects of PCR inhibitors coextracted with pathogen DNA. This assay was specific to *V. dahliae* and high correlation was observed between inoculum densities determined by TaqMan real-time PCR assay and soil plating using a range of field soils with pathogen densities as low as 1–2 MS/g of soil. The results indicated that assaying a minimum of four replicates of each soil would provide reliable estimation of pathogen populations in the soil samples (Bilodeau et al. 2012).

A molecular quantification procedure based on Synergy Brands (SYBR) Green real-time quantitative PCR of wet-sieving samples (wet-sieving qPCR) was developed for quantifying MS of *V. dahliae* present in the soil. This assay was highly sensitive, capable of detecting as low as 0.5 CFU/g soil. By using 40 soil samples, a high correlation (r = 0.98) was observed between the estimates of MS by conventional soil plating method and wet-sieving qPCR assay. More than 400 soil samples taken from the rhizosphere of individual cotton plants with or without visual wilt symptoms in experimental and commercial fields at boll-forming stage were subjected to analysis by wet-sieving qPCR assay. The wilt inoculum estimated by wet-sieving qPCR assay was related to the development of wilt disease. The estimated inoculum threshold varied with the resistance of cultivars. An overall relationship was seen between wilt disease incidence and inoculum density across 31 commercial fields where a single composite soil sample was taken in each field, with an estimated inoculum threshold of 11 CFU/g of soil. The results suggested the possibility of soil inoculum density estimated with wet-sieving qPCR assay might be useful to predict wilt disease risk. Inoculum threshold of 4.0 and 7.0 CFU/g of soil might be used for predicting wilt disease risk in resistant and susceptible cultivars of cotton (Wei et al. 2015).

*V. dahliae*, soilborne fungal pathogen is known to be transmitted through seeds of certain crops like spinach. Seeds of spinach produced in the United States are commonly infected by *V. dahliae* and planting the infected seed not only increases the inoculum density of the soil, but also may introduce the exotic strains of *V. dahliae* that may contribute to Verticillium epidemics on lettuce and other crops grown in rotation with spinach. A sensitive, rapid and reliable method for the detection and quantification of *V. dahliae* in spinach seeds was developed. Quantitative real-time PCR (qPCR) was optimized and employed for the detection and quantification of *V. dahliae* in spinach germplasm and 15 commercial spinach seed lots, using species-specific primer pair, VertBt-F/VertBt-R. An analytic mill was used for grinding spinach seeds which are tough for DNA extraction. This assay had a sensitivity limit of 1 infected seed/100 seed (1.3% infection in a seed lot). With commercial seed lots it was possible to detect *V. dahliae* at 1% infection. A highly significant positive linear relationship was observed between the per cent seed infection and *V. dahliae* DNA copy number (P=0.0001), when assessed on seed samples with an artificial infection gradient of 0.7 to 64%. The relationship between pathogen copy number and per cent seed infection in naturally infected seed lots (15) was statistically less significant (P=0.027), compared with artificially infested seed samples. But the cube root transformation of pathogen copy number and per cent seed infection improved statistical significance (P=0.002) (see Figure 2.6). The overall trend was that the pathogen DNA contents decreased hand in hand with per cent seed infection. The qPCR assay developed in this investigation was sensitive, providing reliable and reproducible results rapidly and showed the potential to be employed as a basis to determine whether the seed lot could be used for planting and for reducing the chances of introducing *V. dahliae* into the soil which might provide favorable conditions for the survival of the pathogen for long periods (see Figure 2.7) (Duressa et al. 2012). *V. longisporum*, causal agent of Verticillium wilt of Chinese cabbage, was detected and identified based on mtSSU rDNA and cytochrome *b* sequence analysis. A real-time PCR assay was developed based on the pathogen species-specific rDNA-ITS sequence. The real-time PCR assay was rapid, sensitive and specific in quantifying *V. longisporum* DNA contents in artificially and field-infected Chinese cabbage seedlings and this procedure has the potential for large-scale screening of host genotypes for resistance to Verticillium wilt disease (Yu et al. 2015).

*V. dahliae*, causes Verticillium wilt disease which is the most serious among the diseases affecting olive trees. The resting structures, MS of the pathogen were quantified using the real-time PCR (qPCR) format. Pathogen DNA contents were correlated with the symptomatology. The relative pathogen DNA contents (molecules/μl) of a defoliating (D) and nondefoliating (ND) strains of *V. dahliae* were quantified in the susceptible cv. Amfissis and tolerant cv. Kalmon and Koroneiki. The viability of *V. dahliae* in plant tissues was indicated by plating method on PDA medium. Symptom severity was correlated with relative DNA content of *V. dahliae* in plant tissues and cultivar susceptibility. The real-time PCR assay showed that D and ND strains were present in significantly higher concentrations in cv. Amfissis than in cv. Kalmon and Koroneiki. Further, the roots had lower pathogen contents than the stems and shoots and decline in pathogen contents occurred over time. The real-time PCR format has the potential for evaluation of the levels of resistance of olive cultivars or olive rootstocks to *V. dahliae* pathotypes (Markakis et al. 2009). Certified pathogen-free olive planting materials are used as an important management practice to

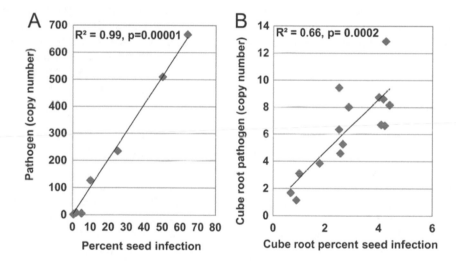

**FIGURE 2.6** Relationship between percent spinach seed infected artificially by *V. dahliae* with grades of infection and quantitative real-time PCR values for pathogen DNA, as determined by copy number of β-tubulin gene of *V. dahliae*.

**[Courtesy of Duressa et al. (2012)) and with kind permission of the American Phytopathological Society, MN, United States]**

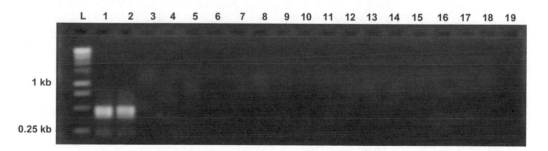

**FIGURE 2.7** Detection of *V. longisporum* in the spinach seed by PCR assay based on the generation of pathogen-specific 340-bp amplicon. Note the presence of *V. longisporum*-specific amplicon from pure culture in lane 1 and 2, but not from that of *V. dahliae*.

**[Courtesy of Duressa et al. (2012)) and with kind permission of the American Phytopathological Society, MN, United States]**

prevent the spread of Verticillium wilt disease caused by *V. dahliae* occurring in two pathotypes (D and ND). A rapid, specific and sensitive method is required for the detection and quantification of *V. dahliae* in plants, especially in the early stages of infection, when plants remain asymptomatic. The accuracy, specificity and efficiency of eight real-time quantitative PCR procedures were compared in detecting *V. dahliae* in plants and soil. DNAs extracted from olive roots, stems and leaves were assayed. The procedures TAQ (based on IGS ribosomal DNA target gene) and SYBR-4 (based on β-tubulin 2 target gene) ranked first in the sensitivity and efficiency for quantification of *V. dahliae* DNA in small quantities and in different types of olive tissues (root and stem) tested. Application of TAQ and SYBR-4 formats enabled accurate quantification of *V. dahliae* DNA, regardless of the background DNA, with a detection limit being fixed at a cycle threshold (Ct) of 36 (= 18 fg for SYBR-4 and 15 fg for TAQ) of *V. dahliae*. Monitoring of colonization of asymptomatic Frantoio plants by *V. dahliae* indicated that the differences in virulence of D and ND pathotypes on susceptible olive cultivars correlated with differences in the amount of *V. dahliae* DNA revealed by the SYBR-4 and TAQ procedures and the percentage of positive pathogen isolation. However, the amount of each pathotype biomass varied with olive tissue assayed. As these two procedures could detect *V. dahliae* in upper tissues of asymptomatic plants, they might be useful for certification programs for the production of disease-free olive plants (Gramaje et al. 2013). Two races, 1 and 2, of *V. dahliae* were differentiated, based on their diffferential pathogenicity on tomato cultivars, carrying resistance gene *Ve1*. Race 2-resistant rootstock cultivars, Aibou and Ganbarume-Karis were bred in Japan, since all commerical tomato cultivars were susceptible. However, some isolates of race 2 could overcome the resistance of the resistant rootstock cultivars. Hence, the current race 2 of *V. dahliae* was proposed to be divided into two races viz., race 2 (nonpathogenic on Aibou) and race 3 (pathogenic on Aibou). Race 3 was detected in 45 of 70 fields examined, indicating that race 3 had already spread in Gifu Prefecture. The presence of race 2 was detected only in 25% of 70 fields (Usami et al. 2017).

Woody plant hosts, ash (*Fraxinus* spp.), sugar maple (*Acer saccharum*) and redbud (*Cercis canadensis*) are severely affected by Verticillium wilt disease caused by *Veticillium dahliae* which has a wide host range, including several crop plant species. As it is very difficult to recognize infected trees by visual symptoms, real-time PCR-based assays were evaluated for their efficacy in detecting and quantifying *V. dahliae* in woody hosts. High quality DNA was extracted in large amounts from presumptive infected woody hosts by collecting drill-press savings from sample tissues, bead beating and extracting using cetylmethylammonium bromide (CTAB) method. Among the six primer sets evaluated, two sets, VertBt-F/VertBt-R and VDS1/VDS2 were selected for evaluation of DNA extracts from field samples. The primer set VertBt-F/VertBt-R amplified a product of 115-bp fragment, while VDS1/VDS2 set amplified a fragment of 540-bp. However, the VertBt primer set was able to detect *V.*

*dahliae* in asymptomatic trees also, indicating its high level of sensitivity. These primers could be employed for pathogen detection and monitoring to restrict disease spread by taking appropriate disease management measures (Aljawasim and Vincelli 2015). The host-specialized *Verticillium longisporum*, infecting oilseed rape (OSR) has a special characteristic of being systemic, nonhomogenous and delayed colonization of host plant xylem, resulting in an extended symptomless period of latency. Assessment of severity of infection in the field by visual examination might not yield reliable estimates, as the infection became apparent only at maturity of crop, when it could be confused with natural senescence of plants. Hence, a qPCR assay was developed to unambiguously differentiate levels of quantitative resistance to *V. longisporum* on OSR genotypes under field conditions. Two primer sets targeting ITS or tubulin loci in *V. longisporum* genome, were evaluated. Tubulin primers exhibited a high specificity to *V. longisporum* isolates. On the other hand, ITS primers showed a significantly higher sensitivity, in detecting fungal DNA in stem tissue (limit of quantification being 0.56 fg DNA) of field-grown presymptomatic plants. The results indicated the effectiveness of the qPCR assay in determining the pathogen DNA content in latently infected OSR plants (Knüfer et al. 2017).

*R. solani* is a species complex comprising of 13 known AGs. *R. solani* AG-3, is the primary incitant of stem canker and black scurf of potato. The pathogen was detected by conventional PCR assay using the primer set, Rs1F2/Rs2R1, designed based on the sequences of nuclear ribosomal ITS1 and ITS2 regions of *R. solani*. A 0.5-kb amplicon was generated by this primer set only from the DNA of all isolates of AG-3, but not from the DNAs of isolates belonging to a range of other *R. solani* anastomosis groups or from other potato pathogens tested, confirming the specificity of the primers for AG-3 only. This PCR format detected *R. solani* AG-3 also in potato tissues with varying intensities of black scurf disease and in soil infested with *R. solani* to a minimum detection level of $5 \times 10^{-4}$ sclerotia/g of soil. Further, the specific primers RsTqF1 (based on the Rs1F2 sequence) and RsTqR1 and a TaqMan™ fluorogenic probe RQP1 were designed for performing real-time quantitative (TaqMan) PCR assay. A direct DNA extraction from soil and a seed-baiting method were combined as pretreatment for conventional PCR and real-time PCR formats for the reliable detection and quantification of *R. solani* AG-3 in both artificially and naturally infested soils. Direct DNA extraction combined with conventional PCR format and real-time qPCR assay detected AG-3. On the other hand, enhancement of *R. solani* population by seed-baiting increased the sensitivity of the assay, compared with direct DNA extraction from soil. *R. solani* AG-3 could be detected more effectively and reliably in artificially and naturally infested soils, when seed-baiting was combined with either conventional PCR or real-time PCR assay (Lees et al. 2002). *R. solani* AG3-PT is the anastomosis group frequently associated with stem canker and scurf disease of potatoes. A real-time PCR assay was developed to target rDNA ITS region of AG3-PT isolates for direct detection in tuber and soil samples infested by the pathogen. The real-time PCR format was highly specific for AG3-PT and no amplification occurred from the DNA from isolates of other AGs or subgroups of AG3. Using the bulk DNA extraction method, capable of extracting from up to 250 g of soil, the assay could detect one individual sclerotium of AG3-PT (weighing 200 µg) in 250 g of soil. The AG3-PT assay was combined with assays for AG2-1, AG5 and AG8 to determine the distribution of the AG present in potato soils and tubers in the United Kingdom. The results showed that AG2-1 and AG3-PT were prevalent predominantly in tubers and soils, although AG3-PT was more frequently associated with tubers (more than 50% of the tuber samples) showing positive results (Woodhall et al. 2013). *R. solani* causes sheath blight disease which is considered as an important yield-limiting factor for rice production. DS scored by visual examination did not provide a reliable assessment of the damage to the rice plant due to *R. solani*. A real-time quantitative PCR (QPCR) format was developed for the detection and quantification of *R. solani* DNA in infected rice plants. Pathogen-specific primer designed based on ITS region of rDNA or *R. solani* AG1-IA was employed for sensitive detection of the pathogen. The detection limit of the QPCR assay was 1 pg of pathogen DNA. This procedure might be useful for the evaluation of rice genotypes and cultivars for resistance to sheath blight disease (Sayler and Yang 2007).

Soilborne *R. solani* AG2-2IIIB infecting sugar beet was detected, using the direct soil DNA extraction method, followed by real-time PCR assay. Detection of *R. solani* AG2-2IIIB by real-time assay was highly specific, as indicated by the absence of positive reaction with strains of other *R. solani* AGs. The assay showed good relationship between cycle threshold and amount AG2-2IIIB sclerotia detected in three spiked field soil samples and also in naturally infested soil samples. Soil IP of *R. solani* AG2-2IIIB was quantified before and after growing a susceptible and resistant sugar beet cultivar, as well as after subsequent growth of a nonhost winter rye crop. Susceptible sugar beet cultivar was severely affected, enabling buildup of high pathogen population that was six-fold greater than in soil planted with resistant sugar beet cultivar. Growing nonhost winter rye plants significantly reduced soil IP to the initial level at sowing, suggesting that susceptibility levels of sugar beet cultivars might accelerate or inhibit the buildup of population of *R. solani* AG2-2-IIIB (Schulze et al. 2016).

*Rhizoctonia cerealis* (teleomorph, *Ceratobasidium cerealis*), causing wheat sharp eyespot disease, is soilborne and it is difficult to identify and quantify this pathogen in soils, using conventional isolation methods. The SYBR Green-based quantitative real-time PCR assay was developed for specific sensitive detection and quantification of *R. cerealis* in soil samples. Using specific primers based on the sequences of β-tubulin gene of pathogen DNA, the assay could detect specifically as low as 100 fg of *R. cerealis* DNA. The real-time PCR format was applied to quantify *R. cerealis* in naturally and artificially infested soil samples (Guo et al. 2012). Rhizoctonia root rot/bare patch disease is one of the important limiting factors in dry land cereal production systems of

the PNW in the United States. *R. solani* and *R. oryzae* are primarily involved in this disease, causing heavy losses in directly seeded (no-till) systems. Although direct plating and bait methods were followed for the detection of the pathogen causing root rot, they were time-consuming and labor intensive. Hence, SYBR Green I-based assay was developed, based on the sequences of ITS1 and ITS2 regions of rDNA of *R. solani* and *R. oryzae*. The QPCR format detected *R. solani* AG2-1, AG-8 and AG-10, three genotypes of *R. oryzae* and an AG-1 like binucleate *Rhizoctonia* sp. Primers Rs2.1/8F and Rs2.1/8R amplified both *R. solani* AG2-1 and AG-8 with equal efficiency. The QPCR assay did not amplify DNA from 26 nontarget soilborne fungi and oomycetes pathogenic to wheat and/ or found in PNW soils. Quantification of the pathogens was reproducible at or below a Ct of 33 or 2 to 10 fg of mycelial DNA from cultured fungi, 200 to 500 fg of pathogen DNA from root extracts and 20 to 50 fg of pathogen DNA from soil extracts. However, it was possible to detect specifically the pathogen in all types of extracts at about 100-fold below the quantification levels. Natural field soils from a direct-seeded farm were analyzed for the detection and quantification of *R. solani* AG-8. Incidence of Rhizoctonia root rot and bare patch had been observed earlier in these fields. The relationship between Ct and pathogen population density was determined, using a soil standard-curve consisting of three *R. solani* isolates. The amplification efficiency of the assay was 0.76. The soil extracts (in triplicates) generally yielded similar Ct values and standard deviations averaged 0.24. The populations of *R. solani* AG-8 in the Ritzville soil in the United States, were estimated, based on the standard-curve. The pathogen was detected in eight of nine soil samples collected in three fields displaying bare patches. Primer specificity was tested by amplifying the target DNA at concentrations near the detection limit in the presence of up to $10^5$ molar excess of all other nontarget *Rhizoctonia* DNA. In these mixtures, *R. solani* target DNA was reliably quantified at about 2 to 17 fg and *R. oryzae* at 30 to 97 fg, depending on the assay. The soil samples contained 0.4 to 19.0 pathogen propagules (ppg) or 9.4 to 780 pg of pathogen DNA/g of soil (see Table 2.2). This QPCR format might be useful in determining whether *R. solani* AG-8 and *R. oryzae* occur in distinct or overlapping locations and also for large-scale analysis of soil samples from infested areas (Okubara et al. 2008).

*A. euteiches*, incitant of root rot of several leguminous crops, including peas, beans, alfalfa and clovers, is widely distributed in North America, Europe, Japan, Australia and New Zealand and it is capable of causing total crop loss under favorable conditions. Conventional PCR using specific primers and TaqMan real-time PCR format with dual-labeled probe were applied for detection and quantification of race 1 and 2 of *A. euteiches* DNA in alfalfa plants exhibiting varying degrees of DS. Based on the DNA contents of *A. euteiches*, alfalfa population could be discriminated for levels of resistance to the root rot disease. The pathogen DNA contents assessed by real-time PCR assay showed positive correlation with DS index ratings. The assessment of race 1 DNA present in resistant check WAPH-1 was significantly less than

**TABLE 2.2**

**Limits of Quantification of Purified Mycelial DNA of *Rhizoctonia* in Real-Time PCR Assays (Okubara et al. 2008)**

| Assay | Purified DNA (fg)[a] | DNA mixture (fg)[b] | Nontarget molar excess[c] |
|---|---|---|---|
| *R. solani* AG-2-1 | 4 | 16 | 39,000–136,000 |
| *R. solani* AG-8 | 3 | 2 | 27,000 |
| *R. solani* AG-10 | 2 | 17 | 515–2,840 |
| Binucleate | 7 | 8 | 190–3,770 |
| *R. oryzae* I | 12 | 30 | 1,500–50,000 |
| *R. oryzae* II | 21 | <48 | 8,900–25,400 |
| *R. oryzae* III | 24 | <97 | 24,000–261,000 |

[a] Minimum quantifiable amount of *Rhizoctonia* target DNA from cultured mycelia

[b] Minimum quantifiable amount of *Rhizoctonia* target DNA that can be quantified in a DNA mixture

[c] Molar excess of nontarget *Rhizoctonia* DNA that could be present and still give reliable quantification of target DNA

that the susceptible check. Saranac. Discrimination between commercial cultivars based on quantitative PCR analysis of bulked plant samples was similar to grouping based on visual assessment of DS (Vandemark et al. 2002). In a later study, real-time fluorescent assay specific for *A. euteiches* was applied to examine the relationship between DS and pathogen DNA contents in infected peas. Interactions between five pea genotypes with varying levels of resistance and five isolates of *A. euteiches* were studied, using real-time PCR assay and visual rating of DS. The susceptible genotypes, Genie, DSP and Bolero tended to show higher levels of disease severity and contained higher amounts of pathogen DNA, than the resistant genotype 902079 and PI180693, which showed less disease severity. Lowest amount of pathogen DNA was present in 90-2079. The relationship between pathogen DNA contents and DS in pea cultivars/genotypes was variable and the real-time PCR format appeared to have limited application for selecting sources of resistance to pea root rot disease caused by *A. euteiches* (Vandemark and Grünwald 2005).

The avoidance of infested soils for cultivation of susceptible crops is considered a desirable option for management of the disease. The development of effective diagnostic methods to identify fields suitable for pea crops was, therefore, considered essential. Determination of IP of soil using baiting plants required a long time, in addition to being less sensitive and inconsistent. The molecular methods available did not enable the detection of less than 200 oospores of *A. euteiches*/ g of soil. Hence, a new, fast and sensitive real-time PCR assay was developed by targeting ITS1 region of rDNA operons, enabling the detection of less than ten oospores of the pathogen/g of soil. Different sets of primers targeting the most dissimilar sequences between pathogenic *Aphanomyces* were designed by varying length and/ location of primers. The selected primers amplified a 129- or 148-bp fragment, depending on the ITS1 sequence consensus considered. The primers amplifying specific sequences only from 40 isolates of *A. euteiches*, but not from *A. cladogamus* or *A. cochlioides* were employed in the assay. The effectiveness of the real-time

assay was proved for different soil-types- sterilized or not, with artificial inoculation of two strains of *A. euteiches* with low or high virulence. *A. euteiches* oospore density in soil was related to the IP based on which suitable soils for pea cultivation may be selected (Gangneux et al. 2014).

*Phialophora gregata* f.sp. *sojae (Pgs)* (syn. *Cephalosporium gregatum* and *Cadophora gregata*), causing brown stem rot (BSR) of soybean, consists of two genotypes A and B which differ in their virulence and type of symptoms induced in soybean. The genotype A isolates are considered more aggressive, causing more severe foliar and stem symptoms than genotype B isolates. The genotypes A and B are distinguished by a 188-bp insertion/detection (INDEL) in the IGS region of ribosomal DNA (Chen et al. 2000). A TaqMan probe based quantitative real-time PCR (qPCR) assay was developed for the specific detection and quantification of A and B genotypes of *P. gregata* in plant and soil samples. The assay had a detection limit of 50 fg of pure pathogen genomic DNA, 100 copies of the target DNA sequence and approximately 400 conidia. The qPCR assay was more sensitive (about 1,000 times) in detecting DNA/conidia of *P. gregata* than standard PCR format. Further, qPCR was more rapid and less sensitive to PCR inhibitors from soybean stem tissues, compared with standard PCR assay. The single-step qPCR format detected the pathogen in stems of soybean at least one to two weeks prior to visible symptom development. This assay was effective in detecting *P. gregata* in field-grown plants and naturally infested soils. The qPCR format might be useful for monitoring the pathogen spread in epidemiological studies (Malvick and Impullitti 2007). In a later study, dilution plating and real-time quantitative PCR (qPCR) assays were employed to quantify the contents of *P. gregata* in cultivars resistant or susceptible to BSR disease of soybean. BSR-susceptible cultivars Comsoy 79 and Century 84 expressed more than 73% foliar and stem symptom severity and they had the highest population density ranging from $\log_{10}$ 4.3 to 4.7 CFU/g of stem tissue. On the other hand, the resistant cultivar Bell expressed less than 10% of foliar symptom severity, but had a pathogen population density that was on par with susceptible accessions. The cvs. Dwight and L84-5873 consistently contained lower population levels than in susceptible and several resistant accessions. PI437833 and IA2008R showed mild symptoms, but contained high pathogen DNA contents. In contrast, cv. Pellar 86 showed severe symptoms, but had lower pathogen DNA contents. The results suggested that some accessions might possess resistance to both pathogen colonization and symptom development, whereas some accessions might be resistant to symptom development, but not to the pathogen colonization, enabling build up of pathogen population in plant tissues (172Impullitti et al. 2009). In order to distinguish populations of genotype A and B of *Pgs*, a real-time quantitative PCR (qPCR) assay was developed for the specific detection and quantification of genotype A. The primer/probe set PgsAspF-PgsAspR-PgsAspPR was specific to *Pgs* genotype A and failed to amplify DNA from genotype B. Further, the genotype A-specific primer/probe did not amplify DNA from other fungi, oomycetes and bacteria associated with soybean. Multiplexing the genotype A-specific primer/probe set with pimer/probe set designed by Malvick and Impulitti (2007) enabled simultaneous amplification of total DNA and genotype A DNA from pure culture, artificially infested soil and inoculated soybean plants and naturally infested soil and field-grown plants. As a separate primer set/probe specific for genotype could not be developed, the DNA contents of genotype B could be determined indirectly by subtracting the amount of DNA quantified by genotype A-specific primer/probe set from the amount of DNA determined with general primer set/probe (total DNA). The multiplex assay was rapid and robust in providing estimates of genotypes A and B of *P. gregata* f.sp. *sojae* (Hughes et al. 2009).

*Sclerotinia sclerotiorum*, causal agent of Sclerotinia stem rot of canola, with a wide host range, can infect different plant parts, and has multiple modes of dissemination. Petal infestation in canola by *S. sclerotiorum* is an important stage in the disease cycle in canola. Quantitative PCR (qPCR) assay was applied to have more rapid and accurate assessment of petal infestation levels. Primers and a hydrolysis probe were designed to amplify a 70-bp region of *S. sclerotiorum*-specific gene SS1G_00263. The hydrolysis-based qPCR assay had a detection limit of $8.0 \times 10^4$ ng of *S. sclerotiorum* DNA and the *S. sclerotiorum* DNA was exclusively amplified. Assessment of petal infestation by *S. sclerotiorum* in ten commercial canola fields at two sampling dates, corresponding to 20–30% bloom was carried out. *S. sclerotiorum* infestation levels, as represented by pathogen DNA content, were 0 to $3.3 - 10^{-1}$ ng/petal. The qPCR procedure efficiently quantified petal infestation levels of *S. sclerotiorum* (Ziesman et al. 2016). *Sclerotinia trifoliorum* and *S. sclerotiorum* cause similar rotting symptoms in clover and it is difficult to differentiate infections caused by these pathogens. PCR amplifications and nucleotide sequencing were used as a basis for differentiation of 40 isolates of *S. trifoliorum* (29 from Poland, 11 from the United States) and 55 isolates of *S. sclerotiorum* (26 from Poland, 29 from the United States). Amplification of ß-tubulin and calmodulin genes with TU1/TU2/TU3 and ScadF1/ScadR1

PCR primers and the presence of introns and SNPs within the ribosomal DNA (rDNA) as detected with NS1/NS8 and ITS1/ITS4 PCR primers, were effective in rapidly and precisely differentiating the isolates of *S. trifoliorum* and *S. sclerotiorum*. No intraspecies variation among isolates of *S. sclerotiorum* could be detected, whereas a high degree of intraspecies variability could be seen among the isolates of *S. trifoliorum* from the United States and Poland (Baturo-Ciesniewska et al. 2017).

Real-time SYBR Green PCR format was developed for the detection and quantification of *Monosporascus cannonballus* infecting roots of melons by targeting the ITS1 region of rDNA conserved between different strains of the pathogen. The assay was shown to be specific by employing the assay against fungi taxonomically and ecologically related to *M. cannonballus*. The real-time PCR assay detected the pathogen in the roots of susceptible cv. Piel de Sapo at two DAI, when no symptoms could be seen. Conventional PCR assay also detected the pathogen specifically. But real-time PCR assay provided reliable quantification of pathogen DNA with a range of four orders of magnitude (from 5 ng to 0.3 pg). This protocol enabled quantitative monitoring of pathogen development from the early stages of infection and also for screening of melon genotypes for resistance to root rot disease. Further, the real-time PCR format provided more accurate estimates of pathogen DNA to reflect the DS on roots, compared with visual scoring of root lesions in root biomass (Picó et al. 2008). Field peas are infected by the soilborne pathogens *Phoma koolunga*, *Didymella pinodes* and *P. medicaginis* var. *pinodella* in four states of Australia. Real-time PCR assay was applied for the detection and quantification of these pathogens in soils. The amounts of DNA of the pathogens in soils prior to planting field pea was positively correlated with Ascochyta blight lesions on field pea subsequently grown on infested soil in pot assay and also on field pea in naturally infested fields. After harvest of field pea crop in the infested fields, the populations of the soilborne pathogens reached the maximum based on DNA contents, followed by a decline in the following three years. The DNA contents in field pea plants were also estimated by real-time PCR assay along with soil plating assay. The results showed that *P. koolunga* and *D. pinodes* were the primary pathogens of blight disease of field pea, whereas *P. medicaginis* var. *pinodella* had a minor role in the disease complex (Davidson et al. 2011).

*R. necatrix*, causal agent of WRR disease of avocado could be detected using several primers based on ITS regions of the rDNA genes for amplification of specific fragments. Screening the primers against two isolates of *R. necatrix* and six other *Rosellinia* spp. resulted in the amplification of a single specific product from the genomic DNA of *R. necatrix* for most of the primer pairs employed in the PCR assay. Two primer pairs R2-R8 and R10-R7 were highly specific in amplifying the product only from 72 isolates of *R. necatrix*, but not from 93 other fungi from different hosts and geographic locations. The R10-R7 primer was modified to obtain a Scorpion primer for detecting a specific 112-bp product by fluorescence emitted from a fluorophore in a self-probing PCR assay. This assay specifically recognized the target sequence of *R. necatrix* over a large number of other fungal species. Conventional PCR format, with primer pairs R2-R8 and R10-R7, had a detection limit of 10 pg/µl using a single set of primers and the nested PCR format was more sensitive with a detection limit of 10 fg/µl. On the other hand, Scorpion-PCR assay could detect 10 pg/µl and 1 fg/µl as nested Scorpion-PCR assay, indicating an increase in sensitivity of detection limit (ten times) over conventional PCR format. A simple and rapid procedure for DNA extraction directly from soil was modified and developed to yield DNA of purity and quality suitable for PCR assays. By combining this protocol with the nested Scorpion-PCR procedure, *R. necatrix* could be detected in artificially infested soils in about 6 hours (Schena et al. 2002). In the further study, quantification of *R. necatrix* in soil samples collected from the base of 47 plants showing different symptoms of canopy decline was taken up, using real-time Scorpion-PCR assay. Comparison of efficiency of detection of *R. necatrix* showed that the pathogen could be isolated from 24 samples, whereas the real-time Scorpion-PCR assay could detect the pathogen in 37 soil samples (based on the cycle threshold values 25.8 to 47.1), demonstrating higher sensitivity and reliability of real-time PCR format. A single real-time PCR amplification was sufficient to detect *R. necatrix* in naturally infested soils. The results indicated that the nested PCR assay has to be avoided, because of risk of cross-contamination. The real-time PCR provides faster and sensitive detection and quantification of *R. necatrix* in roots of infected avocado plants and infested soils around infected plants that might have low population of the pathogen (Ruano-Rosa et al. 2007). Apple replant disease is due to a disease complex and *Cylindrocarpon* spp. form an important component of this complex. *C. macrodidymum* was the most prevalent among the four *Cylindrocarpon* spp. associated with apple replant disease in South Africa. Quantitative real-time PCR (qPCR) assay was developed for the simultaneous detection of all *Cylindrocarpon* spp. from inoculated apple roots. The amount of *Cylindrocarpon* DNA in roots was not correlated to seedling growth reduction (weight/ height) or root rot severity. The qPCR assay was useful for rapid identification of *Cylindrocarpon* spp. in apple roots (Tewoldemedhin et al. 2011).

Petri disease caused *P. chlamydospora* poses serious problems for the establishment of vines in newly planted vineyards. The possibility of nursery procedures being responsible for the dissemination of *P. chlamydospora* was examined. The nucleic acid-based techniques, single PCR, nested PCR, SYBR® Green quantitative PCR and TaqMan® PCR procedures were applied for the detection and quantification of pathogen propagules in water samples. *P. chlamydospora* could be detected in water used to wash grapevine cuttings sampled at a collection point in the field. All samples from the rain water storage facility and cool down tanks, and water used to soak buds during grafting contained detectable populations of *P. chlamydospora* in all sites. The quantitative TaqMan® real-time PCR format had a detection limit of 10 spores/ml of water. The nested PCR was the most sensitive, followed by quantitative TaqMan®real-time PCR format for detecting the pathogen in water samples. Conventional (single) PCR assay was found to be the least efficient in detecting

*P. chlamydospora* in water samples. However, this format required only half the volume of DNA extract that was needed for quantitative PCR assay formats. Quantitative PCR might reduce the risk of carry-over contamination, as the accumulation of PCR products is monitored during cycling and very little handling of the post-amplification products would be required. Further, quantitative PCR could indicate how much of the desired target is present in the sample rather than merely its presence in the sample (Edwards et al. 2007). Black foot disease caused by *Ilyonectria* spp. is a serious problem affecting grapevines in nurseries and rootstock mother fields. A multiplex nested PCR assay was applied to detect the black foot pathogen in soil samples, using species-specific primer pairs, Mac1/MaPa2, Lir1/Lir2 and Pau1/Pau2. Among 180 soil DNA samples analyzed, *Ilyonectria* spp. were detected in 172 samples. *I. macrodidyma* complex was more frequently detected than other species by the assay. *I. liriodendra* was present in 16 soil samples and in all open-root field nurseries and in two rootstock mother fields. In addition, quantitative PCR (qPCR) assay was employed to assess the populations of *I. macrodidyma* and *I. liriodendra* complex in the soil samples. Quantification of *Ilyonectria* spp. DNA by qPCR assay correlated with field soil samples showing positive reaction with nested multiplex PCR format. DNA contents of *Ilynonectria* spp. ranged from 0.004 to 1904.8 pg/µl. Generally, rootstock mother fields contained higher DNA contents of the pathogen species. The results indicated that soils of nurseries and rootstock mother fields could be important sources of inoculum for black foot pathogens (Agusti-Brisach et al. 2014).

Tobacco black root rot disease caused by *T. basicola*, a soilborne pathogen is responsible for significant losses in tobacco crops in several countries. The pathogen can infect a wide range of plant species including tobacco and tomato. Early detection and precise identification of *T. basicola* is essential to prevent disease spread and to reduce the losses effectively. TaqMan real-time PCR assay was developed, based on nucleotide sequences of the ITS regions of rDNA genes of genomic DNA of *T. basicola*. In addition, an inexpensive method of extracting soil DNA was also developed for providing suitable soil DNA of high quality for amplification by TaqMan PCR format. According to the amplicon size, primer pair Tb1/Tb2 was developed for SYBR Green I PCR and Tb3/Tb4 was used for TaqMan PCR with probe Tbp. Real-time quantitative PCR (qPCR) assays were more rapid, specific and sensitive with a detection limit of 100 fg/µl of genomic DNA of *T. basicola*. By combining the qPCR assays with the soil DNA extraction procedure, real-time detection of *T. basicola* in soil could be completed in 4–5 h and the detection limit of 3 conidia/reaction in qPCR assay was achieved. During the survey, the presence of *T. basicola* in naturally infested soils was detected, indicating the usefulness and reliability of the procedure developed in this investigation (Huang and Kang 2010). Onion white rot is due to *Sclerotium cepivorum*, a soilborne pathogen which is capable of causing considerable losses in quantity and quality of produce. A real-time PCR assay was developed to detect and quantify *S. cepivorum* in soils. In combination with a new method of extraction of DNA from soil samples, the real-time PCR assay was highly sensitive and specific in detecting and quantifying *S. cepivorum* in soil samples. Assay specificity was proved by testing against 24 isolates representative of 14 closely related species and other pathogens of onion. A high correlation was observed between Ct and the number of sclerotia detected. *S. cepivorum* was detected in four of 29 samples using the real-time PCR format and the positive samples had a bearing on the current or past outbreaks of onion white rot disease. The real-time PCR procedure was found to be effective in assessing pathogen DNA content of soils and also for detecting other onion pathogens *R. solani* AG-8 and *Botrytis aclada* in the same soil samples (Woodhall et al. 2012).

*Sclerosopora graminicola*, incitant of pearl millet downy mildew and green ear disease, is responsible for heavy losses in hybrids and the pathogen is present in seed, plant and in soil as oospores which are resistant to adverse environmental conditions. Conventional and TaqMan real-time PCR formats were employed for the detection and quantification of *S. graminicola* by targeting 28S region of the nuclear large subunit (nuLSU) of the pathogen. TaqMan real-time PCR format specifically amplified a fragment of 86-bp from diverse DNA samples of *S. graminicola* present in seed, root, leaves and infested soil. Both conventional and real-time PCR formats amplified pathogen-specific product or produced fluorescent signal only from *S. graminicola* infected samples, but not from other microorganisms or healthy pearl millet leaves. The absolute quantity of target molecules (28S) in the DNA sample extracted from sporangia ($6 \times 10^8$/ml) was estimated as 25 pg/µl and the copy number was $2.69 \times 10^8$ molecules/µl. The copy number of target molecules in a diploid nucleus (2n) of *S. graminicola* was 27 molecules/ nucleus. The procedure developed in this study has the potential for diagnosis of *S. graminicola* infection of seeds, facilitating exchange of disease-free seed lots between countries (Kishore Babu and Sharma 2015).*G. graminis* Ggt, incitant of take-all disease of wheat, is soilborne and survives saprophytically in crop debris and by infecting susceptible grass weeds and cereal volunteer plants. A quantitative real-time PCR (qPCR) assay was developed to estimate the quantity of pathogen DNA in plants. The qPCR assay detected DNA of *Ggt* in both symptomatic and asymptomatic roots of wheat plants. The increase in DNA contents of *Ggt* was strongly correlated to increase in intensity of take-all symptoms (Keenan et al. 2015). A real-time PCR-based seedling assay was developed for the detection of *Sphacelotheca reliana*, causing head smut disease in maize (corn) plants. The DNA contents of *S. reliana* in the roots and mesocotyls of maize plants growing from seeds treated with fungicides were assessed. Treatment of seeds with tebuconazole, fludioxonil, sedaxane and mefenoxam + azoxystrobin + thiabendazole provided protection to fungicide-treated plants and reduced infection of roots by *S. reliana*. This real-time PCR procedure was useful for the detection of pathogen in maize plants and also for assessment of infection levels following seed treatment with fungicides (Anderson et al. 2015).

Sugar beet root rot caused by *Aphanomyces cochlioides* was diagnosed, using time-consuming greenhouse bioassay employing bait plants. In order to obtain rapid and reliable results for disease risk assessment, a real-time quantitative PCR (qPCR) assay was developed for determining *A. cochlioides* DNA in field-infested soil samples and validated using the standard bioassay. The qPCR assay was species-specific and was optimized to provide high amplification efficiency suitable for copy quantification. A high correlation (R > 0.98, P < 0.001) with pathogen inoculum was observed, demonstrating the suitability for monitoring pathogen presence in soil samples. The limit of detection by the qPCR format varied between 1 and 50 oospores/g of soil, depending on clay content of the soil sample. Varying levels of *A. cochlioides* target sequence were detected in 20 of the 61 naturally infested soil samples. The results of the qPCR provided reliable data for disease risk assessment enabling the growers to identify high-risk fields for avoiding such fields for sugar beet cultivation (Almquist et al. 2016).

### 2.1.4.3.7 Applications of Nanotechnology

Nanotechnology facilitates pathogen detection more rapidly and precisely, compared to other methods of detection of microbial plant pathogens. Nanotechnology enables development of a new generation of biosensors and imaging techniques with higher level of sensitivity and reliability. Highly sensitive fluorescent dye doped nanoparticles increase the signals by the magnitude of $10^5$ to $10^6$ times as well as tagging pathogens, enabling the equipment to detect targets at very low concentrations. Nanoparticles (about 1–100 nm in diameter) display unique properties over bulk-sized materials and they have been used in diagnostics and environmental monitorimg of microorganisms, pollutants, pathogens and their toxins. Incorporating the nanoparticles into nanosensors provides the advantages of rapid and high throughput detecting ability on a portable device. The nanoparticles are considered as potential sensing materials, due to strong physical confinement of electrons at nanoscale. Their tiny size corresponds to high surface-to-volume ratios. Surface-modified nanocolloids such as gold nanoparticles (GNPs), magnetic nanoparticles (MNPs), quantum dots (QDs) and carbon nanotubes exhibit specific target-binding properties (Koedrith et al. 2015). Conventional molecular detection techniques used to identify microbial plant pathogens, cannot be employed in the field, since they require complex instrumentation and expert operator. Further, they need extensive sample preparation and have long readout periods, resulting in delayed response and disease containment. Hence, taking advantage of the unique properties of nanoparticles (electrical, magnetic, luminescent and catalytic capacity), cost-effective, highspeed and sensitive methods can be developed to detect and monitor microbial pathogens. Nanoparticles when acting as signal reporter, may increase signal significantly and thus, reduce or eliminate the time to increase target cells to detectable levels. Furthermore, nanotechnology-based systems have reasonable reproducibility, cost-effectiveness, robustness and user-friendly features, enabling their effective applications in field tests. Nanomolecular techniques are applied to monitor microbial interactions and gene transfer between pathogens and even the host. Further, nanoparticles such as nanosized silica-silver have been applied as antimicrobial and antifungal agents.

Nanotechnology may be useful for the management of crop diseases either for developing diagnostic tools or controlling delivery of functional molecules. Nanostructured platform is used to detect the microbial pathogens and their mycotoxins. Nanoparticles remain bound to the cell wall of pathogen, resulting in deformity of pathogen propagules, due to high energy transfer, leading to its death. A briefcase-sized kit was used for detecting potential pathogens infecting crops rapidly. The 4-mycosensor is a tetraplex competitive antibody-based assay in a dipstick format for the real-time detection of zearalenone (ZEA), T-2/HT-2, DON and FB1.FB2 mycotoxins on the same single strip in corn, wheat, oat and barley samples at or below their respective European maximum residue limits (MRLs) (Lattanzio et al. 2012). The main role of nanosensors is to decrease the time for fungal pathogen detection (Baeummer 2004). Various types of nanostructures have been evaluated as platforms for the immobilization of biorecognition element to construct a biosensor. The nanomaterials used for biosensor construction include metal and metaloxide nanoparticles, QDs and carbon materials. QDs are a group of semiconducting nanomaterials with unique optical and electronic properties. They provide distinct advantages over traditional fluorescent organic dyes in chemical and biological investigations. QDs have been employed for the construction of biosensor for disease-causing microbial pathogen detection. Due to their unique and advantageous optical properties, they have been used for plant disease diagnosis, using FRET mechanism which facilitates energy transfer between two light-reactive molecules (Fang and Ramasamy 2015). Nanosensors can be linked to a GPS for real-time monitoring of disease incidence and distributed throughout the field to monitor soil conditions and crop health. Nanosensors enable the identification of plant diseases before symptom expression and facilitate the early application of control measures (Rai and Ingle 2012).

The nanosensor is the product of a combination of biological and nanotechnological approaches. The nanobiosensors have the potential for increased sensitivity and offer a significantly reduced response-time to sense potential disease occurrence in crops (Small et al. 2001). A biosensor system was developed for the rapid diagnosis of soilborne diseases, consisting of two biosensors. The system was constructed using equal quantities of two different microbial pathogens, each individually immobilized on an electrode (Hashimoto et al. 2008). An optical DNA biosensor based on FRET utilizing QD as sensor was developed. The modified QD was conjugated with single-stranded DNA (ss-DNA) probe. The target DNA was sandwiched with conjugated QD-ss-DNA and reporter probe labeled with Cy5. Hybridization of the sandwich hybond, enabled detection of related sequences of *Ganoderma boninense* gene by constructed FRET signals which bind between those DNAs bringing them closer for fluorescence emission. The biosensor could detect target DNA at a concentration as low as 1 nM in 10-min hybridization. This method is simple, rapid and sensitive for early detection of specific DNA of fungal pathogen (Bakhori et al. 2012).

A QDs FRET-based biosensor technique was developed for the efficient detection of *P. betae*, the vector of *Beet necrotic yellow vein virus* (BNYVV), causative agent of the rhizomania disease that seriously impacted sugar beet production. QDs were biofunctionalized with a specific antibody against *P. betae*. GST was employed for generating anti-GST antibody, which was effectively conjugated to Tioglicoli acid-modified Cadmium-Telluride QDS (CsTe-QDs) synthesized in an aqueous solution via electrostatic interaction. The dye (rhodomine) molecules were attached to the GST. Donor-acceptor complexes (QDs-Ab-GST-Rhodomine) were then formed, based on the antigen-antibody interaction. The mutual affinity of the antigen and the antibody brought the CsTe-QDs and rhodomine together close enough to enable resonance dipole-dipole coupling required for FRET to occur. The immunosensor showed high sensitivity and specificity of 100%, and acceptable stability and it could be used for real sample detection with consistent results (Safarpour et al. 2012).

### 2.1.4.3.8   Razor Ex BioDetection System

Development of on-site accurate and sensitive detection of soilborne fungal plant pathogens faces formidable challenges, because of the requirement for expensive laboratory equipments and expertise to the needed level. The Razor Ex BioDetection System was originally designed for military application to identify biological threat organisms on-site. This system offers ready-to-use, freeze-dried reagent pouches and barcode-based PCR cycling program upload. For the detection of *Phymatotrichopsis ominovora*, the soilborne pathogen, causing Phymatotrichopsis root rot disease of cotton, alfalfa and many dicotyledonous crops, the TaqMan PCR assay, using the field-deployable Razor Ex BioDetection System was developed. The protocol was found to be reliable, sensitive and precise for the detection of *P. omnivora* using three target genes of the pathogen. This assay required about 30 min, including a modified magnetic bead-based method on-site DNA extraction (~ 10 min), from the infected plant roots. Specific primers and probes were designed based on *P. ominvora* nucleotide sequences of the genes encoding rRNA internal transcribed spacers, β-tubulin and the second largest subunit of RNA polymerase II (RPB2). This multigene qPCR format with maximized reliability, specificity and broad-range detection within *P. ominvora* variants was effective by minimizing the risk of false positives and negative, because each gene acted as an internal control for two other genes. No nonspecific cross-reactivity was obtained for the three pairs of primers and probe sets, when tested against plant and microbial exclusivity panels. Positive PCR results with genomic DNA of 13 diverse *P. omnivora* isolates from different locations confirmed the broad-range detection of the pathogen. All 14 symptomatic and asymptomatic and dead plant roots were PCR-positive, when tested by all three primer and probe sets. In addition, symptomatic and asymptomatic plant and soil samples (52) from alfalfa and cotton fields showed pathogen presence, when tested using the PO4 primer and probe.

The assay could detect as little as 1 fg of plasmid DNA (positive control) and 10 fg (PO4) to 100 fg (Pobt1 and PORB2-2) of genomic DNA of *P. ominivora*. As *P. ominvora* is on multiple lists of regulated organisms, early detection by the qPCR assay using Razor Ex BioDetection System would be effective in preventing dissemination of the pathogen through plant materials during interstate or international commerce (Arif et al. 2013).

### 2.1.4.3.9 Nested Time-Release Fluorescent (NTRF)-PCR Assay

*P. sclerotioides*, a soilborne fungus, causes black rot disease of cucumber and other cucurbit crops. In order to improve the sensitivity of detection and accuracy of quantification, nested PCR, time-release PCR using two different DNA polymerases (recombinant Taq and AmpliTaq Gold) and fluorescent PCR to obtain fluorescent-labeled PCR products that can be analyzed by capillary electrophoresis were combined and termed nested time-release fluorescent (NTRF)-PCR assay. The minimum pathogen DNA content required for detection was 50 fg/µg. Using NTRF-PCR assay, *P. sclerotioides* could be detected at a density of 10 CFU/g of sandy soil that was artificially infested with the pathogen. The NTRF-PCR format was more reliable than the real-time PCR format for the detection and quantification of the pathogen DNA in environmental samples, such as soil, plant debris or compost. The NTRF-PCR assay was rapid and less labor intensive, because the post-PCR analysis could be automated by capillary electrophoresis in the genetic analyzer which utilizes a fluorescent signal for detection. Capillary electrophoresis provides information on the concentration of amplicon and also on the nucleotide number of the amplicon. The pathogen could be detected in most soil samples collected from commercial cucumber fields in which plants showed symptoms or remained asymptomatic, indicating the possibility of assessing the population of *P. sclerotioides* as early as possible to make required disease management decisions. The results showed that *P. sclerotioides* was widely distributed in the melon-growing areas in Japan (Ito et al. 2012).

### 2.1.4.3.10 Loop-Mediated Isothermal Amplification Technique

PCR-based methods require relatively complex and expensive thermocycling equipment. In real-time PCR assays, fluorescence detection is performed concurrently with thermal cycling, increasing the complexity and cost further, although quantitative assessment of the pathogen can be achieved. In contrast to PCR-based assays, isothermal amplification procedure avoids the use of thermocycler and enables reactions to be incubated in a water bath or simple heated block, thus simplifying the procedure to a great extent and the procedure can be completed in less than one hour. Loop-mediated isothermal amplification (LAMP) is a an amplification procedure performed using two sets of primers and a DNA polymerase with strand-displacing activity to produce amplification products containing loop regions to which further primers can bind, enabling amplification to continue without thermal cycling. LAMP functions at isothermal temperature for the synthesis of large amounts of DNA in a short time. LAMP uses large fragments of the *Bst* DNA polymerase which has strand displacement $5' \rightarrow 3'$ polymerase activity, but lacks $5' \rightarrow 3'$ polymerase activity. In most LAMP procedures, sets of two primers (FIP, Forward Inner Primer and BIP, Backward Inner Primer) outer primers (F3 and B3) are designed targeting a specific genomic region. Initially LAMP reaction utilizes all four primers, but later only the primers are used for strand displacement DNA synthesis (Notomi et al. 2000 Nagamine et al. 2001; Tomlinson et al. 2010). Positive LAMP reactions result in the accumulation of a byproduct visible as white precipitate, which can be assessed visually. However, visual assessment could be ambiguous and therefore, quantitative readouts are needed to improve interpretation and in-field application of LAMP methods. Integration of LAMP with (LDFs) which are more efficient, less time-consuming and have potential for in-field application has been carried out in some investigations. Amplification is accelerated by the use of an additional set of primers (loop primers) that bind to those loops which are of incorrect orientation for the internal primers to bind (Nagamine et al. 2002). The requirement for the primers to bind up to eight regions of the target sequence results in high levels of specificity of detection of target sequences. LAMP products can be detected by agarose gel electrophoresis in real-time using intercalation of turbidity or color changes. For unambiguous detection of LAMP products from DNA of microbial plant pathogens *in vivo*, LFDs have been employed. LFDs detect the labels incorporated into the amplification products. Tests in LFD formats have been applied under field conditions and specific LFD immunoassays have been highly effective in detecting the plant pathogens. However, the availability of specific antibodies may be a constraint limiting its large-scale application.

*P. ramorum*, causal agent of sudden oak death disease of tanoak (*L. densiflorus*) and *Quercus* spp. has destroyed large numbers of trees in the forests on the west coast of the United States. A simple and rapid method of extracting DNA on the nitrocellulose membranes of LFD was used. LAMP of target DNA, using labeled primers and detection of the generically labeled amplification products by a sandwich immunoassay in a lateral-flow-device format were applied. Each step in the procedure could be performed without special equipment and applied for on-site testing. The results were obtained in about one hour. The LAMP assay for the detection of plant DNA (cytochrome oxidase, *COX* gene) was employed in conjunction with pathogen-specific assays to confirm negative results. Labeled LAMP could be used to increase the specificity of pathogen detection, when sufficiently specific antibodies are not available, as in the case of *P. ramorum* and *P. kernoviae*, but LFDs are not available for detecting all *Phytophthora* spp. The lowest amount of DNA of *P. kernoviae* and *P. ramorum* detected by LAMP assay was ~ 17 pg. The *P. ramorum* LAMP assay was used in multiplex with COX LAMP assay to test CTAB DNA extracts from healthy and *P. ramorum*-infected rhododendron and an extract from *P. ramorum* culture. The multiplex products were run on DIG and FITC

LFDs, demonstrating the detection of single products (*P. ramorum* or *COX*) and mixed products. This procedure was sufficiently sensitive to detect *Phytophthora* spp. in symptomatic rhododendron (mixed 1:10 or 1:5 infected/healthy material (Tomlinson et al. 2010).

Isothermal amplification methods that are tolerant to inhibitors present in many plant extracts, have been developed for the detection and quantification of fungal plant pathogens. These assays may reduce the need for obtaining purified DNAs of target pathogens for performing diagnostic assays. Recombinase polymerase amplification (RPA) has similarities with real-time PCR format for designing primers and labeled probes. The RPA technique was applied for the detection of *Phytophthora* spp.: one genus-specific assay multiplexed with a plant internal control and other species-specific assays for *P. ramorum* and *P. kernoviae*. The lower limit of linear detection, using purified DNA was 200 to 300 fg of DNA in all pathogen RPA assays. Specificity of the assays was demonstrated by testing against 136 *Phytophthora* taxa, 21 *Pythium* spp., one *Phytopythium* sp. and a wide range of plant species whose DNAs were not amplified by the primers selected for *P. ramorum* and *P. kernoviae*. Of the 222 field-collected plant samples tested, only 56 samples were culture-positive for *Phytophthora* spp., whereas RPA test and TaqMan real-time PCR format detected the pathogen in 91 samples. TaqMan PCR and RPA assays had similar limits of detection. The *Phytophthora* genus-specific (targeting *trnM-trnP-trnM*) and *Phytophthora* species-specific (targeting *atp9-nad9*) TaqMan assays were able to detect purified *P. ramorum* DNA at an initial quantity of 200 fg and detection remained linear at increasing concentration using Ct values ($R^2 = 0.997$ and 0.996), respectively (see Figure 2.8). The *Phytophthora* genus-specific (targeting *trnM-trnP-trnM*) and *Phytophthora* species-specific (targeting *atp9-nad9*) RPA assay were also able to detect purified *P. ramorum* DNA at an initial quantity of 200 fg and detection remained linear at increasing concentrations, using the log of the onset of amplification (OT) values ($R^2 = 0.980$ and 0.974, respectively). Plant extract had a limited effect on both TaqMan and RPA amplification. For TaqMan assays, the standard APHIS-*P. ramorum* DNA

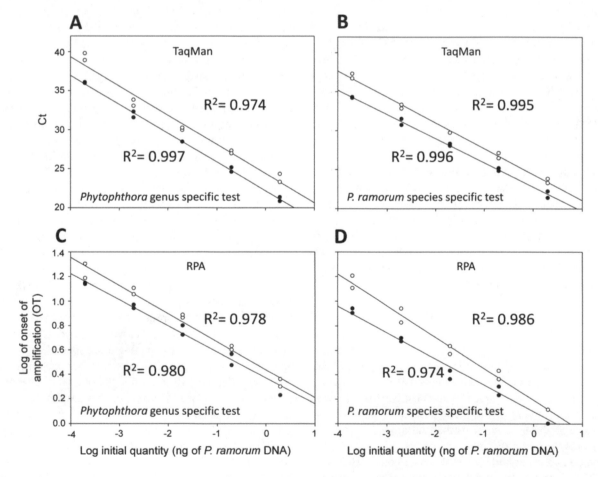

**FIGURE 2.8** Differentiation of *P. ramorum* by employing genus- and species-specific primers by RPA technique multiplexed with plant internal control. A and B: TaqMan *Phytophthora* genus-specific and *P. ramorum* species-specific assays with the log of initial DNA content of *P. ramorum* against Ct values, where the absence or pressure of *U. californica* (Bay laurel) DNA is represented by closed and open circles, respectively. C and D: *Phytophthora* genus-specific and *P. ramorum*-species-specific RPA assays with the log of the initial DNA quantity of *P. ramorum* against the log of onset of amplification minus the agitation step (OT), where the absence of a crude leaf extract of *U. californica* DNA is denoted by closed and open circles, respectively.

**[Courtesy of Miles et al. (2015) and with kind permission of the American Phytopathological Society, MN, United States]**

extraction protocol for genus- and species-specific detection was tested with healthy *U. californica* samples spiked with a serial dilution of purified *P. ramorum* DNA. Addition of plant DNA to the pathogen dilution series increased the Ct values between 1 and 2 across various dilutions and it had a limited effect by delaying onset of amplification across dilutions, with a greater effect observed for the lowest concentration of pathogen DNA. However, *P. ramorum* could be detected at all concentrations in both genus- and species-specific assays. The RPA assays provided additional advantages over conventional methods, since they are rapid (giving results within 15 min); do not require DNA extraction and use less expensive portable equipments, thus facilitating the application of field-deployable capability for pathogen detection that will enable point-of-sample collection processing (Miles et al. 2015).

The comparative efficacy of nested PCR and LAMP assays was assessed for the detection of *Phtyophthora nicotianae*, causal agent of black shank disease of tobacco and other economically important diseases. Primers based on the Ras-related protein gene (*Ypt1*) of *P. nicotianae* were evaluated for the specific detection of the target pathogen, using nested PCR and LAMP assays. Detection limits of the nested PCR and LAMP assays were 100 fg and 10 fg, respectively of genomic DNA per 25 µl reaction volume. The specificity of the primers was indicated by positive reactions only with 47 isolates of *P. nicotianae*, but not with 45 isolates of 16 different oomycetes and 25 isolates of other fungal species. Further, there was no cross-reaction with other pathogens tested. LAMP assay detected *P. nicotianae* in naturally infected tobacco tissues and soil. The results indicated the potential of LAMP assay for obtaining specific detection of *P. nicotianae* in regions that are at risk of contracting black shank disease (Li et al. 2015).

LAMP assay targeting the Ssox5 sequence was applied for the visual detection of *S. sclerotiorum* infecting canola and several other crops. The LAMP reaction was optimal at 63°C for 45 min. Hydroxynaphthol blue (HNB) was added before amplification for easy recognition of amplicon which appeared blue. Samples with pathogen DNA developed characteristic sky blue color, while the samples without pathogen DNA or DNA of nine other fungal pathogens did not develop color. Results obtained with LAMP and HNB were confirmed by electrophoretic analysis of LAMP products. The detection limit of LAMP assay for detection of *S. sclerotiorum* was 0.1 fg/ µl of genomic DNA per reaction. On the other hand, the detection limit of conventional PCR assay was 100 fg/µl. Identical results were obtained by testing 13 samples of canola tissues infected by *S. sclerotiorum* by LAMP, conventional PCR and isolation procedures. The LAMP procedure developed in this study, was found to be simple, rapid, sensitive and specific for the detection of *S. sclerotiorum* in canola plants (Duan et al. 2014). In a later investigation, the detection of mutant genotypes of *S. sclerotiorum* resistant to carbendazim was achieved mainly by determining mean inhibitory concentration (MIC) which required three to five days. A LAMP assay was developed for the rapid detection of mutant genotype F200Y showing resistance to carbendazim, because of point mutation at codon 200 (TTC → TAC). Specific LAMP

primers were designed and concentrations of LAMP components were optimized. The optimal reaction conditions were 62–63°C for 45 min. There was no requirement for any special equipment for the highly senisitve and specific LAMP assay for detection of the F200Y mutant genotype. Inclusion of the loop backward (LB) primer reduced the reaction time to 15 min. The results of LAMP assay corroborated that of MIC determinations. The advantages of the LB-accelerated LAMP assay for the detection of the F200Y mutant genotype were revealed by assaying sclerotia produced on rape stems artificially inoculated in the field. The LAMP assay could be employed for specific detection of target genotype of *S. sclerotiorum* (Duan et al. 2016).

*R. solani* causes root rot diseases of several crops and it persists in soil for many years, surviving as a saprophyte on plant debris or other organic matter. Various kinds of techniques have been applied for the detection and quantification of *R. solani* populations in plants and soils. LAMP procedure can be performed inexpensively, using a water bath or heating block. LAMP primers based on ITS DNA sequence were designed for detecting most AGs of *R. solani*. The LAMP reactions were performed for the detection of four isolates of *R. solani* from tomato and other host plants, with 5 mM $Mg^{2+}$ at 65°C for 60 min. The LAMP test detected very low levels of DNA (10 fg, equivalent to approximately one copy of a 87.1 Mb of *R. solani* genome) for *R. solani* and *R. zeae*, but not from *R. oryzae*. The LAMP products were simultaneously assayed using a generic LFD, turbidity and CYBR green staining. These analyses also showed that 10 fg DNA was the lowest limit that could be detected using LFD, but not with turbidity and CYBR Green staining. The detection limit for *R. zeae* was 1.0 pg which was two orders of magnitude higher than the *R. solani* detection limit. The LAMP procedure was specific for the detection of *R. solani*, as no amplicon was generated from DNA of *F. oxysporum* and *P. parasitica* and several host plant species. The LAMP had a detection limit of 5 µg of *R. solani* DNA/g of soil. For on-site application in the field, the LAMP procedure was performed with a generic anti-biotin and anti-fluorescein antibody-based LFD. The LAMP reactions were performed using biotin-labeled primers which were hybridized with a fluorescein amidite (FAM)-labeled hybridization probe and detected with the LFD. This LAMP-LFD procedure could detect *R. solani* in infected plant tissues. The LAMP-LFD assay and real-time quantitative PCR (qPCR) format had similar detection limits of 10 fg DNA of *R. solani* with no false positive or negative results. But LAMP-LFD procedure is simple, rapid and equally sensitive and specific as qPCR format for the detection and quantification of *R. solani* in different plant species and soil with the distinct possibility of being amenable for on-site testing (Patel et al. 2015).

*R. solani*, causal agent of seedling blight and *M. phaseolina*, incitant of charcoal rot diseases of soybean, were directly detected by LAMP assay. The primers were designed and screened using ITS sequences as targets of both pathogens. An ITS-Rs-LAMP assay for *R. solani* and an ITS-Mp-LAMP assay for *M. phaseolina* were employed to detect these pathogens in diseased soybean plant tissues in the field. Both LAMP assays effectively amplified the target DNA fragments over 60 min at 62°C. Appearance of yellow-green color (visible to the naked eye) for *R. solani* or intense green fluorescence (visible under ultraviolet light) for *M. phaseolina*, after addition of SYBR Green I, indicated the presence of these pathogens. The detection limits of ITS-Rs-LAMP and ITS-Mp-LAMP assays were 10 pg/μl and 100 pg/μl of genomic DNA of these pathogens, respectively. By applying the LAMP assays, both pathogens could be detected in soybean plants under field conditions in China (Lu et al. 2015). A rapid, visual, and sensitive LAMP assay was developed for the detection of *M. phaseolina*, causing root rot in common bean (*P. vulgaris*) seeds, based on SCAR (GU018142.1). The assay was performed providing amplification conditions of 65°C for 45 min. The results of LAMP assay could be visually recorded by the addition of HNB dye. Sky blue color developed to indicate positive reaction, while controls remained violet under ambient light, indicating negative reaction. The assay had a detection limit of 1 pg of *M. phaseolina* genomic DNA per reaction. No cross-reaction was noted with DNA isolated from five other pathogens infecting bean seeds. *M. phaseolina* was detected both in naturally infected and artificially contaminated bean seeds. The detection limit was one infected seed per seedlot of 400 seeds, without the need for isolation of the pathogen DNA prior to LAMP assay. The results indicated the applicability of LAMP assay for sensitive detection of *M. phaseolina* in seeds from which the pathogen could reach the soil (Rocha et al. 2017).

*P. aphanidermatum* is a common soilborne pathogen causing damping-off diseases in nurseries of vegetable crops. A LAMP reaction with primer set designed from the rDNA ITS sequence of *P. aphanidermatum* was developed. The LAMP assay was specific to *P. aphanidermatum* as revealed by the absence of amplification from the DNA of 30 *Pythium* spp., 11 strains of *Phytophthora* spp. and eight other soilborne pathogens. The LAMP assay was ten times more sensitive than conventional PCR format. The detection limit of LAMP assay was 10 fg of genomic DNA of *P. aphanidermatum*. The LAMP assay efficiently detected *P. aphanidermatum* from hydroponic solution samples from tomato plants. In addition, *P. aphanidermatum* could be detected directly in the roots of tomato without extraction of DNA. The results showed that LAMP assay was a simple, sensitive and rapid diagnostic procedure for detection of *P. aphanidermatum* from different substrates (Fukuta et al. 2013). Pythium stalk rot of maize (corn) caused by *Pythium inflatum* is one of the most destructive diseases affecting production seriously. The detection of *P. inflatum*, a soilborne pathogen, early and precisely is the basic requirement for the application of effective disease management practices. A real-time fluorescence LAMP (Real Amp) assay was developed for rapid, quantitative detection of *P. inflatum* in soil. The detection limit of the Real Amp assay was approximately 0.1 pg/μl plasmid DNA when mixed with extracted soil DNA or $10^3$ spores/g of artificially infested soil. No cross-reaction was seen with other related pathogens. There was no significant difference in amounts of soilborne

pathogen DNA determined by Real Amp assays and real-time quantitative PCR format. Further, the Real Amp assay could be employed via an improved closed-tube visual detection system by adding SYBR Green I fluorescent dye to the inside of the lid prior to amplification. Consequently, the inhibitory effects of the dye on DNA amplification could be avoided. The results show that Real Amp assay has the potential for application as a simple, rapid and effective technique for the detection of *P. inflatum* in the field soils (Cao et al. 2016).

*P. ultimum* causes damping-off and root rot in many plant species. The LAMP technique was applied for the rapid and sensitive detection of *P. ultimum* by targeting the gene encoding spore wall protein from *P. ultimum* genome, using a comparative genomics approach. The target DNA fragment was efficiently amplified at 64°C for 60 min. The pathogen-specific amplicon was generated only from the DNA of the isolates of *P. ultimum*, but not from other pathogens tested, indicating the specificity of the LAMP assay. The limit of detection of the assay was 1 pg/μl DNA, indicating that LAMP assay was 1,000 times more sensitive than the conventional PCR assay. The LAMP technique provided positive results with the DNA extract from plant tissues infected by *P. ultimum*. The results showed that the LAMP assay could be employed to detect *P. ultimum* in pure cultures, as well as in infected plants, even when the pathogen population was low (Shen et al. 2017). The LAMP assay and a bait-trap method were combined for the detection of *P. helicoides*, infecting roses, miniature roses and poinsettias grown in hydroponic culture systems. In the 'Bait-LAMP' procedure, the crude extract from perilla seeds used as bait was employed in the LAMP assay, whereas in the 'Bait culture-LAMP' method, the crude extract of mycelia grown out of perilla seeds onto *Pythium*-selective medium was used for the LAMP assay. Both procedures were rapid and simple and they could be employed for monitoring infections by *P. helicoides* in hydroponic culture systems (Miyake et al. 2017).

*T. indica*, a quarantine pathogen, causes Karnal bunt disease of wheat. Specific and rapid detection of this pathogen is essential to prevent its introduction and subsequent spread to new locations. Standard PCR assay encountered specificity issues, due to high DNA homology of *T. indica* with other *Tilletia* spp. A specific and rapid detection of *T. indica* was shown to be possible by employing the LAMP assay at 62°C. Alignment of the mitochondrial DNA of *T. indica* and *T. walkeri* revealed four major unique regions in the DNA of *T. indica*. Six LAMP primers designed based on the sequences of the unique regions, amplified *T. indica* DNA. The amplification could be completed in 30 min and the sensitivity level was equal to that of standard PCR format. Amplification of *T. indica* DNA only occurred during screening of 17 isolates of *T. indica*, *T. walkeri*, *T. horrida*, *T. ehrarte* and *T. caries*. By using fluorescent calcein, end point detection of *T. indica* was possible with the naked eye. The specificity, rapidity and visualization without any equipment are the distinct advantages of LAMP assay over the standard PCR format (Gao et al. 2016). *Sporisorium scitamineum* causes one of the most destructive diseases of sugarcane. Species-specific LAMP assay was developed for the rapid and precise detection of *S*.

*scitamineum.* A set of four species-specific primers F3, B3, FIP and BIP were designed based on the differences in the ITS sequences of *S. scitamineum.* A panel of fungal and bacterial species was included as controls. After optimization of the reaction conditions, the detection limit of LAMP assay was about 2 fg of the pathogen DNA in 25 µl reaction solution, 100-fold lower than that of conventional PCR assay. High specificity of the LAMP assay was revealed by the absence of positive reaction with DNA of nine other fungal and bacterial pathogens tested. Presence of *S. scitamineum* could be detected by the LAMP assay in young sugarcane leaves with no visible disease symptoms. The results showed that the LAMP assay was a single, highly sensitive, rapid and reliable procedure for early detection of *S. scitamineum*, useful for the production of smut-free seedcanes (Shen et al. 2016).

### 2.1.4.3.11   DNA Array Technology

DNA array technology, developed for screening for human genetic disorders, has been adapted and applied to detect and identify microbial plant pathogens infecting various crops. DNA arrays and chips are powerful tools used to study gene expression and their effectiveness in rapidly identifying plant pathogenic oomycetes, fungi, bacteria and viruses. In this technology, detector oligonucleotides hybridize with homologous amplified, prelabeled (with a fluorescent dye, a chemiluminescent reporter or a radioactive isotope) target DNA or RNA which can then be detected. Generally two types of array platforms, the glass slide (microarray) and the nylon or nitrocellulose membrane (macroarray) have been used. The microarray (with spot size less than 200 µm in diameter) enables higher oligonucleotide density where thousands of detector oligonucleotides can be spotted on a single slide. Thus, one microarray can detect hundreds of different pathogens. But high cost and low sensitivity are limitations associated with microarray production, and consequently its application (Bodrossy and Sessitsch 2004). Macroarray, using membranes (with spot size of above 200 mµ in diameter) requires only a pin-tool for array production. Although macroarray has lower throughput than the microarray, 96-well microtiter plate-size membrane can contain over 1,000 different detector oligonucleotides and each array can be washed and reused many times. DNA array technology is essentially a reverse dot blot technique useful for characterization of several DNA fragments of different pathogens simultaneously. High discriminating potential associated with immobilized detector oligonucleotides has been found to be crucial for diagnostic applications, since closely related pathogen species may differ in only a single base pair (single nucleotide polymorphism, SNP) for a target gene. The SNP profiles of fungal pathogens have been used as the basis of detection of a large number of pathogens simultaneously. The capability of detecting a much greater number of samples compared with other molecular diagnostic methods is a distinct advantage provided by macroarrays. Due to its relatively low cost and high throughput, the macroarray has been preferred for diagnosing microbial plant pathogens in different substrates, such as plants, soil and its environments (Lievens et al. 2003, 2005; Agindotan and Perry 2007).

Pure cultures of *P. nicotianae, P. ultimum, F. oxysporum* f.sp. *lycopersici* and *V. dahliae* were obtained and the perfect match oligonucleotides were selected from either ITS1 or ITS2 sequences from which detector sequences with melting temperature of 55°C ± 5°C were derived by adjusting the length of these oligonucleotides, when only a single nucleotide was substituted, mismatched at the fifth nucleotide were the most selective for *P. ulitmum*, enabling SNP discrimination irrespective of amplicon amounts of nucleotide used in the substitution. Similar selectivity of a specific SNP oligonucleotide, depending on its sequence was observed in *P. nicotianae, F. oxysporum* f.sp. *lycopersici* and *V. dahliae*. The results indicated that high specificity can be obtained with DNA arrays that enable the discrimination of single nucleotide sequences and therefore, closely related microorganisms can be differentiated (236Lievens et al. 2006). In another investigation, DNA array technology was applied for the detection of more than 100 species of *Pythium*, using an array containing 172 nucleotides complementary to specific diagnostic regions of the ITS. Positive hybridization reaction with at least one corresponding species-specific oligonucleotide was recognized in all, except *P. ostracodes*. Hybridization patterns were distinct for each species and also for strains of each species tested. Hybridization patterns were consistent with the identification of the isolates based on morphological characteristics and ITS sequence analysis. With DNA array technology, 13 species of *Pythium* were identified and the results were similar to that of with the SD plating method. The results from DNA array technology might form the basis for advancements in complex epidemiological and ecological studies (Tambong et al. 2006).

A membrane-based macroarray technique was applied for the detection of different fungal pathogens infecting solanaceous crops. In the array at least two specific oligonucleotides were employed for each target pathogen. Members of *Fusarium solani* species complex (FSSC) were detected using 33 oligonucleotides (17–27-mers) designed from ITS of rRNA genes of 17 FSSC isolates belonging to 12 phylogenetically closely related species. Of the 33 oligonucleotides, 21 were useful for discriminating all 12 species. The high specificity of the macroarray system was achieved by optimizing the hybridization temperature and oligo probe length. The array was validated by testing inoculated greenhouse samples and diseased field samples (Zhang et al. 2007). *F. oxysporum* f.sp. *cucumerinum* (FOC) and *F. oxysporum* f.sp. *radicis-cucumerinum* (FORC) were detected using robust SCAR markers developed based on RAPD markers. By employing these markers in a DNA macroarray, multiple pathogens belonging to the genus *Fusarium* and the *formae speciales* of *F. oxysporum* could be detected simultaneously with a high level of sensitivity. Three of five plant samples tested were diagnosed with FORC and two with FOC. Plating on selective medium and examining the identities of the pathogens, as determined by DNA macroarray system have been employed in many crop-pathogen systems (Lievens et al. 2007).

Based on the ITS sequences of the rRNA genes, 105 oligonucleotides (17 to 27-bases long) specific for 25 pathogens of solanaceous crops were designed and spotted on a nylon membrane. The array was tested against the 25 target pathogen species, 46 infected field samples and some nontarget fungal species. Target pathogens (23), except members of *FSSC*, were detected accurately and they were differentiated by macroarray without cross-hybridization. The detector oligonucleotides included on the array were highly specific and SNP were clearly differentiated between related fungal species. Two or more pathogens were detected from 23 of the 46 field and greenhouse samples. When ITS amplicons from root, stem and leaf tissues of tomato samples were combined and hybridized with the array, simultaneous detection of *Fusarium solani*, *F. oxysporum*, *Phoma destructiva* and *Alternaria alternata* was accomplished. The macroarray detected *S. subterranea* f.sp *subterranea*, *A. alternata* and *Colletotrichum* sp. in potato tubers, with only powdery scab symptoms at the time of sample collection. The macroarray detected unambiguously as low as 0.04 pg of genomic DNA for *F. sambucinum* from potato tubers. The array was able to detect target DNA, when the PCR amplicon were not visible by gel electrophoresis, indicating a higher level of sensitivity of the array system. The macroarray detected all target microorganisms in a sample, some of which may be pathogens and others may be simply present in the environment samples. The most distinct advantage of the array is that it could detect multiple pathogens in a single assay and it utilizes inexpensive reagents such as an unmodified oligonucleotide and standard Taq DNA polymerase to be cost-effective. Further, the array could be washed and reused for more than ten times and stored at 4°C for more than one year. This investigation clearly demonstrated the potential of macroarrays for the detection of multiple microbial plant pathogens simultaneously in various crops (Zhang et al. 2008).

Simultaneous detection of multiple pathogens and optimization for quantification over at least three orders of magnitude of *Veriticillium dahliae* and *V. albo-atrum* were accomplished using the DNA macroarray technique. Hybridization signals and pathogen concentration from different origins and also infected samples were strongly correlated. Further, it was possible to determine the DNA contents of other tomato pathogens *F. oxysporum*, *F. solani*, *P. ultimum* and *R. solani* in the environmental samples by using a DNA array procedure. The results of real-time PCR assay and DNA array techniques corroborated each other, indicating the robustness, reliability and rapidity of the quantitative estimation by the DNA macroarray procedure (Lievens et al. 2005). DNA /oligonucleotide array was developed based on a PCR and membrane-based DNA hybridization procedure to improve the genus-wide diagnosis for plant pathogens belonging to the genus *Phytophthora*. Sequences of three DNA regions, ITS, the 5′ end of cytochrome oxidase 1 gene (*cox 1*) and intergenic region between cytochrome *c* oxidase 2 (*cox 2*) and *cox 1* (cox2-1 spacer CS) were used to design oligonucleotides. Each sequence data set contained ~ 250 sequences representing 98 described and 15 undescribed species of *Phytophthora*. The nucleotides were grouped into four categories, 'best', 'good', 'acceptable' and 'rejected', based on their performance in hybridization tests. The macroarray was revalidated with 143

pure cultures, representing 96 of 98 described and nine of 15 undescribed *Phytophthora* spp., as well as four *Pythium* isolates. Oligonucleotides designed for the detection of *P. ramorum* and *P. sojae* were sensitive and specific, when tested with field samples. The DNA macroarray procedure targeted multiple genomic loci [ITS, cox 2-1 spacer (CS)], designing signature oligonucleotides to provide more accurate detection of *Phytophthora* spp. than PCR-based assays that target only a single region of pathogen DNA. The hyper variable CS region was used to differentiate *P. ramorum*, *P. pseudosyringae* and *P. nemorosa* through PCR-based assays. Several oligonucleotides designed in this investigation showed high specificity for the detection of economically important pathogens, such as *P. ramorum*, *P. sojae* and *P. cinnamomi* which have quarantine importance. The ability of DNA macroarrays to detect multiple *Phytophthora* spp. in one assay is very important for the disease management and detection of quarantine pathogens in local nurseries and forests where they coexist in container mixes, soils on the same site or different plant parts. The macroarrays provide a high throughput and effective multiplex detection tool for monitoring plant pathogens from complex environmental samples (Chen et al. 2013).

As several microorganisms are associated with rot diseases, molecular tools allowing simultaneous detection of many different targets are required to resolve the identities of the pathogens involved. A new microarray technique, ArrayTube system was employed to enhance the effectiveness of diagnosis of sugar beet root rot disease. Based on three marker genes viz., ITS, translation elongation factor 1-α and 16S ribosomal DNA, 42-well-performing probes enabled the identification of prevalent field pathogens (such as *A. cochlioides*), storage pathogens (like *Botrytis cinerea*) and ubiquitous spoilage fungi (like *Penicillium expansum*). All probes were proven for specificity with pure cultures from 73 microorganism species as well as for in planta detection of their target species, using inoculated sugar beet tissues. Microarray-based identification of root rot pathogens in diseased field beets was confirmed by traditional detection methods. The high discrimination potential of the assay was revealed by differentiation of *Fusarium* species, based on a SNP. The results showed that the ArrayTube procedure was an innovative tool enabling rapid and reliable detection of plant pathogens, especially when multiple microorganisms are involved in the disease (Liebe et al. 2016).

### 2.1.4.3.12  Padlock Probe Technology

Multiplex assays, that can detect several plant pathogens simultaneously, have been developed to increase detection efficiency and to reduce the time required for detecting the pathogens individually by different tests. Multiplex strategies involve amplification with generic primers that target a genomic region containing unique sequences. Targeting a conserved region limits the analysis to a taxonomically defined group of pathogens. Universal amplification coupled with microarray analysis overcomes some problems encountered in the multiplex approach. But the sensitivity is not sufficient for diagnostic purposes in complex environmental samples. Padlock probes (PLPs) enable combination of pathogen-specific molecular recognition and universal amplification, thereby increasing sensitivity and multiplexing capabilities without limiting the range of target organisms. PLPs are long oligonucleotides of ~ 100 bases, containing target complementary regions at both their 5′and 3′ ends. These regions recognize adjacent sequences on the target DNA and between the segments lie universal primer sites and unique sequence identities called ZipCode. Following hybridization, the ends of the probe get into an adjacent position and can be joined by enzymatic ligation. This ligation and the resulting circular molecule will be possible, only if both end segments recognize their target sequences correctly. Non-circularized probes are removed by treatment with exonuclease. The circularized ones may be amplified by using universal primers. Subsequently, the target-specific products are detected by a universal complementary ZipCode (cZipCode) microarray (Nilsson et al. 1994; Shoemaker et al. 1996).

A detection procedure was developed based on PLP and microarray for *P. nicotianae*, *P. cactorum*, *P. infestans*, *P. sojae*, *P. ultimum*, *R. solani*, *F. oxysporum* f.sp. *radicis-lycopersici*, *V. dahliae* and *V. albo-atrum*. Using PLP set, the genomic DNAs from a panel of well-characterized isolates of plant pathogens were tested. All *Phytophthora* spp. were correctly recognized by PLPPPhyt-spp, including *P. cactorum*. The efficacy of the PLP and microarray system for the detection of multiple pathogens simultaneously was assessed using mixtures of equal amounts of genomic DNAs of *P. cactorum*, *R. solani* AG4-1 and *V. dahliae*. The components of the mixture were correctly identified by using species-specific probes. The sensitivity of the assay system was 5 pg genomic DNA and a dynamic range of detection of 100 was observed (Szemes et al. 2005). Ligation-based probe assay employing Plant Research International (PRI)-lock probes was developed for the quantification of pathogen DNA in multiplex format. The circularized probes were amplified by using probe-unique primer pairs via real-time PCR, enabling accurate quantification in a highly multiplex format. High-throughput real-time PCR amplification was achieved by adaptation to OpenArray™ which could detect simultaneously and quantify pathogens such as *Phytophthora* spp., *P. infestans*, *R. solani* (AG2-2, AG4-1 and AG4-2), *F. oxysporum*, *V. dahliae*, *V. albo-atrum*, *Erwinia carotovora*, *Agrobacterium tumefaciens* and the nematode *Meloidogyne hapla*. The sensitivities of detection of pathogens were between $10^3$ and $10^4$ target copies/μl of initial ligation mixture, depending on the PRI-lock probe. The OpenArray™ system employing PRI-lock probes has the potential for highly specific, high-throughput, quantitative detection of multiple pathogens belonging to oomycetes, fungi, bacteria as well as a nematode (van Doorn et al. 2007).

## 2.2 ASSESSMENT OF VARIABILITY IN SOILBORNE FUNGAL PATHOGENS

All microbial plant pathogens, like other organisms, vary in their morphological, physiological, biochemical and genetic characteristics which may influence their pathogenic potential (virulence), resulting in differences in their ability or inability to infect various plant species/cultivars. Variations in other characteristics may affect the survival, and/ or reproduction. Assessment of variability in different characteristics of pathogen is essential to plan and apply effective measures to reduce the incidence and subsequent spread of the disease(s) in different agroecosystems. The pathogenic characteristics of fungal pathogens may be significantly altered due to selection pressure exerted by the introduction of new resistance gene(s) through breeding or indiscriminate application of chemicals for suppressing the development of the pathogens. Fungal pathogens produce new strains/races to overcome the adverse environments created by resistance genes and chemicals applied on host plants. Variations in the virulence (aggressiveness) of and tolerance to chemicals of fungal pathogens have been determined by biological and molecular techniques (Narayanasamy 2011).

### 2.2.1 VARIABILITY IN BIOLOGICAL CHARACTERISTICS

The soilborne fungal pathogens may exhibit variations in their adaptation to host plant species/genotypes, resulting in different degrees of DS. Depending on the interactions between pathogen isolates and the host plant genotypes (differentials), races, strains or pathotypes of a pathogen species are differentiated. By inoculating the pathogen isolates onto a set of host varieties with different specific resistances, it is possible to identify the virulence/avirulence factors present in the pathogen isolates under investigation. The expression of specific resistance may be influenced by environmental factors. A differential interaction between host resistance gene and pathogen virulence gene is a characteristic feature of gene-for-gene relationship. During coevolution of host plants and pathogens, gene-for-gene interactions occur naturally, resulting in a natural balance in an ecosystem. When a gene for resistance is introduced into a cultivar to improve its resistance level, the natural balance is upset. The pathogen reacts by producing populations of individuals (spores) with a virulence gene to match the newly introduced resistance gene for its survival and again, the natural balance between the host and the pathogen may be restored (Parlevliet 1993).

The soilborne fungal pathogens may have a wide host range that includes hundreds of plant species or they may be able to infect only a few plant species belonging to a single family. *Verticilliium dahliae*, *R. solani*, *M. phaseolina*, *T. basicola* and *P. cinnamomi* are some of the fungal pathogens capable of infecting a large number of plant species. *Sclerotium rolfsii*, causal agent of southern blight disease of peanut can infect over 500 plant species and it is widely distributed worldwide (Aycock 1966). In contrast, *Sss*, *P. graminis* and *Ggt* have a restricted host range. The host range of *P. ramorum* was investigated by collecting samples of ornamental plants and inoculating different species present in Serbia. *P. ramorum* could infect ten plant species, including *Rhododendron* and *Robinia pseudoacacia* which are widely distributed. In addition, *Cotoneaster horizontales* and *C. dammeri* were found to be new experimental hosts of *P. ramorum* (Bulajić et al. 2010). Variations in pathogenicity of soilborne fungal pathogens may be differentiated based on the severity of disease symptoms. Determination of races of *P. nicotianae* infecting tobacco was accomplished by employing a laboratory technique and its effectiveness was compared with the conventional greenhouse method. The laboratory technique involved the production and inoculation of tobacco seedlings in tissue culture. *P. nicotianae* isolates (identified as race 0 and 1) and four tobacco cultivars and breeding lines, K-326, K-346, NC-71, NC-1071, L8 and Ky14 × L8, with different types of resistance were evaluated at 7 and 14 DAI. The cultivars were found to be sufficient for differentiating *P. nicotianae* races 0 and 1, as the breeding lines were ineffective as differentials. The laboratory procedure provided reliable results, when the reactions were recorded at 14 DAI. The laboratory technique was validated with 21 isolates of *P. nicotianae* obtained from four counties in North Carolina and the results could be provided earlier than by the conventional greenhouse method. In addition, the use of a large number of seedlings for testing in the laboratory technique enhanced the robustness of the results, especially for isolates for which race identification was doubtful by greenhouse method. The results indicated the potential of the laboratory technique for application for race identification in other pathosystems also (Gutiérrez and Mila 2007).

Isolates of *V. dahliae* may be grouped as race 1 and race 2, based on the responses of differential cultivars of tomato and lettuce or as defoliating or nondefoliating, based on symptom expression in cotton. The frequency and distribution of races and defoliation phenotypes of cotton-associated *V. dahliae* were investigated. Three cotton cultivars FM2484B2F1 (highly resistant), 98M-2983 (highly susceptible) and CA 4002 (partially resistant) were used as differentials. All defoliating/race 2 isolates, except for Ls17, caused defoliation on 98M-2983 and/ CA4002. Isolate Ls17 caused defoliation on 98M-2983 only. The nondefoliating/race 1 isolates caused Verticillium wilt symptoms devoid of defoliation on 98M-2983. The results indicated that the conventional differentiation of nondefoliating and defoliating population structure corresponded to *V. dahliae* race 1 and 2, respectively (Hu et al. 2015). *V. longisporum* is a diploid hybrid, consisting of three different lineages formed due to hybridization between two different sets of parental species. Pathogenicity tests were performed on 20 isolates, representing three *V. longisporum* lineages and the relative *V. dahliae*. These isolates were inoculated on 11 different hosts, including artichoke, cabbage, cauliflower, cotton, eggplant, horseradish, lettuce, linseed, OSR (canola), tomato and watermelon. In general, *V. longisporum* was more virulent on the Brassicaceae crops than *V. dahliae* which was more virulent across non-Brassicaceae crops than *V. longisporum*. Differences in virulence among three *V. longisporum* lineages were studied. *V. longisporum* lineage A1/D1 was the most virulent lineage on OSR and *V. longisporum* lineage A1/D2 was the most virulent on cabbage and horseradish. Further, *V. longisporum* was found to be equally as virulent as or more virulent than *V. dahliae* on eggplant, tomato, lettuce and watermelon, suggesting that *V. longisporum* might have a wider host range than *V. dahliae* (Novakazi et al. 2015).

*Tilletia caries* (= *T. tritici*) and *T. foetida*, causing common bunt and *T. controversa*, inducing dwarf bunt diseases of wheat have a classic gene-for-gene relationship with their host, wheat. Teliospores of *T. controversa* are primarily soilborne, germinating at soil surface to infect the emerging wheat seedlings, whereas seedborne teliospores are essentially inconsequential to the development of dwarf bunt disease (Goates and Peterson 1999). On the other hand, common bunt pathogen is primarily seedborne, but soilborne inoculum may also be important as the teliospores remain viable for two to three years and initiate infection in subsequent wheat crops. Pathogenic races of *T. caries* and *T. foetida* were identified based on their reaction on ten differential wheat lines, each containing single bunt resistance gene *Bt 1* through *Bt 10*, revealing the operation of the classic gene-for-gene system in the wheat-bunt pathosystems. The pathogens are closely related and resistance to both diseases is controlled by the same genes of wheat. Six additional wheat lines containing the genes, *Bt 11* through *Bt 15* and a wheat line with a resistance factor designated *Btp* were added to the differentials used earlier. These differentials were inoculated with all named US races and new isolates from field collections. Six new races of *T. caries*, five new races of *T. foetida* and two new races of *T. controversa* were identified. The new races of common bunt and dwarf bunt pathogens showed unique patterns of virulence that allowed specific targeting and elucidation of bunt resistance genes in wheat facilitating the development of wheat cultivars resistant to bunt disease (Goates 2012).

### 2.2.2 Variability in Vegetative Compatibility

Vegetative or heterokaryon compatibility analysis has been shown to be a useful approach to differentiate strains of some fungal pathogens, such as *V. dahliae* and *F. oxysporum*. Fungal strains that anastomose and form heterokaryons with one another are considered to be vegetatively compatible and are assigned to a single VCG. In contrast, vegetatively incompatible strains fail to form heterokaryons. Auxotrophic, nitrate-nonutilizing (*nit*) mutants were first employed to establish a VC system in *F. oxysporum* by Puhalla (1985). *Nit* mutants can be readily recovered by selecting for chlorate resistance. These mutants are usually recovered at high frequencies, easily generated without a mutagen and they grow on unsupplemented media. Hence, this technique has been used for identifying VCGs in fungal pathogens affecting various crops. The VCGs were considered as good predictors of genetic relatedness and the isolates of one VCG were shown to have similar levels of pathogenic potential (virulence) for infecting a host plant species/variety. Nitrate nonutilizing (*nit*) mutants were recovered from seven strains of *F. oxysporum* cultured on two media, PDA or minimal agar amended with 1.5% potassium chlorate. The mutants were grouped into three phenotypic classes by their growth on supplemented minimal agar medium. These classes presumably reflected mutation at a nitrate reductase structural locus (nit), a nitrate-assimilation pathway-specific regulatory locus (nit 3) and loci (at least five) that affect the assembly of a molybdenum-containing cofactor necessary for nitrate reductase activity (NitM). The majority (59 to 96%) of *nit* mutants recovered, belonged to the class *nit 1* mutants. *nit 3* and *NitM* mutants were recovered at a much higher frequency from minimal agar medium amended with chlorate than from PDA amended with chlorate. The *nit* mutants useful for recognizing VCGs, coupled with virulence tests, may provide valuable information on the genetic diversity of natural populations of soilborne organisms like *F. oxysporum* (Correll et al. 1987).

Strains of *V. dahliae* were tested for VC, using complementary, auxotrophic nitrate nonutilizing (*nit*) mutants. These mutants were generated from wild-type strains of *V. dahliae* by selecting chlorate-resistant sectors on corn meal agar with dextrose, amended with potassium chlorate (15–25 g/l). Complementation tests between *nit* mutants derived from strains of *V. dahliae* from Ohio and tester strains revealed the existence of four VCGs, designated VCG1, VCG2, VCG3 and VCG4. All strains within a VCG were strongly compatible with at least one of the selected tester strains, but were not always completely incompatible with strains of other VCGs. The results showed that VC analysis of pathogen strains might be useful for differentiating strain groups of *V. dahliae* which may have applications in ecological and epidemiological investigations (Joaquim and Rowe 1990). In a later study, strains (187) of *V. dahliae* isolated from potato plants and soil samples from 20 fields in Ohio were assigned to VCG1 (2), VCG2 (53) and VCG4 (128), based on pairings of complementary nitrate, nonutilizing (*nit*) mutants induced on a chlorate-amended medium. VCG4 was subdivided into VCG4A and VCG4B to accommodate differential vegetative compatibility reactions of strains from each subgroup, when paired against tester strains of VCG3. Strains of VCG4A were more virulent than strains of VCG2 and VCG4B (Joaquim and Row 1991). Vegetative or mycelial compatibility groups (MCGs) have been recognized in other soilborne fungal pathogens, such as *A. flavus* (Bayman and Cotty 1991), *C. parasitica* (Anagnostakis 1977), *F. sambucinum* (O'Donell 1992), *R. solani* (Anderson 1982) and *S. sclerotiorum* (Kohn et al. 1991).

Isolates (365) of *S. rolfsii*, incitant of southern blight disease of peanuts were classified into 25 vegetative (mycelial) compatibility groups, based on the formation of an antagonism zone between incompatible mycelia of paired isolates. The same VCG was detected in several geographically isolated counties in Texas, United States. Individual fields contained one to five VCGs, whereas individual plants were infected by one VCG. DNA amplification patterns resulting from the use of 18-base oligonucleotide primer NK2 and from restriction digest of ITS region of rDNA were used to determine the genetic similarity of isolates within and between 12 VCGs. All isolates within a VCG gave identical patterns for each marker and some VCGs shared the same ITS and NK2 patterns. VCGs sharing the same DNA amplification patterns were often present in the same field (Nalim et al. 1995). In a later investigation, isolates (84) of *S. rolfsii*, collected from Florida, Georgia, Louisiana, Southern Carolina, Texas and Virginia were assigned to 23 MCGs. Isolates within one MCG

typically originated from different hosts and different geographical areas, with the exception of MCG11. Nineteen isolates of MCGs (1–10 and 13–18) were inoculated on tomato, pepper and peanut. All isolates were pathogenic to all hosts tested, but there was significant difference in virulence on these hosts. Peanut isolates were most virulent on three host plants compared with isolates from tomato and pepper. The results indicated the existence of variations in virulence of *S. rolfsii* isolates which showed host specificity to some extent (Xie et al. 2014).

*V. dahliae*, causative agent of cotton Verticillium wilt disease, occurs in the form of several VCGs, differing in their virulence and host specificity. Isolates of *V. dahliae* obtained from several hosts growing in Africa, Asia, Europe and the United States were tested by inoculating cultivars of *Gossypium hirsutum*, *G. barbadense* and *G. arboretum*. The strains of VCG1 were both cotton defoliating pathotype and race 3 on cotton and tomato (Daayf et al. 1995). Isolates (40) of *V. dahliae* obtained from potato tubers and plants from various regions in North America were assigned to VCG 4A or VCG 4B. These isolates were characterized using molecular markers. The VCG 4A isolates were of highly virulent pathotypes and interacted synergistically with root-lesion nematode *Pratylenchus penetrans* to induce PED symptoms. The isolates could be differentiated using a RFLP technique (Dobinson et al. 2000). In a later investigation, *V. dahliae* isolates from cotton crops grown in Israel, were tested for vegetative compatibility, using nitrate-nonutilizing (*nit*) mutants and the VCG1, VCG2B and VCG4B were identified. The presence of VCG1 was recorded for the first time. VCG1 isolates belonged to D pathotypes, capable of inducing defoliating symptoms, whereas VCG2B and VCG4B were considered to be defoliating-like (DL) and nondefoliating (ND) pathotypes, based on the types of symptoms induced by them. The D isolates were more virulent than DL isolates on all plants, including okra, cotton, watermelon and tomato tested. The virulence pattern of ND isolates showed variations when compared with that of D and DL isolates. The D isolates were highly virulent on eggplant, while cotton showed less severe symptoms. Tomato was resistant to all cotton isolates of *V. dahliae* tested (Korolev et al. 2008).

Isolates (77) of *V. dahliae* obtained from 87 potato fields were analyzed for the occurrence of VCG. The majority of potato fields (93%) contained VCG4A, whereas VCG4B group isolates were present in 23% of the fields tested. The results suggested that preplant assessment of the nature of *V. dahliae* populations might be useful for applying effective disease management measures (Omer et al. 2008). A diverse collection of 27 isolates of *V. dahliae*, including representatives of all VCGs, both MTs and heterokaryon self-incompatible isolates was investigated to study the genetics underlying complementation tests between *nit* mutants of fungal isolates, resulting in heterokaryon formation. A protocol for rapid generation of *nit* mutants of *V. dahliae* isolates was developed, using UV-irradiation. Further, a reproducible high throughput method was developed for complementation tests between *nit* mutants in liquid cultures, using 96-well microplates. Genetic analyses of selected heterokaryons demonstrated that the frequently observed 'weak' cross-reactions between VCGs and their subgroups might be actually heterokaryotic, implying the absence of strict genetic barriers between VCGs. The high throughput method for VCG assignment of *V. dahliae* populations developed in this investigation might be useful for genetic analysis of heterokaryons formed in VCGs of *V. dahliae* (Papioannou and Typas 2015). In another investigation, *V. dahliae* isolates (106) collected from 85 fields in Idaho, United States were found to be of the same *MAT1-2* mating type. The VCGs were assigned for 93 of the 106 isolates. Of the isolates, 95%, 3%, 1% and 1% belonged to VCG4A, VCG2B, VCG4B and noncompatible, respectively. All the VCG4A isolates had the same mitochondrial haplotype, based on sequencing of *cox3* to *nad6* and *cox1* to *rnl* loci, while the VCG2B isolates had two haplotypes. Pathogenicity tests on sugar beet cv. Monohikari showed that the VCG4A isolates could produce more severe foliar symptoms than VCG1, 1A, 2A, 2B and 4B isolates, but the root and foliage weights were not affected consistently by any of the VCG isolates. VCG4A isolates were pathogenic to potato also and hence, rotation of sugar beet with potato might be avoided to reduce the disease incidence (Strausbaugh et al. 2016).

*F. oxysporum* f.sp. *radicis-cucumerinum*, causes severe root rot and stem rot disease in melon. The pathogen isolates (43) obtained from greenhouse-grown melon plants in Greece, were characterized by pathogenicity and VC tests. In addition, the majority of the isolates was fingerprinted via amplified fragment length polymorphism (AFLP) analysis. Based on pathogenicity, 22 isolates were identified as *F. oxysporum* f.sp. *melonis* and 20 isolates as *F. oxysporum* f.sp. *radicis-cucumerinum*. One isolate was nonpathogenic on cucumber, melon and sponge gourd and pumpkin. All isolates (22) of *F. oxysporum* f.sp. *melonis* were assigned to VCG0134 and all isolates (20) of *F. oxysporum* f.sp. *radicis-cucumerinum* to VCG0260. Isolates of *F. oxysporum* f.sp. *radicis-cucumerinum* were incompatible with isolates of *F. oxysporum* f.sp. *melonis* (Vakalounakis et al. 2005). The relative number of VCGs may be higher in sexual population than in an asexual population. In asexual populations, VCG and pathogenicity may be correlated, enabling VCG to be used as a surrogate for pathogenicity, if the population build up is primarily through asexual reproduction. Results of VC analysis of populations of *F. oxysporum* were used for identifying sub-species-specific groups that could be occasionally correlated with *formae specalis* and/ or pathogenicity (Wang et al. 2006). *F. oxysporum* f.sp. *melongenae (Fom)* isolates (374) collected from different regions of Turkey were analyzed for their genetic diversity, based on pathogenicity and VCG tests. The isolates were grouped as highly virulent, virulent, moderately virulent and low in virulence and they did not exhibit any variations in virulence due to geographic origin. Nitrate nonutilizing mutants (*nit*) were generated as *nit 1*, *nit 3* and *Nit M*, based on phenotyping of cultural characteristics on diagnostic media. The majority of *nit* mutants (39.4%) belonged to *nit 1* and most of the isolates were identified as heterokaryon self-compatible (HSC), based on their ability to form a stable heterokaryon.

Four isolates were identified as heterokaryon self-incompatible (HIS). About 96.3% of *F. oxysporum* f.sp. *melongenae* isolates were assigned to VCG0320, while the remaining isolates (3.7%) were classified as vegetative incompatible group (Altinok et al. 2013).

VCG analysis of isolates of *R. solani* from lettuce, broccoli, spinach, melon and tomato was carried out. All lettuce isolates anastomosed with both AG1-IA and IB subgroups, whereas all isolates from broccoli, spinach, melon and tomato anastomosed with AG4 subgroup HG-1, as well as with subgroups HG-II and HG-III. DNA sequence analysis of ribosomal ITS showed that isolates from lettuce were AG1-IB, isolates from tomato and melon were AG4 HG-I and isolates from broccoli and spinach were AG4 HG-III. Differences in the symptom types induced by the isolates in respective host plants were observed. Tomato isolates caused stem rot and isolates from spinach, broccoli and melon induced hypocotyls and root rot symptoms on their respective hosts. Bottom rot of lettuce was due to isolates from this plant species. *R. solani* AG4 HG-I in tomato and melon, AG4 HG-III in broccoli and spinach and AG1-IB in lettuce were recorded for the first time in Brazil (Kuramae et al. 2003). The genetic diversity of 90 isolates of *S. sclerotiorum* collected from a single sunflower field in Xingjiang province of China was studied, based on MCG, oxalic acid (OA) secretion ability, polygalacturonase (PG) activity and pathogenicity of the isolates within the same and between different MCGs. The distribution of inversion minus (Inv −) and inversion plus (Inv +) isolates of *S. sclerotiorum* at MAT locus within the same MCGs and also in the whole population were identified using specific PCR primers. The sensitivity of *S. sclerotiorum* isolates to both carbendazim and dimethachlon fungicides were also investigated. The results showed that 90 isolates from a single sunflower field could be grouped into 15 MCGs. The MCG1, the biggest group contained 34 isolates (37.8%); six MCGs contained only a single isolate; eight MCGs included two to 13 isolates each. High levels of variability in OA secretion, PG enzyme activities and pathogenicity could be observed among isolates from the same and between different MCGs and no obvious relationship could be inferred between OA, PG and pathogenicity. The fungicide sensitivity tests within MCG1 and MCG2 suggested that 84% of isolates were more sensitive to dimethachlon than to carbendazim. Three isolates, one of MCG1 and two of MCG2 exhibited same sensitivity to both fungicides. The results indicated the existence of high levels of genetic diversity in strains of *S. sclerotiorum* occurring in the same sunflower field, based on different characteristics considered as a basis for differentiation (Li et al. 2016).

### 2.2.3 Variability in Genomic Characteristics

All activities of fungal pathogens are governed by different gene(s). The expression of genes varies depending on the availability of susceptible host plant species and other substrates that support the saprophytic phase in their life cycles. Nucleotide sequences of genes governing pathogenicity and production of enzymes or toxins involved in pathogenesis (process of disease development) may vary in different strains/races, resulting in variations in their levels of virulence (pathogenic potential) on different host plant species/crop cultivars which differ in their susceptibility/resistance levels. Various kinds of nucleic acid-based techniques have been applied to determine the variability in the genomic characteristics of the strains/races of a morphologic species.

### 2.2.3.1 Nucleic Acid Hybridization Techniques

Isolates of *Ggt* showed differences in their virulence on wheat cultivars. Slot blot hybridization technique was applied to differentiate the pathogenic isolates of *Ggt*, using a specific DNA probe pG158. This probe hybridized strongly to isolates of *Ggt* pathogenic to wheat, but moderately to *G. graminis* var. *avenae (Gga)* isolates infecting barley, but not to any of the nonpathogenic isolates of *Ggt* and other soil fungi tested. The probe pG158 efficiently detected isolates of *Ggt* both in the soil and root tissues of wheat. As the isolates of *Ggt* were morphologically indistinguishable, the slot blot hybridization procedure could be useful in determining the population of pathogenic isolates of *Ggt* which might form the basis for predicting the incidence of wheat take-all disease (Harvey and Ophel-Keller 1996). *R. solani*, capable of infecting a wide range of crop plant species and weed hosts, is known to occur in different VCGs. A plasmid DNA fragment (PE-42) was employed as a probe to identify and differentiate isolates of *R. solani* AG2-2-IV, causative agent of large patch disease of *Zoysia* grass. Hybridization of the probe occurred only with *R. solani* isolates infecting *Zoysia* grass, but not with other pathogens infecting the grass host. With the probe PE-42 employed as a marker in Southern hybridization procedure, *R. solani* AG2-2-IV could be differentiated from other intraspecific groups of *R. solani* effectively (Takamatsu et al. 1998).

### 2.2.3.2 Polymerase Chain Reaction-Based Methods

PCR has been the most commonly applied technique either alone or in combination with other nucleic acid-based procedures for the detection of soilborne fungal pathogens and also for differentiation of their variants, such as strains, races or biotypes with differing virulence on their host plant species. *P. ramorum* causes the devastating sudden oak death (SOD) disease, killing large number of native oak trees and also infects several ornamental plants such as *Rhododendron* and *Viburnum* in Europe and the United States of America. Almost all European isolates belong to mating type A1, whereas those from California and Oregon belong to A2 mating type. Presence of both MTs in the same geographical region is likely to result in a population capable of sexual recombination and new source of genetic diversity might be generated. In order to prevent such an eventuality, rapid, reliable and discriminating diagnostic assay had to be developed to easily differentiate the populations of these MTs of *P. ramorum*. Based on a DNA sequence difference in the mitochondrial cytochrome *c* oxidase subunit (*Cox1*) gene, a SNP procedure was applied to distinguish the isolates of *P. ramorum* originating in Europe and those originating from the United States. Isolates of *P. ramorum* from Europe (83)

and from the United States (51) were screened and all the isolates could be consistently and accurately assigned to either the European or the United States populations, using this SNP procedure (Kroon et al. 2004). During a survey of nurseries, greenhouses and landscapes, 121 isolates of *Phytophthora* spp. from 1,657 samples of host plant species belonging to 32 genera, were obtained to assess the genetic diversity of *Phytophthora* spp. Based on the sequence of the ITS region of rDNA, 11 *Phytophthora* species and two hybrid species were identified. *P. citricola* constituted 63.9% and *P. citrophthora* 27.4% of the isolates. Six isolates were found to be hybrids (four of *P. cactorum* × *hedriandra* and two of *P. nicotianae* × *cactorum*) by cloning and sequencing the ITS regions. Three *P. cactorum* × *hedriandra* isolates were obtained from the same site from three *Rhododendron* spp., the known hosts of parental *Phytophthora* spp. The fourth isolate was recovered from *Dicentra* sp. which was not a host to either parental species, suggesting an expansion of host range of the hybrid isolate as compared with either parental species (Leonberger et al. 2013).

The genus *Phytophthora* encloses several highly destructive plant pathogens such as *P. infestans* (potato late blight), *P. ramorum* (sudden oak death) and *P. cinnamomi* (capable of infecting more than 3,000 plant species). Molecular identification of *Phytophthora* spp. has been achieved by PCR-based assays employing primers to amplify sequences of the ITS region, followed by either restriction analysis or direct sequencing and a BLAST search against GenBank or other databases. A *Phytophthora* database was established to provide a tool for identification and monitoring of *Phytophthora* spp. (accessible at http://www.phytophthoradb.org/). During validation of *Phytophthora*-ID, *Phytophthora* isolates collected from four commercial nurseries could be identified with certainty to species level, where species were known. Nurseries were dominated by a few species, including *P. plurivora*, *P. cinnamomi*, *P. syringae*, *P. citrophthora* and *P. cryptogea* (see Figure 2.9) (Grünwald et al. 2011). Several molecular investigations have been carried out to analyze intraspecific variability within *P. nicotianae* which is distributed worldwide and has a wide range of herbaceous and perennial plant species. Genetic variation within the heterothallic *P. nicotianae* was studied, using 96 isolates from different host plant species and geographic locations by characterizing four mitochondrial and three nuclear loci. The combined analysis of nuclear and mitochondrial genomes provides more comprehensive insight into the evolutionary forces acting on natural populations. The mitochondrial and nuclear loci containing SNPs were analyzed to investigate the structure of *P. nicotianae* populations in relation to geographic origin and hosts. A total of 52 SNPs (an average of one in every 58-bp) and 313 sites with gaps representing 5,450 bases, enabled the identification of 50 different multilocus mitochondrial haplotypes. Likewise, 24 SNPs (an average of one in every 69-bp) with heterozygosity observed at each locus, were found in three nuclear regions (*hrp*, *scp* and *β-tub*), differentiating 40 multilocus nuclear genotypes. Both mitochondrial and nuclear markers revealed a high level of dispersal of isolates

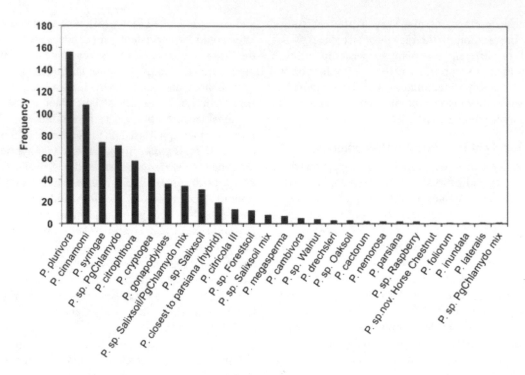

**FIGURE 2.9** Identification of *Phytophthora* spp. and determination of their frequency of occurrence in commercial nurseries in Oregon between 2006 and 2009 based on *Phytophthora*-ID procedure.

**[Courtesy of Grünwald et al. (2011)) and with kind permission of the American Phytopathological Society, MN, United States]**

and inconsistent geographic structuring of populations. The worldwide distribution of *P. nicotianae* and its wide host range might play an important role in facilitating the high level of recombination. However, specific association could be seen for host origin and genetic grouping with both nuclear and mitochondrial sequences. The majority of citrus isolates from Italy, California, Florida, Syria, Albania and the Philippines clustered in the same mitochondrial group and shared at least one nuclear allele. The results suggested an important role of nursery populations in increasing genetic recombination within species and the existence of extensive phenomena of migration of isolates that could have been spread via infected plant materials to various countries in different ecosystems (Mammella et al. 2013).

F. solani, a soilborne fungal pathogen infects many host plant species which differ in their response to different isolates of the pathogen. Based on host specialization, different *formae speciales* have been recognized. *F. solani* f.sp. *glycines (Fsg)* (renamed as *F. virguliforme*) causes sudden death syndrome (SDS) in soybean. In order to differentiate *Fsg* from other *F. solani* isolates, a partial sequence of the mitochondrial small subunit (mtSSU) rRNA gene was amplified by PCR. The amplification products from 14 *Fsg* isolates and 24 *F. solani* isolates from various hosts were sequenced. All *Fsg* isolates contained identical sequences. A single unique insertion of cytosine was observed in all *F. solani* isolates, but not in the isolates of *Fsg*. The sequence data suggested that *F. solani* f.sp. *glycines* had a shorter evolutionary history than the *F. solani* isolates that have either single or multiple nucleotide insertions. The differences in nucleotide insertions in

part of the mtSSU rRNA gene between *F. solani* f.sp. *glycines* and other *F. solani* might be used as a basis for differentiating isolates of *F. solani* (Li et al. 2000). Infection of chickpea by *Fusarium redolens*, causing wilt-like symptoms was observed in Lebanon, Morocco, Pakistan and Spain. Analysis of the sequences of the translation elongation factor (TEF-1α) gene grouped all *F. redolens* isolates from chickpea in the same main clade. Pathogenicity tests on three chickpea cultivars and isolates from different geographic origins indicated that *F. redolens* was mildly virulent. Further, *F. redolens* induced syndromes similar to those caused by the yellowing pathotypes of *F. oxysporum* f.sp. *ciceris*, including leaf yellowing and necrosis that develop upward from the stem base and premature senescence of plants. Application of PCR-based assays for amplification of pathogen-specific sequences facilitated the identification and differentiation of *F. redolens* and *F. oxysporum* f.sp. *ciceris* (Jiménez-Fernández et al. 2011b). In a later investigation, the distribution and frequency of occurrence of *Fusarium* species associated with soybean roots were studied. *Fusarium* colonies were isolated from symptomatic and asymptomatic roots of soybean in different fields and identified based on morphological characteristics. Amplification and sequencing of the translation elongation factor (EF1-α) gene were performed for confirmation of species identity. *F. oxysporum* was the most predominant (> 30%) among the isolates. *F. acuminatum, F. graminearum* and *F. solani* were also frequently recorded. No consistent trend was seen in geographic distribution of *Fusarium* spp. Variability in species frequency was noted between soybean growth stages. *F. oxysporum* was recovered at higher frequency during vegetative stages (40%) than reproductive stages (22%). In contrast, *F. acuminatum, F. graminearum* and *F. solani* were recovered more frequently during reproductive stages of soybean plants under field conditions (Arias et al. 2013).

Banana production is seriously hampered by the Fusarium wilt disease caused by *F. oxysporum* f.sp. *cubense (Foc)*. In order to distinguish strains, races, genomic variations in the translation elongation factor 1 α (TEF-1α) and the IGS region of the nuclear ribosomal operons of *Foc* were analyzed using strains comprising 20 VCGs. Based on two SNPs present in the IGS region, a PCR-based assay was developed to detect isolates of VCG 01213, known as tropical race 4 (TR4) which was highly aggressive on bananas grown in tropical countries. The assay procedure was validated by testing TR4 isolates, *Foc* isolates from 19 other VCGs, other fungal plant pathogens and DNA samples from infected tissues of Cavendish banana cv. Grand Nain (AAA). The PCR protocol was rapid and reliable for detection and differentiation of TR4 isolates and for monitoring the pathogen movement to apply eradication and quarantine programs successfully to reduce incidence and spread of banana wilt disease (Dita et al. 2010).

Isolates of *F. oxysporum* from Gerbera (*Gerbera jamesonii*), chrysanthemum (*Chrysanthemum morifolium*), Paris daisy (*Argyranthemum frutescens*) and African daisy (*Osteospermum* sp.) belonging to Asteraceae, were tested for their pathogenicity on different cultivars of these hosts. The responses of the differentials to these isolates were compared to those induced by *F. oxysporum* f.sp. *chrysanthemi* and f.sp. *tracheiphilum* obtained from American Type Culture Collection. Isolates of *F. oxysporum* f.sp.*chrysanthemi* could be differentiated into three physiologic races based on their reactions on differentials. Sequencing of the IGS region of rDNA and phylogenetic analysis showed that the races of *Fusarium* could be classified into three phylogenetic groups, which were similar to the groups recognized, based on pathogenicity tests. Comparison of IGS sequences revealed a high degree of similarity among strains from Italy and Spain from different host species, suggesting the possibility of introduction of pathogen strains through infected nursery material from a common origin being responsible for outbreaks of wilt diseases in the ornamental plant species (Troisi et al. 2013). Genetic diversity of isolates of *F. oxysporum* f.sp. *dianthi (Fod)* infecting carnation worldwide was investigated. Sequence analyses of partial β-tubulin, translation elongation factor 1 α genes and the full length rDNA IGS region were performed to determine the phylogenetic relationships among isolates of *Fod* from Spanish collection along with some isolates from Italy. The IGS sequence was found to be most effective for resolving phylogenetic relationship among *Fod* isolates and it showed clustering of isolates according to the molecular group (virulence grouping) and vegetative compatibility (VCG). A region in IGS of approximately 300-bp that accumulated enough informative changes to resolve intraspecific relationships and to determine pathogenic variants in *Fod*, was identified. This condensed alignment of the short IGS region showing only the informative positions revealed the existence of virulence group-discriminating positions.

The results indicated that the practical tool developed in this investigation could be applied for pathogen population analyses (Cañizares et al. 2015).

*Fusarium* spp. infecting various crops causing wilt and root rot diseases, produce cell wall-degrading enzymes (CWDEs) such as polygalacturonase (PG) that have a role in pathogenesis (disease development). Variability in the sequences of PG genes of different isolates of *Fusarium* spp. has been observed and this variability is likely to be related to the pathogenic potential (virulence) of *Fusarium* spp. and *formae speciales*. A PCR-based method targeting PG genes was applied to differentiate pathotypes of *F. oxysporum* , causing wilt disease of tomato under field conditions. The partial nucleotide sequences of endoPG (*pg1*) and exoPG (*pg4*) genes of *F. oxysporum* f.sp. *lycopesici* (FOL) and f.sp. *radicis-lycopersici* (FORL) were compared. These pathotypes pathogenic to tomato in Japan could be differentiated by using specific primers in PCR assay (Hirano and Arie 2006). Horseradish roots exhibiting internal discoloration were collected from commercial fields and *Fusarium* and *Verticillium* were isolated from infected root tissues. Isolates (11) identified as *F. oxysporum* based on morphological properties were characterized by DNA sequencing of nuclear translation elongation factor-1α (EF-1α) and mitochondrial small subunit ribosomal DNA (mtSSU rDNA). Analyses of DNA sequences from these two regions and the combined data set showed that six isolates were clearly separated into a common clade that contained *F. commune*, with the remaining five isolates being grouped into a common clade with *F. oxysporum*. Pathogenicity tests showed that isolates of *F. commune* were able to cause internal discoloration in roots, followed by root rot symptoms in the greenhouse, whereas *F. oxysporum* induced internal root discoloration, but not root rot symptoms. The results showed that the PCR assay could be used for differentiating *Fusarium* spp. and that *F. commune* was the cause of internal root discoloration and root rot disease of horseradish (437Yu and Babadoost 2013).

Fusarium basal rot of common onion and Fusarium wilt of Welsh onion caused by *F. oxysporum* f.sp. *cepae* were observed in several areas in Japan. The variability in pathogenicity and genome characteristics was investigated using 55 isolates of *F. oxysporum* f.sp. *cepae* from onion (27 isolates) and Welsh onion (28 isolates). They were characterized based on their rDNA IGS and translation elongation factor-1α (EF-1α) nucleotide sequences, VCGs and the presence of SIX homologs. Sequence analysis of IGS showed that these isolates could be grouped into eight clades (A to H). Pathogenicity tests using onion seedlings indicated that all isolates with high virulence belonged to the IGS clade H. The results suggested that isolates of *F. oxysporum* f.sp. *cepae* included in the IGS clade H might be genetically and pathogenitically different from those placed in other IGS clades (Sasaki et al. 2015). The slow-growing strains (16) of *Fusarium* strains isolated from the roots of soybean plants showing sudden death syndrome (SDS) from South Africa were studied. Molecular phylogenetic analyses of a portion of translation elongation factor 1-α (*TEF1*) and the nuclear ribosomal IGS region rDNA indicated that the etiological agents were *Fusarium brasiliense* and a novel, undescribed *Fusarium* sp. The causal nature of *F. brasiliense* and *Fusarium* sp. for soybean SDS was proved by applying Koch's postulates on seven commercially available soybean cultivars. The virulence of strains of *F. brasiliense* and *Fusarium* sp. showed variations, as reflected by differences in foliar symptoms, root rot and reduction in root rot and reduction in shoot length. Cell-free culture filtrates of these pathogens and two positive control strains of *F. virguliforme* from the United States, caused typical SDS symptoms on susceptible cultivars in whole-seedling assay, indicating the involvement of phytotoxins in symptom induction (Tewoldemedhin et al. 2017).

Verticillium wilt disease of artichoke (*Cynara cardunculus* var. *scolymus)* caused by the soilborne pathogen *V. dahliae*, is an important constraint adversely impacting artichoke production levels. As the infected stumps form the primary sources of inoculum, the presence of different VCGs of *V. dahliae* had to be detected and identified rapidly and precisely. A multiplex nested PCR assay was developed for in planta detection of *V. dahliae* VCGs. PCR markers were identified for the VCGs of *V. dahliae* infecting artichoke: (i) a 334-bp marker amplifying from VCG1A or VCG2B isolates; (ii) a 688-bp marker amplifying from VCG2A or VCG4B isolates and (iii) a 688-bp and a 964-bp marker amplifying from VCG2B isolates. The specific patterns of amplified markers after two rounds of PCR indicated the identity of VCGs present in artichoke tissues. The PCR-based 'molecular tool box' was first optimized using DNA extracted from artichoke plants artificially inoculated with isolates representing known VCGs. The efficiency of the procedure was evaluated using DNA extracted from naturally infected artichoke plants showing different DS levels including asymptomatic plants. This multiplex nested PCR was more efficient in detecting *V. dahliae* than the conventional isolation methods and provided the advantage of establishing the identity of the VCG present in naturally infected symptomatic and asymptomatic plants. In addition, it was possible to detect and differentiate the VCGs involved in multiple infections of artichoke plants in the Communidad Valencia Region of Spain. The multiplex nested PCR format using specific markers provided information on the genetic diversity of *V. dahliae* population infesting a specific artichoke field soil, which might be useful in determining pathogen population structure without the need for applying time-consuming isolation methods (Collado-Romero et al. 2009).

*V. dahliae*, causing vascular wilt diseases in several crops, shows host specificity and two races of *V. dahliae* infecting tomato and lettuce have been identified. Isolates (101) of *V. dahliae* from a range of hosts present in California and ten exotic isolates were characterized to study the genetic variability and race structure. Analysis of 16 simple sequence repeat (SSR) markers showed that tomato populations from central California were distinct relative to the marigold subpopulations. In contrast, cotton and olive isolates were found to be admixture with tomato isolates. Analyses of both rDNA IGS regions and SSR markers showed high genetic variability among isolates, but were unable to delineate races of *V. dahliae*. Results of PCR assay for amplification of race 1-specific product from isolates of *V. dahliae* from several hosts and pathogenicity assays showed 100% concordance and race 1 could be differentiated from 48 isolates of race 2 from tomato. This study indicated the usefulness of PCR assay to distinguish two races of *V. dahliae* and to screen for resistance of tomato genotypes to both races of *V. dahliae* (Maruthachalam et al. 2010). *V. dahliae* and *V. longisporum* are associated with Chinese cabbage yellows disease, one of the most economically important diseases in Japan. Molecular genotyping procedures using group I intron of 18S rDNA, mitochondrial-SSU rDNA and *Cob* gene were applied to differentiate isolates of *V. dahliae* and *V. longisporum*. The results showed that *V. longisporum* was more aggressive on Chinese cabbage than *V. dahliae* under field conditions, although no variation in their virulence could be recognized in the greenhouse tests. Among 67 *Verticillium* isolates from Chinese cabbage, 57 isolates were identified as *V. longisporum*, revealing its greater prevalence with an isolation frequency of 98% during autumn to winter in Ibaraki. But *V. dahliae* was recovered more frequently during cool summer and highland conditions in Nagano. The results indicated the influence of geographical location and season on the frequency of recovery of *Verticillium* spp. in Chinese cabbage (Ikeda et al. 2012). In a later investigation, PCR-based analysis of 18S rDNA, 5.8S rDNA-ITS region, mtSSU-rDNA, cytochrome *b* gene and mating type gene as well as RAPD analysis were applied. Verticillium wilt of cabbage in Gunma Prefecture in Japan was caused by two groups of *V. longisporum* – A1/D1 and A1/D3 and *V. dahliae*. The A1/D1 and A1/D3 lineage strains were equally distributed in Chinese cabbage fields, whereas *V. dahliae* strains were most frequently isolated. A1/D3 type strains produced larger conidia than A1/D1 type strains and produced MS distinguishable from those of A1/D1 type strains of *V. longisporum* and *V. dahliae*. Both types were equally pathogenic on Chinese cabbage (Banno et al. 2015).

*F. oxysporum* f.sp. *cubense* (Foc), incitant of banana wilt disease affects banana production seriously. The pathogen occurs in the form of different races and hence, it has to be differentiated from other races. The mutant W2987 was derived through T-DNA insertional mutagenesis of a wild-type TR4 isolate. The strain W2987 had slower growth rate, produced fewer conidia and it was less virulent than the wild-type strain. Thermal asymmetric interlaced (TAIL)-PCR analysis showed that T-DNA was inserted in the 3′untranslated regions (UTR) of an open reading frame (ORF) in W2987 mutant. Phylogenetic analysis of the gene sequences from various *F. oxysporum* f.sp. *cubense* TR4 isolates was applied. One pair of primers based on the 3′UTR sequence of the putative virulence gene amplified a specific PCR product from TR4 isolates, but not from other races and *Fusarium* species. The results indicated that the genetic locus possibly associated with virulence was unique to TR4 isolates, facilitating specific and reliable detection of *F. oxysporum* sp. *cubense* TR4 (Li et al. 2013).

*R. solani* causes wirestem disease of *Brassica* crops, reducing marketable yield levels significantly. Real-time PCR protocols were developed to detect and discriminate 11 AGs of *R. solani*, using ribosomal ITS regions of AG-1-IA, AG-1-IC, AG-2-1, AG-2-2, AG-4 HGI+HGII, AG-4 HG III, AG-8 or ß-tubulin (AG-3, AG-4 HGII, AG-5 and AG-9) sequences. All real-time assays were target-specific, except for AG-2-2 which showed a weak reaction with AG-2[tabac]. The real-time PCR format could be employed to detect and quantify *R. solani* AG-2-1 (which was the most frequently recovered pathogenic group from cauliflower, cabbage and Brussels sprout) in soil and compost. The DNA extract method developed in this investigation proved to be robust and eliminated the PCR inhibitory materials. The extraction method allowed detection of $10^{-7}$ g sclerotia/g of soil and improved the sensitivity and effectiveness of detection of *R. solani* in soil and compost samples (Budge et al. 2007). Incidence of root rot disease caused by *R. solani* in dry field pea (*Pisum sativum*) increased progressively with increase in area of cultivation in North Dakota, United States. The isolates of *R. solani* were subjected to PCR assay for amplification of ITS region of rDNA. Based on the sequence analysis of the amplicons, 17 *R. solani* pea isolates were considered to belong to AG-4 homogenous group (HG)-II and two isolates to AG-5. Pathogenicity tests on field pea and two rotation hosts, soybean and dry bean showed that the selected isolates were pathogenic to all hosts tested. But the median disease ratings were higher on green pea, dry bean and soybean cultivars, when inoculated with AG-4 HG-II (Mathew et al. 2012).

Sheath blight disease caused by *R. solani* is a major yield-limiting factor in most rice-growing countries. *R. solani* is a complex species composed of 14 identified anastomosis groups (AG1 to AG13 and AG-BI), based on hyphal fusion, morphology, pathogenicity and physiology. *R. solani* AG1-IA is highly aggressive on rice. All isolates of AG1-IA (170) collected from five regions in China were virulent to five rice cultivars with different levels of resistance at rice seedling stage in the greenhouse. These isolates were classified into three pathotypes [weakly virulent (WV), moderately virulent (MV) and highly virulent (HV)], based on DS induced by the isolates. Based on the sequences of ITS region, 43 haplotypes (H1 to H43) were identified and 39 haplotypes were distinct among isolates. High levels of diversity of haplotypes and nucleotides were observed within populations of *R. solani* AG1-IA. Although genetic and pathogenic variation exists among the isolates of *R. solani* AG1-IA, no relationship between genetic diversity and level of aggressiveness was evident. Five populations were divided into two distinct clusters by the unweighted pair group method with arithmetic mean (UPGMA) and no spatial population differentiation was observed. The majority of genetic diversity (97.8%) was distributed among isolates within population with only 2.2% of the genetic diversity attributed to differences among populations. The results indicated that assessment of population characteristics of *R. solani* AG-IA would be useful for evaluation of resistance of rice germplasm to sheath blight disease (Wang et al. 2015b).

Isolates (129) of *R. solani* causal agent of black scurf disease, obtained from potato tubers from different regions in New Zealand were subjected to PCR amplification using primers designed based on sequences of ITS of nuclear rDNA regions. Three AGs, AG-3PT, AG-2-1 and AG-3 were identified among the *R. solani* isolates. AG-3PT isolates were widely distributed, whereas AG-2-1 and AG-5 were confined to distinct areas. Sequence heterogeneity was identified in the ITS regions of 100 AG-3PT and AG-2-1 isolates. AG-2-1 isolates were consistently more virulent (aggressive) than those of AG-3PT, inducing delayed emergence, severe infection of stolons, formation of aerial tubers and severe tuber malformation, but negligible black scurf on tubers. In contrast, AG-3PT isolates caused black scurf on progeny tubers, but variable effects on stem emergence and stolons. Tuber malformation was not associated with infection by all isolates, except AG-2-1 isolates (Das et al. 2014). Multinucleate isolates (128) of *R. solani* obtained from canola and wheat-growing regions of Canada were identified by AG tests and PCR amplification using primers based on the sequences of ITS regions. *R. solani* isolates were grouped into eight distinct clades through phylogenetic comparisons. Each clade corresponded to a specific AG with the exception of two distinct clades that included isolates classified as AG2-1 by anastomosis testing. Most of the isolates of AG5 clustered together according to ITS sequences, whereas three isolates placed in AG5 grouped with AG2-1, AG4 and a binucleate *Rhizoctonia* sp., in the phylogenetic analysis. The results of AG tests were consistent with the classification of isolates of *R. solani* using ITS sequence, although there were some inconsistencies. The results tend to support the use of ITS region sequences as a basis for rapid identification of *R. solani* isolates to their respective AGs (Borders et al. 2014). Isolates (179) of *R. solani* and *Rhizoctonia*-like spp. isolated from soil and plant samples were grouped into three species, subspecies and AGs, based on sequences of ITS regions of rDNA. *R. solani* comprised of 76% of all isolates and included AG4, AG2-1, AG3, AG8, AG5, AG10 and AG9 arranged in the descending order of frequency of recovery. Repeated pathogenicity tests of the most frequently recovered AGs and subspecies on Serge pea at 15°C, showed that *R. solani* AG2-1 caused most adverse effects on seedling emergence and reduction in pea seedling biomass. The AG4 and AG2-1 most frequently detected in pea fields in the Columbia Basin were also the most virulent among the AGs detected. In the seed crop cv. Prevail, a greater frequency of *R. solani* AG8 was detected than AG2-1 or AG4, indicating that AG8 isolates might be associated with root rot complex in some pea crops in the Columbia Basin (Sharma-Poudyal et al. 2015).

*Cylindrocarpon* species cause black foot disease of grapevine which is responsible for considerable losses in Spain. Isolates (82) of *Cylindrocarpon* were characterized by DNA analyses and pathogenicity tests, in addition to phenotypic characteristics. Partial sequences of the ß-tubulin (BT) gene BT1 were amplified, using primers BT1a and BT1b. A unique and conserved 51-bp insertion in the BT1 sequence which was a specific marker for *C. macrodidymum* and this marker was detected in 56 isolates. Other isolates (2) were identified as *C. liriodendri*. BT phylogeny placed all isolates into two well-supported clades. Phenotypic data showed that all isolates of *C. macrodidymum* could be differentiated by production of fewer conidia, longer and wider macroconidia and lower growth rate at 5°C and 10°C, compared with *C. liriodendri*. Rooted cuttings of grapevine rootstock cv. 110R inoculated with selected isolates of both species developed typical black foot disease symptoms, showing that all isolates were pathogenic to grapevines (Alaniz et al. 2007). A combination of species-specific PCR and DNA sequencing procedure was applied for the identification and differentiation of 57 isolates of *Ilyonectria liriodendri*, causal agent of black foot disease of grapevine occurring in New Zealand. The genetic diversity of these isolates was compared with those of Australia and South Africa, using universally primed (UP)-PCR assay. A total of six informative loci distinguished 52 genotypes, of which five contained up to four clonal isolates. The isolates of Australia and South Africa were placed in a clade which did not include New Zealand isolates. A high level of intra- and inter- vineyard genetic variation was observed, indicating the free movement of isolates between regions. Pathogenicity tests revealed variability in virulence, although no correlations could be established (Pathrose et al. 2014).

*S. sclerotiorum*, capable of infecting a large number of plant species, causes stem and crown rot of chickpea. The isolates of *Sclerotinia* were characterized, using morphological characteristics, variations in group I introns and ITS sequences. The isolates were grouped into two types as, fast-growing and slow-growing at 22°C. All fast-growing isolates induced stronger color change in pH-indicating medium at 22°C than slow-growing isolates. The slow-growing isolates contained at least one group I intron in the nuclear small subunit DNA, whereas all fast-growing isolates lacked group I introns in the same DNA region. The ITS sequences of both slow-growing isolates were identical to sequences of *S. trifoliorum*. The slow-growing isolates exhibited ascospore dimorphism, an important morphological character of *S. trifoliorum*. On the other hand, all fast-growing isolates had ITS sequences identical to that of *S. sclerotiorum* and they did not show ascospore dimorphism. Isolates of slow-growing and fast-growing types were pathogenic to chickpea, inducing symptoms similar to those observed in the field. The involvement of *S. sclerotiorum* and *S. trifoliorum* in stem and crown rot of chickpea disease was demonstrated and the use of other methods in addition to symptomatology for differentiating *Sclerotinia* spp. was required for their reliable identification (Njambere et al. 2008).

*V. dahliae* isolates (84) from tomato were characterized by pathogenicity, VCG and mating type determination. Tomato race 2 supplanted race 1 and it was more aggressive on susceptible tomato cultivar compared with race 1. Based on reactions on hosts, tomato, eggplant, sweet pepper and turnip, 59 isolates were assigned to tomato, 19 to eggplant, one to sweet pepper and five to tomato-sweet pepper pathogenicity groups. All isolates of *V. dahliae* could be assigned to VCG-2A, 2B and 4B. A remarkably high incidence of bridging isolates that were compatible with two or more VCGs was also observed. PCR-based molecular marker Tr1/Tr2 provided reliable race prediction among tomato-pathogenic isolates, except for

members of VCG-4B. On the other hand, molecular markers Tm5/Tm7 and 35-1/35-2 were also found to be reliable markers for tomato-pathogenic isolates. The E10 marker was associated with VCG 2B isolates rather than to pathogenicity groups. A SNP in the ITS region and two novel molecular markers, M1 and M2 could provide rapid and precise determination of major VCGs 2A, 2B and 4B and they showed potential for high throughput pathogen population analysis. The mating type was not related to VCG classification and possibly did not control heterokaryon incompatibility in *V. dahliae* (Papaioannou et al. 2013). A combination of PCR, dot blot hybridization and reverse transcription quantitative PCR format was employed to differentiate pathotypes of *P. brassicae*. The presence or absence of nonhousekeeping *P. brassicae* genes (118) was determined by PCR, analyzing five pathotypes (2, 3, 5, 6 and 8). One gene, *Cr811* was present exclusively in the isolate of pathotype 5. This was confirmed by dot blot hybridization and by PCR using alternative DNA preparations and primers. Reverse transcription quantitative PCR analysis indicated that in plant expression of *Cr811* was upregulated during canola infection, especially at the stage of formation of secondary plasmodia. Primers specific to *Cr811* could distinguish a field isolate of *P. brassicae* belonging to pathotype 5 from two other field isolates representing 3 and 8. The results indicated the usefulness of *Cr811* as a molecular marker for differentiating pathotype 5 from other pathotypes of *P. brassicae* (Zhang et al. 2015).

Characterization of isolates of fungal plant pathogens has been accomplished by combining PCR amplification with other nucleic acid-based methods. Pepper and zucchini plants showing Phytophthora blight symptoms, tomato plants with either late blight or buckeye rot symptoms, strawberry showing crown rot symptoms and declining Clementine trees with root and fruit were investigated for *Phytophthora* infections by employing PCR assays, using primers targeting nuclear rDNA repeat sequences. The pathogens detected in the infected plants were differentiated on the basis of PCR primer specificity as well as through extensive RFLP and sequence analysis of PCR-amplified rDNA. The results showed that pepper and zucchini plants were infected by *P. capsici*; tomato plants with late blight and buckeye symptoms were infected by *P. infestans* and *P. nicotianae*, respectively and strawberry plants were infected by *P. cactorum*. Infection by *P. citrophthora* and *P. nicotianae* were proven to be responsible for Clementine decline. Further, using the molecular techniques, the causal agent of new disease, zucchini Phytophthora blight recorded in southern Italy, was identified as *P. tentaculata* (Camele et al. 2005). In a later study, infections of citrus by either *P. nicotianae* or *P. palmivora* or both were observed in Florida. Polymerase chain reaction-restriction fragment length polymorphism (PCR-RFLP) of ITS regions of rDNA was employed for the detection of *P. nicotianae* and *P. palmivora* in citrus root tissues. The sensitivity and usefulness of the PCR-RFLP procedure were assessed and compared with immunoassay and isolation on a semiselective medium. In a semi-nested PCR format with universal primers ITS4 and ITS6, the detection limit was 1 fg of fungal DNA, indicating that this format was 1,000-fold more sensitive than conventional PCR with primers ITS4 and DC6. The lower level of sensitivity (ten fold) of detection of *P. nicotianae*, compared with *P. palmivora* was observed, revealing the limitation of PCR-RFLP for detection of this pathogen in mixed infection. Recovery of *P. palmivora* in the presence of *P. nicotianae* was considerably lower using plating method. However, PCR amplification of ITS regions of rDNA was found to be a valuable tool for detecting and identifying *Phytophthora* spp. in citrus roots (Bowman et al. 2007).

Incidence of WRR disease in grapevines caused by *Armillaria* spp. was observed in northwestern Spain. Variability in *Armillaria* spp. was investigated using VC and molecular methods. RFLP analysis of the ITS region of nuclear ribosomal RNA gene cluster was used to distinguish *Armillaria* spp. Based on VCG test, *Armillaria mellea* was detected in ten of 12 vineyards and *A. gallica* was identified in two vineyards. RFLP-PCR procedure detected four restriction patterns corresponding to *A. mellea*, *A. gallica* and *A. cepistipes*. By employing RFLP-PCR procedure *A. gallica* and *A. cepistipes* were detected for the first time in vineyards established in cleared forest sites (Aguin-Casal et al. 2004). *P. graminis* f.sp. *temperata* and *P. graminis* f.sp. *tepida* infect barley and wheat and they were distinguished on the basis of the differences in their specific ribosomal DNA sequences. In order to verify the existence of host specialization, *P. graminis* inocula were obtained from soils collected from Belgium and France. Their ribotypes were characterized using molecular tools specific to *P. graminis* f.sp. *temperata* and f.sp. *tepida*, such as RFLP analysis of PCR-amplified rDNA, nested and multiplex PCR formats. The presence of both special forms in both countries and coexistence of both forms in some soils were revealed by these assays. *P. graminis* f.sp. *temperata* was associated more frequently with barley, whereas *P. graminis* f.sp. *tepida* was recovered more often from wheat. Quantification of each form on barley and wheat using real-time quantitative PCR assays confirmed the presence of host-specific forms of *P. graminis* (Vaïanopoulos et al. 2007). Two binucleate *Rhizoctonia* (BNR) isolates obtained from cankered stems of potato plants in China were characterized by cultural characteristics and AG testing. The two isolates could anastomose between them, but failed to do so with reference strains of BNR from AG-A to AG-Q and AG-U. These two strains were subjected to RFLP analysis of ITS region of ribosomal DNA (rDNA). RFLP analysis confirmed that these two isolates were different from the reference strains and they were located in a distinct clade from other BNR. This AG was designated AG-W which induced brown, dry and slightly sunken lesions on potato subterranean stems (Yang et al. 2015).

The genetic diversity and structure of 79 isolates of *Sclerotinia sclertiorum* from Brazilian dry bean populations were studied. The internal transcribed spacer (ITS1-5.8S-ITS2) restriction fragment length polymorphism (PCR-RFLP) and sequencing analysis was applied to determine the genetic relatedness among *S. sclerotiorum* isolates. The PCR-RFLP analysis did not reveal any genetic differences among isolates. However, it was possible to detect variability within the sequence which did not support the hypothesis of clonal populations within each population. High variability within and among populations might indicate the introduction of new genotypes in the regions examined, in addition to the presence of clonal and sexual reproduction in populations of *S. sclerotiorum* in Brazil (Gomes et al. 2011). *S. rolfsii* infects several crops including peanut (groundnut) causing considerable yield losses worldwide. Populations of *S. rolfsii* from different hosts and geographical locations have been classified into 12 MCGs. Molecular techniques were employed to group the isolates of *S. rolfsii* with certainty. *S. rolfsii*, incitant of stem and pod rot of groundnut (peanut) is a major yield-limiting factor in Vietnam. Isolates (103) of *S. rolfsii* obtained from 240 observational field plots were examined for the genetic diversity of field populations of the pathogen. PCR assays using primers based on the ITS regions of rDNA were applied for amplification of the specific sequences. Three distinct groups were identified, with the majority of the isolates (n = 90) forming one ITS group. Isolates of *S. rolfsii* from groundnut, tomato and taro were pathogenic to groundnut and relatively sensitive to tebuconazole. However, they exhibited substantial diversity in various genetic and phenotypic traits, including mycelial compatibility, growth rate and sclerotial characteristics (Le et al. 2012). RFLP patterns of ITS regions of nuclear ribosomal DNA (rDNA) and sequence analysis of two protein-coding genes viz., translation elongation factor 1 α (TEF 1 α ) and RNA polymerase II subunit two (RPB2) were applied for characterizing the isolates of *S. rolfsii*. ITS-RFLP analysis of 238 single sclerotial isolates representing 12 MCGs obtained from sugar beet plants from Chile, Italy, Portugal and Spain showed low degree of variability among MCGs. Only three different restriction profiles (using restriction enzyme *Alu*I, *Hpa*II, *Rsa*I and *Mbo*I) were identified among *S. rolfsii* isolates, with no correlation to MCG or to geographic origin. Based on nucleotide polymorphisms, the RPB2 gene was more variable among MCGs, compared with TEF 1α gene. Ten of 12 MCGs could be differentiated, using the RPB2 region, whereas the TEF 1α region resolved seven MCGs. However, combined analysis of partial sequences of TEF 1α and RPB2 genes allowed discrimination of all 12 MCG isolates. All isolates of one MCG showed identical nucleotide sequence that differed by at least one nucleotide from another MCG. The results indicated the effectiveness of using the combined sequence variation in the protein-coding genes TEF 1α and RPB2, as the basis for differentiating MCGs of *S. rolfsii* reliably (Remesal et al. 2013).

*Fusarium* spp. reported to be associated with soybean seedling diseases, root rot and vascular wilt in soybean constitute the *F. oxysporum* species complex (FOSC). *F. commune*, recognized as a component of FOSC is morphologically indistinguishable from other isolates of FOSC. Isolates of FOSC obtained from soybean widely differ in their pathogenicity ranging from nonpathogenic to a highly aggressive level. Sequence analyses of the translation elongation factor (*tef 1-α*) gene and the mitochondrial subunit (mtSSU), PCR-RFLP analysis of the IGS region and identification of the mating type loci were performed for 170 isolates. VC assays were carried out for 114 isolates. Isolates infecting soybean seedlings and roots differed genotypically and phenotypically and they belonged to five clades within FOSC, including *F. commune*. In general, no association was seen between the genotypic characteristics studied and aggressiveness of an isolate. The phylogenetic analyses of the *tef 1-α* gene and mtSSU showed that isolates within the FOSC from soybean were polyphyletic in origin, with soybean isolates belonging to five clades within FOSC. VC analysis indicated the existence of high genetic diversity among FOSC isolates. The PCR-RFLP fingerprinting of the IGS region provided a rapid and easy diagnostic method to determine the clade grouping of the isolates from soybean. Most IGS haplotypes could be determined, using the restriction enzyme *Xho*I to separate clades 3 and 5 from clades 1, 4 and *F. commune* which had similar restriction pattern for this enzyme. The restriction enzyme *Cla*I could be employed for separating clades 2, 4 and *F. commune*. Isolates in clades 3 and 5 could be separated using the restriction enzyme *Ava*I. Both MTs loci MAT-1 or MAT-2 were present in isolates from all clades. The results suggested that there was potential for sexual reproduction in isolates across Iowa fields resulting in greater genetic diversity in FOSC isolates infecting soybean (Ellis et al. 2014).

The RAPD technique may be applied for differentiation of races and strains within a morphologic species of fungal pathogens. Arbitrary short pieces of DNA with 110 or less number of bases from appropriate source are selected as primers which may find suitable complementary sequences in the DNA of target pathogen isolates to be differentiated, resulting in the formation of a mixture of DNA fragments of different sizes. After gel electrophoresis, unique bands in the patterns specific to species or strains or varieties may be recognized in the gels. The unique bands representing the target species/strains may be sequenced to produce specific primers for more accurate PCR analysis or for preparation of probes for dot hybridization method. The simple, sensitive, reliable and rapid RAPD procedure has been applied for characterization of pathogen isolates and assessing the extent of relatedness between isolates of the fungal pathogens. *F. oxysporum* f.sp. *gladioli (Fog)* race 1and 2 differed in their ability to infect both large- and small-flowered *Gladiolus* cultivars. Race 1 could infect both flowered cultivars, while race 2 could infect only small-flowered cultivars. Arbitrary 10-mer oligonucleotide primers (160) were tested on *Fog* by PCR to select RAPD marker specific for race 1. The RAPD primer G12 amplified two discriminating DNA fragments, AB (609-bp) and EF (1196-bp) in race 1 isolates only. Both fragments were cloned and sequenced. Two pairs of race 1-specific primers for multiplex PCR were designed. With these primers, 112 isolates of *F. oxysporum* tested by PCR showed that in almost all cases, race 1 isolates of VCGs 0340 could be distinguished. Seven putative race 1 isolates did not react in multiplex PCR format. Hybridization tests using labeled AB and EF DNA fragments showed that these isolates belonged to different groups (de Haan et al. 2000). RAPD procedure was employed to distinguish pathogenic and nonpathogenic isolates of

*F. oxysporum* f.sp. *dianthi (Fod)*, incitant of carnation wilt disease. By employing the primer OPA 17 and genetic markers, four amplification groups were identified among the isolates of *Fod*. However, RAPD patterns could not be related to the races identified based on the reactions on differential varieties of carnation (Hernandez et al. 1999). In a later investigation, strains of *F. oxysporum* from common bean fields in Spain that were pathogenic/or nonpathogenic were characterized using RAPD procedure. A RAPD marker (RAPD 4.12) specific to highly virulent strains of the seven races of *F. oxysporum* f.sp. *phaseoli (Fop)* was identified. Sequence analysis of the marker RAPD 4.12 enabled designing of oligonucleotides that amplified a 609-bp SCAR marker SAR-B310 A280. It was possible to detect the pathogen by PCR assay in all root samples (100%) of infected symptomless plants and in stem samples of plants with DS of < 4. The assay procedure was rapid and reliable in detecting all known races of *Fop* in early stages of disease development, when no symptom could be seen (Alves-Santos et al. 2002). RAPD technique was shown to be effective to differentiate pathogenic isolates of *F. oxysporum* f.sp. *phaseoli* also. The oligonucleotide primers (23) amplified 229 polymorphic and seven monomorphic DNA fragments ranging from 234- to 2,590-bp. The pathogenic isolates formed a distinct cluster, whereas nonpathogenic isolates were placed in a separate group (Zanotti et al. 2006).

*F. oxysporum* f.sp. *cepae (Foc)* causes Fusarium basal rot of onion occurring in all countries growing onions. Isolates of *Foc* (19) and *F. oxysporum* (33) nonpathogenic to onion were examined for their VC, virulence, MT and genetic diversity. Six *Foc* vegetative compatibility groups (VCG0421 to VCG0425) were identified. The dominant VCGs in Colorado and South Africa were VCG0421 (47% of isolates), and VCG0425 (74%), respectively. Sequence analyses of IGS of rDNA, elongation factor 1-α and mitochondrial small subunit (mtSSU) confirmed the polyphyletic nature of *Foc* and showed that some *Foc* and nonpathogenic *F. oxysporum* isolates were genetically related. Most of the *Foc* isolates clustered into two distinct, well-supported clades. The largest clade only contained highly virulent isolates, including two main VCGs, 0421 and 0425. On the other hand, the basal clade contained MV isolates. Groupings along with VCG data might provide a dependable basis for selection of *Foc* isolates for screening onion genotypes for their resistance and development of molecular markers to rapidly identify VCGs. MTs genotyping revealed the distribution of both MTs (MAT-1 and MAT-2) idiomorphs across phylogenetic clades (Southwood et al. 2012). *F. oxysporum* f.sp. *ciceris*, causal agent of Fusarium wilt of chickpea, occurs in the form of eight races distinguished based on their reactions on chickpea differentials. The sequences of 32 genomic regions for each of the eight races were analyzed to study the genetic relationships between the races of the pathogen. Twelve of these regions were newly designed microsatellite markers polymorphic for the *F. oxysporum* complex developed from a microsatellite enriched genome library. The translation elongation factor 1-α (TEF), ITS region of the rDNA, five mitochondrial regions, a xylanase gene (*xyl4*) and its trasnscriptional activator (*xlnR*), two SCARs of the pathogen and 11 microsatellites were entirely identical for all races. Only a few polymorphisms were observed between and sometimes within races for β-tubulin gene and IGS of rDNA, endogalacturonase *pg1* and exopolygalacturonase *pgx4* genes and six microsatellite regions. The high degree of similarity among races supported a monophyletic origin of *F. oxysporum* f.sp. *ciceris* and subsequent development of pathogenic races within this linkage (Demers et al. 2014).

The genetic diversity of *F. oxysporum* f.sp. *dianthi (Fod)* soil population was studied for selecting suitable cultivars of carnation resistant to wilt disease. Isolates of *Fod* were obtained from soils and carnation plants in the most important cultivation areas in Spain and some *Fod* strains from Italy were included as a reference. RAPD fragments generated by single-primer PCR were used to compare the relationships between isolates. Separation of *Fod* isolates into three clusters, A, B and C was more related to aggressiveness (virulence) than to the races of the isolates. The results of PCR amplifications using primers specific for race 1 and race 2 and SCAR primers developed in this investigation, correlated with molecular groups determined from RAPD analysis earlier and new molecular markers were identified for accurate identification of isolates. Pathogenicity tests showed that molecular differences between isolates of the same race correlated with differences in virulence. Isolates of races 1 and 2 in cluster A (R11 and R21 isolates) and cluster C (R1-type isolates) were highly virulent. On the other hand, isolates of races 1 and 2 in cluster B (R1II and R2II isolates) showed low levels of aggressiveness. The results indicated a change in the composition of the Spanish *Fod* population over time and this temporal variation was likely to be due to continuous change in commercial carnation cultivars used by growers (Cabanás et al. 2012). Isolates of *F. oxysporum* f.sp. *dianthi (Fod)* obtained from carnation plants and soils were analyzed for vegetative incompatibility by pairing all isolates in all possible combinations. Isolates of race 1 and race 2 were distributed among three VCGs which correlated with molecular groups I and II. Race 1 and race 2 isolates in molecular Group I were included in VCG0021, isolates of race 1 type were in VCG 0022 and isolates of race 1 and race 2 in molecular Group II constituted a new VCG (002-). Isolates in each VCG contained the same mating type gene (MAT-1 or MAT-2) with the exception of the new VCG002- that contained both idiomorphs (Cabanás and Pérez-Artés 2014).

Isolates (27) of *V. dahliae* infecting cotton were characterized by VCG tests and 13 isolates were assigned to VCG1, eight isolates to VCG2B and six isolates to VCG4B. The DNAs were extracted from these isolates for use in RAPD and PCR assays. DNA bands diagnostic of the nondefoliating (ND) pathotypes were produced by VCG2B and VCG4B isolates consistently, whereas VCG1 isolates (except one) yielded RAPD markers specific to defoliating (D) pathotype. As the results were inconsistent, PCR assays employing primers specific to D- and ND- pathotypes were performed. The results showed that all VCG2B and VCG4B isolates belonged to ND pathotypes. In addition, all except three isolates of VCG1

were also found to be ND pathotypes. The three isolates of VCG1 did not amplify products similar to either D- or ND-pathotypes. The isolates of VCG2B and VCG4B from Israel also belonged to ND pathotypes and they were found to be molecularly homogenous, as indicated by RAPD-PCR assay (Korolev et al. 2008). Genetic diversity of isolates of *Ggt*, causal agent of wheat take-all disease was investigated using 98 isolates of *Ggt* from the world collection. Phylogenetic analysis of the DNA sequences of two regions, the coding regions of the RAPD-PCR marker previously selected and used for genotyping G1 and G2 isolates corresponding to the 1,2-dioxygenase-like gene and the rDNA region corresponding to a part of the 18S gene, 5.8S gene, ITS1 and ITS2 DNA sequences was performed. The distribution of *Ggt* genotypes into two main groups was confirmed by the phylogenetic analysis. The use of a universal molecular descriptor of *Ggt* groups was useful for carrying out investigations on the occurrence, distribution and changes in *Ggt* populations in relation to development of epidemics (Daval et al. 2010).

Variability in virulence levels of *Ascochyta pisi* isolates (109) was assessed by excised pea leaf assay technique. Twenty-eight isolates were avirulent, whereas other isolates induced formation of varying sizes of lesions on pea leaves. RAPD technique using PCR with single primers was employed and DNA samples from 86 isolates were amplified. One dominant genotype of *A. pisi* was identified in several locations, but variations in virulence were not clearly differentiated by the RAPD technique (Wang et al. 2000). RAPD analysis of the isolates of *P. parusitica* var. *nicotianae*, causal agent of TBS disease was carried out to differentiate the pathogenic isolates from nonpathogenic isolates. Virulence levels of *P. parasitica* var. *nicotianae* was assessed using zoospore inoculum in the greenhouse. The RAPD results were highly correlated to that of virulence assay. Amplification patterns generated by RAPD reactions, were used to generate phenogram depicting genetically distinct nature of the cluster defined by pathogenic isolates. The nonpathogenic isolates formed a separate group. The results indicated the potential of RAPD procedure for the detection and differentiation of *P. parasitica* var. *nicotianae* infecting tobacco (Zhang et al. 2001). *Ascochyta rabiei* (telemorph- *Didymella rabiei*), causal agent of Ascochyta blight of chickpea, is heterothallic, requiring the presence of two compatible MTs for production of sexual spores (ascospores) that may contribute to genetic diversity in pathogen population. The genetic diversity among isolates of *A. rabiei* from Canada and other countries was assessed by virulence assays and RAPD markers. Eight lines, Kabul lines UC27, Sanford, ILC3856, ILC4421 and Flip 83-48 and desi lines ICC4200, ICC4475 and ICC6328 were selected as differentials for pathogenicity assays and 14 pathotypes were identified among the Canadian isolates. The desi lines were resistant to a wider range of isolates than the Kabuli line. For RAPD analysis, seven isolates of *A. lentis (Didymella lentis)* and two isolates of *A. pinodes (Mycosphaerella pinodes)* were also included. RAPD analysis of 68 isolates using eight primers produced 112 fragments of which 96% were polymorphic. Similarities among *A. rabiei* isolates from Canada ranged from 20 to 100%. A broad genotypic variation among the 39 Canadian isolates of *A. rabiei* was detected. However, most of the isolates clustered together into a large group. Isolates of *A. lentis* formed a group that was moderately similar to one cluster of *A. rabiei* isolates. But *A. pinodes* showed little similarity to *A. rabiei*. The association between RAPD banding patterns and pathotypes was weak. The results indicated that the population of *A. rabiei* in Canada was highly diverse (Chongo et al. 2004).

*Phoma sclerotioides*, the incitant of brown root rot of alfalfa was detected in plants and soil by employing PCR-RAPD analysis. Amplification products obtained from RAPD reactions were cloned and sequenced and two extended primer sets were designed from the resulting data that were used to detect sequence characterized DNA markers. A single 499-bp DNA amplification product was consistently obtained from primer PSB12499 that was specific for 19 isolates of *P. sclerotioides*. No amplification occurred from the DNAs of *Phoma medicaginis*, or *Phoma betae* or from other soilborne pathogens, including *A. euteiches*, *R. solani*, *F. oxysporum*, *P. ultimum* or *P. infestans*. The pathogen-specific amplicon (499-bp) was also produced from alfalfa root tissues infected with *P. sclerotioides* whose presence was visualized by microscopy. This PCR product was generated also from soil samples collected from fields infested with the pathogen. The PCR format provided reliable detection of *P. sclerotioides* in alfalfa root tissues and soil samples within one day, including DNA extraction, whereas the conventional isolation procedure required up to 100 days to obtain results (Larsen et al. 2002).

Genetic structure of the populations of *Sss*, causing potato powdery scab disease, could not be investigated, because of the obligate parasitism of *Sss* and nonavailability of molecular markers for this pathogen. A single cystosorus inoculation technique was employed to produce large amounts of *Sss* plasmodia or zoosporangia in eastern black shade (*Solanum ptycanthum*) roots from which DNA was extracted and cryopreserved for long-term storage of the isolates. *Sss*-specific RFLP markers were developed from RAPD fragments. Cystosori of *Sss* were used for RAPD assays and putative pathogen-specific RAPD fragments were cloned and sequenced. Four polymorphic *Sss*-specific probes containing repetitive elements and one containing single-copy DNA were identified. These RFLP probes were employed for the analysis of 24 single cystosorus isolates derived from eight geographic locations in the United States and Canada. Variations in genotypic characteristics were seen among, but not within geographic locations. Two major groups were recognized based on cluster analysis: group I included isolates from western North America with the exception of those from Colorado and group II contained isolates from eastern North America and from Colorado.

AFLP technique was developed to overcome the limitations of methods available previously which required a long time and were unable to differentiate pathogen populations beyond species level. The AFLP procedure is a versatile and powerful technique that could be applied for fingerprint analysis, mapping and other genetic investigations on fungal pathogens. Amplification of 50-=–100 DNA fragments simultaneously by PCR may be achieved in the AFLP technique. Technically AFLP procedure might be more difficult than RFLP analysis. Yet AFLP analysis has become a very powerful marker system to study the genetic variations among strains/races of fungal pathogens. The AFLP method involves selective amplification of restriction fragments from a digest of total genomic DNA using PCR. The DNA fragments may be identified after electrophoresis and designated AFLP markers. High levels of sensitivity and the reproducibility of results are the distinct advantages of the AFLP procedure for differentiation of isolates/strains of fungal pathogens. Identification of *Phytophthora* spp. up to species level has been difficult, although about 60 described species are included in the genus *Phytophthora*, many of which are under phytosanitary regulations. In order to develop a reliable detection and identification technique, the digests of ITS regions of ribosomal RNA subunit of over 450 *Phytophthora* isolates representing almost all known taxa were examined. Sophisticated band-matching software was used to compare the isolates rapidly with those in the database. Closely related taxa with identical ITS regions could be differentiated at intraspecific level. The database of AFLP fingerprints of *Phytophthora* spp. was established at PRI, Scottish Crop Research Institute, the United Kingdom (79Cooke et al. 2000). Isolates (132) of *P. cactorum* and 15 other *Phytophthora* spp. were characterized by AFLP analysis to determine their variability in their genomic DNA, pathogenicity and sensitivity to the fungicide mefenoxam. *P. cactorum* isolates were collected from almond (30), strawberry (86), walnut (5) and 11 other host plant species for AFLP analysis. All 16 *Phytophthora* species could be differentiated by their AFLP banding patterns. High coefficiency of genetic similarities (> 0.9) was observed among all California isolates of *P. cactorum*. Among the isolates (132) of *P. cactorum*, 30.8% and 24.5% of the AFLP variation was associated with hosts and geographical location of isolates, respectively. Sensitivity of all isolates of *P. cactorum* did not show variation at 1.0 ppm concentration of mefenoxam. Californian populations of *P. cactorum* were apparently sensitive to mefenoxam and exhibited host specificity with relatively minor variation in genomic DNA (Bhat et al. 2006). *Pythium* species cause damping-off and root rot diseases in the nurseries of many vegetable crops. A AFLP fingerprinting technique was applied to characterize pathogenic *Pythium* spp. and intraspecific populations. Species diagnostic AFLP fingerprints of *P. aphanidermatum*, *P. irregulare* and *P. ultimum* and tentative fingerprints for six other species were identified. Intraspecific distance analyses of *P. aphanidermatum*, *P. ultimum* and *P. irregulare* revealed distinct patterns of intraspecific variation among the three species. *P. aphanidermatum* showed the smallest mean distance among the isolates (15%), followed by *P. ultimum* (37%). *P. irregulare* had the largest mean distance among isolates (64%). The results indicated the usefulness of AFLP procedure for assessing the inter- and intra-specific variability in genetic characteristics of the genus *Pythium* (Garzón et al. 2005).

Isolates (153) of *P. nicotianae* collected from a single tobacco field over a four-year period were analyzed by AFLP to assess the effect of host resistance on genetic variability in pathogen population. Race 1 isolates, initially absent, were detected in multiple plots by the end of four-year period. Presence of 102 race '0' isolates and 51 race 1 isolates was detected. Of the 153 isolates, 76 had a unique AFLP profile, whereas the remaining 76 isolates had one of 27 AFLP profiles shared by at least two isolates. Isolates of both races were present in both the unique and shared AFLP profile groups. Race 1 isolates detected over multiple years, were always obtained from the same plot. None of the race 1 profile was detected in more than one plot, indicating that the multiple occurrences of the race throughout the field could be the result of independent events and not pathogen spread. Three identical race 0 AFLP profiles were detected in noncontiguous plots and in each case, the plots were cropped with the same partially resistant cultivar. Estimates of genetic diversity ranged from 0.365 to 0.831 and varied depending on tobacco cultivar planted and pathogen race. Based on the averages of all treatments, diversity in race 1 isolates was lower than in race 0 isolates at the end of each season. Planting tobacco cultivar with single-gene resistance initially decreased the genetic diversity of the population, but diversity increased each year, indicating that *P. nicotianae* was able to adapt to the host genotypes deployed in the field (Sullivan et al. 2010). Genetic variability of isolates (18) of *P. fragariae* var. *fragariae* was assessed using AFLP procedure. The AFLP fingerprint patterns of some of the isolates and their single-spore derivatives were analyzed. The isolates exhibited little difference in race type and the majority of the isolates were genetically identical at 433 AFLP loci. Races inoculated in the 1970s in the site could be recovered during this investigation. The fingerprints of the single variant isolate matched that of an isolate, originally used to inoculate the site in the 1950s. The results revealed clearly that *P. fragariae* var. *fragariae* was highly stable and persisted retaining its genetic integrity and remained dormant in soil for several years, ensuring its survival between epidemiologically favorable conditions which were available erratically (Newton et al. 2010).

The detection and identification of *V. albo-atrum* hop pathotypes PG1 and PG2 occurring in Slovenia was carried out, using PCR assays. Of the 17 pathotype-linked AFLP markers, 11 were successfully cloned and sequenced. Specific primer pairs for PG2 (22) and PG1 (10) were designed from 16 sequences to convert polymorphic AFLP markers into pathotype-specific SCAR markers. Primer specificity was demonstrated by testing on a wide range of *Verticillium* isolates. Ten PG2- and PG1-specific primer pairs showed specificity for amplification of hop isolates of *V. albo-atrum*. But they also amplified sequences in *V. albo-atrum* and *V. dahliae* hop isolates from different hop production areas in Europe as well as some isolates from other hosts. Primer combinations obtained from the AFLP 9-1 marker, were specific only for *V. albo-atrum* PG2 isolates prevalent in Slovenia. The highly specific primers were employed in multiplex- and nested PCR formats for the detection of *V. albo-atrum* PG2 pathotype

in xylem tissues of hop plants. The SCAR markers showed potential for specific detection of PG1 and PG2 pathotypes of *V. albo-atrum* infecting hop crops (Radišek et al. 2004). AFLP procedure was applied to assess genetic variability within the VCGs of *V. dahliae*. Characterization of isolates of *V. dahliae* from artichoke (53), cotton (96), cotton field soil (7) and olive trees (45) by AFLP and PCR procedures showed that clustering of isolates within VCGs occurred and it correlated with respective VCGs, regardless of host plant species and geographical locations. However, VCGs revealed molecular diversity, the variability being highest in VCG2B and VCG2A. Molecular differences were correlated with virulences of isolates of VCG2B pathogenic to artichoke and cotton cultivars (Collado-Romero et al. 2006).

Genetic variability in the isolates (288) of *F. oxysporum* f.sp. *vasinfectum (Fov)*, causal agent of cotton wilt disease, recovered from three consecutive cotton crops (2002–2006) was investigated. Among 288 isolates, 25 distinct AFLP genotypes were distinguished. These genotypes were classified into two main groups corresponding to known VCGs 01111 and 01112. The *Fov* populations were dominated by four genotypes, I-A, I-B, II-A and II-B that accounted for 87.5% of the isolates. Temporal variations in the isolates were significant in the two sampled fields with 6.8% and 10.7% of total genetic variation being attributed to differences in collections in different years. Significant changes in the frequency of the dominant *Fov* genotypes were observed, where genotype I-A declined from 84.8% to 40.0% over a four-year period. In contrast, the genotype I-B increased from 7.6% to 35.4%. Intergenotype competition was strong as revealed by glasshouse bioassays with 93.4% of symptomatic plants sampled from dual inoculation trials being infected by single genotypes. Competition was differentially mediated by cotton cultivars, as the competitive ability of pathogen genotype I-B was enhanced on resistant cultivar Sicot189, relative to the susceptible cultivar Siokra1-4. The results suggested that host-mediated inter-genotype competition might play an important role in temporal variation in the population of *F. oxysporum* f.sp. *vasinfectum* under field conditions (Wang et al. 2010). The AFLP procedure was employed for studying the genetic composition of *F. oxysporum* f.sp. *vasinfectum*, including race 3, 7 and 8, Australian genotype strain and 80 strains collected from China. Based on AFLP analysis, these strains were placed in four groups. Race 3, strain CN8 was included in group B race 7, strain CN7 was grouped with 75 strains from China in group C. The Australian genotype strain ATCC9629 was grouped with five strains from China in group D (Guo et al. 2015).

Isolates of *F. oxysporum* f.sp. *melonis* (FON) (22) and *F. oxysporum* f.sp. *radicis-cucmerinum* (FORC) (20) were characterized based on pathogenicity and VC tests. One of the 42 isolates tested was found to be nonpathogenic. The majority of these isolates were subjected to AFLP analysis. AFLP fingerprinting allowed for clustering of the isolates of FON and FORC along two phenetic groups: FON to AFLP major haplotype I and FORC to AFLP major haplotype II. In general, pathogenicity, vegetative compatibility grouping and AFLP analysis were correlated and effectively differentiated isolates of *F. oxysporum* from melon (Vakalounakis et al. 2005). *F. oxysporum* f.sp. *cubense (Foc)*, incitant of Fusarium wilt of banana occurs in the form of three races (1, 2 and 4) of which race 4 (tropical race, TR 4) is the most destructive reducing the banana production very seriously. The genetic relationship of the 55 isolates of *Foc* race 4, including 18 from Taiwan and 37 from Hainan and Guangdong Provinces, was investigated by using race 4-specific primers in PCR assay and AFLP analysis. One race from Australia, one race from the Philippines and ten non-banana *F. oxysporum* isolates were also included for comparison. The isolates from Taiwan (16) and Southern China (21) were pathogenic to Cavendish banana and identified as race 4 and the remaining isolates were identified as race 1. PCR amplification with tropical race 4-specific primers confirmed their identity as TR 4. AFLP analysis with eight *Eco*RI-*Mse*I primer combinations clustered the 67 *F. oxysporum* isolates analyzed into two major groups, one containing all race 4 isolates and another consisting of race 1 isolates along with non-banana *F. oxysporum* isolates. The results suggested that TR 4 isolates might share the same linkage, irrespective of geographical origin (Li et al. 2011). *F. oxysporum* f.sp. *radicis-lycopersici* (FORL), causing Fusarium crown and root rot is a highly destructive soilborne pathogen of tomato. In order to study the population structure of FORL, 125 isolates were obtained from three counties of Florida and VCG tests and ten microsatellite loci formed the basis for the study. The isolates up to 69.8% could be assigned to one of three VCGs, 0094, 0098 or 0099 with frequencies of 38.6, 24.4 and 6.8%, respectively. A medium level of population differentiation was detected among the three counties and three optimal clusters of populations were differentiated by discriminant analysis of principal components. In addition, each population had some individuals that could have migrated from other populations. Considerable numbers of migrants (>1.33 migrants/generation) were also detected between three counties, resulting in an increase in the effective population size and genetic diversity of FORL. However, mutation and parasexual recombination could not be entirely eliminated as contributing factors for the genetic diversity of *F. oxysporum* f.sp. *radicis-lycopersici* (Huang et al. 2013).

*S. sclerotiorum* has a wide host range that includes several crop plant species and also wild hosts. The population structure of *S. sclerotiorum* in the UK was studied using eight microsatellite markers and sequenced sections of the IGS region of the rRNA gene and the elongation factor 1-α (EF 1-α) gene. A total of 229 microsatellite haplotypes was identified within 384 isolates from 12 *S. sclerotiorum* populations collected from England and Wales. One microsatellite haplotype was generally found at high frequency in each population and it was widely distributed across different hosts, locations and years. Fourteen IGS and five EF haplotypes were detected in the 12 populations, with six IGS haplotypes and one EF haplotypes, exclusive to the wild host buttercup. Three of the IGS haplotypes and one EF haplotype were present also in the isolates from the United States, Canada, New Zealand and Norway, whereas eight IGS haplotypes were detected only in

the UK. Overall, no consistent difference in the populations of *S. sclerotiorum* in crop hosts and buttercup present in the UK could be observed, indicating the multiclonal population structure of *S. sclerotiorum* in the UK (Clarkson et al. 2013). Monitoring clonal populations of *V. dahliae* using mitochondrial genome sequences was examined. Mitochondrial genome sequences of *V. dahliae* were used to design primers for amplifying spacer regions for assessing differences in mitochondrial haplotypes among isolates. Five regions (5,229-bp) were examined for 30 isolates representing different VCGs, host and geographic origin. The differences among isolates were due to SNP, different numbers of bases in specific homopolymorphic regions and copies of subrepeated sequences. The isolates were placed in 28 groups and some groups correlated with VCG. But five isolates of VCG-4 fell into four mitochondrial haplotypes, one of which was identical to the largest VCG-2 grouping. Phylogenetic analysis of the isolates revealed the mitochondrial background of VCG-1 and VCG-2B to be monophyletic, but VCG-2A and VCG-4 could not be separated. The results indicated that there might be variation in mitochondrial haplotypes and this type of analysis might be useful for characterization of isolates (Martin 2010). Isolates (286) of *V. dahliae* were analyzed by mating type, VCG and multilocus microsatellite haplotype assays to determine the genetic structure of populations infecting mint and potato crops. Populations from mint and potato could fit in a clonal reproductive model, with all isolates clustering in a single mating type (MAT1-2) and multiple occurrences of the same haplotypes. Haplotype HO2 represented 88% of mint isolates and was primarily VCG2B, while haplotype HO4 represented 70% of potato isolates and was primarily VCG4A. Haplotypes HO2 and HO4 were highly virulent, inducing severe symptoms on mint and potato, respectively, regardless of host origin. Mint and potato populations were significantly genetically diverged. Migration of isolates between potato plants and seed tubers, infested tare soil and field soils was detected. Genetic differentiation of *V. dahliae* from mint and potato might be due to the presence of a single MT and differences in VCG. Populations of *V. dahliae* in potato and mint were characterized by the presence of aggressive clonally reproducing haplotypes which were widely distributed in commercial mint and potato populations (Dung et al. 2013).

Mitochondrial haplotype analysis was performed for the differentiation of isolates of *P. cinnamomi*. Seven mitochondrial loci spanning a total of 6,961-bp were sequenced for 62 isolates, representing geographically diverse collection of isolates with A1 and A2 MTs, to determine mitochondrial haplotypes of *P. cinnamomi*. Forty-five mitochondrial haplotypes were identified with differences due to SNPs totaling 152-bp) and length mutations. SNPs were the predominant mutations in the four-coding region and their flanking regions, while both SNPs and length mutations were observed in the three primarily intergenic regions. Some of the length mutations were due to the addition or loss of unique sequences, whereas others were due to variable numbers of subrepeats. Network analysis of the haplotypes identified eight primary clades, with the most divergent clade representing primarily

A1 isolates from Papua New Guinea. Among the 62 isolates examined, some isolates recovered from different regions had the same mitochondrial haplotype, suggesting the movement of isolates via plant materials (Martin and Coffey 2012). *P. citrophthora*, incitant of Phytophthora branch canker of Clementines, was responsible for heavy losses in Spain. In order to determine the population structure of *P. citrophthora*, 134 isolates were analyzed genotypically and phenotypically. Genetic variability was studied by analyzing inter-simple sequence repeat (ISSR) markers, sporangial characteristics, sexual behavior and growth rates. Furthe, colony morphology of the isolates at different temperatures was determined. Host specificity and virulence of the selected isolates were assessed by pathogenicity tests on sweet orange and Clementine plants under field conditions. One genotype P-1 contained 88% of the isolates causing the disease in Clementines. This genotype was present even before the first outbreak of the disease in Spain. In addition, 13 other minor genotypes were also distinguished, but most of them contained only one isolate. All *Phytophthora* isolates obtained from infected Clementines were sexually sterile and showed the characteristic petalloid colony pattern, although wide variations in the morphological and physiological characters of the isolates were discernible. Pathogenicity tests showed that Clementines and their hybrids were more susceptible to *P. citrophthora* than sweet orange (Alvarez et al. 2011).

The population structure of *F. solani*, causative agent of soybean sudden death (SDS) syndrome was investigated, using phenotypic and multilocus molecular phylogenetic analyses, as well as pathogenicity assays. *F. solani*, *F. solani* f.sp. *glycines* or *F. solani* f.sp. *phaseoli* were considered to be primarily associated with SDS syndrome. Soybean SDS and mung bean root rot (BRR) were also considered to be due to distinct *Fusarium* sp. To establish the identity of the pathogens involved in these diseases, comparative DNA sequence analyses of the soybean SDS and BRR pathogens showed that highly conserved regions of three loci could be employed in the design of assays. None of them was found to be species-specific, based on the species limits within SDS-BRR clade. Then a high throughput multilocus genotyping (MLGT) assay was employed. The soybean SDS and two closely related *Phaseolus* and mung bean BRR pathogens could be precisely differentiated based on nucleotide polymorphism within the nuclear ribosomal IGS region of rDNA and two anonymous intergenic regions designated locus 51 and 96. The single-well diagnostic assay, using flow cytometry and a fluorescent microsphere array was validated by independent multilocus molecular phylogenetic analyses of a 65-isolate design panel. The MLGT assay was used to reproducibly type a total of 262 and nine isolates, respectively of soybean SDS and BRR pathogens. The MLGT assay has the potential for accurate identification and molecular surveillance of important pathogens (O'Donnell et al. 2010).

Single-strand conformation polymorphism (SSCP) analysis can detect small nucleotide changes in the fungal DNA fragments by electrophoresis facilitating differentiation of fungal pathogenic species within a genus. The SSCP analysis may be applied to detect DNA polymorphism in the ITS region of rDNA of the target pathogen. The genetic diversity of *Fusarium* spp. populations associated with the decline of asparagus (*Asparagus officinalis*) present in field soils and plants in Japan was investigated by employing SSCP procedure. The ITS2 regions of nuclear rDNA genes were amplified by PCR assay using specific primers. Over 651 isolates obtained from roots of asparagus plants showing decline symptoms were analyzed by SSCP technique. *Fusarium* spp. differentiated by SSCP analysis were *F. oxysporum* f.sp. *asparagi*, *F. proliferatum* and *F. solani*. *F. oxysporum* f.sp. *asparagi* was the most predominant in five regions (68%), whereas *F. proliferatum* was recovered from 28% of infected plants and *F. solani* was found only occasionally (2.5%). The frequency of recovery of *F. proliferatum* gradually increased from north to south of Japan, although considerable differences between field in each region could be observed. The results showed the relative importance of *Fusarium* spp. in the development of asparagus decline in Japan (Nahiyan et al. 2011). Over 40 pathotypes of *S. endobioticum* have been reported and the biological methods of differentiating them using infectivity tests was found to be laborious and time-consuming. The five pathotypes 1(D1), 2(G1), 6(O1), 8(F1) and 18(T1) were found to be predominant in Germany and other European countries. Both genomic DNA and cDNA of the German pathotype 18(T1) from infected potato tissue were sequenced and 5,422 expressed sequence tags (ESTs) and 423 genomic contigs were generated. Comparative sequencing of 33 genes and SSCP analysis with PCR fragments of 27 additional genes, as well as the analysis of 41 SSR loci revealed extremely low levels of variation among the five German pathotypes. From the markers, one SCAR marker and five SSR markers revealed polymorphisms among the German pathotypes and an extended set of 11 additional European isolates. Pathotypes 8(F1) and 18(T1) displayed discrete polymorphisms, enabling their differentiation from other pathotypes. The results showed that based on the six markers, 16 isolates could be differentiated into three distinct genotype groups (Busse et al. 2017).

## REFERENCES

Abad ZG, Abad JA, Cofffey MD et al. (2008) *Phytophthora bisheri* sp. nov., a new species identified in isolates from the rosaceous raspberry, rose and strawberry in three continents. *Mycologia* 100: 99–110.

Adams GC Jr., Butler EE (1979) Serological relationships among anastomosis groups of *Rhizoctonia solani*. *Phytopathology* 69: 629–633.

Afouda L, Wolf G, Wydra K (2009) Development of a sensitive serological method for detection of latent infection of *Macrophomina phaseolina* in cowpea. *J Phytopathol* 157: 15–23.

Agindotan B, Perry KL (2007) Macroarray detection of plant RNA viruses using randomly primed and amplified complementary DNAs from infected plants. *Phytopathology* 97: 119–127.

Aguin-Casal O, Sáinz-Osés MJ, Mansilla-Vázquez P (2004) *Armillaria* species infesting vineyards in northwestern Spain. *Eur J Plant Pathol* 110: 683–687.

Agusti-Brisach C, Mostert L, Armengol J (2014) Detection and quantification of *Ilynectria* spp. associated with black foot disease of grapevine in nursery soils using multiplex nested PCR. *Plant Pathol* 63: 316–322.

Alaniz S, León M, Vicent A, García-Jiménez J, Abad-Campos P, Armengol J (2007) Characterization of *Cylindrocarpon* species associated with black foot disease of grapevine in Spain. *Plant Dis* 91: 1187–1193.

Al-Daoud F, Gossen BD, Robson J, McDonald MR (2017) Propidium monoazide improves quantification of resting spores of *Plasmodiophora brassicae* with qPCR. *Plant Dis* 101: 442–447.

Ali-Shtayeh MS, McDonald JD, Kabashima J (1991) A method for using commercial ELISA tests to detect zoospores of *Phytophthora* and *Pythium* species in irrigation water. *Plant Dis* 75: 305–311.

Aljawasim B, Vincelli P (2015) Evaluation of polymerase chain reaction (PCR)-based methods for rapid, accurate detection and monitoring of *Verticillium dahliae* on woody hosts by real-time PCR. *Plant Dis* 99: 866–873.

Almquist C, Persson L, Olsson A, Sundström J, Jonsson A (2016) Disease risk assessment of sugar beet root rot using quantitative real-time PCR analysis of *Aphanomyces cochlioides* in naturally infested soil. *Eur J Plant Pathol* 145: 731–742.

Altinok HA, Can C, Çolak H (2013) Vegetative compatibility, pathogenicity and virulence diversity of *Fusarium oxysporum* f. sp. *melongenae*. *J Phytopathol* 161: 651–660.

Alvarez LA, León M, Abad-Campos P, Garcia-Jaminez J, Vincent A (2011) Genetic variation and host specificity of *Phytophthora citrophthora* isolates causing branch cankers in Clementine trees in Spain. *Eur J Plant Pathol* 129: 103–117.

Alves-Santos FM, Ramos B, García-Sánchez MA, Eslava AP, Díaz-Minguez JM (2002) A DNA-based procedure for in planta detection of *Fusarium oxysporum* f.sp. *phaseoli*. *Phytopathology* 92: 237–244.

Anagnostakis SL (1977) Vegetative compatibility in *Endothia parasitica*. *Exp Mycol* 1: 306–316.

Anderson NA (1982) The genetics and pathology of *Rhizoctonia solani*. *Annu Rev Phytopathol* 20: 329–347.

Anderson SJ, Simmons HE, Munkvold GP (2015) Real-time PCR assay for detection of *Sphacelotheca reiliana*. *Plant Dis* 99: 1847–1852.

Andrade O, Muñoz G, Galdames R, Durán P, Honorato R (2004) Characterization, in vitro culture, and molecular analysis of *Thecaphora solani*, the causal agent of potato smut. *Phytopathology* 94: 875–882.

Arias MMD, Munkvold GP, Ellis ML, Leandro LFS (2013) Distribution and frequency of *Fusarium* species associated with soybean roots in Iowa. *Plant Dis* 97: 1557–1562.

Arie T, Gouthu S, Shimagaki S et al. (1998) Immunological detection of endopolygalacturonase secretion by *Fusarium* in plant tissue and sequencing of its encoding gene. *Ann Phytopathol Soc, Jpn* 64: 7–15.

Arie T, Hayashi Y, Yoneyama K, Nagatani A, Furuya M, Yamaguchi I (1995) Detection of *Fusarium* spp. in plants with monoclonal antibody. *Ann Phytopathol Soc Jpn* 61: 311–317.

Arif M, Fletcher J, Marek SM, Ochoa-Corona FM (2013) Development of a rapid, sensitive and field deployable Razoe Ex Biodetection System and quantitative PCR assay for detection of *Phymatotrichopsis omnivora* using multiple gene targets. *Appl Environ Microbiol* 79: 2312–2320.

Aroca A. Raposo R (2007) PCR-based strategy to detect and identify species of *Phaeoacremonium* causing grapevine disease. *Appl Environ Microbiol* 73: 2911–2918.

Atallah ZK, Bae J, Jansky SH, Rouse DI, Stevenson WR (2007) Multiplex real-time quantitative PCR to detect and quantify *Verticillium dahliae* colonization in potato lines that differ in response to Verticillium wilt. *Phytopathology* 97: 865–872.

Avanzato MA, Rothrock CS (2010) Use of selective media and baiting to detect and quantify the soilborne plant pathogen *Thielaviopsis basicola* on pansy. *The Plant Health Instructor*

Aycock R (1966) Stem rot and other diseases caused by *Sclerotium rolfsii* or the status of Rolf's fungus after 70 years. *NC Agric Exp Stn Tech Bull* 174: 202.

Baeummer A (2004) Nanosensors identify pathogens in food. *Food Technol* 58: 51–55.

Baird RE, Mullinix BG (2000) Determination of whole-cell fatty acid profiles for characterization and differentiation of isolates of *Rhizoctonia solani* AG-4 and AG-7. *Plant Dis* 84: 785–788.

Bakhori NM, Yusof NA, Abdullah AH, Hussein MZ (2012) Development of fluorescence based DNA sensor utilizing quantum dot for early detection of *Ganoderma boninense*. *2nd Internatl Conf Environmental Science and Biotechnology (IPCBEE)* 48: 138–142.

Banno S, Ikeda K, Saito H, Sakai H, Urushibara T, Shiraishi T, Fujimura M (2015) Characterization and distribution of two types of *Verticillium longisporum* from cabbage fields in Japan. *J Gen Plant Pathol* 81: 118–126.

Baturo-Cieniewska A, Groves CL, Albrecht KA, Grau R, Willis DK, Smith DL (2017) Molecular identification of *Sclerotinia trifoliorum* and *Sclerotinia sclerotiorum* isolates from the United States and Poland. *Plant Dis* 101: 192–199.

Bayman P, Cotty PJ (1991) Vegetative compatibility and genetic diversity in *Aspergillus flavus* populations in signle field. *Canad J Bot* 69: 1707–1710.

Bennett RS, Hutmacher RB, Davis RM (2008) Seed transmission of *Fusarium oxysporum* f.sp. *vasinfectum* race 4 in California. *J Cotton Sci* 12: 160–164.

Benson DM (1991) Detection of *Phytophthora cinnamomi* in azalea with commercial serological assay kits. *Plant Dis* 75: 478–482.

Benson DM (1992) Detection by enzyme-linked immunosorbent assay of *Rhizoctonia* species in poinsettia cuttings. *Plant Dis* 76: 578–561.

Bhat RG, Browne GT (2010) Specific detection of *Phytophthora cactorum* in diseased strawberry plants using nested polymerase chain reaction. *Plant Pathol* 59: 121–129.

Bhat RG, Colowit PM, Tai TH, Aradhya MK, Browne GT (2006) Genetic and pathogenic variation in *Phytophthora cactorum* affecting fruit and nut crops in California. *Plant Dis* 90: 161–169.

Bilodeau GJ, Koike ST, Uribe P, Martin FN (2012) Development of an assay for rapid detection and quantification of *Verticillium dahliae* in soil. *Phytopathology* 102: 331–343.

Bilodeau GJ, Lévesque CA, de Cock AWAM et al. (2007) Molecular detection of *Phytophthora ramorum* by real-time polymerase chain reaction using TaqMan, SYBR Green and molecular beacons. *Phytopathology* 97: 632–642.

Bilodeau GJ, Pelletier G, Pelletier F, Hamelin RC, Lévesque CA (2009) Multiplex real-time polymerase chain reaction (PCR) for detection of *Phytophthora ramorum*, the causal agent of sudden oak death. *Canad J Plant Pathol* 31: 195–210.

Bodrossy L, Sessitsch A (2004) Oligonucleotide microarrays in microbial diagnostics. *Curr Opin Microbiol* 7: 245–254.

Böhm J, Hahn A, Schubert R et al. (1999) Real-time quantitative PCR: DNA determination in isolated spores of the mycorrhizal fungus *Glomus mosseae* and monitoring *Phytophthora infestans* and *Phytophthora citricola* in their respective host plants. *J Phytopathol* 147: 404–416.

Bonants PJM, van Gent-Pelzer PE, Hooftman R, Cooke DEL, Guy DC, Duncan JM (2004) A combination of baiting and different PCR formats including measurement of real-time quantitative fluorescence for the detection of *Phytophthora fragariae* var. *fragariae* in strawberry plants. *Eur J Plant Pathol* 110: 689–702.

Bonants PJM, van Gent-Pelzer PEM, Hagenaarde Weerdt M (2000) Characterization and detection of *Phytophthora fragariae* in plant, water and soil by molecular methods. *OEPP Bull* 30: 525–531.

Bonde MR, Nester SE, Schaad NW, Frederick RD, Luster DG (2003) Improved detection of *Tilletia indica* teliospores in seed or soil by elimination of contaminating microorganisms with acid electrolyzed water. *Plant Dis* 87: 712–718.

Bonde MR, Peterson GL, Dowler WM, May B (1984) Isozyme analysis to differentiate species of *Peronosclerospora* causing downy mildews for maize. *Phytopathology* 74: 1278.

Borders KD, Parker ML, Melzer MS, Boland GJ (2014) Phylogenetic diversity of *Rhizoctonia solani* associated with canola and wheat in Alberta, Manitoba and Saskatchewan. *Plant Dis* 98: 1695–1701.

Bouchek-Mechiche K, Montfort F, Merz U (2011) Evaluation of the Sss AgriStrip rapid diagnostic test for the detection of *Spongospora subterranea* on potato tubers. *Eur J Plant Pathol* 131: 277–287.

Bowman KD, Albrecht U, Graham JH, Bright DB (2007) Detection of *Phytophthora nicotianae* and *P. palmivora* in citrus roots using PCR-RFLP in comparison with other methods. *Eur J Plant Pathol* 119: 143–158.

Braun H, Levivier S, Eber F, Renard M, Chevre AM (1997) Electrophoretic analysis of natural populations of *Leptosphaeria maculans* directly from leaf lesions. *Plant Pathol* 46: 147–154.

Budge GE, Shaw MW, Colyer A, Pietravalle S, Boonham N (2007) Molecular tools to investigate *Rhizoctonia solani* distribution in soil. *Plant Pathol* 58: 1071–1080.

Bulajić A, Djekić I, Jović J, Krnjajić S, Vačurivićc A, Krstić B (2010) *Phytophthora ramorum* occurrence in ornamentals in Serbia. *Plant Dis* 94: 703–708.

Bush EA, Hong C, Stromberg EL (2003) Fluctuations of *Phytophthora* and *Pythium* spp. in components of a recycling irrigation system. *Plant Dis* 87: 1500–1506.

Busse F, Bartkiewicz A, Terefe-Ayana D, Niepold F, Schleusner Y, Flath K, Sommerfeld-Impe N, Lubeck J, Strahwald J, Tacke E, Hofferbert H-R, Linde M, Przetakiewicz J, Debener T (2017) Genomic and transcriptomic resources for marker development in *Synchytrium endobioticum*, an elusive but severe potato pathogen. *Phytopathology* 107: 322–328.

Butterfield EJ, DeVay JE (1977) Reassessment of soil assays for *Verticillium dahliae*. *Phytopathology* 67: 1073 – 1078.

Cabanás CG, Pérez-Artés E (2014) New evidence for intra-race diversity in *Fusarium oxysporum* f.sp. *dianthi* populations based on vegetative compatibility groups. *Eur J Plant Pathol* 139: 445–441.

Cabanás CG, Valverde-Corredor A, Pérez-Artés E (2012) Molecular analysis of Spanish populations of *Fusarium oxysporum* f.sp. *dianthi* demonstrates a high genetic diversity and identifies virulence groups in races 1 and 2 of the pathogen. *Eur J Plant Pathol* 132: 561–576.

Cahill DM, Hardham AR (1994) A dipstick immunoassay for the specific detection of *Phytophthora cinnamomi* in soils. *Phytopathology* 84: 1284–1292.

Camele I, Macone C, Cristinzio G (2005) Detection and identification of *Phytophthora* species in southern Italy by RFLP and sequence analysis of PCR-amplified nuclear ribosomal RNA. *Eur J Plant Pathol* 113: 1–14.

Cañizares MC, Gómez-Lama C, García-Pedrajas MD, Pérez-Artés E (2015) Study of phylogenetic relationships among *Fusarium oxysporum* f.sp. *dianthi* isolates: Confirmation of intra-race diversity and development of a practical tool for simple population analyses. *Plant Dis* 99: 780–787.

Cao T, Tewari J, Strelkov SE (2007) Molecular detection of *Plasmodiophora brassicae* causal agent of club root of crucifers in plant and soil. *Plant Dis* 91: 80–87.

Cao Y, Li LJ, Wang L, Chang Z, Wang H, Fan Z, Li H (2016) Rapid and quantitative detection of *Pythium inflatum* by real-time fluorescence loop-mediated isothermal amplification assay. *Eur J Plant Pathol* 144: 83–95.

Chandelier A, Ivors K, Garbelotto M, Zini J, Laurent F, Cavelier M (2006) Validation of a real-time PCR method for detection of *Phytophthora ramorum*. *Bull OEPP*/EPPO 36: 409–414.

Chen W, Grau CR, Adee EA, Meng X-Q (2000) A molecular marker for identifying subspecies specific population of the soybean brown stem rot pathogen *Phytophthora gregata*. *Phytopathology* 90: 875–883.

Chen W, Robleh Djama Z, Coffey MD, Martin FN, Bilodeau GJ, Radmer L, Denton G, Lévesque CA (2013) Membrane-based oligonucleotide array developed from multiple markers for the detection of many *Phytophthora* species. *Phytopathology* 103: 43–54.

Chen X-R, Liu B-B, Xing Y-P et al. (2016) Identification and characterization of *Phytophthora helicoides* causing stem rot of Shatangju manadarin seedlings in China. *Eur J Plant Pathol* 146: 715–727.

Chiocchetti A, Sciaudone L, Durando F, Garibaldi A, Migheli Q (2001) PCR detection of *Fusarium oxysporum* f.sp. *basilici* on basil. *Plant Dis* 85: 607–611.

Chittem K, Mathew FM, Gregoire M et al. (2015) Identification and characterization of *Fusarium* sp. associated with root rots of field pea in North Dakota. *Eur J Plant Pathol* 143: 641–649.

Cho JH, Rupe JC, Cummings MS, Gbur Jr EE (2001) Isolation and identification of *Fusarium solani* f.sp. *glycines* from soil on modified Nash and Snyder's medium. *Plant Dis* 85: 256–260.

Chongo G, Gossen BD, Buchwaldt L, Adhikari T, Rimmer SR (2004) Genetic diversity of *Ascochyta rabiei* in Canada. *Plant Dis* 88: 4–10.

Cipriani MG, Schena L, Sialer MMF, Gallitelli D (2000) Characterization and cloning of a molecular probe for diagnosis of *Verticillium* spp. *Atti Gionrate fitopatologiche* 2: 551–558.

Clarkson JP, Coventry E, Kitchen J, Carter HE, Whipps JM (2013) Population structure of *Sclerotinia sclerotiorum* in crop and wild hosts in the UK. *Plant Pathol* 62: 309–324.

Collado-Romero M, Berbegal M, Jiménez-Díaz RM, Armengol J, Mercado-Blanco J (2009) A PCR-based 'molecular tool box' for in planta differential detection of *Verticillium dahliae* vegetative compatibility groups infecting artichoke. *Plant Pathol* 58: 515–526.

Collado-Romero M, Mercado-Blanco J, Olivares-Garcia C, Valverde-Corredor, Jiménez-Díaz RM (2006) Molecular variability within and among *Verticillium dahliae* vegetative compatibility groups determined by fluorescent amplified fragment length polymorphism and polymerase chain reaction markers. *Phytopathology* 96: 485–495.

Cooke DEL, Duncan JM, Williams NA, Hagenaar-de Weerdt M, Bonants PJM (2000) Identification of *Phytophthora* species on the basis of restriction enzyme fragment analysis of the internal transcribed spacer regions of ribosomal RNA. *EPPO Bull* 30: 519–523.

Cooke DEL, Schena L, Cacciola SO (2007) Tools to detect, identify and monitor *Phytophthora* species in natural ecosystems. *J Plant Pathol* 89: 145–150.

Copes WE, Yang X, Hong CX (2015) *Phytophthora* species recovered from irrigation reservoirs in Mississippi and Alabama nurseries and pathogenicity of three new species. *Plant Dis* 99: 1390–1395.

Correll JC, Klittich CJR, Leslie JF (1987) Nitrate nonutilizing mutants of *Fusarium oxysporum* and their use in vegetative compatibility tests. *Phytopathology* 77: 1640–1646.

Crutcher FK, Doan HK, Bell AB et al. (2016) Evaluation of methods to detect the cotton wilt pathogen *Fusarium oxysporum* f.sp. *vasinfectum* race 4. *Eur J Plant Pathol* 144: 225–230.

Cullen DW, Lees AK, Toth IK, Duncan JM (2001) Conventional PCR and real-time quantitative PCR detection of *Helminthosporium solani* in soil and on potato tubers. *Eur J Plant Pathol* 107: 387–398.

Daayf F, Nicole M, Geiger JP (1995) Differentiation of *Verticillium dahliae* population on the basis of vegetative compatibility and pathogenicity on cotton. *Eur J Plant Pathol* 101: 69–79.

Dai T-T, Meng J, Dong S et al. (2013) A *Phytophthora* conserved transposon-like DNA element as a potential target for soybean root rot disease diagnosis. *Plant Pathol* 62: 719–726.

Das S, Shah FA, Butler RC, Falloon RE, Stewart A, Raikar S, Pitman AR (2014) Genetic variability and pathogenicity of *Rhizoctonia solani* associated with black scurf of potato in New Zealand. *Plant Pathol* 63: 651–666.

Daval S. Lebreton L, Gaznegel K, Guillerm-Erckelboudt AY, Sarniguet A (2010) Genetic evidence for differentiation of *Gaeumannomyces graminis* var. *tritici* into two major groups. *Plant Pathol* 59: 165–178.

Davidson JA, Krysinska-Kaczmarek M, Wilmshurst CJ, McKay A, Scott ES (2011) Distribution and survival of Ascochyta blight pathogens in field pea-cropping soils of Australia. *Plant Dis* 95: 1217–1223.

de Haan LAM, Numansen A, Roebroeck BJA, van Doorn J (2000) PCR detection of *Fusarium oxysporum* f.sp. *gladioli* race 1, causal agent of Gladiolus yellows disease from infected corms. *Plant Pathol* 49: 89–100.

Degola F, Bemi E, Dall'Asta C et al. (2007) A multiplex RT-PCR approach to detect aflatoxigenic strains of *Aspergillus flavus*. *J Appl Microbiol* 103: 409–417.

Delfosse P, Reddy AS, Legreve A et al. (2000) Serological methods for detection of *Polymyxa graminis*, an obligate root parasite and vector of plant viruses. *Phytopathology* 90: 537–545.

Demers JE, Garzon CD, Jiménez-Gasco MDM (2014) Striking similarity between races of *Fusarium oxysporum* f.sp. *ciceris* confirms a monophyletic origin and clonal evolution of the chickpea vascular wilt pathogen. *Eur J Plant Pathol* 139: 309–324.

Deora A, Gossen BD, Amirsadeghi S, McDonald MR (2015) A multiplex qPCR assay for detection and quantification of *Plasmodiophora brassicae* in soil. *Plant Dis* 99: 1002–1009.

De Sousa MV, Machado da, Simmons HE, Munkvold GP (2015) Real-time quantitative PCR assays for the rapid detection and quantification of *Fusarium oxysporum* f.sp. *phaseoli* in *Phaseoulus vulgaris* (common bean) seeds. *Plant Pathol* 64: 478–488.

Dita MA, Waalwijk C, Buddenhagen IW, Souza Jr MT, Kema GHJ (2010) A molecular diagnosis for tropical race 4 of the banana Fusarium wilt pathogen. *Plant Pathol* 59: 348–357.

Dobinson KF, Harrington MA, Omer M, Rowe RC (2000) Molecular characterization of vegetative compatibility group 4A and 4B isolates of *Verticillium dahliae* associated with potato early dying. *Plant Dis* 84: 1241–1245.

Duan Y, Ge C, Zhang XD, Wang J, Zhou M (2014) A rapid detection method for the plant pathogen *Sclerotinia sclerotiorum* based on loop-mediated isothermal amplification (LAMP). *Austr Plan Pathol* 3: 61–66.

Duan Y, Yang Y, Wang Y et al. (2016) Loop-mediated isothermal amplification for the rapid detection of the F200Y mutant genotype of carbendazim resistant isolates of *Sclerotinia sclerotiorum*. *Plant Dis* 100: 976–983.

Duffy BK, Weller DM (1994) A semiselective and diagnostic medium for *Gaeumannomyces graminis* var. *tritici*. *Phytopathology* 84: 1407–1415.

Duncan JM (1980) A technique for detecting red stele (*Phytophthora fragariae*) infection in strawberry stocks before planting. *Plant Dis* 64: 1023–1025.

Duncan JM (1990) *Phytophthora* species attacking strawberry and raspberry. *Bull OEPP/EPPO* 20: 107–115.

Dung JKS, Hamm PB, Eggers JE, Johnson DA (2013) Incidence and impact of *Verticillium dahliae* in soil associated with certified seed lots. *Phytopathology* 103: 55–63.

Dung JKS, Peever TL, Johnson DA (2013) *Verticillium dahliae* populations from mint and potato are genetically divergent with predominant haplotypes. *Phytopathology* 103: 445–459.

Duressa D, Rauscher G, Koike ST et al. (2012) A real-time PCR assay for detection and quantification of *Verticillium dahliae* in spinach seed. *Phytopathology* 102: 443–451.

Duy J, Smith RL, Collins SD, Connell LB (2015) Rapid colorimetric detection of the fungal phytopathogens *Synchytrium endobioticum*, using cyanine dye-indicated PNA hybridization. *Amer J Potato Res* 92: 398–409.

Eden MA, Hill RA, Galpoththage M (2000) An efficient baiting assay for quantification of *Phyhtophthora cinnamomi* in soil. *Plant Pathol* 49: 515–522.

Edwards J, Constable F, Wiechel T, Salib S (2007) Comparison of the molecular tests–single PCR, nested PCR and quantitative PCR (SYBR® Green and TaqMan®) for detection of *Phaeomoniella chlamydospora* during grapevine nursery propagation. *Phytopathol Mediterr* 46: 58–72.

Ellis ML, Cruz Jimenez DA, Leandro LF, Munkvold GP (2014) Genotypic and phenotypic characterization of fungi in the *Fusarium oxysporum* species complex from soybean roots. *Phytopathology* 104: 1329–1339.

Elmer WH, Covert SF, O'Donnell K (2007) Investigation of an outbreak of Fusarium foot and fruit rot of pumpkin within the United States. *Plant Dis* 91: 1142–1146.

Engelbrecht J, Duong TA, van den Berg N (2013) Development of a nested quantitative real-time PCR for detecting *Phytophthora cinnamomi* in *Persea Americana* rootstocks. *Plant Dis* 97: 1012–1017.

Eshraghi L, Aryamanesh N, Anderson JP et al. (2011) A quantitative PCR assay for accurate in planta quantification of the necrotrophic pathogen *Phytophthora cinnamomi*. *Eur J Plant Pathol* 131: 419.

Etebu E, Osborn AM (2010) Molecular quantification of the pea foot rot disease pathogen (*Nectria haematococca*) in agricultural soils. *Phytoparasitica* 38: 447–454.

Faedda R, Cacciola SO, Pane A, et al. (2013) *Phytophthora* × *plegrandis* causes root rot and collar rot of *Lavandula stoechas* in Italy. *Plant Dis* 97: 1091–1096.

Faggian R, Bulman SR, Lawrie AC, Porter IJ (1999) Specific polymerase reaction primers for the detection of *Plasmodiophora brassicae* in soil and water. *Phytopathology* 89: 392–397.

Fang Y, Ramasamy RP (2015) Current and prospective methods for plant disease detection. *Biosensors* 4: 537–551.

Feng J, Jiang J, Feindel D, Strelkov SE, Hwang S-F (2016) The gene *Cr811* is present exclusively in pathotype 5 and newly emerged pathotypes of the club root pathogen *Plasmodiophora brassicae. Eur J Plant Pathol* 145: 615–620.

Ferraris L, Cardinale F, Valentino D, Roggero P, Tamietti G (2004) Immunological discrimination of *Phytophthora cinnamomi* from other *Phytophthora* pathogenic on chestnut. *J Phytopathol* 152: 193–199.

Fichtner EJ, Lynch SC, Rizzo DM (2007) Detection, distribution, sporulation and survival of *Phytophthora ramorum* in a California redwood-tanoak forest soil. *Phytopathology* 97: 1366–1375.

Fouly HM, Wilkinson HT (2000) Detection of *Gaeumannomyces graminis* varieties using polymerase chain reaction with variety-specific primers. *Plant Dis* 84: 947–951.

French-Monar RD, Jones JB, Roberts PD (2006) Characterization of *Phytophthora capsici* associated with roots of weeds on Florida vegetable farms. *Plant Dis* 90: 345–350.

Fukuta S, Takahashi R, Kuroyanagi S, Miyake N, Nagai H, Suzuki H, Hashizume F, Tsuji T, Taguchi H, et al. (2013) Detection of *Pythium aphanidermatum* in tomato using loop-mediated isothermal amplification (LAMP) with species-specific primers. *Eur J Plant Pathol* 136: 689–701.

Gabor BK, O'Gara ET, Philip BA, Horan DP and Hardham AR (1993) Specificities of monoclonal antibodies to *Phytophthora cinnamomi* in two rapid diagnostic assays. *Plant Dis* 77: 1189–1197.

Gagon M-C, Kawachuk L, Tremblay DM, Carisse O, Danies G, Fry WE, Lévesque CA, Bilodeau G (2016) Identification of the dominant genotypes of *Phytophthora infestans* in Canada using real-time PCR with ASO-PCR assays. *Plant Dis* 100: 1482–1491.

Gangneux C, Cannesan M-A, Bressan M et al. (2014) A sensitive assay for rapid detection and quantification of *Aphanomyces euteiches* in soil. *Phytopathology* 104: 1138–1147.

Gao X, Jackson TA, Lambert KN, Li S, Hartman GL, Niblack TL (2004) Detection and quantification of *Fusarium solani* f sp. *glycines* in soybean roots with real-time quantitative polymerase chain reaction. *Plant Dis* 88: 1372–1380.

Gao Y, Tan MK, Zhu YG (2016) Rapid and specific detection of *Tilletia indica* using loop-mediated isothermal amplification. *Austr Plant Pathol* 45: 361–367.

Garzón CD, Geiser DM, Moorman GW (2005) Diagnosis and population analysis of *Pythium* species using AFLP fingerprinting. *Plant Dis* 89: 81–89.

Gayoso C, de Ilarduya OM, Pomar F, de Caceres FM (2007) Assessment of real-time PCR as a method for determining the presence of *Verticillium dahliae* in different Solanaceae cultivars. *Eur J Plant Pathol* 118: 199–209.

Gevens AJ, Donahoo RS, Lamour KH, Hausbeck MK (2007) Characterization of *Phytophthora capsici* from Michigan surface irrigation water. *Phytopathology* 97: 421–428.

Ghimire SR, Richardson PA, Moorman GW, Lea-Cox JD, Ross DS, Hong CX (2009) An in situ baiting bioassay for detecting *Phytophthora* species in irrigation runoff containment basins. *Plant Pathol* 58: 577–583.

Goates BJ (2012) Identification of new pathogenic races of common bunt and dwarf bunt fungi and evaluation of known races using an expanded set of differential wheat lines. *Plant Dis* 96: 361–369.

Goates BJ, Peterson GL (1999) Relationship between soilborne and seedborne inoculum density and incidence of dwarf bunt of wheat. *Plant Dis* 83: 819–824.

Gomes EV, Do Nascimento LB, De Freitas MA, Nasser LCB, Petrofeza S (2011) Microsatellite markers reveal genetic variation within *Sclerotinia sclerotiorum* populations in irrigated dry bean crops in Brazil. *J Phytopathol* 159: 94–99.

Goodwin PH, English JT, Neber DA, Duniway JM, Kirkpatrick BC (1990) Detection of *Phytophthora parasitica* from soil and host tissue with species-specific DNA probe. *Phytopathology* 80: 277.

Gramaje D, Pérez-Serrano V, Montes-Borrego M et al. (2013) A comparison of real-time PCR protocols for the quantitative monitoring of asymptomatic olive infections by *Verticillium dahliae* pathotypes. *Phytopathology* 103: 1058–1068.

Grote D, Olmos A, Kofoet A, Tuset JJ, Bertolini E, Cambra M (2000) Detection of *Phytophthora nicotianae* by PCR. *OEPP Bull* 30: 539–541.

Grote D, Olmos A, Kofoet A, Tuset JJ, Bertolini E, Cambra M (2002) Specific and sensitive detection of *Phytophthora nicotianae* by simple and nested-PCR. *Eur J Plant Pathol* 108: 197–207.

Grünwald NJ, Martin FN, Larsen MM et al. (2011) Phytophthora-ID. org: A sequence-based *Phytophthora* Identification tool. *Plant Dis* 95: 337–342.

Guo Q, Li S, Lu X, Gao H, Wang X, Ma Y, Zhang X, Wang P, Ma P (2015) Identificationof a new genotype of *Fusarium oxysporum* f.sp. *vasinfectum* on cotton in China. *Plant Dis* 99: 1569–1577.

Guo Y, Li W, Sun H, Wang, Yu H, Chen H (2012) Detection and quantification of *Rhizoctonia solani* in soil using real-time PCR. *J Gen Plant Pathol* 78: 247–254.

Gutiérrez WA, Mila AL (2007) A rapid technique for determination of races of *Phytophthora nicotianae* on tobacco. *Plant Dis* 91: 985–999.

Haegi A, Catalano V, Luongo L, Vitale, Scotton M, Ficcadenti N, Belisario A (2013) A newly developed real-time PCR assay for detection and quantification of *Fusarium oxysporum* and its use in compatible and incompatible interactions with grafted melon. *Phytopathology* 103: 802–810.

Hardham AR, Gubler F, Duniec J, Elliott J (1991) A review of methods for the production and use of monoclonal antibodies to study zoosporic plant pathogens. *J Microscopy* 162: 305–318.

Hardham AR, Suzaki E, Perkin JL (1986) Monoclonal antibodies to isolate species- and genus-specific components of the surface of zoospores and cysts of the genus *Phytophthora cinnamomi. Canad J Bot* 64: 311–321.

Harvey B, Ophel-Keller K (1996) Quantification of *Gaeumannnomyces graminis* var. *tritici* in infected roots and arid soil using slot-blot hybridization. *Mycol Res* 100: 962–970.

Hashimoto Y, Nakamura H, Koichi AK, Karube I (2008) A new diagnostic method for soilborne disease using a microbial sensor. *Microbes and Environment* 23: 35–39.

Hayden KJ, Ivors K, Wilkinson C, Garbelotto M (2006) TaqMan chemistry for *Phytophthora ramorum* detection and quantification with comparison of diagnostic methods. *Phytopathology* 96: 846–854.

Hayden KJ, Rizozo D, Tse J, Garbelotto M (2004) Detection and quantification of *Phytophthora ramorum* from California forests using a real-time polymerase chain reaction assay. *Phytopathology* 94: 1075–1083.

Heinz PR, Platt HW (2000) A competitive PCR-based assay to quantify *Verticillium tricorpus* propagules in soil. *Canad J Plant Pathol* 22: 122–130.

Henson JM, French R (1993) The polymerase chain reaction and plant disease diagnosis. *Annu Rev Phytopathol* 31: 81–109.

Henson JM, Goins T, Grey W, Mathre DE, Elliott ML (1993) Use of polymerase chain reaction to detect *Gaeumannomyces graminis* DNA in plants grown in artificially and naturally infested soil. *Phytopathology* 83: 283–287.

Hernandez JF, Posado MA, del Portillo P, Arbelaez G (1999) Identification of molecular markers of *Fusarium oxysporum* f.sp. *dianthi* by RAPD. *Acta Hortic* 482, 123–131.

Hirano Y, Arie T (2006) PCR-based differentiation of *Fusarium oxysporum* f.sp. *lycopersici* and *radicis-lycopersici*. *J Gen Plant Pathol* 72: 273–283.

Holben WE, Jansson JK, Chelm BK, Tiedje JM (1983) DNA probe method for detection of specific microorganisms in the soil bacterial community. *Appl Environ Microbiol* 54: 703–711.

Holtz BA, Karu AF, Weinhold AR (1994) Enzyme-linked immunosorbent assay for detection of *Thielaviopsis basicola*. *Phytopathology* 84: 977–984.

Hong CF, Hseih HY, Chen CT, Huang HC (2013) Development of a semiselective medium for detection of *Nalanthamala psidii*, causal agent of wilt of guava. *Plant Dis* 97: 1132–1136.

Hsieh SPY, Huang RZ, Wang TC (1996) Application of tannic acid in qualitative and quantitative growth assay of *Rhizoctonia* spp. *Plant Pathol Bull* 5: 100–106.

Hu X-P, Gurung S, Short DPG, Sandoya GV, Shang W-J, Hayes RJ, Davis RM, Subbarao KV (2015) Nondefoliating and defoliating strains from cotton correlated with races 1 and of *Verticillium dahliae*. *Plant Dis* 99: 1713–1720.

Huang C-H, Roberts PD, Gale LR, Elmer WH, Datnoff LE (2013) Population structure of *Fusarium oxysporum* f.sp. *radicis-lycopersici* in Florida inferred from vegetative compatibility groups and microsatellites. *Eur J Plant Pathol* 136: 509–521.

Huang C-H, Tsai R-T, Vallad GE (2016) Development of a TaqMan real-time polymerase chain reaction assay for detection and quantification of *Fusarium oxysporum* f.sp. *lycopersici* in soil. *J Phytopathol* 164: 455–463.

Huang J, Kang Z (2010) Detection of *Thielaviopsis basicola* in soil with real-time quantitative PCR assays. *Microbiol Res* 165: 411–417.

Huang J, Wu J, Li C, Xiao C, Wang G (2010) Detection of *Phytophthora nicotianae* with real-time quantitative PCR. *J Phytopathol* 158: 15–21.

Hughes KJD, Tomlinson JA, Griffin RL, Boonham N, Inman AJ, Lane CJ (2006) Development of a one-step real-time polymerase chain reaction assay for diagnosis of *Phytophthora ramorum*. *Phytopathology* 96: 975–981.

Hughes KTD, Inman AJ, Beales PA, Cook RTA, Fulton CE, Mc Reynolds ADK (1998) PCR-based detection of *Phytophthora fragariae* in raspberry and strawberry roots. BCPC *Conference on Pests and Diseases* 2: 687–692.

Hughes TJ, Atallah ZK, Grau CR (2009) Real-time PCR assays for the quantification of *Phialophora gregata* f.sp. *sojae* IGS genotype A and B. *Phytopathology* 99: 1008–1114.

Ikeda K, Banno S, Watanabe K, Fujinaga M, Ogiso H, Sakai H, Tanaka H, Miki S, Shibata S, et al. (2012) Association of *Verticillium dahliae* and *Verticillium longisporum* with Chinese cabbage yellow and their distribution in the main production areas of Japan. *J Gen Plant Pathol* 78: 331–337.

Impullitti AE, Malvick DK, Grau CR (2009) Characterizing reaction of soybean to *Phialophora gregata* using pathogen population density and DNA quantity in stems. *Plant Dis* 93: 734–740.

Ioos R, Laugustin L, Schenk N, Rose S, Husson C, Frey P (2006) Usefulness of single copy genes containing introns in *Phytophthora* for the development of detection tools for the regulated species *P. ramorum* and *P. fragariae*. *Eur J Plant Pathol* 116: 171–176.

Ippolito A, Schena L, Nigro F (2002) Detection of *Phytophthora nicotianae* and *P. citrophthora* in citrus roots and soils by nested PCR. *Eur J Plant Pathol* 108: 855–868.

Ippolito A, Schena L, Nigro, Ligorio VS, Yaseen T (2004) Real-time detection of *Phytophthora nicotianae* and *P. citrophthora* in citrus roots and soil. *Eur J Plant Pathol* 110: 833–843.

Ito S, Maehara T, Maruno E, Tanaka S, Kameya-Iwaki M, Kishi F (1999) Development of a PCR-based assay for the detection of *Plasmodiophora brassicae* in soil. *J Phytopathol* 147: 83–88.

Ito T, Fuji S, Sato E, Iwadate Y, Toda T, Furuya H (2012) Detection of *Phomopsis sclerotioides* in commercial cucurbit field soil by nested time-release PCR. *Plant Dis* 96: 515–521.

Jiménez-Fernández D, Montes-Borrego M, Jiménez-Díaz RM, Navas-Cortes JA, Landa BB (2011) In planta and soil quantification of *Fusarium oxysporum* f.sp. *ciceris* and evaluation of Fusarium wilt resistance in chickpea with a newly developed quantitative polymerase chain reaction assay. *Phytopathology* 101: 250–262.

Jimenéz-Fernández D, Navas-Cortés JA, Montes-Borrego M, Jiménez-Díaz RM (2011) Molecular and pathogenic characterization of *Fusarium redolens*, a new causal agent of Fusarium yellows in chickpea. *Plant Dis* 95: 860–867.

Joaquim TR, Rowe RC (1990) Reassessment of vegetative compatibility relationships among strains of *Verticilliium dahliae* using nitrate-nonutilizing mutants. *Phytopathology* 80: 1160–1166.

Joaquim TR, Rowe RC (1991) Vegetative compatibilty and virulence of strains of *Veticillium dahliae* from soil and potato plants. *Phytopathology* 81: 552–588.

Jordan VWL (1971) Estimation of the distribution of *Verticillium* populations in infected strawberry plants and soil. *Plant Pathol* 20: 21–24.

Judelson HS, Messenger-Routh B (1996) Quantification of *Phytophthora cinnamomi* in avocado roots using a species-specific DNA probe. *Phytopathology* 86: 763–768.

Kabir Z, Bhat RG, Subbarao KV (2004) Comparison of media for recovery of *Verticillium dahliae* from soil. *Plant Dis* 88: 49–55.

Kageyama K, Kobayashi M, Tomita M, Kubota N, Suga H, Hyakumachi M (2002) Production and evaluation of monoclonal antibodies for the detection of *Pythium sulcatum* in soil. *J Phytopathol* 150: 97–104.

Kandel YR, Haudenshield JS, Srour AM et al. (2015) Multilaboratory comparison of quantitative PCR assays for detection and quantification of *Fusarium virguliforme* from soybean roots and soil. *Phytopathology* 105: 1601–1611.

Karajeh MR, Masoud SA (2006) Molecular detection of *Verticillium dahliae* Klet. in asymptomatic olive trees. *J Phytopathol* 154: 496–499.

Kashiwa T, Inami K, Teraoka T, Komatsu K, Arie T (2016) Detection of cabbage yellows fungus *Fusarium oxysporum* f.sp *conglutinans* in soil by PCR and real-time PCR. *J Gen Plant Pathol* 82: 240–247.

Keenan S, Cromey MG, Harrow SA, Bithell SL, Butler RC, Beard SS, Pitman BAR (2015) Quantitative PCR to detect *Gaeumannomyces graminis* var. *tritici* in symptomatic and nonsymptomatic wheat roots. *Austr Plant Pathol* 44: 591–597.

Keer JT, Birceh L (2003) Molecular methods for assessment of bacterial viability. *J Microbiol Meth* 53: 175–183.

Kim HJ, Lee YS (2001) Development of an in planta molecular marker for the detection of Chinese cabbage (*Brassica campestris pekinensis*) clubroot pathogen *Plasmodiophora brassicae*. *J Microbiol* 39: 56–61.

Kingsnorth CS, Asher MJC, Keane GJP, Chwarszczynska DM, Luterbacher MC, Musta-Gottgens ES (2003) Development of a recombinant antibody ELISA test for the detection of *Polymyxa betae* and its use in resistance screening. *Plant Pathol* 52: 673–680.

Kishore Babu B, Sharma R (2015) TaqMan real-time PCR assay for the detection and quantification of *Sclerospora graminicola*, the causal agent of pearl millet downy mildew. *Eur J Plant Pathol* 142: 149–158.

Kitagawa T, Sakamoto Y, Furumi K, Ogura H (1989) Novel enzyme immunoassays for specific detection of *Fusarium oxysporum* f.sp. *cucumerinum* and for general detection of various *Fusarium* species. *Phytopathology* 79: 162–165.

Klassen GR, Blacerzak M, de Cock AWAM (1996) 5S ribosomal RNA gene spacers as species-specific probes for eight species of *Pythium*. *Phytopathology* 86: 581–587.

Klemsdal SS, Herrero ML, Wanner LA, Lund G, Hermansen A (2008) PCR-based identification of *Pythium* spp. causing cavity spot in carrots and sensitive detection in soil samples. *Plant Pathol* 57: 877–886.

Knüfer J, Lopisso DT, Koopmann B, Karlovsky P, von Tiedemann A (2017) Assessment of latent infection with *Verticillium longisporum* in field-grown oilseed rape by qPCR assay. *Eur J Plant Pathol* 147: 819–831.

Koedrith P, Thasiphu T, Weon J-I, Boonprasert R, Tuitemwong K, Tuitemwong P (2015) Recent trends in rapid environmental monitoring of pathogens and toxicants: Potential of nanoparticle-based biosensor and applications. *The Scientific World Journal* 2015, Article ID 510982, 12 p.

Kohler G, Milstein C (1975) Continuous cultures of fused cells secreting antibody of predetermined specificity. *Nature (London)* 256: 495–497.

Kohn LM, Stavoski E, Cartone I, Royer J, Anderson JB (1991) Mycelial compatibility and molecular markers to identify genetic variability in field populations of *Sclerotinia sclerotiorum*. *Phytopathology* 81: 480–485.

Kong P, Hong C, Jeffers SN, Richardson PA (2003) A species-specific polymerase chain reaction assay for rapid detection of *Phytophthora nicotianae* in irrigation water. *Phytopathology* 93: 822–831.

Kong P, Hong CX, Richardson PA (2003) Rapid detection of *Phytophthora cinnamomi* using PCR with primers derived from the *Lpv* putative storage protein genes. *Plant Pathol* 52: 681–693.

Korolev N, Pérez-Artés E, Mercado-Blanco J, Bejarano-Alcázar J, Rodriguez-Jurado D, Jimenéz-Díaz RM, Katan T, Katan J (2008) Vegetative compatibility of cotton defoliating *Verticillium dahliae* in Israel and its pathogenicity to various crop plants. *Eur J Plant Pathol* 122: 603–617.

Kox LFF, van Brouwershaven IR, van de Vossenberg BTLH, van den Beld HE, Bonants PJM, de Gruyter J (2007) Diagnostic values and utility of immunological, morphological and molecular methods for in planta detection of *Phytophthora ramorum*. *Phytopathology* 97: 1119–1129.

Kraft JM, Boge WL (1994) Development of an antiserum to quantify *Aphanomyces euteiches* in resistant pea lines. *Plant Dis* 78: 179–183.

Kroon LPNM, Verstappen ECP, Kox LFF, Flier WG, Bonants PJM (2004) A rapid diagnostic test to distinguish between American and European populations of *Phytophthora ramorum*. *Phytopathology* 94: 613–620.

Kuramae EE, Buzeto AL, Ciampi MB, Souza NL (2003) Identification of *Rhizoctonia solani* AG1-IB in lettuce, AG4 HG-I in tomato and melon and AG4 HG-III in broccoli and spinach. *Eur J Plant Pathol* 109: 391–395.

Lan CZ, Liu PQ, Li BJ, Chen QH, Weng QY (2013) Development of a specific PCR assay for the rapid and sensitive detection of *Phytophthora capsici*. *Austr Plant Pathol* 42: 379–384.

Lane CR, Hobden E, Walker L et al. (2007) Evaluation of a rapid diagnostic field test kit for identification of *Phytophthora* species, including *P. ramorum* and *P. kernoviae* at the point of infection. *Plant Pathol* 56: 828–835.

Larkin RP, Groves CL (2003) Identification and characterization of isolates of *Phytophthora infestans* using fatty acid methyl ester (FAME) profiles. *Plant Dis* 87: 1233–1243.

Larkin RP, Ristaino JB, Campell CL(1995) Detection and quantification of *Phytophthora capsici* in soil. *Phytopathology* 85: 1057–1063.

Larsen RC, Hollingsworth CR, Vandemark GJ, Gritsenko MA, Gray FA (2002) A rapid method using PCR-based SCAR markers for the detection and identification of *Phoma sclerotioides*: The cause of brown root rot disease of alfalfa. *Plant Dis* 86: 928–932.

Lattanzio VMT, Nivarlet N, Lippolis V et al. (2012) Multiplex dispstick immunoassay for semiquantitative determination of *Fusarium* mycotoxins in cereals. *Analyt Chim Acta* 718: 99–108.

Lees AK, Cullen DW, Sullivan L, Nicolson MJ (2002) Development of conventional and quantitative real-time PCR assays for the detection and identification of *Rhizoctonia solani* AG-3 in potato and soil. *Plant Pathol* 51: 293–302.

Lees AK, van de Graaf P, Wale S (2008) The identification and detection of *Spongospora subterranea* and factors affecting infection and disease. *Amer J Potato Res* 85: 247–252.

Leonberger AJ, Speers C, Ruhl G, Creswell T, Beckerman JL (2013) A survey of *Phytophthora* spp. in Midwest nurseries, greenhouses and landscapes. *Plant Dis* 97: 635–640.

Lévesque CA, Harlton CE, de Cock AWAM (1998) Identification of some oomycetes by reverse dot-blot hybridization. *Phytopathology* 88: 213–222.

Lévesque CA, Vrain TC, De Boer SH (1994) Development of a species-specific probe for *Pythium ultimum* using amplified ribosomal DNA. *Phytopathology* 84: 474–478.

Li AY, Crone M, Adams PJ, Fenwick SG, Hardy GESJ, Williami N (2014) The microscopic examination of *Phytophthora cinnamomi* in plant tissues using fluorescent in situ hybridization. *J Phytopathol* 162: 747–757.

Li D, Liu P, Xie S, Yin R, Weng Q, Chen Q (2015) Specific and sensitive detection of *Phytophthora nicotianae* by nested PCR and loop-mediated isothermal amplification assays. *J Phytopathol* 163: 185–193.

Li M, Asano T, Suga H, Kageyama K (2011) A multiplex PCR for the detection of *Phytophthora nicotianae* and *P. cactorum* and a survey of their occurrence in strawberry production areas of Japan. *Plant Dis* 95: 1270–1278.

Li M, Ishiguro Y, Otsubo K et al. (2014) Monitoring by real-time PCR of three water-borne zoosporic *Pythium* species in potted flower and tomato greenhouses under hydroponic culture systems. *Eur J Plant Pathol* 140: 229–242.

Li M, Jia R, Na R et al. (2016) Genetic diversity of *Sclerotinia sclerotiorum* within a single sunflower field in Wenquan, Xinjiang province, China. *J Plant Pathol* 98: 43–53.

Li M, Shi J, Xie X et al. (2013) Identification and application of a unique genetic locus in diagnosis of *Fusarium oxysporum* f.sp. *cubense* tropical race 4. *Canad J Plant Pathol* 35: 482–493.

Li MH, Yang B, Leng Y, Chao CP, Liu JM, He ZF, Jiang ZD, Zhong S (2011) Molecular characterization of *Fusarium oxysporum* f.sp. *cubense* race 1 and 4 isolates from Taiwan and Southern China. *Canad J Plant Pathol* 33: 168–178.

Li S, Harman GL (2003) Molecular detection of *Fusarium solani* f.sp. *glycines* in soybean roots and soil. *Plant Pathol* 62: 74–83.

Li S, Tam YK, Harman GL (2000) Molecular differentiation of mictochondrial subunit rDNA sequences. *Phytopathology* 90: 491–497.

Liebe S, Christ DS, Ehricht R, Varrelmann M (2016) Development of a DNA microarray-based assay for detection of sugar beet root rot pathogens. *Phytopathology* 106: 76–86.

Lievens B, Brouwer M, Vanachter ACRC, Lévesque CA, Cammue BPA, Thomma BPHJ (2005) Quantitative assessment of phytopathogenic fungi in various substrates using DNA microarray. *Environ Microbiol* 7: 1698–1710.

Lievens B, Brouwere M, Vanachter ACRC, Lévesque CA, Cammue BPA, Thomma BPHJ (2003) Design and development of a DNA array for rapid detection and identification of multiple tomato vascular wilt pathogens. *FEMS Microbiol Lett* 223: 113–122.

Lievens B, Claes L, Vakalounakis DJ, Vanachter ACRC, Thomma BPHJ (2007) A robust idenficaion and detection assay to discriminate the cucumber pathogens *Fusarium oxysporum* f.sp. *cucumerinum* and f.sp. *radicis-cucumerinum*. *Environ Microbiol* 9: 2145–2161.

Lievens B, Claes L, Vanachter ACRC, Bruno PA, Cammue BPA, Thomma BPHJ (2006) Detecting single nucleotide polymorphisms using DNA arrays for plant pathogen diagnosis. *FEMS Microbiol Lett* 255: 129–139.

Liew ECY, Maclean DJ, Irwin JAG (1998) Specific PCR-based detection of *Phytophthora medicaginis* using the intergenic spacer region of the ribsosomal DNA. *Mycol Res* 102: 73–80.

Lin Y-H, Chang J-Y, Liu E-T, Chao C-P, Huang J-W, Chang P-FL (2009) Development of a molecular marker for species specific detection of *Fusarium oxysporum* f.sp. *cubense* race 4. *Eur J Plant Pathol* 123: 353–365.

Lin Y-H, Su C-C, Chao C-P et al. (2013) A molecular diagnosis method using real-time PCR for quantification and detection of *Fusarium oxysporum* f.sp. *cubense* race 4. *Eur J Plant Pathol* 135: 395–405.

Loyd AL, Benson DM, Ivors KL (2014) *Phytophthora* populations in nursery irrigation water in relationship to pathogenicity and infection frequency of *Rhododendron* and *Pieris*. *Plant Dis* 98: 1213–1220.

Lu C, Song B, Zhang H-F, Wang Y-C, Zheng X-B (2015) Rapid diagnosis of soybean seedling blight caused by *Macrophomina phaseolina* using LAMP assays. *Phytopathology* 105: 1612–1617.

Lübeck M, Poulsen H (2001) UP-PCR cross blot hybridization as a tool for identification of anastomosis groups in the *Rhizoctonia solani* complex. *FEMS Microbiol Lett* 201: 83–89.

Lyons NF, White JG (1992) Detection of *Pythium violae* and *Pythium sulcatum* in carrots with cavity spots using competition ELISA. *Ann Appl Biol* 120: 235–244.

Mahuku GS, Platt (Bud) HW (2002) Quantifying *Verticillium dahliae* in spoils collected from potato fields using a competitive PCR assay. *Amer J Potato Res* 79: 107–117.

Malvick DK, Impullitti AE (2007) Detection and quantification of *Phialophora gregata* in soybean and soil samples with quantitative, real-time PCR assay. *Plant Dis* 91: 736–742.

Mammella MA, Martin FN, Cacciola SO, Coffey MD, Fedda R, Schena L (2013) Analyses of the population structure in a global collection of *Phytophthora nicotianae* isolates inferred from mitochondrial and nuclear DNA sequences. *Phytopathology* 103: 610–622.

Mansoori B (2011) An improved ethanol medium for efficient recovery and estimation of *Verticillium dahliae* population in soil. *Canad J Plant Pathol* 33: 88–93.

Markakis EA, Tjamos SE, Antoniou PP, Paplomatas EJ, Tjamos EC (2009) Symptom development, pathogen isolation and real-time QPCR qunatificatin as factors for evaluating the resistance of olive cultivars to *Verticillium* pathotypes. *Eur J Plant Pathol* 124: 603–611.

Martin B (1987) Rapid tentative identification of *Rhizoctonia* spp. associated with diseased turf grasses. *Plant Dis* 71: 47–49.

Martin FN (2010) Mitochondrial haplotype analysis as a tool for differentiating isolates of *Verticillium dahliae*. *Phytopathology* 100: 1231–1239.

Martin FN, Coffey MD (2012) Mitochondrial haplotype analysis for differentiation of isolates of *Phytophthora cinnamomi*. *Phytopathology* 102: 229–239.

Martínez-Ferri E, Zumaquero A, Ariza MT, Barcelo A, Pliego C (2016) Nondestructive detection of white root rot disease in avocado rootstocks by leaf chlorophyll fluorescence. *Plant Dis* 100: 49–58.

Maruthachalam K, Atallah ZK, Vallad GE, Klosterman SJ, Hayes RJ, Davis RM, Subbarao KV (2010) Molecular variation among isolates of *Verticillium dahliae* and polymerase chain reaction-based differentiation of races. *Phytopathology* 100: 1222–1230.

Masago H, Yoshikawa M, Fukuda M, Naknishi N (1977) Selective inhibition of *Pythium* spp. on a medium for direct isolation of *Phytophthora* spp. from soils and plants. *Phytopathology* 67: 425–428.

Mathew FM, Lamppa RS, Chittem K, Chang YW, Botschner M, Kinzer K, Goswami RS, Markell SG (2012) Characterization and pathogenicity of *Rhizoctonia solani* isolates affecting *Pisum sativum* in North Dakota. *Plant Dis* 96: 666–672.

Matsumoto M, Matsuyama N (1998) Trials of identification of *Rhizoctonia solani* AG1-IA, the causal agent of rice sheath rot disease using specifically primed PCR analysis in diseased plant tissues. *Bull of Inst Trop Agric Kyushu Univ* 21: 27–32.

Matthew JS, Brooker JD (1991) The isolation and characterization of polyclonal and monoclonal antibodies to anastomosis group 8 of *Rhizoctonia solani*. *Plant Pathol* 40: 67–77.

Mattupalli C, Genger RK, Charkowski AO (2013) Evaluating incidence of *Helminthosporium solani* and *Colletotrichum coccodes* on asymptomatic organic potatoes and screening potato lines for resistance to silver scurf. *Amer J Potato Res* 90: 369–377.

Meadows LM, Zwart DC, Jeffers SN, Waldrop TA, Bridges Jr. WC (2011) Effects of fuel reduction treatments on incidence of *Phytophthora* species in soil of a Southern Appalachian Mountain forest. *Plant Dis* 95: 811–820.

Mehl HL, Epstein L (2007) Identification of *Fusarium solani* f.sp. *cucurbitae* race 1 and race 2 with PCR and production of disease-free pumpkin seeds. *Plant Dis* 91: 1288–1292.

Meng J, Wang Y (2010) Rapid detection of *Phytophthora nicotianae* in infected tobacco tissues and soil samples based on its *Ypt1* gene. *J Phytopathol* 158: 1–7.

Mercado-Blanco J, Rodriguez-Jurado, Parrilla-Araujo S, Jimenez-Díaz RM (2003) Simultaneous detection of the defoliating and nondefoliating *Verticillium dahliae* pathotypes in infected olive plants by duplex, nested PCR. *Plant Dis* 87: 1487–1494.

Merz U, Walsh JA, Bouchek-Mechiche K, Oberhänshi Th, Bitterlin W (2005) Improved immunological detection of *Spongospora subterranea*. *Eur J Plant Pathol* 111: 371–379.

Miles TD, Martin FN, Coffey MD (2015) Development of rapid isothermal amplification assays for detection of *Phytophthora* spp. in plant tissue. *Phytopathology* 105: 265–278.

Miller SA, Bhat RG, Scmitthenner (1994) Detection of *Phytophthora capsici* in pepper and cucurbit crops in Ohio with two commercial immunoassay kits. *Plant Dis* 78: 1042–1046.

Minerdi D, Moretti M, Li Y, Gaggero L, Garibaldi A, Gullino ML (2008) Conventional PCR and real-time quantitative PCR detection of *Phytophthora cryptogea* on *Gerbera jamesonii*. *Eur J Plant Pathol* 122: 227–237.

Miyake N, Nagai H, Kato S, Matsusaki M, Fukuta S, Takahashi R, Suzuki R, Ishiguro Y (2017) Practical method combining loop-mediated isothermal amplification and bait trap to detect *Pythium helicoides* in hydroponic culture solutions. *J Gen Plant Pathol* 83: 1–6.

Moine LM, Labbe C, Louis-Seize G, Seifer KA, Belanger RR (2014) Identification and detection of *Fusarium striatum* as a new record of pathogen to greenhouse tomato in Northeastern America. *Plant Dis* 98: 292–298.

Morita Y, Tojo M (2007) Modifications of PARP medium using fluazinam, miconazole and nystatin for detection of *Pythium* spp. in soil. *Plant Dis* 91: 1591–1599.

Morrica S, Ragazzi A, Kasuga T, Mitchelson KR (1998) Detection of *Fusarium oxysporum* f.sp. *vasinfectum* in cotton tissue by polymerase chain reaction. *Plant Pathol* 47: 486–494.

Moukhamedov R, Hu X, Nazar RN, Robb J (1994) Use of polymerase chain reaction amplified ribosomal intergenic sequences for the diagnosis of *Verticillium tricorpus*. *Phytopathology* 84: 256–259.

Mumford R, Boonham N, Tomlinson J, Barker J (2006) Advances in molecular phytodiagnostics–new solutions for old problems. *Eur J Plant Pathol* 116: 1–19.

Nagamine K, Hase H, Notomi T (2002) Accelerated reaction by loop-mediated isothermal amplification among loop primers. *Mol Cell Probes* 16: 223–229.

Nagamine K, Watanabe K, Ohtsuka K, Hase T, Notomi T (2001) Loop-mediated isothermal amplification reaction using non-denatured template. *Clin Chem* 47: 1742–1743.

Nahiyan ASM, Boyer LR, Jeffries P, Matsubara Y (2011) PCR-SSCP analysis of *Fusarium* diversity in asparagus decline in Japan. *Eur J Plant Pathol* 130: 191–203

Nalim FA, Starr JL, Woodard KE, Segner S, Keller NP (1995) Mycelial compatibility groups in Texas peanut field populations of *Sclerotium rolfsii*. *Phytopathology* 85: 1507–1512.

Nanayakkara UN, Singh M, Al-Mughrabi KI, Peters RD (2009) Detection of *Phytophthora erythroseptica* in above ground potato tissues, progeny tubers, stolons and crop debris using PCR techniques. *Amer J Potato Res* 86: 239

Nanayakkara UN, Singh M, Al-Mughrabi KI, Peters RD (2010) Detection of *Phytophthora erythroseptica* in soil using nightshade as bait combined with PCR techniques. *Amer J Potato Res* 87: 67–77.

Narayanasamy P (2001) *Plant Pathogen Detection and Disease Diagnosis*, Second edition, Marcel Dekker, New York.

Narayanasamy P (2005) *Immunology in Plant Health and Its Impact on Food Safety*. The Haworth Press, New York.

Narayanasamy P (2011) *Microbial Plant Pathogens–Detection and Disease Diagnosis*, Volume 1, Springer Science + Business Media B. V., Heidelberg, Germany.

Newton AC, Duncan JM, Augustin WH, Guy DC, Cooke DEL (2010) Survival, distribution and genetic variability of inoculum of the strawberry red core. *Plant Pathol* 59: 472–479.

Niepold F, Schöber-Butin G (1995) Application of PCR technique to detect *Phytophthora infestans* in potato tubers and leaves. *Microbiol Res* 150: 379–385.

Niepold F, Stachewicz H (2004) PCR-detection of *Synchytrium endobioticum* (Schub) Perc. *J Plant Dis Protect* 111: 313–321.

Nilsson HE (1995) Remote sensing and image analysis in plant pathology. *Annu Rev Phytopathol* 33: 489–527.

Nilsson M, Malmgren H, Samiotaki M, Kwiatkowski M, Chowdhary BP, Landegren U (1994) Padlock probes: Circularizing oligonucleotides for localized DNA detection. *Science* 265: 2085–2088.

Njambere EN, Chen W, Frate C, Wu B-M, Temple SR, Muehlbauer FJ (2008) Stem and crown rot of chickpea in California caused by *Sclerotinia sclerotiorum*. *Plant Dis* 92: 917–922.

Notomi T, Okayama H, Masubuchi H, Yonekawa T, Watanabe K, Amino N, Hase T (2000) Loop-mediated isothermal amplification of DNA. *Nucleic Acids Res* 28: E63.

Novakazi F, Inderbitzin P, Sandoya G, Hayes RJ, von Tiedemann A, Subbarao KV (2015) The three linkages of the diploid hybrid *Verticillium longisporum* differ in virulence and pathogenicity. *Phytopathology* 105: 662–673.

O'Donell K (1992) Ribosomal DNA internal transcribed spacers are highly divergent in the phytopathogenic ascomycete *Fusarium sambucinum*. *Curr Genet* 22: 213–220.

O'Donnell K, Sink S, Scandiani MM, Luque A, Colletto A, Biasoli M, Lenzi L, Salas G, Onzález VG, Daniel Ploper L, Formento N, Pioli RN, Aoki T, Yang XB, Sarver BAJ (2010) Soybean suddet death syndrome species diversity within North and South America revealed by multilocus genotyping. *Phytopathology* 100: 58–71.

Okubara PA, Harrison LA, Gatch EW, Vandemark G, Schroeder KL, du Toit LJ (2013) Development and evaluation of a TaqMan real-time PCR assay for *Fusarium oxysporum* f.sp. *spinaciae*. *Plant Dis* 97: 927–937.

Okubara PA, Schroeder KL, Paulitz TC (2008) Identification and quantification of *Rhizoctonia solani* and *R. oryzae* using real-time polymerase chain reaction. *Phytopathology* 98: 837–847.

Olsson CHB, Heiberg N (1997) Sensitivity of the ELISA test to detect *Phytophthora fragariae* var *rubi* in raspberry roots. *J Phytopathol* 145: 285–288.

Omer MA, Johnson DA, Douhan LI, Hamm PB, Rowe RC (2008) Detection, quantification and vegetative compatibility of *Verticillium dahliae* in potato and mint production soils in the Columbia basin of Oregan and Washington. *Plant Dis* 92: 1127–1131.

Orihara S, Yamammoto T (1998) Detection of resting spores of *Plasmodiophora brassicae* from soil and plant tissues by enzyme-linked immunoassay. *Ann Phytopathol Soc Jpn* 64: 569–573.

Ortiz CS, Bell AA, Magill CW, Liu J (2017) Specific PCR detection of *Fusarium oxysporum* f.sp. *vasinfectum* California race 4 based on a unique *Tfo1* insertion event in the *PHO* gene. *Plant Dis* 101: 34–44.

Oudemans P, Coffey MD (1991a) Isozyme comparison within and among worldwide sources of three morphologically distinct species of *Phytophthora*. *Mycol Res* 95: 19–30.

Oudemans P, Coffey MD (1991b) A revised systematic of twelve papillate *Phytophthora* species based on isozyme analysis. *Mycol Res* 95: 1025–1046.

Palmero D, Iglesias C, de Cara M, Lomas T, Santos M Tello JC (2009) Species of *Fusarium* isolated from river and sea water of southern Spain and pathogenicity on four plant species. *Plant Dis* 93: 377–385.

Papaioannou IA, Typas MA (2015) High-throughput assessment and genetic investigation of vegetative compatibility in *Verticillium dahliae*. *J Phytopathol* 163: 475–485.

Papaioannou K, Ligoxigakis EK, Vakalounakis DJ, Markakis EA, Typas M (2013) Phytopathogenic, morphological, genetic and molecular characterization of a *Verticillium dahliae* population from Crete, Greece. *Eur J Plant Pathology* 136: 577–596.

Parkunan V, Ji P (2013) Isolation of *Pythium literole* from irrigation ponds used for vegetable production and its pathogenicity on squash. *Canad J Plant Pathol* 35: 415–423.

Parlevliet JE (1993) What is durable resistance?–a general outline. In: Jacobs Th, Parlevliet JE (eds.), *Durability of Disease Resistance*, Kluwer Academic Publishers, Dordrecht, The Netherlands, pp. 23–35.

Pasche JS, Mallik, I, Anderson NR, Gudmestad NC (2013) Development and validation of a real-time PCR assay for the quantification of *Verticillium dahliae* in potato. *Plant Dis* 97: 608–618.

Pasquali M, Dematheis F, Gullino ML, Garibaldi A (2007) Identification of race 1 of *Fusarium oxysporum* f.sp. *lactucae* on lettuce by inter-retrotransposon sequence characterized amplified region technique. *Phytopathology* 97: 987–996.

Pasquali M, Piatti P, Gullino ML, Garibaldi A (2006) Development of real-time polymerase chain reaction for detection of *Fusarium oxysporum* f.sp. *basilici* from basil seed and roots. *J Phytopathol* 154: 632–636.

Patel JS, Brennan MS, Khan A, Ali GS (2015) Implementation of loop-mediated isothermal amplification methods in lateral flow devices for the detection of *Rhizoctonia solani*. *Canad J Plant Pathol* 37: 118–129.

Pathrose B, Jones EI, Jaspers MV, Ridgway HJ (2014) High genotypic and virulence diversity in *Ilyonectria liriodendra* associated with black foot disease in New Zealand vineyards. *Plant Pathol* 63: 613–624.

Patzak J (2003) PCR detection of hop fungal pathogen. *Proceedings of* International Hop Growers *Convention*, Dobrna-Zalec, Slovania, pp. 1–16.

Paulitz TC, Schroeder KL (2005) A new method for the quantification of *Rhizoctonia solani* and *R. oryzae* from soil. *Plant Dis* 89: 767–772.

Paulitz TC, Zhang H, Cook RJ (2003) Spatial distribution of *Rhizoctonia oryzae* and Rhizoctonia root rot in direct-seeded cereal. *Canad J Plant Pathol* 25: 295–303.

Pavon CF, Babadoost M, Lambert KN (2008) Quantification of *Phytophthora capsici* oospores in soil by sieving-centrifugation and real-time polymerase chain reaction. *Plant Dis* 92: 143–149.

Pekarova B, Krátka J, Slováček J (2001) Utilization of immunological methods to detect *Phytophthora fragariae* in strawberry plants. *Plant Protect Sci* 37: 57–65.

Pérez-Sierra A, León M, Álvarez LA et al. (2010) Outbreak of a new *Phytophthora* sp. associated with severe decline of almond trees in eastern Spain. *Plant Dis* 94: 534–541.

Pettitt TR, Wainwright MF, Wakeham AJ, White JG (2011) A simple detached leaf assy provides rapid and inexpensive determination of pathogenicity of *Pythium* isolates to 'all year round' (AYR) chrysanthemum roots. *Plant Pathol* 60: 946–956.

Pettitt TR, Wakeham AJ, Wainwright MF, White JG (2002) Comparison of serological, culture and bait methods for detection of *Pythium* and *Phytophthora* zoospores in water. *Plant Pathol* 51: 720–727.

Picó B, Roig C, Fita A, Nuez F (2008) Quantitative detection of *Monosporoascus* in infected melon roots using real-time PCR. *Eur J Plant Pathol* 120: 147–156.

Platiño-Álvarez B, Rodríguez-Cámara MC, Rodríguez-Fernandez T, González-Jaen MT, Vazquez-Estévez (1999) Immunodetection of an exopolygalacturonase in tomato plants infected with *Fusarium oxysporum* f.sp. *radicis-lycopersici*. *BolSandad Vegetal Plagas* 25: 529–536.

Platt (Bud) HW, Mahuku G (2001) Detection methods for *Verticillium* species in naturally infested and inoculated soils. *Amer J Potato Res* 77: 271–274.

Puhalla JE (1985) Classification of strains of *Fusarium oxysporum* on the basis of vegetative compatibility. *Canad J Bot* 63: 179–183.

Qin Z, Zhang M (2000) Detection of rice sheath blight for in-season disease management using multispectral remote sensing. *International Applied Earth Observatory of Geoinformation* 7: 115–128.

Qiu R, Qu D, Hardy GE St. J, Trengove R, Agarwal M, Ren Y (2014) Optimization of head space solid-phase microextraction conditions for the identification of *Phytophthora cinnamomi* Rands. *Plant Dis* 98: 1088–1098.

Qiu RQD, Trengove R, Agarwal M, Hardy E. St, Ren Y (2014) Head-space solid-phase microextraction and gas chromatography-mass spectrometry for analyses of VOCs produced by *Phytophthora cinnamomi*. *Plant Dis* 98: 1099–1015.

Qu X, Christ BJ (2001) Single cystosorus isolate production and restriction fragment length polymorphism characterization of the obligate biotroph *Spongospora subterranea* f.sp. *subterranea*. *Phytopathology* 96: 1157–1163.

Qu X, Kavanagh JA, Egan D, Christ BJ (2006) Detection and quantification of *Spongospora subterranea* f.sp. *subterranea* by PCR in host tissues and naturally infested soils. *Amer J Potato Res* 83: 21–40.

Rachdawong S, Cramer CL, Grabau EA, Stromberg VK, Lacy GH, Stromberg EL (2002) *Gaeumannomyces graminis* vars. *avenae*, *graminis* and *tritici* identified using PCR amplification of avinacinase-like genes. *Plant Dis* 86: 652–660.

Radisĕk S, Jakše J, Javornik B (2004) Development of pathotype-specific SCAR markers for detection of *Verticillium alboatrum* isolates from hop. *Plant Dis* 88: 1115–1122.

Rai M, Ingle A (2012) Role of nanotechnology in agriculture with special referene to management of insect pests. *Appl Microbiol Biotechnol* 94: 287–293.

Remesal E, Land BB, Jiménez-Gasco MDM, Navas-Cortés JA (2013) Sequence variation in two protein-coding genes correlates with mycelia compatibility groupings in *Sclerotinia rolfsii*. *Phytopathology* 103: 479–487.

Reynolds GJ, Windels CE, MacRae IV Laguette S (2012) Remote sensing for assessing Rhizoctonia crown and root rot severity in sugar beet. *Plant Dis* 96: 497–505.

Robold AV, Hardham AR (1998) Production of species-specific antibodies that react with surface components on zoospores and cysts of *Phytophthora nicotianae*. *Canad J Microbiol* 44: 1161–1170.

Rocha DC, Oliveira MB, de Freitas mA, Petrofeza S (2017) Rapid detection of *Macrophomina phaseolina* in common bean seeds, using a visual loop-mediated isothermal amplification assay. *Austr Plant Pathol* 46: 205–217.

Rollins L, Coats K, Elliott M, Chastagner G (2016) Comparison of five detection and quantification methods for *Phytophthora ramorum* in stream and irrigation water. *Plant Dis* 100: 1202–1211.

Ruano-Rosa D, Schena L, Ippolito A, López-Herrera CI (2007) *Rosellinia necatrix* in avocado orchards in souther Spain. *Plant Pathol* 56: 251–256.

Safarpour H, Safarnejad MR, Tabatabaei M et al. (2012) Development of a quantum dots FRET-based biosensor for efficient detection of *Polymyxa betae*. *Canad J Plant Pathol* 34: 507–515.

Salazar O, Julian MC, Rubio V (2000) Primers based on specific rDNA-ITS sequence for PCR detection of *Rhizoctonia solani*, *R. solani* AG-2 subgroups and ecological types and binucleate *Rhizoctonia*. *Mycol Res* 104: 281–285.

Sánchez-Hernández E, Muñoz-Garcia M, Brasier CM, Trapero-Casas A (2001) Identify and pathogenicity of two *Phytophthora* taxa associated with a new root disease of live trees. *Plant Dis* 85: 411–416.

Sasaki K, Nakahara K, Shigyo M, Tanaka S, Ito S-i (2015) Detection and quantification of onion isolates of *Fusarium oxysporum* f.sp. *cepae* in onion plant. *J Gen Plant Pathol* 81: 232–236.

Sasaki K, Nakahara K, Tanaka S, Shigyo M, Shin-ichilto (2015) Genetic and pathogenic variability of *Fusarium oxysporum* f.sp. *cepae* isolated from onion and Welsh onion in Japan. *Phytopathology* 105: 525–532.

Sayler RJ, Yang Y (2007) Detection and quantification of *Rhizoctonia solani* AG-1IA, the rice sheath blight pathogen in rice using real-time PCR. *Plant Dis* 91: 1663–1668.

Scarlett K, Tesoriero L, Daniel R, Guest D (2013) Detection and quantification of *Fusarium oxysporum* f.sp. *cucumerinum* in environmental samples using a specific quantitative PCR assay. *Eur J Plant Pathol* 137: 315–324.

Schaad NW, Frederick RD, Shaw J et al. (2003) Advances in molecular-based diagnostics in meeting crop biosecurity and phytosanitary issues. *Annu Rev Phytopathol* 41: 305–324.

Schena L, Hughes KJ, Cooke DEL (2006) Detection and quantification of *Phytophthora ramorum*, *P. kernoviae*, *P. citricola* and *P. quercina* in symptom leaves by multiplex real-time PCR. *Molec Plant Pathol* 7: 365–379.

Schena L, Nigro F, Ippolito A (2002) Identification and detection of *Rosellinia necatrix* by conventional and real-time Scorpion-PCR. *Eur J Plant Pathol* 108: 355–366.

Schlenzig A, Cooke DEL, Chard JM (2005) Comparison of a baiting method and PCR for the detection of *Phytophthora fragariae* var. *rubi* in certified raspberry stocks. *EPPO Bull* 35: 87–91.

Schroeder KL, Okubara PA, Tambong JT, Lévesque CA, Paulitz TC (2006) Identification and quantification of pathogenic *Pythium* spp. from soils in eastern Washington using real-time polymerase chain reaction. *Phytopathology* 96: 637–647.

Schulze S, Bahnweg G (1998) Critical review of identification techniques for *Armillaria* spp. and *Heterobasidium annosum* root and butt rot diseases. *J Phytopathol* 146: 61–72.

Schulze S, Koch H-J, Märländer B, Varrelmann M (2016) Effect of sugar beet variety and nonhost plant on *Rhizoctonia solani* AG2-2IIIB soil inoculum potential measured in soil DNA extracts. *Phytopathology* 106: 1047–1054.

Sharma-Poudyal D, Paulity TC, Porter LD, du Toit LJ (2015) Characterization and pathogenicity of *Rhizoctonia* and *Rhizoctonia*-like spp. from pea crops in the Columbia Basin of Oregon and Washington. *Plant Dis* 99: 604–613.

Shen D, Li Q, Yu J, Zhao Y, Zhu Y, Xu H, Dou D (2017) Development of a loop-mediated isothermal amplification method for rapid detection of *Pythium ultimum*. *Austr Plant Pathol* 46: 571–576.

Shen W, Xu G, Sun L, Zhang L, Jiang Z (2016) Development of a LAMP assay for rapid and sensitive detection of *Sporisorium scitamineum* in sugarcane. *Ann Appl Biol* 168: 321–327.

Shishido K, Murakami H, Kanda D, Fuji S, Toda T, Furuya H (2016) Effect of soil inoculum density and temperature on the incidence of cucumber black root rot. *Plant Dis* 100: 125–130.

Shoemaker DD, Lashkari DA, Morris D, Mittmann M, Davis RW (1996) Quantitative phenotypic analysis of yeast deletion mutants using highly parallel molecular bar-coding strategy. *Nature Genet* 14: 450–456.

Short DPG, Gurung S, Koike ST, Klosterman SJ, Subbarao KV (2015) Frequency of *Verticillium* species in commercial spinach fields and transmission of *V. dahliae* from spinach subsequent lettuce crops. *Phytopathology* 105: 80–90.

Silvar C, Diaz J, Merino F (2005) Real-time polymerase chain reaction quantification of *Phytophthora capsici* in different pepper genotypes. *Phytopathology* 95: 1423–1429.

Silvar C, Duncan JM, Cooke DEL, Williams NA, Merino F (2005) Development of specific primers for identification and detection of *Phytophthora capsici* León. *Eur J Plant Pathol* 112: 43–52.

Sippel DN, Hall R (1995) Glucose phosphate isomerase polymorphisms distinguishes weakly virulent from highly virulent strains of *Leptoshpaeria maculans*. *Canad J Plant Pathol* 17: 1–6.

Small J, Call DA, Brockman FJ, Straub TM, Chandler DP (2001) Direct detection of 16S rRNA in soil extracts by using oligonucleotide microarrays. *Appl Environ Microbiol* 67: 4708–4716.

Somai BM, Keinath AP, Dean RA (2002) Development of PCR-ELISA detection and differentiation of *Didymella bryoniae* from related *Phoma* species. *Plant Dis* 86: 710–716.

Southwood MJ, Vilojen A, Mostert L, Rose LJ, McLeod A (2012) Phylogenetic and biological characterization of *Fusarium oxysporum* isolates associated with onion in South Africa. *Plant Dis* 96: 1250–1261.

Spies CFJ, Mazzola M, McLeod A (2011) Characterization and detection of *Pythium* and *Phytophthora* species associated with grapevines in South Africa. *Eur J Plant Pathol* 131: 103.

Spurlock TN, Rothrock CS, Monfort WS (2015) Evaluation of methods to quantify populations of *Rhizoctonia solani* in soil. *Plant Dis* 99: 836–841.

Stevens Johnk J, Jones RK (2001) Differentiation of three homogeneous groups of *Rhizoctonia solani* anastomosis group 4 by analysis of fatty acids. *Phytopathology* 91: 821–830.

Strausbaugh CA, Eujayl IA, Martin FN (2016) Pathogenicity, vegetative compatibility and genetic diversity of *Verticillium dahliae* isolates from suar beet. *Canad J Plant Pathol* 38: 492–505.

Suga H, Hirayama Y, Morishima H, Suzuki T, Kageyama K (2013) Development of PCR primers to identify *Fusarium oxysporum* f.sp. *fragariae*. *Plant Dis* 97: 619–625.

Sullivan MJ, Parks EJ, Cubeta MA, Gallup CA, Melton TA, Moyer JW, Shew HD (2010) An assessment of the genetic diversity in a field population of *Phytophthora nicotianae* with a changing race structure. *Plant Dis* 94: 455–460.

Sundelin T, Christensen CB, Larsesn J, Moller K, Lübeck M, Bodker L, Jensen B (2010) In planta quantification of *Plasmodiophora brassicae* using signature fatty acids and real-time PCR. *Plant Dis* 94: 432–438.

Sundaram S, Plascencia S, Banttari EE (1991) Enzyme-linked immunosorbent assay for detection of *Verticillium* spp., using antisera produced to V. *dahliae* from potato. *Phytopathology* 81: 1485–1489.

Suo B, Xiao C, Lin He et al. (2015) Development of quantification bioassays for live propagules of *Phytophthora sojae* in soil. *J Gen Plant Pathol* 81: 271–278.

Sutton W, Hansen EM, Reeser PW (2009) Stream monitoring for detection for *Phytophthora ramorum* in Oregon tanoak forests. *Plant Dis* 93: 1182–1186.

Szemes M, Bonants P, de Weerdt M, Baner J, Landegren U, Schoen CD (2005) Diagnostic application of padlock probes—multiplex detection of plant pathogens using universal microarrays. *Nucleic Acids Res* 33: e70, pp 1–13.

Takamatsu S, Nakano M, Yokoto H, Kumoh (1998) Detection of *Rhizoctonia solani* AG2-2-IV, the causal agent of large patch of *Zoysia* grass using plasmid DNA as a probe. *Ann Phytopathol Soc Jpn* 64: 451–457.

Tambong JT, de Cock AWAM, Tinker NA, Lévesque CA (2006) Oligonucleotide array for identification and detection of *Pythium* species. *Appl Environ Microbiol* 72: 2691–2706.

Tardi-Ovadia R, Linker R, Tsror L (Lakhim) (2017) Direct estimation of local pH change at infection sites of fungi in potato tubers. *Phytopathology* 107: 132–137.

Tewoldemedhin YT, Lamprecht SC, Vaughan MM, Doehring G, O'Donnell K (2017) Soybean SDS in South Africa is caused by *Fusarium brasiliense* and a novel undescribed *Fusarium* sp. *Plant Dis* 101: 150–157.

Tewoldemedhin YT, Mazzola M, Mostert L, McLeod A (2011) *Cylindrocarpon* species associated with apple tree roots in South Africa and their quantification using real-time PCR. *Eur J Plant Pathol* 129: 637–651.

Themann K, Werres S, Diener HA, Luttmann R (2002) Comparison of different methods to detect *Phytophhora* spp. in recycling water from nurseries. *J Plant Pathol* 84: 41–50.

Thompson JP, Clewatt TG, Jennings RE, Sheady JG, Owen KJ, Persley DM (2011) Detection of *Polymyxa graminis* in a barley crop in Australia. *Austr Plant Pathol* 40: 66–75.

Thornton CR, Dewey FM, Gilligan CA (1993) Development of monoclonal antibody-based immunological assays for the detection of live propagules of *Rhizoctonia solani* in soils. *Plant Pathol* 42: 763–773.

Thornton CR, Groenhof AC, Forrest R, Lamotte R (2004) A one-step immunochromatographic lateral flow device specific to *Rhizoctonia solani* and certain related species and its use to detect and quantify *R. solani* in soil. *Phytopathology* 94: 280–288.

Thornton CR, O'Neill TM, Hilton G, Gilligan CA (1999) Detection and recovery of *Rhizoctonia solani* in naturally infested field soils using a combined baiting double monoclonal antibody ELISA. *Plant Pathol* 48: 627–634.

Timmer LW, Menge JA, Zitko SE, Pond E, Miller SA, Johnson EL (1993) Comparison of ELISA techniques and standard isolation methods for *Phytophthora* detection in citrus orchards in Florida and California. *Plant Dis* 77: 791–796.

Tomlinson JA, Dickinson MJ, Boonham N (2010) Rapid detection of *Phytophthora ramorum* and *P. kernoviae* by two-minute DNA extraction followed by isothermal application and amplification detection by generic lateral flow device. *Phytopathology* 100: 143–149.

Tooley PW, Carras MM, Lamber DH (1998) Application of a PCR-based test for detection of potato late blight and pink rot in tubers. *Amer J Potato Res* 75: 187–194.

Tooley PW, Martin FN, Carras MM, Frederick RD (2006) Real-time fluorescent polymerase chain reaction detection of *Phytophthora ramorum* and *Phytophthora pseudosyringae* using mitochondrial gene regions. *Phytopathology* 96: 336–345.

Troisi M, Bertetti D, Gullino ML, Garibaldi A (2013) Race differentiation in *Fusarium oxyporum* f.sp. *chrysanthemi*. *J Phytopathol* 161: 675–688.

Usami T, Momma N, Kikuchi S et al. (2017) Race 2 of *Verticillium dahliae* infecting tomato in Japan can be split into two races with differential pathogenicity on resistant rootstocks. *Plant Pathol* 66: 230–238.

Vaïanopoulos CV, Bragard C, Moreau V, Maraite H, Legréve A (2007) Identification and quantification of *Polymyxa graminis* f.sp. *temperata* and *P. graminis* f.sp. *tepida* on barley and wheat. *Plant Dis* 91: 857–864.

Vakalounakis DJ, Doulis AG, Klironomou E (2005) Characterization of *Fusarium oxysporum* f.sp. *radicis-cucumerinum* attacking melon under natural conditions in Greece. *Plant Pathol* 54: 339–346.

van de Graaf P, Lees AK, Cullen DW, Duncan JM (2003) Detection and quantification of *Spongospora subterranea* in soil, water and plant tissue samples using real-time PCR. *Eur J Plant Pathol* 109: 589–597.

Van de Graaf P, Lees AK, Wales SJ, Duncan JM (2005) Effect of soil inoculum level and environmental factors on potato powdery scab caused by *Spongospora subterranea*. *Plant Pathol* 54: 22–28.

van den Boogert PHJF, van Gen-Pelzer MPE, Bonants PJM et al. (2005) Development of PCR-based detection methods for quarantine phytopathogen *Synchytrium endobioticum*, causal agent of potato wart disease. *Eur J Plant Pathol* 113: 47–57.

van Doorn R, Szemes M, Bonants P, Kowalchuk GA, Salles JF, Ortenberg E, Schoen CD (2007) Quantitative multiplex detection of plant pathogens using a novel ligation-based system coupled with universal high-throughput real-time PCR on Open ArraysTM. *BMC Genomics* 8: 276

van Gent-Pelzer MPE, Krijger M, Bonants PJM (2010) Improved real-time PCR assay for detection of the quarantine potato pathogen *Synchytrium endobioticum* in zonal centrifuge extracts from soil and in plants. *Eur J Plant Pathol* 126: 129–133.

Van Regenmortel MHV (1982) *Serology and Immunochemistry of Plant Viruses*. Academic Press, New York.

Vandemark GJ, Barker BM (2003) Quantifying *Phytophthora medicaginis* in susceptible and resistant alfalfa with real-time fluorescent assay. *J Phytopathol* 151: 577–583.

Vandemark GJ, Barker, Gristeno MA (2002) Quantifying *Aphanomyces euteiches* in alfalfa with a fulorescent polymerase chain reaction assay. *Phytopathology* 92: 265–272.

Vandemark GJ, Grünwald NJ (2005) Use of real-time PCR to examine the relationship between disease severity in pea and *Aphanomyces euteichus* DNA concentration in roots. *Eur J Plant Pathol* 111: 309–316.

Vettraino AM, Sukno S, Vannini A, Garbelotto M (2010) Diagnostic sensitivity and specificity of different methods used by two laboratories for the detection of *Phytophthora ramorum* on natural hosts. *Plant Pathol* 59: 289–300.

Volossiouk J, Robb EJ, Nazar RN (1995) Direct DNA extraction for PCR-mediated assays of soil organisms. *Appl Environ Microbiol* 61: 3972–3976.

Wander JGN, van den Berg W, van den Boogert PHJF et al. (2007) A novel technique using the Hendrickx centrifuge for extracting winter sporangia of *Synchytrium endobioticum* from soil. *Eur J Plant Pathol* 119: 165–174.

Wang B, Brubaker CL, Tate W, Woods MJ, Matheson BA, Burdon JJ (2006) Genetic variation and population structure of *Fusarium oxysporum* f.sp. *vasinfectum* in Australia. *Plant Pathol* 55: 746–755.

Wang B, Brubaker CL, Thrall PH, Burdon JJ (2010) Host resistance mediated inter-genotype competition and temporal variation in *Fusarium oxysporum* f.sp. *vasinfectum*. *Eur J Plant Pathol* 128: 541–551.

Wang H, Hwang SF, Chang KF, Turnbull GD, Howard RJ (2000) Characterization of *Ascochyta* isolates and susceptibility of pea cultivars to the Ascochyta complex in Alberta. *Plant Pathol* 49: 540–545.

Wang J, Jacobs JL, Byrne JM, Chilvers MI (2015a) Improved diagnoses and quantification of *Fusarium virguliforme*, causal agent of soybean sudden death syndrome. *Phytopathology* 105: 378–387.

Wang L, Liu LM, Hou YX, Li L, Huang SW (2015b) Pathotypic and genetic diversity in the population of *Rhizoctonia solani* AG1-IA causing rice sheath blight in China. *Plant Pathol* 64: 718–728.

Wang PH, Chang CW (2003) Detection of low-germination rate resting spores of *Pythium myriotylum* from soil by PCR. *Lett Appl Microbiol* 36: 157–161.

Wang PH, Chung CY, Lin YS, Yeh Y (2003a) Use of polymerase chain reaction to detect the soft rot pathogen *Pythium myriotylum* in infected ginger rhizomes. *Lett Appl Microbiol* 36: 116–120.

Wang PH, Wang YT, White JG (2003b) Species-specific PCR primers for *Pythium* developed from ribosomal ITS1 region. *Lett Appl Microbiol* 37: 127–132.

Wang PH, White JG (1996) Development of a species-specific primer for *Pythium violae*. *Proc BCPC Symp* 65: 205–210.

Wang Y, Ren Z, Zheng X, Wang Y (2007) Detection of *Phytophthora melonis* in samples of soil, water and plant tissue with polymerase chain reaction. *Canad J Plant Pathol* 29: 172–181.

Wang Y, Zhang W, Wang Y, Zheng X (2006) Rapid and sensitive detection of *Phytophthora sojae* in soil and infected soybean by species- specific polymerase chain reaction assays. *Phytopathology* 96: 1315–1321.

Ward LI, Beales PK, Barnes AV, Lane CR (2004) A real-time PCR assay based method for routine diagnosis of *Spongospora subterranea* on potato tubers. *J Phytopathol* 152: 633–638.

Watanabe H, Kageyama Ki, Taguchi Y, Horinouchi H, Hyakumachi MC (2008) Bait method to detect *Pythium* species that grow at high temperatures in hydroponic solutions. *J Gen Plant Pathol* 74: 417–424.

Watson WT, Kenerley CM, Appel DM (2000) Visual and infrared assessment of root colonization of apple trees by *Phymatotrichopsis omnivora*. *Plant Dis* 84: 539–543.

Wei F, Fan R, Dong H et al. (2015) Threshold microsclerotia inoculum for cotton Verticillium wilt determined through wet-sieving and real-time quantitative PCR. *Phytopathology* 105: 220–229.

Weier HU, Gray JW (1988) A programmable system to perform the polymerase chain reaction. *DNA* 7: 441–447.

Weiland JL (2011) Influence of isolation method on recovery of *Pythium* species from forest nursery soils in Oregon and Washington. *Plant Dis* 95: 547–553.

Whisson DL, Herdina A, Francis L (1995) Detection of *Rhizoctonia solani* AG-8 in soil using a specific DNA probe. *Mycol Res* 99: 1299–1302.

Woodhall IW, Adams IP, Peters JC, Harper G, Boonham N (2013) A new quantitative real-time PCR assay for *Rhizoctonia solani* AG3-PT and detection of AGs of *Rhizoctonia solani* associated with potato in soil and tuber samples in Great Britain. *Eur J Plant Pathol* 136: 273–280.

Woodhall JW, Webb KM, Giltrap PM et al. (2012) A new large scale soil DNA extraction procedure and real-time PCR assay for the detection of *Sclerotium* in soil. *Eur J Plant Pathol* 134: 467–473.

Wright J, Lees AK, Van der Waals JE (2012) Detection and eradication of *Spongospora subterranea* in mini-tuber production tunnels. *South Afr J Sci* 108 (5/6) Article No. 614, 4p.

Xie C, Huang C-H, Vallad GE (2014) Mycelial compatibility and pathogenic diversity among *Sclerotium rolfsii* isolates in Southern United States. *Plant Dis* 98: 1685–1694.

Xu ML, Melchinger AE, Luberstdet T (1999) Species-specific detection of the maize pathogen *Sporisporium reiliana* and *Ustilago maydis* by dot-blot hybridization and PCR-based assays. *Plant Dis* 83: 390–395.

Yang YG, Zhao C, Guo ZJ,Wu XH (2015) Characterization of a new anastomosis group (AG-W) of binucleate *Rhizoctonia*, causal agent of potato stem canker. *Plant Dis* 99: 1757–1763.

Yu JM, Babadoost M (2013) Occurrence of *Fusarium commune* and *F. oxysporum* in horseradish roots. *Plant Dis* 97: 453–460.

Yu S, Su T, Chen T et al. (2015) Real-time PCR assay as a diagnostic tool for evaluating the resistance of Chinese cabbage cultivars to Verticillium wilt. *Eur J Plant Pathol* 143: 549–557.

Yuen GY, Craig ML, Avila F (1993) Detection of *Pythium ultimum* with a species-specific monoclonal antibody. *Plant Dis* 77: 692–698.

Yuen GY, Xia JQ, Sutula (1998) A sensitive ELISA for *Pythium ultimum* using polyclonal and species-specific monoclonal antibodies. *Plant Dis* 82: 1029–1032.

Zambounis AG, Paplomatas E, Tsfataris AS (2007) Intergenic spacer-RFLP analysis and direct quantification of Australian *Fusarium oxysporum* f.sp. *vasinfectum* isolates from soil and infected cotton tissues. *Plant Dis* 91: 1564–1573.

Zanotti MGS, de Queiroz MV, dos Santos JK, Rocha RB, de Barros EG, Arujo EF (2006) Analysis of genetic diversity of *Fusarium oxysporum* f.sp. *phaseoli* pathogenic and nonpathogenic to common bean (*Phaseolus vulgaris* L.). *J Phytopathol* 154: 549–559.

Zhang H, Feng J, Manoli VP, Strelkov SE, Hwang S-F (2015) Characterization of gene identified in pathotype 5 of the clubroot pathogen *Plasmodiophora brassicae*. *Phytopathology* 105: 764–770.

Zhang N, Geiser DM, Smart CD (2007) Macroarray detection of solanaceous plant pathogens in Fusarium solani species complex. *Plant Dis* 91: 1612–1620.

Zhang N, McCarthy ML, Smart CD (2008) A macroarray system for the detection of fungal and oomycete pathogens of solanaceous crops. *Plant Dis* 92: 953–960.

Zhang XG, Zheng GS, Han HY, Han W, Shi CK, Chang CJ (2001) RAPD PCR for diagnosis of *Phytophthora parasitica* var. *nicotianae* isolates which cause black shank in tobacco. *J Phytopathol* 149: 569–574.

Zhang Z, Zhang J, Wang Y, Zheng X (2005) Molecular detection of *Fusarium oxysporum* f.sp. *niveum* and *Mycosphaerella melonis* in infected plant tissues and soil. *FEMS Microbiol Newslett* 249: 39–47.

Zhang ZG, Li YQ, Fan H, Wang YC, Zheng XB (2006) Molecular detection of *Phytophthora capsici* in infected plant tissues, soil and water. *Plant Pathol* 55: 770–775.

Zhu ZX, Zheng L, Hsiang T et al. (2016) Detection and quantification of *Fusarium commune* in host tissue and infested soil using real-time PCR. *Plant Pathol* 65: 218–226.

Ziems AD, Giesler LJ, Yuen GY (2006) First report of sudden death syndrome of soybean caused by *Fusarium solani* f.sp. *glycines* in Nebraska. *Plant Dis* 90: 109.

Ziesman BR, Turkington TK, Strelkov SE (2016) A quantitative PCR system for measuring *Sclerotinia sclerotiorum* in canola (*Brassica napus*). *Plant Dis* 100: 984–900.

Zitnick-Anderson KK, Nelson Jr. BD (2015) Identification and pathogenicity of *Pythium* on soybean in North Dakota. *Plant Dis* 99: 31–38.

# 3 Biology of Soilborne Fungal Plant Pathogens

Soilborne fungus-like oomycetes and fungal pathogens have to interact with both biotic and abiotic factors that significantly influence their development and ability to infect susceptible host plant species. The fungal pathogens have innate capacity to either adapt to the adverse environmental conditions, or to produce resistant propagules for their survival and subsequent perpetuation, until the return of favorable conditions. Successful cultivation of crops demands effective integration of several practices/operations. The knowledge on the distribution and economic importance of the diseases, biological characteristics including disease cycles, nutritional requirements, temperature and pH optima for growth and reproduction of the fungal pathogens, host range, existence of races/biotypes with varying virulence, variations in the phenotypic and genotypic characteristics and factors that contribute to the development of epidemics is essential to developing effective management systems for containing the soilborne diseases occurring in different agroecosystems

## 3.1 BIOLOGICAL CHARACTERISTICS OF FUNGAL PATHOGENS

### 3.1.1 DISEASE INCIDENCE AND DISTRIBUTION

Occurrence of soilborne diseases of various crops has been observed in all countries to different extent, depending on the availability of susceptible cultivar, virulent strain(s) of the pathogen and the favorable environmental conditions. The oomycetes *Phytophthora* spp. and *Pythium* spp. have wide distribution, as they are capable of infecting a large number of plant species and once they are introduced into a new location, they can survive for a long time (extending to several years) in the soil, in plant debris or as oospores which are highly resistant to adverse environmental conditions. One or more species belonging to the genus *Phytophthora* may cause diseases such as crown and root rot of apple. Infection by *P. cactorum* in crown and root tissues of apple was observed previously. Later on, *P. cryptogea*, *P. gonapodyides* and *P. megasperma* were isolated from necrotic tissues of roots and crown or rhizospheres of apple trees. *P. cactorum* was the most predominant and virulent on apple trees, indicating the differential pathogenic potential of different *Phytophthora* spp. on apple (Latorre et al. 2001). The soilborne pathogen with multiple modes of dispersal may get dispersed to several geographically separated locations rapidly. *Fusarium oxysporum* f.sp. *nicotianae* was found to be seedborne, as well as soilborne. This pathogen was introduced into Greece through imported tobacco seed. In addition, *F. oxysporum* f.sp. *vasinfectum*, present already in Greece, also caused vascular wilt in

tobacco. The results indicated the possibility of *F. oxysporum* f.sp. *nicotianae* getting rapidly established in Greece because of the seedborne nature of the pathogen (Tjamos et al. 2006). Distribution of Fusarium crown rot (FCR) and common root rot (CRR) pathogens associated with wheat (*Triticum aestivum*) was investigated in 91 fields in Montana between 2008 and 2009 crop seasons. A conventional isolation method was employed for the detection of *Bipolaris sorokiniana*, *Fusarium culmorum* and *F. pseudograminearum* associated with FCR and CRR diseases. Percentages of incidences of *B. sorokiniana*, *F. culmorum* and *F. pseudograminearum* were 15%, 13% and 8%, respectively, as determined by isolation method. Incidence of FCR disease was observed in 57% of fields, whereas 93% of the fields had CRR incidence, indicating the predominance of root rot disease in wheat crops (Moya-Elizondo et al. 2011).

Constant monitoring of the incidence of diseases in a geographic location may be useful to recognizing the newly introduced or already occurring pathogens about which required information has to be gathered for effective management of diseases caused by these pathogens. When the cause of strawberry decline in Latvia and Sweden was investigated, infection of strawberry by *Gnomonia fragariae* was observed. The pathogen was repeatedly recovered from discolored roots and crown tissues of severely stunted plants and its pathogenicity was proved. *G. fragariae* was shown to be one of the serious pathogens involved in the root rot complex of strawberry in Latvia and Sweden (Morocko et al. 2006). The frequency and incidence of *Pyrenochaeta terrestris* on the root internodes of different maize hybrids were assessed at 70 days after sowing. The pathogen developed in the roots of all well developed internodes (from the primary to sixth or seventh internodes). The average frequency and incidence of *P. terrestris* in the roots of late and medium early maturity hybrids ranged from 29.5 to 55.2% and from 11.8 to 22.7%, respectively. The highest frequency of pathogen recovery was at the second root internode (93.3%) and its greatest incidence was detected in the mesocotyls of the medium early hybrid H-1 (56.9%). The pathogen produced a distinctive red pigment on carnation leaf agar (CLA) medium in a light regime. Presence of this red pigment could be used as a reliable marker for the identification of *P. terrestris*, even when pycnidia were not produced. The results revealed the distribution of the pathogen in the root internodes of maize hybrids (Levic et al. 2011). *Fusarium circinatum*, causal agent of pitch canker of *Pinus* spp. occurs in many parts of the world. Investigation on the population biology of *F. circinatum* carried out in Western Cape Province indicated that the pathogen population was highly differentiated and pitch canker might have originated from one or more separate introductions of the pathogen and its movement in West Cape Province of South Africa was not restricted, necessitating more effective strategies for limiting the spread of the pathogen (Steenkamp et al. 2014).

Some fungal pathogens are able to cause disease on susceptible host. In addition, they function as vectors of viruses that induce diseases, resulting in a complex. Incidence and association of *Spongospora subterranea* f.sp. *subterranea (Sss)*, (causal agent of potato powdery scab) with *Potato mop-top virus* (PMTV, genus *Pomovirus*) was investigated in 39 potato fields in Costa Rica. Paired samples (633) of leaf tissue and corresponding tubers were collected, assessed visually for disease and subsequently assayed by enzyme-linked immunosorbent assay (ELISA). Presence of *Sss* in tuber tissue and PMTV in leaf and tuber tissues was tested by double antibody sandwich (DAS)-ELISA and triple antibody sandwich (TAS)-ELISA, respectively. Further, the field soil samples were assayed for both pathogens by ELISA and bioassay. Incidence of both diseases ranged from 0 to 100% within individual fields, with incidences lower than 40% occurring in more than 70% of the fields. Of the 633 paired samples, 179 and 146 were positive for both PMTV and *Sss*, respectively, as indicated by ELISA tests with either foliage or tuber tissues. The ELISA tests detected PMTV in leaf samples of symptomatic plants (14/81) and asymptomatic plants (76/81). *Sss* was detected in four of ten soil samples by ELISA tests. Soilborne PMTV was detected by ELISA test in roots of bait plants, raised in field soil samples. Occurrence of both *Sss* and PMTV was detected in 64 samples. A significant but low degree of association for vector and virus was revealed in this investigation (Montero-Astúa et al. 2008).

The interaction between *Rhizoctonia solani* and *Rhizopus stolonifer* isolated from sugar beet plants showing severe root rot in commercial fields in Michigan, United States, was investigated. The susceptible sugar beet cv. USH20 and resistant cv. FC716 were inoculated with *R. solani* anastomosis group (AG) 2-2 IIIB with and without *R. stolonifer*. No clear rot symptoms were induced in plants inoculated with spore suspension of *R. stolonifer*. But some surface scarring was observed on the crowns. Lesions typical of Rhizoctonia crown and root rot (RCRR) were observed on susceptible cultivar, when inoculated with *R. solani* alone. However, more extensive root discoloration was noted on sugar beet plants inoculated with both pathogens in both susceptible and resistant cultivars. The susceptible cultivar showed higher disease severity rating when exposed to both pathogens, than the resistant cultivar. Plants inoculated with *R. stolonifer* alone showed little or no internal discoloration in susceptible USH20 plants. Susceptible plants inoculated with *R. solani* exhibited shallow, dark lesions that penetrated less than 1 cm into the interior of the roots. On the other hand, several plants inoculated with both pathogens had extensive internal discoloration. The results indicated the existence of synergism under conditions that were not conducive to infection of sugar beet by *R. stolonifer* (Hanson 2010). The mechanisms of plant root defenses against soilborne necrotrophic pathogens are unclear, especially in perennial fruit crops like apple. Apple replant disease

(ARD), considered as a complex, includes *Pythium ultimum* as a component. Responses of apple rootstocks, with different levels of tolerance to ARD, were evaluated, using tissue-cultured apple plants, following inoculation with *P. ultimum*, at biological, cellular and molecular levels. ARD tolerant G.935 and susceptible B.9 apple rootstocks were assessed for their responses to inoculation with *P. ultimum*, based on differences in survival rate, root biomass, shoot biomass and maximum root length. Microscopic examination revealed contrasting dynamics in symptom development. The susceptible B.9 rootstock was unable to arrest pathogen invasion, resulting in rapid progress of necrosis across the entire root system. In contrast, roots of tolerant G.935 plants limited the pathogen invasion, leading to localized root necrosis. Apple genes, including *MdCHs*, *MdCHIB*, *MdBIS* and *MdGLUC* were activated as defense responses, resulting in differential expression profiles, during infection of tolerant and susceptible apple rootstocks. The results suggested the existence of discrete molecular and biochemical processes in apple genotypes in response to infection by *P. ultimum* (Zhu et al. 2016).

Various factors influencing incidence of the soybean sudden death (SDS) caused by *Fusarium virguliforme* were investigated under greenhouse conditions with soybean cultivars susceptible and partially resistant to SDS, four pathogen population levels and six crop residue treatments. Population level increases and crop residues showed significant interacting effects in increasing foliar disease severity and root rot. Disease severity was positively correlated with pathogen population and biomass was negatively correlated. Disease severity reached its peak in corn and sorghum seed. The results showed that pathogen population levels and crop-derived nutrients in soil interact and influence severity of SDS (Freed et al. 2017). Incidence of soybean suddent death syndrome (SDS) caused by *Fusarium* spp. was observed for the first time between 2013 and 2014 in South Africa. Slow growing *Fusarium* strains (16) were isolated from roots of symptomatic plants. Molecular phylogenetic analyses of a portion of translation elongation factor 1-α (TEF1) and the nuclear ribsosomal intergenic spacer region (IGS rDNA) indicated the involvement of *F. brasiliense* and an undescribed *Fusarium* sp. The causal nature of *Fusarium* spp. was established by following Koch's postulates. The strains of *F. brasiliense* showed variations in their virulence to soybean. Cell-free culture filtrates of the SDS pathogens present in South Africa and two positive control strains of *F. virguliforme* from the United States induced typical SDS symptoms on susceptible soybean cultivars in the whole-seedling assay, indicating the role of toxins produced by the pathogens in symptom induction (Tewoldemedhin et al. 2017). The possibility of enhancement of susceptibility to fungal pathogen in plants with associated virus infection of plants has been observed in some pathosystems. Potato tubers from plants infected by *Potato leafroll virus* (PLRV) were found to be more densely covered by *R. solani*, causing black scurf disease under field conditions in Hokkaido, Japan. In order to determine the influence of PLRV infection on the susceptibility of potato tubers to *R. solani*, PLRV-infected mother tubers were inoculated with *R. solani* under pot culture conditions. PLRV-infected plants produced significantly fewer and smaller tubers compared to tubers from virus-free plants. The results suggested that PLRV-infected plants might be more susceptible to infection by *R. solani* and that virus-free seed tubers needed to be planted to reduce incidence of blackscurf disease (Ito et al. 2017).

## 3.1.2 Disease Cycles of Fungal Pathogens

The soilborne fungal pathogens produce various types of spores during asexual and sexual phases of their life cycles for dispersal to other plants in the same location or to other locations and for perpetuation through subsequent generations. In general, asexual spores such as sporangia, and zoospores or conidia are produced by oomycetes (fungus-like) and eukaryotic (true) fungi, respectively. The asexual spores may be enclosed in sporangia or spore-bearing structures such as pycnidia, acervuli or synnemata.

### 3.1.2.1 Production of Pathogen Propagules

The fungus-like members of Phycomycotina have coencytic (nonseptate) mycelia, lacking septa (cross-walls) and they are like a continuous, tubular, branched or unbranched multinucleate cell or multinucleate hyphae. Septa are formed later, when spore-bearing structures (sporangiophores) are produced. The mycelium is formed by elongation of hyphal tips, following germination of spores/cysts. Members of Plasmodiophorales do not have a true mycelium, but form naked, amoeboid, multinucleate plasmodia. On the other hand, Chytridiales produce a system of strands that vary in diameter, known as rhizomycelium. In true fungi (eukaryotes), the mycelium produced following germination of conidia is septate (with cross-walls) and after required growth, when nutrient supply is sufficient, repeats the asexual cycle indefinitely without any change in the genetic characteristics, unless mutations occur. In contrast, during sexual phase, two physiologically different gametangia (+ and –) come in contact with each other, resulting in the formation of sexual spores (ascospores or basidiospores). The phenomenon of heterothallism requires the union between thalli of two different (self-incompatible) isolates (+ and –) for formation of sexual spores, as observed in most Ascomycotina.

Root infection by soilborne fungal pathogens can be inferred by loss of turgidity and yellowing of foliage. These symptoms are not a reliable indication of infection of roots of plants by fungal pathogens. However, at this stage, the pathogen would have well established in the vascular tissues of infected plants, making it almost impossible to save such infected plants. Soilborne fungal pathogens induce lesions on root tissues which may be seen only after destructive sampling. Lesions may extend in size involving many secondary roots affecting the absorption of water and nutrients and rotting of roots and collar tissues may result in collapse and death of infected plants eventually. It has not been possible to visualize formation of pathogen spores from infected root tissues through examination under light microscope. But after the collapse of plants, the formation of sclerotia or chlamydospores may occur in the infected tissues and they may

be released into the soil, when plant tissues are degraded by saprophytic organisms present in the soil. The commonly adopted procedure is to isolate the fungal pathogens from the root/collar tissues on appropriate media. Cultural characteristics of the pathogen, such as the colony development, and temperature and pH optima for growth and sporulation and production of enzymes and/ or toxins are determined. Few investigations have been successful in determining the conditions required for *in situ* production of spores in root tissues. *Monosporascus cannonballus*, a soilborne pathogen, causes destructive vine decline disease of melons, occurring in all countries around the world. The pathogen produces ascospores (primary inoculum on the roots primarily at the end of cropping season). High soil temperatures (20 to 30°C) form a crucial factor affecting the rate of production of ascospores by the pathogen. Under *in vitro* conditions, the optimal temperature for sexual reproduction ranged from 25 to 30°C. In addition, the root system of a single mature cantaloupe plant was able to support the production of about 400,000 ascospores. This population of ascospores, if incorporated uniformly into 0.03 m2 (1 ft3) of soil, would be equivalent to 10 ascospores/g of soil. Naturally infested soil might contain as few as 2 ascospores/g of soil. The results suggested that strategies aimed at reducing production of ascospores might be effective in containing the disease incidence and spread in melons (Waugh et al. 2003).

Among the four fusaria involved in soybean sudden death (SDS) disease, *Fusarium tucumaniae* was the most important in Argentina. Formation of perithecia on soybean by *F. tucumaniae* was observed for the first time in Argentina. Ascospores derived from perithecia formed colonies that produced sporodochial conidia, characteristic of *F. tucumaniae*. Conidia had an acuate apical cell and a distinctly foot shaped basal cell. This study appears to be the first authentic report on sexual reproduction by *F. tucumaniae* involved in SDS in soybean in nature (Scandiani et al. 2010). The effects of pathogen population levels, seed exudates, crop residues and their interactions on development of sudden death syndrome (SDS) of soybean and growth of the pathogen, *F. virguliforme* were investigated under greenhouse conditions, using susceptible and partially resistant soybean cultivars. Severity of symptoms in roots was assessed at 15 and 50 days after inoculation (*dai*). Increases in pathogen population levels had significant interacting effects on increasing foliar disease severity and root rot and on host biomass reduction. Disease severity was positively correlated with pathogen population and biomass was negatively correlated. Addition of corn and sorghum seed residues added to the soil increased disease severity to the maximum extent. Low or no root rot or foliar disease was observed in plots receiving no crop residue. Exudates collected from germinating soybean and corn seed enhanced growth of *F. virguliforme* and *F. solani* significantly. Growth of the pathogen was increased more effectively by exudates collected during soybean radicle emergence than those sampled at other stages of seed germination (Freed et al. 2017).

*Phytophthora ramorum* has a distinct soil phase inoculum, consisting of infected organic debris, infected roots and persistent chlamydospores and this is important epidemiologically, because of the potential for movement of the pathogen after rain and in waterways such as streams and rivers (Sutton et al. 2009). *P. ramorum* infects rhododendron, by either producing germ tube or indirectly by forming zoospores that become encysted zoospores on contact with plant surface (root or leaf). The impact of propagule type on disease development was investigated. Rhododendron leaf disks were inoculated with *P. ramorum* zoospores (2,400/disk), sporangia (75/disk) or sporangia + trifluoperazine hypochloride (TFP) (75/disk) which inhibited zoospore formation. The higher concentration of zoospores (2,400/disk) induced a significantly higher percentage of necrotic leaf disk area (96.9%) than sporangia (87.6%) The sporangia + TFP induced the lowest intensity of necrosis (47.5%). Rooted rhododendron cuttings had a higher percentage of necrotic leaves/plant, when inoculated with zoospores (3,000 or 5,000/ml) or cysts (50,000/ml) than with sporangia (3,000/ml) with or without TFP and zoospores at 3,000/ml. Full inoculum potential could be reached when zoospores were used as inoculum (Widmer 2009). Under controlled conditions, the exposure time required to infect Quercus (oak) species by *P. ramorum*, zoospore concentration and inoculum produced from the host plant roots was determined. Sprouted acorns (fruits of oaks) were exposed to zoospores (3,000/ml) for different periods and transplanted in potting soil. The roots were infected within one hour after exposure to zoospore suspension. Roots of *Q. prinus* seedlings showed 0.5 to 3.2% colonization of total root mass after five months. Radicle infection occurred even at a concentration of 1 zoospore/ml. To assess inoculum production, roots were inoculated with sporangia and transplanted into pots. Periodically samples of runoff were collected and plated on selective medium. Then, root segments were plated to assess percent colonization by the pathogen. At 16 and 35 *dai*, root colonization and inoculum production from oak was lower than that of *Viburnum tinus*, a positive control. The results showed that *P. ramorum* could infect sprouted oak acorns and produce secondary inoculum which might be of edpidemiological importance (Widmer et al. 2012).

A procedure was developed to produce inocula of five *Phytophthora* spp. *in situ* that would be similar to the inocula found in nursery effluent so that wetland plants could be exposed to continuous supply of zoospores in an aqueous environment. The V8 agar plugs from actively growing cultures of three or four isolates of *P. cinnamomi*, *P. citrophthora*, *P. cryptogea*, *P. nicotianae* and *P. palmivora* were used to produce inocula. For laboratory tests, agar plugs were kept in plastic cups and covered with 1.5 nonsterile soil extract solution (SES) for 29 days and presence and activity of zoospores in the solution were monitored at two- or three-day intervals, using a rhododendron leaf disk baiting assay. For greenhouse assays, agar plugs of *Phytophthora* spp. were placed in plastic pots and covered with either SES or Milli-Q water (ultrafiltered water) for 13 days during both summer and winter months. Baiting bioassay and filtration methods were employed for determining the zoospore presence and activity. The mean percentage of the perimeter of leaf disks colonized

was used as a measure of zoospore activity in the respective solutions containing agar plugs, because zoospores are both negatively geotrophic and chemotactic in the laboratory tests. The number of zoospores in solutions was closely related to the amount of leaf bait colonized. Leaf bait colonization was considered as a better measure of zoospore activity, because it took into account both the number of zoospores present in the solution and the aggressiveness/virulence of individual isolates. Zoospores were present in solutions throughout the 29- and 13-day experimental periods, but consistency of zoospore release differed with pathogen species. Colonization of leaf baits decreased over time between species and varied also among isolates within a species under *in vitro* conditions. Bait colonization by some *Phytophthora* spp. decreased over time in both summer and winter months under greenhouse conditions. But zoospore densities of all *Phytophthora* spp. were lower in winter experiments than in summer experiments. The type of treatment solution appeared to influence the number of zoospores present in these solutions, with greater densities of zoospores estimated in Milli-Q water than in SE. The protocol developed in this investigation might be useful for assessing susceptibility/resistance of plants to different species of *Phytophthora* spp. infecting different aquatic plant species (Ridge et al. 2014).

*Phytophthora capsici* causes Phytophthora blight disease in several crops such as pepper, tomato, cucumber, pumpkin and watermelon. Viability of the spores is the important factor for determining the pathogenicity of *P. capsici*. Germination tests are performed to confirm the viability of spores, but they are time-consuming and subject to variations due to the environmental conditions to which they are exposed. The procedure involving the use of vital stains requires less time and has the capability to provide quantitative data without affecting biological activity of the spores. The fluorescent vital dye FIN-1R [2-choloro-4 (2,3-dihydro-3-methyl-(benzo-1,3-thiazol-2-yl-methylidene)-1-phenylquinolinum iodide] was used to stain mycelia, sporangia and zoospores from cultures of *P. capsici*. The FUN-1 dye was useful to quantitatively assess the live sporangia and other spores and differentiate them from dead ones. The dye had the potential for monitoring the changes in the metabolic activity of sporangia over time (Lewis Ivey and Miller 2014). *Phytophthora sojae* (syn. *P. megasperma* f.sp. *glycines*), causal agent of Phytophthora stem and root rot (PSR), has a restricted host range, with soybean as its primary host (Tyler 2007). A method of inoculating *P. sojae* zoospores directly into the hydroponic solution was developed to obtain reproducible infections by *P. sojae*. Cultures of *P. sojae* race 3 and race 7 were grown on clarified V8 agar medium for one week at 25°C in the dark. In order to induce sporangium production, active mycelial plugs were transferred into 15 ml of soil extract placed in 9-cm petridishes and incubated for five days at 25°C under constant fluorescent light. The fungal suspension was then placed in the dark at 15°C for 48 h and then returned to previous environment of 25°C under fluorescent light for two days, followed by one day at 15°C in the dark. Zoospores were released by placing the suspension at 4°C for 1 h, followed by another hour at 25°C. Zoospore concentration of each race was determined using a hemacytometer. Inoculation of roots of plants in a hydroponics system was performed by adding 200 ml of a suspension ($2 \times 10^5$ zoospores/ml) of each race directly into the tank containing the hydroponic solution at two weeks after the seedlings had been transferred into the hydroponic system. This method of inoculation allowed inoculation of different soybean cultivars with different pathogen races and assessment of the resistance of host plants to *P. sojae* (Guérin et al. 2014).

*Leptosphaeria maculans* and *L. biglobosa* 'brassicae' are involved in blackleg (Phoma stem canker) disease, which has significant adverse impact on production of oilseed rape and several cruciferous crops. The effect of flooding on the survival of *L. biglobosa* in the stubbles of winter oilseed rape (*Brassica napus*) was assessed. Infected pieces of basal stems were submerged in water at different temperatures (16–40°C). Under field conditions, the infected stem pieces were placed on the soil surface. The pathogen was isolated after different intervals (one to eight weeks) after flooding the soil. Isolates of *L. biglobosa* 'brassicae'were identified, based on the amplification of a pathogen-specific 444-bp product in the polymerase chain reaction (PCR) assay. Flooding for one to two weeks resulted in negligible recovery of *L. globosa* 'brassicae'. The results indicated that flooding resulted in rapid decomposition of the infected stem pieces. After flooding for eight weeks, the dry weight of stem pieces was reduced by 28 to 42% and 26 to 36%, respectively under the laboratory and field conditions. As the host tissue decomposed rapidly under flooded conditions, the survival of the pathogen in host stem tissues of oilseed rape could be expected to be reduced significantly (Cai et al. 2015). Potato tubers are infected by *Verticillium dahliae* and *V. tricorpus*, inducing Verticillium wilt disease of potato plants. The role of tuber infection by *Verticillium* spp. in colonization of stems, symptom development, tuber yield, progeny tuber infection, quantity of soilborne inoculum and contribution of soil inoculum to subsequent disease development was investigated under greenhouse conditions, using naturally infected and uninfected seed tubers of seven potato cultivars. Internal seed tuber infection of all cultivars did not contribute to severity of disease, although *V. dahliae* could be isolated from symptomless plants. Susceptible cultivars showed highest petiole infection, while cultivar with moderate resistance showed lower disease incidence at 65 days after planting. Planting infected tubers contributed to *V. dahliae* in the potting soil, with Russet Burbank (susceptible), showing highest pathogen population (70 pg DNA/g of soil), whereas the Ranger Russet (moderately resistant) had the least soil inoculum (11 pg DNA/g of soil). Soil inoculum originating from naturally infected tubers did not contribute to subsequent foliar disease development in Russet Burbank grown in infested soil. However, *V. dahliae* was isolated from the petiole of symptomless plants. Progeny tubers were entirely free from *V. dahliae*. Tubers of cv. Nicola infected with *V. tricorpus* produced very few wilt symptoms, but the pathogen was reisolated from petiole of symptomless plants. Progeny tuber infection by *V. tricorpus* was limited to 4% and soil inoculum levels were low (1 CFU/g of dry soil). No significant reduction

in tuber yield was recorded due to tuber infection by *V. tricorpus* (Nair et al. 2016).

*V. dahliae* is a soilborne fungal pathogen and its transmission via spinach seed is of epidemiological significance. Verticillium wilt was found to be endemic in spinach seed production areas in northern Europe and Washington State in the United States. Spinach seeds infested with *V. dahliae* can disseminate the pathogen to long distances, as well as to subsequent generations of crops. Once *V. dahliae* is introduced into the soil from infested seed, microsclerotia are formed and they can survive in the soil for more than ten years, making the elimination of the pathogen from the soil very difficult. Commercial seed lots from Denmark, New Zealand, Netherlands and the United States were infested up to an extent of 92%, while transmission of *V. dahliae* across spinach generations ranged from 0 to 100% (Butterfield and DeVay 1977). *V. dahliae* colonizes the pericarp, seed coat, cotyledons and radicle, forming about 250 microsclerotia per seed. Seedborne infection by *V. dahliae* results in systemic invasion of spinach seedlings and subsequently the flowers and seeds (Maruthachalam et al. 2013). The green fluorescent protein (GFP)-tagged strain of *V. dahliae* VdSo925-316 colonized the lettuce root surface at five weeks after planting lettuce in soil, following two prior plantings of infested spinach seeds and incorporation of residue. Another portion of the same lettuce root system showed the presence of GFP-expressing *V. dahliae* strain. The GFP-tagged strain could also penetrate spinach root surface, as revealed by confocal laser scanning microscope (CLSM) (Short et al. 2015). The potential of mustards, grasses and Austrian winter pea to serve as sources of inoculum was assessed by inoculating them with *V. dahliae*. Typical wilt symptoms were not observed in any of the rotation crops, but plant biomass was reduced or not affected or increased by infection by specific isolates of *V. dahliae*. Each isolate was host-specific and infected a subset of the rotation crop tested, but microsclerotia from at least one isolate were observed on each rotation crop. Some isolates were host-adapted and differentially altered plant biomass or produced differential amounts of inoculum on rotation crops like arugula and Austrian winter pea, which supported more inoculum of specific isolates on potato. Asymptomatic and symptomatic infection and differential inoculum production on rotation crops by *V. dahliae* may be crucial factors influencing incidence of *V. dahliae* (Wheeler and Johnson 2016).

*Macrophomina phaseolina* causes charcoal rot disease of soybean that is widely distributed. A reliable method was developed for producing large amounts of inoculum sufficient for performing high throughput resistance-screening of soybean genotypes. Seven semi-defined media were evaluated for their usefulness for production of pycnidia in culture. The number of pycnidia produced by eight isolates of *M. phaseolina* was dependent on induction medium. Peanut butter extract-saturated filter paper placed over soynut butter extract agar (PESEA) induced formation of greater number of pycnidia and conidia than other media tested. Production of pycnidia on PESEA ranged from 269 to 1,082 per plate. No difference was observed in conidial germination among isolates on PESEA with an average of 83 + 2% germination. Using the conidial inoculum produced on PESEA, the soybean genotype DT 97-4290 was identified as partially resistant to charcoal rot. The results indicated that the PESEA medium could be used for mass production of pycnidia and conidia for screening soybean genotypes for identification of sources of resistance to characoal rot disease (Ma et al. 2010). A foliar blight disease of tomato of unknown etiology was observed in North Carolina (NC) between 2005 and 2006 and the cause of the disease was identified as *R. solani*, based on the morphological characteristics. The basidiospores were generated in planta. DNA sequence analysis, hyphal anastomosis and somatic hyphal interactions of the isolates were studied. Phylogenetic analysis and hyphal anastomosis placed the pathogen in *R. solani* AG-3. Tomato foliar blight isolates occurring in NC formed a single phylogenetic group with tomato isolates of *R. solani* AG-3 from Japan and were more closely related to *R. solani* AG-3 isolates from potato than tobacco (Bartz et al. 2010).

The genetic structure of three populations of *R. solani* AG-1 IA from three locations in Iran was analyzed. All three populations exhibited a mixed reproductive mode, including both sexual and asexual reproduction. No inbreeding was detected, suggesting that the pathogen was random mating. The results indicated that a recombining population structure with some clonality in all three field populations at the early growth stage. Asexual propagules (mycelia and sclerotia) were probably the main source of primary infection and the evidence for recombination at early stages of the epidemic reflected sexual recombination that occurred in these populations in earlier years. It was considered that basidiospores might be a significant source of secondary inoculum during the development of rice sheath blight epidemics in northern Iran. These characteristics indicated that *R. solani* AG-1 IA had high evolutionary potential and hence, control based on systemic fungicides and major resistance genes has to be implemented carefully to avoid the emergence of isolates with greater levels of fungicide resistance or virulence (Padasht-Dehkaei et al. 2013). *R. solani*, infecting common bean causes web blight (WB) in the tropics and root rot (RR) in all countries around the world, since the pathogen is soilborne. *R. solani* is a species complex, grouped into 14 AGs. The AG1-IA, AG1-IB, AG1-IE, AG1-IF, AG2-2 and AG4 induce WB on aerial plant parts, whereas AG4 and AG2-2 may induce RR also. Nine isolates of *R. solani*, including three AG1 and three AG4 WB isolates and three AG4 WB isolates and three AG RR isolates obtained from both leaves and roots, respectively, of common bean in Puerto Rico, were evaluated for their interactions with 12 common bean genotypes. WB isolates were inoculated on detached leaves and suspensions of mycelia of RR isolates were used to inoculate the roots in the greenhouse. All *R. solani* isolates induced both RR and WB symptoms. RR intensities were generally more severe than WB severity ratings. The RR isolate RR1 (AG4) induced the highest RR severity scores. All bean genotypes had RR mean scores of ≥ 5.0 on a 1 to 9-point scale, when inoculated with isolates RR1, RR2 and RR3, belonging to AG4 group. Significant

line-pathogen isolate interactions were recorded for WB and RR inoculations for three planting dates, suggesting a differential response to the common bean lines to *R. solani* due to variations in growing conditions (Torres et al. 2016).

*Plasmodiophora brassicae*, an obligate pathogen, causing club root disease in cabbage and other cruciferous crops, has three phases in its life cycle: survival in soil as resting spores, primary infection of root hairs and secondary infection of and development within root cortex. Each resting spore germinates, releasing one primary zoospore which infects root hairs by penetrating the cell wall. Primary plasmodia develop within the root hairs and then cleave into zoosporangia, each containing 4–16 secondary zoospores. Visible symptoms are not expressed, following primary infections. The secondary zoospores are either released into the rhizosphere and then infect the root cortex from outside or they infect neighboring cortical cells directly from inside the root hair. Secondary infection is followed by the development of secondary plasmodia within root cortex, resulting in formation of characteristic club roots in the root system. Each secondary plasmodium is later cleaved into large numbers of resting spores within the clubroot. When the infected root tissues disintegrate, resting spores release into the soil to complete the life cycle (Ingram and Tommerup 1972; Kageyama and Asano 2009; Howard et al. 2010). As it is difficult to generate single spore-derived isolates of the biotrophic pathogens, two methods of producing pathotype isolates of *P. brassicae* were compared. Single root hairs of *Brassica* seedlings were inoculated with *P. brassicae* during the first infection cycle in root hairs. Ten single spore-derived isolates from 328 single sporangiosorus inoculations were produced, using different populations of *P. brassicae*. From 125 inoculations with single resting spores only one single spore-derived isolate was achieved. The root hair inoculation method was more practicable than the methods available earlier and it had reasonable efficacy for the production of single spore-derived isolates (Diederichsen et al. 2016).

*Thecaphora solani* causes potato smut disease in the South American countries and it induces formation of characteristic galls in stems, stolons and tubers. Galls contain oval to irregular locular sori of variable sizes with reddish-dark, granular mass of teliospores which are in the form of spore balls composed of two to eight firmly united ustilospores (Mordue 1988). In solid and liquid media, teliospores produce two kinds of vegetative tissues, depending on the media and nutrient availability. Slow-growing hyaline and septate mycelium in water agar + rifampicin (WA-R) and sponge-like mycelia mass in potato dextrose agar (PDA) + streptomycin sulfate (PDA+St) and Holliday's complete medium (HCM) + activated charcoal, were produced, after 40 to 50 days, under laboratory conditions (see Figure 3.1). Teliospore-like structures within the sponge-like mycelial mass were observed and after 90 to 100 days many teliospores could be seen. In liquid medium, formation of chlamydospore-like structures and teliospores was much faster. The teliospores were firmly attached to the sponge-like mycelial mass. The sexual cycle of the pathogen was visualized under laboratory conditions. The identity of the teliospores and sponge-like mycelial mass was

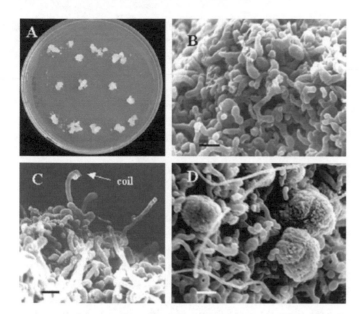

**FIGURE 3.1** Morphological characteristics of *T. solani* observed under scanning electron microscope (SEM) A: sponge-like mycelial mass developed in petriplate; B: mycelial growth as seen under SEM; C: coiling of hyphal tips and D: formation of teliospores with characteristic reticulated spore balls developing in the mycelial mass (Bar = 10 μm).

[Courtesy of Andrade et al. (2004) and with kind permission of the American Phytopathological Society, MN, United States]

established as that of *T. solani* by DNA-fingerprinting and partial sequencing of the large subunit (LSU) rDNA region of the pathogen (Andrade et al. 2004). Peanut smut caused by *Thecophora frezii*, occurring in all regions of Argentina, accounts for heavy losses. *T. frezii* forms thick-walled structures for overwintering in soil and crop residues. Teliospores survive in a dormant metabolic state in the soil in the absence of crop plants. When peanut pegs penetrate into the soil, their exudates disrupt the dormancy of teliospores, inducing them to germinate and initiating local infections. During teliospore germination, a probasidium and basidium are formed, followed by generation of basidiospores through meiosis. The basidisopres fuse to form dikaryotic mycelium that penetrates the peanut gynophores in the soil; colonizes the tissues and replaces cells with reddish brown teliospores. The teliospores are brown and echinulated. During harvest, the teliospores are dispersed by wind to adjacent fields. Long-distance dispersal may occur through infected seeds. Infested machinery may also be involved in pathogen dispersal to other fields. Teliospors of *T. frezii* could be detected and quantified in seeds and soil by PCR assay (Rago et al. 2017).

Potato tubers are affected by black dot disease caused by *Colletotrichum coccodes*. Effects of crop production on black dot disease severity in potato crops grown for different durations (days from 50% emergence to harvest) in soils that posed low, medium and high risk of disease incidence, were investigated. In field trials for over four growing seasons (2005–2008), black dot severity at harvest increased with increasing crop duration, within the range of 103–146 days

from 50% emergence to harvest (P < 0.05). The influence of storage temperature on black dot disease was also assessed between 2005 and 2006. In 2005, no difference in black dot severity was recorded on tubers stored for 20 weeks at 2.5°C and 3.5°C. But in 2006, increasing the duration of curing at harvest from four to 14 days, increased black dot severity on tubers from 8.9 to 11.2% (P < 0.01) in long duration crops (> 131 days after 50% emergence) grown under high (>1,000 pg pathogen DNA/g of soil) soil pathogen inoculum. The number of days of curing did not affect disease severity for shorter duration crops grown at high soil inoculum on crops grown at medium or low (100–1,000- and < 100 pg pathogen DNA/g of soil, respectively) soil inoculum concentrations. Soil inoculum and crop duration provided a reasonable prediction of black dot severity at harvest and after a 20-week storage period (Peters et al. 2016).

*Aspergillus flavus* and *A. parasiticus* infecting cereals and grains produce highly potent carcinogens with toxic and immunosuppressive properties. *A. flavus* shows high genetic variations, possibly due to its capacity for sexual reproduction in sclerotia naturally formed in crops. Sclerotia of *A. flavus* are survival structures and germinate sporogenically in soil by producing aerial conidiophores. *A. flavus* and *A. parasiticus* belong to *Aspergillus* section Flavi and both species commonly produce sclerotia in culture. Two morphotypes of *A. flavus*, based on sclerotial size have been differentiated: large (L) strain with sclerotia > 400 μm in diameter and small (S) strain with numerous sclerotia of < 400 μm in diameter. *A. flavus* is heterothallic and laboratory crosses produce ascospore-bearing ascocarps embedded within sclerotia. The sclerotia of *A. flavus* were produced by single strains as well as by crossed pairs of sexually compatible strains and sclerotia from both sources germinated sporogenically. Corn was grown for three years under different levels of drought stress and sclerotia were recovered from 146 ears and sclerotia of *A. flavus* strain L were dominant in 2010 and 2011, while strain S was dominant in 2012. Ascospores were not present in sclerotia at harvest. However, sclerotia buried in soil for 16 weeks formed less numbers of ascospores than the sclerotia incubated on the soil surface in both *A. flavus* L and S strains (see Figure 3.2). *Aspergillus alliaceous*, a homothallic pathogen also produced ascospores. Differences were significant for sclerotia from all crosses (< 0.05) in both experiments. When the sclerotia of section Flavi were incubated on the surface of nonsterile soil under laboratory conditions, ascospore formation was observed in 6.1% of the 6,022 sclerotia tested in 2010, 0.1% of the 2,846 sclerotia in 2011 and 0.5% of the 3,106 sclerotia in 2012. The majority of sclerotia that formed ascospores belonged to *A. flavus* L strain. Sporogenic germination of sclerotia incubated on soil surface depended on the cross from which the sclerotia were derived. Burial of *A. flavus* sclerotia inhibited ascocarp formation significantly. Formation of ascocarps with viable ascospores in *A. flavus* sclerotia that were collected from corn and incubated on nonsterile soil in the laboratory revealed the potential for sexual reproduction in the field. Variability of buried sclerotia without ascospores among crosses was 1.1 to 31.8% in experiment 1 and 90 to

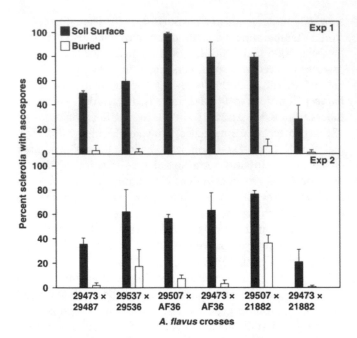

**FIGURE 3.2** Ascospore formation (%) from sclerotia of *A. flavus* incubated for 16 weeks on the surface of nonsterile soil or when buried in soil determined in two experiments (1 and 2). Bar represents means ± standard deviation of three soil cups.

**[Courtesy of Horn et al. (2014) and with kind permission of the American Phytopathological Society, MN, United States]**

100% in experiment 2. The role of soil microbial populations during sexual development and the environmental conditions conducive for sexual reproduction in nature have to be investigated (Horn et al. 2014).

Sugarcane smut pathogen *Sporisorium scitamineum* has a multistep process in its life cycle. Haploid sporidia of compatible mating types (MAT-1 versus MAT-2) fuse to generate dikaryotic hyphae that can infect susceptible sugarcane plant. Within the host tissues, diploid teliospores are formed and induce characteristic long, black whip-like sorus enclosing millions of teliospores. The diploid teliospores germinate to form haploid sporidia by meiosis. The pathogen development was followed throughout the entire life cycle, by using sporidia of *S. scitamineum* MAT-1 and MAT-2 expressing the GFP and red fluorescent protein (RFP), respectively. Epifluorescence microscopic observations showed that conjugation tube formation and sporidia fusion occurred at 4 to 8 h and formation of dikaryotic filaments could be observed at 12 h after mating. The resultant teliospores with diffused GFP and RFP, underwent meiosis as revealed by septate hypha with single fluorescent signal. The results clearly indicated the potential of using GFP- and RFP-tagged stains to investigate the process of sugarcane smut pathogen development in host tissues. This dual-color imaging system is a valuable tool for studying biotic and abiotic factors affecting the progress of life cycle of fungal pathogens and pathogenesis (Yan et al. 2016).

*Synchytrium endobioticum*, incitant of potato wart disease, is a biotrophic, soilborne, quanrantine pathogen and infection of potato tubers may result in unmarketable tubers and even

complete loss in yield. Infested fields are forbidden for cultivation, until the soil is shown to be free of pathogen propagules (Obidiegwu et al. 2014). The sporangia of *S. endobioticum* can survive in the soil for over 30 years. After 45 years, under favorable conditions, infection may occur even from a single spore of the pathogen (Przetakiewicz 2015). *S. endobioticum* has distinct phases in its life cycle, commencing with a haploid sexual reproduction phase, which may change under stress conditions, into sexual reproduction with a diploid phase. In spring, uniflagellate zoospores are released from infected tissue and may penetrate into the meristematic tissue of young tubers or stolons of potato, leaving their flagella outside. After infection, the potato host cells enlarge remarkably, forming a tumor-like tissue of the wart. The zoospore inside develops into a haploid, thick-walled sorus, containing several prosorium or zoosporangia. The summer sporangia release hundreds of haploid zoospores, which can infect new tissues. The zoospores, under favorable conditions, may function as isogametes and fuse to form a diploid, biflagellate zygote. The zygotes penetrate into host plant tissues, like the zoospores. The infected cells are buried deep in the tissue forming thick-walled resting or winter sporangia, which are released into the soil. The diploid nuclei divide repeatedly, with a first reduction division forming uninucleate haploid zoospores, which are released after rupturing the cell wall (Lange and Olson 1981; Busse et al. 2017).

### 3.1.2.2 Survival and Dispersal of Fungal Pathogens

Fungal pathogens produce thin-walled spores in large numbers during the asexual phase in their life cycle. These spores are short-lived and they have to find suitable susceptible plants for successful infection. If susceptible plants are available continuously, the pathogens can survive by producing several crops of asexual spores that can infect same or different host plant species in one crop season. At the end of season, when the plants reach harvest stage, the pathogens produce thick-walled sexual spores for overwintering/ survival during the absence of crop plant species. The sexual spores such as oospores, ascocarps or teliospores may remain in the residues or fall on the soil surface from the disintegrating infected plant tissues. These spores have different periods of survivability.

Some of the soilborne fungal pathogens can survive in the seeds, as in the case of *Peronosclerospora sorghi*, causing downy mildew of sorghum. The oospores released from infected plant debris left after harvest of sorghum germinate when the seeds are sown in the next crop season and infect the emerging seedlings and remain dormant in the seeds. The seedlings growing from infected seeds develop disease symptoms in the next season (Ahamad and Majumder 1987). *Fusarium oxysporum* f.sp. *vasinfectum (Fov)*, incitant of cotton Fusarium wilt disease, is soilborne and has worldwide distribution. The seedborne nature of *Fov* in cotton was investigated, because of wilt disease occurrence in isolated spots in uninfested field plots. The presence of *Fov* race 4, highly virulent on most commercial cotton cultivars was detected in delinted Prima cv. DP744 seeds by isolating the pathogen in Komada's agar or a selective liquid medium and also by

PCR assay using race 4-specific primers. The results clearly revealed that cotton wilt pathogen could survive in seeds and become soilborne in locations where incidence of race 4 of *F. oxysporum* f.sp. *vasinfectum* was not known previously (Bennett et al. 2008). *V. dahliae* is seedborne in spinach and when infected spinach seeds are sown, the pathogen becomes soilborne. Lettuce plants raised in the subsequent seasons are infected by *V. dahliae* present in the soil. Thus, *V. dahliae* could survive in the spinach seeds and later become soilborne to infect lettuce crops (Short et al. 2015).

*Tilletia tritici*, causal agent of wheat Karnal bunt or partial bunt disease, is soilborne and the teliospores of this pathogen may fall on the soil surface during harvest of wheat heads. The teliospores with thick walls are resistant to adverse conditions existing in the soil and they persist in the soil for a long time. When favorable conditions are available, the teliospores germinate, produce primary and secondary sporidia which are carried by wind currents and deposited on wheat head at the time of flowering, resulting in infection of florets of susceptible wheat cultivars. Survival of teliospores of *T. tritici* in soils with different texture, and other soil characteristics was studied. The teliospores enclosed in polyester bags were kept in polyvinyl chloride tubes and they were buried vertically in the ground. The percentages of teliospores recovered from soil samples decreased progressively from 90.2 to 13.3%, as the interval after burial of teliospores increased from one day to 32 months. The soil type and soil temperature had significant influence of the teliospore survival. The loamy soil was less inhibitory to the teliospores than silt loamy soil. A negative correlation was observed between soil temperature (22°C to – 18°C) and percentages of teliospore germination after 37 months of incubation, the maximum germination occurring at – 5°C. Microscopic examination revealed disintegration of teliospores after breakdown of the sheath covering teliospores (Babadoost et al. 2004).

Survival of *Didymella bryoniae*, causal agent of gummy blight disease of watermelon was studied by placing naturally infected dried vines in nylon mesh bags covered with soil and buried in the field soil at depths of 0-, 12.5-, or 25.0 cm in November and December of 1997 to 1999. The percentages of 1-cm vine segments that yielded *D. bryoniae* and the number of segments retrieved intact declined over time. Vine segments cultured on semiselective medium yielded the pathogen for 30, 24 and 21 weeks, respectively for three years. A hypocotyl-infection assay, using watermelon seedlings was developed to detect *D. bryoniae* conidia produced on recovered vine sections. The percentage of seedlings exhibiting disease symptoms increased with the logarithm of inoculum density. Viable conidia were produced on retrieved vine sections placed on semiselective agar for up to 32 weeks after burial, but conidia induced disease for only 16 weeks after burial in the hypocotyl-infection assay. Conidia of *D. bryoniae* were recovered for eight weeks longer at 0 cm (on soil surface) than at lower depths in 1999, but not in 1998. The results indicated that incorporation of infested crop residues into the soil reduced the number of infective propagules and reduced the survival of *D. bryoniae* (Keinath 2002). In a further study,

the survival of *D. bryoniae* in infested crowns of muskmelon (*Cucumis melo* subsp. *melo*) on the soil surface and buried at a depth of 12.5 cm was investigated. Dried crowns with cankers were buried or placed on soil surface or placed on top of raised beds covered with white-on-black polyethylene film mulch at different months between 2002 and 2007. At regular intervals, crowns or crown debris were retrieved, washed, cut into pieces and cultured in semiselective medium to recover *D. bryoniae*. The pathogen was not recovered from crowns buried for 35 and 45 weeks in 2003 and 2004, but it could be recovered from 2.5% of crowns buried 66 weeks in 2005. In contrast, *D. bryoniae* was recovered for 48, 45, 66 and 103 weeks from 6.6, 6.3, 2.5 and 10% of crowns placed on the soil surface in 2003, 2004, 2005 and 2007, respectively. *D. bryoniae* was recovered after 66 and 103 weeks from 12.5 and 8% of crowns on mulched beds in 2005 and 2007. The results indicated that the survival of *D bryoniae* might be reduced by incorporating the infected debris promptly into the soil after harvest (Keinath 2008).

Survival and infectivity of oospores of *Phytophthora infestans*, causing potato late blight disease, were investigated in central Mexico. Sporangia were selectively eliminated from soil by two cycles of wetting and drying the soil to determine the infectivity attributable to the presence of oospores. Oospore concentration, viability and infectivity varied among soils collected during winter fallow in different locations of central Mexico. In some soils, oospores were infective regardless of the time at which they were collected during the winter fallow. However, the oospore viability and infectivity decreased, following two years of intercropping. The number of stem lesions and initial disease severity were significantly higher in soils with moderate oospore infestation (20 to 39 oospores/g of soil), compared to soils with low (0 to 19 oospores/g of soil) infestation. The results confirmed that oospores could survive winter fallow and serve as a source of primary inoculum in the central highlands of Mexico. Oospore survival appeared lower in the Toluca valley soil which might be an indication of soil suppressiveness (Fernández-Pavía et al. 2004). Persistence of relatively new *P. infestans* clonal lineages US-22, US-23 and US-24 during winter season in Wisconsin, United States, suggested investigation to determine the ability of the late blight pathogen under freezing conditions. Tomato seed was used as a culture medium to determine the survival of *P. infestans* isolates of the three lineages under temperatures of 18, 4, 0, – 3 and – 5°C for 11 time points (1 to 112 days postincubation). Survival of clonal lineages varied interactively with temperature and duration of exposure. US-22, US-23 and U-24 isolates survived for 112 days at 18°C and 4°C, 84 days at 0°C and 14 days at – 3°C. US-23 isolates survived longer at – 3 °C and – 5°C than did US-22 or US-24. The vigor of US-22 and US-24 isolates decreased with increasing exposure time to cold temperatures and this trend was not seen in US-23 isolates. Based on the length of time needed to kill the lineage isolates on infested tomato seed at five temperatures, it was possible to predict that *P. infestans* would survive in 5% of tomato seed for 99, 25 and 16 days at 0, – 3 and – 5°C, respectively.

The results suggested that *P. infestans* might survive over the winter season by asexual means in infested tomato seed in Wisconsin and other Northern latitudes (Frost et al. 2016).

Most *Phytophthora* spp. have a soilborne phase in their life cycle which is crucial for infection of roots and survival during the absence of the host plants. The soil phase of *P. ramorum* infecting several ornamental plant species and forest trees was investigated. Colonies of *P. ramorum* could be recovered from moist potting mix or sand for many months, whether buried as infected plant leaf tissue or as mycelium bearing chlamydospores. The buried material was resistant to treatment with acidic electrolyzed water (AEW, a product of electrolysis) used for sterilizing the soil. No significant difference could be seen in the recovery over time among plant tissue or mycelium. Approximately after one year, colonies could be recovered (0.8 to 14.3%), when excised roots inoculated with *P. ramorum* sporangia buried in mesh bags in potting mix. The pathogen was recovered from buried roots for at least eight to 11 months. But it was not clearly known, whether pathogen survived as mycelium or chlamydospores. When chlamydospores were placed near roots and observed directly, they germinated forming sporangia. Most of the roots were infected, the tips of roots being covered with sporangia. The results indicated the existence of a soil phase of *P. ramorum* at least under greenhouse and nursery conditions (Shishkoff 2007). In another investigation, *P. ramorum* could be recovered from symptomless roots under laboratory or greenhouse conditions. Survival of *P. ramorum* in the root ball of *Rhodendron* container plants as well as in different rootless forest substrates and a horticultural potting medium was investigated. A novel nondestructive baiting assay was developed for the recovery of *P. ramorum* from inoculated root balls of *Rhododendron*, the aerial plant parts of which remained asymptomatic. The leaf bait technique involved a combination of partial flooding and baiting and plating onto semiselective medium. The baiting technique was nondestructive, sensitive and applicable to large numbers of samples (*ex situ*) and it outperformed baiting of roots and direct microscopic detection of cleared roots. The pathogen could be recovered until at least eight months postinoculation. Plating of surface-sterilized roots and direct microscopic observations confirmed the presence of *P. ramorum* in the roots. Inoculum of *P. ramorum* could be baited from the root balls. Baiting experiments with plants from commercial nurseries also showed survival of the pathogen in plants with latent infection, since *P. ramorum* was detected in root balls of symptomless and noninoculated *Rhododendron* plants. The pathogen was baited from these root balls up to two years (between 2008 and 2010) after the plants were obtained from the nurseries. Survival of *P. ramorum* in rootless media was assessed by burying disks of infected leaf material below soil surface in columns filled with four different undisturbed forest substrates, or a potting medium. *P. ramorum* could be recovered at least 33 months after burial from all substrates, with a significant increase in recovery after the winter season. The results suggested that *P. ramorum* might be present in root balls which remained asymptomatic surviving as latent infection in root balls as well as in potting medium and different forest substrates under western European climatic conditions. *P. ramorum* might overwinter outdoors in Belgium, in forest soils as well as in commercial potting medium contributing to the survival and spread of *P. ramorum* to other locations (Vercauteren et al. 2013).

Sclerotial development is a basic requirement for development of diseases induced by *Sclerotinia scleortiorum*, since sclerotia are important for the survival of the pathogen during the absence of host plant species. A highly conserved homolog of ERK-type mitogen-activated protein kinases (MAPKs) was required for sclerotial development in *S. sclerotiorum* (*Smk1*). During sclerogenesis, the *smk1* transcription and MAPK enzyme activity were induced substantially. Sclerotial maturation was impaired by applying inhibitors of MAPK activation. Addition of cAMP inhibited *smk1* transcription, MAPK activation and sclerotial development. The results showed that *S. sclerotiorum* could coordinate environmental signals (such as pH changes) to trigger a signaling pathway mediated by SMK1 to induce sclerotial production and this pathway was negatively regulated by cAMP (Chen et al. 2004). *L. maculans*, causal agent of Phoma stem canker or blackleg disease of oilseed rape (canola) overwintered in crop residues in South Australia. Presence of *L. maculans* in soil surface and deeper layers of soil was detected by a quantitative PCR assay. As the size of the organic matter particles in the soil decreased, the population of *L. maculans* also decreased. In a survey of 49 commercial fields, most *L. maculans* population was detected in the fields one year after oilseed rape had been cultivated. Two years after harvest of oilseed rape, the population of *L. maculans* detected was much less and was negligible after three or more years after oilseed rape cropping. The diagnostic DNA-based assay applied in this study, reduced the time and cost of studying the survival of *L. maculans* in soil and increased the sensitivity and accuracy of results, compared with estimates of propagule number or CFU on a semiselective medium (Sosnowski et al. 2006).

*R. solani* and *R. oryzae-sativae* cause bordered sheath spot and brown sclerotium diseases of rice, respectively. Their survival in soil and stubble during the preplanting period and the effect on disease development during the maturation period of rice were investigated. All field isolates of these pathogens from soil, stubble, rice sclerotial disease lesions (infected tissues) and weeds (belonging to 17 families) were assigned to mycelial compatibility groups (MCGs). Members belonging to a single MCG from diseased rice tissues were detected from maximally five weeds growing in the neighboring fields. The pathogen causing sclerotial diseases at the maturation stage of rice plants survived on and in soil and stubble until the preplanting period of the next year, followed by wide dispersal in and out of fields by infection and disease development on rice plants and different weed species present in adjacent fields (Guo et al. 2006). *M. phaseolina*, causal agent of crown and RR of strawberry, is a cosmopolitan pathogen, capable of infecting > 500 botanical plant species (Mihail 1992). Survival of *M. phaseolina* in soil was studied, using 151 isolates of the pathogen obtained from infected strawberry plants of commercially grown cultivars in Israel. Sclerotial viability declined more rapidly in soil maintained at 25°C or at soil temperatures fluctuating from 18°C to 32°C under greenhouse conditions, compared to sclerotial viability in soil at 30°C. After 30 to 40 weeks of exposure in soil, inocula maintained at 25°C or 30°C or at fluctuating temperatures in greenhouse declined to negligible levels. A significant increase in plant mortality was observed in infested soils maintained at 30°C versus 25°C, whereas water stress at 25°C or 30°C did not affect plant mortality in *M. phaseolina*-infested soils. Infection of various crops grown in close proximity by *M. phaseolina* was detected. No significant host specialization was observed among isolates of the pathogen obtained from almond, aralia, protea, melon, watermelon or strawberry. It is possible that *M. phaseolina* might survive on these crops in the absence of strawberry plants, suggesting that these crops have to be avoided as rotation crop in strawberry fields (Zveibil et al. 2012). Survival of *R. solani* AG-1IA, incitant of rice sheath blight disease, in diseased rice straw as sclerotia and mycelia was assessed. Sclerotia placed in the desiccators, soaked in sterile water or immersed in wet paddy soil were viable, after storage for ten months at 4°C, 25°C, and non-aircondtioned natural room temperature (NRT, 6 to 35°C). In contrast, only 15% of sclerotia in dry paddy soil were viable. Temperature and humidity reduced the viability of mycelia drastically. After ten months in the desiccators at 4°C, 55% of the mycelia survived, whereas at 25°C and NRT, mycelia could survive only for seven and five months, respectively. However, at 4°C or 25°C, mycelia stored in sterile water at constant temperatures (4°C/25°C) could survive for ten months. The survival rates of *R. solani* in diseased straw stored for 16 months were 100% at 4°C or 50% at 25°C and 35% at NRT. The survival rates of the pathogen in diseased rice straw buried in dry, wet and flooded paddy soils were 75%, 100% and 100%, respectively, after storage for ten months at NRT. The results indicated that *R. solani* could survive in diseased rice straw, till next planting with susceptible rice crop (Feng et al. 2017).

Survival, persistence and efficiency of *V. dahliae* passed through the digestive tract of sheep were determined. Pathogenicity of and disease intensity induced by 32 isolates of *V. dahliae* were assessed by inoculating eggplant, turnip, tomato and potato. The infected plant material was used to feed four one-year-old sheep. Presence of *V. dahliae* in the fecal samples from the test animal was detected using PCR assays up to five days after feeding the sheep with *V. dahliae*-infected plant materials and the pathogen could not be detected at zreo, six and seven days after feeding. Pathogenicity tests were performed by planting eggplant seedling in soil substrate amended with 20% decomposted manure collected from the four test animals. Verticillium wilt disease symptoms were observed at 52 days after transplanting eggplant seedlings. Disease incidence, disease severity and percentage of pathogen isolations from stem tissues of eggplants at 60 *dai* were 58.3, 30.7 and 48.3%, respectively. All control plants did not show any symptoms of infection by *V. dahliae*. The results revealed the important role of *V. dahliae* passed through animal systems as a source of inoculum

and the relative persistence and transmission via sheep digestive system (Markakis et al. 2014).

*P. brassicae*, an obligate endoparasite, causes one of the most devastating diseases of cabbage and *Brassica* crops. When infected roots decay, a large number of resting spores are dispersed into the soil, where they can survive for about 20 years (Wallenhammer 1996). Soil transmission is the principal mode of dispersal of *P. brassicae*. The possibility of pathogen dispersal through livestock manure was suggested by Karling (1968). A quantitative PCR (qPCR) assay was developed to detect and quantify resting spores of *P. brassicae* in manure samples from naturally and artificially infested chickens and pigs. Resting spores of *P. brassicae* could be reliably and unequivocally quantified in seven of 28 naturally infested manure samples, with an infection rate of 25%. The highest level of resting spore infestation of $8.3 \times 10^7$ spores/g of pig manure was detected, whereas the average infestation level of chicken manure was $3.6 \times 10^7$ resting spores/g of manure. The levels of resting spore infestation decreased gradually in chicken manure samples collected from two to 48 h after feeding (HAF). Resting spores of *P. brassicae* were not detected until 6 HAF in manure samples from pigs fed on artificially infested feed. The resting spores could be detected in pig manure samples for six to 72 HAF. Resting spores could stay in the digestive tracts of pigs for 48 h and the manure excreted as long as two days, following ingestion might carry resting spores of *P. brassicae*. The viability of *P. brassicae* resting spores after passage through the highly acidic livestock gut environments was assessed by double staining with Hoechst 33342 and propidium iodide (PI). The dead spores showed intense red fluorescence, and the live spores emitted pale blue fluorescence. The assay showed that 99–100% of resting spores were viable after passage through the digestive tracts of chicken and pigs. In addition, cabbage plants were used as bait plants. Severe symptoms of club root developed in bait plants at infestation levels of $10^6$–$10^7$ resting spores/g of manure. A positive linear relationship was observed between the numbers of *P. brassicae* resting spores (determined by qPCR assay) and disease severity indices (DSI) (assessed by bioassay). The DSI increased as the infestation levels of manure increased from $10^3$ to $10^8$ resting spores/g of manure. The results indicated that the extent of spread of clubroot through contaminated manure might be much more widespread than that was realized, suggesting that feeding livestock on clubroot-infested feed or applying infested manure should be avoided for reducing incidence and spread of club root disease in cruciferous crops (Chai et al. 2016).

*A. flavus* produces aflatoxins which contaminate cereal grains, cotton seeds and peanuts which cause serious health problems in humans and animals. Corn-cotton rotations commonly adopted in Texas, United States, resulted in long-term residence of corncobs on soil surface in fields where reduced tillage practice was followed. Corncobs were colonized by *A. flavus* either prior to harvest or while in the soil. *A. flavus* communities in corncobs and soil samples were obtained from 29 fields between 2001 and 2003. Persistence of *A. flavus* in corncobs was assessed by extracting *A. flavus* communities in corncobs and soil every two to five months intervals from four fields for a period of three years. Corncobs were found to be major sources of *A. flavus* inoculum. Corncobs from the previous season contained, on an average, over 190 times more *A. flavus* propagules than soil samples from the same field and two-year old corncobs still retained 45 times more propagules than the soil samples. The populations of *A. flavus* in corncobs decreased with increase in the age of corncobs (Jaime-Garcia and Cotty 2004). Soil environmental conditions exert significant influence on the survival of soilborne fungal pathogens. The influence of irrigation on the survival of teliospores of wheat Karnal bunt pathogen *Tilletia indica* in field plots in Arizona, United States was assessed. Two methods were employed to test the viability of teliospores during a 48-month period. Polyester mesh bags (21 μm pore size) containing teliospore-infested soil were buried in irrigated and nonirrigated field plots at two sites. The total number of viable teliospores and percentages of germination of teliospores extracted from the soil samples by sucrose density gradient centrifugation were determined. The total number of viable teliospores decreased from 55.7% at time zero to 9.7% and 6.7% for nonirrigated and irrigated field soils, respectively in 48 months. The total number of viable teliospores in soil in the laboratory decreased from 55.7 to 34.0%, after 48 months. Germination of teliospores decreased significantly over time. The rate of decrease in germination was significantly greater for teliospores from irrigated field plots than from nonirrigated plots and the laboratory soil. The field site and soil depth at which the teliospores were buried did not have any effect on total number of viable teliospores or their germination percentages (Bonde et al. 2004).

The survival of asexual propagules, sporangia and zoospores, of *P. infestans* in soil was studied. Coverless petridishes containing water suspensions of sporangia and zoospores of *P. infestans* were embedded in sandy soil in eastern Washington in July and October 2001 and July 2002 to determine the longevity of the propagules in water under natural conditions. Effects of solar radiation intensity, presence of soil in petridishes (15 g/dish) and a two-h chill period on survival of isolates of clonal lineages US-8 and US-11 were also assessed. Spores in water suspensions survived 0 to 16 days under nonshaded conditions and two to 20 days under shaded conditions. Mean spore survival significantly increased from 1.7 to 5.8 days, when soil was added to the water. Maximum survival time of spores in water without soil exposed to sunlight was two to three days in July and six to eight days in October. Mean duration of survival of spores did not vary significantly between chilled and nonchilled spores that survived for extended periods than that of nonchilled spores. Spores of US-11 and US-8 isolates did not show variation in the mean duration of survival, but significantly greater number of sporangia of US-8 survived than did sporangia of US-11 in one of three trials conducted (Porter and Johnson 2004).

Survival of *P. ramorum* in recirculating irrigation water was investigated, using an open air simulation system with nine separate container stands, each connected to its own water collection system. The water in these reservoirs was inoculated with *P. ramorum* and then used for overhead irrigation over the course of the season to study the spread of P. *ramorum* and development of disease symptoms of blight in *Rhododendron* and *Viburnum* spp. The maximum number of *Rhododendron* plants (19.0%) showed infection by *P. ramorum* applied through contaminated irrigation water. In the two years of experimentation, symptom development began at eight and 16 *dai* with contaminated water. Presence of *P. ramorum* could be detected in the water reservoirs over the course of growing season and the percentage of plants getting infected varied with year and season (Werres et al. 2007). Survival of *Phytophthora* spp. as zoospores or sporangia was investigated in response to an important water quality parameter, electrical conductivity (EC) at its range in irrigation water reservoirs and irrigated cropping systems. Hoagland's solution at different strengths were tested. *P. ramorum*, *P. kernoviae* and *P. alni* survived at a broad range of EC levels for at least three days and these pathogens were stimulated to grow and sporulate at ECs >1.89 dSm–1. Recovery of initial populations after a 14-day exposure was over 20% for *P. alni* subsp. *alni* and *P. kernoviae* and 61.3% and 130% for zoospores and sporangia of *P. Ramorum,* respectively. Zoospore survival of these pathogens at ECs < 0.41 dSm–1 was poor, barely beyond three days in pure water. The results indicated that these pathogens could survive better in cropping systems than in irrigation water. Containment of run-off and reduction in EC levels might be nonchemical management options to reduce the risk of pathogen through natural waterways and irrigation systems (Kong et al. 2012).

The effects of temperature on the survival of oospores of *Aphanomyces cochlioides* isolate C22 were assessed. The oospores were exposed in water to 35, 40, 45 or 50°C for predetermined duration and their viability was assessed. Lethal doses of temperature for 50% germination of oospores were 25, 49.8, 9.8 and 1.9 h at 35, 40, 45 and 50°C, respectively. The effects of alternating high and low temperatures were also investigated. The oospores were examined after each of four 24-h cycles at 45°C for 4 h and 21°C for 20 h. Significant variability in heat tolerance among five isolates was observed. The results indicated that under wet conditions, there could be predictable patterns of mortality of *A. cochlioides* oospores exposed to continuous fluctuating high temperatures (Dyer et al. 2007).

### 3.1.2.3 Transmission of Soilborne Fungal Pathogens through Vectors

Soilborne fungal pathogens are known to be transmitted through seeds, propagules and irrigation water. Some of them have been shown to be transmitted via biological vectors. Involvement of fungus gnats in the transmission of *Pythium* spp. in the greenhouse was studied. The ability of *Bradysia impatiens* larvae to ingest propagules of two strains of P. *aphanidermatum* and *P. ultimum* and to transmit the pathogens to healthy geranium seedlings on a filter paper substrate in petridishes was assessed. *Pythium* spp. transmission by larval fungus gnats varied greatly with the assay substrate and also with the number and nature of ingested propagules.

Transmission was highest (65%) in petridish assays, when larvae fed on *P. aphanidermatum* K-13, a strain that produced abundant oospores, were used. Transmission of strain K-13 was much lower (< 6%) in plug cells with potting mix. Larvae were less efficient at vectoring *P. ultimum* strain PSN-1 which produced few oospores. No transmission occurred of two non-oospore-producing strains of *P. aphanidermatum* Pa558 and *P. ultimum* P4. Passage of *P. aphanidermatum* K-13 through larval guts significantly increased oospore germination. On the other hand, decreased germination of hyphal swellings was recorded following larval gut passage for strains of *P. ultimum*. The results suggested that larval fungus gnats may function as vector of *Pythium* spp. (Braun et al. 2012). Mango sudden decline disease caused by *Ceratocystis manginecans* is responsible for considerable losses due to the death of infected trees. The role of bark beetle *Hypocryphalus mangiferae* in the spread of sudden decline disease in Oman was investigated. *H. mangiferae* were fed on mango trees with sudden decline symptoms and then they were fed on mango seedlings which developed disease symptoms after six weeks. *C. manginecans* was isolated from the wilted mango seedlings. The control seedlings remained healthy. The results indicated that the bark beetles could vector the mango sudden decline pathogen (Al Adawi et al. 2013). Forest soils may contain many soilborne fungi which may be parasitic, mutualistic (ectomycorrhizal) or hyperparasitic (antagonists) in their behavior. Two earthworm species, *Lumbricus terrestris* and *Aporrectodea caliginosa* were able to feed on roots infected by *Phytophthora cactorum*, *Nectria radicicola*, *Fusarium reticulatum* and *V. dahliae*. Casts were collected from each earthworm species fed with different species. The casts were analyzed by molecular methods and the presence of viable pathogenic fungal species was detected in the casts, indicating that earthworms might be involved in the spread of fungal pathogens in forests (Montecchio et al. 2015).

### 3.1.2.4  Host Range of Fungal Pathogens

The ability of a soilborne fungal pathogen to infect large number of plant species other than the primary host plant, is a crucial epidemiological factor significantly contributing to high levels of disease incidence which in turn determines the rate of spread of the disease induced by the pathogen concerned. Fungal pathogens with wide host range may be able to overcome the limitation due to crop discontinuity, because of availability of alternative sources of inoculum which may be the crop hosts and/ or weed hosts growing adjacent to the cropped area. As the number of alternative host plant species increases, the buildup of inoculum may be more rapid, leading to epidemics and consequent greater economic losses. *P. infestans*, causal agent of potato late blight, survives as oospores in soil and is able to infect 47 *Solanum* spp. that can serve as alternative hosts, most of which are weeds present in adjacent fields. Plants belonging to other genera *Datura*, *Geranium* and *Ipomoea* were also reported as hosts of *P. infestans* (Erwin and Ribeiro 1996).

*P. capsici* is another soilborne pathogen capable of infecting many crop hosts as well as weed plant species. *P. capsici* causes significant losses in bell pepper (*Capsicun annuum*), eggplant (*Solanum melongena*), squash (*Cucurbita pepo* var *condensa*), and watermelon (*Citrullus lanatus*). The weed hosts affected by *P. capsici* were American black nightshade (*Solanum americanum*), *Portulaca oleracea* and *Geranium carolinianum*. These weeds may contribute to the survival of *P. capsici* in the absence of crop plant species or when propagules may not readily survive in soil or plant debris (French-Monar et al. 2006). In a later study, isolates of *P. capsici* from four species of *Capsicum*, *C. annuum*, *C. baccatum*, *C. chinense* and *C. pubescens* and tomato (*Solanum lycopersicum*) collected from 33 field sites were found to belong predominantly to one genotype, indicating the inoculum from one host species could infect pepper and tomato (Hurtado-Gonzales et al. 2008).

*P. brassicae*, causal agent of club root diseases of cruciferous crops, has assumed importance increasingly in China, because of an extended host range that includes several economically important crops. In the field, presence of visible galls in 17 species including radish, *Capsella bursa-pastoris*, *Orychophragmus violaceus*, *Sinaphis alba* and 13 *Brassica* crops was observed. Under pot culture conditions, an additional 13 plant species in 11 genera were found to be hosts of *P. brassicae*. Five weed species viz., *C. bursa-pastoris*, *Lepidium apetalum*, *Descurainia sophia*, *S. alba* and *Thellungiella salsuginea* were infected by *P. brassicae*. Infection of these plants was confirmed by PCR assay with pathogen-specific primers. Presence of *P. brassicae* could be detected in asymptomatic roots of *Matthiola incana*. Microscopic examination showed infection only in root hairs of *M. incana*, indicating its resistance to *P. brassicae*. Of the 297 accessions of oilseed rape tested in the field, three accessions of *B. napus* and one accession of *B. juncea* showed high levels of resistance to *P. brassicae* (Ren et al. 2016).

*Phytophthora cinnamomi* causes extensive RR in many crops, ultimately killing them, when fine roots and lower stem tissues are destroyed. It has been estimated that *P. cinnamomi* causes decay of fine surface roots, resulting in host vulnerability to seasonal drought stress in over 3,000 plant species worldwide, including agricultural, ornamental and forest tree species (Erwin and Ribeiro 1996; Hardham 2005). Infection of major oak species *Quercus glaucoides*, *Q. peduncularis* and *Q. salicifolia* by *P. cinnamomi* was diagnosed in a 300-ha of mixed oak trees in a native forest in southern Mexico by isolating and proving the pathogenicity of the isolates of *P. cinnamomi* (Tainter et al. 2000). Sudden oak death (SOD) disease caused by *P. ramorum* affects some members of Fagaceae, Ramorum shoot dieback on some members of Ericaceae and conifers and Ramorum leaf blight on diverse hosts. Infection of Pacific *Rhododendron*, salmon-berry, cascara and poison oak by *P. ramorum* was confirmed by applying Koch's postulates. Douglas-fir was found to be most susceptible to shoot dieback, shortly after budburst, with infection being initiated in the buds (Hansen et al. 2005). *Rhododendron* spp., *Viburnum* and *Pieris japonica* are commonly infected and they are used as trap plants for detecting *P. ramorum* in soil and irrigation water. Under inoculated conditions, *Acer macrophyllum, Camellia oleifera, C. sinensis, C. sasanqua, Taxus baccata, Umbellularia californica, Vaccinium macrocarpum, Viburnum davidii* and *Syringa vulgaris* were infected by *P. ramorum*, indicating possible further expansion of host range of this pathogen (Shishkoff 2007; Loyd et al. 2014). *Phtyophthora nicotianae* has a broad host range infecting herbaceous and perennial host plants which exhibit RR and crown rot symptoms. Tobacco black shank and citrus RR and gummosis diseases are responsible for heavy losses. Some of the other plants species infected by *P. nicotianae* are *Chrysanthemum* spp., *Cyclamen* sp. *Dodonaea viscosa, Diffenbachia maculata, Lavendula* sp., *Rhamnus alaternus, Catharanthus roseus, Carthamus tinctorius, Citrus jambiri, C. clementine* and *C. aurantium*. Presence of *P. nicotianae* in these plant species was confirmed by molecular technique (Mammella et al. 2013).

*R. solani*, capable of infecting large numbers of plant species, is a complex species composed of 64 AGs differing in their pathogenicity and aggressiveness (virulence). The isolates of AG1 are placed in three subgroups, IA, IB and IC and the subgroup AG1-IA causes rice sheath blight disease (González-Vera et al. 2010). Other isolates of *R. solani* AG1-IB, AG1-IC and *R. oryzae-sativae* are associated with sheath blight symptoms to a lesser extent (Chaijuckam et al. 2010). *R. solani* AG1-IA also causes banded leaf and sheath blight on maize, aerial blight and stem blight on soybean, mungbean and cowpea, sheath blight on sorghum and foliar blight on durian and coffee (Padasht-Dehkaei et al. 2013). *R. solani* AG2-2 is the causal agent of RCRR disease of sugar beet, accounting for significant losses (Reynolds et al. 2012). *R. solani* isolates, infecting dry field pea (*Pisum sativum*) were assigned to AG4 homogenous group (HG)-II (17 isolates) and AG5 (two isolates), indicating the variability in populations of *R. solani* associated with RR disease (Matthew et al. 2012). The phylogenetic diversity of isolates of *R. solani* infecting wheat and canola (oilseed rape) was investigated. The isolates occurring in major canola- and wheat-growing regions belonged to AG2-1, AG4 or AG5. The isolates of *R. solani* could be grouped based on the internal transcribed spacer (ITS) region of the pathogen DNA (Borders et al. 2014). Isolates of *R. solani* collected from black scurf on potato tubers from different potato-growing regions in New Zealand, were classified into three anastomosis groups, AG-3PT, AG-2-1 and AG-5. Isolates of AG-3PT were widely distributed, whereas AG-2-1 and AG-5 were confined to distinct locations. The isolates of AG 2-1 were consistently more aggressive than those of AG-3PT. Delayed emergence, severe infection on stolons, formation of aerial tubers and high yield losses were associated with AG-2-1 isolates, but they caused negligible black scurf. In contrast, AG-3PT isolates induced black scurf on progeny tubers, but variable effects on stem emergence and stolons. Further, AG-2-1 isolates caused severe tuber malformation, but isolates of other AGs did not. The results revealed the variations in the biological characteristics of isolates of *R. solani* from potato occurring in different locations

(Das et al. 2014). Two binucleate *Rhizoctonia* (BNR) isolates were obtained from cankered stems of potato in China. Based on the lack of anastomosis with reference strains of BNR from AG-A to AG-Q and AG-U, the BNR isolates in China were considered as a new BNR AG and designated as AG-W. These two BNR isolates induced brown, dry and slightly sunken lesions on potato subterranean stems. Various investigations have shown that *R. solani* infecting large number of crops and other plant species may be classified into different groups/classes, based on anastomosis and molecular assays (Yang et al. 2015).

*V. dahliae* is another soilborne pathogen with wide host range, including several crops that are cultivated the world over. Two races of *V. dahliae* infecting tomato and lettuce have been identified. Genetic variability and race structure of 101 isolates of *V. dahliae* from various host plant species were investigated. Tomato subpopulations of *V. dahliae* were distinct from marigold subpopulations. In contrast, cotton and olive isolates showed admixture with tomato isolates (Maruthachalam et al. 2010). *V. dahliae* and *V. longisporum* were involved in Chinese cabbage yellows disease in Japan. Under greenhouse conditions, both pathogen species were equally virulent to Chinese cabbage. However, *V. longisporum* was more aggressive in the field compared to *V. dahliae*. Frequency of occurrence of *Verticillium* isolates was determined. Among the 67 isolates tested, 53 were *V. longisporum* and hence, it was considered to be the major pathogen causing yellows disease. These two pathogens showed different geographical distribution during autumn-winter and cool summer seasons (Ikeda et al. 2012). Populations of *V. dahliae* from potato and mint were characterized for mating types and vegetative compatibility groups (VCGs). The potato isolates were primarily VCG4A, whereas mint isolates were primarily VCG2B. Mint and potato populations of *V. dahliae* were significantly genetically diverged. Populations of *V. dahliae* in potato and mint were characterized by the presence of aggressive, clonally reproducing haplotypes which were widely distributed in commercial mint and potato production (Dung et al. 2013). Of the 84 isolates of *V. dahliae*, 59 isolates were assigned to tomato, 19 to eggplant, one to sweet pepper and five to tomato-sweet pepper pathogenicity groups. The tomato-sweet pepper pathogenicity group was morphologically quite distinct from other groups (Pappaionnou et al. 2013). *Verticillium longisporum* is an economically important vascular pathogen infecting many members of Brassicaceae in different countries. *V. longisporum* is a diploid hybrid that consists of three different lineages, each of which originated from a separate hybridization event between two different sets of parental species. Pathogenicity tests were performed on 11 different hosts, using 20 isolates representing the three *V. longisporum* lineages and closely related *V. dahliae*. Overall, *V. longisporum* was more virulent on the Brassicaceae crops than *V. dahliae*, which was more aggressive than *V. longisporum* across the non-Brassicaceae crops. Differences in virulence of *V. longisporum* lineages were observed. *V. longisporum* lineage A1/D1 was the most aggressive on oilseed rape, whereas the lineage A1/D2 was the most aggressive on cabbage and horseradish. In addition, V. *longisporum* was equally or more virulent than *V. dahliae* on non-Brassicaceae hosts, eggplant, tomato, lettuce and watermelon, suggesting that *V. longisporum* might have a wider host range than that which is currently known (Novakazi et al. 2015).

*Thielaviopsis basicola*, a soilborne pathogen, causing black RR of tobacco, has worldwide distribution with a wide host range of plants included under 137 genera (Koike and Henderson 1998; Maria et al. 2006). Cotton Phytmatotrichum RR, caused by *Phymatotrichopsis ominvora* induces RR diseases in over 2,000 dicotyledonous plant species, including alfalfa, vegetable crops and fruit and nut trees, accounting for annual losses of up to $100 million to the cotton crop alone in the United States (Streets and Bloss 1973; Marek et al. 2009). The soilborne necrotrophic pathogen *Sclerotinia sclerotiorum* causing stem rot disease of sunflower, infects over 500 plant species, including several economically important field and fruit crops grown around the world. The majority of these hosts are dicotyledonous plants, although a number of agriculturally important monocotyledonous plants such as onions and tulips are also infected by *S. sclerotiorum* (Boland and Hall 1994; Saharan and Mehta 2008).

### 3.1.2.5 Disease Complexes of Soilborne Fungal Pathogens and Nematodes

Disease complexes affecting various crops, involve frequently soilborne fungal pathogens and nematodes. The role of nematodes in the development of diseases caused by fungal pathogens has been investigated. Some of them include *Fusarium solani* with *Heterodera glycines* in soybean (Sugawara et al. 1997), *R. solani* with *Pratylenchus thornei* in chickpea (Bhatt and Vadhera 1997), *Fusarium oxysporum* f.sp. *pisi* with *Rotylenchulus reniformis* in pea (Vats and Dalal 1997), *R. solani* with *Meloidogyne incognita* in peanut (Abdel-Momen and Star 1998), *V. dahliae* with *Pratylenchus neglectus* in potato (Hafez et al. 1999), *T. basicola* with *M. incognita* in cotton (Walker et al. 2000), *R. solani* with *M. incognita* in tomato (Arya and Saxena 1999), *P. sojae* with *H. glycines* in soybean (Kaitany et al. 2000), *R. solani* with *Globodera rostochiensis* in potato (Back et al. 2000), *V. dahliae* with *Pratylenchus penetrans* in mint (Johnson and Santo 2001) and *Fusarium incarnatum* with *Pratylenchus delattrei* in crossandra (Mallaiah et al. 2014).

A multitude of microorganisms exists in soil under natural conditions. Different populations of bacteria (106–108), actinomycetes (106–107), fungi (5 × 104 – 5 × 106) CFUs, protozoa (105–106) and algae (104 – 5 × 105) were estimated to be present in a gram of field surface soil (Gottlieb 1976). On the other hand, the population of nematodes in 1 m2 was ca. 1 × 107 (Richards 1976). The moist soil conditions favored the development of both plant parasitic fungi and nematodes. A range of relationships between fungal pathogens and plant parasitic nematodes (PPNs) has been observed. A disease complex is the resultant of synergistic interaction between these two soilborne pathogenic organisms, leading to plant damage exceeding the sum of individual damage likely to be induced by them separately. Antagonistic relationship

between fungus and nematode may result in less plant damage than expected. The interaction between Fov and root-knot nematode *Meloidogyne* sp. was first reported by Atkinson (1892). When soil sterilant like ethylyne dibromide or dichloropropene was applied, the incidence of wilt disease was significantly reduced, because the chemical reduced nematode population, indirectly indicating the synergistic association between the fungal pathogen and the nematode (Newson and Martin 1953). The endoparasites *Globodera*, *Heterodera*, *Meloidogyne*, *Rotylenchulus* and *Pratylenchus* are more commonly reported to be involved in complexes with fungal pathogens such as *Fusarium* and *Verticillium* (causing wilt diseases), *Pythium*, *Phytophthora* and *Rhizoctonia* (causing RR diseases) (Back et al 2002). *M. incognita* was found associated with *T. basicola*, causing black rot of cotton (Walker et al. 1998). Both, in combination, consistently increased seedling mortality, increased root necrosis, suppressed early seedling growth and subsequently reduced the number of bolls formed (Wheeler et al. 2000). Guava decline, accounting for heavy losses in Brazil, is a disease complex involving the fungal pathogen *F. solani* and the nematode *Meloidogyne enterolobii*. Guava plants, that were immune to *F. solani*, were predisposed to infection by the fungal pathogen, following infection with *M. enterolobii*. *F. solani* induced extensive root deterioration. As such the nematode was not highly aggressive to guava, when present alone (Gomes et al. 2011). M. *enterolobii* exerted a localized effect that favored *F. solani* infection of guava plants. Infection by the nematode induced morphophysiological changes in the roots of guava plants that enhanced the level of susceptibility to *F. solani*. The enhanced virulence of *F. solani* required the presence of *M. enterolobii* at the same space and at the same time for the guava decline to be initiated and further invasion of guava plants by *F. solani* (Gomes et al. 2014). It is essential to study the nature of association of the components of the disease complex, since even a low incidence of infection of fungal pathogen or nematode may lead to a synergistic interaction, resulting in considerable increase in disease severity. Studies on the interaction of fungal pathogen and nematode under field conditions were few. Hence, the spatial correlations between nematode species and *R. solani*, causative agent of stem canker and black scurf disease of potato were investigated. In the potato fields in Norway, the poor growth of potato in patches was most likely caused by *R. solani*. There was a correlation between plants with high disease index (DI) and high abundance of stubby root nematode (Trichodoridae) and potato cyst nematode (*Globodera* spp.). No difference in the number of nematodes within the patches of diseased plants could be seen for any nematode taxa tested. But the severity of stem canker was higher on plants graded in the center of the patches, compared to those graded in the margins. The results indicated no spatial correlation between *R. solani* and root-lesion neamtodes (*P. penetrans*) in the severity of the disease complex in potato (Björsell et al. 2017).

*Globodera-Verticillium dahliae* and *Pratylenchus-V. dahliae* disease complexes have been a serious limiting factor in potato production in several countries. Early senescence or early dying of potato plants due to infection by *V. dahliae* and *V. albo-atrum* was accentuated by high populations of *Pratylenchus spp.* or *G. rostochiensis/G. pallida* (Storey and Evans 1987; Hafez et al. 1999). The severity of SDS in potato induced by *F. solani*, was enhanced by *H. glycines*, the soybean cyst nematode (SCN). Incidence of SDS symptoms in plots containing both *F. solani* and *H. glycines* was higher (35%) than in plots inoculated with the fungal pathogen alone (18%) (McLean and Lawrence 1993). Plants infected by soilborne fungal pathogens may produce enlarged root system, releasing greater amounts of root exudates which may attract more populations of parasitic nematodes. Clover (*Trifolium subterraneum*) seedling roots infected by *Fusarium avenaceum* attracted greater population of *Heterodera daverti*, because of the presence of diffusates of *F. avenaceum* in the root system, compared with uninfected control plants (Nordmeyer and Sikora 1983). *Verticilliium dahliae* produced root growth-promoting substances such as indole-3-acetic acid, resulting in enlarged root system in peppermint. The root system, in turn, released greater volumes of root exudates, thereby attracting greater populations of *Pratylenchus minyus* (Clarke and Hennessy 1987; Rolfe et al. 2000).

Soybean SDS caused by *F. virguliforme (F. solani* f.sp. *glycines)*, when combined with SCN *H. glycines*, becomes highly destructive, accounting for heavy yield losses. The severity of foliar symptoms of SDS was intensified, when *H. glycines* also infected soybean plants synergistically. The ecology of the disease complex was investigated under field conditions from 2000 to 2007. Initially, susceptible soybean cv. Spencer was planted, while inoculating *F. virguliforme* into nonfumigated or preseason-fumigated plots (methylbromide, MB at 450 kg/ha) and incidence of SCN and SDS was monitored. In one field, SCN population densities declined in nonfumigated plots, but increased in fumigated plots. After years of limited SDS incidence in 2003 and 2004, SDS developed later in nonfumigated plots. In 2006 in the greenhouse, nondisturbed or disturbed soil cores (10-cm diameter, 30 cm depth) from field plots were tested. At harvest, cores from nonfumigated plots had fewer nematodes, less SDS, regardless of disturbance or inoculation with fungal pathogen than the corresponding cores from fumigated plots (1,070 kg/ha, MB). In the second field, SCN became detectable after 2003 during the monoculture in nonfumigated plots and lagged in fumigated plots. Incidence of SDS was at low levels in both treatments (Westphal and Xing 2011).

## 3.2 GENETIC CHARACTERISTICS OF FUNGAL PATHOGENS

Microorganisms present in the environment differ distinctly in their ability or inability to infect and cause disease(s) on their host plants or animals. Microorganisms capable of infecting plants are pathogens endowed with genes that produce compounds that facilitate the pathogens to initiate infection and induce symptoms characteristic of the disease. As the first step, the isolates of a fungus obtained from infected plant tissues are cultured on suitable medium and they are then inoculated onto the healthy plants under optimal conditions required for disease development (pathogenesis). If the fungus is pathogenic, after a certain period after inoculation, symptoms similar to those observed under natural conditions will develop, indicating the pathogenicity of the fungus isolated from naturally infected plants. The fungus from the inoculated plant is reisolated and the characteristics of the fungus are compared with those of the fungus originally isolated from naturally infected plant. These steps of proving pathogenicity of the fungus were enunciated as Koch's postulates. In contrast, the microorganisms, lacking pathogenicity genes, fail to infect plants, are termed as saprophytes which are involved in the degradation of complex organic matter and they greatly help in the cleansing of environment, soil and water bodies.

### 3.2.1 PATHOGENICITY OF FUNGAL PATHOGENS

Soilborne fungi require some conditions that favor initiation of infection and establishment in the susceptible plants. Presence of injuries/wounds is known to favor entry of the pathogen that cannot penetrate unwounded root/ crown tissues. In order to determine progression of Phytophthora blight disease caused by *P. capsici* on wounded and unwounded pepper plants and to assess whether susceptibility to the pathogen decreases with wound aging, two isolates of *P. capisci* with differing levels of virulence were used for inoculation. Trimming roots followed by immediate inoculation with either isolates increased susceptibility significantly ($P \leq 0.05$), compared with plants that were not trimmed as indicated by higher disease severity and early appearance of disease symptoms, following trimming of roots. Isolate NM6011 with higher virulence induced greater disease severity (three to four times) than less virulent isolate NM6040. Resistance to *P. capsici* increased with increase in interval after trimming roots, as the wounds aged resulting in lower disease severity. Inoculation of roots with zoospores, soon after trimming roots, led to significantly higher levels of attachment than, when inoculated at 48-h after trimming. The results suggested that wound repair mechanisms might have a role in decreasing infection by *P. capsici* (Adorada et al. 2000). *P. capsici* infects cucurbits, causing RR and fruit rot disease. Summer squash and winter squash seedlings were evaluated for their resistance to *P. capsici* under greenhouse conditions by placing pathogen-infested millet seed (approximately 1 g) on surface of soilless potting media and assessing disease severity every two days up to 14 *dai*. Crop type, pathogen isolate, or crop-pathogen isolate interaction significantly influenced symptom appearance and area under disease progress curve (AUDPC) values. Butternut squash and zucchini cultivars, when inoculated with solanaceous isolate 13351 showed differences in their levels of susceptibility to *P. capsici* (Enzenbacher and Hausbeck 2012). Seasonal variations in disease incidence may be observed due to their effect on pathogen development. Excised feeder roots from mature citrus trees located in climatically different regions were infected with zoospores of *Phytophthora citrophthora* and *P. nicotianae* var. *parasitica* at different times of the year under identical laboratory conditions. Zoospores encysted on and caused infection in roots from all locations year-round. Both pathogens had the most encysted zoospores on roots from November to January and the least from March to May. Infection by *P. nicotianae* var. *parasitica* was consistently higher than *P. citrophthora* in excised roots in summer (May to September) and lower in January in 1998 and November 1997. Infection by both species of *Phytophthora* dropped to the minimum in March, when carbohydrate levels in the roots were the lowest. However, there was no correlation between root carbohydrate contents and seasonal infection fluctuations (Dirac et al. 2003).

Races of fungal pathogens are differentiated by inoculating a set of differentials that have distinct genes of resistance or tolerance to the target pathogen. Race 3 of *Phytophthora nicotianae* occurring in NC was identified from a tobacco field with a history of tobacco varieties with *Phl* gene for resistance and numerous field sites with no known deployment of varieties with *Phl* gene. Race 3 was defined as overcoming the *Phl* gene from *Nicotiana longiflora*, but not the *Php* gene from *N. plumbaginifolia*. Stem inoculation was unable to differentiate race 3 from race 0 of *P. nicotianae* and could not be relied upon for identifying virulence types of the pathogen. However, race 1 gave a unique phenotype, using stem inoculation, whereas root inoculation could be employed for distinguishing races 0 and 3 of the pathogen. The results showed that race 3 was able to damage seedlings following root inoculation and to infect plants containing the *Phl* gene in naturally infested soil (Gallup and Shew 2010). *P. infestans*, incitant of potato late blight disease of historical importance, may be disseminated through airborne propagules, tuberborne inoculum and soilborne oospores. Percentage of infection of potato sprouts by *P. infestans* was investigated by examining preemerged sprouts which were removed prior to emergence from the field with potato plants severely infected by late blight. Percentages of sprouts infected by *P. infestans* were 49.5 and 43%, respectively in three experiments. Infection potential of potato sprouts was evaluated in the greenhouse by applying 10-ml sporangial suspensions (50 and 250 sporangia/ml) daily for ten days to the soil surface of pots planted with sprouted potato tubers. The daily inoculation rate of 50 sporangia/ml ($15.9 \times 10^3$ sporangia/m2) resulted in 100% infection of sprouts. The results revealed the potential for preemergence infection of potato sprouts by *P. infestans* in the highlands of Ecuador, where year-round aerial inoculum was available, indicating the need for applying appropriate disease management options for restricting the incidence and spread of late blight disease of potato (Komann et al. 2008).

Pathogenicity traits of *P. sojae*, causal agent of soybean root and stem rot were investigated. Isolates (121) of soybean-infecting *Phytophthora* spp. were baited using susceptible cv. Sloan seedlings from soybean field soils with history of seedling diseases in 24 counties across Illinois. The pathotype and race of isolates of *P. sojae* were characterized, using 11 differential soybean cultivars in the greenhouse assays, employing the hypocotyl inoculation procedure. The majority of isolates (96%) studied were identified as *P. sojae* based on phenotypic and genotypic traits. Based on eight *Rps* gene differentials (*Rps1a, 1b, 1c, 1d, 1k, 3a, 6* and *7*), 22 virulence pathotypes of *P. sojae* were identified and 88% of all isolates were characterized to a defined race. Predominant constituents of the population belonged to race 1 (21%), 4 (5%), 33 (12%) and 28 (10%). Based on 11 differentials (*Rps* 2, 4 and in addition to eight mentioned above), 31 virulence pathotypes were differentiated. The results showed that the populations of *P. sojae* in Ilionois, were diverse and composed of multiple pathotypes and races (Malvick and Grunden 2004). In a later investigation, the population structure of *P. sojae* present in two provinces of China was studied. *P. sojae* was baited and isolated from 258 soil samples using a soybean leaf bait method. Using 13 differential cultivars, the pathotypes of all isolates were characterized, employing a hypocotyl slit inoculation procedure. Thirty-five new pathotypes were identified. Less than 5% of the isolates were virulent on cultivars with individual *Rps* genes *1a, 1c* or *1k*. The results indicated that cultivation of soybean cultivars with *Rps* genes might be effective in managing soybean root and stem rot caused by *P. sojae*, supplemented by application of metalaxyl (Cui et al. 2010). Fungal pathogens differ in their pathogenic potential and different pathotypes have been differentiated based on their aggressiveness on their host plant species. Pathotype classification of 49 populations of *P. brassicae* was performed on two sets of differentials – the European Clubroot Differential set and the set of differentials of Somé. In addition, the levels of virulence of the isolates of *P. brassicae* were determined, using the clubroot-resistant oilseed rape cv. Mendel. Variations in pathotype distribution in Germany were oberved. The majority of isolates were pathotypes 1 and 3 of Somé, respectively, with pathotypes 2 and 5 in the minority. For all populations tested for virulence on cv. Mendel, 15 isolates were found to be moderately or highly virulent. The virulent populations of *P. brassicae* were not restricted to a small geographical area in Germany (Zamani-Noor 2017).

The effects of amino acids and sugars in the root exudates of soybean on chemotaxis, encystment, cyst germination and germ tube growth of *P. sojae* zoospores were assessed. All eight L-amino acids (glutamate, asparate, asparagine, alanine, serine, valine, histidine and hydroxyproline) and 5 L-sugars (glucose, sucrose, fructose, arabinose and galactose) stimulated zoospore chemotaxis, encystment and cyst germination, but did not stimulate subsequent growth of the germ tube in *P. sojae*. Germ tube growth was significantly retarded in response to each of the compounds tested. The results suggested that chemotaxis, encystment and cyst germination in *P. sojae* zoospores might be receptor-mediated and the zoospores could not utilize the sugars and amino acids present in the root exudates directly. Promotion or inhibition of zoospore development occurred at specific concentration of each compound (Suo et al. 2016). The mechanism of chemotaxis of zoospores of *P. sojae* toward soybean isoflavones was found to be essential in the early stages of infection of soybean plants. A G- protein α subunit encoded by PsGPA1 required for regulating the chemotaxis and pathogenicity of *P. sojae* was previously identified. Affinity purification procedure was applied to identify PsGPA-interacting proteins, including PsHint1, a histidine triad (HIT) domain-containing protein orthologous to human HIT nucleotide-binding protein (HINT1). PsHint1 interacted with both the guanosine triphosphate (GTP)- and guanosine diphosphate (GDP)-bound forms of PsGPA1. Analysis of the gene-silenced transformants revealed that PsHint1 was involved in the chemotropic response of zoo-spores to the isoflavones daidzein. During interaction with susceptible soybean cultivar, PsHint1-silenced transformants displayed significantly reduced infectious hyphal extension and caused a strong cell death in plants. In addition, the transformants showed defective cyst germination and formation of abnormal highly branched germ tubes with apical swelling. The results suggested that PsHint1 might be involved in regulation of chemotaxis by interacting with PsGPA1 as well as in participation in a G α-independent pathway involved in the pathogenicity of *P. sojae* in soybean (Zhang et al. 2016).

*P. ramorum*, causal agent of sudden oak death (SOD) disease, infects different forest tree species and other large number of diverse plant species. Variations in the symptomatology due to differential inoculation methods were studied. None of the inoculation methods induced the full range of symptoms observed under natural conditions, but whole plant dip procedure produced symptoms closest to natural infection. Detached leaf-dip inoculation provided a rapid assay and allowed a reasonable assessment of susceptibility to leaf blight. Stem-wound inoculation of seedlings correlated with field symptoms observed on several host plant species. Log inoculation resulted in a realistic test of susceptibility to SOD, but it was cumbersome and subject to seasonal variability (Hansen et al. 2005). The potential of some soilborne inoculum of *P. ramorum* for root infection of rhododendron was investigated. Rhododendron 'Nova Zembla' plants grown from rooted cuttings and native Pacific Rhododendron (*Rhododendron macrophyllum*) plants grown from seed were planted into potting medium artificially infested with *P. ramorum* (cultures, chopped infected leaves or zoospores). Plants were watered from the bottom to prevent splash dispersal inoculum into stems and foliage. Both infested amendments and zoospore inoculum resulted in mortality of plants within three to seven weeks. *P. ramorum* could be isolated from hairy roots, large roots and stems above and below the potting medium surface. Observations under epifluorescence microscopy on inoculated tissue culture plantlets revealed the attraction of zoospores to wounds and root primordia and colonization of the cortex and vascular tissues of roots and stems including xylem. The results indicated the need for monitoring freedom of potting media from the pathogen to prevent spread of *P. ramorum* on nursery stock (Parke and Lewis 2007).

Several species of *Pythium* are soilborne and are present commonly in field soils in the United States, but their pathogenic potential has not been assessed. Pathogenicity and virulence of *Pythium* spp. associated with corn and soybean seed and seedling diseases in 42 production fields in Ohio were studied. Eleven species and two distinct morphological groups of *Pythium* were identified, of which six species were moderately to highly pathogenic to corn seed and nine species were highly pathogenic on soybean seed (Borders et al. 2007). In a later study, pathogenicity and virulence of 44 *Pythium* isolates representing 16 species recovered from three forest nursery soils in Washington and Oregon were assessed. Responses of Douglas-fir to seedlings to inoculation with infested soil, varied significantly depending on *Pythium* spp. and isolate. *P. dissotocum*, *P. irregulare*, *P. aff. macrocarpum*, *P. mamillatum*, *P. aff.oopapillum*, *P. rostratifingens*, *P. sylvaticum* and *P. ultimum var. ultimum* significantly reduced seedling survival, compared to the control. However, all other *Pythium* spp. induced root lesions. The results showed that *P. irregulare*, *P. mamillatum* and *P. ultimum* var. *ultimum* caused damping-off of Douglas-fir seedlings predominantly and other *Pythium* spp. might be responsible for seedling loss to different extent (Weiland et al. 2013). Pathogenicity of *Pythium* isolates to cut-flower chrysanthemum roots was determined using detached leaf assay, involving insertion of mycelial plug into the slits cut in the excised leaf petiole. Induction of necrosis and rate of necrosis spread indicated pathogenicity and virulence of the *Pythium* isolates. The isolates of *P. sylvaticum* and *P. ultimum* were more virulent, with a mean rate of spread of 14.6 mm per day compared to other less virulent isolates with 1.6 mm per day, belonging to *P. irregulare*, *P. oligandrum* and *P. aphanidermatum*. However, P. *aphanidermatum* had been reported to cause epidemics of Pythium RR in chrysanthemum elsewhere. The detached leaf assay was found to be useful during the survey of commercial chrysanthemum beds. *P. sylvaticum* was considered to be the most likely pathogen spreading through contaminated soil. Detached leaf assay demonstrated a relationship between pathogenic inoculum concentration in soil and the expression of RR symptoms (Pettitt et al. 2011).

Different species of *Pythium* spp. cause damping-off of soybean and corn (maize) and the disease development is favored by cool temperatures and wet soil conditions. The effects of temperature on aggressiveness (virulence) and fungicide sensitivity of *Pythium* spp. were assessed. The most prevalent *Pythium* spp. in Iowa were identified as *P. lutarium*, *P. oopapillum*, *P. sylvaticum* and *P. torulosum*. Seed and seedling assays were used to quantify the aggressiveness of the *Pythium* spp. on soybean and corn at 13, 18 and 23°C. Isolates recovered from soybean or corn were equally pathogenic on both hosts. *P. torulosum* was more aggressive at 13°C, compared with 18°C and 23°C. On the other hand, *P. sylvaticum* was more aggressive at 18°C and 23°C rather than at 13°C. The plate assay to determine the effects of temperature on sensitivity to seven fungicides showed that $EC_{50}$ values for *P. torulosum* were higher for all fungicides tested at 13°C rather than at 18°C or 23°C, whereas $EC_{50}$ values for *P. sylvaticum* were higher for all fungicides at 18°C and 23°C, rather than at 13°C. The results showed the influence of temperature on *Pythium* spp. development and their sensitivity may have to be taken into consideration for developing effective disease management measures (Matthiesen et al. 2016). The soilborne oomycetes, causing seed decay, damping-off and RR in soybean and corn were evaluated for their relative aggressiveness (virulence). Isolates collcted from seedlings and soils belonged to 30 oomycete species. The aggressiveness of the isolates of

21 species was determined, using seed decay and seedling methods. Seven *Pythium* spp. were pathogenic on soybean and corn, while two *Pythium* spp. and one *Phytopythium* sp. were pathogenic only on soybean. Aggressiveness of many isolates increased with increase in temperatures from 15°C to 25°C. *Pythium* spp. in Minnesota, United States were diverse and capable of causing significant adverse effects on growth and yield of soybean and corn (Radmer et al. 2017).

Plant hormones play important roles in defense against infection by microbial plant pathogens. The role of ethylene and strigolactones (SLs) in response to infection of pea plants by *Pythium irregulare* was studied at molecular and whole-plant levels, using a set of well-characterized hormone mutants, including an ethylene-insensitive ein2 mutant and SL-deficient and insensitive mutants. A key role for ethylene signaling was identified in specific cell types that reduced pathogen invasion. No evidence was found indicating that SL biosynthesis or response influences the interaction of pea with *P. irregulare* or that synthetic SL influenced the growth or hyphal branching of the oomycete *in vitro* (Blake et al. 2016). The role of indole-3-acetic acid (IAA) on infection of rapeseed by *P. brassicae* during early stages was studied. Treatment of infected oilseed rape (*B. napus*) seedlings with exogenous application of IAA promoted the development of club root disease, including an increase in the number and size of galls, whereas treatment with N-1- naphthalamic acid (NPA) attenuated the disease severity and altered the location and size of root galls, resulting in a reduced DIs. The level of free IAA was increased in the roots at three and seven *dai*, compared with the control, suggesting activation of IAA signaling in the early stages of *P. brassicae* infection. Treatment with IAA did not alter the rate of germination of resting spores, but resulted in enhancement of root hair infection. RT-PCR and quantitative real-time PCR (qPCR) assay showed that five IAA biosynthesis-related genes were induced in the roots after inoculation with *P. brassicae* but not in the leaves. The rapid induction of BnAAO4 expression at three days after inoculation might be responsible for overproduction of IAA during early stages of infection. The results suggested that IAA might act as a signaling molecule, resulting in putative stimulation of root hair infection as the early response of rapeseed to *P. brassicae* infection (Xu et al. 2016).

Soilborne pathogenic fungi may vary in their virulence on different plant hosts, although they may show cross-pathogenicity on two or more host plant species. The cross-pathogenicity of *V. dahliae* between potato and sunflower was studied. Four-week-old potato and sunflower seedlings were inoculated with ten isolates from each of two host species. Kennebec (susceptible) and Ranger Russet (moderately resistant) potato cultivars and sunflower hybrids IS8048 (susceptible) and IS6946 (moderately resistant) were inoculated with *V. dahliae* and the extent of vascular discoloration at two, three, four, five and six weeks after inoculation was determined. Most *V. dahliae* isolates were highly aggressive on both host species and susceptible genotypes were more severely affected than resistant ones. The pathogen isolates caused higher disease severity (vascular discoloration) in their original host than in the alternative host. However, isolates from sunflower caused less infection and disease severity on both hosts, compared to potato isolates. The results showed that *V. dahliae* was cross-pathogenic to both host species and hence, rotations involving these crops should be avoided, as a disease management strategy (Alkher et al. 2009). Differentiation of races 1 and 2 of *V. dahliae* infecting tomato could be achieved, based on the differential pathogenicity on tomato cultivars carrying resistance gene *Ve1*. The tomato cultivar Aibou and Ganbarune-Karis bred in Japan were resistant to race 2 and the resistance was controlled by a single dominant locus, but their reistance appeared to be unstable in commercial tomato fields. Some isolates of race 2 could overcome this reistance. It was, therefore, proposed that the current race 2 of *V. dahliae* should be divided into two races, ie., race 2 (nonpathogenic on cv. Aibou) and race 3 (pathogenic on cv. Aibou). Race 3 was detected in 45 fields, indicating that race 3 had already spread well throughout production areas in Hida, Gifu Prefecture, Japan. Presence of race 2 in 25 fields was also observed and the race 2-resistant rootstocks could be grown in those fields. Southern hybridization and race-specific PCR assay were not effective in differentiating the races 2 and 3, that could be identified based on the reactions on cv. Aibou (Usami et al. 2017).

Development of wheat take-all disease, caused by *Gaeumannomyces graminis* var. *tritici* (*Ggt*) is favored by factors associated with low manganese (Mn) availability such as high soil pH, $NO_3-$ -N sources or low soil Mn content. *Ggt* is capable of oxidizing Mn from soluble, plant-available $Mn2+$ oxidation. Mn oxidation was correlated with pathogen virulence. *Ggt* isolates with low Mn oxidation capacity could cause only fewer and less extensive lesions than isolates characterized as strong oxidizers (Peddler et al. 1996). In a later investigation, manganese-oxidizing isolate 1158-1 and presumed nonoxidizing isolate 1079-1 were found to be pathogenic and produced similar lesions. There was no detectable difference in the virulence between the isolates. Treatment of cell-free agar medium containing Mn oxidizing factor (MOF) with heat or proteinase K eliminated all Mn oxidation activity, confirming the involvement of an enzyme in Mn oxidation by *Ggt*. The results indicated that MOF produced by *Ggt* is a laccase-like multicopper oxidase with an estimated MW of 50 to 100 kDa. Although the isolate 1079-1 was unable to oxidize Mn in culture, its ability to oxidize Mn in the presence of host tissue might account for it being as virulent as isolate 1058-1 which oxidized Mn in culture. This supports the hypothesis of a link between Mn oxidization and virulence of the pathogen (Thompson et al. 2006). Strains of *Fusarium oxysporum* f.sp. *melonis (Fom)*, causal agent of Fusarium wilt disease of melon, were characterized for their pathogenicity on melon. Of the 44 strains of *Fom*, 37 were pathogenic and belonged to known races 0, 1, 2 and 1,2 of the pathogen, while two strains were nonpathogenic. Beauvericin was produced by 36 strains in the range of 1 to 310 μg/g and eight isolates of race 1,2 did not produce the mycotoxin. One nonpathogenic strain was also able to produce beauvericin at a concentration of 290 μg/g. Eleven strains belonging to all three races produced enniatin B up to 60 μg/g. Melon fruit inoculated with strain

ITEM 3464 contained beauvericin. The results suggested that production of beauvericin and enniatin B was not related to the pathogenicity of *Fom* and it was not specific to any one race of the pathogen. Beauvericin was a common metabolite of phytopathogenic *Fusarium* spp. (Moretti et al. 2002).

Pathogencity tests were performed to determine the virulence of newly emerged populations of *V. longisporum* in the United Kingdom and to compare the British isolates with reference isolates characterized earlier. *Arabidopsis thaliana*, four cultivars of three *Brassica* spp., viz., oilseed rape (cvs. Quartz and Incentive), cauliflower (cv. Clapton), and Chinese cabbage (cv. Hilton) were inoculated with four British isolates. Vascular discoloration of the roots, plant biomass accumulation and fungal stem colonization upon inoculation with test isolates were the parameters used for aggressiveness of the isolates of the pathogen. British isolates appeared to be remarkably aggressive based on the reduction in plant biomass and severity of vascular discoloration. The British isolates were efficient stem colonizers and there was a negative correlation between the extent of fungal colonization and cauliflower and Quartz oilseed rape plant biomass. The A1/D1 isolates formed due to hybridization between different *Verticillium* spp., including British isolates were aggressive on oilseed rape, in spite of levels of colonization being limited, compared to a virulent isolate of A1/D3 lineage of *V. longisporum* (Depotter et al. 2017).

Biotrophic smut fungi have a restricted host range generally. *Sporisorium reilianum* exists in the form of two host-adapted *formae speciales*: *S. reilianum* f.sp. *reilianum* (SRS), causing sorghum head smut and *S. reilianum* f.sp. *zeae* (SRZ), causing maize head smut disease. The mechanism of host specificity of these smut fungi was investigated. By fungal DNA quantification and fluorescence microscopy of stained plant samples, colonization behavior of both SRS and SRZ was followed in sorghum and maize, respectively. Both pathogens were able to penetrate and multiply in the leaves of both hosts. In sorghum, the hyphae of SRS reached the apical meristems, whereas the hyphae of SRZ did not. SRZ strongly induced several defense responses in sorghum, such as generation of $H_2O_2$, callose and phytoalexin, whereas the hyphae of SRS did not induce such responses. In maize, both SRS and SRZ were able to spread in the plant up to the apical meristem. Transcriptome analysis of colonized maize leaves revealed more genes induced by SRZ than by SRS, with many of them being involved in defense responses. Among the maize genes specifically induced by SRS were 11 pentatricopeptide repeat proteins. The results indicated that SRZ might be inhibited by plant defense after sorghum penetration, whereas SRS proliferates in a relatively undisturbed manner, but not efficiently, in maize, explaining the phenomenon of host specificity operating in these smut pathogens (Poloni and Schirawski 2016).

### 3.2.2 Production of Toxic Metabolites and Enzymes by Fungal Pathogens

Soilborne fungal pathogens may produce toxic compounds and/or enzymes that may aid in establishment of infection or degradation of complex substances into simpler

ones required for their development. Different species of *Fusarium* involved in Fusarium head blight (FHB) disease of wheat and other cereals have been extensively investigated for their ability to produce different types of mycotoxins in grains (Narayanasamy 2017). *Fusarium graminearum*, a known producer of trichothecene mycotoxins in cereal hosts, causes dry rot of potato tubers. In order to determine accumulation and diffusion of trichothecenes in potato tubers, following infection by *F. graminearum*, potato tubers of cv. Russet Burbank were inoculated with 14 isolates of F. *graminearum* from potato, sugar beet and wheat. Twelve isolates of *F. graminearum* produced deoxynivalenol (DON), whereas two isolates produced nivalenol (NIV). Trichothecenes were detected in rotted potato tissues. DON genotypes of *F. graminearum* produced up to 39.68 μg/g in rotted potato tissues, whereas NIV accumulated up to 18.28 μg/g in potato tubers infected by NIV genotypes. However, DON genotypes produced NIV also in addition to DON in potato tuber tissues with dry rot lesions. Diffusion of DON was assayed by inoculating tubers with two isolates of DON chemotype and incubating for seven weeks at 10 to 12°C. *F. graminearum* was recovered from > 53% of the lesions from inoculated tubers at 3 cm distal to the rotted tissue, after seven weeks of incubation, but DON was not detected in the surrounding tissues. The results indicated that accumulation of trichothecenes in asymptomatic tissue surrounding dry rot lesions induced by *F. graminearum* was minimal at customary processing storage conditions (Delgado et al. 2010). Cell wall-degrading enzymes (CWDEs) secreted by *F. graminearum* have been found to be important pathogenicity factors for infection of susceptible host plants. *FgGpmK1* regulates the activities of extracellular endoglucanase and xylanolytic and proteolytic enzymes (Jenczmionka and Schafer 2005). MAPK, homologous to the *Saccharomyces cerevisiae* mating/filamentation MAPKs, Fus3/Kss1 has been shown to be involved in pathogenicity of several fungal pathogens infecting plants. The transcription factor Ste12, which functions downstream of the MAPKs, Fus3/Kss1, binds to pheromone and filamentation response elements in the promoters of target genes involved in invasive growth of *S. cerevisiae* (Madhani et al. 1997; Qi and Elion 2005). A whole-genome search revealed that *F. graminearum* contained an ortholog of Ste12, named FgSte12 which might play an important role in its pathogenicity. A conserved MAPK cascade homologous to the yeast Fus3/Kss1 was involved in the regulation of vegetative development and pathogenicity in *F. graminearum*. The transcription factor FgSte12 was characterized. The FgSte12 deletion mutant *ΔFgSte12* was impaired in virulence and secretion of cellulase and protease, although it did not show recognizable phenotype changes in hyphal growth condition or DON biosynthesis. The results indicated that FgSte12, a transcription factor, had an important role in pathogenicity of *F. graminearum* (Gu et al. 2015).

Plant roots respond to infection by soilborne fungal pathogens through activation of general and systemic resistance, including lignifications of cell walls and increased release of phenolic compounds in root exudates. Degradation of lignin by pathogens has been demonstrated by producing lignolytic extracellular peroxidases and laccases. Aromatic lignin breakdown products may be further catabolized via ß-ketoadipate pathway. The role of 3-carboxy-cis,cis-muconate lactonizing enzyme (CMLE), an enzyme of the ß-ketoadipate pathway in the pathogenicity of *F. oxysporum* f.sp. *lycopersici* to tomato, was investigated. The *cmle* deletion mutant could not catabolize phenolic compounds known to be degraded via β-ketoadipate pathway. Further, the mutant was impaired in root infection and remained nonpathogenic to tomato, although it was able to colonize tomato roots superficially. The results suggested that operation of β-ketoadipate pathway in soilborne fungal pathogens might be necessary to degrade phenol compounds in root exudates and/ or inside root tissues for successful establishment of infection and disease development (Michielese et al. 2012). In silico search and biochemical enzyme activity analyses showed that 25 structural secreted lipases could be predicted in *Fusarium oxysporum* f.sp. *lycoperisic (Foc)*, based on the conserved pentapeptide Gly-X-Ser-X-Gly-, characteristic of fungal lipases and secretion signal sequences. Further, a predicted lipase regulatory gene was also identified, in addition to *cft1*. Transcription profile of 13 lipase genes during tomato plant colonization by *Foc*, revealed that *lip1*, *lip3*, and *lip22* were highly induced between 21 and 96 h after inoculation. Deletion mutants in five lipase genes (*lip1*, *lip2*, *lip3*, *lip5* and *lip22*) and in the regulatory genes *ctf1* and *ctf2* as well as Δ*cft1*, Δ*cft2*, double mutant were generated. Quantitative RT-PCR expression analyses of structural lipase genes in the Δ*ctf1*, Δ*ctf2* and Δ*ctf1*Δ*ctf2* mutants indicated the existence of a complex lipase regulation network in *F. oxysporum*. The important role of the lipolytic system of *F. oxysporum* f.sp. *lycopersici* in pathogenicity to tomato was indicated by the reduced virulence of mutants (Bravo-Ruiz et al. 2013). The genome of *Fusarium oxysporum* f.sp. *lycopersici* (Foc) encodes four endopolygalacturonases (endoPGs) and exoPGs. Quantitative real-time reverse transcription-polymerase chain reaction (RT-PCR) assay showed that endoPGs *pg1* and *pg5* and exoPGs *pgx4* and *pgx6* were expressed at significant levels, during growth on citrus pectin, polygalacturonic acid or the monomer galacturonic acid as well as during infection of tomato plants. The remaining PG genes exhibited low expression levels under cell conditions tested. Secreted PG activity was reduced significantly during growth on pectin in the single deletion mutants lacking either *pg1* or *pgx6*, as well as in the double mutant. Although the single deletion mutants did not display a significant virulence reduction in tomato plants, the Δ*pg1*Δ*pgx6* double mutant was significantly attenuated in virulence. The combined action of exoPGs and endoPGs was found to be essential for plant infection by vascular wilt pathogen infecting tomato (Ruiz et al. 2016).

*V. dahliae* isolates have different levels of virulence (aggressiveness) on different host plant species. Two isolates differing in their aggressiveness were evaluated for their ability to produce cellulase in media with different carbon sources. *V. dahliae* could degrade crystalline celluloses (Avicel), because it could produce three enzymes required for its hydrolysis. Two isolates behaved similarly in the presence of soluble cellulose, but most aggressive isolate had greater β-1,4-glucosidae (EC.3.2.1.2.1) and endo-β-1-glucanase (EC.3.2.1.4) activity. The less virulent isolate required more time to degrade crystalline cellulose. The results indicated that cellulases might have a role in penetration of host tissues (Novo et al. 2006). The role of hydrolytic CWDEs in virulence of *V. dahliae* which has a wide host range, was investigated. The sucrose nonfermenting 1 gene (*VdSNF1*) which regulates catabolic repression, was disrupted in *V. dahliae* race 1 infecting tomato. Expression of CWDE in the mutants was not induced in inductive medium and in simulated xylene fluid medium. Growth of the mutants was significantly reduced in media containing pectin or galactose as carbon source, whereas the mutants grew normally as wild-type strain in media with glucose, sucrose and xylose. The mutants were severely impaired in virulence on tomato and eggplant. Microscopic observations showed that the infection behavior of GFP-labeled *VdSNF1* mutant (*70ΔSF-gfp1*) was significantly altered. The mutant was defective in initial colonization of roots. Cross-sections of tomato stem at the cotyledonary level showed that *70ΔSF-gfp1* mutant colonized xylem vessels at an intensity considerably less than the wild-type strain. The wild-type strains heavily colonized xylem vessels and adjacent parenchyma cells. Quantification of fungal biomass in plant tissues further confirmed reduced colonization of roots, stems and cotyledons by *70ΔSF-gfp1* mutant, relative to that of wild-type strains of *V. dahliae* in tomato plants (Tzima et al. 2011). Expression of eight CWDE genes of 15 isolates of *V. dahliae* with different hosts of origin and virulence levels was studied. The test isolates had similar levels of gene expression after 12, 24, 48 and 72 h in the presence of host plant root extracts. However, correlation analyses indicated a clear association between pathogenicity and expression of genes controlling synthesis of pectinase, cutinase, 1,4-β-glucosidase and SNF protein kinase. On the other hand, xylanase and endoglucanase G1 showed negative relationship with pathogenicity. Over time, a sequential production of CWDEs by *V. dahliae* was observed. Pectinases were produced earlier than xylanase and endoglucanase G1. Correlation of gene expression results with respective host of origin of *V. dahliae* isolates, depended on the cell wall composition of host plants. Exclusive expression of acetylxylan esterase and ornithine decarboxylase was observed only in isolates of *V. dahliae* from sunflower plants, but not in isolates from potato or olive (Gharbi et al. 2015).

Hemibiotrophic pathogens such as *V. dahliae* require induction of plant cell death (PCD) to provide saprophytic nutrition for the transition from biotrophic to necrotrophic stage and successful colonization of host tissues. A necrosis-inducing *Phytophthora* protein (NPP1) domain-containing protein family, enclosing nine genes was identified in a virulent defoliating isolate of *V. dahliae* (V592), named as the *VdNLP* genes. Only two of these genes, VdNLP1 and *VdNLP2* encoded proteins capable of inducing necrotic lesions and triggering defense responses in cotton, *Nicotiana benthamiana* and *Arabidopsis* sp. Both VdNLP1 and VdNLP2 induced wilting of cotton seedling cotyledons. However, gene-deletion mutants targeted by VdNLP1, VdNLP2, or both did not affect pathogenicity of *V. dahliae* V592 on cotton. Similar expression and induction patterns were observed in seven of nine *VdNLP* transcripts. By comparing conserved amino acid residues of VdNLP with different necrosis-inducing activities, several novel conserved amino acids that are indispensable for necrosis-inducing activity of VdNLP2 protein were identified (Zhou et al. 2012). *V. longisporum*, showing host specificity infects members of Brassicaceae, including canola (oilseed rape) causing wilt diseases. Interaction of *V. longisporum* with oilseed rape was investigated. The pathogen reacted to the presence of canola xylem sap with production of six different upregulated and eight downregulated proteins visualized by two-dimensional gel electrophoresis. All upregulated proteins were involved in oxidative stress response. The pathogen catalase peroxidase (VlCPEA) was the most upregulated protein and it was encoded by two isogenes, *Vlcpe A-1* and *Vlcpe A-2*. Both genes were identical (98%). Knockdown mutants of both *VlcpeA* genes had reduced protein expression by 80% and resulted in sensitivity against reactive oxygen species (ROS). The mutants had low level of pathogenic activity at late phase of disease development, although their saprophytic growth and activity in the initial phase of disease development were not affected. The results suggested that catalase and peroxidase might play a role in protecting the pathogen against oxidative stress generated by the host plant at advanced stages of disease development (Singh et al. 2012). *Verticillium* spp. infect susceptible host plant species through roots and colonize xylem vessels of the plant. The xylem fluid provides an environment with limited carbon sources and unbalanced amino acid supply, necessitating the pathogen to induce the cross-pathway control of amino acid biosynthesis. RNA-mediated gene silencing reduced the expression of the two *cpc1* isogenes (VlCPC 1-1 and VlCPC 1-2) of the allodiploid *V. longisporum* up to 85%. *VlCPC1* encodes the conserved transcription factor of the cross-pathway control. The silenced mutant was highly sensitive to amino acid starvation and the infected plants exhibited significantly fewer symptoms such as stunting or senescence in oilseed rape (canola) plant infection assays. Deletion of single *CPC1* of the haploid *V. dahliae* resulted in strains sensitive to amino acid starvation and they induced strongly reduced severity of symptoms in tomato. The allodiploid *V. longisporum* and the haploid *V. dahliae* were shown to require *CPC1* for infection and colonization of their respective host plants – oilseed rape and tomato (Timpner et al. 2013).

The role of *V. dahliae* homolog of Sge1, a transcriptional regulator implicated in pathogenicity and effector gene expression in *Fusarium oxysporum* was investigated. The results showed that *V. dahliae Sge1 (VdSge1)* was required for radial growth and production of conidia, since *VdSge1* deletion mutants displayed reduced radial growth and conidia production. Further, *VdSge1* deletion strains lost pathogenicity on tomato. *VdSge1* was not required for induction of Ave1, an effector of *V. dahliae* that activated resistance mediated by the Ve1 immunity receptor in tomato. Assessment of the role of *VdSge1* in the induction of nine most highly in-planta-induced genes encoding putative effectors revealed differential activity. *VdSge1* appeared to be required for the expression of six putative effectors, whereas two of the putative effector genes were found to be negatively regulated by *VdSge1*. The results suggested that *VdSge1* might differentially regulate *V. dahliae* effector gene expression (Santhanam and Thomma 2013). *V. dahliae* relies on the virulence and the amount of microsclerotia present in the soil for its pathogenicity and extent of disease incidence and severity. Genes functional in pathogenicity and microsclerotial formation were screened. A uracil-DNA glycosylate VdUDG was identified in *V. dahliae* and it was found to be involved in pathogenicity as well as in microsclerotial formation, based on screening a T-DNA insertion library. Transcription analysis of wild-type strain and *VdUDG* gene knockout mutants, suggested that *VdUDG* might regulate transcription of a series of genes involved in microsclerotial formation and pathogenesis of *V. dahliae* (Zhang et al. 2015). The initial interaction between fungal pathogen and host plant involves numerous metabolic pathways and regulatory proteins. A random genetic screen enabled identification of 58 novel candidate genes that are involved in the pathogenic potential of *V. dahlia*. One of the candidate genes identified, was a putative biosynthetic gene involved in the production of nucleotide sugar precursor, as it encodes a putative nucleotide-rhamnose synthase/epimerase-reductase (NRS/ER). Rhamnose is a minor cell wall glycan in fungi. However, this investigation showed that deletion of the *VdNRS/ER* gene from the *V. dahliae* genome resulted in the loss of pathogenicity entirely on tomato and *N. benthamiana* plants, whereas mycelia growth and sporulation were not affected. The results demonstrated that VdNRS/ER was a functional enzyme in the biosynthesis of uridine diphosphate (UDP)-rhamnose. Further, *VdNRS/ER* deletion mutants were impaired in the colonization of tomato roots. Rhamnose, although only a minor cell wall component, was found to be essential for pathogenicity of *V. dahliae* (Santhanam et al. 2017).

The phytotoxin produced by *R. solani*, incitant of rice sheath blight disease, was used to investigate the genetics of sheath blight susceptibility in rice genotypes. Infiltration of the toxin preparation into the plant leaves caused necrosis in rice, maize and tomato. Rice cultivars (17) differing in their levels of resistance to sheath blight disease exposed to the toxin were grouped into toxin-sensitive (tox-S) and toxin-insensitive (tox-I) genotypes. A correlation between the toxin sensitivity and disease susceptibility (r = 0.66) was observed among rice genotypes tested. A total of 154 $F_2$ progenies from a cross between Cypress (tox-S) and Jasmine (tox-I) segregated in a 9:7 ratio for tox-S/tox-I, indicating an epistatic interaction between two genes controlling sensitivity to the toxin in rice. The protocol was useful to genetically map toxin sensitivity genes and to eliminate susceptible genotypes during screening of rice genotypes for their resistance to rice sheath blight pathogen (Brooks 2007). Phytotoxins produced by *R. solani* AG-3 in culture were extracted and purified by preparative thin layer chromatography (TLC) technique. The toxin capable of inducing canker on potato stems contained eight toxin fractions separated by column layer chromatography, whereas only four fractions (spots) could be recognized by TLC method. Four fractions (1, 4, 6 and 7) showed phytotoxic activity, indicating that many related compounds might be involved in the development of disease symptoms caused by *R. solani* in potato (Kankam et al. 2016). The black pigment produced by *R. solani* was characterized. Transmission electron microscopic observations showed that cell walls of the rind tissues from black sclerotia of *R. solani* were heavily pigmented, designated Rs-melanin, based on physical and chemical properties, including its ultraviolet (UV) and infrared (IR) spectra. Rs-melanin was greatly affected by pH values and bleached by strong oxidant. Catechol, an intermediate of the catechol pathway, induced Rs-melanin deposition in the hyphal cell walls of *R. solani*, when pathogen was grown on catechol-amended medium. The results suggested that Rs-melanin was a kind of catechol melanin (Chen et al. 2015).

*A. flavus*, a soilborne fungal pathogen, infects maize (corn), peanut (groundnut) and several other crops. Aflatoxin, the mycotoxin produced by the pathogen in grains and kernels leads to reduced crop value and adverse health hazards to human beings and animals, when contaminated food/feed are consumed. Two morphotypes, L and S strains, differentiated based on morphological characteristics, were investigated for their ability to produce aflatoxins. Isolates of L strain produced few large sclerotia and highly variable amounts of aflatoxins with some isolates entirely lacking the ability to produce aflatoxins (atoxigenic). In contrast, isolates of S strains produced a large number of small sclerotia and higher concentrations of aflatoxins (Cotty 1989). The atoxigenic isolates (lacking ability to produce aflatoxins) were evaluated for their potential to reduce aflatoxin concentrations in maize kernels. Atoxigenic isolates (12) were able to reduce aflatoxin levels by about 80%, indicating the possibility of utilizing the atoxigenic isolates for reducing the aflatoxin contamination of grains and kernels (Probst et al. 2011). The involvement of proteins in the process of infection of pepper by *P. capsici* was investigated, using secretome analysis. Bioassay-guided fractionation of secretome of *P. capsici* resulted in purification of a phytotoxic protein fraction designated p47f, capable of inducing wilting and necrosis on leaves of *Capsicum chinense*. This protein p47f had a 47 kDa polypeptide with proteolytic activity as the major component. The p47f fraction induced DNA degradation and decreased survival of *C. chinense* cell suspension culture. Sequencing of p47f showed the presence of 15 proteins belonging to seven classes, including a protease group, cell wall remodeling proteins and the transglutaminase elicitor M81D. The results indicated that *P. capsici* secreted proteins that could modulate cell responses mediated by ROS in the host plant (Flores-Giubi et al. 2014).

Fungal pathogens produce different kinds of enzymes required for their growth and reproduction in cultures as well as in planta during infection and development of disease symptoms (pathogenesis). *S. sclerotiorum* produces sclerotia which are pigmented, multihyphal structures that have a central role in the survival and infection cycles of this pathogen. Sclerotial formation is affected by intracellular cyclic AMP (cAMP) levels. The cAMP is a key modulator of cAMP-dependent protein kinase A (PKA). Changes in relative PKA activity levels were monitored during sclerotial development. Relative PKA activity levels increased during the white-sclerotium stage in the wild-type strain, whereas low levels of PKA activity were maintained in non-sclerotium-producing mutants. Application of caffeine, an inducer of PKA activity, resulted in increased relative PKA activity levels and it was correlated with production of sclerotial initial-like aggregates by non-sclerotium-producing mutants. The changes in PKA activity and abundance of phosphorylated MAPKs that accompanied sclerotial development appeared to represent a potential target for restricting pathogen development (Harel et al. 2005). The roles of cutinase A (SsCUTA) and polygalacturonase 1 (SsPG1) in pathogenesis of *S. sclerotiorum* was investigated. *SsCutA* transcripts appeared within one hour postinoculation of leaves and its expression was primarily governed by contact of mycelia with solid surfaces. Expression of *SsPg1* was moderately induced by contact with leaf surfaces and its expression was restricted to the expanding margin of the lesion, as the infection progressed. Supply of glucose supported a basal level of *SsPg1* expression, but accentuated expression, when provided to mycelia used as inoculum. Disruption of calcium signaling influenced *SsCutA* and *SsPg1* expression and decreased the virulence of *S. sclerotiorum*, whereas cAMP levels reduced virulence without affecting gene expression (Bashi et al. 2012).

*S. sclerotiorum* produces white mycelial mass on soybean leaf surface and later reaches stem via vascular tissue, resulting in induction of stem rot symptoms. The pathogen forms sclerotia, resting bodies in the harvested seeds or they may fall onto the ground to overwinter. Under favorable conditions, the propagules germinate to produce stipitate, spore-forming apothecia. The liberated ascospores land on petals or wounded/necrotic tissue, germinate and infect the plant, leading to a new disease cycle. The infectivity of *S. sclerotiorum* largely depends on its major virulence factor, oxalic acid (OA). Virulence of the pathogen was abolished in transgenic plants that express OA-degrading enzymes like oxalate oxidase (OxO). The interactions between *S. sclerotiorum* and susceptible and resistant soybean plants were studied, using histological methods. The infection processs of *S. sclerotiorum* and OA accumulation were investigated, using transgenic line overexpressing OxO (OxO-OE) and its isogenic parent (WT). *In situ* flower inoculation showed that the OxO-OE plants were highly resistant to the pathogen, whereas WT parents were susceptible. The difference in resistance was not apparent in floral tissues, as aggressive hyphal activity was similar on both hosts, indicating that high OxO activity and low OA accumulation in OxO-OE was not a deterrent. However, the process of infection by the pathogen on excised leaf tissue differed on susceptible and resistant plants. Primary lesions developed and showed similar severe structural derangement on both types of hosts, but rapid lesion expansion (colonization) proceeded only on WT plants, concomitant with OA accumulation. OA content was enhanced in OxO-OE plants at one day postinoculation (dpi) and did not change during the next three days, indicating that colonization could be blocked by maintaining low levels of OA. However, OxO degradation of OA did not inhibit initial host penetration and primary lesion formation. The results showed that OA was the major virulence factor of *S. sclerotiorum*, as indicated by previous investigations, and it was critical for host colonization, but may not be required, during primary lesion formation, suggesting that other factors might also contribute to the formation of primary lesions (Davidson et al. 2016).

The germinating resting spores of *P. brassicae*, causal agent of clubroot disease of crucifers, were transformed by two fungal expression vectors, containing either a green fluorescent protein (GFP) gene or a hygromycin resistance (*hph*) gene. Putative transformants were produced from both transformations, with ~ 50% of the galls containing resting spores with transforming DNA could be detected by PCR and qPCR formats. The transforming DNA, which was integrated into *P. brassicae* genome. Transcript of *hph* but not *gfp* was detected by RT-qPCR assay from selected transformants. From all galls produced by transformants, no GFP actitivity could be identified. New galls were formed on canola inoculated with verified transformants. The transforming DNA was detected in the new galls induced by transformants. The results indicated the possibility of applying the genetic transformation approach for investigations on canola-*P. brassicae* interactions (Feng et al. 2013a).

Soilborne fungal pathogens produce diverse compounds that influence pathogen development. *Ustilago hordei*, causing barley covered smut, produces mating pheromones that breakdown to smaller peptide compounds that act as potent inhibitors of mating and germination of some fungi. The pheromones are members of farnesylated family of proteins. Synthetic peptide analogs of pheromone derivatives were farnesylated, methyl esterified, or both and tested for their effects on mating or teliospore germination of *U. hordei* or *T. tritici*. Glasshouse experiments showed that some selected antagonists, applied as seed treatments inhibited covered smut of barley and common bunt of wheat, although the level of inhibition was inconsistent. Use of pheromone-related compounds appeared to be a novel strategy for control of smut and bunt pathogens (Kosted et al. 2002). *Sclerotium cepivorum*, incitant of onion white rot disease produces secondary metabolites with inhibitory effects. The culture filtrate (CF) of *S. cepivorum* at 50% concentration inhibited growth to the maximum extent and reduction in the number of sclerotia by 98.6%. Sclertoial germination was entirely inhibited even at 10% CF concentration. Application of ethyl acetate extract of CF on onion seedlings suppressed disease development completely under greenhouse conditions. The main constituent of ethyl acetate extract was 5-hydroxymethyl-furfural (HMF). The results indicated the presence of secondary metabolites of *S. cepivorum* capable of suppressing pathogen development and consequently disease progression (Elsherbiny et al. 2015).

## 3.3 PROCESS OF INFECTION OF FUNGAL PATHOGENS

Soon after a fungal pathogen reaches the infection court, through different dispersal mechanisms, a physiological contact is established between the pathogen and the host plant at the initial site of contact. In a compatible interaction between the pathogen and susceptible host plant, the surveillance mechanism of the plant does not recognize the arrival of the intruder (pathogen). In contrast, the surveillance mechanism in an incompatible interaction (resistant/nonhost) the plant rapidly recognizes the presence of the pathogen and triggers plant defense mechanisms that produce a range of compounds that may interfere with the different phases of disease development (pathogenesis). Under optimal conditions, five different phases of pathogenesis (disease development) viz., attachment of pathogen to plant surface, germination of spore or pathogenic unit, penetration of the host, colonization of host tissues and expression of symptoms characteristic of the disease under study, have been recognized. The progress of these phases may not be clearly distinguishable, as the phenomenon of pathogenesis is continuous and may overlap with each other. Several factors may influence the progress of disease development, such as host resistance/susceptibility levels and environmental conditions, exerting significant favorable or adverse effects (Narayanasamy 2002).

### 3.3.1 GERMINATION OF SPORES

Investigations on soilborne pathogens have been hampered by the inherent problems associated with visualization of process of infection in the soil environment. Comparatively, it has been easier to study the process of infection of pathogens infecting aerial plant parts. Initiation of infection by soilborne fungal pathogens depends on the availability of optimal inoculum density in the rhizosphere/rhizoplane of the susceptible host plant species under optimal environmental conditions. The effects of inoculum density of *Fusarium oxysporum* f.sp. *ciceris* races 0 (*Foc-0*) and 5 (*Foc-5*) and susceptibility of chickpea cultivars P-2245 and PV-61 on development of Fusarium wilt disease were investigated. *Foc-5* was much more virulent than *Foc-0*. With cultivar P-2245 and race *Foc-5* combination, the highest disease intensity could be attained with 6 chlamydospores/g of soil, the lowest inoculum density required for infection. On the other hand, 1,000 chlamydospores were needed to induce maximum disease intensity in cv. PV-61. In the case of *Foc-0*, 20,000 chlamydospores were necessary to induce maximum disease intensity in cv. P-2245. The results indicated the critical requirement of necessary inoculum density to infect plants with different levels of resistance (Navas-Cortés et al. 2000).

The genus *Phytophthora* comprises about 100 species of destructive plant pathogens capable of infecting a wide range of economically important agricultural and horticultural crops as well as forest trees species (Gallegly and Hong 2008). *Phytophthora* spp. produce flagellate, single-celled zoospores as their primary dispersal and infection units. Many of the zoospore behaviors are density-dependent. Plant infection by *Phytophthora* spp. requires certain densities of inoculum, although the threshold is much lower compared with that required for aggregation and this density may vary depending on pathogen species. *P. nicotianae* (syn. P. *parasitica*) was used as a model to investigate regulation of zoospore communal behaviors to demonstrate autoregulation of some zoospore behaviors, using signal molecules that release zoospores into the environment. Zoospore aggregation, plant targeting and infection required were enhanced by threshold concentrations of these signal molecules. Below the threshold concentration, zoospores did not aggregate and move toward a cauline leaf of *A. thaliana* (Col-O) and failed to individually attack annual vinca (*C. roseus* cv. Little Bright Eye). These processes were reversed, when supplemented with zoospore-free fluid (ZFF) prepared from a zoospore suspension above threshold densities, but not with calcium chloride at a concentration equivalent to extracellular Ca2+ ZFF. Zoospores coordinated their communal behaviors by releasing, detecting and responding to signal molecules. This chemical communication raises the possibility that *Phytophthora* infection of plants may not depend entirely on zoospore number in reality. Single zoospore infection may be possible, if it is signaled by a common molecule available on the environment which contributes to the aggressiveness of *Phytophthora* spp. (Kong and Hong 2010).

Soilborne fungal pathogens have to interact with host roots as well as other microorganisms that affect their development significantly. Root exudation from host plants have marked influence on the spore germination and further development of the pathogen, resulting in success/failure of infection. The influence of root exudates and roots of various plant species, including pea, bean, alfalfa, oat, soybean, corn and tomato on germination of *Aphanomyces euteiches* oospores was studied. A procedure was developed for prolific production, easy collection and assessing germination frequency of single oospore of *A. euteiches*. Root exudates of several host plant species could slightly stimulate oospore germination, ranging from 0 to 11.1%. In contrast, oospores placed directly on plant roots germinated at higher frequencies. Oospores of pea pathotype isolates P30 and P46 germinated at a greater frequency (30.6 to 61.1%) on pea, bean and oat roots than on roots of other plant species tested. Oospores of bean pathotype isolates GB33 and GB71 showed higher germination frequency (47.2 to 52.8%) on bean roots than on roots of other plant species tested. Oospore germination was greater, when they were placed on lateral roots than on tap roots of pea and bean. Roots of young plants (ten-day old) seemed to stimulate oospore germination more efficiently than the roots of older plants (20-day old) of bean. The results indicated that roots of bean and pea could be used to germinate oospores of *A. euteiches* more effectively (Shang et al. 2000).

The bioactivity of exudates from tomato was assessed, using different membrane filters. The effects of root exudates from tomato on germination of macroconidia of *Fusarium oxysporum* f.sp *lycopersici* were determined. Membrane filtration of unsterile root exudates with filters of different membrane materials and filter brands, resulted in enhancement of germination of macroconidia. The enhanced conidial germination varied, depending on the membrane filter used, but lacking when sterile root exudates were used. Bioactivity associated with root exudates appeared to be due to the presence of microbial contaminants. Variants in the effects of different filter brands might be due to their differential potential of retaining inhibitory compounds (Steinkellner et al. 2008). Peanut brown RR disease caused by *F. oxysporum* and *F. solani*, is a major constraint, limiting production in China. The effects of root exudates of peanut plants of susceptible and resistant cultivars on the pathogen development were investigated. The root exudates from susceptible cv. Ganhua-5 (GH) and moderately resistant cv. Quanhua-7 (QH) enhanced spore germination, sporulation and mycelial growth of *F. oxysporum* and *F. solani*, were compared with a control. Strains of the pathogen species responded differently to the root exudates, the extent of stimulation increasing with increase in concentrations of root exudates. Root exudates of cv. GH contained higher contents of sugars, alanine, total amino acids, compared with that of cv. QH which contained higher amounts of p-hydroxybenzoic acid, benzoic acid, p-coumaric acid and total phenolic acids than that of cv. GH, as determined by high performance liquid chromatography (HPLC) analysis. The results suggested that the phenolic compounds might have a role in the host resistance at later preinfectional stages (Li et al. 2013).

Different *Phytophthora spp., P. sojae, P. vigne, P. pisi* and *P. niederhauserii* infect various leguminous crops. P. *sojae* and *P. vigne* cause Phytophthora blight in soybean and cowpea, respectively, whereas *P. pisi* is pathogenic to pea and faba bean. *P. niederhauserii* has a wider host range. Zoospores of *Phytophthora* spp. are chemotactically attracted to the isoflavones that are secreted by their respective host plant species. The chemotaxic behavior of zoospores of closely related *Phytophthora* spp. infecting legume roots was investigated. *P. sojae* and *P. vigne* were attracted to the non-soybean isoflavones prunetin as well as to the soybean flavones genistein and daidzein which was in contrast with their host specificity on soybean and cowpea, respectively. On the other hand, *P. pisi* and *P. niederhauserii* were only attracted to prunetin, reported to be produced by pea, not to isoflavones associated with the nonhost bean. The lack of responsiveness to genistein and daidzein in *P. pisi* might represent an adaptation to host specialization toward pea. However, the affinity of *P. niederhauserii* to prunetin indicated that this trait could also be present in taxa not specifically associated with legume hosts (Hosseini et al. 2014).

Grafting watermelon (*C. lanatus*) onto bottle gourd (*Lagenaria siceraria*) is a common practice to enhance resistance to wilt pathogen, *Fusarium oxysporum* f.sp. *niveum* (FON) in China. The effects of root exudates from own-root bottle gourd, graft-root watermelon and own-root watermelon on FON conidial germination were investigated. Regardless of the type of plant, the root exudates showed stimulatory effect on FON conidial germination to varying extents. However, the highest germination impact index (diameter of colony in treatment – control/control) was obtained by addition of root exudates from own-root watermelon plants into the 2% water agar. On the other hand, the lowest germination impact index was observed in root exudates from own-root bottle gourd plants (see Figure 3.3). The germination impact index of the root exudates from the grafted-root watermelon plants was

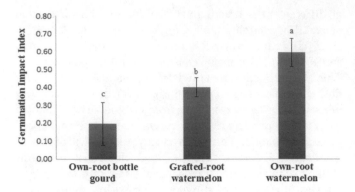

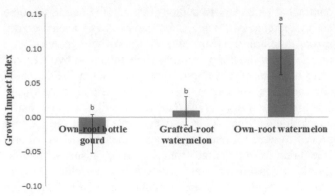

**FIGURE 3.3** Influence of root exudates from own-root bottle gourd, grafted-root watermelon and own-root watermelon on germination of conidia of FON, represented by germination impact index. Bars indicate the standard error for three replicates and different letters represent the significant differences between pairs of mean values at P ≤ 0.005 according to LSD test.

**[Courtesy of Ling et al. (2013) and with kind permission of *PLoS ONE* Open Access Journal]**

**FIGURE 3.4** Effect of root exudates from own-root bottle gourd, grafted-root watermelon and own- root watermelon on growth of FON. Bars represent the standard error for three replicates and significant differences are indicated by different letters based on LSD test (P ≤ 0.005).

**[Courtesy of Ling et al. (2013) and with kind permission of *PLoS ONE* Open Access Journal]**

approximately 67% and higher (200%) than that from own-root watermelon plants and own-root bottle gourd plants, respectively. Likewise, root exudates from own-root watermelon stimulated growth of FON to the maximum extent (see Figure 3.4), whereas treatment with root exudates from own-root bottle gourd had little or negative effect on pathogen growth. HPLC analysis showed that composition of root exudates released from graft-root watermelon differed not only from the own-root watermelon, but also from the bottle gourd plants. Salicylic acid was detected in all root exudates, whereas chlorogenic acid and caffeic acid were present in graft-root watermelon and bottle gourd root exudates. Cinnamic acid was present abundantly in own-root watermelon root exudates only. Chlorogenic acid and caffeic acid facilitated enhancement of resistance in graft-root watermelon to FON. Both chlorogenic acid and caffeic acid inhibited conidial germination and growth of FON in a dose-dependent manner. FON was more sensitive to chlorogenic acid than to caffeic acid. The results indicated that root exudates of watermelon could be altered by grafting onto bottle gourd rootstock and resistance to Fusarium wilt in watermelon was mainly because of grafting on a rootstock of a nonhost of the pathogen. Secretion of chlorogenic acid and caffeic acid might reduce secretion of cinnamic acid that increased susceptibility of watermelon to *F. oxysporum* f.sp *niveum* (Ling et al. 2013).

*L. maculans*, causing Phoma stem canker of oilseed rape (canola) produced two types (A and B group) of ascospores. A-group ascospores produced highly branched hyphae that grew tortuously, whereas B-group ascospores produced long, straight hyphae. Effects of temperature on germination and penetration by A and B group ascospores were investigated. Ascospores of both groups germinated at temperatures from 5 to 20°C on leaves of oilseed rape (canola) at 2 h after inoculation (hai), reaching the maximum by about 14 hai. Both percentage of A-/B- group ascospores that had germinated after 24 h incubation and germ tube length increased with

increasing temperature. Germ tubes from B-group ascospores were longer than those from A-group ascospores at all temperatures, with the greatest difference at 20°C. Hyphae from ascospores of both groups penetrated the leaves predominantly through stomata at temperatures from 5 to 20°C. The percentage of germinated ascospores that penetrated stomata at 5 to 20°C was greater for A-group ascospores of *L. maculans*, after a 40-h incubation period (Huang et al. 2003).

Variation in the requirements of temperature and moisture for germination and infection of chickpea by ascospores and conidia of *Didymella rabiei* were studied. Germination of ascospores and conidia on cover glasses coated with agar began after 2 h with maximum germination (> 95%), occurring after 6 h at 20°C. Germination was entirely inhibited at 0 and 35°C. Ascospores germinated at a faster rate than conidia at temperatures tested. Germination declined rapidly as the water potential varied from 0 to – 4 MPa, although some germination occurred at – 6 MPa at 20 and 25°C. Ascospores could germinate over a wider range of water potentials than conidia and their germ tubes were longer than those formed from conidia at most water potentials and temperatures. The optimum temperature for infection and disease development by both ascospores and conidia was around 20°C. Disease severity was greater when ascospores were discharged directly onto plant surfaces from naturally infected chickpea debris, compared with aqueous suspensions of ascospores and conidia sprayed onto plants. Disease severity increased, as the period of wetness increased. Dry periods of 6 to 48 h immediately after inoculation decreased disease severity, which was higher in plants inoculated with ascospores than with conidia as inoculum. The results showed that ascospores were more effective in initiating infection and enhancing disease severity compared with conidia of *D. rabiei* (Trapero-Casas and Kaiser 2007). The type of inoculum viz., soilborne or tuberborne, was shown to have marked influence on disease incidence and severity. *C. coccodes*, causal agent of potato black dot disease, is both soilborne and tuberborne. Plants grown in infested soil had more sclerotia on roots and greater reduction in yield than plants developing from infected tubers. Increase in concentrations of soilborne inoculum showed nonlinear association with disease development. Soilborne inoculum induced more disease than tuberborne inoculum and disease severity did not increase above a threshold of soilborne inoculum (Nitzan et al. 2008).

### 3.3.2 PENETRATION AND COLONIZATION OF HOST TISSUES

Soilborne fungal pathogens may form more complex infection structures, after spore germination. *R. solani*, produces both lobed appressoria and more complex cushions consisting of branched hyphae. *R. solani* can infect both roots and aerial plant parts. Generally, appressoria are formed on aerial plant parts and infection cushions on root tissues. The infection cushions produce multiple infection hyphae. In the case of *T. basicola*, the vegetative apex, coming into contact with the tobacco root, rapidly differentiates to form infection structures from which thread-like infection hyphae are produced. These infection hyphae penetrate root hair and sickle-shaped intercellular hyphae are then produced (Hood and Shaw 1997). *Fusarium verticillioides* (syn: *F. moniliforme*, teleomorph: *Gibberella moniliformis*), causing ear rot of maize (corn) may be present commonly on plant residues after harvest, but the disease symptoms vary widely and range from asymptomatic infections to severe rotting of all plant parts. Often diseased and asymptomatic plants may occur in the same field planted with a genetically uniform maize cultivar. Infection of maize by *F. verticillioides* may result via different routes. The most common method of kernel infection is through airborne conidia that infect silks. Another route is through systemic infection spreading from infected seeds (Headrick and Pataky 1991; Munkvold and Carlton 1997). Continuous monitoring of the pathogen has been taken up by using transgenic pathogen lines expressing reporter genes. β-glucuronidase- expressing isolate was employed to study the maize-*F. verticillioides* interaction. Tissue sectioning and other experimental manipulation required to monitor β-glucuronidase activity, limited the applicability of this procedure (Brown et al. 2001).

Pathogen-plant interactions have been investigated commonly using the GFP which provides certain advantages. Fungal spores and hyphae of GFP-expressing fungal isolates can be identified by fluorescence microscopy in intact tissues or tissue sections without the need for extensive manipulations, facilitating visualization of the process of colonization of plant tissues by pathogens. Transgenic isolates of *F. verticillioides* expressing GFP were generated for visualization of early events in the *F. verticillioides*-maize interactions. Plants grown in *F. verticillioides*-infested soil were smaller and chlorotic. The pathogen colonized all underground plant parts, but was found primarily in the lateral roots and mesocotyl tissue. In some mesocotyl cells, conidia were produced within 14 to 21 *dai*. A greater amount of mycelium was seen in the cell wall spaces and contained thick organelles that looked like conidiophores or aerial branching. At 21 *dai*,

infected cells were filled with mature conidia. A large number of hyphae that developed along the root axis and green fluorescing macroconidia were detected on root hairs. External symptoms were not expressed at this stage. At 25 to 30 days after planting, the mesocotyls and main roots were heavily infected and rotting developed in those tissues. Other tissues including adventitious roots and stem contained only few hyphae. Asymptomatic systemic infection seemed to be characterized by a mode of fungal development that resulted in infection of certain tissues, intercellular growth of a limited number of fungal hyphae and reproduction of the pathogen in a few cells without invasion of other cells. Rotted tissue with massive production of mycelium and retarded plant growth reflect severity of symptoms induced by *F. verticillium* in maize (Oren et al. 2003).

*Fusarium oxysporum* f.sp. *cubense* race 1 (*Foc1*) and race 4 (*Foc4*) are widely distributed in the banana-growing countries. Differences in the infection process of *Foc1* and *Foc4* were studied, using GFP-tagged strains. Tropical race 4 (*Foc4*) and *Foc1* were inoculated onto Brazil Cavendish banana. At one day postinoculation (dpi), spores and hyphae got attached to the root hairs and root epidermis. At 3 dpi, the hyphae of both *Foc1* and *Foc4* were detected in the vascular tissue of roots. However, *Foc4* was present in the parenchyma cells of banana root, whereas *Foc1* could not be detected in parenchyma cells at 7 dpi. Only few *Foc1* hyphae could be seen in a few xylem tissues, whereas more number of hyphae of *Foc4* was present in many xylem and phloem tissues. The presence of *Foc4* in the vascular tissues of banana rhizomes was observed. In contrast, *Foc1* was not detected in rhizomes at two months after inoculation. Scanning electron microscopic observations showed that *Foc4* was able to penetrate into banana roots from intercellular spaces of the epidermis and wounds. On the other hand, *Foc1*, mainly penetrated through wounds, but not from the intercellular space of the epidermis. The results indicated that direct root penetration and rhizome vascular colonization by *F. oxysporum* f.sp. *cubense* might be the critical stages influencing the aggressiveness of the races of this pathogen (Li et al. 2017).

Interactions between *C. coccodes* and *V. dahliae* , major incitants of potato early dying (PED) syndrome were studied, using aseptic plantlets of cvs. Nicola Desiree, Alpha and Cara with differing susceptibility to PED syndrome. The plantlets were inoculated with identical concentrations of each pathogen or with a mixture of these pathogens. Coinoculation of Nicola with both pathogens caused more severe foliar symptoms and crown rot and greater colonization by *C. coccodes*, than when pathogens were inoculated separately. In Desiree, more roots were covered by *C. coccodes* sclerotia and the disease symptoms were significantly more severe in plants inoculated with both pathogens. Amounts of sclerotia covering the roots were not affected by simultaneous inoculation with both pathogens. In Cara, plants inoculated with pathogen mixture or either pathogen separately showed similar symptom severity and sclerotial production. The differential responses of potato cultivars to PED syndrome might reflect their levels of susceptibility to these pathogens (Tsror (Lakhim) and

Hazanovsky 2001). Colonization of roots, stolons as well as above and below ground stems by *C. coccodes*, following inoculation of seed tubers and foliage was studied during two growing seasons. The black dot pathogen was detected in potato stems as early as 14 days prior to emergence. Colonization of above and below ground stems occurred at a higher frequency than in roots and stolons, resulting in significantly higher relative area under colonization progress curves (RAUCPCs). Although fungal colonization and disease incidence were higher in inoculated and/ or infested treatments, sufficient natural inoculum was present to cause substantial levels of disease in noninoculated and noninfested plots, but they displayed the lowest RAUCPC values. Treatments with more than one inoculation and / or infestation event, tended to have higher disease incidence, indicating the effect of inoculum density on disease incidence/spread under field conditions (Pasche et al. 2010). The fungal pathogens have the ability to modify the pH in or around the infected site through alkalanization or acidification. Investigations to monitor pH changes may be useful in understanding the dynamics of host-pathogen interactions. The ability of the fungal pathogens *C. coccodes* and *Helminthosporium solani* to modify the pH of the potato tuber tissue, following artificial inoculation *in situ* was assessed. Direct visualization and assessment of pH changes near the inoculated area were accomplished, using pH indicators and image analysis. Both pathogens increased the potato native pH of approximately 6.0 to 7.4–8.0. Further, the growth curve of each fungus could be derived by performing simple image analysis. The lag phase of radial growth was estimated to be ten days for *C. coccodes* and 17 days for *H. solani*. A distinctive halo (edge area with higher pH) was observed only during the lag phase of *H. solani* infection. This simple method could be employed to monitor pH modulation, a major factor in the host-pathogen interactions (Tardi-Ovadia et al. 2017).

*V. dahliae* and *V. longisporum* are closely related soilborne pathogens, causing vascular diseases of several cruciferous crops. They produce microsclerotia which accumulate in the soil and are resistant to adverse environmental conditions. Germination of the microsclerotia is triggered by root exudates and infection hypha penetrates the root epidermal cells near the root tips (Zhou et al. 2006). The pathogens traverse the root cortex, inter- and intracellularly, followed by colonization of xylem vessels. *V. longisporum* is confined to the vascular system, a nutrient-poor environment to which the pathogen is well adapted (Pegg 1985). The pathogen spreads with growing hyphae and /or conidia carried with the transpiration stream into the upper parts of the plant vascular system. As the host tissues become senescent, the pathogen enters a final saprophytic growth stage in which microsclerotia are abundantly produced in deteriorating stem parenchyma (Pegg 1985). Systemic spread of *V. longisporum* and vascular responses were investigated, using a susceptible cultivar Falcon and resistant genotype SEM 06-500526 of oilseed rape (canola, *B. napus*) for biochemical and histochemical characterization. Colonization of both genotypes by the pathogen, after dip-inoculation of roots was followed by quantitative

PCR (qPCR) assay. Real-time PCR analysis revealed successful root penetration and invasion of lower plant parts in both genotypes. However, significant differences were observed in pathogen spread further into the shoots. In susceptible Falcon cultivar, the pathogen could spread readily into upper plant parts, until 79 days postinoculation (dpi). On the other hand, colonization of the resistant genotype SEM 05-500256 was confined to the basal parts of infected plants, indicating that resistance to *V. longisporum* was internally expressed during stages later than root penetration and primary invasion of the vascular system. The qPCR data provided evidence that the hypocotyl tissue might play a crucial role in expression of resistance to *V. longisporum*. The morphological and biochemical nature of barriers induced in the hypocotyl tissue, following infection, was studied with histochemical methods followed by biochemical analyses. The buildup of vascular occlusions and the reinforcements of trecheary elements through the deposition of cell wall bound phenolics and lignins were observed. Accumulation of phenolics occurred with a significantly higher intensity in the resistant genotypes, compared with susceptible cultivar, was detected and it corresponded with disease phenotype. In the resistant genotypes, phenols were differentially expressed in a time-dependent manner with preformed soluble and cell wall-bound phenolics at earlier time points and *de novo* formation of lignin and lignin-like polymers in the later stages of infection. The results indicated a crucial role of phenol metabolism in internal defense of *B. napus* against *V. longisporum* and defense responses in the hypocotyl tissues of host plants being a crucial factor for limiting pathogen spread (Eynck et al. 2009).

*V. dahliae*, produces microsclerotia at late stages of development of cotton Verticillium wilt disease in dying aerial plant tissues and degradation of infected tissues releases the microsclerotia into the soil. Microsclerotia have an important role as inoculum source and long-term survival structures. A MAPK signaling pathway plays a role in the microsclerotial development. Mutation in the MAPKinase gene *VMK1* affected formation of microsclerotia and pathogenicity (Rauyaree et al. 2005). The disruption of the gene *VDH1*, encoding a hydrophobin led to reduction in microsclerotial production, but did not affect pathogenicity (Klimes et al. 2008). However, VdPKAC1 (cAMP-dependent PKA catalytic subunit gene) mutants possessed the ability for enhanced microsclerotial production with reduced virulence (Tzima et al. 2010). VdGARP1 (glutamic acid-rich protein) and VGB (the G protein B subunit) were also shown to have a role in microsclerotial production and pathogenicity (Gao et al. 2010; Tzima et al. 2012). Cerevisin, a vacuolar protease, included in serine protease group, has been implicated to have a role in pathogenicity of fungal pathogens. Cerevisin gene-interrupted mutant of *V. dahliae* (Vd991) was obtained through *Agrobacterium tumefaciens*-mediated transformation (ATMT). The T-DNA insertion mutants that produced significantly fewer microsclerotia were selected for further analyses. Mutant T0065 was identified as having the cerevisin gene interrupted by T-DNA and it was named cerevisin mutant. Two-week-old cotton seedlings were root-dip inoculated with conidia from the wild-type strain Vd991 and the cerevisin mutant to assess the effects of mutation on virulence. The cerevisin protein showed a high amino acid sequence similarity with vacuolar protease B. The mutant strain cerevisin displayed significantly decreased microsclerotial and conidial production, growth rate and virulence, compared to wild-type strain. Cotton seedlings inoculated with cerevisin mutant did not exhibit noticeable symptoms, indicating cerevisin was essential for the virulence of *V. dahliae*. The composition of secreted proteins differed between the cerevisin mutant and the wild-type strain. Loss of function of cerevisin decreased the secretion of proteins of low-molecular weight (21–45 kDa). Treatment with the secreted proteins of the mutant resulted in a decrease in the degree of leaf wilting, indicating that cerevisin was involved in the production of these proteins which were putative pathogenicity factors of *V. dahliae*. The results suggested that cerevisin might be involved in controlling multiple processes of development and metabolism and the effects of cerevisin disruption on vegetative growth as well as on mitochondrial formation and the secreted proteases might synergistically affect virulence of V. dahliae (He et al. 2015).

The role of plasma membrane transport proteins in nutrient acquisition by *V. dahliae* was studied. A plasma-membrane protein, the *V. dahliae* thiamine transporter protein VdThit, was characterized functionally by deletion of the *VdThit* gene in *V. dahliae*. The deletion mutants were viable, but growth and conidial germination and production were reduced and virulence of the *VdΔThit* mutants was partially restored. Stress tolerance assays showed that the *VdΔThit* mutant strains were markedly more susceptible to oxidative stress and UV damage. High performance liquid chromatography-mass spectrometry (HPLC-MS) and gas chromatography-mass spectrometry (GC-MS) analyses showed low levels of pyruvate metabolism intermediates acetoin and acetyl coenzyme A (acetyl-CoA) in *VdΔThit* mutant strains, suggesting that pyruvate metabolism was suppressed. Expression analysis of *VdΔThit* confirmed the importance of VdThit in vegetative growth, reproduction, and invasive hyphal growth. In addition, the GFP-labeled *VdΔThit* mutant (VdΔThit-7-GFP) was suppressed in initial infection and root colonization, as viewed with light microscopy. The results revealed the indispensable role of VdThit in the pathogenicity of *V. dahliae* (Qi et al. 2016).

Different proteins are secreted by microbial plant pathogens to suppress host immunity for successfully colonizing host tissues. Identification and characterization of pathogen-secreted proteins will be useful to understand the mechanism of pathogenicity of fungal pathogens. Proteomic search was applied for proteins secreted into xylem by *Verticillium nonalfalfae* during colonization of hop plants. Three highly abundant fungal proteins viz., two enzymes a-N-arabinofuraosidase (VnaAbf4.216) and peroxidase (VnaPRX1.1277) and one small secreted hypothetical proteins (VnaSSP4.2) were identified. These were the first secreted proteins identified so far in xylem sap following infection with *Verticillium* spp. VnaPRX1.1277 classified as a hema-containing peroxidase from class II, similar to other *Verticillium* spp.

lignin-degrading peroxidases and VnaSSP4.2, a 14-kDa cysteine-containing protein with unknown function and with a close homolog in related *V. alfalfae* strains were examined further. In planta expression of VnaPRX1.127 and VnaSSP4.2 genes increased with the progression of colonization, implicating their role in fungal virulence. *V. nonalfalfae* deletion mutants of both genes showed attenuated virulence on hop plants which returned to the level of wild-type pathogenicity in the knockout complementation lines. The results indicated that VnaPRX1.1277 and VnaSSP4.2 might function as a virulence factor facilitating pathogen colonization (Flajsman et al. 2016).

The infection process of *Phytophthora pisi* in pea was studied by inoculating roots of pea plants. Zoospores of *P. pisi* were attracted to the tips of pea seedling roots and they encysted on the roots within 30 min. At 6 h after inoculation, the pathogen had invaded five cortical cell layers by inter- and intra-cellular hyphae. Water-soaked lesions appeared at 20 h after inoculation. *P. pisi* was detected using quantitative PCR format. Gradual accumulation of *P. pisi* DNA in pea root tips continued up to 48 h postinoculation. Induction of genes encoding putative enzyme inhibitors occurred during early stage of infection during 2-6 h postinoculation. Genes encoding putative cysteine protease, glucanase, pleiotropic drug transporter and crinkler proteins were induced, during late stage of infection (Hosseini et al. 2012). The role of pectolytic enzymes, polygalacturonase (PG), pectin methyl esterase (PME) and pectate lyase (PL) produced by *P. capsici*, during pathogenesis, has been investigated. These enzymes are known to degrade the pectin in plant tissues, resulting in their maceration and the activities of these enzymes are increased under conditions favorable for rapid development of disease. *P. capsici* produced high levels of PG, PME and PL in planta. A high correlation between enzymatic activity and pathogenicity of isolates of *P. capsici* was observed (Jia et al. 2009). *P. capsici* secretes a series of pectinase, including pectatelyase (PEL) which have a role in the infection process. A pectate lyase gene (*Pcpel1*) was identified in a genomic library of a highly virulent *P. capsici* strain SD33. *Pcpel1* was identified, as an open reading frame (ORF) of 1,233-bp encoding a protein of 140 amino acids with a predicted amino-terminal signal sequence of 21 amino acids. Western blotting and northern blotting analyses showed that *Pcpel1* was strongly expressed during interaction of *P. capsici* with pepper plant, suggesting its involvement in the infection process of *P. capsici* (Fu et al. 2013).

Pepper-*P. capsici* interaction was visualized by using *P. capsici* transformed with green (*pgfp*) or red (*tdTomato*) fluorescent protein genes. All transformants emitted fluorescence at all stages of the life cycle, but the intensity of fluorescence differed. The transformants contained one, two or five copies of *pgfp* or *tdTomato*, as revealed by Southern blot tests. One of the transformants expressing *pgfp* had reduced growth on artificial medium, produced smaller lesions on detached pepper fruit and showed reduced virulence on pepper seedlings, compared with wild-type strain. However, four transformants that had similar growth rate, as the wild-type strain was useful for monitoring pathogen spread in pepper tissues at different intervals after inoculation (Dunn et al. 2013). Interactions of *P. capsici* with roots, crowns and stems of resistant and susceptible bell pepper plants were studied, using an isolate of the pathogen constitutively expressing the gene for GFP. The susceptible bell pepper cv. Red Knight, resistant cv. Paladin and resistant landrace Crillos de Morelos 334 (CM-334) were inoculated with zoospore suspensions. The same number of zoospores got attached to the roots and germinated on root of all cultivars at 30 and 120 min postinoculation, respectively. The secondary roots of Red Knight showed a higher number of necrotic lesions than Paladin and CM-334 plants at three days postinoculation (dpi). Presence of necrotic lesions on tap roots of Red Knight could be detected at 4 dpi, but not on Paladin or CM-334 plants. All Red Knight plants showed the presence of pathogen hyphae in crown and stems at 4 dpi. On the other hand, only a few plants of Paladin had crown infection and none of CM-334 plants showed crown and stem infections. The differential responses of bell pepper plants susceptible and resistant to *P. capsici* indicated the operation of an efficient defense system in resistant bell pepper plants (Dunn and Smart 2015).

*Tilletia horrida, T. indica* and *T. walkeri* induce rice kernel smut, wheat Karnal bunt and ryegrass bunt, respectively. Thick-walled teliiospores of these smut pathogens persist for several years in the field soil. Teliospores present at or near soil surface, on germination, develop filiform primary sporidia (basidiospores) which produce hyphae or allantoid secondary sporidia. Additional secondary sporidia or hyphae are formed from secondary sporidia. The secondary sporidia are forcibly discharged and they germinate via hyphae, forming the primary inoculum. Initial infection by *T. indica* occurs through airborne sporidia deposited by air on glumes and produce hyphae which penetrate stomata and eventually enter the caryopsis where the pathogen proliferates and sporulates, producing a sorus filled with millions of teliospores (Goates 1988). *Tilletia* spp. induce similar symptoms, when the grains mature, converting the seed into sori containing a dark-colored mass of teliospores. As the secondary sporidia were thin-walled spores, they were considered to be short-lived. Hence, the longevity of secondary sporidia and their role in initiating infection of floral parts of cereals were investigated under laboratory and field conditions. Sporidia naturally discharged onto petridishes, air-dried and maintained at 10 to 20% relative humidity (RH) for 31 to 49 days and at 40 to 50% for 56 to 88 days regenerated rapidly. No difference in the viability of sporidia that were initially dried rapidly or dried slowly over ten hours was observed (see Figure 3.5). Sporidia of *T. horrida* or *T. indica* dried on petridish lids placed in the lower canopy of barley or wheat fields in Idaho and Arizona during early flag leaf to soft dough stages and held until crops were near or beyond maturity, regenerated rapidly, despite temperatures of up to 46°C and several days of RH < 20% (see Figure 3.6). The results showed that the secondary sporidia were highy durable and tolerant to fairly extreme natural and laboratory condtions and had the potential to function as infecting agents in the absence of germinating teliospores . The persistence of

secondary sporidia might have a significant role in the epidemiology of diseases caused by *Tilletia* spp. (Goates 2010).

*P. cinnamomi* causes extensive root rot in many susceptible plant species, including several crops and forest trees. Decay

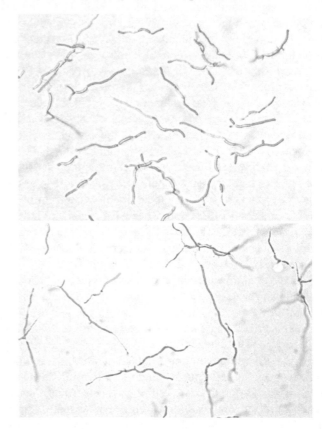

**FIGURE 3.5** Formation of secondary sporidia of *T. horrida* on agar surface in petridish. Sporidia discharged from regenerated dried sporidia on petridish lid dried for 25 days at 10% to 20% (RH) and hyphae produced after 18 hours.

**[Courtesy of Goates (2010) and with kind permission of the American Phytopathological Society, MN, United States]**

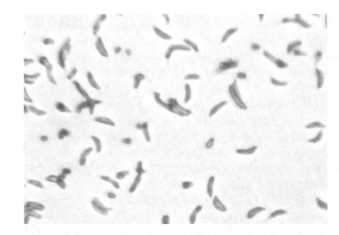

**FIGURE 3.6** Formation of secondary sporidia of *T. horrida* on petridish lid placed in the field and incubated for 49 days Sporidia were similar to those formed under *in vitro* conditions.

**[Courtesy of Goates (2010) and with kind permission of the American Phytopathological Society, MN, United States]**

of surface of fine roots, subsequently leading to vulnerability of seasonal drought stress is a common syndrome. The pathogen, a soilborne oomycete, infects host plants primarily via motile zoospores that are attracted to roots. The infected plants are killed, following decay of fine roots and lower stem tissues, restricting the ability of plants to absorb water and nutrients from soil. It is difficult to identify and differentiate *P. cinnamomi* from other *Phytophthora* spp. and *Pythium* spp. present in the roots of the same infected plants. Hence, fluorescent in situ hybridization (FISH) technique was developed to visualize *P. cinnamomi* in naturally infected plant materials. The FISH procedure was useful to identify different structures of *P. cinnamomi*, such as hyphae and chlamydospores in plant roots containing structures of other fungi or oomycetes. The species-specific probe did not hybridize to other species of *Phytophthora* or *Pythium*. The use of multiple species-specific probes allowed direct comparison of the competitive/synergistic interaction between different *Phytophthora* species in field samples. The widely employed PCR formats did not allow microscopic examination of the pathogen directly within the host tissues. On the other hand, detection of *P. cinnamomi* within host tissue by FISH technique was found to be a distinct advantage over other nucleic acid-based procedures. The probe developed in this investigation was able to penetrate several layers of plant cell walls and hybridize to *P. cinnamomi* embedded within all plant materials analyzed. The FISH assay detected *P. cinnamomi* within the woody tissue of *Stylidium diuroides*, while the pathogen could not be viewed under bright field magnification. The major limitation of FISH was the autofluorescence produced by many plant cells without quenching the fluorescence of P. *cinnamomi*-specific probe conjugated with AlexFluoro350. Treatment with toluidine blue completely quenched the autofluorescence across all plant samples analyzed, thus overcoming the limitation of FISH technique in visualization of P. *cinnamomi* in infected root tissues unambiguously. The FISH procedure showed potential for use in the microscopic examination of *P. cinnamomi* even in deeper woody tissues of plants (Li et al. 2014).

The role of proteolytic enzymes produced by the soilborne fungal pathogens, A. *flavus* and A. *parasiticus*, causal agents of peanut (groundnut) aflaroot disease was studied. *A. parasiticus* produced a protease before infection of seed, when the pathogen was inoculated on seeds of resistant and susceptible peanut cultivars. *A. paraiticus* grown on casein medium (inductive medium) induced greater internal and external infection and the seeds had higher fungal protease content than those induced by the pathogen grown on PDA or sucrose media (non-inductive media). Higher fungal colonization and aflatoxin contamination in seeds of resistant cultivar preincubated with *Aspergillus* extracellular protease was observed than in those incubated with proteases. A. *flavus* and A. *parasiticus* produced metallo and serine proteases, the greatest activity being associated with the latter enzyme (Asis et al. 2009). In a later investigation, aflatoxins were detected in husks and kernels of peanuts inoculated with A. *flavus* and A. *parasiticus*, using HPLC procedure. However, the results indicated that the presence of these pathogens need not necessarily result in production of aflatoxins in the substrate. The main route of peanut infection and aflatoxin contamination was through soilborne inoculum (Atayde et al. 2012). The pathways of invasion of kernels of maize by A. *flavus*, localization, morphology and transcriptional profile of the pathogen were investigated during internal seed colonization. A. *flavus* was found to be capable of infecting all tissues of the immature kernel by 96 h after inoculation. Presence of mycelia in and around the point of inoculation in the endosperm was observed and the hyphae advanced down the germ. At the endosperm-germ interface, hyphae seemed to differentiate and form a biofilm-like structure that surrounded the germ. A custom-designed Afymetrix GeneChipR microarray system was applied to monitor genome-wide transcription during pathogenicity. A total of 5,061 genes were differentially expressed. Genes encoding secreted enzymes, transcription factors and secondary metabolite gene cluster were upregulated and considered to be potential effector molecules responsible for disease in kernel. The results indicated the possibility of preventing or slowing down of A. *flavus* colonization, using suitable disease management strategies (Dolezal et al. 2013).

*P. brassicae*, causal agent of club root disease of canola (*B. napus*) and other cruciferous crops, has two phases in its life cycle: a primary phase restricted to root hairs and occasionally epidermal cells and a secondary phase involving infection and proliferation within root and subsequent symptom development. The roles of primary and secondary stages were investigated separately, using resting spores (RS, source of primary zoospores) and secondary zoospores (SZ) formed from multinucleate zoosporangia produced by primary plasmodia (McDonald et al. 2014). The development of plasmodia of *P. brassicae*, causal agent of club root disease of cruciferous crops, in dual culture and turnip suspension cells was studied. Suspension culture of *P. brassicae*-infected turnip cells was developed using *P. brassicae*-infected callus in Murashige-Skoog medium supplemented with 0.1 mg of 2,3-D/l and 0.02 mg of kinetin/l. Spherical to subspherical secondary plasmodia were formed in the suspension cell culture. A few young plasmodia divided, resulting in numerous spherical, small plasmodia, which eventually formed a plasmodial cluster. Secondary plasmodia could move within transformed suspension host cells by cytoplasmic streaming of host cells. Secondary plasmodia divided synchronously with transformed turnip cells (see Figure 3.6) (Asano and Kageyama 2006). In a later study, the responses of *Brassica* hosts susceptible and resistant to *P. brassicae* were assessed. Primary (root hair) and secondary (cortical) infections occurred in both resistant and susceptible hosts and the symptoms included cell wall breaks, presence of vesicles or inclusion bodies within cell walls, cell wall thickening in association with plasmodesmata and enlarged and / or disorganized nuclei. Degradation of the secondary thickening and cell walls of the xylem was observed in susceptible hosts, but not in resistant plants. The results suggested that amoeboid form of the pathogen might be present and the reduced number of cell wall breakages in resistant hosts might restrict movement of amoeboid form, but not its development (Donald et al. 2008).

Primary and secondary infections of canola cv. Westar by *P. brassicae* were investigated, using different concentrations of RS or SZ. Inoculation with 1 × 105 resting spores/ml, resulted in primary (root hair) infection at 12 h after inoculation (*hai*). Secondary (cortical) infection was commenced at 72 *hai*. When inoculated onto plants at a concentration of 1 × 104/ml, SZ produced primary infections similar to those induced by RS. Primary and SZ could produce primary infections synchronously that were similar in morphology, supporting the hypothesis that primary and SZ may essentially be the same in their identities. On the other hand, in the early stages of infection, when both spores are able to cause primary infection, recognition of each other as non-self would result in competition between the two kinds of spores. After inoculation of healthy plants with SZ, primary and secondary infections could be observed simultaneously. Primary zoospores can also cause directly secondary infection, when the host is under primary infection. This makes it difficult to determine the requirement of a primary infection phase for the production of SZ to cause secondary infections (Feng et al. 2013b).

Interactions between *P. brassicae* and compatible and incompatible canola cultivars were studied. Canola cv. Zephyr was resistant to pathotype 6 (P6), but susceptible to pathotype 3 (P3) of *P. brassicae*. SZ of virulent and avirulent pathotypes were used to inoculate compatible (susceptible) and incompatible (resistant) cultivars. The pattern of symptom development and the extent and development of cortical infection were evaluated at 52 days after seeding. In both compatible and incompatible interactions, inoculation with RS elicited a pattern of response different from that of inoculation with SZ. Inoculation with RS-P3 resulted in the highest values for both cortical infection (33%) and number of cells with RS (32 cells) and 100% clubroot incidence and severity. SZ-P3 produced less cortical infection (12%) and lower incidence and severity (78 and 67% DSI) than RS-P3. In the incompatible interaction, the opposite pattern was observed. No clubs developed on plants inoculated with RS-P6 or on the noninoculated control plants. The results showed that primary infection played a role in subsequent cortical infection in both compatible and incompatible reactions. Cortical infection was almost completely suppressed after primary infection with an avirulent pathotype (RS-P6), but the root cortex was infected and colonized to a substantial extent with SZ of an avirulent pathotype (SZ-P6). This indicated that primary infection with P6 induced a resistance response that was expressed at high levels in the root cortex, although there was a small effect on pathogen development in root hairs. In contrast, cortical infection following primary infection with a virulent pathotype (RS-P3) was more extensive than with SZ of a virulent pathotype (SZ-P3) alone, indicating that primary infection induced susceptibility or suppressed resistance in the cortex. Recognition of the pathogen as incompatible or compatible occurred in the root cortex, but the resistance reaction developed more rapidly and was expressed more strongly in response to primary infection (McDonald et al. 2014).

*P. brassicae* forms two types of galls, club-shaped and spindle-shaped in appearance in different genotypes of *Brassica* spp. Spindle gall formation was coincident with expansion of stele of secondary tissues by *P. brassicae*, whereas spheroid galls had a region of proliferating tissue that corresponded to the secondary cortex and periderm of healthy plants, with the outer proliferating tissue less infected than the inner portions. The underlying tissue showing limited secondary tissue development, was largely uninfected and where infection occurred, a continuous stele was maintained. An active host defense reaction, in the form of cell lignification or hypersensitive reaction (HR), was not observed, whereas RS were visible in a spheroid gall. The results suggested that restriction of proliferation of *P. brassicae* in spheroid galls might not be indicative of resistance to club root disease (Rennie et al. 2013). The development of *P. brassicae* pathotype in susceptible cv. Bronco, resistant cv. Kilaherb and intermediate resistant cv. B-2819 cabbage cultivars was investigated under controlled conditions. The infected roots were examined by root sectioning, staining with methylene blue and bright field microscopy. The presence of *P. brassicae* plasmodia was observed in root tissues of all cultivars. Extensive cortical colonization and some resting spore formation in susceptible cultivar were seen. In contrast, resistant cultivar showed lower level of colonization without formation of RS and absence of disease symptoms (Gludovacz et al. 2014).

*S. sclerotiorum*, a soilborne pathogen, causing stem rot disease, can infect cotyledons, apical buds, base of stems, leaves and heads of sunflower. Under favorable conditions, sclerotia of *S. sclerotiorum* germinated directly to initiate new infections or germinated carpogenetically to release huge numbers of ascospores which are the primary source of inoculum. As a typical necrotrophic pathogen, *S. sclerotiorum* penetrates and kills susceptible tissues to absorb nutrients from dead cells. The pathogen can enter into plant cells through direct penetration, actions of enzymes and toxins and mechanical pressure or indirectly through wounds or natural openings such as lenticels or stomata. In most cases, *Sclerotinia* penetrates cuticle directly by appressoria and not through stomata. Enzymatic degradation of cuticle may also aid the penetration process. Secretion of CWDEs and OA has been shown to have a significant role in pathogen penetration into host cells (Lumsden 1969; Cessna et al. 2000). The process of infection of four economically important crops viz., dry bean, canola, soybean and sunflower by *S. sclerotiorum* was studied, using transformants expressing *gfp* gene *in vitro* and in planta. Isolates of *S. sclerotiorum* ND30 and ND21 were transformed using pcT74 and gFGP constructs, both containing genes for GFP and hygromycin β-phosphotransferase putative transformants were obtained using polyethylene glycol-mediated transformation of protoplasts. Seven stable *gfp* transformants were evaluated for fluorescence *in vitro* and in planta, pathogenicity and colonization of host tissues. Real-time quantitative PCR assay detected a single copy of the *gfp* gene in transformants. Seven transformants (four from ND30 and three from ND2) were pathogenic on all four crop plant species tested. The pathogen directly penetrated tissues through the cut end of petioles, but also grew down the outside of the petiole. Mycelia and individual hyphae fluoresced on and in petiole tissue of sunflower, dry bean, soybean and canola. Hyphae began to ramify through cortex and parenchyma tissues at four to five *dai*. Hyphal branches penetrated the parenchyma cells and ran parallel to and around vessels. No definitive penetration of xylem tissue could be confirmed. Fluorescing hyphae could be clearly distinguished from plant cells. Infection and colonization of tissues could be readily visualized, using a fluorescent microscope. The transformants differed significantly in the length of lesions induced, compared to the wild-type strain, depending on the host plant species (de Silva et al. 2009). The tips of hyphae of *S. sclerotiorum*, prior to penetration, become swollen and extensively branched and then produce modified, multicellular, melanin-rich hyphal structures that are known as compound appressoria or infection cushions. Sticky mucilage covering the surface of compound appressoria, might facilitate firm adhesion of *S. sclerotiorum* to the plant surface and consequently effective penetration into the host tissues. Further, the plant cuticle may be perforated by thin penetration pegs that originate from flattened hyphae of the compound appressorium and some infection holes may be seen on host plant surface, after removal of compound appressoria (Huang et al. 2008).

Sunflower-*S. sclerotiorum* interactions were investigated at plant surface and cellular levels of a susceptible genotype (C146) at 12, 24 and 48-h postinoculation. Appressoria were formed and the hyphal strands branched upon contact of pathogen with the host plant surface. In addition, a direct penetration by fungal hyphae through cuticle within 12 h of inoculation was observed. Microscopic examination of inoculated tissues after 24 h revealed that the pathogen had formed both inter- and intra-cellular growth. The hyphal development was incremental among and inside host cells. The host cells were entirely colonized by pathogen mycelium at 48 h after inoculation, resulting in tissue collapse (Davar et al. 2012). In a later investigation on soybean-*S. sclerotioum* interaction, leaf, flower, and stem tissues of wild type (WT) and transgenic lines of soybean with high oxalate oxidase activity overexpressing (OxO-OE) plants inoculated with *S. sclerotiorum* were stained with 4-chloro-1-naphthol at 20, 30 and 40 days after sowing. Histological examination using light microscope revealed similar patterns of staining in unifoliate and the first, second and third trifoliates. Leaves of transgenic OxO-OE plants stained purple at torn edges (see Figure 3.7a) within 1 h in assay solution. After overnight incubation, entire leaf tissue was stained and this reaction was specific for OxO activity, since the WT plant tissues were not stained (see Figure 3.7d). Cell walls of stomata, epidermis and vascular tissues were mainly stained (see Figure 3.7b,c). Newly emerged senescing and dead flowers in OxO-OE plants cut longitudinally stained along the cut surface of peduncle, sepals, petals, stamen and pistil (see Figure 3.7e). In contrast, flowers of WT plants, even after overnight incubation, were not stained (see Figure 3.7f). Tissues stained with control assay solution lacking OA did not exhibit the purple color due to staining, indicating the specificity of the stain, indicating OxO activity

in plant tissues. Changes in the infection process of *S. sclerotiorum* was followed in the flowers of OxO-OE and WT soybean plants at 8, 12, 18, 24, 48 and 72 h postinoculation (hpi), after staining with lactophenol blue under light microscope. The flowers were inoculated with a spore suspension containing 5,000 ascosposres of *S. sclerotiorum*/10 µl and incubated under humid conditions in petriplates. Most of the ascospores germinated by 8–24 hpi on both OxO-OE and WT plants. Long straight hyphae could be observed. Mass of mycelia was formed after 24 hpi. Mycelial growth of *S. sclerotiorum* was not inhibited in OxO-OE plants, although the growth rate was reduced, compared to the WT plants (Davidson et al. 2016) (see Figure 3.8).

Necrotrophic fungal pathogens induce PCD to obtain nutrients required for the establishment of infection and subsequent disease development. Two genes of *S. sclerotiorum*, *SsNep1* and *SsNep2* encoding necrosis-inducing peptides, were cloned. The peptides encoded by these genes induced necrosis, when expressed transiently in tobacco leaves. *SsNep1* was expressed at a very low level, relative to *SsNep2* during infection. The expression of *SsNep2* was induced by contact with solid surfaces and occurred in both the necrotic zone and at the leading margin of infection zone. *SsNep2* expression was dependent on calcium and cyclic adenosine monophosphate signaling, as compounds affecting these pathways reduced or abolished *SsNep2* expression coincident with a partial or total loss of virulence (Bashi et al. 2010). In another investigation, *S. sclerotiorum*, a model necrotrophic fungal pathogen, capable of producing OA and CWDEs to kill plant cells and to draw nutrition from the dead host cells was studied. OA is a key pathogenic factor and the pathogen may regulate secretion of OA. A pathogenicity-defective mutant, *Sunf-MT6* was isolated from T-DNA insertional library. *Sunf-MT6* could not form compound appressorium and failed to induce lesions on leaves of oil seed rape, though it could produce more OA than the WT strain. However, it could enter into host tissues via wounds and cause typical necrotic lesions. Further, *Sunf-MT6* produced fewer, but larger sclerotia than the WT strain Sunf-M. A gene, *Ss-caf1* was disrupted by T-DNA insertion in *Sunf-MT6*. Gene complementation and knockdown experiments confirmed that the disruption of *Ss-caf1* was responsible for the phenotypic changes of *Sunf-MT6*. *Ss-caf1* encodes a novel secretory protein (SsCaf1) with a putative Ca2+ binding EF-hand motif. High expression levels of *Ss-caf1* were observed at an early stage of compound appressorium formation and in immature sclerotia. Expression of *Ss-caf1* without signal peptides in *N. benthamiana* via *Tobacco rattle virus* (TRV)-based vectors elicited cell death. The results showed that *Ss-caf1* has a secretory property and *Ss-caf1* showed strong expression at 4 h postinoculation, after contacting host leaves. Further, expression of signal peptide-deleted *Ss-caf1* in *N. benthamiana* led to cell death in plants. This implied that *Ss-caf1* had the potential to enter host cells to interact with host proteins or other substances and trigger host cell death to create a favorable environment for pathogen development and thus play an important role in pathogenicity of *S. sclerotiorum*. The results suggested that blocking expression

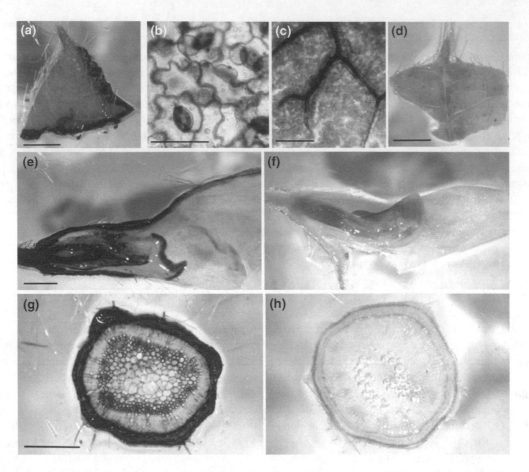

**FIGURE 3.7**    Visualization of OxO activity in leaf, stem and floral tissues of transgenic OxO-OE and WT soybean plants inoculated with *S. sclerotiorum* and stained with 4-chloro- 1-naphthol (a): leaf segment of OxO-OE plant showing OxO activity at torn edges; (b): transdermal section of OxO-OE leaf exhibiting OxO activity in cell walls, stomata, and (c): vascular tissue; (d): leaf segment of WT plant lacking OxO activity without any discoloration; (e): floral tissue of OxO-OE plants showing OxO activity in longitudinal sections; and (f): absence of OxO activity in floral tissues of WT plants In transverse sections of OxO-OE stem tissues, OxO activity specifically was detected in the cell walls (g), but non-specific staining of phloem and cambium in sections of WT plant stems (h) was seen.

[Courtesy of Davidson et al. 2016 and with kind permission of Her Majesty the Queen in Right of Canada, open access article published in *Plant Pathology*]

of *Ss-caf1* could be a feasible strategy for suppressing pathogen development and consequent reduction in disease incidence (Xiao et al. 2014).

During sclerotium, apothecium and compound appressorium development in *S. sclerotiorum*, transcripts encoding γ-glutamyl transpeptidase (Ss-Ggt1) accumulated. Gene deletion mutants of *Ss-ggt* were generated and five independent homokaryotic *ΔSs-ggt1* mutants were characterized. All deletion mutants over-produced sclerotial initials and their further development was arrested or eventually produced sclerotia with aberrant rind layers. When incubated for carpogenic germination, the sclerotia of mutants decayed and failed to produce apothecia. Accumulation of total glutathione (tenfold) and $H_2O_2$ was observed in *ΔSs-ggt1* sclerotia, compared with WT strain. Production of compound appressoria significantly reduced. These mutants showed a defect in infection efficiency and a delay in initial symptom development, unless the host tissue was wounded prior to inoculation. The results suggested that Ss-ggt1 might be the primary enzyme involved in glutathione recycling, during these key developmental

stages in the life cycle of *S. sclerotiorum*, but Ss-Ggt was not required for host colonization and symptom development (Li et al. 2012).

Potential virulence factors in *Sclertoinia sclerotiorum* were identified by generating *Agrobacterium*-mediated transformants (AMTs). Two mutants showing significant reduction in virulence were recognized by AMTs that exhibited similar growth rate, sclerotial formation and oxalate production, as the WT strain. The mutation was due to a single T-DNA insertion at 212-bp downstream of the Cu/Zn superoxide dismutase (SOD) gene (*SsSOD1*). Expression levels of *SsSOD1* were significantly increased under oxidative stresses or during plant infection in the WT strain, but not in the mutant. The SOD mutant showed higher level of sensitivity to heavy metal toxicity and oxidative stress in culture and reduced ability to detoxify superoxide in infected leaves. The mutant also showed reduced expression levels of other known pathogenicity genes such as endopolygalacturonases *sspg1* and *sspg3*. The functions of *SsSOD1* were further confirmed by *SsSOD1*-deletion mutation. The *SsSOD1* deletion

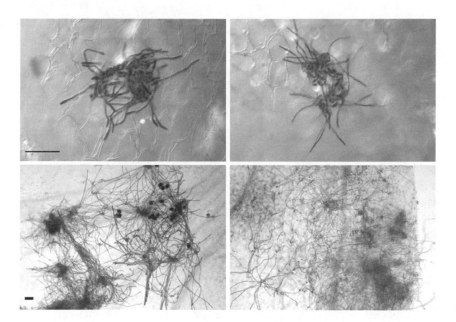

**FIGURE 3.8** Differential growth of *S. sclerotiorum* on petal tissues of transgenic OxO-OE and WT soybean plants inoculated with ascospores of the pathogen. Top left: hyphal growth on petals of OxO-OE plants at eight hours postinoculation (*hpi*); top right: hyphal growth on petals of WT plants at 8 *hpi*; bottom left: pathogen development at 24 *hpi* on petals of OxO-OE plants; and bottom right: pathogen development at 24 *hpi* on petals of WT plants.

**[Courtesy of Davidson et al. 2016 and with kind permission of Her Majesty the Queen in Right of Canada, open access article published in *Plant Pathology*]**

mutants showed characteristics similar to that of AMT insertion mutants. The results suggested that *SsSOD1* might play critical roles in detoxification of ROS during host-pathogen interactions and as an important virulence factor of *S. sclerotiorum* (Xu and Chen 2013). The ability of fungal pathogens to tolerate/dampen oxidative environment is a crucial factor during growth and pathogenesis. Regulatory components of oxidative stress include both enzymatic and non-enzymatic antioxidants. Catalases play an important role in enzymatic detoxification of ROS by converting hydrogen peroxide ($H_2O_2$) into water and molecular oxygen. The type A catalase (Scat1) in *S. sclerotiorum* was evaluated and functionally characterized, as this enzyme was highly induced during infection of susceptible host plant. Insertional inactivation of Scat1 (*ΔScat1*) resulted in hyperbranching of hyphae, accompanied by slower growth and smaller sclerotia. *ΔScat1* strains were attenuated in pathogenicity and rendered the pathogen hypersensitive to sodium dodecylsulfate (SDS), as well as to osmotic and salt stresses. On the other hand, *ΔScat1* exhibited increased tolerance to $H_2O_2$, suggesting that ΔScat1 might not be required for detoxification of ROS. *ΔScat1* strains had a two-fold decrease in ergosterol content and overall lower sterol levels, compared to the WT strain. The results suggested that Scat1 might be involved in modulation of ROS in a manner that deviates from detoxification of $H_2O_2$, alters membrane integrity and contributes to the pathogenicity of *S. sclerotiorum* (Yarden et al. 2014). Accumulation of high levels of OA is essential for *S. sclerotiorum* pathogenesis. Factors influencing OA accumulation were, therefore, investigated by characterizing two putative oxalate decarboxylase (OxDC)

genes (*Ss-odc1 and Ss-odc2*). *Ss-odc1* transcripts accumulated significantly in vegetative hyphae, apothecia, early stages of compound appressorium development and during plant colonization. In contrast, *Ss-odc2* transcripts accumulated significantly only during mid to late stages of compound appressorium development. Neither gene was induced by low pH or exogenous OA in vegetative hyphae. A loss-of-function mutant for *Ss-odc1(Δss-odc1)* showed WT growth, morphogenesis and virulence. Δss-odc2 mutants hyper-accumulated OA *in vitro*, were less efficient in compound appressorium differentiation and defective in virulence which could be fully bypassed by wounding plant surface prior to inoculation. All Δdss-odc2 phenotypes were restored to the WT by ectopic complementation. One *S. sclerotiorum* strain overexpressing *Ss-odc2* exhibited strong OxDC activity, but no OxO activity. The results indicated different roles for *S. sclerotiorum* OxDCs with Odc2 playing a significant role in plant infection related compound appressorium formation and function (Liang et al. 2015).

A peroxisomal carnitine acetyl transferase (CAT)-encoding gene *Ss-pth2*, was required for virulence-associated host colonization by *S. sclerotiorum*. Null (*ss-pth2*) mutants obtained by *in vitro* transposon mutagenesis, failed to utilize fatty acids, acetate, or glycerol, as sole carbon sources for growth. Gene expression analysis of these mutants showed altered levels of transcript accumulation for glyoxylate cycle enzymes. Ss-pth2 disruption also affected sclerotial, apothecial and appressorial development and morphology, as well as OA accumulation, when cultured with acetate or oleic acid as sole carbon sources. Mutants could penetrate and initially colonize host tissue, but subsequent colonization was impaired. Genetic complementation with the WT *Ss- pth2* restored WT virulence phenotypes. The results suggested an essential role for peroxisomal metabolic pathways in *S. sclerotiorum* for OA synthesis and host colonization (Liberti et al. 2013). Bax inhibitor-1 protein functions as a suppressor of programmed cell death (PCD) and it is involved in the response to biotic/abiotic stress in plants. A putative Bax inhibitor-1 protein, *Ss-Bi1* of *S. sclerotiorum* was functionally characterized. High expression levels of *Ss-Bi1* was observed in hyphae under various stresses. Targeted silencing of *Ss-Bi1* resulted in reduced virulence in host plants. *Ss-Bi1* gene-silenced strains were more sensitive to heat stress than the WT strain. The results suggested that *Ss-Bi1* might encode a putative BAX inhibitor-1 protein that is required for full virulence of *S. sclerotiorum* (Yu et al. 2015). The mechanism of sexual reproduction of *S. sclerotiorum,* causing white mold, stem rot, stalk rot and head rot in different crops was investigated. Forkhead-box transcription factors (FOXTFs) was reported to play key regulatory roles in the sexual reproduction of some fungi. *Ss-FOXE2* is one of the four *FOXTF* family member genes identified in *S. sclerotiorum*. Based on ortholog function in other fungi, it was hypothesized to function in sexual reproduction of *S. sclerotiorum*. The role of *Ss-FOXE2* in *S. sclerotiorum* was identified with a gene knockout strategy. Following transformation and screening, strains having undergone homologous recombination in which a *hph* gene replaced the *Ss-FOXE2* gene from the genomic DNA were identified. No variation in cultural characteristics and virulence was observed among the knockout mutant and genetically complemented mutant. However, following the induction of sclerotia for sexual development, apothecia were not formed in the Ss-FOXE2 knockout mutant. The *Ss-FOXE2* gene expressed significantly higher in the apothecial stages than in other developmental stages. The results indicated that *Ss-FOXE2* gene might be necessary for the regulation of sexual reproduction, but vegetative development and pathogenicity were not apparently affected significantly in *S. sclerotiorum* (Wang et al. 2016).

Host-induced gene silencing (HIGS) methods involving host expression of ds-RNA generating constructs directed against genes in *S. sclerotiorum* were examined. The target gene chosen was chitin synthase (*chs*) which controls chitin synthesis. Polysaccharides form a crucial structural component of cell walls of many fungi. Tobacco plants were transformed with an interfering intron-containing hair pin RNA construct for silencing the fungal *chs* gene. Five transgenic lines of tobacco showed a reduction in disease severity ranging from 55.5 to 86.7%, compared with the non-transgenic lines at 72 h after inoculation (*hai*). The lesion area did not show extensive progress up to 120 *hai*. Disease resistance and silencing of the fungal *chs* gene was positively correlated with the presence of detectable siRNA in the transgenic lines. Expression of endogenous genes from the aggressive necrotrophic pathogen *S. sclertiorum* could be prevented by HIGS. The HIGS of the fungal *chs* gene could generate white mold-tolerant plants, demonstrating the possibility of development of cultivars resistant to necrotrophic plant pathogens (Andrade et al. 2016).

Penetration of pansy roots by *T. basicola* was studied, using a transmission electron microscope (TEM). The sequence of events observed in roots of 7–10-day old plants produced on moist filter paper differed slightly from those in roots of 4-week old plants inoculated after washing roots free of potting media. By 3 h postinoculation (PI), epidermal cells of roots produced on filter paper showed aggregated cytoplasm and papilla formation in response to germ tube tips. Presence of callose in papillae was detected by immunogold labeling technique. Papilla formation was not effective in preventing host cell penetration. A slender infection hypha emerged from a germ tube tip and grew through a papilla. The tip of infection hypha expanded to form a globose infection vesicle. At 6 h PI, infection hypha had grown into cortical cells, in spite of papilla formation in these cells. After 24 h PI, distinctive intracellular hyphae were present in necrotic cortical cells. In washed roots, most epidermal cells failed to respond to penetration by *T. basicola*. Hyphae readily penetrated through epidermal cells and contacted cortical cells which showed aggregated cytoplasm and papillae formation. Infection structures similar to those produced in epidermal cells from roots grown on filter paper were formed also in cortical cells of washed roots. *T. basicola* formed infection structures only in cells that responded to invasion, suggesting that the pathogen might have more complex relationship with its host plant. *T. basicola* may be considered as a necrotrophic hemibiotrophic pathogen, as it differs from typical necrotrophic fungal pathogens (Mims et al. 2000). The infection process of *Ceratocystis fimbriata*, causing mango wilt disease, was studied. The extent of disease progress on inoculated stem tissues of mango cultivars was evaluated. Stem sections obtained from the site of inoculation were examined under light microscope. The pathogen isolates apparently heavily colonized stem tissues in susceptible cultivars, Espada, Haden and Palmer, starting from collenchyma and moving toward cortical parenchyma, xylem vessels and pith parenchyma. On the other hand, in resistant cultivars Tommy Atkins and Ubrá, most of the cells reacted to pathogen presence by accumulating amorphous material. The results indicated the significant role of phenolic compounds in incompatible interactions of *C. fimbriata* with resistant mango cultivars (Araujo et al. 2014).

Penetration of woody stem tissues of tree species by *P. cinnamomi* was investigated. Jarrah seedlings (4–6- and 18-month old) were inoculated with a suspension of motile zoospores on the green stem/young periderm region. Light microscopy and fluorescent microscopy procedures were employed to examine sites of taxis and penetration sites on seedlings with intact periderm. The emerging axillary shoot and the region of stem immediately surrounding an axillary shoot were identified as infection courts. Invasion also occurred at sites where the developing shoot had not yet emerged, but was at the stem surface. At these sites also, *P. cinnamomi* directly invaded through the thin-walled phellem of the periderm surrounding the shoot. Zoospores were not attracted toward stomata on mature leaves or green stems. Penetration of the epidermal cell layer of the axillary bud leaf primordia was inter- and intra-cellular, whereas growth of hyphae in the periderm surrounding the shoot occurred intercellularly. In collenchyma, the pathogen growth was both inter- and intra-cellular, while intercellular growth was seen between polyphenol-rich cells. Exposed stem collenchyma was also directly invaded immediately adjacent to the young axillary shoot. Zoospores exhibited taxis to sites of discontinuous periderm, similar to wounded tissues, where the outer protective layers of the plant were already breached. The results revealed the ability of *P. cinnamomi* to penetrate suberized periderm intercellularly (O'Gara et al. 2015).

Fungal pathogens produce enzymes that facilitate penetration of host plant surface. Cutinase produced by fungal pathogens may have a role in the initiation of infection. Plant pathogenic *Fusarium* spp. perceive environmental signals via cell receptors and signal transduction cascades which regulate the expression of effector genes that are directly involved in pathogenesis. *F. solani* and *F. oxysporum* have been used as models to study cellular mechanisms controlling early signaling between plant hosts and fungal pathogens. Plant signal molecule identified as an inducer of fungal pathogenicity response was identified as the monomer of cutin. The cutin monomer released from the plant cuticle triggered rapid upregulation of *cut1*, a gene encoding an extracellular cutinase in the pathogen *F. solani* f.sp. *pisi* (Woloshuk and Kolattukudy 1986). Targeted disruption of the *cut1* gene drastically reduced virulence of *F. solani* f.sp. *pisi* on intact host surface, whereas heterologous expression of *cut1* in a wound infecting *Mycosphaerella* sp. conferred the ability to infect the intact plant without wounding, indicating that *cut1* acted as a virulence factor (Dickman et al. 1989; Rogers et al. 1994). A constitutively expressed and starvation-upregulated cutinase gene (cut1) was identified and its expression was regulated by Cys6Zn6 motif, containing factor, CTF1β (Li and Kolattukudy 1997). Cutinase activity of mutant of *Nectria haematococca* with reduced pathogenicity showed marked reduction in cutinase activity (80–90%), compared with WT strain. *Mycosphaerella* sp., a wound pathogen of papaya fruits, transformed with the cutinase gene of *N. haematococca*, could infect unwounded papaya fruits, indicating an increase in the virulence of *Mycosphaerella* sp. (Dickman et al. 1989). However, the mutant of *N. haematococca* produced by transformation-mediated gene disruption, deficient in cutinase activity, was pathogenic, as the WT strain despite the loss of cutinase activity (Stahl and Schafer 1992). Likewise, loss of cutinase activity in the isolates of *F. solani* f.sp. *cucurbitae* did not affect their pathogenicity on cucurbit fruits (Crowhurst et al. 1997). The role of cutinase may vary depending on the host-pathogen combinations. *Fusarium oxysporum* f.sp. *lycopersici*, causing wilt disease in tomato produces endopolygalacturonase encoded by the gene *pg1*. Expression of *pg1* in roots and lower stems of infected plants was detected by applying RT-PCR assay. The *pg1* gene was cloned and sequenced. The transformed isolates of *F. oxysporum* f.sp. *melonis* deficient in PG1 were as pathogenic as the WT strain, indicating the endoPG was not essential for infection of plants (Pietro and Roncero 1998). The role of polygalacturonase and proteolytic activities in the development

of potato dry rot disease caused by *F. solani* f.sp. *eumartii* was studied. Fungal enzymatic activities were determined at different periods after inoculation either with pathogenic *F. solani* f.sp *eumarti*i isolate 3122 or nonpathogenic *F. solani* isolate 1042. After inoculation with isolate 3122, proteolytic and polygalacturonase activities were higher in susceptible interaction than in resistant interaction. Further, a correlation was observed between levels of proteolytic activity detected in the intracellular washing fluids and size of lesions area caused by *F. solani* f.sp. *eumartii* in susceptible Hinkul tubers. An extracellular potato protein with homology to proteinase inhibitors was identified as a substrate of the proteolytic activity in the susceptible cultivar. Microscopic examination revealed differences between the potato genotypes in the rate of response to infection by *F. solani* f.sp. *eumartii* (Olivieri et al. 2004).

*Ggt*, incitant of take-all disease of wheat, is responsible for significant losses in several countries. The presence of cell wall components, cellulose, xylan, pectin and lignin in roots of healthy and infected wheat plants was detected by applying enzyme-gold and immune-gold labeling techniques. Labeling densities for cellulose, xylan and pectin in the cell walls of infected roots were significantly reduced, compared to those of healthy tissues. Degradation of the cell wall components used as substrate for the activities of the CWDEs was indicated by labeling intensities. Quantification of labeling densities of lignins showed only slight increase in the cell walls of infected roots, whereas greater deposition of lignin was observed in the cell walls of healthy roots, indicating the successful initiation of infection by *G. graminis* var. *Tritici* (*Ggt*) (Kang et al. 2000). In a later study, the distribution of extracellular β-1,3-glucanase secreted by*Ggt* was determined *in situ* in inoculated wheat roots, by employing immunogold labeling and TEM. Antiserum specific to β-1,3-glucanase secreted by *Ggt* was prepared in rabbits. A specific antibody of β-1,3-glucanase, anti-GluGgt, was purified and characterized. Double immunodiffusion tests showed that the antiserum was specific to *Ggt* β-1,3-glucanase, but not to β-1,3-glucanase from wheat plants. The antibody was monospecific for β-1,3-glucanase in fungal extracellular protein concentrations. After incubation of ultrathin sections of *Ggt*-infected wheat roots with anti-β-1,3-glucanase antibody and secondary antibody, deposition of gold particles occurred over hyphal cells and host tissue. Hyphal cell walls and septa as well as membranous structures showed regular labeling with gold particles, whereas a few gold particles were detected over the cytoplasms and other organelles such as mitochondria and vacuoles. In host tissues, cell walls in contact with pathogen usually showed a few gold particles, whereas host cytoplasm and cell walls at a distance from the hyphae were free of labeling. In addition, labeling with gold particles over lignitubers in the infected host cells, was detected. Noninoculated wheat roots were devoid of any gold labeling. The results revealed that β-1,3-glucanase secreted by *Ggt* might be involved in pathogenesis of *G. graminis* var. *tritici*, through degradation of callose in postinfectionally formed cell wall appositions, such as lignitubers (Yu et al. 2010).

The dynamics of infection of wheat (*T. aestivum*), rye (*Secale cereale*), barley (*Hordeum vulgare*) and triticale (× *Triticosecale*) b*Ggt*, incitant of wheat take-all disease was investigated at different growth stages of the host plant species. Under field conditions, *Ggt* spreads along the rows of plants from inoculum sources. At harvest, the pathogen DNA was detected up to 60 cm away from inoculum sources in all species, except rye, although most of the take-all lesions occurred in roots less than 30 cm away in all species, with rye being the least affected by *Ggt*. The populations of fluorescent *Pseudomonas* spp. were in highest numbers in the rhizosphere of cereal roots sampled near the points of *Ggt* inoculation, prior to the booting growth stage in triticale and rye. However, *Psuedomonas* populations were not related to the concentration of pathogen DNA in the roots of affected plants. In the greenhouse pot experiment, *Ggt* colonized seedling wheat roots to a concentration of 103 ng DNA/ mg dried roots and caused 14% take-all severity in roots during plant development. In rye, seedling roots contained *Ggt* DNA at 15 ng/mg dried root, which decreased to negligible concentrations until heading, then increased rapidly to 280 ng DNA ng/mg dried roots at kernel development. Take-all severity in rye increased from 1 to 50% over that period. In another greenhouse experiment, inoculation of roots of host plants at various growth stages with actively growing hyphae of *Ggt*, showed that the pathogen could overcome resistance in rye plants, after an establishment phase (van Toor et al. 2015).

CWDEs have an important role in at least two key phases of development of vascular wilt diseases: during penetration of different layers of the root cortex to gain access to the vascular system and during colonization of the host by spreading upward through xylem vessels. *F. oxysporum* secreted a defined sequence of extracellular CWDEs, when brought in contact with the host cell wall and the production of these enzymes was regulated via substrate induction and catabolic repression (Jones et al. 1972; Cooper and Wood 1973). The enzymes depolymerizing pectin, a major component of primary plant cell wall and middle lamella have been investigated. PLs that catalyze the trans-elimination of pectate were suggested to be involved in vascular wilt diseases. PL activity was detected inside tomato root and stem tissues infected by *F. oxysporum* f.sp. *lycopersici*. Further, transcription of the *pl1* gene, encoding an endopectate lyase was observed, during different stages of disease cycle (Di Pietro and Roncero 1996; Huertas-Gonzalez et al. 1999). Polygalacturonases (PGs) efficiently macerate plant tissues by depolymerizing homogalacturan, a major component of plant cell wall. PG1 is the major endoPG produced by *F. oxysporum* and PG2 and PG3 are also produced by this pathogen. These enzymes were secreted during the pathogen growth on tomato vascular tissue liberating required compounds for use as carbon sources. In planta, only PG1 activity could be detected in stems of infected tomato plants (Di Pietro and Roncero 1996; Garcia-Maceira et al. 1997). As there was no difference in the virulence patterns of PG1-overproducing transformants of this isolate and PG1-lacking WT strain, PG1 was considered to be essential for pathogenicity on muskmelon (Di Pietro

and Roncero 1998). Xylan, the main polymer within hemicellulose, is the major component of the cell wall. The role of xylanolytic enzymes in pathogenesis was studied. The *xyl 2* and *xyl 3* genes encoding family of ten endoxylanases, as well as the *xyl 4* and *xyl 5* genes encoding two families of 11 endoxylanases were cloned. The gene *xyl 2* was expressed only during the final stages of disease development, whereas *xyl 5* was expressed only in the roots, during early infection stages. As there was no reduction in the virulence of knockout mutants, these enzymes might not be essential for pathogenicity (Ruiz-Roldan et al. 1999; Gomez-Gomez et al. 2001, 2002). Plant cell walls contain structural proteins, in addition to the polysaccharides as the major component. Extracellular proteases of the pathogen origin may have a role in successful colonization of host tissues. *Fusarium oxysporum* f.sp. *lycopersici* (Fol), causal agent of tomato wilt disease, has a subtilase Prt1 (a protease) which was expressed at low levels, during the entire infection process. However, targeted inactivation did not affect virulence of the pathogen on tomato plants, indicating its nonrequirement for pathogenicity (Di Pietro et al. 2001).

Chitinases are an important component of plant defense, in response to pathogen invasion. Chitinases in host plants may have a role adversely affecting pathogenesis (disease development) by degrading fungal cell walls containing chitin. Melon cultivars Galia (susceptible) and Bredor (resistant) to *Fom* were examined at different intervals after inoculation with the pathogen. By employing an antibody against a class III chitinases, the presence of this enzyme was traced by both Western blotting analysis and immunolocalization technique. Western blotting analysis detected constitutive expression of chitinase III in the stem base tissues of susceptible Galia. Three bands (26-, 30- and 35-kDa) were detected prior to infection. Two bands (30- and 35-kDa) became faint and the third one was markedly reduced, following infection by *Fom*. Immunolocalization observations revealed the presence of chitinase in the intercellular spaces surrounding a few isolated cells or cell groups dispersed through the cortical region in both susceptible and resistant cultivars. Chitinase became undetectable in susceptible cultivar after eleven days after infection, whereas chitinase could be detected till 18 days after infections in resistant Bredor melon (Baldé et al. 2006). A cDNA library was constructed to identify specific chitinase in tomato by using total RNA from a resistant tomato genotype inoculated with conidia of isolates of race 2 of *F. oxysporum* f.sp. *lycopersici (Fol)*. One chitinase (*SolChi*) clone was isolated. The sequence analysis showed that the cDNA clone *SolChi* encoded an acidic form of class III chitinase. Southern blotting indicated that the expression of this gene was induced upon infection with *Fol* and the accumulation of transcripts for this R protein was rapid in the resistant genotype, during the first 24 h after inoculation. The results suggested a putative role for chitinase in the development of resistance in tomato to *F. oxysporum* f.sp. *lycopersici* (do Amaral et al. 2012).

Infection of watermelon by *Fusarium oxysporum* f.sp *niveum (Fon)* and the development of wilt disease were studied by inoculating resistant and susceptible cultivars with GFP-tagged isolate of *Fon*. Confocal laser scanning microscopy was employed for visualization of colonization of watermelon tissues by *Fon*. The pathogen penetrated and colonized roots of resistant watermelon plants. But further invasion by the pathogen did not progress, as the tap root was not invaded. However, tertiary and secondary lateral roots of resistant plants were colonized, although not as extensively as in susceptible plants. The hyphae penetrated into the central cylinder of lateral roots forming a dense fungal mat, which was followed by a sequential collapse of the lateral roots. The initial infection zone for roots of both susceptible and resistant plants appeared to be the epidermal cells within the root hair zone, which the pathogen penetrated directly after forming appressoria. Preferential penetration of areas where secondary roots emerged and wounded root tissues was also observed (Lu et al. 2014). In a later investigation, colonization dynamics of watermelon cultivar, Sumi 1 (SM1) susceptible to wilt disease was followed, using a GFP-expressing strain of *Fon* race 1. Pathogen hyphae adhered to and grew on epidermal cells of roots at two days postinoculation (*dpi*). Hyphae crossed the epidermis, penetrated into the cortex and progressed through parenchymatous cell borders reaching the xylem vessels at 7 *dpi*. Growth of hyphae was restricted to xylem vessels, even in the colonized lower hypocotyl. In the roots, infected xylem vessels were plugged with gum and tyloses and some occluded vessels were separated from root vascular tissue. Parenchyma cells surrounding the vascular cylinder showed wall thickening and cell disintegration forming cavities eventually. Vascular tissues were densely colonized by hyphae until the vascular bundle became hollow and plants withered ultimately (Zhang et al. 2015).

Many fungal pathogens are known to produce different extracellular enzymes that may be involved in the degradation of cell wall elements for penetration and colonization of host tissues, during pathogenesis. Interactions between fungal endopolygalacturonases (endoPG) and polygalacturonase-inhibiting proteins (PGIPs) present in plant cell walls have been investigated in soybean-*S. sclerotiorum* pathosystems. The pathogen secretes OA and endoPGa and endoPGb, being acidic and basic respectively in nature. The activity of PGb3 was stimulated by OA (oxalates) and PGa activity was maintained at pH 3.6 also. PGa escaped inhibition by soybean PGIP, possibly at this pH and PGIP present in the plant tissues was inactive. RT-PCR assay showed that during infection of soybean by *S. sclerotiorum*, the expression of putative *pga* gene was delayed in comparison to the basic one. The different temporal expression of PGa and PGb and their differential responses to pH, oxalate and PGIP appeared to be consistent with the possible maximization of fungal PG activity in the soybean tissues (Favaron et al. 2004).

The role of lignin degradation in infection, colonization and survival of *Fusarium solani* f.sp. *glycines (Fsg)*, causative agent of soybean SDS, in the root tissues of soybean was investigated. Lignin degradation by *Fsg* occurred by the catalyzed release of 14CO$_2$ from purified 14C-labeled Klason lignin (a method used to test lignolytic capacity of microorganisms in culture) and the production of laccase and lignin peroxidase (major fungal lignin degrading enzymes). The laccase and lignin peroxidase activities and the amount of decolorization or aromatic polymeric dyes by *Fsg* were determined. Studies on lignin synthesis from (14C) phenylalanine with soybean hairy root cultures showed that *F. solani* f.sp. *glycines* treatment stimulated lignin synthesis at 2 h and by 24 h after inoculation, some lignin degradation could be observed. The results indicated that the pathogen could degrade lignin and this activity might facilitate progress of different stages of disease development (Lozovaya et al. 2006). Differential colonization of melon tissues by *Fom* and defense responses of resistant and susceptible melon lines were investigated, using GFP-expressing *Fom* mutant. In susceptible line, pathogen mycelium crossed the cortex and endodermis through narrow pores in cell walls, initially growing on root epidermis and adhering to epidermal cell borders. After reaching the xylem vessels where sporulation commenced, secondary hyphae grew upwards. Although *Fom* colonized the vascular system of the resistant line, seedling infection was lower, compared to the susceptible line. Infection of the vascular system of the resistant line was comparatively slower, suggesting stronger defense responses in resistant lines, expressed at prexylem stage of infection. Analysis of expression patterns of three defense genes, using real-time PCR assay, showed that transcript levels of phenylalanine-ammonia lyase (PAL), chitinase (CHI) and hydrogen peroxide lyase (HPL) were induced to a greater extent in the resistant line than in the susceptible line. In addition, constitutive 2–4-fold difference between resistant and susceptible lines in basal levels of all three transcripts was also apparent. The results indicated that both constitutive and inducible defense responses could contribute to reduced vascular colonization of resistant genotype by F. oxysporum f.sp. *melonis* (Zvirin et al . 2010).

Colonization by *Fusarium oxysporum* f.sp. *fragariae* of strawberry cultivars susceptible and resistant to wilt disease, during 4 h to 7 days postinoculation (dpi) was investigated, using light and scanning electron microscopy techniques. Resistant cultivar Festival arrested spore germination and penetration from 4 to 12 hpi and the pathogen growth subsequently. On the other hand, fungal colonization in susceptible cv. Camarosa was similar to that in Festival until 7 *dpi*. Fungal colonization at 7 *dpi* in cv. Festival remained confined to the epidermal layer of the root, while in cv. Camarosa, the fungal hyphae had not only heavily colonized the cortical tissue throughout, but had also colonized vascular tissues. The results indicated that resistance in strawberry cultivar to wilt pathogen was dependent on inhibition of growth and colonization of pathogen, both on plant surface and within host tissues (Fang et al. 2012). Inoculum density of *Fusarium oxysporum* f.sp. *lactucae*, incitant of lettuce wilt disease, is an important factor affecting disease severity. Crop rotation, considered as a disease management strategy was investigated for its influence on the frequency of infection of lettuce by wilt pathogen. The extent of colonization of roots of broccoli, cauliflower and spinach by *F. oxysporum* f.sp. *lactucae* was assessed. Frequency of infection was significantly lower on all rotation crops than on lettuce cultivar. The pathogen was restricted to the cortex of roots of broccoli. But the pathogen could be isolated from root vascular stele of 7.4% of cauliflower plants and 50% of spinach plants, indicating potential for colonization and production of inoculum on these rotation crops. Root vascular stele of Fusarium wilt-resistant lettuce cultivars showed the presence of the pathogen. Cultivars showing indistinguishable symptoms on aerial plant parts differed significantly in the extent to which they were colonized by *F. oxysporum* f.sp. *lactucae*. The extent of root colonization by the pathogen may be employed as a criterion for the selection of sources of resistance for breeding programs (Scott et al. 2014).

An isolate of *Phoma macdonaldii*, causal agent of sunflower black stem disease, was transformed with a constitutively expressing reporter gene *gfp*. Colonization of sunflower roots by the transformed strain was followed by confocal and scanning microscopy methods. Penetration of the pathogen into the roots of sunflower occurred through natural fissures or through epidermis on partially resistant and susceptible lines. But the colonization rate of stele was reduced in the partially resistant line and the morphology of pathogen hyphae was altered. The effect on hyphal morphology was strongest in the stele, indicating localized production of defense compounds in the partially resistant sunflower genotype (Al Fadil et al. 2009). Colonization patterns of *Foc*, causing banana wilt disease, were monitored, using two races of Foc NT320 (race 1) and B2-gfp (race 4) tagged with GFP. Penetration and colonization of banana cvs. Pisang Awak and Brazil by both isolates were revealed within six days postinoculation (*dpi*), but sporulation in xylem vessels could not be seen until 30 *dpi*. The isolate B2-gfp could penetrate into xylem vessels of Pisang Awak banana roots more rapidly than isolate NT320, suggesting that race 4 isolate was more virulent than race 1. Disease severity induced by these isolates also indicated such difference in virulence of these races of *Foc*. Results of quantitative real-time PCR assay showed that some pathogenicity-associated genes, including *Fga1*, *Fhk1*, *Fow2* and *Ste12* were upregulated by B2-gfp isolate during infection of Brazil bananas, while they were either downregulated by NT320 or not significantly altered (Guo et al. 2015).

CWDEs in fungal pathogens are often subject to carbon catabolite repression at the transcriptional level, such that when glucose is available, CWDE-encoding genes, along with many other genes, are repressed. The gene *SNF1* in *S. cerevisiae* (yeast) control this process and encodes a protein kinase. Snf1 is required to release glucose repression, when this sugar is depleted from growth medium. A reverse genetic approach was employed to explore the role of the *SNF1* ortholog as a potential regulator of CWDE gene expression in *Ustilago maydis*. The gene *snf1* was identified and deleted it from the genome of *U. maydis*. The relative expression of an endoglucanase and a pectinase was higher in WT strain than in the Δ*snf1* mutant strain, when glucose was depleted from the growth medium. However, when cells were grown in depressive conditions, the relative expression of two xylanase genes was unexpectedly higher in the Δ*snf1* strain than in the WT strain, indicating that *snf1* negatively regulated the expression of these genes. In addition, contrary to several

other fungal species, *U. maydis* Snf1 was not required for utilization of alternative carbon sources. Deletion of *snf1* did not significantly affect virulence in *U. maydis* (Nadal et al. 2010).

Colonization of tissues of tomato by *Fusarium oxysporum* f.sp. *radicis-lycopersici* (FORL), incitant of wilt disease, was studied. The β-glucuronidase (*gus*) reporter gene was integrated into FORL in a co-transformation experiment, using *hph* gene, as a selective marker, which resulted in the generation of ten mitotically stable transformants. Using the selected transformant F30, a strong positive correlation was observed between GUS activity and accumulated biomass of *in vitro* grown pathogen. Hence, the GUS activity was used to measure the pathogen growth in tomato lines. A parallel increase in lesion development and GUS activity was recorded in both tomato lines. However, a correlation between GUS activity and disease progression was not seen always. The levels of GUS activity obtained for the more resistant line were higher than those obtained for the susceptible line, indicating that increase in biomass might not be the only factor influencing disease development in tomato infected by *F. oxysporum* f.sp. *radicis-lycopersici* (Papadopoulou et al. 2005). Tomato plants contain α-tomatines, a steroidal glycoalkaloid saponin which has been implicated as a preformed antimicrobial compound effective against some potential fungal pathogens. *Fusarium oxysporum* f.sp. *lycopersici* has been shown to degrade α-tomatinase by the activities of extracellular enzymes (Ito et al. 2002). *F. oxysporum* f.sp. *lycopersici* (FOC) and *F. oxysporum* f.sp. *radicis-lycopersici* (FORL) produced tomatinases with the same mechanisms involving hydrolysis of a-tomatine. But the gene encoding tomatinase seemed to be different between these two pathogens (Ito et al. 2004). The tomatinase gene *FoTom1* capable of degrading α-tomatine to less toxic metabolites was present in FOL and also in some strains of *F. oxysporum*, belonging to *formae speciales* nonpathogenic to tomato. Four FoFom1-positive strains (2, 3, 22 and 17) of *F. oxysporum* nonpathogenic to tomato colonized not only the roots and basal stems, but also the upper stems of tomato plants. The strain #3 transformed with β-glucuronidase (GUS) reporter gene colonized vascular vessels of all tomato cultivars tested. The *FoTom1* gene was expressed in roots and stems of plants, as revealed by RT-PCR assays, suggesting that *FomTom1* was important for the four *FomTom1*-positive strains to survive in tomato plants. Prior to inoculation of tomato plant with either #2 or #3 strain resulted in the suppression of vascular wilt of tomato caused by FOL, due to activation of defense genes, as indicated by accumulation of transcripts of acidic chitinase gene (*Chi3*) (Ito et al. 2005).

Interaction of *Fusarium oxysporum* f.sp. *cubense*, race 4 (*Foc 4*), incitant of banana wilt disease, was studied. Foatf1, a basic leucine zipper (bZIP) transcription factor in *Foc4* was characterized and a deletion mutant could induce oxidative burst of banana seedlings in early stages of infection and the mutant exhibited higher sensitivity to hydrogen peroxide ($H_2O_2$) than the WT strain *Foc4*. The deletion mutant (lacking Foatf1) was significantly less virulent on Cavendish banana (*Musa* spp.). The results indicated that Foatf1 might be required to induce the full virulence in *Foc4* by regulating the transcription of catalase to impair plant defenses mediated by reactive oxygen (ROS) (Qi et al. 2013). Differences in the virulence of *Fusarium sulpherum* and *F. sambucinum* and changes in the ROS production and metabolism in inoculated slices of potato tubers were compared. *Fusarium* infection induced significant production of ROS, lipid peroxidation and loss of cell membrane integrity, but low activity of superoxide dismutase (SOD) and ascorbate peroxidase (APX). *F. sulpherum* induced larger lesion diameters on potato tubers and slices, resulting in greater superoxide anion (O–2) and earlier peak of $H_2O_2$, but lower catalase activity (CAT) and APX, resulting in higher malondialdehyde (MDA) content and lower cell membrane activity, compared with effects induced by *F. sambucinum*. The results suggested that overproduction of ROS was primarily involved in the pathogenicity of *Fusarium* spp. in potato tubers (Bao et al. 2016).

Soilborne fungal pathogens are known to produce toxic metabolites that have a role in the induction of symptoms in susceptible host plants. *V. dahliae*, incitant of cotton Verticillium wilt disease, produces a glycoprotein in CF, contributing to the development of disease symptoms. The crude toxin, VD-toxin, present in CF induced damage to plasma membrane and cell wall in the susceptible callus cells, as revealed by transmission electron micrographs. Invaginated plasmalemma, distinctive plasmolysis and extensive development of membrane-bound vesicles and ultimately agglutinated cytoplasm and disintegration of plasma membrane were the adverse effects induced by VD-toxin at different periods after treatment (Zhen and Li 2004). *V. dahliae* causes PED disease. Colonization dynamics of *V. dahliae* in two potato cultivars with varying responses to PED were monitored. QPCR assay was applied to assess colonization and spatial progression of *V. dahliae* in moderately resistant potato cv. Ranger Russet and highly susceptible cv. Russet Norkotah. *V. dahliae* was detected in samples from plant bases by week 2 and from apical parts at week 4 in highly susceptible cultivar. Pathogen could be detected in apical samples from moderately resistant cultivar only by week 6. Consistent increase was observed in colonization of apical and basal samples of highly susceptible potato during ten weeks, whereas it plateaued after week 6 in moderately resistant potato. Differences in response to PED seemed to be associated with the speed of colonization and establishment of a high population density of *V. dahliae* in plants (Bae et al. 2007). Interaction between *Verticillium albo-atrum* and potato plants was studied. *V. albo-atrum*, infecting potato produces phytotoxins in susceptible potato plants. A low-molecular mass toxin was separated from the CF of *V. albo-atrum*. Petioles of susceptible cv. Desirée, when treated with the phytotoxin, produced greater quantities of ethylene, compared to the petioles of tolerant cv. Home Guard. Pretreatment of leaflets of cv. Desirée with silver thiosulfate (capable of inhibiting perception of ethylene) prevented chlorosis and necrosis associated with toxin activity. Likewise, application of aminoethoxyvinylglycine (AVG), an inhibitor of aminocyclopropane-1-carboxylic acid (ACC) synthetase, to petioles of cv. Desirée reduced toxin-induced ethylene synthesis and symptom development. The results indicated that

Verticillium toxin, in part, could act through induction of ethylene biosynthesis in the host tissues and different responses of potato cultivars susceptible and tolerant to *V. albo-atrum*, might be due to differential production of ethylene (Mansoori and Smith 2005). Colonization of olive inflorescences by *V. dahliae* was investigated. Presence of viable propagules of *V. dahliae* in dried inflorescences was detected by isolation on modified sodium pectate agar medium. The microsclerotia of the pathogen were present inside peduncles and flowers, indicating the possibility of infected olive inflorescences being sources of inoculum of Verticillium wilt epidemics in olive (Trapero et al. 2011).

The process of infection of *L. maculans* (anamorph: *Phoma lingam*), the incitant of canola Phoma stem canker, was studied, using the cultivar Surpass 400, containing single dominant gene-based resistance (SDGBR). Two strains of *L. maculans*, UWA192 and UWAP11 were inoculated on cultivars Surpass 400 and Westar. The prepenetration and penetration behavior of both strains were similar. Variations in further stages of infection could be observed. When UWAP strain infected Surpass through stomata, guard cells were killed rapidly within a few hours and necrosis appeared in the surrounding mesophyll cells, resulting in hypersensitive reaction (HR). Hyphal growth continued slowly, through the intercellular palisade, mesophyll and spongy spaces, but hyphae rarely spread on the HR region and did not sporulate. Accumulation of phenolic compounds occurred in the area bordering the HR. In contrast, when UWA192 infected through stomata, symptoms appeared at 10–12 *dai*. Pale to tan/gray circular lesions with abundant pycnidia were observed. Hyphae extensively spread beyond the lesion border, reaching the veins and progressing down the petiole toward the stem. HR reaction to UWAP strain restricted its spread to a few palisade cells, although the presence of hyphae in the lower tissue layers of cotyledons could be noted. In the absence of SDGBR, as in UWA192 strain infection, extensive death of epidermal and upper and lower cells occurred throughout infected areas of the cotyledon abundantly. Accumulation of phenolic compounds characteristic of HR did not occur in the interaction with UWA192 strain (Li et al. 2007). In a later study, the cytological responses in the HR induced in the cotyledon and stem tissues of *B. napus* inoculated with an avirulent strain of *L. maculans* were investigated. At cotyledon or the sixth true leaf stage, plants were inoculated with pycnidiospores of *L. maculans* UWAP11. Condensation of cytoplasm, shrinkage in cell size and nuclear DNA fragmentation in the cotyledon and stem tissues, shrinkage and condensation of cytoplasm, chromatin fragmentation and lobing of nucleus were the cytological responses induced by the avirulent strain of *L. maculans* in Surpass 400 canola (Li et al. 2008).

### 3.3.3 In Planta Expression of Fungal Pathogen Genes

The genes of fungal pathogens expressed in planta, form a group of genes involved in the establishment and the maintenance of infection, but they are not directly associated with acquisition of nutrients from host plant tissues. The elements that control in planta gene expression and genes that influence disease development, after initial contact with the plant surface, have been characterized. Various sets of fungal genes operate during pathogenesis (disease development). Following penetration and invasion of the first few cells, new genes are required for further disease progression. In necrotrophic fungal pathogens, activation of toxin and CDWEs is accelerated. On the other hand, in hemibiotrophic and biotrophic fungal pathogens, genes are required for regulation of the formation of infection structures, such as infection vesicles and haustoria, and prevention of elicitation of host defenses (Tudzynski and Sharon 2003). The complexity of gene expression during interaction between pathogens and host plant species is analyzed by using cDNA microarrays developed from expressed sequence tags (ESTs). The ESTs are partial sequences of cDNA clones in an expressed cDNA library and they can be used to identify all of the unique sequences (genes) in order to study their functions. The identified unique cDNA sequences can be used to fabricate a cDNA microarray for functional study. Microarrays have been used to analyze samples for the presence of gene variations or mutations (genotypes) or for patterns of gene expression (Aharoni and Vorst 2001).

Expression of genes of *P. brassicae*, causal agent of cabbage club root disease, during disease development was studied, using differential display analysis of total RNA extracted from roots of Chinese cabbage inoculated with the pathogen. Differentially expressed bands (30) of cDNA were detected. Expression of the clone *PbSTKL1* in diseased roots was confirmed and this clone was a single-copy gene in the pathogen genome, as revealed by Southern blot analysis. Expression of *PbSTKL1* increased strongly beginning 30 *dai* and coincident with resting spore production (Ando et al. 2006). In a later study, the role of serine protease gene (*PRO1*) was investigated. *PRO1* was indicated to be a single-copy gene by Southern blot and PCR assays. Expression of *PRO1* was induced during plant infection and the quantity of transcripts fluctuated according to the stage of pathogenesis. The conditions required for expression of *PRO1* coincided with the temperature and pH conditions favorable for germination of RS of *P. brassicae*. The results indicated that PRO1 might play a role during club root pathogenesis by stimulating resting spore germination through its proteolytic activity (Feng et al. 2010).

Infection of resistant and susceptible canola at different intervals (5, 7, 14, 21, 28 and 35 days) after inoculation with *P. brassicae* causing clubroot of canola (*B. napus*) disease was compared to assess the differences in their defense responses. At each time interval, the primary infections were at similar levels in both resistant and susceptible cultivars. At seven *dai* all root hairs were infected at 14 *dai* and thereafter, primary infections declined to less than 20%, most of the newly formed root hairs being free of infection. Secondary infections were common at 14 to 28 *dai*, and the level of secondary infection was greater in resistant cultivar than in susceptible cultivar. The in planta expression of 12 selected *P. brassicae* genes was investigated by quantitative real-time PCR assay. In the resistant cultivar, a single expression peak was observed for all 12 genes at five or seven *dai*, indicating the involvement of those genes in the primary infection, production of SZ, or early secondary infection in the resistant cultivar. In the susceptible cultivar, the 12 genes could be classified into three groups, according to their expression patterns: *G75* and *G116* had an expression peak at 14 *dai*; *G60*, *G117* and *G10* had two expression peaks at 14 and 35 *dai*, whereas all other genes had an expression peak at 35 *dai*. Expression of all pathogen genes, except *G75* and *G116*, reaching the peak at 35 *dai* in the susceptible cultivar indicated that they are important during the process of resting spore formation. Expression of all 12 genes of the pathogen in the resistant cultivar was downregulated after seven *dai*, suggesting a restriction on their expression as a result of the operation of a resistance mechanism. The genes specifically upregulated at five or seven *dai* in the resistant cultivar, might participate in the early resistance interaction and likely contribute to the ability of the pathogen to cause primary infections in the resistant cultivar, resulting in a level of primary infection equal to that observed in the susceptible cultivar. Microscopic investigations on infection indicated that the resistance was expressed phenotypically, after secondary infection. The results could be useful to having a better understanding of clubroot pathogenesis and provide gene candidates for gene functional studies, facilitating breeding programs for development of canola cultivars resistant to clubroot disease (Fei et al. 2016).

The patterns of gene expression in soybean-*P. sojae* pathosystem, were determined by constructing a 4,896-gene microarray of host and pathogen cDNA transcripts. The rRNAs from soybean and *P. sojae* were used to calculate the ratio of host and pathogen RNA present in the mixed samples. This ratio showed significant changes between 12 and 24 h after infection, indicating the rapid growth and multiplication of pathogen within host tissues. The genes encoding enzymes of phytoalexin biosynthesis and defense and PR-proteins were strongly upregulated during infection as shown by microarray analysis. The number of pathogen genes expressed during infection reached the maximum at 24 h after infection. This investigation indicated the possibility of using a single microarray to simultaneously probe gene expression in two interacting organisms (Moy et al. 2004). The genes of *P. sojae* (infecting soybean) and *P. ramorum* (causing sudden oak death) have been fully sequenced. The *P. sojae* genome contains 499 single nucleotide polymorphisms (SNPs), whereas ~ 13,643 SNPs have been identified in the genome of *P. ramorum*. The nature of secreted proteins may primarily determine the host-pathogen interactions. The predicted secretomes for *P. sojae* and *P. ramorum* are 1,464 and 1,188 proteins, respectively. A comparison of genomes of these two *Phytophthora* spp. reveals a rapid expansion and diversification of many protein families associated with pathogenesis such as hydrolases, ABC transporters, protein toxins, proteinase inhibitors and in especially, a superfamily of 700 proteins with similarity to known oomycete avirulence genes (Tyler et al. 2006). The genetic basis of avirulence toward *Rps1c* gene in soybean was studied, using the $F_2$ population derived from a cross between two isolates pRx (Avr1c-Avr1c) and ps1 (avr1c-avr1c) of *Phyophthora sojae*, causing stem and root rot of soybean. This avirulcnce was found to be dominant and controlled by a single locus, as expected for a simple gene-for-gene model. Four of 80 AFLP primers could efficiently distinguish the avirulent pRx from virulent ps1 isolate. Among the five specific markers, band C was amplified from the avirulent pRx by the primer set EGC/MAT and then cloned. This AFLP marker was transferred to a SCAR marker through scanning, primer design and specific amplification of the DNA of the avirulent pRx. This SCAR marker (616-bp fragment) could be specifically amplified from *P. sojae* isolates that had *Avr1c* gene (Wen et al. 2014).

P. cinnamomi, infecting cork oak (*Quercus suber*), causes decline in the Mediterranean ecosystems. During infection, fungal pathogens secrete effector proteins that reprogram host physiology and immunity for their advantage. The pathogen may find its way to counteract the basic defense responses. Host-specific interactions between cork oak (*Q. suber*) and *P. cinnamomi* was investigated by studying the genes expressed as a cellular response to the infection and their role in disease development. The genes involved in the *Q. suber*-*P. cinnamomi* pathosystem, were studied, using cDNA-AFLP analysis to identify differential gene expression and defense-related transcripts in the roots of micropropagated clonal *Q. suber* during infection. Selective amplification with 20 primer combinations allowed visualization of 53 transcript-derived fragments (TDFs) with 47 to 485-bp size. The fragments were sequenced and the sequences were further analyzed by blastn and blastx. A homology search showed that among the determined sequences, 66% were homologous to known sequences, while 26.5% were homologous to genes with unknown function and 7.5% had no matches with database. Among these TDFs, 92.5% were related to Q. *suber* and 7.5% were attributed to *P. cinnamomi* (see Figure 3.9). Six candidate genes were selected, based on their expression patterns and homology to genes known to play a role in defense. They encoded a cinnamyl alcohol dehydrogenase 2(QsCAD2), a protein disulfide isomerase (QsPD1), a CC-NBS-LRR resistance protein (QsRPC), a thaumatin-like protein (QsTLP), a chitinase (QsCH1) and a 1,3-β-glucanase (QsGlu). The qPCR analysis of gene expression revealed that transcript levels of QsRPC, QsCH1, QsCAD2 and QsPD1 increased during the first 24 h postinoculation, whereas those of QsTLP decreased. 1,3-β-glucanase level did not show differential expression. The results suggested that *P. cinnamomi* might suppress host defense processes (reduction in QsTLP production) to facilitate invasion or production of others may be reduced to undetectable levels within 24h postinoculation (Ebadzad and Cravador 2014).

Changes in different gene expression profiles of cotton root and hypocotyl tissues inoculated with Fov were investigated using microarray technique. The microarray analysis revealed large scale temporal and tissue-specific plant gene expression changes during a compatible interaction. Enhanced expression of defense-related genes was observed in hypocotyl tissues, whereas few changes in the expression levels of defense-related

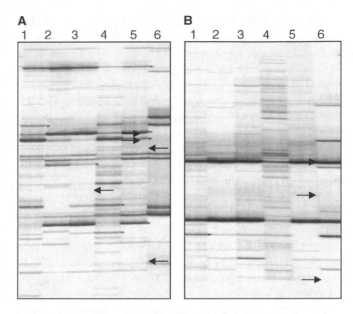

**FIGURE 3.9** Selective amplification of cDNA AFLP fragments derived from micropropagated *Q. suber* roots before and after inoculation with zoospores of *P. cinnamomi* Selective amplification with primer combinations *Eco*RI-Acc/*Mse*I-CTA (A) and with primer combination *Eco*RI-ATG/*Mse*I-CAA (B) in host root cells at 8, 8, 14, 20, 26 and 32 h post-challenge inoculation respectively in lanes 1 to 6. Arrows indicate differentially expressed fragments.

**[Courtesy of Ebadzad and Cravador (2014) and with kind permission of *SpringerPlus* Journal]**

genes could be seen in the root tissues. In infected roots, more host plant genes were repressed than induced especially at the earlier stages of infection. Many known cotton defense responses including induction of PR-genes and gossypol biosynthesis genes have been reported. Potential new defense responses, such as the biosynthesis of lignins were identified. The involvement of genes coding for phytohormones, ethylene and auxin in pathogenesis was implicated by gene expression results (Dowd et al. 2004). Strains of *Fusarium oxysporum* f.sp. *phaseoli*, causing Fusarium wilt disease of runner bean were classified into three groups as super virulent, highly virulent and weakly virulent, whereas they were all highly virulent on common bean plants. Expression of the gene encoding the Fusarium transcription factor *ftf1*, specific to virulent strains of *F. oxysporum* strain was analyzed. This gene was highly upregulated during plant infection. In planta expression analysis based on real-time qPCR assay showed that expression of *ftf1* was correlated with the degree of virulence (de Vega-Bartol et al. 2011).

Fungal pathogens, causing vascular wilt diseases, have to breach the highly structured and rigid xylem walls to enter the vessels. In addition, the pit membranes are major barriers for vascular pathogens, since they are too large to pass through pit membrane pores. Xylem provides a nutritionally poor environment that limits pathogen development and survival. Vascular wilt pathogens derive their nutritional requirements by efficiently acquiring the scarce nutrients available in the xylem sap, by enzymatic degradation of host cell walls, by invading neighboring cells or by inducing nutrient leakage from surrounding tissues (Divon et al. 2005; Klosterman et al. 2011). Among the nutrients, the availability of nitrogen in the xylem sap is one of the limiting constraints for the development of vascular wilt pathogens (Divon et al. 2005). *Fusarium oxysporum* f.sp *lycopersici* global nitrogen regulator (FNR1) regulates the utilization of secondary nitrogen sources, as the primary nitrogen sources such as ammonia and glutamine are scarce in xylem sap. Disruption of FNR1 affected virulence of the pathogen as well as regulation of three nitrogen acquisition genes *Gap1, Mtd1* and *uricase* during growth in planta (Divon et al. 2005). Analysis of the whole genome sequences of *F. oxysporum* f.sp. *lycopersici, V. dahliae* and *V. albo-atrum* showed that these genomes are enriched in genes that encode CWDEs that may be useful for degrading xylem walls and pit membranes (Ma et al. 2010; Klosterman et al. 2011). The sugars liberated by enzymatic activity may be used by the pathogens as nutrient sources. Vascular wilt pathogens have been reported to produce high- and low-molecular weight phytotoxins, during colonization. The phytotoxins disturb plant cell membrane integrity, resulting in leakage of nutrients from cells surrounding xylem vessels that may be utilized by the pathogens. Two *Verticillium* necrosis- and ethylene inducing-like proteins, NLP1 and NLP2 were shown to possess cytotoxic activity and differentially contribute to virulence on various host plant species (Zhou et al. 2012; Santhanam et al. 2013). The NLP gene family is expanded in *V. dahliae* genome. In addition to the NLP1 and NLP2, none of the other *V. dahliae* NLPs possessed cytotoxic activity (Zhou et al. 2012; Santhanam et al. 2013).

*F. oxysporum*, causing wilt diseases of different crops, interacts with roots and hypocotyls of seedlings and adult plants. It produces necrosis and ethylene-inducing 124-kDa peptide (Nep1) (Bailey et al. 1997). The response of *Arabidopsis* (model plant) to Nep1 was assessed. Nep1 from *F. oxysporum*, following application on *Arabidopsis*, inhibited both root and cotyledon growth and triggered cell death, resulting in necrotic spots. Nep1 was localized to the cell wall and cytosol, as revealed by immunolocalization observations. The contents of most water-soluble metabolites were reduced at 6 h after Nep1 treatment, indicating loss of cellular membrane integrity. The results of quantitative PCR confirmed that of microarray analysis. Induction of genes localized in the chloroplast, mitochondria and plasma membrane and genes responsive to calcium/calmodulin complexes, ethylene, jasmonate, ethylene biosynthesis and cell death was attributed to Nep1 treatment. Nep1 can facilitate death as a component of diseases induced by necrotrophic pathogens of plants (Bae et al. 2006). Expression of defense-related genes in tomato following inoculation with two strains of FORL 27.2 and usaFORL of *Fusarium oxysporum* f.sp. *radicis-lycopersici* were investigated. RT-PCR analysis of tomato roots inoculated with strains of FORL showed that PR1, PR6 and CH9 expressions were upregulated by usaFORL. PR4 expression in plants infected with both strains was downregulated, indicating that systemic acquired resistance (SAR) was not associated with PR4 activation. No transcripts for *CH3* were

detected after inoculation with both strains, suggesting lack of participation of CH3 in defense response of tomato against FORL. Ethylene (ET)-regulated genes (ETR1, ACO1, ACO2 and ACO3) expressions were increased only by usaFORL strain, indicating the relationship of gene expression with pathogen virulence and in turn, reduced ethylene sensitivity and subsequent necrosis. However, both strains downregulated ACO4 transcripts. Furthermore, both strains induced *Pal* expression, indicating gene involvement in FORL, induction of SAR in tomato seedlings (Çakir et al. 2014).

Effector genes are required for disease development, but not necessary for the completion of the life cycle of the pathogen. Fungal effector genes may be grouped into several classes based on their functions, including coding for infection structures, cuticle and CWDEs, ability to respond to environmental stimuli, fungal toxins, signal cascade components and suppression of plant resistance. *F. oxysporum* may utilize a wide array of secreted molecules, often called elicitors or virulence and avirulence factors (Idnurm and Howlett 2001; van der Does and Rep 2007). *Fusarium oxysporum* f.sp. *betae (Fob)* causes Fusarium yellows in sugar beet crops and the strains are highly variable in growth, production of conidia and pigmentation and virulence. A large number of isolates of *F. oxysporum* recovered from symptomatic sugar beet plants was found to be nonpathogenic. Hence, identification of pathogenicity factors and their diversity in *Fob* was considered essential to differentiating pathogenic strains from nonpathogenic strains. Conserved gene markers have been employed to characterize the populations of *Fob* and group them into phylogenetic clades, but they were found to be ineffective to differentiating pathogenic from nonpathogenic isolates. The genes reported to contribute to fungal pathogenicity were utilized to describe a population of *F. oxysporum* from sugar beet into phylogenetic classes, and potentially be used to identify pathogenic isolates of *Fob*. Six genes viz., *Fmk1, Fow1, PelA, Rho1, Snf1* and *Ste12* were present in both pathogenic and nonpathogenic isolates from sugar beet, whereas genes, *Pda1, PelD, Pep1, Prt1, Sge1, Six1* and *Six6* were found to be dispersed within the population. Of these *Fmk1, Fow1, PelA, Rho1, Sge1, Snf1 and Ste12* were useful for relating clade designations. The genes *PelD* and *Prt1* were associated with pathogenicity of *F. oxysporum* f.sp *betae*. The results indicated that the multilocus approach utilizing putative effector genes might be useful for differentiation of pathogenic isolates causing Fusarium yellows in sugar beet (Covey et al. 2014).

Soybean is affected by root rot, seedling blight and wilt diseases induced by *F. oxysporum* species complex. In order to identify genetic markers for detecting aggressive soybean wilt isolates, 80 isolates collected from soybean primarily were tested in the greenhouse for their ability to produce wilt symptoms, using susceptible soybean cv. Jack. Further, these isolates were assessed for the presence of fungal effector genes *Fmk1, Fow1, Pda1, PelA, PelD, Pep1, Prt1, Rho1, Sge1, Six6* and *Snf1*. All PCR amplicons were sequenced; phylogenies were inferred and analysis of molecular variance (AMOVA) was performed for ten of the 12 genes. Three isolates induced high intensity of vascular discoloration of roots or stems, whereas 25 isolates caused moderate to low levels of vascular discoloration in roots or stems. Fungal effector genes *Fmk1*, *Fow1*, *PelA*, *Rho1*, *Sge1* and *Snf1* were detected in all isolates tested, whereas *Pda1*, *PelD*, *Pep1*, *Prt1*, *Six1* and *Six6* were dispersed among isolates. None of the genes had a significant association with wilt symptoms on soybean. *Six6* was present only in three wilt isolates existing earlier in soybean, common bean and tomato. The soybean and common bean isolates produced high levels of disease severity (Ellis et al. 2016). In the further investigation, a collection of *F. oxysporum* isolates from soybean roots and *F. oxysporum* f.sp. *phaseoli* isolates from common bean, was screened for the presence of *Six6* gene. All isolates from which the *Six6* amplicon was generated, caused wilt symptoms on soybean and two-thirds of isolates showed high levels of aggressiveness (virulence), indicating a positive association between the presence of the effector gene *Six6* and induction of wilt symptoms. The expression profile of the *Six6* gene analyzed by quantitative reverse-transcription-PCR (qPCR) revealed an enhanced expression for the isolates that caused more severe wilt symptoms on soybean, as revealed by greenhouse tests. The results suggested that *Six6* gene might be suitable as a possible locus for pathogenicity-based molecular diagnostics across the various *formae speciales* of *F. oxysporum* (Lanubile et al. 2016).

*F. oxysporum*, a species complex, is classified as *formae speciales*, based on their host specificity. Lineage-specific (LS) genomic regions in *F. oxysporum* have been identified. LS regions are mostly organized into supernumerary chromosomes, which contain genes that are not required for basic metabolic activities. The LS regions of *F. oxysporum* f.sp. *lycopersici* (FOL) include four entire chromosomes (3, 6, 14 and 15), parts of chromosome 1 and 2 and some small scaffolds. Transfer of some LS chromosomes between strains of *F. oxysporum* might result in nonpathogenic strains acquiring the property of pathogenicity. These regions show enrichment of transposable elements and genes predicted to encode secreted proteins, some specifically expressed during plant infections, such as the *Secreted In Xylem (SIX)* genes. The *Fusarium transcription factor 1* (FTF1) is one of the expanded gene families in the LS genome of *F. oxysporum*. The *FTF* gene family comprises a single copy gene *FTF2*, which is present in all filamentous ascomycetes. Several copies of *FTF1* are exclusively present in *F. oxysporum*. An RNA-mediated gene silencing system was developed to target mRNA produced by all the *FTF* genes and tested in *F. oxysporum* f.sp. *phaseoli* infecting common bean and *F. oxysporum* f.sp. *lycopersici* infecting tomato. Quantification of the mRNA levels showed breakdown of *FTF1* and *FTF2* in randomly isolated transformants of both pathogens. Attenuation of *FTF* expression resulted in a marked reduction in virulence, a reduced expression of several *SIX* genes and lower levels of *SIX Gene Expression 1* (SGE1) mRNA, the presumptive regulator of *SIX* gene expression. Further, the knockdown mutants had a pattern of colonization of host plant similar to that displayed by strains devoid of *FTF1* copies (weakly virulent strains). Gene knockout of *FTF2* also resulted in reduction in virulence, but

to a lesser extent. The results indicated the role of the *FTF* gene expansion as a regulator of virulence of *F. oxysporum* (Niño-Sánchez et al. 2016).

The LS, mobile pathogenicity chromosomes which contain *SIX* and other pathogenicity-related genes have been identified in *Fusarium oxysporum* f.sp. *lycopersici* (FOL). These small effector proteins are secreted by FOL, during colonization of tomato plants. Fourteen *SIX* genes have been identified in FOL, which did not show any homology with each other or with any other sequenced gene. *SIX1* (also known as *Avr3*), *SIX3* (*AvrZ*), *SIX4* (*Avr1*) and *SIX5* were recognized by host plant resistance genes which have been introgressed into tomato, whereas gene knockouts demonstrated that *SIX1*, *SIX3*, *SIX5* and *SIX6* could contribute directly to virulence ( Rep 2005; Houterman et al. 2008, 2009; Takken and Rep 2010; Gawehns et al. 2014; Ma et al. 2015). *Fusarium oxysporum* f.sp. *cepae* (FOC) isolates (31) collected from onions in the UK were characterized, using pathogenicity tests, sequencing of housekeeping genes and identification of effectors. A concatenated tree separated the *F. oxysporum* isolates into six classes, but did not distinguish pathogenic isolates from nonpathogenic isolates. Ten putative effectors were identified within FOC, including seven *SIX* genes first reported in *F. oxysporum* f.sp. *lycopersici* (FOL). Two proteins highly homologous with signal peptides and RxLR motifs (*CRX1/CRXZ*) and a gene with no previously characterized domains (C5) were also identified. The presence/absence of nine genes was strongly related to pathogenicity against onion and all of them were expressed in planta. Different *SIX* gene complements were identified in other *Fusarium* spp., but none was identified in three other *Fusarium* spp. from onion. Although the FOC *SIX* genes showed high homology with other *Fusarium* spp., there were clear differences in sequences which were unique to FOC, whereas *CRX1* and *C5* genes appeared to be largely FOC-specific (Taylor et al. 2016).

The key genes involved in the pathogenicity of *V. dahliae* were investigated to develop measures for reducing losses due to Verticillium wilt diseases affecting several crops. *Sho1* encodes a conserved tetraspan transmembrane protein which is a key element of two upstream branches of the HOG-MAPK pathway in fungi. *Sho1* is required for full virulence in a wide range of fungal pathogens. The *sho1* mutant of *V. dahliae* (*ΔVdsho1*) was generated by *Agrobacterium tumefaciens* (At)-mediated transformation. The *ΔVdsho1* strain was highly sensitive to menadione (at a concentratin of 120 μM) and hydrogen peroxide (250 μm), which delayed spore germination and reduced sporulation in relation to the WT strain and the complemented strain. During infection of cotton plants, *ΔVdsho1* exhibited impaired ability of root attachment and invasive growth. The results suggested that *Vdsho1* might control external sensing, virulence and multiple growth-related traits in *V. dahliae* and hence, this gene could be a potential target directed to arrest pathogen development and adverse effects of wilt disease on cotton and other susceptible crops (Qi et al. 2016). Fungal plant pathogens secrete effector molecules that are capable of modifying host physiology, including immune responses that are triggered, when plants recognize the presence of potential pathogen(s). Effectors are small secreted proteins that are species-specific or even LS. But some effectors, such as the necrosis- and ethylene-inducing-like proteins that are phytotoxic, are more broadly conserved. Another group of conserved fungal effectors are the lysin motif (LysM) effectors produced by several fungal pathogens, as well as saprophytes. Production of LysM effectors by *V. dahliae* was studied. Comparative genomics showed that a collection of *V. dahliae* strains were able to produce LysM effectors. Expression of core LysM effectors was not detected in planta in a taxonomically diverse panel of host plants. In addition, targeted deletion of the individual LysM effector genes in *V. dahliae* strain JR2 did not compromise virulence in infections in *Arabidopsis*, tomato or *N. benthamiana*. An additional LS LysM effector was encoded in the genome of *V. dahliae* strain VdLs17, but not in any other *V. dahliae* strain sequenced. This LS effector was expressed in planta and contributed to the virulence of VdLs17 strain on tomato, but not on *Arabidopsis* or *N. benthamiana*. Functional analysis showed that this LysM effector bound chitin, suppressed chitin-induced immune responses and protected pathogen hyphae against hydrolysis by host plant hydrolytic enzymes. The results indicated that in contrast to the core LysM effectors of *V. dahliae*, the LS LysM effector of VdLs17 strain, could contribute to its virulence in planta (Kombrink et al. 2016).

Plants secrete chitinases that degrade chitin, the major structural component of fungal cell walls, as part of their defense strategy against fungal pathogens. Due to secretion of chitin-binding effector proteins that protect the cell walls of some fungal pathogens, they are not sensitive to host plant chitinases. However, some fungal pathogens that lack chitin-binding effectors are able to overcome the plant defense barrier. The ability of tomato pathogens to cleave chitin-binding (CBD)-containing chitinases and its effect on fungal virulence was studied. Four tomato CBD chitinases were produced in *Pichia pastoris* and they were incubated with secreted proteins isolated from seven fungal tomato pathogens. Of these, *F. oxysporum* f.sp. *lycopersici*, *V. dahliae* and *Botrytis cinerea* were able to cleave the extracellular tomato chitinases SlChi1 and SlChi13. Cleavage by *F. oxysporum* removed the CBD from N-terminus, as shown by mass spectrometry and significantly reduced the chitinase and antifungal activity of both chitinases. Both secreted metallo-protease FeMep1 and serine protease FoSep1 were responsible for this cleavage. Double deletion mutants *FeMep1* and *FoSep1* of *F. oxysporum* lacked cleavage activity on SlChi1 and SlChi13 and showed reduced virulence on tomato. The importance of plant chitinase cleavage in fungal pathogen virulence was clearly revealed by this investigation (Jashni et al. 2015).

The initial interaction between fungal pathogen and the host plant is complex and numerous metabolic pathways and regulatory proteins are involved in the host-pathogen interactions. Contribution of molecules such as glycans, other than proteins (effector molecules) is not clear still. A random genetic screen was applied to identify 58 novel candidate genes that were involved in the virulence of *V. dahliae*, incitant of vascular

wilt disease in several crops. A putative biosynthetic gene involved in nucleotide sugar precursor formation was identified and this gene encoded a putative nucleotide-rhamnose synthase/epinorase-reductase (NRS/ER). This enzyme is homologous to bacterial enzymes involved in the biosynthesis of the nucleotide sugar deoxy-thymidine diphosphate (dTDP)-rhamnose, a precursor of L-rhamnose, which was required for virulence in several human pathogenic bacteria. Deletion of the *VdNRS/ER* gene from the *V. dahliae* genome results in complete loss of pathogenicity on tomato and *N. benthamiana* plants, whereas vegetative growth and sporulation were not affected. VdNRS/ER was found to be a functional enzyme in the biosynthesis of UDP-rhamnose. Further, *VdNRS/ER* deletion strains were impaired in the colonization of tomato roots. The results showed that rhamnose, although only a minor cell wall component, was essential for the pathogenicity of *V. dahliae* (Santhanam et al. 2017).

*R. solani* AG2-2IIIB isolates are most aggressive on sugar beet. The genome of the pathogen was sequenced using Illumina technology. The genome of *R. solani* AG2-2IIIB was predicted to harbor 11,897 genes, of which 4,908 genes were found to be isolate-specific. The pathogen was predicted to contain 1,142 putatively secreted proteins and 473 of them were unique for this isolate. The genome encoded a high number of carbohydrate-active enzymes. The highest numbers were observed for the polysaccharide lyases family 1 (PL-1), glycoside hydrolase family 43 (GH-43) and carbohydrate esterase family 12 (CE-12). Transcription analysis of selected genes representing different enzyme clades revealed a mixed pattern of up- and down-regulation at six days after infection of sugar beets, featuring variable levels of resistance, compared to mycelia of the pathogen (Wibberg et al. 2016). Early infection and establishment of *R. solani*, causal agent of rice sheath blight disease, on tolerant cv. Swarnadhaan and susceptible cv. Swarna was followed, using a detached leaf assay method. Disease severity induced on Swarna was higher (50%) than in Swarnadhaan. *R. solani* showed different growth behavior on tolerant and susceptible cultivars, the hyphal growth being more in susceptible cultivar. The pathogen produced profuse branching, making intimate contact with host surface to form more inter- and intra-cellular structures and greater sclerotial development in susceptible host compared to the tolerant one. The pathogen could intercept host surface structures and use these for anchorage or penetration, as revealed by light and electron microscopy. Using confocal laser scanning microscopy to investigate pathogen behavior in transformed *R. solani* isolate, expressing GFP, formation of infection cushions and subsequent colonization of the host tissues were observed (Basu et al. 2016). The process of infection of potato by *R. solani*, causing potato stem canker and black scurf disease, was investigated by histological observations using light and electron microscopy. Observations were made at the invasion site, for the initial infection time and for the presence of infection structures such as appressoria and infection cushions in inoculated potato cv. Desiree (resistant) and cv. Atlantic (susceptible). In the aerial stems of potato, the invasion of *R. solani* was mainly limited to intercellular spaces. In the underground stems and tubers, primary invasion sites were epidermal cracks and lenticels. Initial infection time was 8 to 12 h postinoculation (*hpi*) in above ground stems, 8 *hpi* in underground stems and 4 *hpi* in tubers. The hyphae of *R. solani* produced various infection structures with a few infection cushions and a large number of appressoria with different morphologies. In above ground stems, more infection structures were produced than in underground stems and tubers. Desiree had fewer numbers and smaller lenticels, thicker cuticle and periderm and fewer epidermal cracks compared to Atlantic. Fewer numbers of infection structures of *R. solani* were present in Desiree than in Atlantic. The results suggested that the epidermal and perithecial structures on potato might be related to resistance of the cultivar to *R. solani* (Zhang et al. 2016).

*U. maydis*, incitant of maize smut disease, is a model to study biotrophic interactions of fungal pathogens with their host plants. *U. maydis* encodes several hundred putative effector proteins involved in the modification of metabolic functions of susceptible maize plant. The role of ApB73 (apathogenic in B73) in colonization of maize by *U. maydis* was investigated. The gene *apB73* was transcriptionally induced during the biotrophic stages of the life cycle of *U. maydis*. Deletion of *apB73* gene resulted in cultivar-specific loss of gall formation, the characteristic symptom of smut, in maize. The ApB73 protein was conserved among closely related smut fungi. However, virulence assay showed that only the ortholog of *S. reilianum*, causal agent of maize head smut, could complement the mutant phenotype of *U. maydis*. Microscopic observations revealed that ApB73 was secreted into the biotrophic interface and it appeared to be associated with fungal cell wall components or fungal plasma membrane. The results indicated that ApB73 was a conserved protein, functioning as an important virulence factor of *U. maydis* that could localize in the interface between pathogen and maize plant (Stirnberg and Djamei 2016).

### 3.3.4 Molecular Basis of Interactions between Host Plants and Fungal Pathogens

Fungal pathogens derive all nutrients from their hosts for growth and reproduction, when infection has been successfully established. Under the stress imposed due to the damage done by the pathogen(s), the host plants express symptoms characteristic of the disease and the severity of disease depends on the levels of resistance of the cultivars, virulence of the pathogen isolate(s) and availability of favorable environmental conditions for disease development. *V. dahliae* causes gradual wilting, senescence, defoliation and ultimate death of plants, depending on the levels of virulence of the pathogen strain and host resistance. The physiological and biochemical changes induced by *V. dahliae* include reduced water potential, stomatal conductance, rate of carbon assimilation and accumulation of proline, soluble sugars and abscisic acid (Goicoechea et al. 2000; Sadras et al. 2000). The protein lipopolysaccharide complexes and small peptides (considered as pathogenicity factors) released from *V. dahliae*, induced

leaf chlorosis and necrosis and inhibition of root growth. *V. dahliae-Arabidopsis thaliana* (model plant) interaction was investigated. Symptoms induced by the pathogen included chlorosis, stunting, anthocyanin accumulation, induction of early flowering and finally, mortality of plants. A single dominant locus of *V. dahliae*-tolerance (VET1) was identified and it may function as a negative regulator of the transition to flowering. VET1 was mapped on chromosome IV. The results indicated that the ability of *V. dahliae* to induce disease symptom might be connected to the genetic control of development and life span in *Arabidopsis* (Veronese et al. 2003). In another investigation, a random insertional mutant library of *V. dahliae* was constructed, using *ATMT* to investigate pathogenicity and regulatory mechanisms of the pathogen. The pathogen-specific transcription factor-encoding gene *Vdpf* was associated with vegetative growth phase and virulence, with transcription expression reaching the peak during conidial production in V991 strain. The deletion mutants (Δ*Vdpf*) and insertion mutants (*IM*Δ*Vdpf*) produced fewer conidia, compared with WT strain, resulting in reduction in virulence of the mutants. In contrast to WT strain, the complimented strains *IM*Δ*Vdpf* and Δ*Vdpf* produced swollen, thick-walled and hyaline mycelium rather than melanized microsclerotia. The Δ*Vdpf* mutants were melanin deficient, with undetectable expression of melanin biosynthesis-related genes (*Brn1*, *Brn2* and *Scd1*). The Δ*Vdpf* and *IM*Δ*Vpdf* mutants were also successfully able to penetrate into cotton and tobacco roots like the WT and complemented strains, but had reduced virulence, because of lower biomass in plant roots and significantly reduced the expression of pathogenicity-related genes in *V. dahliae*. The results revealed the role of *Vdpf* in the production of melanized microsclerotia and conidia as well as in pathogenicity (Luo et al. 2016).

*A. flavus* and *A. parasiticus*, infecting peanut are ubiquitous, soilborne pathogens, occurring all over the world. The aflatoxin produced by *Aspergillus* spp. contaminate seeds, which when consumed, become serious health hazards for humans and animals. As development of peanut genotypes with resistance to *Aspergillus* spp. is the most effective disease management strategy, the mechanisms of development of resistance was studied. The expression of pathogenesis-related (PR) proteins, ß-1,3-glucanase and isoform patterns of peanut seed inoculated with *A. flavus* was determined. The activities of b-1,3-glucanase in susceptible genotypes, Georgia Green and A100 and resistant genotypes, GT-YY0 and GT-YY20 were colorimetrically assayed in extracts from infected and uninfected (control) seeds. The ß-1,3-glucanase activity was stimulated by infection with *A. flavus* in both resistant and susceptible genotypes, the increase in activity being remarkable in resistant genotypes at 24 hours after inoculation (*hai*), reaching the maximum on the third day after inoculation. The susceptible genotypes showed gradual increase in enzyme activity, reaching the maximum at four *dai*. The enzyme activity remained at a very low level in both resistant and susceptible controls (unifected seeds). The ß-1,3-glucanase activity in resistant genotypes was three–four-fold higher than in susceptible genotypes. Polyacrylamide gel electrophoresis

(PAGE) analysis revealed the presence of more protein bands corresponding to ß-1,3-glucanase isoforms in extracts of seeds of resistant genotypes than in susceptible genotypes (see Figure 3.10). TLC analysis showed that individual bands corresponding to the bands of ß-1,3-glucanase isoforms Glu 1 to 5 could be identified in the plates, after separation using sodium dodecylsulfate (SDS)-PAGE technique. The results indicated the association between ß-1,3-glucanase activity and resistance to colonization of peanut seed by *A. flavus* (Liang et al. 2005)

*F. virguliforme (*syn. *F. solani f.sp. glycines)* causes SDS, affecting the soybean production heavily in the United States. *F. tucumaniae*, closely related to *F. virguliforme*, induces SDS in South America. Two components of SDS, root necrosis and foliar SDS have been recognized. *F. virguliforme* has not been isolated from diseased foliar tissues, suggesting the involvement of toxin(s), capable of causing foliar chlorosis and necrosis. *F. virguliforme* secreted one or more toxins in culture media, that could produce SDS-like symptoms in three-week old cut soybean seedlings (Li et al. 1999). Cell-free *F. virguliforme* CFs containing a toxin caused foliar SDS in soybean. A low-molecular weight protein of approximately 13.5 kDa (FvTox1), purified from Fv CFs induced foliar SDS-like symptoms in cut soybean seedlings. The candidate *FvTox1* gene was expressed in baculovirus-infected insect cell line Sf21. In the presence of light, the recombinant FvTox1 isolated from baculovirus-infected insect cells were infiltrated into soybean leaf disks of soybean cv. Essex. Anti-FvTox1 monoclonal antibodies raised against the purified toxin were used in identifying FvTox1 gene. SDS-susceptible, but not SDS-resistant soybean lines were sensitive to the baculovirus-expressed toxin. The requirement of light for foliar SDS-like symptom development indicated that FvTox1 induced foliar SDS in soybean, possibly through production of free radicals by interrupting photosynthesis. RT-PCR assays showed that *FvTox1* gene was transcribed in infected soybean roots. However, FvTox1 could not be detected in infected roots by using Western blot procedure, probably because of the low concentration of the toxin in the infected roots. The results suggested that *F. virguliforme* remained in the infected soybean roots and yet it might induce soybean SDS foliar symptoms by one or more toxins produced by the pathogen (Brar et al. 2011).

Affymetrix analysis was applied for measuring transcript abundances in soybean cultivars resistant (PI567.374) and susceptible (Essex) to soybean SDS caused by *F. virguliforme* at five and seven *dai*. Many of the genes with increased expression were common between resistant and susceptible plants. Assessment of changes in small (sm) RNA levels between inoculated and mock-treated samples indicated a role for these molecules in the soybean-*F. virguliforme* interaction, impacting functioning of a total of 2,467 genes of which 1,694 showing changes in response to *F. virguliforme* infection. The smRNA (93) and microRNA (42) with putative soybean gene targets were identified from infected tissue. Comparison of genotypes showed that 247 genes were uniquely modulating in the resistant genotype, whereas 378 genes were uniquely modulating in the susceptible genotype (Radwan et al. 2011).

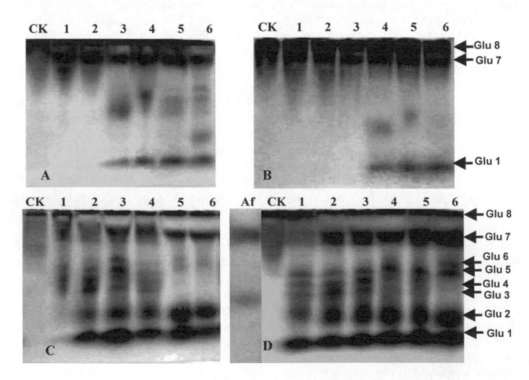

**FIGURE 3.10**    Isoform patterns of β-1-3-glucanase detected in seeds of peanut genotypes A100 (A), Georgia (B), GT-YY9 (C) and GT-YY20 (D) infected by *A. flavus* after one to six days (lanes 1 to 6) and uninoculated seeds (lane CK) and protein extract from *A. flavus* loaded as control (lane Af).

**[Courtesy of Liang et al. (2005) and with kind permission of the American Phytopathological Society, MN, United States]**

*F. virguliforme* produces phytotoxins that are translocated from roots to leaves in which they cause SDS foliar symptoms. Additional putative phytotoxins of *F. virguliforme*, including three secondary metabolites and 11 effectors were identified. Citrin, fusaric acid and radiciol induced foliar chlorosis and wilting. Soybean mosaic virus (SMV)-mediated overexpression of *F. virguliforme* necrosis-inducing secreted protein1 (FvNIS1) induced SDS foliar symptoms that mimicked foliar symptoms observed in the field. The expression level of FvNIS1 remained steady over time, although foliar symptoms were delayed, compared with the expression levels. SMV::FvNIS1 also displayed genotype-specific toxicity to which 75 of 80 soybean cultivars showed susceptibility. CFs of FvNIS1 knockout mutants displayed a mild reduction in phytotoxicity, indicating that FvNIS1 could be one of the phytotoxins responsible for SDS foliar symptoms and might contribute to the quantitative susceptibility of soybean (Chang et al. 2016).

The roles of G-protein a and b subunit genes *fga2* and *fgb1* in the development of pathogenicity of *Fusarium oxysporum* f.sp. *cubense* were investigated. Deletion of either or both genes resulted in increased heat resistance, lower cAMP levels and enhanced pigmentation, whereas phenotype changes in colony morphology and condition were seen in *Δfgb1* and *Δfga2/Δfgb1* deletion mutants, but not in *Δfga2* mutant. Conversely, *Δfgb1* retained greater virulence in banana, suggesting that FGA2 might regulate full virulence, whereas FGB1 might modulate both development and virulence, potentially via the cAMP-dependent PKA pathway (Guo et al. 2016). LS chromosomes have been characterized in some plant pathogenic filamentous fungi, including *F. oxysporum*. There is a link between pathogenicity and LS chromosomes. In *F. oxysporum* f.sp. *lycopersici* (FOL) strain Fol 4287, the genes encoding small proteins (SIX proteins or effectors) are located on an LS chromosome that could be horizontally transferred from one strain to another. One LS chromosome of FOL was found to harbor effector genes that contribute to pathogenicity. The rate of spontaneous loss of the 'effector' LS chromosome *in vitro* was at around 1 in 35,000 spores. Further, a viable strain of FOL lacked chromosome 12, which was considered to be a part of the core genome. Whole-genome sequencing revealed that some of the sites at which deletions occurred were the same in several independent strains, indicating the existence of deletion hotspots. For the core chromosome, the deletion hotspot was the site of insertion of marker used to select for loss of events. Loss of core chromosome did not affect pathogenicity, whereas loss of effector chromosome led to loss of pathogenicity entirely (Vlaardingerbroek et al. 2016).

Metabolomic, biochemical and gene expression analyses were performed to have an insight into the underlying responses of resistant and susceptible *Musa* species, during early stages of infection with *Fusarium oxysporum* f.sp *cubense (Foc)*, causing Fusarium wilt disease of banana. The susceptible cv. BX and a wild banana relative *Musa yunnanensis* (YN) were inoculated with *Foc* strain T4. Numerous metabolomic compounds, including defense-responsive signaling molecules, phytohormones, phenolics and antioxidants were identified through metabolomic analysis. Changes in salicylic acid (SA), methyl-jasmonic acid, abscisic acid (ABA), cytokinin, 3-indole acetic acid, gibberellic acid and total phenolic contents were determined, using liquid chromatography and mass spectrometry and the Folin–Ciocalteu method. The quantitative real-time PCR assay was applied to assess the expression levels of genes involved in the biosynthesis of some defense-responsive molecules. The resistant YN showed greater changes in SA content and lower level of ABA throughout the stage of infection, compared with those in susceptible BX. Further, the susceptible BX contained lower amounts of phenolic compounds. The resistant YN showed higher expression levels of PR genes, particularly *PR1*, *PR4*, *PR5-1* and *PDF2.2* than in susceptible BX plants. The results indicated differential dynamic changes in metabolic and gene-expression profiles in susceptible and resistant genotypes of banana during early stages of infection by *F. oxysporum* f.sp. *cubense* (Li et al. 2017).

Development of symptoms of infection by fungal pathogens depends on compatible or incompatible interactions with host plants. Changes in physiology of host plants induced by pathogens indicate the extent of derangements in the functioning of physiological processes. Defense responses of host plants show the effectiveness of synthetic machinery of host plants in inhibiting/blocking invasion of the pathogen from the site of infection. Elicitins, small proteins secreted by *Phytophthora* spp. and *Pythium* spp. display the ability to induce resistance to pathogens in plants. Formation of a calcium pectate gel in intercellular spaces of parenchymatous tissues, impregnation of pectin by phenolic compounds in intercellular spaces of phloem bundles and accumulation of phloem proteins (P proteins) in the lumen of leaf sieve elements were found to be ultrastructural changes induced by treatment of tobacco plants with cryptogein. These cytological modifications resulted in enhancement of physical barriers that may prevent pathogen ingress and restrict host tissue colonization, when cryptogein-treated tobacco plants were challenged with *Phytophthora parasitica*. Wall appositions also were observed at most sites of penetration by pathogen hyphae. Further, hyphae exhibited severe morphological damages, suggesting a modified toxic environment. Similar induction of P proteins in mature sieve tubes of tobacco leaves was obtained with oligandrin treatment, another elicitin. In addition, the presence of P protein plugs and occlusion of pore sites by callose could be seen in sieve elements of treated plants, resulting in prevention of in planta migration of phloem-restricted plant pathogens (Lherminier et al. 2003).

Regulation of defense responses in roots of avocado seedlings inoculated with *P. cinnamomi* was studied. A burst of ROS occurred at four *dai*. The higher physiological concentration of $H_2O_2$ induced by P. *cinnamomi* in avocado roots did not affect its growth on artificial media. Total phenols and epicatechin content showed a significant decrease, but lignin and pyocianidins did not show any change after inoculation. Nitric oxide (NO) production was also enhanced at 72 h after treatment. The results indicated an important role for NO production during avocado-*P. cinnamomi* interaction. The extent of defense responses activated in avocado, following infection by P. *cinnamomi* did not seem to be sufficient to avoid pathogen development in avocado tissues (Garcia-Pineda et al. 2010). The role of elicitins in the interaction of *P. cinnamomi* with another host plant species, cork oak was investigated, using WT strain and two genetically modified strains. The results indicated that the elicitin, β-cinnamomin might be either directly or indirectly associated with the infection process of *Phytophthora* spp. (Horta et al. 2010). Metabolic responses elicited by white root rot of avocado caused by *Rosellinia necatrix* in the aerial parts of the plant were assessed. Leaf metabolism of infected avocado plants was significantly affected, only when symptoms began to appear, possibly related to the loss of root functionality (Granum et al. 2015).

The role of pathogen gene expression during different stages of pathogenesis has been studied in different host plant-fungal pathogen interactions. *S. sclerotiorum* secretes the nonhost-specific phytotoxin, OA, required for effective colonization of host plant tissues and OA is known to suppress generation of reactive oxygen intermediates (ROI). A full-length cDNA coding for an oxalate decarboxylase (ToxDC) which converts OA into $CO_2$ and formate, was isolated from the basidiomycete *Trametes versicolor*. It was overexpressed in tobacco plants to investigate the role of ROI and OA in the interaction between tobacco and *S. sclerotiorum*. The transgenic plants contained less OA and showed delayed colonization of host tissues by *S. sclerotiorum*. The results indicated that OA might support the infection process of *S. sclerotiorum* (Walz et al. 2008). MAPK pathways are universal and evolutionarily conserved signal transduction molecules in all eukaryotic cells. In *P. sojae* (causal agent of root and stem rot disease of soybean), PsSAK1 encoding a stress-activated MAPK was identified. PsSAK1 is highly conserved in oomycetes. RT-PCR analysis showed that PsSAK1 expression was upregulated in zoospores and cysts and during the early infection stage. In addition, its expression was induced by osmotic and oxidative stress mediated by NaCl and $H_2O_2$, respectively. In order to elucidate the function, expression of PsSAK1 was silenced using stable transformation of *P. sojae*. Silencing of PsSAK1 did not impair fungal growth, sporulation or oospore production, but severely affected zoospore development. The silenced strains exhibited more rapid encystment and a lower germination ratio than the WT. PsSAK1-silenced mutants produced much longer germ tubes and could not colonize either wounded or unwounded soybean leaves. The results indicated that PsSAK1 is an important regulator of zoospore development and pathogenicity in *P. sojae* (Li et al. 2010).

Sensing of stress signals and their transduction into appropriate responses are crucial for adaptation and survival and infection by microbial plant pathogens. MAPK cascades function as key signal transducers that use phosphorylation to convey information. A gene *PsMAPK7*, one of 14 predicted genes encoding MAPKs in *P. sojae* was identified. *PsMAPK7* was highly transcribed in each test stage, but it was upregulated in the zoospore, cyst and cyst germination

stages. Silencing *PsMAPK7* affected the growth of germinated cyst, oospore production and pathogenicity of soybean. Transformants in which *PsMAPK7* expression was silenced (*PsMAPK7*-silenced) were significantly more sensitive to osmotic and oxidative stress. Staining with aniline blue and diaminobenzidine showed that silenced lines did not suppress the host ROS burst, indicating that either the inoculated plants activated stronger defense responses to the transformants and or the *PsMAPK7*-silenced transformants failed to overcome plant defenses. Further, extracellular secretion of laccase decreased in the silenced lines. The results indicated that the *PsMAPK7* gene encodes a stress-associated MAPK in *P. sojae* that is required for responses to various stresses, as well as for ROS detoxification, cyst germination, sexual oospore production and pathogenicity to soybean (Gao et al. 2015).

Induction of PCD occurs during host-pathogen interactions to varying extent, depending on susceptibility/resistance levels of host genotype. Conserved Tat D protein in *P. sojae* was identified. The purified protein showed DNase activity *in vitro*. Its expression was upregulated in sporangia and later infective stages, but downregulated in cysts and during the early infection stage. Functional analysis showed that the gene was required for sporulation and zoospore production and expression levels were associated with the numbers of $H_2O_2$-induced terminal dUTP nick end-labeling-positive cells. In addition, overexpression of *PsTatD4* gene reduced the virulence in susceptible soybean cultivar. The results suggested that apoptosis might play different roles in *P. sojae* and that *PsTatD4* might be a key regulator of infection (Chen et al. 2014). In *P. sojae*, at least 12 avirulence genes have been genetically identified and mapped. The avirulence genes *Avr4* and *Avr6* were found to be a single gene, designated *Avr4/6*, located near the 24-kb region of the pathogen genome. *Avr4/6* encodes a secreted protein of 123 amino acids. Transient expression of *Avr4/6* in soybean leaves revealed that its gene product could trigger a hypersensitive response (HR) in the presence of *Rps4* or *Rps6*. Silencing of *Avr4/6* in *P. sojae* stable transformants abolished the avirulence phenotype exhibited on both Rps4 and Rps6 soybean cultivars. The N-terminus of *Avr4/6* was sufficient to trigger *Rps4*-dependent HR, while its C-terminus was enough to trigger *Rps6*-mediated HR, compared with alleles from avirulent races. Alleles of *Avr4/6* from virulent races possessed nucleotide substitutions in the 5′ untranslated region of the gene, but not in the protein-coding region (Dou et al. 2010). Another effector gene *Avr1b-1* of *P. sojae* determined the efficacy of resistance gene *Rps1b* in soybean. Sequences of *Avr1b-1* locus in 34 Chinese isolates were analyzed by PCR and inverse PCR assay. Four different alleles and a complete deletion mutant of *Avr1b-1* gene were identified. Transcription of *Avr1b-1* was analyzed in virulent isolates containing four alleles of *Avr1-b-1* by real-time RT-PCR assay. In all virulent isolates, only those isolates containing the second allele, transcribed *Avr1b-1* (Cui et al. 2012).

The *Rps* genes have been widely employed to improve resistance of soybean cultivars to *P. sojae*, causing root and stem rot disease. *Rps1k* has been frequently incorporated into soybean cultivars. Products of two distinct, but closely linked RxLR effector genes were detected by *Rps1k*-containing plants, resulting in disease resistance. One of the genes *Avr1b-1* conferred avirulence in the presence of *Rps1b*. Overexpression and gene silencing of *Avr1b-1* in stable *P. sojae* transformants and transient expression of this gene in soybean indicated that Avr1b could trigger an *Rps1k*-mediated defense response. Some isolates of *P. sojae* that did not express Avr1b were unable to infect *Rps1k* plants. In such isolates, a second RxLR effector gene (*Avr1k*) was identified. Silencing or over-expression of *Avr1k* in *P. sojae* stable transformants resulted in the loss or gain, respectively of the avirulence phenotype in the presence of *Rps1k*. Only isolates of *P. sojae* with mutant alleles of both *Avr1b-1* and *Avr1k* could evade perception by the soybean plants carrying *Rps1k* (Song et al. 2013). Based on bioinformatic and global transcription analysis, *P. sojae* RxLR effector, Avr1d with 125 amino acids, was identified. Transient expression of the *Avr1d* gene triggered a HR in the presence of *Rps1d*. Sequencing of *Avr1d* genes in different *P. sojae* isolates showed two *Avr1d* alleles. Although polymorphic, the two *Avr1d* alleles could trigger *Rps1d*-mediated HR. *P. sojae* strains carrying either of the alleles were avirulent on *Rps1d* soybean lines. *Avr1d* was upregulated during the germination of cyst and early infection stages. In addition, transient expression of *Avr1d* in *N. benthamiana* enhanced *P. capsici* infection. Avr1d also suppressed effector-triggered induction of immunity by associating with Avr1b and Rps1b (Yin et al. 2013).

Zoospore chemotaxis to soybean isoflavones is crucial during early stages in pathogenesis for the establishment of *P. sojae* infection in soybean, resulting in root and stem rot disease. A G-protein α subunit encoded by *PsGPA1* was identified and it regulated the chemotaxis and pathogenicity of *P. sojae*. The PsGPA1 interacting proteins were identified, following affinity purification procedure. PsHint1 interacted with both GTP- and guanosine diphosophate (GDP)-bound forms of PsGPA1. Analysis of gene-silenced transformants revealed that *PsHint1* was involved in the chemotropic response of zoospores to the isoflavones daidzein. During interaction with susceptible soybean cultivar, *PsHint1*-silenced transformants displayed significantly reduced infectious hyphal extension and caused strong cell necrosis in plants. Further, the transformant showed defective cyst germination, forming abnormal germ tubes with several branches showing apical swelling. The results suggested that PsHint1 might regulate chemotaxis by interacting with PsGPA1, in addition to participating in a Gα-independent pathway involved in the pathogenicity of *P. sojae* (Zhang et al. 2016). Resistance to diseases may be enhanced by using pathogen-induced promoters to control expression of functional gene in transgenic plants. Soybean GmaSKT136 gene was strongly induced following infection by *P. sojae*. Functional analysis revealed that its promoter could mediate rapid and strong induction of GUS expression in *N. benthamiana* leaves and soybean hairy roots infected by *P. sojae*. A 122-bp fragment, required for activity, was identified by progressive 5′ deletion analysis. A synthetic promoter by tetramerizing this fragment could confer strong *P. sojae* induction activities. The results suggested

that GmaSKT136 promoter, the 122-bp fragment and the synthetic promoter could be effective pathogen inducible promoters (Chai et al. 2016). *P. sojae* encodes several hundreds of RXLR effectors capable of manipulating host immunity responses. Phosphatidylinositol-3-phosphate [PtdIns(3)P], an essential lipid may be produced by *P. sojae* for binding the effectors to be transported into the host cells. Transgenic *P. sojae* expressing PtdIns(3)P-specific fluorescent probes were used for determining PtdIns(3)P distribution in the pathogen. Fluorescence was associated mainly with the ER, where effectors might bind PtdIns(3)P for transport. In addition, PtdIns(P)3 biosensor was found enriched in haustoria during infection. The results indicated that PtdIns(P)3 generated by *P. sojae* might aid in the infection process operating in soybean (Chen et al. 2016).

Primary defense against microbial pathogens infecting plants is mostly inducible and associated with cell wall modification and defense-related gene expression, including several secreted proteins. A yeast-based signal sequence trap screening procedure was applied to investigate the role of secreted proteins, using RNA from *P. capsici*-inoculated root of pepper plants. *Capsicum annuum* (CaS) clones (101) were identified and of these, 92 clones were predicted to have a secretory signal sequence at their N-terminus. Reverse northern blots and RT-PCR assays were performed with RNA samples isolated at different intervals after inoculation with *P. capsici* to differentiate expressed *CaS* genes in resistant and susceptible pepper cultivars. Knockdown assays were performed in planta, using TRV-based gene-silencing method to assign biological functions to *CaS* genes of pepper. Silencing of eight *CaS* genes in pepper led to suppression of cell death induced by the nonhost bacterial pathogen *Pseudomonas syringae* pv. *tomato* T1. Three *CaS* genes induced phenotype abnormalities in silenced plants and one *CaS* gene CaS259 (PR4-1) induced both cell death suppression and perturbed phenotypes. The results indicated important roles of *CaS* genes in pathogen defense and also developmental processes (Yeom et al. 2011).

The spectrum of genes expressed during preinfection stages of *P. cinnamomi*, capable of infecting a wide range of plant species (> 3,000) was investigated by analyzing an RNA-Seq library of cysts and germinating cysts of a pathogen isolate, originating from *Persea americana*. Over 70,000 transcripts were identified from 225,049 contigs, assembled from 13 million Illumina paired-end reads. A total of 1,394 transcripts were found to have a putative role in pathogenesis. Genes aiding detoxification and metabolite transport (cytochrome P450 and ABC transporters) and protection against oxidative stress were most abundant, followed by the genes coding CWDEs. The transcript set included 44 putative RXLR effector genes and genes encoding elicitin and necrosis-inducing proteins. Expression patterns of seven putative pathogenicity genes [encoding RXLR-necrosis-inducing Phytophthora protein (NPP1)], elicitin, polygalacturonase, cellulose binding and elicitor lectin (CBEL), mucin and adhesion proteins were assessed across four *in vitro* developmental stages of *P. cinnamomi*. High expression of these genes in zoospores suggested their functional importance in the subsequent developmental stages, germination of cysts, implying a role in preinfection and revealing the molecular basis of infection stages by *P. cinnamomi* (Reitmann et al. 2017).

During interactions between *Phytophthora* spp. and their host plant species, exchange of complex molecular signals occurs between them. Various appoplastic elicitors produced by the pathogens known to trigger plant immunity have been identified. Variations in elicitin production by isolates of *P. ramorum* were determined in detached leaves of *Rhododendron* sp. Real-time PCR assays detected expression of class I elicitins (ram-α1 and ram-α2) in all isolates tested. The ram-α2 differed between three clonal lineages of *P. ramorum*. Ram-α2 expression showed a positive relationship with isolate virulence or lesion size induced. A significant positive, linear relationship was observed between ram-α2 expression and sporulation, but it was not strong. Isolates belonging to clonal lineages EV1 and NA2 were generally more virulent and produced a higher number of sporangia and greater amounts of ram-α2 elicitin *in vitro* than isolates belonging to lineage NA1 of *P. ramorum* (Manter et al. 2010). *P. parasitica* produces a protein encoded by *OPEL* during interaction with tobacco plants. Quantitative RT-PCR assay showed that *OPEL* was expressed throughout the development of *P. parasitica* and was especially highly induced after plant infection. Infiltration of OPEL recombinant protein from *Escherichia coli* into leaves of tobacco cv. Samsun NN, resulted in cell death, callose deposition, production of ROS and induced expression of pathogen-associated molecular pattern (PAMP)-triggered immunity markers and SA-responsive defense genes. In addition, infiltration conferred systemic resistance against a broad spectrum of pathogens, including *Tobacco mosaic virus*, *Ralstonia solanacearum* and *P. parasitica* (Chang et al. 2015).

*Phytophthora* spp. and solanaceous plants produce peptides and small molecules in the appoplastic space, which mediated the relationship between the interplaying pathogen-host plant. Various *Phytophthora* appoplastic effectors, including small cysteine-rich (SCR) secretory proteins have been identified, but their roles during interaction remains unclear. The SCR-effector encoded by *scr96*, one of the three novel genes, encoding SCR proteins in *P. cactorum* with similarity to the *P. cactorum* phytotoxic protein PcF, was identified. The *scr96* gene, along with other two genes, was transcriptionallly induced throughout the developmental and infection stages of the pathogen. These genes triggered PCD in the Solanaceae, including tomato and *N. benthamiana*. The *scr96* gene did not show SNPs in a collection of *P. cactorum* isolates from different countries and host plants, suggesting that its role is essential and non-redundant during infection. Homologs of SCR were present only in oomycetes, but not in fungi and other organisms. A stable protoplast transformation method was adapted for *P. cactorum*, using GFP as a marker. The silencing of *scr96* in *P. cactorum* caused gene-silenced transformants to lose their pathogenicity on host plants and these transformants were significantly more sensitive to oxidative stress. Transient expression of *scr96* partially recovered virulence of gene-silenced transformants. The results indicated that the *scr96* gene in *P. cactorum* encodes an important virulence factor that not only causes PCD, but is also important for pathogenicity and oxidative stress tolerance (Chen et al. 2016).

The vegetative (asexual) phase of the fungal pathogens has a pivotal role in pathogen proliferation and spread of the pathogen and development of disease(s) induced by them. A global proteomics investigation was taken up for comparing two key asexual propagative structures of *P. capsici* viz., mycelium and cysts, to identify stage-specific biochemical processes. Using qualitative and quantitative proteomics, 1200 proteins were identified. Transcript abundance of some of the enriched proteins was analyzed by quantitative real-time PCR assay. Different levels of abundance were estimated in 73 proteins between the mycelium and cysts. Proteins present in high concentrations were associated with glycolysis, the tricarboxylic acid (or citric acid) cycle and the pentose phosphate pathway, providing energy required for the biosynthesis of cellular building blocks and hyphal growth. On the other hand, the predominant proteins of cysts were essentially involved in fatty acid degradation, suggesting that the early infection stage of *P. capsici* relied primarily on fatty acid degradation for energy production. The results revealed the nature of potential metabolic targets at two different developmental stages that could be exploited for effective management of diseases induced by *P. capsici* (Pang et al. 2017).

Two types of immunity (resistance) to microbial plant pathogens involved in suppression of disease development have been recognized: (i) PAMP, also known as microbe-associated molecular pattern, (MAMP)-triggered immunity (PTI), and (ii) effector-triggered immunity (ETI). PAMPs derived from microbial pathogens are conserved molecular structures that are recognized by pattern recognition receptors (PRRs) and elicit the first layer of plant immunity (PTI). Plant pathogens, in their turn, can produce effectors to suppress PTI, resulting in progress of disease development. Plants have evolved resistance (R) proteins to recognize cognate effectors or effector-mediated changes of host targets and consequently activate ETI (Jones and Dangl 2006). Most plant PRRs are members of the receptor-like kinase (RLR) or receptor-like protein (RLP) family implicated in many physiological functions of host plant, including immunity to microbial plant pathogens. The molecular events leading to PTI induced by *P. parasitica* in tomato were studied. Tomato (*S. lycopersicum*) *SlSOB1R1* and *SlSOB1R1*-like genes were shown to be involved in defense responses to *P. parasitica*. Silencing of *SlSOB1R1* and *SlSOB1R1*-like genes enhanced susceptibility to *P. parasitica* in tomato. Callose deposition, ROS production, and PTI marker gene expression were compromised in *SlSOB1R1* and *SlSOB1R1*-silenced plants. *P. parasitica* infection and elicitin (ParA1) treatment induced the relocalization of SlSOB1R1 from plasma membrane to endosomal compartments and silencing of NbSOB1R1 compromised ParA1-mediated cell death on *N. benthamiana*. Further, the SlSOB1R1 kinase domain was indispensable for ParA1 to trigger SlSOB1R1 internalization and plant cell death (Peng et al. 2015).

In spite of intensive breeding efforts, development of potato cultivar(s) with durable resistance to the late blight disease caused by *P. infestans* still remains an elusive problem, since newly formed pathogen strains could overcome major resistance genes rapidly. The RB protein, from the diploid wild potato relative *Solanum bulbocastanum*, conferred resistance to most strains of *P. infestans* through its recognition of members of the corresponding pathogen effector protein family IPI-0 which comprises a multigene family. Some variants could be recognized by RB to elicit host resistance like IPI-01 and IPI-02, whereas others like IPI-04 might be able to elude detection. IPI-01 was detectable in almost all *P. infestans* strains and IPI-04 was present only in a few strains. In planta expression of both IPI-01 and IPI-04 increased the aggressiveness of *P. infestans*, resulting in production of enlarged lesions on potato leaflets. IPI-04 showed the ability to suppress hypersensitive response induced by IPI-01 in the presence of RB. The gain-of-function of IPI-04 did not compromise its effect on pathogen virulence, since IPI-04 expression resulted in formation of lesions larger than that formed by IPI-01. Higher expression of IPI-0 correlated with enlarged lesions, indicating that IPI-0 could contribute to virulence quantitatively. The results demonstrated the effect of IPI-0 on virulence of *P. infestans* on potato (Chen and Halterman 2017). PAMPs are conserved in all classes of microorganisms and PAMP-triggered immunity (PTI) is considered to be durable in plants. Elicitins are conserved extracellular proteins in oomycete species and they have been well characterized as having characteristics of PAMPs. *P. infestans*, incitant of potato late blight disease, secretes an elicitin protein designated INF1. A cell surface RLP, that mediates INF1 response, was cloned in potato. Further, some other genes also were reported to be involved in INF1-triggered immune responses. Isobaric tags for relative and absolute quantification-based quantitative proteomics were applied to analyze proteins involved in INF1-triggered cell death responses in *N. benthamiana*. Of the 2,964 proteins, 32 were significantly altered in abundance, following induction of INF1. Two of the eight selected, upregulated proteins viz., ATP-dependent transporters and 60S ribosomal protein L15, were found to be essential in INF1-triggered cell death responses by virus-induced gene silencing analysis (Du et al. 2017).

Sclerotia of *S. sclerotiorum*, survival structures, are pigmented multicellular structures formed by aggregation of hyphae. Sclerotia have a pivotal role in the successful completion of life and infection cycles of *S. sclerotiorum*. Atypical forkhead (FKH)-box containing protein, SdFKH1, involved in sclerotial development and virulence, was characterized. Partial sequence of *SsFkh1* was cloned to investigate the role of *SsFkh1*. RNA-silenced mutants with significantly reduced *SsFkh1* RNA levels exhibited slow hyphal growth and defective sclerotial development. Further, the expression levels of a set of putative melanin biosynthesis-rleated laccase genes and a polyketide synthase-encoding gene were significantly downregulated in silenced strains. Pathogenicity of RNAi-silenced mutants was significantly reduced, as indicated by smaller lesions produced by them on tomato leaves.

The results suggested the involvement of *SsFkh1* in hyphal growth, virulence and the formation of sclerotia in *S. sclerotiorum* (Fan et al. 2017).

In *L. maculans*, causal agent of black leg disease of *Brassica* spp., an avirulence gene *AvrLmJ1*, conferring avirulence toward three cultivars of *B. juncea*, was identified. Analysis of RNA sequence data showed that AvrLmJ1 was in a region of *L. maculans* genome that contained only one gene that was highly expressed in planta. Like other avirulence genes, *AvrLmJ1* was located within an AT-rich, gene-poor region of the pathogen genome. The protein encoded by *AvrLmJ1* contained 141 amino acids, had a predicted signal peptide and was cysteine-rich. Two virulent isolates had a premature stop codon in *AvrLmJ1*. Complementation of an isolate that induced lesions on cotyledons of *B. juncea* with the WT allele of *AvrLmJ1* conferred avirulence toward all three *B. juncea* cultivars tested. The results suggested that *AvrLmJ1* might confer species-specific avirulence activity in *Brassica-L. maculans* interactions (Van de Wouw et al. 2014). The APSES transcription factor StuA is a key developmental regulator of fungi involved in morphogenesis, conidia production and effector gene expression in fungal pathogens. The involvement of the ortholog of *StuA* in *L. maculans*, causing stem canker of oilseed rape, in morphogenesis, pathogenicity and effector gene expression was studied. *LmStuA* was induced during mycelial growth and at 14 days after infection, corresponding to the development of pycnidia on oilseed rape leaves, consistent with the function of *StuA*. RNA interference approaches applied for functional characterization of *LmStuA*. Silenced *LmStuA* transformants showed typical phenotypic defects of *StuA* mutants with altered typical phenotypic defects of *StuA* mutants with altered growth in culture and impaired conidia production and perithecia formation. Silencing *LmStuA* abolished the pathogenicity of *L. maculans* on oilseed rape leaves and also resulted in significant reduction in expression of at least three effector genes during in planta infection. The results suggested that either *LmStuA* might regulate directly or indirectly, the expression of several effector genes in *L. maculans* (Soyer et al. 2015). Pathogen-secreted effectors interfere with a range of physiological processes of the host plant. The hemibiotrophic fungal pathogen *L. maculans* infecting rapeseed (*B. napus*) produces an effector AvrLm4-7 which is one of the cloned effectors produced by the pathogen. AvrLm4-7 is strongly involved in the virulence of *L. maculans* and its effect on plant defense responses in a susceptible rapeseed cultivar was investigated. Using two isogenic *L. maculans* isolates differing in the presence of a functional *AvrLm4-7* allele (absence of *a4a7* and presence of *A4A7* of the allele), the plant hormone concentrations, defense-related gene transcription and ROS accumulation were analyzed in infected *B. napus* cotyledons. Various components of the plant immune system were affected. Infection with the A4A7 isolate induced suppression of SA-and ethylene (ET)-dependent signaling.These pathways regulated an effective defense against *L. maculans* infection. In addition, ROS accumulation was reduced in cotyledons infected with the *A4A7* isolate. Treatment with an antioxidant agent, ascorbic acid, increased the aggressiveness of the *a4a7*, but not that of *A4A7* isolate. The results suggested that the increased aggressiveness of the *A4A7* isolate of *L. maculans* could be due to defects in ROS-dependent defense and/or linked to suppressed SA and ET signaling (Nováková et al. 2016).

Infection of potato sprouts by *R. solani*, causing black scurf disease, results in lesions that may be lethal to the sprouts or sprout tips, followed by compensatory growth of new sprouts. Such infection may induce defense against pathogen in sprouts that recover and emerge successfully. The mechanism underlying this phenomenon of recovery of potato sprouts was studied. The basal portion of the sprout was isolated from the apical portion with a soft plastic collar and inoculated with a highly virulent isolate of *R. solani*. Microarray and quantitative PCR assay were applied to monitor induction of defense-related responses in the apical portion of the sprouts at 48 and 120 h postinoculation (*hpi*). Differential expression of 122 and 779 genes, including many well-characterized defense-related genes was detected at 48 and 120 *hpi*, respectively. In addition, the apical portion of the sprout expressed resistance which inhibited secondary infection of the sprouts. The results showed that systemic induction of resistance in sprouts following infection with virulent *R. solani* strain revealed the nature of defense in potato operating in potato prior to emergence of plant and development of photosynthetic activity (Lehtonen et al. 2008). The induction of defense genes of grapevine by *Armillaria mellea*, causative agent of destructive Armillaria root rot disease was studied, using suppression subtractive hybridization approach during early stages of infection. The genes (24) upregulated in the roots of inoculated rootstock Kober 5BB were identified at 24 h, after challenge inoculation. Real-time RT-PCR analysis confirmed the induction of genes, encoding protease inhibitors, thaumatins, glutathione-S-transferase and aminocyclopropane carboxylate oxidase, as well as phase-change related, tumor-related and proline-rich proteins and gene markers of the ethylene and jasmonate signaling pathway. Gene modulation was generally stronger in Kober 5BB than in Pinot Noir plants and *in vitro* inoculation induced higher modulation than in greenhouse *Armillaria* spp. treatments. The grapevine homolog of the *Quercus* spp. phase-change-related protein inhibited the growth of *A. mellea* mycelia *in vitro*, suggesting that this protein might have a role in the defense response of grapevine to *A. mellea* (Perazzoli et al. 2010).

## 3.3.5 Influence of Environment on Disease Development

The process of infection of plants by soilborne fungal pathogens may be significantly influenced by environmental conditions prevailing in the soil. The effect of temperature on infection and development of *P. brassicae* in root hairs of Shanghai pak choi (*Brassica rapa* subsp. *chinensis*) and on initiation of clubroot symptoms was studied. Seedlings (tenday old) were grown in liquid-sand culture, inoculated and maintained in growth chambers at 10, 15, 20, 25 and 30°C.

Roots of plants harvested at two-day intervals from two *dai* were assessed for root hair infection (RHI), stage of development of infection (primary plasmodia, zoosporangia, release of zoospores and secondary plasmodia), symptom development and for clubroot severity at 24 *dai*. Temperature affected every stage of clubroot development. RHI reached its peak and visual symptoms initiated earliest at 25°C, intermediately at 20°C and 30°C and lowest and latest at 15°C and 10°C. RHI was observed at every temperature, but clubroot symptom developed only at 15°C and above. Marked delay in pathogen development was seen at 10°C and 15°C. None of the seedlings showed symptoms grown at 10°C at 28 *dai*. Swelling of tap root was visible at 28 *dai* in plants grown at 15°C, 14 *dai* at 20°C and 30°C and ten *dai* at 25°C. The results indicated that cool temperatures slowed down development of symptoms of club root disease in *Brassica* crops (Sharma et al. 2011). In another investigation, the effects of temperature on infection of canola (*B. napus*), in addition to Shanghai pak choi and the development of *P. brassicae* were determined under controlled conditions. Clubroot severity was affected by temperatures (14–26°C) during both infection and vegetative development of the plants. Little or no clubroot developed at temperatures at or below 17°C and disease development was slower at 26°C and above than at 23–26°C in both crops throughout the three-week period of experimentation. High levels of inoculum used resulted in high disease incidence in canola, regardless of temperature treatments. A similar trend in the temperature treatments was noted in Shanghai pak choi also. The results indicated that temperature might have a significant effect on infection and development of *P. brassicae*, during vegetative growth phase of these two crops (Gossen et al. 2012).

Potted strawberry trees (*Arbutus unedo*) were affected seriously by leaf and twig blights caused by different *Phytophthora* spp. in Spain. Based on morphological characteristics and sequences of ITS regions of rDNA genes, the isolates were identified as *P. syringae*, *P. citrophthora*, *P. ramorum*, *P. tropicalis* and *P. nicotianae*. The frequency of isolation of the pathogens showed that *P. syringae* occurred most frequently in late autumn and winter, whereas *P. citrophthora* was dominant during late summer and autumn. Variations in pathogenicity of *Phytophthora* spp. were determined by inoculating detached leaves of *A. unedo* with zoospores and twigs with mycelial plugs. *In vitro* sporangial production was assessed by inoculating excised leaves and on agar plugs at 12, 15 and 20°C. *P. citrophthora* produced the largest lesions both on leaves and twigs at all temperatures tested. The size of lesions produced by *P. ramorum* was comparable to those induced by *P. syringae*, but *P. ramorum* produced significantly more sporangia on excised leaves and agar plugs. Log inoculation assay showed that *P. syringae* caused large lesions in the inner bark, whereas *P. ramorum* induced moderate sized lesions (Moralejo et al. 2008). *P. ramorum*, causes sudden oak death disease and it is heterothallic, requiring compatible isolates of mating type A 1 and A2 (MAT-1 and MAT-2) for sexual reproduction. Temperature had a significant effect on lesion size, but that response varied depending on the host. At 20°C, the optimal temperature for growth of *P. ramorum in vitro*, the susceptibility of both rhododendron and bay laurel was very similar. There was some variability among isolates used to inoculate rhododendron, but generally the clonal lineage NA2 (restricted to North America) produced larger lesions at 20°C. At day and night temperatures of 24°C and 15°C, respectively, commonly existing in coastal California in late spring, lesion sizes were reduced in both hosts equally. But at 12°C, there was a significantly decreased quantitative susceptibility in rhododendron, compared with bay laurel leaves, as shown by the smaller lesions on rhododendron. Phenotypic differences between lineages appeared to be heavily influenced by the representation of isolates used, host and temperature (Eyre et al. 2014).

Fungal pathogens may produce compounds that affect host plant development. *P. ultimum* produces IAA in liquid culture amended with L-tryptophan, tryptamine or tryptophol and also in unamended culture. The influence of IAA on the development of symptoms induced by *P. ultimum* on tomato plants was investigated, using different bioassays. Application of IAA (5 µg/ml) on tomato seedlings inoculated with *P. ultimum*, did not affect their emergence, suggesting that IAA had no effect on development of Pythium damping-off disease. But IAA increased the symptom severity on tomato plantlets, when applied (at 0.01 µg/ml) within the rhizosphere of plantlets. On the other hand, the application of higher concentration (10 µg/ml) applied either by drenching to the growing medium or by spraying on the shoot, reduced the symptom severity induced by the pathogen (Gravel et al. 2007). The influence of environment on the biochemical and physiological changes in plants induced by root-infecting fungal pathogens has not been well elucidated. Changes in phenolic compounds were assessed in pepper plants grown hydroponically and inoculated with zoospores of *Pythium aphanidermatum*, causing Pythium root rot. Biotrophic colonization of roots (remaining white) and necrotrophic colonization of roots (turning brown) by *P. aphanidermatum* showed distinct differences. Colonization of roots remained biotrophic in a temperature regime of 22:19°C (temperature reduced at two *dai*), but became necrotrophic, when the temperature was raised from 22°C to 28°C and when the plants were kept continuously at 28°C. Concentrations of free and cell wall-bound phenolics, respectively, increased by two–three- fold and up to six–fold in inoculated roots, compared with uninoculated roots maintained at 28°C and after the temperature was raised in the 22:28°C regime; no differences were found in the 22:19 regime. Accumulation of phenolics in inoculated roots coincided with and paralleled root browning. Increases in the bound and free phenolics in the shoots coincided with phenolic accumulation and browning in the roots, but the leaves remained green. Phenolic compounds accumulated in the nutrient solutions of all treatments, but levels were higher for inoculated plants with brown roots, compared with plants with white roots and uninoculated controls. The development of browning symptoms in roots resulted in marked increases in phenolics contents of roots and decline in the dry mass of infected plants (Owen-Going et al. 2008).

# REFERENCES

Abdel-Momen SM, Starr JL (1998) *Meloidogyne javanica-Rhizoctonia solani* disease complex of peanut. *Fundament Appl Nematol* 21: 611–616.

Adorada DL, Biles CL, Liddell CM, Fernandez-Pavia S, Waugh KO, Waugh ME (2000) Disease development and enhanced susceptibility of wounded pepper roots to *Phytophthora capsici*. *Plant Pathol* 49: 719–726.

Ahamad R, Majumder SK (1987) Seedborne nature, detection and seed transmission of downy mildew of sorghum. In: *Plant Quarantine and Phytosanitary Barriers to Trade in the ASEAN*. ASEAN Plant Quarantine Center Training Institute, Malaysia, p. 173.

Aharoni A, Vorst O (2001) DNA microarrays for functional plant genomics. *Plant Mol Biol* 48: 99–118.

Al Adawi AO, Al Jabri RM, Deadman ML, Barnes I, Wingfield B, Wingfield MJ (2013) The mango sudden decline pathogen *Ceratocystis manginecans* is vectored by *Hypocryphalus mangiferae* (Coleopter: Scolytinae) in Oman. *Eur J Plant Pathol* 135: 243–251.

Al Fadil TA, Jauneau A, Martinez Y, Rickauer M, Dechamp-Guillaume G (2009) Characterisation of sunflower root colonization by *Phoma macdonaldii*. *Eur J Plant Pathol* 124: 93–103.

Alkher H, El-Hadrami A, Rashid KY, Adam R, Daayf F (2009) Cross-pathogenicity of *Verticillium dahliae* between potato and sunflower. *Eur J Plant Pathol* 124: 505–519.

Ando S, Yamada T, Asano T et al. (2006) Molecular cloning of *PbSTKL1* gene from *Plasmodiophora brassicae* expressed during club root development. *J Phytopathol* 154: 185–189.

Andrade CM, Tinoco MLP, Rieth AF, Maia FCO, Aragão FJL (2016) Host-induced gene silencing in the necrotrophic fungal pathogen *Sclerotinia sclerotiorum*. *Plant Pathol* 65: 626–632.

Andrade O, Munoz G, Galdames R, Duran P, Honorato R (2004) Characterization, in vitro culture, and molecular analysis of *Thecaphora solani*, the causal agent of potato smut. *Phytopathology* 94: 875–882.

Araujo L, Silva Bispo WM, Cacique IS, Cruz MFA, Rodrigues FA (2014) Histopathological aspects of mango resistance to the infection process of *Ceratocystis fimbriata*. *Plant Pathol* 63: 1282–1295.

Arya R, Saxena SK (1999) Influence of certain rhizosphere fungi together with *Rhizoctonia solani* and *Meloidogyne incognita* on germination of Pusa Ruby tomato seeds. *Ind Phytopathol* 52: 121–126.

Asano T, Kageyama K (2006) Growth and movement of secondary plasmodia of *Plasmodiophora brassicae* in turnip suspension-culture cells. *Plant Pathol* 55: 145–151.

Asis R, Muller V, Barrionuveo DL, Araujo SA, Alado MA (2009) Analysis of protease activity in *Aspergillus flavus* and *A. parasiticus* on peanut seed infection and aflatoxin contamination. *Eur J Plant Pathol* 124: 391–403.

Atayde DD, Reis TA, Godoy IJ, Zorzete P, Reis GM, Correa B (2012) Mycobiota and aflatoxins in a peanut variety grown in different regions in the stage of Sao Paulo, Brazil. *Crop Protect* 33: 7–12.

Atkinson GF (1892) Some diseases of cotton. *Alabama Polytech Inst Agric Exp Sta Bull* 41: 61–65.

Babadoost M, Mathre DE, Johnston RH, Bonde MR (2004) Survival of teliospores of *Tilletia indica* in soil. *Plant Dis* 88: 56–62.

Back MA, Haydock PPJ, Jenkinson P (2002) Disease complexes involving plant parasitic nematodes and soilborne pathogens. *Plant Pathol* 51: 683–697.

Back MA, Jenkinson P, Haydock PPJ (2000) The interaction between potato cyst nematodes and *Rhizoctonia solani* disease in potato. In: *Proc BCPC Pests and Diseases*, British Crop Protection Council, Farnham, UK, pp. 503–506.

Bae H, Kim MS, Sicher RC, Bae HJ, Bailey BA (2006) Necrosis- and ethylene-inducing peptide from *Fusarium oxysporum* induces a complex cascade of transcripts associated with signal transduction and cell death in *Arabidopsis*. *Plant Physiol* 141: 1056–1067.

Bae T, Atallah ZK, Jansky SH, Rouse DI, Stevenson WR (2007) Colonization dynamics and spatial progress of *Verticillium dahliae* in individual stems of two potato cultivars with differing responses to potato early dying. *Plant Dis* 91: 1137–1141.

Bailey BA, Jennings JC, Anderson JD (1997) The 24-kDa protein from Fusarium oxysporum f.sp. *erthroxyli*: occurrence on related fungi and the effect of growth medium on its production. *Canad J Microbiol* 43: 45–55.

Baldé JA, Francisco R, Queiroz A, Regaldo AP, Ricardo CP, Veloso M (2006) Immunolocalization of class III chitinase in two muskmelon cultivars reacting differently to *Fusarium oxysporum* f.sp. *melonis*. *J Plant Physiol* 163: 19–25.

Bao G, Ni Y, Li Y et al. (2016) Overproduction of reactive oxygen species involved in the pathogenicity of *Fusarium* in potato tubers. *Physiol Molec Plant Pathol* 86: 35–42.

Bartz FE, Cubeta MA, Toda T, Naito S, Ivors KL (2010) An in planta method for assessing the role of basidiospores in Rhizoctonia foliar disease of tomato. *Plant Dis* 94: 515–520.

Bashi ZD, Hegedus DD, Buchwaldt L, Rimmer R, Borhan MH (2010) Expression and regulation of *Sclerotinia sclerotiorum* necrosis and ethylene-inducing peptides (NEPs). *Molec Plant Pathol* 11: 43–53.

Bashi ZD, Rimmer SR, Khachatourians GG, Hegedus DD (2012) Factors governing the regulations of *Sclerotinia sclerotiorum* cutinase A and polygalacturonase 1 during different stages of infection. *Canad J Microbiol* 58: 605–616.

Basu A, Chowdhury S, Raychaudhuri T, Kundu S (2016) Differential behavior of sheath blight pathogen *Rhizoctonia solani* in tolerant and susceptible rice varieties before and during infection. *Plant Pathol* 65: 1333–1346.

Bennett RS, Hutmacheer RB, Davis RM (2008) Seed transmission of *Fusarium oxysporum* f.sp. *vasinfectum* race 4 in California. *J Cotton Sci* 12: 160–164.

Bhatt J, Vadhera I (1997) Histopathological studies in cohabitation of *Pratylenchus thornei* and *Rhizoctonia bataticola* on chickpea (*Cicer arietinum* L.) *Adv Plant Sci* 10: 33–38.

Björsell P, Edin E, Viketoft M (2017) Interactins between some plant-parasitic nematodes and *Rhizoctonia solani* in potato fields. *Appl Soil Ecol* 113: 151–154.

Blake SN, Barry KM, Gill WM, Reid JB, Foo E (2016) The role of strigolactones and ethylene in disease caused by *Pythium irregulare*. *Molec Plant Pathol* 17: 680–690.

Boland GJ, Hall R (1994) Index of plant hosts of *Sclerotinia sclerotiorum*. *Canad J Plant Pathol* 16: 93–108.

Bonde MR, Nester SE, Olsen MW, Berner DK (2004) Survival of teliospores of *Tilletia indica* in Arizona field soils. *Plant Dis* 88: 804–810.

Borders KD, Lipps PE, Paul PA, Dorrance AE (2007) Characterization of *Pythium* spp. associated with corn and soybean seed and seedling disease in Ohio. *Plant Dis* 91: 727–735.

Borders KD, Parker ML, Melzer MS, Boland GJ (2014) Phylogenetic diversity of *Rhizoctonia solani* associated with canola and wheat in Alberta, Manitoba and Sasketchewan. *Plant Dis* 98: 1695–1701.

Brar HK, Swaminathan S, Bhattacharyya MK (2011) The *Fusarium virguliforme* toxin FvTox1 causes foliar sudden death syndrome-like symptoms in soybean. *Molec Plant-Microbe Interact* 24: 1179–1188.

Braun SE, Sanderson JP, Wraight SP (2012) Larval *Bradysia impatiens* (Diptera: Sciaridae) potential for vectoring Pythium root rot pathogens. *Phytopathology* 102: 283–289.

Bravo-Ruiz G, Ruiz-Roldán C, Roncero MIG (2013) Lipolytic system of the tomato pathogen *Fusarium oxysporum* f.sp. *lycopersici*. *Molec Plant-Microbe Interact* 26: 1054–1067.

Brooks SA (2007) Sensitivity to a phytotoxin from *Rhizoctonia solani*. *Phytopathology* 97: 1207–1212.

Brown R, Cleveland T, Woloshuk C, Payne A, Bhatnagar D (2001) Growth inhibition of a *Fusarium verticillioides* GUS strain in corn kernels of aflatoxin-resistant genotype. *Appl Biotechnol Microbiol* 57: 708–711.

Busse F, Bartkiewicz A, Terefe-Ayana D, Niepold F, Schleusner Y, Flath K, Sommerfeld-Impe N, Lubeck J, Strahwald J, Tacke E, Hofferbert H-R, Linde M, Przetakiewicz J, Debener T (2017) Genomic and transcriptomic resources for marker development in *Synchytrium endobioticum*, an elusive but severe potato pathogen. *Phytopathology* 107: 322–328.

Butterfield E, DeVay J (1977) Reassessment of soil assays for *Verticillium dahliae*. *Phytopathology* 67: 1073–1078.

Cai X, Zhang J, Wu M, Jiang D, Li G, Yang L (2015) Effect of water flooding on survival of *Leptosphaeria biglobosa* 'brassicae' in stubble of oilseed rape (*Brassica napus*) in central China. *Plant Dis* 99: 1426–1433.

Çakir B, Gül A, Yolageldi L, Ozaktan H (2014) Response to *Fusarium oxysporum* f.sp. *radicis-lycopersici* in tomato roots involves regulation of SA- and ET- responsive gene expressions. *Eur J Plant Pathol* 139: 379–391.

Cessna SG, Sears VE, Dickman MB, Low PS (2000) Oxalic acid, a pathogenicity factor for *Sclerotinia sclerotiorum*, suppresses the oxidative bursts of the host plant. *Plant Cell* 12: 2191–2200.

Chai AL, Li JP, Xie XW, Shi YX, Li BJ (2016) Dissemination of *Plasmodiophora brassicae* in livestock manure detected by qPCR. *Plant Pathol* 65: 137–144.

Chai C, Zeng W, Lin Y, Shen D, Zhang M, Dou D (2016) Synthetic tetramer of a *Phytophthora sojae*-inducible fragment from soybean *GmaSKT136* promoter improves its pathogen induction of activities. *Physiol Molec Plant Pathol* 93: 49–57.

Chaijuckam P, Baek JM, Greer CA, Webster RK, Davis R (2010) Population structure of *Rhizoctonia oryzae-sativae* in California rice fields. *Phytopathology* 100: 502–510.

Chang H-X, Domier LL, Radwan O, Yendrek CR, Hudson ME, Hartman GL (2016) Identification of multiple phytotoxin produced by *Fusarium virguliforme* including a phytotoxic effector (FvNIS1) associated with sudden death syndrome foliar symptoms. *Molec Plant-Microbe Interact* 29: 96–108.

Chang Y-H, Yan H-Z, Liou R-F (2015) A novel elicitor protein from *Phytophthora capsici* induces plant basal immunity and systemic acquired resistance. *Molec Plant Pathol* 16: 123–126.

Chen C, Harel A, Gorovoits R, Yarden O, Dickman MB (2004) MAPK regulation of sclerotial development in *Sclerotinia sclerotiorum* is linked with pH and cAMP sensing. *Molec Plant-Microbe Interact* 17: 404–413.

Chen J, Wang C, Shu C, Zhu M, Zhou E (2015) Isolation and characterization of a melanin from *Rhizoctonia solani*, the causal agent of rice sheath blight. *Eur J Plant Pathol* 142: 281–290.

Chen L, Shen D, Sun N, Xu J, Wang W, Dou D (2014) *Phytophthora sojae* TatD nuclease positively regulates sporulation and negatively regulates pathogenesis. *Molec Plant-Microbe Interact* 27: 1070–1080.

Chen L, Wang W, Hou Y, Wu Y, Li H, Dou D (2016a) Phosphatidylinositol 3-phosphate, an essential lip in *Phytophthora sojae* enriches in the haustoria during infection. *Austr Plant Pathol* 45: 435–441.

Chen X-R, Li Y-P, Li Q-Y, et al. (2016b) SCR96, a small cysteine-rich secretory protein of *Phytophthora cactorum*, can trigger cell death in the Solanaceae and is important for pathogenicity and oxidative stress tolerance. *Molec Plant Pathol* 17: 577–587.

Chen Y, Halterman DA (2017) *Phytophthora infestans* effectors IPI-01 and IPI-04 each contribute to pathogen virulence. *Phytopathology* 107: 600–606.

Clarke AJ, Hennessy J (1987) Hatching agents as stimulants of movement of *Globodera rostochiensis* juveniles. *Revue de Nematolgie* 10: 471–476.

Cooper RM, Woods RKS (1973) Induction of synthesis of extra-cellular cell wall-degrading enzymes in vascular wilt fungi. *Nature* 246: 309–311.

Cotty PJ (1989) Virulence and cultural characteristics of two *Aspergillus flavus* strains pathogenic on cotton. *Phytopathology* 79: 808–814.

Covey PA, Kuwitzky B, Hanson M, Webb KM (2014) Multilocus analysis using putative fungal effectors to describe a population of *Fusarium oxysporum* from sugar beet. *Phytpathology* 104: 886–896.

Crowhurst RN, Binnie SI, Bowen JK et al. (1997) Effect of disruption of a cutinase gene (*cutA*) on virulence and tissue specificity of *Fusarium solani* f.sp *cucurbita* race 2 toward *Cucurbita maxchta* and *Cucurbita moschata*. *Molec Plant-Microbe Interact* 10: 355–368.

Cui L, Yin W, Dong S, Wang Y (2012) Analysis of polymorphism and transcription of the effector gene *Avr1b* in *Phytophthora sojae* isolates from China virulent to *Rsp1b*. *Molec Plant Pathol* 13: 114–122.

Cui L, Yin W, Tang W et al. (2010) Distribution, pathotypes and metalaxyl sensitivity of *Phytophthora sojae* from Heilongjiang and Fujian Provinces of China. *Plant Dis* 94: 881–884.

Das S, Shah FA, Butler RC et al. (2014) Genetic variability and pathogenicity of *Rhizoctonia solani* associated with black scurf of potato in New Zealand. *Plant Pathology* 63: 651–666.

Davar R, Darvishzadeh R, Majd A, Kharabian-Masouleh A, Ghosta Y (2012) The infection process of *Sclerotinia sclerotiorum* in basal stem tissue of a susceptible genotype of *Helianthus annus* L. *Notlac Botanicae HortiAgrobotanici* 40: 143–149.

Davidson AL, Blahut-Beatty L, Itaya A, Zhang Y, Simmonds D (2016) Histopathology of *Sclerotinia sclerotiorum* infection and oxalic acid function in susceptible and resistant soybean. *Plant Pathol* 65: 878–887.

Depotter JRL, Rodriguez-Moreno L, Thomma BPHJ, Wood TA (2017) The emerging British *Verticillium longisporum* population consists of aggressive *Brassica* pathogens. *Phytopathology* 107: 1399–1405.

de Silva AP, Bolton MD, Nelson BD (2009) Transformation of *Sclerotinia sclerotiorum* with the green fluorescent protein gene and fluorescence of hyphae in four inoculated hosts. *Plant Pathol* 58: 487–496.

de Vega-Bartol JJ, Martín-Dominguez R, Ramos B, García-Sanchez M, Días-Mínguez M (2011) New virulence groups in *Fusarium oxysporum* f.sp. *phaseoli*: The expression of the gene coding for the transcription factor ftf1 correlates with virulence. *Phytopathology* 101: 470–479.

Delgado JA, Schwarz PB, Gillespie J, Rivera-Varas VV, Secor GA (2010) Trichothecene mycotoxins associated with potato dry rot caused by *Fusarium graminearum*. *Phytopathology* 100: 290–296.

Di Pietro A, Huertas-Gonzalez MD, Gutierrez-Corona JF, Martinez-Cadena G, Meglcaz E, Roncero MIG (2001) Molecular characterization of a subtilase from the vascular wilt fungus *Fusarium oxysporum*. *Molec Plant-Microbe Interact* 14: 653–662.

Di Pietro A, Roncero MIG (1996) Endopolygalacturonase from *Fusarium oxysporum* f.sp. *lycopersici*: purification, characterization and production during infection of tomato plants. *Phytopathology* 86: 1324–1330.

Di Pietro A, Roncero MIG (1998) Cloning, expression and role in pathogenicity of *pg1* encoding the major extracellular endo-polygalacturonase of the vascular wilt pathogen *Fusarium oxysporum*. *Molec Plant-Microbe Interact* 11: 91–98.

Dickman MB, Podilal GK. Kolattukudy PE (1989) Insertion of cutinase gene into wound pathogen enables it to infect host. *Nature (London)* 342: 446–448.

Diederichsen E, Wagenblatt B, Schallehn V (2016) Production of pure genotype isolates of *Plasmodiophora brassicae* Wor.–Comparison of inoculations with root hairs containing single sporangiosori or with single resting spores. *Eur J Plant Pathol* 145: 621–627.

Dirac MF, Mange JA, Madore MA (2003) Comparison of seasonal infection of citrus roots by *Phytophthora citrophthora* and *P. nicotianae var. parasitica*. *Plant Dis* 87: 493–501.

Divon HH, Rothan-Denoyes B, Davydov O, Di Pietro A, Fluhr R (2005) Nitrogen-responsive genes are differentially regulated in planta during *Fusarium oxysporum* f.sp. *lycopersici* infection. *Molec Plant Pathol* 6: 459–470.

do Amaral DOJ, de Almeida CMA, dos Santos Correia MT, de Menzes Lima VL, da Silva MV (2012) Isolation and characterization of chitinase from tomato infected by *Fusarium oxysporum* f.sp. *lycopersici*. *J Phytopathol* 160: 741–744.

Dolezal AL, Obrian GR, Nielsen DM, Woloshuk CP, Boston RS, Payne GA (2013) Localization, morphology and transcriptional profile of *Aspergillus flavus* during seed colonization. *Molec Plant Pathol* 14: 898–909.

Donald EC, Jaudzems G, Porter IJ (2008) Pathology of cortical invasion by *Plasmodiophora brassicae* in club root resistant and susceptible *Brassica oleracea* hosts. *Plant Pathol* 57: 201–209.

Dou D, Kale SD, Liu T et al. (2010) Different domains of *Phytophthora sojae* effector Avr 4/6 recognized by soybean resistance genes *Rsp4* and *Rsp6*. *Molec Plant-Microbe Interact* 23: 425–435.

Dowd C, Wilson IW, McFadden H (2004) Gene expression profile changes in cotton root and hypocotyl tissues in response to infection with *Fusarium oxysporum* f.sp. *vasinfectum*. *Molec Plant-Microbe Interact* 17: 654–667.

Du J, Guo X, Chen L, Xie C, Liu J (2017) Proteomic analysis of differentially expressed proteins of *Nicotiana benthamiana* triggered by INF1 elicitin from *Phytophthora infestans*. *J Gen Plant Pathol* 83: 66–77.

Dung JKS, Peever TL, Johnson DA (2013) *Verticillium dahliae* populations from mint and potato are genetically divergent with predominant haplotypes. *Phytopathology* 103: 445–449.

Dunn AR, Fry BA, Lee TY et al. (2013) Transformation of *Phytophthora capsici* with genes for green and red fluorescent protein for use in visualizing plant-pathogen interactions. *Austr Plant Pathol* 42: 583–593.

Dunn AR, Smart CD (2015) Interactions of *Phytophthora capsici* with resistant and susceptible pepper roots and stems. *Phytopathology* 105: 1355–1361.

Dyer AT, Windels CE, Cook RO, Leonard KJ (2007) Survival dynamics of *Aphanomyces cochlioides* oospores exposed to heat stress. *Phytopathology* 97: 484–491.

Ebadzad G, Cravador A (2004) Qunatitative RT-PCR analysis of differentially expressed genes in *Quercus suber* in response to *Phytophthora cinnamomi* infection. *SpringerPlus* 2014, 3: 613, (13p). www.springerplus.com/content/3/1/613

Ellis ML, Lanubile A, Garcia C, Munkvold GP (2016) Association of putative fungal effectors in *Fusarium oxysporum* with wilt symptoms in soybean. *Phytopathology* 106: 762–773.

Elsherbiny EA, Saad AS, Zaghloul MG, El-Sheshtawi MA (2015) Efficiency assessment of the antifungal metabolites from *Sclerotinia cepivorum* against onion white rot disease. *Eur J Plant Pathol* 142: 843–854.

Enzenbacher TB, Hausbeck (2012) An evaluation of cucurbits for susceptibility to cucurbitaceous and solanaceous *Phytophthora capsici* isolates. *Plant Dis* 96: 1404–1414.

Erwin DC, Ribeiro OK (1996) *Phytophthora Diseases Worldwide*. The American Phytopathological Society, St Paul, MN, USA.

Eynck C, Koopmann B, Karlovsky P, von Tiedemann A (2009) Internal resistance in winter oilseed rape inhibits systemic spread of the vascular pathogen *Verticillium longisporum*. *Phytopathology* 99: 802–811.

Eyre CA, Hayden KJ, Kozanitas M, Grünwald NJ, Garbelotto M (2014) Lineage, temperature and host species have interacting effects on lesion development in *Phytophthora ramorum*. *Plant Dis* 98: 1717–1727.

Fan H, Yu G, Liu Y et al. (2017) An atypical forkhead-containing transcription factor SsFKH1 is involved in sclerotial formation and is essential for pathogenicity in *Sclerotinia sclerotiorum*. *Molec Plant Pathol* 18: 963–975.

Fang X, Kuo J, You MP, Finnegan PM, Barbetti MJ (2012) Comparative root colonization of strawberry cultivars Camarosa and Festival by *Fusarium oxysporum* f.sp. *fragariae*. *Plant and Soil* 358: 75–89.

Favaron F, Sella L, D'Ovidio R (2004) Relationships among endopolygalacturonase, oxalate, pH and plant polygalacturonase-inhibiting protein (PGIP) in the interaction between Sclerotinia sclerotiorum and soybean. *Molec Plant-Microbe Interact* 17: 1402–1409.

Fei W, Feng J, Rong S, Strelkov SE, Gao Z, Hwang S-F (2016) Infection and gene expression of the club root pathogen *Plasmodiophora brassicae* in resistant and susceptible canola cultivars. *Plant Dis* 100: 824–828.

Feng J, Hwang S-H, Strelkov SE (2013a) Genetic transformation of the obligate parasite *Plasmodiophora brassicae*. *Phytopathology* 103: 1052–1057.

Feng J, Hwang S-H, Strelkov SE (2013b) Studies on primary and secondary infection processes by *Plasmodiophora brassicae* on canola. *Plant Pathol* 62: 177–183.

Feng J, Hwang R, Hwang S-F et al. (2010) Molecular characterization of a serine protease Pro1 from *Plasmodiophora brassicae* that stimulates resting spore germination. *Molec Plant Pathol* 11: 503–512.

Feng J, Xiao Q, Hwang S-H, Strelkov SE, Gossen BD (2012) Infection of canola by secondary zoospores of *Plasmodiophora brassicae* produced on a nonhost. *Eur J Plant Pathol* 132: 309–315.

Feng S, Shu C, Wang C, Jiang S, Zhou E (2017) Survival of *Rhizoctonia solani* AG-1IA, the causal agent of rice sheath blight, under different environmental conditions. *J. Phytopathol* 165: 44–52.

Fernández-Pavía SP, Grünwald NJ, Díaz-Valasis M, Cadena-Hinojosa M, Frey WE (2004) Soilborne oospores of *Phytophthora infestans* in Central Mexico survive winter fallow and infect potato plants in the field. *Plant Dis* 88: 29–33.

Flajsman M, Mandelc S, Radisek S, Stajner N, Jakse J, Kosmelj K, Javornik B (2016) Identification of novel virulence-associated proteins secreted to xylem by *Verticillium nonalfalfae* during colonization of hop plants. *Molec Plant-Microbe Interact* 29: 362–373.

Flores-Giubi MG, Diaz-Brito MA, Brito-Aregaez L et al. (2014) Purification of p47f, a necrosis-inducing protein fraction from the secretome of *Phytophthora capsici*. *J Phytopathol* 162: 788–800.

Freed GM, Floyd CM, Malvick DK (2017) Effects of pathogen population levels and crop-derived nutrients on development of soybean sudden death syndrome and growth of *Fusarium virguliforme*. *Plant Dis* 101: 431–441.

French-Monar RD, Jones JB, Roberts PD (2006) Characterization of *Phytophthora capsici* associated with roots of weeds on Florida vegetable farms. *Plant Dis* 90: 345–350.

Frost KE, Johnson ACS, Grevens A (2016) Survival of isolates of the US-22, US-23 and US-24 clonal lineages of *Phytophthora infestans* by asexual means in tomato seed at cold temperatures. *Plant Dis* 100: 180–187.

Fu L, Wang HZ, Feng BZ, Zhang XG (2013) Cloning, expression, purification and initial analysis of a novel pectate lyase Pcpel1 from *Phytophthora capsici*. *J Phytopathol* 161: 230–238.

Gallegly ME, Hong C (2008) *Phytophthora: Identifying Species by Morphology and DNA Fingerprints*. The American Phytopathological Society, St Paul, MN, USA.

Gallup CA, Shew HD (2010) Occurrence of race 3 of *Phytophthora nicotianae* in North Carolina, the causal agent of black shank of tobacco. *Plant Dis* 94: 557–562.

Gao F, Zhou BJ, Li GY et al. (2010) A glutamic acid-rich protein identified in *Verticillium dahliae* from an insertional mutagenesis affects microsclerotial formation and pathogenicity. *PLOS ONE* 5, e15319.

Gao J, Cao M, Ye W et al. (2015) PsMPK7, a stress-associated mitogen-activated protein kinase (MAPK) in *Phytophthora sojae*, is required for stress tolerance, reactive oxygenated species detoxification, cyst germination, sexual reproduction and infection of soybean. *Molec Plant Pathol* 16: 61–70.

Garcia-Maceira FI, Di Pietro A, Roncero MIG (1997) Purification and characterization of a novel exopolygalacturonase from *Fusarium oxysporum* f.sp. *lycopersici*. *FEMS Microbiol Lett* 154: 37–43.

Garica-Pineda E, Benezer-Benezer M, Gutiérrez- Segundo A, Rangel-Sánchez G, Arreola-Cortés A, Castro-Mercado E (2010) Regulation of defense responses in avocado roots infected with *Phytophthora cinnamomi* (Rands). *Plant and Soil* 331: 45–56.

Gawehns F, Houterman PM, Ichou FA et al. (2014) The *Fusarium oxysporum* effector Six6 contributes to virulence and suppresses I-2-mediated cell death. *Mol. Plant-Microbe Interact* 27: 336–348.

Gharbi Y, Alkher H, Triki MA et al. (2015) Comparative expression of genes controlling cell wall-degrading enzymes in *Verticillium dahliae* isolates from olive, potato and sunflower. *Physiol Molec Plant Pathol* 91: 56–65.

Gludovacz TV, Deora A, Mc Donald MR, Gossen BD, (2014) Cortical colonization by *Plasmodiophora brassicae* in susceptible and resistant cabbage cultivars. *Eur J Plant Pathol* 140: 859–862.

Goates BJ (1988) Histology of infection of wheat by *Tilletia indica*, the Karnal bunt pathogen. *Phytopathology* 78: 1434–1441.

Goates BJ (2010) Survival of secondary sporidia of floret-infecting *Tilletia* species: Implications for epidemiology. *Phytopathology* 100: 655–662.

Goicoechea N, Aguirreolea J, Cenoz S, Garcia-Mina JM (2000) *Verticillium dahliae* modifies the concentration of proline, soluble sugars, starch, soluble protein and abscisic acid in pepper plants. *Eur J Plant Pathol* 106:19–25.

Gomes VM, Souza RM, Almeida AM, Dolinski C (2014) Relationship between *M. enterolobii* and *F. solani*: Spatial and temporal dynamics in the occurrence of guava decline. *Nematoda* 1: e01014

Gomes VM, Souza RM, Mussi-Dias V, Silveira SF, Dolinski C (2011) Guava decline: A complex disease involving *Meloidogyne mayaguensis* and *Fusarium solani*. *J Phytopathol* 159: 45–50.

Gomez-Gomez E, Isabel M, Roncero MIG, Di Pietro A, Hera C (2001) Molecular characterization of a novel endo-beta-1,4-xylanase gene from the vascular wilt fungus *Fusarium oxysporum*. *Curr Genet* 40: 268–275.

Gomez-Gomez E, Ruiz-Roldan MC, Roncero MIG, Di Pietro A, Hera C (2002) Role in pathogenesis of two endo-beta-1,4-xylanase genes from the vascular wilt fungus *Fusarium oxysporum*. *Fungal Genet Biol* 35: 213–222.

Gonzáles-Vera AD, Bernardes-de-Assis J, Zala M et al. (2010) Divergence between symparic rice- and maize-infecting populations of *Rhizoctonia solani* AG1-IA from Latin America. *Phytopathology* 100: 172 182.

Gossen BD, Adhikari KKC, McDonald MR (2012) Effect of temperature on infection and subsequent development of club root under controlled conditions. *Plant Pathol* 61: 593–599.

Gottlieb D (1976) Production and role of antibiotics in soil. *J Antibiotics* 29: 987–1000.

Granum E, Pérez-Bueno, Calderón CE et al. (2015) Metabolic responses of avocado plants to stress induced by *Rosellinia necatrix* analyzed by fluorescence and thermal imaging. *Eur J Plant Pathol* 142: 625–632.

Gravel V, Antoun H, Tweddell RJ (2007) Effect of indole-acetic acid (IAA) on the development of symptoms caused by *Pythium ultimum* on tomato plants. *Eur J Plant Pathol* 119: 457–462.

Gu Q, Zhang C, Liu X, Ma Z (2015) A transcription factor FgSte12 is required for pathogenicity in *Fusarium graminearum*. *Molec Plant Pathol* 16: 1–13.

Guérin V, Lebreton A, Cogliati EE et al. (2014) A zoospore inoculation method with *Phytophthora sojae* to assess the prophylactic role of silicon on soybean cultivars. *Plant Dis* 98: 1632–1638.

Guo L, Yang L, Liang C, Wang G, Dai Q, Huang J (2015) Differential colonization patterns of bananas (*Musa* spp.) by physiological races 1 and race 4 isolates of *Fusarium oxysporum* f.sp. *cubense*. *J Phytopathol* 163: 807–817.

Guo L, Yang L, Liang C, Wang J, Liu L, Huang J (2016) The G-protein subunits FGA2 and FGB1 play distinct roles in development and pathogenicity in banana fungal pathogen *Fusarium oxysporum* f.sp *cubense*. *Physiol Molec Plant Pathol* 93: 29–38.

Guo Q, Kamio A, Sharma BS, Sagara Y, Arkawa M, Inagaki K (2006) Survival and subsequent dispersal of rice sclerotial disease fungi, *Rhizoctonia oryzae* and *Rhizoctonia oryzae-sativae* in paddy fields. *Plant Dis* 90: 615–622.

Hafez SI, Al-Rehiayani S, Thornton M, Sundararaj P (1999) Differentiation of two geographically isolated populations of *Pratylenchus neglectus*, based on their parasitism of potato and interaction with *Verticillium dahliae*. *Nematropica* 29: 25–36.

Hansen EM, Parke JL, Sutton W (2005) Susceptibility of Oregon forest trees and shrubs to *Phytophthora ramorum*: A comparison of artificial inoculation and natural infection. *Plant Dis* 89: 63–70.

Hanson LE (2010) Interaction of *Rhizoctonia solani* and *Rhizopus stolonifer* causing root rot of sugar beet. *Plant Dis* 94: 504–509.

Hardham AR (2005) *Phytophthora cinnamomi*. *Molec Plant Pathol* 6: 589–604.

Harel A, Gorovits R, Yarden O (2005) Changes in protein kinase A activity accompany sclerotial development in *Sclerotinia sclerotiorum*. *Phytopathology* 95: 397–404.

He X-J, Li X-L, Li Y-Z (2015) Disruption of cerevisin via *Agrobacterium tumefaciens*-mediated transformation affects microsclerotia formation and virulence of *Verticillium dahliae*. *Plant Pathol* 64: 1157–1167.

Headrick J, Pataky J (1991) Maternal influence on the resistance of sweet corn lines to kernel infection by *Fusarium moniliforme*. *Phytopathology* 81: 268–274.

Hood ME, and Shaw HD (1997) Initial cellular interactions between *Thielaviopsis basicola* and tobacco root hairs. *Phytopathology* 87: 228–235.

Horn BW, Sorensen RB, Lamb MC et al. (2014) Sexual reproduction in *Aspergillus flavus* sclerotia naturally produced in corn. *Phytopathology* 104: 75–85.

Horta M, Caetano P, Medeira C, Maia I, Cravador A (2010) Involvement of the ß-cinnamomin elicitin in infection and colonization of cork oak roots by *Phytophthora cinnamomi*. *Eur J Plant Pathol* 127: 427–436.

Hosseini S, Heyman F, Olsson U, Broberg A, Jensen DF, Karlsson M (2014) Zoospore chemotaxis of closely related legume root-infecting *Phytophthora* species towards host isoflavones. *Plant Pathol* 63: 708–714.

Hosseini S, Karlsson M, Jensen DF, Heyman F (2012) Quantification of *Phytophthora pisi* DNA and RNA transcripts during in planta infection of pea. *Eur J Plant Pathol* 132: 455–468.

Houterman PM, Cornelissen BJC, Rep M (2008) Suppression of plant resistance gene-based immunity by a fungal effector. *PLoS Pathog*. 4: e1000061.

Houterman PM, Van Ooijen G, De Vroomen MJ, Cornelissen BJC, Takken FLW, Rep M (2009) The effector protein Avr2 of the xylem-colonizing fungus *Fusarium oxysporum* activates the tomato resistance protein I-2 intracellularly. *Plant J* 58: 970–978.

Howard RJ, Strelkov SE, Handing MW (2010) Club root of cruciferous crops–news perspectives on old disease. *Canad J Plant Pathol* 32: 43–57.

Huang L, Buchenauer H, Han Q, Zhang X, Kang Z (2008) Ultrastructural and cytochemical studies on infection process of *Sclerotinia sclerotiorum* in oilseed rape. *J Plant Dis Protect* 115: 9–16.

Huang YJ, Toscano-Underwood C, Fitt BD, Hu XJ, Hall AM (2003) Effects of temperature on ascospore germination and penetration of oilseed rape (*Brassica napus*) leaves by A- or B-groups of *Leptosphaeria maculans* (Phoma stem canker). *Plant Pathol* 52: 245–255.

Huertas-Gonzalez MD, Ruiz-Roldan MC, Garcia Maceira FI, Roncero MIG, Di Pietro A (1999) Cloning and characterization of *pl1* encoding an in planta secreted pectatelyase of *Fusarium oxysporum*. *Curr Genet* 35: 36–40.

Hurtado-Gonzáles O, Aragon-Caballero L, Apaza-Tapia W, Donahoo R, Lamour K (2008) Survival and spread of *Phytophthora capsici* in coastal Peru. *Phytopathology* 98: 688–694.

Idnurm A, Howlett BJ (2001) Pathogenicity genes of phytopathogenic fungi. *Molec Plant Pathol* 2: 241–255.

Ikeda K, Banno S, Watanabe K et al. (2012) Association of *Verticillium dahliae* and *V. longisporum* with Chinese cabbage yellows and their distribution in the main production areas of Japan. *J Gen Plant Pathol* 78: 331–337.

Ingram DS, Tommerup IC (1972) The life history of *Plasmodiophora brassicae* Woron. *Proc Royal Soc B Biol Sci* 180: 103–112.

Ito M, Meguro-Maoka A, Maoka T, Akino S, Masuta C (2017) Increased susceptibility of potato to Rhizoctonia solani disease in Potato leafroll virus-infected plants. *J Gen Plant Pathol* 83: 169–172.

Ito S, Kawaguchi T, Nagata A et al. (2004) Distribution of the *FoTom1* gene encoding tomatinase in *formae speciales* of *Fusarium oxysporum* and identification of a novel tomatinase from *Fusarium oxysporum* f.sp. *radicis-lycopersici*, the causal agent of Fusarium crown and root rot of tomato. *J Gen Plant Pathol* 70: 195–201.

Ito S, Nagata A, Kai T, Takahara H, Tanaka S (2005) Symptomless infection of tomato plants by tomatinase-producing *Fusarium oxysporum formae speciales* nonpathogenic on tomato plants. *Physiol Molec Plant Pathol* 66: 183–191.

Ito S, Takahara H, Kawaguchi T, Tanaka S, Kameya-Iwaki M (2002) Post-transcriptional silencing of the tomatinase gene in *Fusarium oxysporum* f.sp. *lycopersici*. *J Phytopathol* 150: 474–480.

Jaime-Garcia R, Cotty PJ (2004) *Aspergillus flavus* in soils and corncobs in South Texas: Implications for management of aflatoxins in corn-cotton rotations. *Plant Dis* 88: 1366–1371.

Jashni MK, Dols IHM, Iida Y et al. (2015) Synergistic action of a metallprotease and a serine protease from *Fusarium oxysporum* f.sp. *lycopersici* cleaves chitin-binding tomato chitinases, reduces their antifungal activity and enhances fungal virulence. *Molec Plant-Microbe Interact* 28: 996–1008.

Jenczmionka NJ, Schafer W (2005) The Gpmk1 MAP kinase of *Fusarium graminearum* regulates the induction of specific secreted enzymes. *Curr Genet* 47: 29–36.

Jia, Feng BZ, Sun WX, Zhang XG (2009) Polygalacturonase, pectatelyase and pectin methylesterase activity in pathogenic strains of *Phytophthora capsici* incubated at different concentrations. *J Phytopathol* 157: 581–591.

Johnson DA, Santo GS (2001) Development of wilt in mint in response to infection by two pathotypes of *Verticillium dahliae* and coinfection by *Pratylenchus penetrans*. *Plant Dis* 85: 1189–1192.

Jones TM, Anderson AJ, Albersheim PC (1972) Host-pathogen interactions. IV. Studies on the polysaccharide degrading enzymes secreted by *Fusarium oxysporum* f.sp. *lycopersici*. *Physiol Plant Pathol* 2: 153–166.

Jones JD, Dangl JL (2006) The plant immune system. *Nature* 444: 323–329.

Kageyama K, Asano T (2009) Life cycle of *Plasmodiophora brassicae*. *J Plant Growth Regulation* 28: 203–211.

Kaitany R, Melakeberhan H, Bird GW, Safir G (2000) Association of *Phytophthora sojae* with *Heterodera glycines* and nutrient stressed soybeans. *Nematropica* 30: 193–199.

Kang Z, Huang L, Buchenauer H (2000) Cytochemistry of cell wall component alteration in wheat roots infected by *Gaeumannomyces graminis* var. *tritici*. *Z. Pflanzenkrank Pflanensch* 107: 337–351.

Kankam F, Qiu H, Pu L et al. (2016) Isolation, purification and characterization of phytotoxins produced by *Rhizoctonia solani* AG-3, the causal agent of potato stem canker. *Amer J Potato Res* 93: 321–330.

Karling JS (1968) *The Plasmodiophorales*, Second edition, Hafner, New York, USA.

Keinath AP (2002) Survival of *Didymella bryoniae* in buried watermelon vines in South Carolina. *Plan Dis* 86: 32–38.

Keinath AP (2008) Survival of *Didymella bryoniae* in infested muskmelon crowns in South Carolina. *Plant Dis* 92: 1223–1228.

Klimes A, Amyotte SG, Grant S, Kang S, Dobinson KF (2008) Microsclerotia development in *Verticillium dahliae*: Regulation and differential expression of the hydrophobin gene *VDH1*. *Fungal Genet Biol* 45: 1525–1532.

Klosterman SJ, Subbarao KV, Kang S et al. (2011) Comparative genomics yields insights into niche adaptation of plant vascular wilt pathogens. *PLOS Pathol* 7: e 1002137

Koike ST, Henderson DM (1998) Black root rot caused by *Thielaviopsis basicola* on tomato transplants in California. *Plant Dis* 82: 447 (Abs.).

Komann P, Andrade-Piedra ATJL, Forbes GA (2008) Preemergence infection of potato sprouts by *Phytophthora infestans* in the highland Tropics of Ecuador. *Plant Dis* 92: 569–574.

Kombrink A, Rovenich H, Shi-Kunne X et al. (2016) *Verticillium dahliae* LysM effectors differentially contribute to virulence on plant hosts. *Molec Plant Pathol* 18: 596–608.

Kong P, Hong C (2010) Zoospore density-dependent behaviors of *Phytophthora nicotianae* are autoregulated by extracellular products. *Phytopathology* 100: 632–637.

Kong P, Lea-Cox JD, Hong CX (2012) Effect of electrical conductivity on survival of *Phytophthora alni*, *P. kernoviae* and *P. ramorum* in a simulated aquatic environment. *Plant Pathol* 61: 1179–1186.

Kosted PJ, Gerhardt SA, Sherwood JE (2002) Pheromone-related inhibitors of *Ustilago hordei* mating and *Tilletia tritici* teliospore germination. *Phytopathology* 92: 210–216.

Lange L, Olson L W (1981) Germination and parasitation of the resting sporangia of *Synchytrium endobioticum*. *Protoplasma* 166: 69–82.

Lanubile A, Ellis ML, Marocco A, Munkvold GP (2016) Association of effector Six6 with vascular wilt symptoms caused by *Fusarium oxysporum* on soybean. *Phytopathology* 106: 1404–1412.

Latore BA, Rioja ME, Wicox WF (2001) *Phytophthora* species associated with crown and root rot of apple in Chile. *Plant Dis* 85: 603–606.

Lehtonen MJ, Somervuo P, Valionen JPT (2008) Infection with *Rhizoctonia solani* induces defense genes and systemic resistance in potato sprouts grown without light. *Phytopathology* 98: 1190–1198.

Levic J, Petrovic T, Stankovic S, Krnjaja V, Stankovic G (2011) Frequency and incidence of *Pyrnochaeta terrestris* in root internodes of different maize hybrids. *J Phytopathol* 159: 424–428.

Lewis Ivey ML, Miller SA (2014) Use of vital stain FUN-1 indicates ability of *Phytophthora capsici* propagules can be used to predict maximum zoospore production. *Mycologia* 106: 362–367.

Lherminier J, Benhamou N, Larrue J et al. (2003) Cytological characterization of elicitin-induced protection in tobacco plants infected by *Phytophthora parasitica* or phytoplasma. *Phytopathology* 93: 1308–1319.

Li A, Wang Y, Tao K et al. (2010) PsSAK1, a stress-activated MAPK kinase of *Phytophthora sojae*, is required for zoospore viability and infection of soybean. *Molec Plant-Microbe Interact* 23: 1022–1031.

Li AY, Crone M, Adams PJ, Fenwick SG, Hardy GESJ, Williams N (2014) The microscopic examination of *Phytophthora cinnamomi* in plant tissues using fluorescent in situ hybridization. *J. Phytopathol* 162:747–757.

Li C, Yang J, Li W, Sun J, Peng M (2017) Direct root penetration and rhizome vascular colonization by *Fusarium oxysporum* f.sp. *cubense* are the key steps in the successful infection of Brazil Cavendish. *Plant Dis* 101: 2073–2087.

Li D, Kolattukudy PE (1997) Cloning cutinase transcription factor 1, a transactivating protein containing Cys6Zn2 binuclear cluster DNA-binding motif. *J Biol Chem* 272: 12462–12467.

Li D, Sirakova T, Rogers L, Ettinger WF, Kolattukudy PE (2001) Regulation of constitutively expressed and induced cutinase genes by different zinc finger transcription factors in *Fusarium solani f.sp. pisi (Nectria hematococca)*. *J Biol Chem* 277: 7905–7912.

Li H, Sivasithamparam K, Barbetti MJ, Wylie SJ, Kuo J (2008) Cytological responses in the hypersensitive reaction in cotyledon and stem tissues of *Brassica napus* after infection by *Leptosphaeria maculans*. *J Gen Plant Pathol* 74: 120–124.

Li H, Stone V, Dean N, Sivasithamparam K, Barbetti MJ (2007) Breaching by a new strain of *Leptosphaeria maculans* of anatomical barriers in cotyledons of *Brassica napus* cultivar Surpass 400 with resistance based on single dominant gene. *J Gen Plant Pathol* 73: 297–303.

Li M, Liang X, Rollins JA (2012) *Sclerotinia sclerotiorum* γ-gluatamyl transpeptidase (Ss-Ggt) is required for regulating glutathione accumulation and development of sclerotia and compound appressoria. *Molec Plant-Microbe Interact* 25: 412–420.

Li S, Hartman GL, Widholm JM (1999) Viability staining of soybean suspension cultured cells and stem-cutting assay to evaluate phytotoxicity of *Fusarium solani* f.sp. *glycines* culture filtrates. *Plant Cell Rep* 18: 375–380.

Li W, Li C, Sun J, Peng M (2017) Metabolomic, biochemical and gene expression analyses reveal the underlying responses of resistant and susceptible banana species during early infection with *Fusarium oxysporum* f.ap. *cubense*. *Plant Dis* 101: 534–543.

Li X-G, Zhan T-L, Wang X-W, Hua K, Zhao L, Han Z-M (2013) The composition of root exudates from two different resistant peanut cultivars and their effects on the growth of soilborne pathogen. *Internatl J Biol Sci* 9: 164–173.

Liang X, Moomaw EW, Rollins JA (2015) Fungal oxalate decarboxylase activity contributes to *Sclerotinia sclerotiorum* early infection by affecting both compound appressoria development and function. *Molec Plant Pathol* 16: 825–836.

Liang XQ, Holbrook CC, Lynch RE, Guo BZ (2005) β-1,3-gluacanse activity in peanut seed (*Arachis hypogaea*) is induced by inoculation with *Aspergillus flavus* and copurfies with a conglutin-like protein. *Phytopathology* 95: 506–511.

Liberti D, Rollins JA, Dobinson KF (2013) Peroxisomal carnitine acetyl transferase influences host colonization capacity of *Sclerotinia sclerotiorum*. *Molec Plant-Microbe Interact* 26: 768–780.

Ling N, Zhang W, Wang D, Mao J, Huang Q, Guo S (2013) Root exudates from grafted-root watermelon showed a certain contribution in inhibiting *Fusarium oxysporum f.sp niveum*. *PLOS ONE* 8 (5): e63383.

Loyd AL, Benson DM, Ivors KL (2014) *Phytophthora* populations in nursery irrigation water in relationship to pathogenicity and infection frequency of *Rhododendron* and Pieris. *Plant Dis* 98: 1213–1220.

Lozovaya VV, Lygin AV, Zernova OV, Li S, Widholm JM, Hartman GL (2006) Lignin degradation by *Fusarium solani* f.sp. *glycines*. *Plant Dis* 90: 77–82.

Lu G, Guo S, Zhang H, Geng L, Martyn RD, Xu Y (2014) Colonization of Fusarium wilt-resistant and -susceptible watermelon roots by a green fluorescent protein-tagged isolate of *Fusarium oxysporum* f.sp. *niveum*. *J Phytopathol* 162: 228–237.

Lumsden RD (1969) *Sclerotinia sclerotiorum* infection of bean and the production of cellulase. *Phytopathology* 59: 653–657.

Luo X, Mao H, Wei Y et al. (2016) The fungal-specific transcription factor *Vdpf* influences condia production, mealanized microsclerotia formation and pathogenicity in *Verticillium dahliae*. *Molec Plant Pathol* 17: 1364–1381.

Ma J, Hill CB, Hartman GL (2010) Production of *Macrophomina phaseolina* conidia by multiple soybean isolates in culture. *Plant Dis* 94: 1088–1072.

Ma L, Houterman PM Gawehns F et al. (2015) The AVR2-*SIX5* gene pair is required to activate I-2-mediated immunity in tomato. *New Phytol* 208: 507–518.

Madhani HD, Styles CA, Fink GR (1997) MAP kinases with distinct inhibitory functions impart signaling specificity during yeast differentiation. *Cell* 91: 673–684.

Mallaiah B, Muthamilan M, Prabhu S, Ananthan R (2014) Studies on interaction of nematode, *Pratylenchus delattrei* and fungal pathogen, *Fusarium incarnatum* associated with crossandra wilt in Tamil Nadu, India. *Curr Biotica* 8: 157–164.

Malvick DK, Grunden E (2004) Traits of soybean-infecting *Phytophthora* populations from Illinois agricultural fields. *Plant Dis* 88: 1139–1145.

Mammella MA, Martin FN, Cacciola SO, Coffey MD, Faedda R, Schena L (2013) Analyses of the population structure in a global collection of *Phytophthora nicotianae* isolates inferred from mitochondrial and nuclear DNA sequences. *Phytopathology* 103: 610–622.

Mansoori B, Smith CJ (2005) Elicitation of ethylene by *Verticillium albo-atrum* phytotoxins in potato. *J Phytopathol* 153: 143–149.

Manter DK, Kolodny EH, Hansen EM, Parke JL (2010) Virulence, sporulation and elicitin production in three clonal lineages of *Phytophthora ramorum*. *Physiol Molec Plant Pathol* 74: 317–322.

Marek SM, Hansen K, Romanish M, Thorn RG (2009) Molecular systematic of the cotton root rot pathogen, *Phytmatotrichum omnivora*. *Persoonia* 22: 63–74.

Maria MG, Jolanda R, Andre JC, Brenda DW, Michae JW (2006) Clonality in South African isolates and evidence for a European origin of root pathogen *Thielaviopsis basicola*. *Mycol Res* 161: 306–311.

Markakis EA, Ligoxigakis EK, Avramidou EV, Tzanidakis N (2014) Survival, persistence and efficiency of *Verticillium dahliae* passed through the digestive system of sheep. *Plant Dis* 98: 1235–1240.

Maruthachalam K, Atallah ZK, Vallad GE et al. (2010) Molecular variation among isolates of *Verticillium dahliae* and polymerase chain reaction-based differentiation of races. *Phytopathology* 100: 1222–1230.

Maruthachalam K, Klosterman SJ, Anchieta A. Mou B, Subbarao KV (2013) Colonization of spinach by *Verticillium dahliae* and effects pathogen localization on the efficacy of seed transmission. *Phytopathology* 103: 268–280.

Matthew FM, Lamppa RS, Chittem K, Chang YW, Botschner M, Kinzer K, Goswami RS, Markell SG (2012) Characterization and pathogenicity of *Rhizoctonia solani* isolates affecting *Pisum sativum* in North Dakota. *Plant Dis* 96: 666–672.

Matthiesen RL, Ahmad AA, Robertson AE (2016) Temperature affects aggressiveness and fungicide sensitivity of four *Pythium* spp. that cause soybean and corn damping-off in Iowa. *Plant Dis* 100: 583–591.

McDonald MR, Sharma K, Gossen BD, Deora A, Feng J, Hwang S-F (2014) The role of primary and secondary infection in host resistance to *Plasmodiophora brassicae*. *Phytopathology* 104: 1078–1087.

McLean KS, Lawrence GW (1993) Interrelationship of *Heterodera glycines* and *Fusarium solani* in sudden death syndrome of soybean. *J Nematol* 25: 434–439.

Michielese C, Reijnen L, Olivian C, Alabouvette C, Rep M (2012) Degradation of aromatic compounds through the β-ketoadipate pathway is required for pathogenicity of the tomato wilt pathogen *Fusarium oxysporum* f.sp. *lycopersici*. *Molec Plant Pathol* 13: 1089–1100.

Mihail JD (1992) *Macrophomina*. In: Singleton LS, Mihail JD, Rush CM (eds.), *Methods for Research on Soilborne Phytopathogenic Fungi*, The American Phytopathological Society, St Paul, MN, USA, pp. 134–136.

Mims CM, Copes WE, Richardson EA (2000) Ultrastructure of the penetration and infection of pansy roots by *Thielaviopsis basicola*. *Phytopathology* 90: 843–850.

Montecchio L, Scattolin L, Squartini A, Butt KR (2015) Potential spread of forest soilborne fungi through earthworm consumption and casting. *Forest* 8: 295–301.

Montero-Astúa M, Vasquéz V, Turechek WW, Merz U, Rivera C (2008) Incidence, distribution and association of *Spongospora subterranea* and *Potato mop-top virus* in Costa Rica. *Plant Dis* 92: 1171–1176.

Moralejo E, Belbahri L, Calmin G, Garcia- Muñoz JA, Lefort F, Descals E (2008) Strawberry tree blight in Spain, a new disease caused by various *Phytophthora* species. *J Phytopathol* 156: 577–587.

Mordue JEM (1988) *Thecaphora solani*. CMI Descroption of Pathogenic Fungi and Bacteria No. 966. Mycophytopathologia 103: 177–178.

Moretti A, Belisario A, Tafuri A, Ritieni A, Corazza L, Logrieco A (2002) Production of beauverin by different races of *Fusarium oxysporum* f.sp. *melonis*, the Fusarium wilt agent of muskmelon. *Eur J Plant Pathol* 108: 661–666.

Morocko I, Fatchi J, Gerhardson B (2006) *Gnomonia fragariae*, a cause of strawberry root rot and petiole blight. *Eur J Plant Pathol* 114: 235–244.

Moy P, Qutob D, Chapman BP, Atkinson I, Gijzen M (2004) Patterns of gene expression upon infections of soybean plants by *Phytophthora sojae*. *Molec Plant-Microbe Interact* 17: 1051–1062.

Moya-Elizondo EA, Rew LJ, Jacobsen BJ, Hogg AC, Dyer AT (2011) Distribution and prevalence of Fusarium crown rot and common root rot pathogens of wheat in Montana. *Plant Dis* 95: 1099–1108.

Munkvold GP, Carlton WM (1997) Influence of inoculation method on systemic *Fusarium moniliforme* infection of maize plants grown from infected seeds. *Plant Dis* 81: 211–216.

Nadal M, Garcia-Perajas MD, Gold SE (2010) The *snf1* gene of *Ustilago maydis* acts as a dual regulation of cell wall-degrading enzymes. *Phytopathology* 100: 1364–1372.

Nair PVR, Wiechel TJ, Crump NS, Taylor PWJ (2016) Role of *Verticillium dahliae* and *V. tricorpus* naturally infected tubers in causing Verticillium wilt disease, contribution to soil pathogen inoculum and subsequent tuber infection. *Austr Plant Pathol* 45: 517–525.

Narayanasamy P (2002) *Microbial Plant Pathogens and Crop Disease Management*, Science Publishers, Enfield, USA.

Narayanasamy P (2017) *Microbial Plant Pathogens: Detection and Management in Seeds and Propagules*, Volume 1, Wiley, UK.

Navas-Cortés JA, Alcalá-Jiménez AR, Hau B, Jiménez-Diaz RM (2000) Influence of inoculum density of races 0 and 5 of *Fusarium oxsporum* f.sp. *cicereis* on development of Fusarium wilt in chickpea cultivars. *Eur J Plant Pathol* 106: 135–146.

Newson LD, Martin DWJ (1953) Effects of soil fumigation on populations of plant parasitic nematodes, incidence of Fuarium wilt and yield of cotton. *Phytopathology* 43: 292–293.

Niño-Sánchez J, Castillo VC, Tello V et al. (2016) The FTF genes family regulates virulence and expression of SIX effectors in *Fusarium oxysporum*. *Molec Plant Pathol* 17: 1124–1139.

Nitzan N, Cummings TF, Johnson DA (2008) Disease potential of soil- and tuber-borne inocula of *Colletotrichum coccodes* and black dot severity on potato. *Plant Dis* 92: 1497–1502.

Nordmeyer D, Sikora RA (1983) Studies on the interaction between *Heterodera daverti*, *Fusarium avenaceum* and *F. oxysporum* on *Trifolium subterraneum*. *Revue de Nematologie* 6: 193–198.

Novakazi F, Inderbitzin P, Sandoya G, Hayes RJ, von Tiedemann A, Subbaro KV (2015) The three lineages of the diploid hybrid *Verticillium longisporum* differ in virulence and pathogenicity. *Phytopathology* 105: 662–673.

Nováková M, Šašek V, Trdá L et al. (2016) *Leptosphaeria maculans* effect AvrLm4-7 affects salicylic acid (SA) and ethylene (ET) signaling and hydrogen peroxide (H2O2) accumulation in *Brassica napus*. *Molec Plant Pathol* 17: 813–831.

Novo M, Pomar F, Gayoso C, Merino F (2006) Cellulase activity in isolates of *Verticillium dahliae* differing in aggressiveness. *Plant Dis* 90: 155–160.

O'Gara E, Howard K, McComb J, Colquhoun IJ, Hardy EStJ (2015) Penetration of suberized periderm of a woody host by *Phytophthora cinnamomi*. *Plant Pathol* 64: 207–215.

Obidiegwu J. E, Flath K, Gebhardt C (2014) Managing potato wart: A review of present research status and future perspective. *Theor Appl Genet* 127: 763–780.

Olivieri FP, Maldonado S, Tondón CV, Casalongué CA (2004) Hydrolytic activities of *Fusarium solani* and *Fusarium solani* f.sp. *eumartii* associated with infection process of potato tubers. *J Phytopathol* 152: 337–344.

Oren L, Ezrati S, Cohen D, Sharon A (2003) Early events in the *Fusarium verticillioides*-maize interaction characterized by using a green fluorescent protein expression transgenic isolate. *Appl Environ Microbiol* 69: 1695–1701.

Owen-Going TN, Beninger CW, Sutton JC, Hall JC (2008) Accumulation of phenolic compounds in plants and nutrient solution of hydroponically grown peppers inoculated with *Pythium aphanidermatum*. *Canad J Plant Pathol* 30: 214–225.

Padasht-Dehkaei F, Ceresini PC, Zala M, Okhovvat SM, Nikkhah MJ, McDonald BA (2013) Population genetic evidence that basidiospores play an important role in the disease cycle of rice-infecting populations of *Rhizoctonia solani* AG1-IA in Iran. *Plant Pathol* 62: 49–58.

Pang Z, Srivastava V, Liu X, Bulone V (2017) Quantitative proteomics links metabolic pathways to specific developmental stages of the plant-pathogenic oomycete *Phytophthora capsici*. *Molec Plant Pathol* 18: 378–390.

Papadopoulou KK, Kavroulakis N, Tourn M, Aggelou I (2005) Use of β-glucuronidase activity to quantify the growth of *Fusarium oxysporum* f.sp. *radicis-lycopersici* during infection of tomato. *J Phytopathol* 153: 325–332.

Pappaionnou I, Ligoxigakis EK, Vakalounakis D, Markakis EA, Typas MA (2013) Phytopathogenic, morphological, genetic and molecular characterization of a *Verticillium dahliae* population from Crete, Greece. *Eur J Plant Pathol* 136: 577–596.

Parke JL, Lewis C (2007) Root and stem infection of rhododendron from potting medium infested with *Phytophthora ramorum*. *Plant Dis* 91: 1265–1270.

Pasche JS, Taylor RJ, Gudmesad NC (2010) Colonization of potato by *Colletotrichum coccodes*: Effect of soil infestation and seed tuber and foliar inoculation. *Plant Dis* 94: 905–914.

Peddler JF, Webb MJ, Buchhorn SC, Graham RD (1996) Manganese oxidizing ability of isolates of take-all fungus is correlated with virulence. *Biol Fert Soils* 22: 272–278.

Pegg GF (1985) Life in a black hole–The microenvironment of the vascular pathogen. *Trans Brit Mycol Soc* 85: 1–20.

Peng K-C, Wang C-W, Wu C-H, Huang C-T, Liu R-F (2015) Tomato SOB1R1/EVR homologs are involved in elicitin perception and plant defense against the oomycete pathogen *Phytophthora parasitica*. *Molec Plant-Microbe Interact* 28: 913–926.

Perazzolli M, Bampi F, Faccin S et al. (2010) *Armillaria mellea* induces a set of defense genes in grapevine roots and one of them codifies a protein with antifungal activity. *Molec Plant-Microbe Interact* 23: 485–496.

Peters JC, Harper G, Brierly JL et al. (2016) The effect of post-harvest storage conditions on the development of black dot (*Colletotrichum coccodes*) on potato crops grown for different durations. *Plant Pathol* 65: 1484–1491.

Pettitt TR, Wainwright MF, Wakeham AJ, White JG (2011) A simple detached leaf assay provides rapid and inexpensive determination of pathogenicity of *Pythium* isolates to 'all year round' (AYR) chrysanthemum roots. *Plant Pathol* 60: 946–956.

Pietro AD, Roncero MI (1998) Cloning, expression and role in pathogenicity of *pg1* encoding the major extracellular endopolygalacturonase of the vascular wilt pathogen *Fusarium oxysporum*. *Molec Plant-Microbe Interact* 11: 91–98.

Poloni A, Schirawski J (2016) Host specificity in *Sporisorium reilianum* is determined by distinct mechanisms in maize and sorghum. *Molec Plant Pathol* 17: 741–754.

Porter LD, Johnson DA (2004) Survival of *Phytophthora infestans* in surface water. *Phytopathology* 94: 380–387.

Probst C, Bandyopadhyay R, Price LE, Cotty PJ (2011) Identification of atoxigenic *Aspergillus falvus* isolates to reduce aflatoxin contamination of maize in Kenya. *Plant Dis* 95: 212–218.

Przetakiewicz J (2015) The viability of winter sporangia of *Synchytrium endobioticum* (Schilb.) Perc. from Poland. *Amer J Potato Res* 92: 704–708.

Qi M, Elison EA (2005) MAP kinase pathways. *J Cell Sci* 118: 3569–3572.

Qi X, Guo L, Yang L, Huang J (2013) Foatf1, bZIP transcription factor of *Fusarium oxysporum* f.sp *cubense* is involved in pathogenesis by regulating the oxidative stress responses of Cavendish banana (*Musa* spp.). *Physiol Molec Plant Pathol* 84: 76–85.

Qi X, Su X, Guo H, Qi J, Cheng H (2016) VdThit, a thiamine transport protein is required for pathogenicity of the vascular pathogen *Verticillium dahliae*. *Molec Plant-Microbe Interact* 29: 545–559.

Qi X, Zhou S, Shang X, Wang X (2016) VdSho1 regulates growth, oxidant adaptation and virulence in *Verticillium dahliae*. *J Phytopathol* 164: 1064–1074.

Radmer L, Anderson G, Malvick DM, Kurle JE, Rendahl A, Mallik A (2017) *Pythium*, *Phytophthora* and *Phytopythium* spp. with soybean in Minnesota, their relative aggressiveness on soybean and corn and their sensitivity to seed treatment fungicides. *Plant Dis* 101: 62–72.

Radwan O, Liu Y, Clough SJ (2011) Transcriptional analysis of soybean root response to *Fusarium virguliforme*, the causal agent of sudden death syndrome. *Molec Plant-Microbe Interact* 24: 958–972.

Rago AM, Cazón LI, Paredes JA et al. (2017) Peanut smut: From an emerging disease to an actual threat to Argentine peanut production. *Plant Dis* 101: 400–408.

Rauyaree P, Ospina-Giraldo MD, Kang S et al. (2005) Mutations in *VMK1*, a mitogen-activated protein kinase gene, affect microsclerotia formation and pathogenicity in *Verticillium dahliae*. *Curr Genet* 48: 109–116.

Reitmann A, Berger DK, van den Berg N (2017) Putative pathogenicity genes of *Phytophthora cinnamomi* identified via RNA-Seq analysis of pre-infection structures. *Eur J Plant Pathol* 147: 211–228.

Ren L, Xu L, Liu F, Chen K, Sun C, Li J, Fang X (2016) Host range of *Plasmodiophora brassicae* on cruciferous crops and weeds in China. *Plant Dis* 100: 933–939.

Rennie DC, Manolii VP, Plishka M, Strelkov SE (2013) Histopathological analysis of spindle and spheroid galls caused by *Plasmodiophora brassicae*. *Eur J Plant Pathol* 135: 771–781.

Rep M (2005) Small proteins of plant-pathogenic fungi secreted during host colonization. *FEMS Microbiol Lett* 253: 19–27.

Reynolds GJ, Windels CE, MacRae IV, Laguette S (2012) Remote sensing for assessing Rhizoctonia crown and root rot severity in sugar beet. *Plant Dis* 96: 497–505.

Richards BN (1976) *Introduction to Soil Ecosystem*, Longman, London, UK.

Ridge GA, Jeffers SN, Bridges WC Jr., White SA (2014) In situ production of zoospores of five species of *Phytophthora* in aqueous environments for use as inocula. *Plant Dis* 98: 551–558.

Rogers LM, Flaishman MA, Kolattukudy PE (1994) Cutinase gene disruption in *Fusarium solani* f.sp. *pisi* decreases its virulence on pea. *Plant Cell* 6: 935–945.

Rolfe RN, Barret J, Perry RN (2000) Analyses of chemosensory responses of second stage juveniles of *Globodera rostochiensis* using electrophysical techniques. *Nematology* 2: 523–533.

Ruiz GB, Di Pietro A, Roncero MIG (2016) Combined action of the major secreted exo- and endo-polygalacturonases is required for full virulence of *Fusarium oxysporum*. *Molec Plant Pathol* 17: 339–353.

Ruiz Roldan MC, Di Pietro A, Huertas-Gomez MD, Roncero MIG (1999) Two xylanase genes of the vascular wilt pathogen *Fusarium oxysporum* are differentially expressed during infection of tomato plants. *Molec Gen Genet* 261: 530–536.

Sadras VO, Quiroz F, Echarte L, Escande A, Pereyra VR (2000) Effect of *Verticillium dahliae* on photosynthesis, leaf expansion and senescence of field grown sunflower. *Ann Bot* 86: 1007–1015.

Saharan GS, Mehta N (2008) *Sclerotinia Diseases of Crop Plants: Biology, Ecology and Disease Management*, Springer-Verlag GmbH, Heidelberg, Germany.

Santhanam P, Boshoven JC, Salas O et al. (2017) Rhamnose synthase activity is required for pathogenicity of the vascular wilt fungus *Verticillium dahliae*. *Molec Plant Pathol* 18: 347–362.

Santhanam P, Thomma BPHJ (2013) *Verticillium dahliae* Sge1 differentially regulates expression of candidate effector genes. *Molec Plant-Microbe Interact* 26: 249–256.

Santhanam P, van Esse HP, Albert I, Faino L, Nurenberger T, Thomma BPHJ (2013) Evidence for functional diversification within a fungal NEP1-like protein family. *Molec Plant-Microbe Interact* 26: 278–286.

Scandiani MM, Aoki T, Luque AG, Carmona MA, O'Donnell K (2010) First report of sexual reproduction by the soybean sudden death syndrome pathogen *Fusarium tucumaniae* in nature. *Plant Dis* 94: 1411–1416.

Scott JC, McRoberts DN, Gordon TR (2014) Colonization of lettuce cultivars and rotation crops by *Fusarium oxysporum* f.sp. *lactucae*, the cause of Fusarium wilt of lettuce. *Plant Pathol* 63: 548–553.

Shang H, Grau CK, Peters RD (2000) Oospore germination of *Aphanomyces euteiches* in root exudates and on rhizoplanes of crop plants. *Plant Dis* 84: 994–998.

Sharma K, Gossen BD, McDonald MR (2011) Effect of temperature on primary infection by *Plasmodiophora brassicae* and initiation of club root symptoms. *Plant Pathol* 60: 830–838.

Shishkoff N (2007) Persistence of *Phytophthora ramorum* in soil mix and roots of nursery ornamentals. *Plant Dis* 91: 1245–1249.

Short DPG, Gurung S, Koike ST, Klosterman SJ, Subbarao KV (2015) Frequency of *Verticillium* species in commercial spinach fields and transmission of *V. dahliae* to subsequent lettuce crops. *Phytopathology* 105: 80–90.

Singh S, Braus-Stromeyer SA, Timpner C et al. (2012) The plant host *Brassica napus*, induces in the pathogen *Verticillium longisporum*, the expression of functional catalase peroxidase which is required for the late phase of disease. *Molec Plant-Microbe Interact* 25: 569–581.

Song T, Kale SD, Arredondo FD, Shen D, Su L, Liliu, Wu Y, Wang Y, Dou D, Tyler BM (2013) Two *RxLR* avirulence genes in *Phytophthora sojae* determine soybean *Rps1k*-mediated disease resistance. *Molec Plant-Microbe Interact* 26: 711–720.

Sosnowski MR, Scott ES, Ramsey MD (2006) Survival of *Leptosphaeria maculans* in soil on residues of *Brassica napus* in South Australia. *Plant Pathol* 55: 200–206.

Soyer JL, Hamiot A, Ollivier B, Balesdent M-H, Rouxel T, Fudal I (2015) The APSES transcription factor LmStuA is required for sporulation, pathogenic development and effector gene expression in *Leptosphaeria maculans*. *Molec Plant Pathol* 16: 1000–1015.

Stahl DJ, Schäfer W (1992) Cutinase is not required for fungal pathogenicity on pear. *Plant Cell* 4: 621–629.

Steenkamp ET, Makhari OM, Coutinho TA, Wingfield BD, Wingfield MJ (2014) Evidence for a new introduction of the pitch canker fungus *Fusarium circinatum* in South Africa. *Plant Pathol* 63: 538–548.

Steinkellner S, Mammerler IR, Vierheilig H (2008) Effects of membrane filtering of tomato root exudates on conidial germination of *Fusarium oxysporum* f.sp. *lycopersici*. *J Phytopathol* 156: 489–492.

Stirnberg A, Djamei A (2016) Characterization of ApB73, a virulence factor important for colonization of *Zea mays* by the smut pathogen *Ustilago maydis*. *Molec Plant Pathol* 17: 1467–1479.

Storey GW, Evans K (1987) Interactions between *Globodera palida* juveniles, *Verticillium dahliae* and three potato cultivars with descriptions of associated histopathologies. *Plant Pathol* 36: 192–200.

Streets RB, Bloss HE (1973) Phymatotrichum root rot. *Monographs of the American Phytopathological Society*, Volume 8, The American Phytopathological Society, St Paul, MN, USA.

Sugawara K, Kobayashi K, Ogoshi A (1997) Influence of the soybean cyst nematode (*Heterodera glycines*) on the incidence of brown stem rot in soybean and adzuki bean. *Soil Biol Biochem* 29: 1491–1498.

Suo B, Chen Q, Wu D et al. (2016) Chemotactic responses of *Phytophthora sojae* zoospores to amino acids and sugars in root exudates. *J Gen Plant Pathol* 82: 142–148.

Sutton W, Hansen EM, Reeser PW, Kanaskie A (2009) Stream monitoring for detection of *Phytophthora ramorum* in Oregon tanoak forests. *Plant Dis* 93: 1182–1186.

Tainter FH, O'Brien JG, Hernandez A, Orozco F, Rebolledo O (2000) *Phytophthora cinnamomi* as a cause of oak mortality in the State of Colima, Mexico. *Plant Dis* 84: 394–398.

Takken F, Rep M (2010) The arms race between tomato *and Fusarium oxysporum*. *Molec Plant Pathol* 11: 309–314.

Tardi-Ovadia R, Linker R, Tsror (Lahkim) L (2017) Direct estimation of local pH change at infection sites of fungi in potato tubers. *Phytopathology* 107: 132–137.

Taylor A, Vágány V, Jackson AC, Harrison RJ, Rainoni A, Clarkson JP (2016) Identification of pathogenicity-related genes in *Fusarium oxysporum* f.sp. *cepae*. *Molec Plant Pathol* 17: 1032–1047.

Tewoldemedhin YT, Lamprecht SC, Vaugh MM, Doehring G, O'Donnell K (2017) Soybean SDS in South Africa is caused by *Fusarium brasiliense* and a novel undescribed *Fusarium* sp. *Plant Dis* 101: 150–157.

Thompson IA, Huber DM, Schulze DG (2006) Evidence of a multi-copper oxidase in Mn oxidation by *Gaeumannomyces graminis* var. *tritici*. *Phytopathology* 96: 130–136.

Timpner C, Braus-Stromeyer SA, Tran VT, Braus GH (2013) The Cpc1 regulator of the cross-pathway control of amino acid biosynthesis is required for pathogenicity of the vascular pathogen *Verticillium longisporum*. *Molec Plant-Microbe Interaction* 26: 1312–1324.

Tjamos SE, Markakis EA, Antoniou P, Paplomatas EJ (2006) First record of Fusarium wilt of tobacco in Greece imported as seedborne inoculum. *J Phytopathol* 154: 193–196.

Torres SV, Vargas MM, Godoy-Lutz G, Porsch TG, Beaver JS (2016) Isolates of *Rhizoctonia solani*, can produce both web blight and root rot symptoms in common bean (*Phaseolus vulgaris* L.). *Plant Dis* 100: 1351–1357.

Trapero C, Roca LF, Alcántara E, Lopez-Escudero FJ (2011) Colonization of olive inflorescence by *Verticillium dahliae* and its significance for pathogen spread. *J Phytopathol* 159: 638–649.

Trapero-Casas A, Kaiser WJ (2007) Differences between ascospores and conidia of *Didymella rabiei* in spore germination and infection of chickpea. *Phytopathology* 97: 1600–1607.

Tsror (Lakhim) L, Hazanovsky M (2001) Effect of combination of *Verticillium dahliae* and *Colletotrichum coccodes* on disease symptoms and fungal colonization in four potato cultivars. *Plant Pathol* 50: 483–488.

Tudzynski P, Sharon A (2003) Fungal pathogenicity genes. In: Arora DK, Khachatourians GG (eds.), *Applied Mycology and Biotechnology*, Volume 3, Fungal Genetics, Elsevier Sciences, BV, Amsterdam, pp. 187–212.

Tyler BM (2007) *Phytophthora sojae*: Root rot pathogen of soybean and model oomycete. *Molec Plant Pathol* 8: 1–8.

Tyler BM, Tripathy S, Zhang X, Dehal et al. (2006) *Phytophthora* genome sequences uncover evolutionary origins and mechanisms of pathogenesis. *Science* 313: 1261–1266.

Tzima AK, Paplomatas EJ, Rauyaree P, Kang S (2010) Roles of the catalytic subunit of cAMP-dependent protein kinase A in virulence and development of the soilborne plant pathogen *Verticillium dahliae*. *Fungal Genet Biol* 47: 406–415.

Tzima AK, Paplomatas EJ, Rauyaree P, Ospina-Giraldo MD, Kang S (2011) *VdSNF1*, sucrose nonfermenting protein kinase gene of *Verticillium dahliae*, is required for virulence and expression of genes involved in cell-wall degradation. *Molec Plant-Microbe Interact* 24: 129–142.

Tzima AK, Paplomatas EJ, Tsitsigiannis DJ, Kang S (2012) The G protein β-subunit control virulence and multiple growth- and development-related traits in *Verticillium dahliae*. *Fungal Genet Biol* 49: 271–283.

Usami T, Momma N, Kikuchi S et al. (2017) Race 2 of *Verticillium dahliae* infecting tomato in Japan can be split into two races with differential pathogenicity on resistant rootstocks. *Plant Pathol* 66: 230–238.

Van de Wouw AP, Lowe RGT, Elliott CE, Dubois DJ, Howlett BJ (2014) An avirulence gene *AvrLmJ1*, from the blackleg fungus, *Leptosphaeria maculans* confers avirulence to *Brassica juncea* cultivars. *Molec Plant Pathol* 15: 523–530.

van der Does HC, Rep M (2007) Virulence genes and the evolution of host specificity in plant pathogenic fungi. *Molec Plant-Microbe Interact* 20: 1175–1182.

van Toor RF, Chang SF, Warren RM, Butler RC, Cromey MG (2015) The influence of growth stage of different cereal species on host susceptibility to *Gaeumannomyces graminis* var. *tritici* on *Pseudomonas* populations in the rhizosphere. *Austr Plant Pathol* 44: 57–70.

Vats R, Dalal MR (1997) Interaction between *Rotylenchulus reniformis* and *Fusarium oxysporum* f.sp. *pisi* on pea (*Pisum sativum* L). *Ann Biol* 13: 239–242.

Vercauteren A, Riedel M, Maes M, Werres S, Heungens K (2013) Survival of *Phytophthora ramorum* in *Rhododendron* root balls and in rootless substrates. *Plant Pathol* 62: 166–176.

Veronese P, Narasimhan ML, Stevenson RA et al. (2003) Identification of a locus controlling Verticillium disease symptom response in *Arabidopsis thaliana*. *Plant J* 35: 574–587.

Vlaardingerbroek I, Beerens B, Schmidt SM, Cornelissen BJC, Rep M (2016) Dispensable chromosomes in *Fusarium oxysporum* f.sp. *lycopersici*. *Molec Plant Pathol* 17: 1455–1466.

Walker NR, Kirkpatrick TL, Rothrock CS (1998) Interaction between *Meloidogyne incognita* and *Thielaviopsis basicola* on cotton (*Gossypium hirsutum*). *J Nematol* 30: 415–422.

Walker NR, Kirkpatrick TL, Rothrock CS (2000) Influence of *Meloidogyne incognita* and *Thielaviopsis basicola* populations on early-season disease development and cotton growth. *Plant Dis* 84: 449–453.

Wallenhammer AC (1996) Prevalence of *Plasmodiophora brassicae* in a spring oilseed rape growing area in central Sweden and factors influencing soil infestation levels. *Plant Pathol* 45: 710–719.

Walz A, Zingen-Sell I, Theisen S, Kotekamp A (2008) Reactive oxygen intermediates and oxalic acid in the pathogenesis of the necrotrophic fungus *Sclerotinia sclerotiorum*. *Eur J Plant Pathol* 120: 317–330.

Wang L, Liu Y, Liu J, Zhang Y, Zhang X, Pan H (2016) The *Sclerotinia sclerotiorum FOXE2* gene is required for apothecial development. *Phytopathology* 106: 484–490.

Waugh MM, Kim DH, Ferrin DM, Stanghellini ME (2003) Reproductive potential of *Monosporascus cannonballus*. *Plant Dis* 87: 45–50.

Weiland JE, Beck BR, Davis A (2013) Pathogenicity and virulence of *Pythium* species obtained from forest nursery soils on Douglas-fir seedlings. *Plant Dis* 97: 744–748.

Wen J, Chen Q, Sun L, Zhao L, Suo B, Tian M (2014) A SCAR marker specific for rapid detection of the avirulence gene *Avr1c* in *Phytophthora sojae*. *J Gen Plant Pathol* 80: 415–422.

Werres S, Wagner S, Brand T, Kaminski K, Seipp D (2007) Survival of *Phytophthora ramorum* in recirculating irrigation water and subsequent infection of *Rhodendron* and *Viburnum*. *Plant Dis* 91: 1034–1044.

Westphal A, Xing L (2011) Soil suppressiveness against the disease complex of soybean cyst nematode and sudden death syndrome of soybean. *Phytopathology* 101: 878–886.

Wheeler DL, Johnson DA (2016) *Verticillium dahliae* infects, alters plant biomass and produces inoculum on rotation crops. *Phytopathology* 106: 602–613.

Wheeler TA, Hake KD, Dever JK (2000) Survey of *Meloidogyne incognita* and *Thielaviopsis basicola*: their impact on cotton fruiting and producers management choices in infested fields. *J Nematol* 32: 576–581.

Wibberg D, Anderson L, Tzelepis G et al. (2016) Genome analysis of the sugar beet pathogen *Rhizoctonia solani* AG2-2IIIB revealed high numbers in secreted proteins and cell wall degrading enzymes. *BMC Genomics* 17: 245 12864-016.251-1

Widmer TL (2009) Infective potential of sporangia and zoospores of *Phytophthora ramorum*. *Plant Dis* 93: 30–35.

Widmer TL, Shishkoff N, Dodge SC (2012) Infectivity and inoculum production of *Phytophthora ramorum* on roots of Eastern United States oak species. *Plant Dis* 96: 1675–1682.

Woloshuk CP, Koalttukudy PE (1986) Mechanism by which contact with plant cuticle triggers cutinase gene expression in the spores of *Fusarium solani* f.sp. *pisi*. *Proc Natl Acad Sci USA* 83: 1704–1708.

Xiao X, Xie J, Chang J, Li G, Yi X, Jiang D (2014) Novel secretory protein Ss-Caf1 of plant pathogenic fungus *Sclerotinia sclerotiorum* is required for host penetration and normal sclerotial development. *Molec Plant-Microbe Interact* 27: 40–55.

Xu L, Chen W (2013) Random T-DNA mutagenesis identifies a Cu/Zn superoxide dismutase gene as a virulence factor of *Sclerotinia sclerotiorum*. *Molec Plant-Microbe Interact* 26: 431–441.

Xu L, Ren L, Chen K, Liu F, Fang X (2016) Putative role of IAA during the early response of *Brassica napus* L to *Plasmodiophora brassicae*. *Eur J Plant Pathol* 145: 601–613.

Yan M, Cai E, Zhou J et al. (2016) A dual-color imaging system for sugarcane smut fungus *Sporisorium scitamineum*. *Plant Dis* 100: 2357–2362.

Yang YG, Zhao C, Guo ZJ, Wu XH (2015) Characterization of a new anastomosis group (AG-W) of binucleate *Rhizoctonia*, causal agent of potato stem canker. *Plant Dis* 99: 1757–1763.

Yarden O, Veluchamy S, Dickman MB, Kabbage M (2014) *Sclerotinia sclerotiorum* catalase SCAT1 affects oxidative stress tolerance, regulates ergosterol levels and controls pathogenic development. *Physiol Molec Plant Pathol* 85: 34–41.

Yeom S-I, Baek H-K, Oh S-K ET et al. (2011) Use of a secretion trap screen in pepper following *Phytophthora capsici* infection reveals novel functions of secreted plant proteins in modulating cell death. *Molec Plant-Microbe Interact* 24: 671–684.

Yin W, Dong S, Zhai L, Lin Y, Zheng X, Wang Y (2013) The *Phytophthora sojae Avr1d* gene encodes an RxLR dEER effector with presence and absence of polymorphisms among pathogen strains. *Molec Plant-Microbe Interact* 26: 958–968.

Yu Y, Kang Z, Han Q, Buchenauer H, Huang L (2010) Immunolocalization of 1,3-β-glucanases secreted by *Gaeumannomyces graminis* var. *tritici* in infected wheat roots. *J Phytopathology* 158: 344–350.

Yu Y, Xiao J, Yang Y, Bi C, Qing L, Tan W (2015) *Ss-Bi* encodes a putative BAX inhibitor-1 protein that is required for full virulence of *Sclerotinia sclerotiorum*. *Physiol Molec Plant Pathol* 90: 115–122.

Zamani-Noor N (2017) Variation in pathotypes and virulence of *Plasmodiophora brassicae* populations in Germany. *Plant Pathol* 66: 316–324.

Zhang M, Xu JH, Liu G, Yao XF, Li PF, Yang XP (2015) Characterization of the watermelon seedling infection process by *Fusarium oxysporum* f.sp. *niveum*. *Plant Pathol* 64: 1076–1084.

Zhang X, Zhai C, Hua C et al. (2016) PsHint1, associated with the G-protein α-subunit PsGPA1, is required for the chemotaxis and pathogenicity of *Phytophthora sojae*. *Molec Plant Pathol* 7: 272–285.

Zhang XY, Huo HL, Xi XM, Liu LL, Yu Z, Hao JJ (2016) Histological observation of potato in response to *Rhizoctonia solani* infection. *Eur J Plant Pathol* 145: 289–303.

Zhang Y-L, Mao J-C, Huang J-F, Meng P, Gao F (2015) A uracil-DNA glycosylase functions in spore development and pathogenicity of *Verticillium dahliae*. *Physiol Molec Plant Pathol* 92: 148–153.

Zhen XH, Li YZ (2004) Ultrastructural changes and location of β-1,3-glucanase in resistant and susceptible cotton callus cells in response to treatment with toxin of *Verticillium dahliae* and salicylic acid. *J Plant Physiol* 161: 1367–1377.

Zhou B-J, Jia P-S, Gao F, Gao H-S (2012) Molecular characterization and functional analysis of a necrosis- and ethylene-inducing protein encoding gene family from *Verticillium dahliae*. *Molec Plant-Microbe Interact* 25: 964–975.

Zhou L, Hu Q, Johannson A, Dixelius C (2006) *Verticillium longisporum* and *Verticillium dahliae*: Infection and disease in *Brassica napus*. *Plant Pathol* 55: 137–144.

Zhu Y, Shi S, Mazzola M (2016) Genotype responses of two apple rootstocks to infection by *Pythium ultimum* causing apple replant disease. *Canad J Plant Pathol* 38: 483–489.

Zveibil A, Mor N, Gnayem N, Freeman S (2012) Survival, host-pathogen interaction and management of *Macrophomina phaseolina* on strawberry in Israel. *Plant Dis* 96: 265–272.

Zvirin T, Herman R, Brotman Y et al. (2010) Differential colonization and defence responses of resistant and susceptible melon lines infected by *Fusarium oxysporum* race 1– 2. *Plant Pathol* 59: 576–585.

# 4 Detection and Identification of Soilborne Bacterial Plant Pathogens

Among the microbial plant pathogens, bacterial pathogens are mostly unicellular organisms, much smaller in size, compared to fungal pathogens. They are classified as prokaryotes which have less defined structural characteristics compared with eukaryotes which include fungal pathogens. The presence of fossil prokaryotes was found on rocks which were estimated to be about 3.5 billion years old, indicating their existence as the first recorded forms of life on Earth (Schumann 1991). Bacterial species pathogenic to plants have been estimated to be about 100 of the 1,600 species reported to be present in the world (Agrios 2005). The nuclear material present in bacterial cells is not separated by a membrane as in eurkaryotes. The genome of the bacteria is in the form of a single chromosome with double-stranded (ds) DNA as a closed circular form. Plasmids capable of self replication are present in bacterial cells and they are extrachromosmal DNA that determines certain characteristics, including pathogenicity, resistance to chemicals and antibiotics, and tumor production. The plasmids may be exchanged between bacterial species/strains, resulting in variations in characteristics governed by plasmids. The bacteria lack other organelles present in eukaryotic cells. Several bacterial species are strictly saprophytic and are involved in the decomposition of organic wastes produced from industries or by dead animals and plants. Several bacterial species are useful in crop production, because of their ability to fix atmospheric nitrogen and function as biological control agents that may be applied as an effective disease management strategy against crop diseases.

The bacterial pathogens have simple structure with different shapes like spherical, ellipsoidal, rod-shaped, spiral, filamentous or comma-shaped. Bacterial cells generally retain their shape and size under favorable environmental conditions. All plant pathogenic bacterial species are rod-shaped, except filamentous *Streptomyces* spp. The rod-shaped bacterial cells are short and cylindrical, measuring $0.3-1.0 \times 0.5-3.5$ μm. A thin or thick slime layer made of gummy materials may envelop pairs of cells or short chains of cells and this slimy layer functions as a capsule protecting the bacterial cells from desiccation. Slime layer may be found as a larger mass surrounding groups of bacterial cells. Most of the plant pathogenic bacterial species possess flagella, useful for locomotion, and they may be present either singly or in groups at one or both ends of bacterial cell or distributed over the entire cell surface. *Streptomyces* spp. have nonseptate, branched threads which cut off conidia in chains on aerial hyphae. Morphological characteristics of bacterial colonies, growing on artificial media, such as shape, size, color, elevations, and form of edges may be useful for identifying certain bacterial genera. The bacterial colonies may be circular, oval or irregular with smooth, wavy,

or angular edges. The elevation of colonies may be flat, raised dome-like or gray. Gram stain, developed by Hans Christian Gram, has been used to differentiate the bacterial species which differ in their ability to retain the Gram stain. The bacterial species capable of retaining the Gram stain, after washing, are named as gram-positive, whereas the bacterial species, that lose the stain after washing, are designated as gram-negative. The gram-positive bacteria have relatively thick uniform cell walls. On the other hand, gram-negative bacteria have thinner cell walls with an additional outer layer of polysaccharides and lipids. The presence of lipids in the cell walls of gram-negative bacteria prevents absorption of some substances like penicillin making them insensitive/resistant to such substances. *Bacillus* spp. form endospores, protecting them against desiccation, whereas bacteria belonging to other genera are sensitive to dry, low moisture environments. Bacterial pathogens do not have complex life cycles of producing different spore forms like fungal pathogens. Asexual reproduction of bacteria known as binary fission is the most common and predominant method for increase in the population size of the bacterial species. The binary fission process continues indefinitely until a threshold level is reached to initiate new infections or until nutrients are available in the substrate (culture medium/plant tissues). The phenomenon of conjugation, considered to be similar to sexual reproduction in eukaryotic organisms, involves side by side contact between two physiologically opposite cells. A small fragment of DNA from the male cell (donor) is transferred to the female cell (receptor), followed by multiplication of female cell by binary fission, resulting in production of a population of cells possessing the characteristics encoded by the DNA fragment of the donor.

Variability in genetic characteristics observed in different isolates of bacterial species may be due to three phenomena: mutation, transformation and transduction. Mutation in specific gene(s) may occur, leading to recognizable variations in the isolate undergoing mutation. The variations may remain stable through several generations. Variability in genetic characteristics may occur, when the genetic material is liberated from one cell either by secretion or rupture of the cell wall. A fragment of the released DNA gains entry into a genetically compatible bacterial cell of the same or closely related species. The recipient cells incorporated with a foreign DNA fragment become genetically different. This cell formed by transformation, then multiplies by binary fission, giving rise to a population of cells genetically different from the parent cell. The phenomenon of transduction requires the involvement of a bacteriophage as a vector for transfer of a DNA fragment from the donor cell to a receptor cell. The phage acquires a fragment of DNA from the infected bacterial cell which is lysed later, liberating the phage. When another bacterial cell is infected by the phage, the DNA fragment is integrated into the genome of the recipient. Presence of the DNA fragment of the donor leads to changes in the genetic characteristics of the bacterial cells that may be recognized, when the population of cells reaches required level. Thus, the genetic characteristics of the bacterial pathogens may show variations due to mutation, transformation or transduction. The ability of bacterial pathogens to have genetic diversity because of rapid multiplication to reach high population levels is an important factor to be reckoned with in all ecosystems in general and with pathogens of plants in particular (Narayanasamy 2001, 2011).

## 4.1　DETECTION OF BACTERIAL PATHOGENS

Soilborne bacterial pathogens have been responsible for various diseases, affecting different crops resulting in significant quantitative and qualitative losses. Infection of plants initiated through roots by bacterial pathogens can be recognized by the symptoms exhibited in aerial plants parts. At this stage, the pathogen would have well established in internal tissues of affected plants. Furthermore, the symptoms expressed in aerial plant parts may be due to the toxins produced by the pathogen confined to the roots making the disease diagnosis difficult. The bacterial pathogens may also be seedborne and the bacterial pathogen may migrate from the infected plants growing from infected seeds to the rhizosphere soil around the root system and proliferate in the organic matter or plant debris present in the soil, when the crop is harvested. Likewise, the bacterial pathogens present in plant residues may also reach the soil and survive in the soil. Soilborne bacterial pathogens have to be detected at early stages of infection prior to symptom expression by employing sensitive technique(s) that can provide results rapidly. The precise identity of the pathogen has to be established precisely, since the pathogen may exist in the form of different pathovars, varying in their virulence (pathogenic potential). Detection techniques have been developed, based on the biological, biochemical, immunological and nucleic acid properties of the bacterial pathogen concerned. The effectiveness and limitations of techniques applied for detection and differentiation of bacterial pathogens present in plants, soil and water are discussed hereunder.

### 4.1.1　Isolation-Dependent Methods

Bacterial pathogens present in the soil, plants growing on infested soil and water in lakes/ponds and irrigation channels may be isolated on appropriate media that support the development of the target pathogens. The isolates growing on the culture media are then tested for their pathogenicity on plant species from which they were isolated. The type of symptoms induced and severity of disease caused by the isolates may vary depending on their pathogenic potential (virulence). The pathogenic isolates are reisolated from artificially inoculated plants and compared with the characteristics of the isolates obtained earlier from naturally infected plants. This process of proving Koch's postulates has to be completed before identifying the pathogenic isolates. Initially the taxonomic characteristics and guidelines described in *Bergey's Manual of Systematic Bacteriology* (Krieg and Holt 1984) are used for identification of pathogenic isolates and naming them in accordance with the *International Code of Nomenclature of Bacteria*. As variations in morphologic characteristics of the bacteria in axenic cultures are insufficient to distinguish different bacterial species, the results of several physiological and biochemical tests are taken into consideration for classification of bacterial species. Polyphasic tests, including nucleic acid analyses, chemotaxonomic comparisons such as cell wall composition, lipid profiles, soluble and total proteins, fatty acid profiles, enzyme characterization, in addition to determination of nutritional requirements are used for differentiation of bacterial species (Young et al. 1992). The International Society of Plant Pathology formulated criteria for using the term 'pathovar' that is applied at the infrasub-specific level for distinguishing bacteria mainly based on the differences in pathogenicity determined, using a set of plant species/cultivars. The taxonomy and systems of nomenclature of bacteria are less stable, compared to that of fungi, because of a lack of distinctive morphological characteristics of most of bacterial species. Bacterial strains that share some phenotypic and genotypic similarities are placed in a bacterial species, the predominant strain being named as type strain. Other strains of the same species may differ to varying degrees from the type strain in morphological, cultural and physiological properties.

#### 4.1.1.1　Use of Indicator Plants

Bacterial pathogens present in the soil may be detected directly by growing susceptible plants as indicator (bait) on the infested soil or after isolating them on specific media that favor their development differentially. *Ralstonia solanacearum* with its wide host range could be directly detected by growing sensitive plants like potato on infested soil and the indicator plants show typical wilt symptoms and brown rot symptoms in tubers (Graham and Lloyd 1978). Tomato bioassay method was very effective by consistently detecting *R. solanacearum* in soil samples with populations greater than $7.5 \times 10^5$ CFU/g of soil (Pradhanang et al. 2000). A simple rapid and reliable bioassay technique was developed for the detection of *Agrobacterium tumefaciens* capable of infecting large number of plant species. Detached leaves of *Kalanchoe tubiflora* were employed as bait for trapping the pathogen present in soil samples. The leaf bait selectively retrieved *A. tumefaciens* even in the presence of large and heterogeneous microflora in the soil. The pathogen could be isolated from leaf bait on culture medium (Romeiro et al. 1999). The bioassays require large volumes of soil samples, greenhouse space and a long time to provide results. However, estimates of viable pathogen populations could be obtained from the bioassays and this information would be very useful for epidemiological investigations.

#### 4.1.1.2　Use of Selective Media

Soil microflora consist of high populations of saprophytic bacteria and fungi that may overgrow the pathogenic bacteria, making it difficult to recover them from soil samples.

Semiselective and selective media are employed for isolating plant pathogenic bacterial species from soil. A modified semiselective medium (SMSA) was used to detect *R. solanacearum* by dilution plating at populations fewer than $10^2$ CFU/g of soil. By combining an enrichment step in SMSA broth with enzyme-linked immunosorbent assay (ELISA) or nested polymerase chain reaction (PCR) assay, it was possible to detect *R. solanacearum* at a concentration of $10^2$ CFU/g of naturally infested soil. It was possible to enhance the sensitivity of pathogen detection by using the modified SMSA medium. Soilborne *R. solanacearum* populations could be observed on the semiselective medium up to 18 months after harvest of infected potato crop, indicating that *R. solanacearum* could survive in the soil during the absence of potato plants in the field. The lowest detectable population of the pathogen using SMSA medium was $5 \times 10^2$ cells/g of soil (Pradhanang et al. 2000).

*R. solanacearum (Rs)* biovar 2 was reported to survive in waterways at low temperatures in the United Kingdom and the Netherlands (Elphinstone et al. 1998; Janse and Schans 1998) and later in other European countries. Survival of *R. solanacearum* cells in water for variable periods depended on the temperature and inoculum density. The pathogen was recovered at low densities (10–80 CFU/ml) by direct plating on modified SMSA agar from water samples at 14°C or higher. Waterways could be major dissemination routes of *R. solanacearum* which was able to survive for long periods in sterilized water, but its survival in natural water along with other microorganisms was unclear. The fate of a Spanish strain of *R. solanacearum* inoculated in water microcosms from Spanish river containing different microbiota fractions at 24°C and 14°C was investigated by plating the samples on yeast extract peptone glucose agar (YPGA) and SMSA media. Presence of lytic phages and protozoa adversely affected the *R. solanacearum* population buildup. In water microcosm, the temperature of 14°C was more favorable for the survival of *R. solanacearum* than at 24°C, since biotic interactions were slower at lower temperatures (Álvarez et al. 2007). In a later investigation, surveys were undertaken for over three seasons to assess the population levels of *R. solanacearum* (Rs) biovar 2 race 3 (phylotype II sequevar 1) in irrigation, drainage and artesan well water throughout the major potato-growing areas in Egypt. The presence of *R. solanacearum* was limited to the canals of the traditional potato-growing regions in the Nile Delta. The presence of *R. solanacearum* was detected more commonly in the network of smaller irrigation canals flowing through potato-growing areas. Populations of *R. solanacearum* in the canals of the Delta (~ 100–200 CFU/l) were generally variable throughout the year with presence linked to potato cultivation in the immediate areas. The pathogen was not detected in irrigation or drainage water associated with potato cultivation in the newly reclaimed desert areas (designated as Pest-Free Areas, PFAs) or in the main branches of the Nile upstream from these areas. Temperature and microbial activity were principal factors affecting the survival of *R. solanacearum* in canal water, as indicated by *in vitro* studies. The pathogen survival was the longest at 15°C, whereas

high temperature (35°C) reduced the survival most seriously. Aeration, solarization and pH variation (4 to 9) had little effect on survival of *R. solanacearum*. The maximum survival time in nonsterile Egyptian canal water at high inoculum pressure was estimated to be up to 300 days at optimum temperature for survival (15–30°C), suggesting the potential for the long-distance spread of *R. solanacearum* in Egyptian surface waters to form sources of contamination (Tomlinson et al. 2009).

The medium NGM was developed for the isolation of *Erwinia chrysanthemi* and to distinguish the pathogen from other *Erwinia* spp. based on the production of specific blue-pigmented indigoidine. The NGM medium consisted of nutrient agar supplemented with 1% glycerol that induced pigment synthesis and 2 mM $MnCl_2.4H_2O$ that further enhanced color development. All *E. chrysanthemi* strains (> 50 strains) from six different host plant species grew well on NGM medium and formed dark brown to blue colonies easily distinguishable from other *Erwinia* spp. The results indicated that pigment production on the NGM medium was a stable characteristic useful for differentiation of *E. chrysanthemi* from other *Erwinia* spp. All *E. chrysanthemi* strains tested contained *indC* gene, as determined by PCR amplification, employing a specific oligonucleotide primer set for detection of *indC* involved in indigoidine biosynthesis. Pigment production by E. chrysanthemi on NGM medium was correlatable with the presence of *indC* sequences. The NGM medium was used to identify the isolates of *E. chrysanthemi*, causing soft rot of *Phalaenopsis* orchids in Taiwan. The results indicated the usefulness of NGM medium in identifying and differentiating *E. chrysanthemi* from other *Erwinia* spp. (Lee and Yu 2006). The presence of *Erwinia* spp., causal agent of potato soft rot disease, was detected in wash water used for washing tubers at various sites in the washing plants and from ponded recycled water. The water samples, after serial dilutions, were plated on crystal violet pectate medium in petridishes (Hyman et al. 2001). The pathogen population and soft rot disease incidence and severity were assessed. Populations of *Erwinia* spp. around $10^4$ CFU/ml and occasionally $10^6$ CFU/ml were present in wash water at various sites. The highest disease severity was induced, when tubers were immersed in water containing a concentration of $10^4$ CFU/ml of *Erwinia* spp. or greater. The results suggested that potato wash water should be replaced frequently with clean water to reduce pathogen population and consequently to reduce disease incidence in potato tubers (Wicks et al. 2007).

*Xanthomonas campestris* pv. *musacearum (Xcm)*, a soilborne pathogen, causes the highly destructive Xanthomonas wilt disease of banana in eastern Africa. A selective culture medium was developed for isolation of *Xcm* in soil samples from infested fields. Various antibiotics (29) were tested for their efficiency in inhibiting the saprophytes present in the soil. Cephalexin inhibited most of the saprophytes, whereas cycloheximide inhibited the fungal contaminants. The semiselective medium effective for isolation of *Xcm*, YTSA-CC contained yeast extract (1%), tryptone (1%), sucrose (1%), agar (1.5%), cephalexin (50 mg/l) and cycloheximide (150 mg/l), pH 7.0. The pathogen developed on YTSA medium as yellowish mucoid and circular colonies which could be easily identified. This semiselective medium was useful for isolating *X. campestris* pv. *musacearum* from infected banana plant tissues and soil (Tripathi et al. 2007). In a later investigation, a semiselective medium was developed for isolating *Xcm* from infected banana plants, soil and insect vectors. The cellobiose-cephalexin agar (CCA) medium was evaluated for its efficacy using 21 bacterial isolates and or plating efficiency for *Xcm*. The plating efficiency of *Xcm* on CCA medium was lower, when compared with nonselective YPGA medium, but its selectivity was significantly higher, averaging 60 and 82% for isolation from banana fruits and soil samples, respectively. The CCA medium was more efficient in recovering *Xcm* from insect vectors with selectivity of 48–75%, compared with 8–17% on YPGA medium. Of the 33 suspected *Xcm* strains, 29 strains could be isolated from plants, soil and insects using CCA medium and these strains proved to be pathogenic to banana plants, indicating the reliability of CCA medium for specific isolation and detection of *Xcm* in samples from suspected banana fields (Mwangi et al. 2007).

### 4.1.1.3 Use of Bacteriophages

Bacteriophages are bacterial viruses capable of infecting plant bacterial pathogens under natural conditions with potential for use as biological control agents (BCAs) for the management of bacterial diseases of crops. Bacteriophages show specificity to certain bacterial species/strains and hence, they have been employed for phage typing for identification and detection of bacteria in plants, soil and irrigation water. *Xanthomonas oryzae* pv. *oryzae (Xoo)* is susceptible to several tadpole-shaped bacteriophages (with polyhedral head and a tail) and filamentous phages Xf (Kuo et al. 1967). By employing OP1 phage, the possibility of detecting and quantifying *Xoo* was indicated by Wakimoto (1957). *Xoo* isolates could be divided into 15 lysotypes, based on their sensitivity to phages. The lysotypes differ in formation of plaques (bacteria-free areas in the bacterial colonies developing on nutrient medium, due to lysis of bacterial cells induced by phage), indicating the level of susceptibility to the phage employed (Wakimoto 1960). Sensitivity of *Xoo* strains to phages did not show any relationship with virulence or immunological properties of the pathogen strains (Ou 1985). The specificity of phages to *R. solanacearum*, causal agent of bacterial wilt (BW) diseases of several crops, has been useful for detection and quantification of the pathogen in plant and soil samples. *R. solanacearum* is capable of surviving in the soil for several years, during the absence of crop hosts. Four phages ØRSL, ØRSA, ØRDM and ØRSS, capable of infecting *R. solanacearum* were isolated. These phages had relatively wide host ranges and induced large, clear plaques in the susceptible bacteria grown on culture media. The phage ØRSA1 could infect all 15 strains of *R. solanacearum* of different races or biovars (bvs). Three host strains contained ØRSA1-related sequences in the genomic DNAs, suggesting a lysogenic cycle of ØRSA1. The results indicated the possibility of employing bacteriophages for detection of bacterial pathogens and management of diseases caused by them in different crops (Yamada et al. 2007).

Bacteriophages, a component of water microbiota form a constant source of mortality of bacterial populations in water systems. A strong effect of bacteriophages on the survival of *R. solanacearum* in agricultural drainage water microcosms has been reported (van Elsas et al. 2001). Lytic phages of *R. solanacearum* were detected and isolated from water samples. Initial concentration of the lytic phages in the river water samples were estimated between $10^2$ and $>10^3$ lytic virus particles/ml. Phage enumeration on YPGA medium inoculated with *R. solanacearum* yielded plaques which were visible after 36–48 h of incubation at 29°C and continued expanding on plates up to 72 h. Characterization of lytic activity by selected river water phages against *R. solanacearum* strains IVIA-1602.1 and IPO-1609 in modified Wilbrink broth (MWB) confirmed phage activity at 29°C, but not at 35°C. The time course of the interaction between *R. solanacearum* and one selected phage strain was monitored at 24°C and 14°C in sterile river water for 24 and 48 h, respectively. At 24°C the initial population [$(2 \times 10^2$–$2 \times 10^6$ plaque forming unit (PFU)/ml)], bacterial populations decreased from $10^6$ CFU/ml to values ranging from $10^4$ to $10^1$ CFU/ml within 12 h. The subsequent increase in bacterial population was due to development of phage-resistant variants. Phage populations increased from $10^3$ to about $10^4$ PFU/ml. In sterile water (control), *R. solanacearum* and phage populations remained at levels similar to the initial population levels (Álvarez et al. 2007).

### 4.1.1.4 Biochemical and Physical Methods

Based on the patterns of utilization of over 100 carbon sources by the target pathogen, a technique was developed by API (API System, Monatiew-Vercelu, France). This system depending on the visualization of growth of the target pathogen on a set of selected carbon sources may require about one week for providing the results. API 50CH and APIZYM systems were employed to characterize 53 strains of *Clavibacter michiganensis* subsp. *sepedonicus (Cms)*, causing ring rot disease of potato tubers from different geographic locations and several reference strains of the same and different species, including other potato pathogens. Strains of *Cms* displayed a high level of homogeneity, both in carbohydrate utilization and enzymatic activity. It was possible to differentiate strains of *Cms*, by employing API 50CH and API ZYM systems from other taxa tested. These systems showed potential for use in characterization of strains of *C. michiganensis* subsp. *sepedonicus*. The results obtained with API systems agreed with current taxonomic classification of *C. michiganensis*, clearly separating *sepedonicus* from other subspecies belonging to this species (Palomo et al. 2006). Biochemical and physiological tests involving the determination of different enzymatic activities or fatty acid analysis are time-consuming and labor-intensive. These tests are, therefore, not useful for large scale application.

A two-dimensional polyacrylamide gel electrophoresis (2-D PAGE) analysis of acid ribosome-enriched proteins was employed to detect and differentiate *Erwinia* spp. and soft rot-causing strains. Electrophoretic profiles of acid ribosome-enriched proteins of selected strains of *Erwinia carotovora* subsp. *atroseptica (Eca)*, *E. carotovora* subsp. *carotovora (Ecc)* and *E. chrysanthemi* were compared. Seven major clusters were identified. Soft-rotting bacterial strains could be readily identified and differentiated by the 2-D PAGE procedure. *Ecc* and *Eca* produced protein profiles that were consistent and distinct enough to be separated into two species. Other unidentified soft-rotting isolates also could be reliably identified (Moline 1985). The pulsed-field gel electrophoresis (PFGE) technique is capable of resolving extremely large DNA, raising the upper size limit of DNA separation from 30–50 kb to well over 10,000 kb. Geranium plants in north Florida, United States showed symptoms of infection by an exotic strain of *R. solanacearum* biovar 1. The PFGE technique was used to identify the isolates of *R. solanacearum* from irrigation ponds and aquatic weeds, the potential hosts of the pathogen between 2002 and 2003. PFGE analysis revealed different haplotypes upon comparison of the collected pond strains and Floridian strains. Based on PFGE polymorphism, endoglucanase gene (*egl*) sequencing and phylogenetic analysis, the Caribbean strains of *R. solanacearum* were found to be identical to the strain isolated from infected geranium plants. The presence of *R. solanacearum* in irrigation ponds and associated weeds was monitored. *R. solanacearum* was detected in surface-disinfested common aquatic weeds growing in the irrigation ponds, including *Hydrocotyle ranunculoides* (dollar weed) and *Polygonum pennsylvaticum* (smart weed). Both weed species were latently infected and did not exhibit wilt symptoms. Under greenhouse conditions two different *Hydrocotyle* spp. were inoculated with the pathogen which was recovered from plants showing symptoms at 14 days postinoculation. A positive correlation was observed between ambient temperature and population of R. solanacearum in irrigation water (Hong et al. 2008).

Plant pathogenic bacteria may be detected and differentiated by employing a thin layer chromatography (TLC) technique, using chloroform-methanol-NaCl or 2- propanol solvent system. Comparison of TLC profiles of aminolipids extracted from 315 plant pathogenic and nonpathogenic bacterial species showed that TLC profiles were characteristic at the genus or species level and the results were highly reproducible. For most of the gram-negative bacterial species, the upper most spot had an $R_f$ of 0.70 on the chromatogram and this spot was absent on the chromatograms of gram-positive bacteria like *Clavibacter michiganensis*. The profiles of 96 isolates of *R. solanacearum* were identical, irrespective of their sources. In the case of *Erwinia carotovora*, an intensive bench mark spot appeared at $R_f$ of 0.64 which was absent on chromatograms of pathovars of *E. chrysanthemi*. The profiles of *Xoo* showed significant variations, reflecting the genetic diversity of isolates from different sources. The TLC profiles of aminolipids may be prepared easily and used for presumptive identification and differentiation of bacterial pathogens (Matsuyama et al. 2009).

## 4.1.2 Isolation-Independent Methods

### 4.1.2.1 Immunoassays

Bacterial plant pathogens are less complex, compared with fungal pathogens which produce different kinds of spore forms at different stages of life cycle. These fungal spores have different surface proteins that are antigenically distinct. On the other hand, bacteria may have various immunodeterminants of capsular polysaccharide antigens, lipopolysaccharides (LPS) O and K antigens and murine lipoproteins. The membrane proteins of *R. solanacearum* were found to be useful for detection and identification of the pathogen (Schaad et al. 1978). Both polyclonal antibodies (PAbs) and monoclonal antibodies (MAbs) have been produced against bacterial plant pathogens. PAbs have been reported to provide inconsistent results. In addition, when PAbs are employed, cross-reactions with unrelated bacterial species are commonly observed. The development of MAbs has been useful to overcoming some problems associated with the use of PAbs. The MAbs secreted by hybridoma lines exhibit defined specificity to a single epitope of the target bacterial species/strains and they provide reliable and reproducible results, although generation of MAbs is expensive and requires high technical expertise. After the characterization of MAbs, panels of MAbs may be combined to form a reactive group that can be employed to detect and identify up to genus, species, subspecies and pathovars of bacterial pathogens (Álvarez et al. 1996). The MAbs produced against certain cellular or extracellular fractions may be used to differentiate serogroups of bacterial pathogens, as in the case of *Eca* (later named as *Pectobacterium carotovorum* subsp. *atrosepticum*) or *E. chrysanthemi* (De Boer and Mc Naughton 1987). The MAbs generated against the European potato strains of *E. chrysanthemi* reacted with a fimbrial antigen present in all except two strains of *E. chrysanthemi* isolated from potato, demonstrating the effectiveness of MAbs in detecting the pathogen in potato, as well as in other host plant species (Singh et al. 2000). Gram-negative bacterial plant pathogens produce LPSs that constitute structural components of the outer membrane of bacterial cells. Antisera specific to LPSs were efficient in detecting *Eca* (De Boer and Mc Naughton 1987) and *R. solanacearum* (Mc Garvey et al. 1999). MAbs provide greater specificity and sensitivity to immunoassays. Commercial test kits for detection and identification of plant pathogenic bacterial pathogens are available for application under different conditions. Soilborne plant pathogenic bacterial pathogens have been detected and differentiated pathogens by employing different immunoassays that vary in their sensitivity.

### 4.1.2.1.1 Immunofluorescence Test

Immunofluorescence colony (IFC)-staining testing which combines bacterial colony growth and immunoassay, allows sensitive and quantitative detection of several phytopathogenic bacteria in complex backgrounds. Both MAbs and PAbs generated against *R. solanacearum (Rs)* showed cross-reaction with taxonomically related bacteria such as *R. szygii*, causal agent of banana blood disease as well as more distantly related soil saprophytes. The results indicated the need for careful interpretation, since populations of *R. solanacearum* may be low along with high numbers of nontarget soil bacteria (Griep et al. 1998). A specific and sensitive quantitative technique for detection of *R. solanacearum* biovar 2 (race 3) in soil based on IFC procedure, was developed. The results of

IFC were confirmed by PCR amplification or dilution plating on semiselective SMSA medium. The addition of sucrose and antibiotic cycloheximide and crystal violet to the nonselective trypticase soybroth agar resulted in increased colony size and staining intensity of *R. solanacearum* in IFC test. It was possible to detect consistently ca. 100 CFU/g of soil, a detection threshold being similar, but less laborious, whereas the tomato leaf bioassay had a detection limit of $10^4$ to $10^5$ CFU/g of soil (van der Wolf et al. 2000).

### 4.1.2.1.2 Enzyme-Linked Immunosorbent Assay

The specificity of monoclonal antibodies (MAbs) generated against several bacterial pathogens have been screened initially using ELISA tests. In quality assurance programs, the ELISA procedure has been demonstrated in several countries to be useful for testing large numbers of samples. The ELISA test is amenable for automation and reproducible results could be obtained in large–scale testing programs. Commercial detection kits for the detection of bacterial pathogens are available. *E. chrysanthemi*, a gram-negative bacterial pathogen, causes rot diseases in agricultural and ornamental crops, including potato and carnation. *E. chrysanthemi* persists in and spreads through surface water sources and is primarily a tuber disease, inducing decay of seed and progeny potato tubers. Some strains of *E. chrysanthemi* may be relatively specific and they may be differentiated into BVs, based on physiological and biochemical characteristics (Dickey 1979). The comparative effectiveness of ELISA or PCR was compared. The MAb secreted by murine hybridoma cell line 6A6 reacted with a fimbrial antigen and fibrillin protein of *E. chrysanthemi*, as revealed by immunoelectron microscopy and Western blotting, respectively. The MAb 6A6 , typed out as an IgG2b reacted with all strains of *E. chrysanthemi* from potato, except two strains, one of which was isolated from potato in Australia and the other was of unknown origin. The MAb also reacted with 20 of the 36 *E. chrysanthemi* strains. Reactivity of strains was unrelated to either biovar or host origin. Sensitivity of ELISA was about $10^7$ CFU/ml, when dilutions of pure cultures of *E. chrysanthemi* and spiked tuber tissue were tested. On the other hand, PCR was more sensitive, with the limit of detection (LOD) being $10^2$ CFU/ml for pure cultures and $10^3$ CFU/ml for spiked tuber tissue. Further, of the 14 stems grown from *E. chrysanthemi*-infected tubers, only four were positive in ELISA, but 12 stem samples were positive in PCR assay. The absence of positive reaction in *Pectobacterium carotovorum* subsp. *atrosepticum*-infected stem and tuber samples, indicated the specificity of the MAb. The negatively stained bacterial cell preparations were examined under electron microscope. Filaments of variable length with a diameter of about 14 nm with typical fimbrial morphology were observed. Selective attachment of gold beads to fimbria-like structures in *E. chrysanthemi* could be seen in preparations treated with 6A6 MAb and antimouse IgG antibodies conjugated to gold beads (see Figure 4.1) (Singh et al. 2000).

*R. solanacearum (Rs)*, the causal agent of BW diseases of several crops, infects peanut plants. Latent infection of peanut by *R. solanacearum* could be detected in the main roots, hypocotyls and stems of infected peanut plants, using the ELISA procedure (Shan et al. 1997). In another study, PAbs were raised against virulent strains of *R. solanacearum* which have cells encapsulated with mucin. The ELISA test using the PAbs detected *R. solanacearum* efficiently at a concentration of 100 cells/ml. The PAbs specifically reacted with tomato isolate of *R. solanacearum*, but not with isolates from pepper (chilli) or aubergine (eggplant) (Rajeshwari et al. 1998).

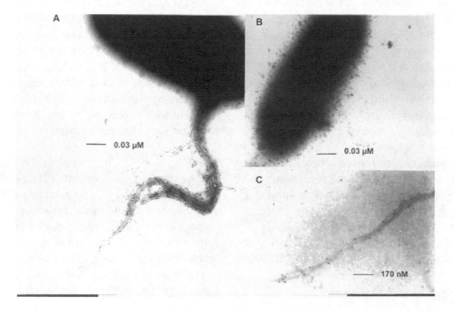

**FIGURE 4.1**   Detection of pathogen-specific fimbrial protein in *E. chrysanthemi*, by electron microscopy using the microbial antibody 6A6 A: whole cell showing attached bundle of fimbriae; B: whole cell showing single attached fimbrial and C: fimbrial with selective attachment of gold beads (under high magnification).

[Courtesy of (Singh et al. (2000) and with kind permission of the American Phytopathological Society, MN, United States]

In order to overcome the problem of cross-reaction with the use of PAbs with other bacterial species, recombinant single-chain antibodies (scFv) were prepared against the LPS of *R. solanacearum* biovar 2, race 3. The scFv antibodies were successfully selected by phage display from a large combinatorial antibody library. Antibodies with improved efficiency compared with PAbs were used for detection of *R. solanacearum*, causal agent of brown rot disease of potato tubers. Only a few cross-reactions with saprophytic bacteria were observed. It was possible to detect as few as $5 \times 10^3$ bacteria in potato tuber extracts using the ELISA test (Griep et al. 1998). *R. solanacearum (Rs)* is soilborne as well as tuber-borne in potato and it can survive in soil for a long time in the absence of potato plants or other susceptible host plant species. It is, therefore, essential to estimate the population of *R. solanacearum* in the field soil, in order to reduce the incidence of brown rot disease by avoiding highly infested fields. Double antibody sandwich (DAS)-ELISA format was applied after selective enrichment for the detection of R. solanacearum in the soil. The pathogen was efficiently detected by post enrichment DAS-ELISA test, even when the population levels of *R. solanacearum* were low in inoculated soil. The DAS-ELISA format had a satisfactory level of sensitivity and no cross-reaction was observed with soil saprophytes. Likewise, reliable results in tests were obtained using extracts of naturally infested soils and an enrichment step for concentrating pathogen populations (Priou et al. 1999). In a later investigation, the bacterial suspension obtained from the soil sample was incubated in a modified selective medium for enrichment of pathogen population prior to performing the indirect ELISA format and the test was effective in detecting *R. solanacearum* at a concentration of as few as $10^4$ CFU/g of soil (Pradhanang et al. 2000).

Sensitive and specific routine detection of *R. solanacearum* in symptomless potato tubers is essential to prevent the movement of the pathogen from tubers to the soil in new locations. The optimized protocol included an initial enrichment step consisting of shaking the samples in MWB for 72 h at 29°C and employing a specific MAb 8B-IVIA which reacted with 168 typical strains of *R. solanacearum* and did not recognize 174 other pathogenic or unidentified bacteria isolated from potato. The enrichment step enabled specific detection by DAS-indirect (DASI)-ELISA of 1 to 10 CFU of *R. solanacearum*/ml of potato extract. Analysis of three commercial potato lots by DASI-ELISA format provided results similar to that of conventional methods. The protocol combined the high sensitivity of post enrichment ELISA with the specificity and high affinity for *R. solanacearum* by Mab 8B-IVIA. This procedure had the potential for use in epidemiological investigations and certification programs (Caruso et al. 2002). A new semiselective broth containing potato tuber infusion which differentially supported the development of *R. Solanacearum (Rs)* was formulated at the International Potato Center (CIP), Peru. Potato isolates of *R. solanacearum* (273) belonging to five different bvs originating from 33 countries were incubated on this broth prior to detection of the pathogen by DAS-ELISA format. Presence of *R. solanacearum* in the soil samples was detected by employing specific antibodies at low levels of the pathogen population, after incubation of soil suspensions for 48 h at 30°C in the semiselective broth. Detection thresholds for biovars bv1 and bv2 were 20 and 200 CFU/g of inoculated soil, respectively (Priou et al. 2006).

The detection of infection by *R. solanacearum* in asymptomatic potato stem before harvest would be useful to avoid the selection of seed tubers from potato plants with latent infection. The possibility of detecting BW latent infection in stem pieces at about three weeks before harvest was explored by testing plant materials from 57 fields of the Andean Highlands of Peru. DAS-ELISA and indirect ELISA formats were applied after enrichment of plant extracts in semiselective broth. Optimum sample sizes of stems and tubers were evaluated for 37 potato crops showing between 0 and 0.1% BW incidence, using a binomial distribution model to calculate the detection probabilities. The results of two ELISA formats were in agreement in all samples (100%). However, detection probabilities were higher using the DAS-ELISA format for all plant parts tested. Probabilities of detection of *R. solanacearum* in tubers were greater than in stems. The high percentage (63%) of symptomless infection of tubers by BW confirmed that infected tubers coming from infected fields were responsible for contamination of seed potato-producing areas in the Highlands of Peru. Detection of BW infection in areas where BW was not reported earlier was a consequence of the flow of contaminated seed from endemic areas. Detection of BW infection in symptomless plants at 20 days before harvest, using post enrichment DAS-ELISA was found to be a reliable and user-friendly procedure that had potential for use by national plant protection services and seed programs operating in various countries (Priou et al. 2010). Levels of contamination of certified visually healthy potato seed tubers imported from Europe to Israel by *Dickeya* spp. was assessed using ELISA and PCR assay. The seed tubers of 277 lots were grown in commercial potato fields which were inspected twice in a season. Stem samples were tested for the presence of *Dickeya* spp. The results of PCR assay and ELISA test from seed lot testing correlated with symptom expression in 74% and 83.8% of seed lots, respectively. Maximum disease incidence as well as the number of cultivars expressing disease symptoms increased over years, indicating an increase in the prevalence of the disease. Latent infection of seed tubers by *Dickeya* spp. was maximum in seed lots imported from the Netherlands, followed by Germany and France. None of the seed lots from Scotland showed latent infection by *Dickeya* spp. (Tsror et al. 2012).

*X. campestris* pv. *musacearum (Xcm)*, causal agent of a destructive banana Xanthomonas wilt disease might be transmitted by a broad range of mechanisms. A specific and sensitive method of detection of the pathogen was considered essential for restricting the incidence and spread of the disease. A polyclonal antibody (PAb) was raised against *Xcm* and deployed in a lateral flow device (LFD) format to allow rapid in-field detection of the pathogen. Pure cultures of *Xcm* were used to immunize the rabbit and the IgG antibodies purified from the antiserum were employed for testing by ELISA and LFD techniques. The PAb effectively detected all strains of *Xcm*, representing isolates from seven countries and the known genetic diversity of *Xcm*. But the PAb cross-reacted with closely related *X. axonopodis* pv. *vasculorum*, primarily a sugarcane pathogen which has not been recorded in banana. LFD assay detected *Xcm* in both naturally and artificially inoculated banana plants with the LOD of $10^5$ cells/ml. The results indicated that the LFD method might be used as a first line screening tool to detect *Xcm* in the field. LFD testing is cost-effective and required no equipment and it could be performed by nontechnical personnel, proving to be an efficient tool for the management of banana *Xanthomonas* wilt disease of banana (Hodgetts et al. 2015).

Infection of potato plants by *R. solanacearum*, causal agent of BW disease, occurred frequently after irrigation with contaminated water. In aquatic ecosystems, outbreaks of BW disease were related to watercourses in which infected *Solanum dulcamara* plants were present. *R. solanacearum* cell survival in water for variable periods depended on the temperature and inoculum density. The presence of *R. solanacearum* bv 2 in the watercourses of European countries was found to increase. The population density of *R. solanacearum* was assessed at different locations on Spanish river over a period of three years. The pathogen was recovered at low densities (10 to 80 CFU/ml) by direct plating on modified SMSA agar from water samples at 14°C or higher, but isolation of the pathogen was usually unsuccessful at temperatures below 9°C. The abundance of *R. solanacearum* was monitored in winter, using two liquid selective media for enrichment (at 29°C and 35°C) and compared them by using spiked river water samples. MWB was more efficient than modified SMSA broth for detection of *R. solanacearum* DAS-indirect ELISA format. Enrichment in MWB at both temperatures allowed recovery of *R. solanacearum* cells that were nonculturable on solid media up to 25 days after their entry into the viable, but nonculturable state. This technique was effective in detecting *R. solanacearum* in water samples taken during cold months of 2001 and 2002. The enrichment protocol was combined with DASI-ELISA and validated by Co-PCR to detect both naturally and artificially starved and cold-stressed cells in water which was still infective. The results showed that the effects of temperature varied on population and culturability of *R. solanacearum* cells on solid media and their survival at low temperatures (Caruso et al. 2005).

### 4.1.2.1.3 Immunolabeling and Electron Microscopy

Localization and distribution of various proteins of pathogen/host plant origin in different pathosystems have been investigated by immunolabeling and electron microscopy. Enzyme-linked antibodies are applied after embedding and thin sectioning of plant tissues or they may be allowed to diffuse inside fixed cells and to interact with antigenic sites prior to sectioning. The MAbs used for detection of *Xoo* could be characterized by employing immunosorbent electron microscopy (ISEM) and immunofluorescence (IF) techniques (Benedict et al. 1989). Some of the pectolytic *Erwinia* spp. possess nonflagellar appendages known as fimbriae that are

considered to play a role in the adhesion of bacterial cells to the plant surface/substrates. By employing fimbriae-specific MAbs, the presence of fimbriae in bundles in *E. chrysanthemi* was revealed by observations under electron microscope. A specific immunogold labeled MAb targeted a fimbrial epitope. All European potato isolates of *E. chrysanthemi* could be detected by using MAbs specific to fimbrial antigens (Singh et al. 2000).

### 4.1.2.1.4 Immunomagnetic Separation Technique

The immunomagnetic separation (IMS)/fishing procedure improves the recovery ratio between the target and nontarget bacteria present together in samples. The plant samples, especially seeds and soils are contaminated with many saprophytic bacteria which often predominate, making the isolation of bacterial pathogen difficult. The IMS procedure involves selective separation of target pathogen species and cultivation of the separated pathogen in a selective medium that supports rapid prolixferation of the pathogen. The IMS method was applied for the detection of *Eca*, causing soft rot of potato tubers. *Eca* was separated from the potato peel extract by employing a combination of advanced magnetic (AM)-protein A particles-antirabbit IgG particles and the selective medium crystal violet pectate medium supplemented with 100 μg/ml of streptomycin. A streptomycin-tolerant strain of *Eca* was included to indicate the adverse effect of streptomycin on *Eca* strain. The pathogen could be enumerated in potato extract consistently. The detection limit of the test was 100 *Eca* cells/ml (van der Wolf et al. 1996).

### 4.1.2.1.5 Lateral Flow Immunoassay

Lateral flow immunoassay (LFIA), also known as immune chromatographic assay, combines the lateral flow principles that allow rapid analysis under out-of-laboratory conditions and use of antibodies, making the assay sensitive and specific for the detection of target pathogen. All reactants are applied on the test strip before the assay and contact between the test strip and the sample initiates specific interactions that lead to recognition of the positive result visually. The presence or absence of coloration on certain areas of the strip indicates the positive or negative result of the assay. A LFIA was developed for rapid detection of *Cms*, incitant of potato ring rot disease. Multimembrane composites (test stripes) containing PAbs against *Cms* and gold nanoparticles-antibody conjugates were employed for analysis. The test strips were found to be suitable for the analysis of potato tuber tissues and leaf extracts within ten min. The LOD of *Cms* was $4 \times 10^5$ cells/ml. No cross-reactivity was observed with strains of *C. michiganensis* subsp. *michiganensis*, *Pectobacterium carotovorum* susbsp. *carotovorum* and saprophytes associated with healthy potato plants. The results of LFIA, ELISA and PCR techniques were compared. The results of LFIA could be confirmed in 96.2% of tests using the ELISA and PCR procedures, indicating high correlations among the techniques employed (Safenkova et al. 2014).

*Pectobacterium atrosepticum* causes blackleg and soft rot disease in potato plants and tubers. Early detection of *P. atrosepticum* is vital for production of healthy seed tuber stocks.

A method for rapid detection of the pathogen was developed, based on the LFIA technique. Rabbit PAbs specific to various strains of *P. atrosepticum* were generated. Conjugates of the PAbs specific to the pathogen with gold nanoparticles (GNPs, 20nm diameter) were synthesized. Optimal concentrations of antibodies and conjugates deposited on membranes of the test strips were used. The conjugate concentrations were determined using OD at 520 nm, taking into consideration that immobilization of the antibodies led to an approximately 10% decrease in GNP adsorption. The maximum color intensity in the test and control zones was obtained by depositing antibodies from a solution at a concentration of 10 mg/ml (at a deposition density of 0.2 µl/mm). Antirabbit antibodies for control zones were applied to the membrane from a solution with a concentration of 0.5 mg/ml to give the maximum color intensity of the control line. The OD of the IgG-GNP conjugate solution used for application in LFIA was 10.0. The intensity of color of the test lines increased gradually with increasing analyte concentrations in the sample (up to $10^7$ bacterial cells/ml). *P. atrosepticum* could be reliably detected by LFIA at concentrations of $2 \times 10^4$ cells/ml for the strain Pa204, $3.32 \times 10^4$ for strain Pa393 cells/ml, and $2 \times 10^3$ cells /ml for strain Pa18077. The LOD of LFIA was similar to that of ELISA tests. LFIA results were confirmed by ELISA (100%) and by PCR [(87.5%) for positive samples and 95.5% for negative samples)]. Diagnosis of potato blackleg and soft rot by LFIA could be undertaken without any equipment by personnel without training. Further, LFIA is cost-effective and it could be used under field conditions to monitor incidence and spread of *P. atrosepticum* (Safenkova et al. 2015).

### 4.1.2.2 Nucleic Acid-Based Techniques

Generally, the sensitivity of detection of bacterial pathogens by nucleic acid-based techniques has been reported to be at a higher level, when compared to immunoassays and isolation-based methods. Closely related bacterial pathogens share greater nucleotide similarity than those that are distantly related. Techniques based on nucleic acid hybridization and PCR assay or its variants have been applied for detection, identification and quantification of bacterial pathogens in planta, soil and water bodies supplying irrigation water. Single and/ or multiple infections by bacterial pathogens have been detected efficiently using nucleic acid-based techniques. The presence of inhibitors of PCR amplification in plant tissues and soils and the inability to differentiate living and dead bacterial cells are the major limitations of nucleic acid-based methods.

#### 4.1.2.2.1 Nucleic Acid Hybridization Methods

The use of specific DNA probes has been shown to be effective for the detection, identification and quantification of bacterial plant pathogens belonging to different species, subspecies, pathovars or strains. The diagnostic probes may exhibit specificity at the genus, species, pathovar or race level. A probe that can be employed for taxonomic comparisons is designed by identifying a DNA fragment that is present only in the target bacterial species, but not in other closely related species. The DNA sequences of 16S ribosomal RNA of different bacterial species have been compared for selecting probes at genus level. A DNA probe for comparing partial sequences of 16S rRNA from 52 strains of bacteria, including *Xoo* was designed. The rRNA molecule is present in large numbers (>10,000 copies/bacterial cells) in actively multiplying cells of *Xoo*. The sensitivity of the probe derived from unique 16S rRNA sequences could be greatly improved, since a large number of copies/bacterial cells was present. A DNA sequence based on the 16S rRNA which hybridized only with pathovars of *Xoo* could be identified (De Parasis and Roth 1990). *R. solanacearum* race 3 biovar 2 could be detected in potato tissues by using a fluorescent *in situ* hybridization (FISH) procedure. The probe RSLOB had the potential for specific detection of *R. solanacearum* in pure cultures as well as in potato tissues exhibiting a strong fluorescence signal (Wullings et al. 1998). Detection of *Cms*, the causal agent of potato ring rot disease was possible by using radiolabeled probes obtained from unique three single copy DNA fragments designated Cms50, Cms72 and Cms85 isolated from CS3 strains of *Cms* by subtraction hybridization. These probes specifically hybridized to all North American strains of the pathogen tested by Southern hybridization. All strains including plasmidless and mucoid strains could be detected reliably (Mills et al. 1997).

#### 4.1.2.2.2 Polymerase Chain Reaction-Based Assays

If the target bacterial pathogen is contaminated with saprophytic bacteria, it may be difficult to detect the bacterial pathogens using hybridization methods. Under such conditions, PCR-based assay may be employed for reliable detection of the target bacterial pathogen(s). The sensitivity of detection may also be enhanced by different PCR formats. The PCR-based assays, using pathogen-specific primers, are effective for the detection, identification and quantification of a large number of bacterial pathogens infecting a wide range of crop plants. PCR assay has been employed for the detection of *A. tumefaciens* in infected plants, pure cultures and soil. PCR assay was applied to detect 34 strains of *A. tumefaciens* infecting *Vitis* spp. The results of PCR assay were corroborated by those of pathogenicity tests and a DNA slot blot hybridization method (Dong et al. 1992). By employing two primers designed from the sequences of *virD2* and *ipt* genes, a wide spectrum of *A. tumefaciens* strains could be detected. The T-DNA-borne cytokinin synthesis gene was detected using primers corresponding to sequences of *ipt* gene in *A. tumefaciens*, but not in *A. rhizogenes* (Haas et al. 1995). The *tms2* gene in T-DNA coding for indole acetamide amidohydrolase was shown to be essential for the pathogenicity of *A. tumefaciens*. Primers flanking a 220-bp fragment of one of the conserved regions of *tms2* gene were designed for PCR amplification which revealed the presence of T-DNA in infected plants and in infested soils (Sachadyn and Kur 1997). The PCR assay efficiently detected *A. tumefaciens* in more than 200 samples, including naturally infected almond, apricot, chrysanthemum, grapevine, peach, raspberry, rose, tobacco and tomato. Detection of *A. tumefaciens* by PCR assay was more effective in crown and root tumors than in aerial tumors (Cubero et al. 1999). In a later study, use of an internal control (IC) along with a set of validated primers was shown to be effective in overcoming the adverse effects of PCR inhibitors. An IC was generated from a fragment of 172-bp amplified from *A. tumefaciens* strain 58. PCR assay was employed to detect *A. tumefaciens* in tumor samples of naturally infected plants after enrichment in selective liquid medium. The PCR products from the IC and the target sequence were obtained from most of the samples analyzed. Coamplification of the IC and target sequence from *A. tumefaciens* ensured amplification of at least one product in every PCR reaction, even in the uninfected plant material. The detection limit of PCR procedure was $10^2$–$10^3$ CFU/ml of *A. tumefaciens* (Cubero et al. 2002).

*Eucalyptus occidentalis* plants were infected by crown gall disease in Algeria and the pathogen isolated from the galls was identified, based on cultural and biochemical characteristics as *A. tumefaciens* biovar 1. PCR assay analysis of the DNA extracted from virulent strains yielded an amplification product of 247-bp located within the virulence region of nopaline-type Ti (tumor-inducing principle) plasmid. The opine, nopaline was detected in the tumors induced on test plants, but not in Eucalyptus plants. Nopaline was degraded by 20 pathogenic isolates that were sensitive to agrocine 84, indicating the presence of a nopaline-type *pTi* in these strains. The chromosomal region encoding the 16S rRNA was analyzed in a subpopulation of the pathogenic agrobacterial isolates. The strains causing crown galls in *E. occidentalis* belonged to the ribogroup of the reference strain B6. Other species of *Eucalyptus,* E. *camaldulensis* and *E. cladocalyx* grown in the same nursery and in the same soil substrate did not develop galls, suggesting that these two species were not susceptible to *A. tumefaciens* (Krimi et al. 2006). The PCR assay, using the primers based on sequences of intercistronic region between *virB* and *virG* of the *vir* region of pTi (tumor-inducing plasmid) was highly sensitive for the detection of *A. tumefaciens*. The population of *A. tumefaciens* at a concentration of $10^5$ CFU/g of fresh tumor tissue could be detected by the PCR assay. The PCR assay could detect *A. tumefaciens* in root and stem tissues of symptomless plants also. In addition, the PCR procedure allowed differentiation of pathogenic forms from the nonpathogenic forms, since primers specific for the *vir* region of Ti-plasmid of *A. tumefaciens* were employed (Puopolo et al. 2007). *Agrobacterium* spp. were isolated from 23 soil samples and assigned to the genomospecies G4, G7 and G9 by specific primers. The majority of samples yielded strains belonging to more than one genomospecies. G4 and G7 members were present predominantly. Partial *recA* gene sequencing revealed new alleles and a high allelic diversity at both local and country (Tunisia) scales. BOX-PCR fingerprinting of non-tumorigenic strains from dominant alleles did not show a clear correlation with geographic origin/soil plantation type, or a clear clonal spread of *Agrobacterium* strains in soils. Ti-plasmid-containing strains were only recovered from soils of fields with evidence of crown gall disease occurrence and they were exclusively placed in genomospecies

G4. Tumorigenic strains isolated from soils with galled grapevines were distinct from tumorigenic strains isolated from soils with galled stone fruit trees, based on *recA* gene sequences, Ti-plasmid type, and L-tartarate utilization (Bouri et al. 2016).

The PCR assay was applied for confirmation of the results of detection of *R. solanacearum* by using the IFC staining procedure. PCR amplification based on primer pair D2/B, allowed specific and rapid confirmation of *R. solanacearum* biovar 2 taken from IFC-positive colonies. The primers D2/B reacted with all strains of *R. solanacearum* biovar 2. The success rates of rapid verification of IFC-positive detection by PCR assay were 86% and 96% with spiked and naturally contaminated soil samples, respectively (van der Wolf et al. 2000). The primers PS96H and PS96I were able to detect 28 strains of *R. solanacearum* specifically. Sequence analysis revealed the presence of six different sequence groups in the region between these primers (Hartung et al. 1998). By employing the PCR assay for amplifying pathogen-specific primers followed by electrophoretic analysis of amplicons enabled the interception of distinct pathotype of *R. solanacearum* race 1, biovar 1 infecting pothos (*Epipremnum aureum*) at the entry point (Norman and Yuen 1998). A DNA fragment of 0.7-kb was amplified by random amplified polymorphic DNA (RAPD) protocol from the total DNA extract of *R. solanacearum* and it was cloned and evaluated as a specific DNA probe. This DNA fragment hybridized to a 2.7-kb fragment in the *Eco*RI-digested total DNA of *R. solanacearum*. Specific oligonucleotide primers were designed based on the sequences of the 2.7-kb fragment for PCR amplification. The primers specifically amplified the expected product from all strains of *R. solancearum* and no amplification occurred from the genomic DNA of other plant pathogenic bacterial strains tested. The LOD of *R. solanacearum* by the PCR assay was about 20 bacterial cells, providing sensitive detection of the pathogen in soil samples (Young and Chichung 2000).

The development of DNA probes specific for *R. solanacearum* is complicated by the genetic diversity of this species complex. A 0.7-kb fragment, amplified by RAPD procedure from total pathogen DNA was cloned for its evaluation as a specific DNA probe. This 0.7-kb DNA fragment hybridized to a 2.7-kb *Eco*RI fragment in the *Eco*RI-digested *R. solanacearum* total DNA. This 2.7-kb fragment hybridized with genomic DNAs of all *R. solanacearum* strains tested. By constructing specific primers BP4-R and BP4-L based on the sequence of 2.7-kb fragment, *R. solanacearum* could be detected at a concentration between 10 and 50 cells in suspensions, without the need for prior enrichment or cultivation. The primers amplified 1.1-kb PCR product from all *R. solanacearum* strains tested, but not from any other plant pathogenic bacterial strains tested. The sensitivity of PCR was about 20 cells of *R. solanacearum*. In order to detect *R. solanacearum* in soil, the DNA from the soil samples was extracted by a simple procedure developed in this investigation. The primers BP4-R and BP4-L were able to detect *R. solanacearum* in soil extracts. The sensitivity of detection of *R. solanacearum* in soil samples was $2 \times 10^3$ cells (Lee and

Wang 2000). Nine strains of *R. solanacearum* isolated from infected eggplant (brinjal), pepper (chilli) and tomato were analyzed using the RAPD technique. All ten primers tested showed high polymorphism in different isolates of *R. solanacearum*. The number of bands ranged from 0 to 10 and amplicon size varied from 0.1- to 5.0-kb. DNA amplification with different primers indicated a high level of genetic diversity among populations of *R. solanacearum* belonging to bvs 3 and 3A. None of the primers generated bands specific to any of the two bvs tested (Deepa et al. 2003).

Isolates of *R. solanacearum*, causal agent of pepper (chilli) BW disease, were detected and identified by employing a species-specific PCR format. PCR primers were designed by alignment of *hrpB* gene sequences and sequences specific for *R. solanacearum* at their 3′ ends were selected. The primers were specific for *R. solanacearum* and no amplification occurred when closely related species were tested. The primers RshrpBF and RshrpBR amplified the predicted product of 819-bp from 20 isolates of *R. solanacearum* infecting pepper. Indian isolates showed homology to standard reference isolates from other countries (Chandrasekhar and Umesha 2015). *R. solanacearum* race 3 biovar 2 (R3bv2) subgroup is a high-concern quarantine pathogen, whereas the related sequevar 7 group is endemic in the southern United States. In order to prevent introduction of R3bv2 strain in asymptomatic geranium cuttings, a sensitive and rapid method had to be developed. The sensitivity, cost and technical complexity of available detection methods were compared for detecting and differentiating R3bv2 and sequevar 7 strains of *R. solanacearum* in geranium, tomato and potato tissues in the laboratory and field samples of tomato. The sensitivity of the PCR-based methods was significantly improved by using kapa3G plant, a polymerase with enhanced performance in the presence of inhibitors. R3bv2 cells were killed within 60 min of application to Whatman FTA(R) nucleic acid binding cards, suggesting that samples on FTA cards could be safely transported for diagnosis. Overall, the culture enrichment followed by dilution plating was the most sensitive detection method (10 CFU/ml), but it was also the most laborious method. Conducting PCR assay from Plat-trapped antigen (PTA) cards was faster, easier and sensitive enough to detect approximately $10^4$ CFU/ml levels similar to those present in latently infected geranium plants (Tran et al. 2016).

The PCR assays may be developed based on primers designed on the sequences of specific genes involved in the process of disease development (pathogenesis). A segment of *necl* gene governing necrosis of tissues infected by *Streptomyces scabies*, causing potato scab disease of tubers, is involved in the pathogenicity. Amplification of the segment of *necl* gene using PCR assay was useful for reliable and sensitive detection of *S. scabies* in potato tissues (Bukhalid et al. 1998; Joshi 2007). A single pair of primers was designed, based on sequences of the spacer region between 16S and 23S rRNA genes for five different *Clavibacter* subspecies. This primer pair specifically amplified a 215-bp fragment of *C. michiganensis* subsp. *sepedonicus*. This PCR format was more sensitive than ELISA and IF tests in detecting the pathogen in naturally infected potato

tubers (Li and De Boer 1995). Standard PCR assay was performed with primer sets Cms50 and Cms72 obtained through subtractive hybridization for the detection of C. *michiganensis* subsp. *sepedonicus (Cms)*. Both primers amplified *Cms* DNA consistently from bacterial suspensions. Postharvest samples of potato tubers collected from ring rot-infected plants were tested. The primers Cms50 and Cms72 amplified the products of 195- and 164-bp, respectively from tuber tissues. In infected tuber tissue processed by macerating, *Cms* was detected by PCR with a sensitivity of 85%. Detection sensitivity could be enhanced by hybridizing PCR amplicons with specific digoxigenin (DIG)-labeled DNA probes in an enzyme-linked oligonucleosorbent assay (ELOSA). In naturally infected tuber samples representing three potato cultivars, the diagnostic sensitivity of PCR/ELOSA was 96%, while specificity of detection exceeded 99%. The PCR/ELOSA procedure detected C. *michiganensis* subsp. *sepedonicus* in tuber samples with equal sensitivity, regardless of colony morphology, potato cultivar or primer sets (Baer et al. 2001).

The PCR-based procedure was developed for rapid and cost-effective detection of *S. scabies* and *S. turgidiscabies*, causal agents of potato scab disease. The PCR assay was specific and detected a collection of previously characterized strains of *Streptomyces* isolated from potato scab lesions. The scab lesions (1,245) from potato cvs. Matilda and Sabina grown in two geographic regions of Finland were tested. Freshly harvested or stored potato tubers were incubated at room temperature (18 to 21°C) under humid conditions for a few days. DNA was isolated from the bacterial growth developed on the culture medium and analyzed by PCR assay. Both scab pathogens were detected in the same potato fields, tubers and scab lesions. The relative incidence of *S. scabies* was high in freshly harvested tubers, but was much lower than that of *S. turgidiscabies* after storage. Both pathogens were transmitted via tubers in the potato cultivars Matilda and Sabina after storing at 4°C for a period of 24 weeks (Lehtonen et al. 2004).

The incidence of common scab of potato caused by *S. scabies* may vary from region to region, year to year and field to field, due to pathogen load, environmental conditions and appearance of new strains or species that have acquired different virulence characteristics. The spread of plant pathogenicity determinants in soil streptomycetes by horizontal transfer of a large chromosomal region containing pathogenicity might be possible. Phylogenetically diverse soil streptomycetes may presumably acquire a set of genes for pathogenicity harbored in a pathogenicity island (PAI) to become plant pathogens. Presence of a conserved gene region with characteristics of a PAI has been observed in *S. scabies*, *S. acidiscabies* and *S. turgidiscabies* (Kers et al. 2005). Isolates of a new common scab-causing streptomycete were isolated from scabby potatoes from southeastern Idaho, United States. These isolates were morphologicaly distinct from the type strain of *S. scabies* ATCC49173, producing a reddish-brown substrate mycelium with gray aerial spores and a red-brown diffusible pigment on yeast-malt extract (YME) agar. They produced an inky black diffusible melanoid pigment like the type strain. These isolates were pathogenic to radish and potato and they

**TABLE 4.1**

**Virulence Characteristics of New Isolates of *Streptomyces* Infecting Radish in Relation to the Type Strain *S. scabiei* ATCC49173 (Wanner 2007)**

| Isolate | Initial inoculum density[a] | Scab lesion score[b] | Seedling emergence (%)[c] | Plant survival after 5 weeks (%) |
|---|---|---|---|---|
| ID01-12c | 7.03 | DD[d] | 15 | 0 |
|  | 6.65 | 4.43(D)[e] | 51.8 | 0 |
| ID01-6.2a | 7.47 | DD | 14 | 4 |
|  | 7.06 | DD | 4 | 0 |
| ID03-1A | 7.49 | 4.0 (D) | 95 | 14 |
| ID03-2A | 7.06 | 4.1 (D) | 95 | 36 |
| ID03-3A | 7.0 | 3.5(DD) | 90 | 7 |
| *S. scabiei* type strain ATCC49173 | 6.75[f] | 3.48[f] | 88[f] | 82[f] |
| Range | 6.18–7.27 | 1.78–4.43 | 63–100 | 52–100 |
| Independent experiments | 14 | 14 | 12 | 12 |

[a]  Log CFU of *Streptomyces* isolate or strain/cc of soil

[b]  Severity × area, 0–5 scale; scores are averages of three pots in an independent experiment (9 plants/pot) or for the number of experiments given in parenthesis for type strain

[c]  Percentages-averages (details as in 'b' above)

[d]  All plants dead ( no radish plant for scoring)

[e]  Many plants dead (few radishes available for scoring)

[f]  Average in multiple independent experiments

were consistently more virulent than the type strain on radish as indicated by scab lesion scores, reduced seedling emergence and reduced long-term plant survival (see Table 4.1). The new strain showed hallmarks of the *Streptomyces* PAI and it had genes encoding the synthetase for the pathogenicity determinant thaxtomin A (*txtAB* gene cluster) and for a second pathogenicity factor tomatinase (*tomA*), although it lacked a third gene characteristic of *Streptomyces* PAI, *nec1* gene (Wanner 2007).

Potato soft rot disease is widespread in all potato cultivating countries around the world. *Eca,* infecting potato was detected in stem and tuber tissues, using primers capable of specifically amplifying the fragment of pathogen DNA. The PCR assay proved to be more sensitive than the ELISA test using MAbs in detecting the pathogen (De Boer and Ward 1995). A one-step PCR-based protocol was applied for detection of all five species of *Erwinia* (causing soft rot diseases), including subspecies *atroseptica* and *carotovora* and all pathovars/bvs of *E. chrysanthemi* in micorpropagated potato plants. The primers SR3F1 and SRIcR based on conserved region of the 16S rRNA gene amplified a DNA fragment of 119-bp from all the 65 strains tested, when an enrichment step was used prior to PCR amplification. The sensitivity of the assay was enhanced by about 200 times (Toth et al. 1999). Pathogenic enterobacteria in the genera *Pectobacterium* and *Dickeya* (earlier known as *Erwinia*) were isolated from diseased stems and tubers. PCR assay were applied to establish the identity of the bacteria which were identified as *P. atrosepticum, P. carotovorum* and *Dickeya* spp. In addition, *Dickeya* strains were isolated from river water samples in Finland.

Phylogenetic analysis with 16S–23S rDNA intergenic spacer sequences suggested that *Dickeya* strains could be divided into three groups, two of which were isolated from potato samples. Phylogenetic analysis with 16S rDNA sequences and growth at 39°C suggested that one of the groups corresponded to *D. dianthicola*, a quarantine pathogen in greenhouse cultivation of ornamental plants, whereas two of the groups did not clearly resemble any of the previously characterized *Dickeya* spp. Field trials with strains indicated that *D. dianthicola*-like strains isolated from river water samples caused the highest incidence of rotting and necrosis of potato stems, but some of the *Dickeya* strains isolated from potato samples also induced symptoms. The results showed that although *P. atrosepticum* remained as the major cause of blackleg in Finland, virulent *Dickeya* strains were frequently detected in potato stocks and rivers (Laurila et al. 2008).

Species- and subspecies-specific PCR assays were applied for the detection of *P. atrosepticum, P. carotovorum* subsp. *brasiliensis (Pcb), P. carotovorum* subsp. *carotovorum* in potato stems with blackleg symptoms. Another PCR assay was formulated for the detection of *P. wasabiae*, based on the phytase gene sequence. Identification of isolates from diseased potato stems by biochemical or physiological characterization, PCR and multilocus sequence typing (MLST) largely confirmed the results of PCR assays, using potato stem samples. *P. atrosepticum* was predominantly detected, but it was the only pathogen detected in 52% of infected stem samples. *P. wasabiae* was most frequently found along with *P. atrosepticum* and it was detected exclusively in 13% of stem samples. *P. wasabiae* was shown to be pathogenic to potato by following standard procedure. In addition, *P. wasabiae* was isolated as the sole pathogen from field-grown potato plants produced from diseased potato tubers. Incidence of *Pcb* was low in diseased stem. Canadian strains of *Pcb* differed from Brazilian isolates in diagnostic biochemical tests, but conformed to the subspecies in PCR specificity and typing MLST (De Boer et al. 2012). The bacterial soft rot disease affected cucumber stems and leaves in China, causing serious economic losses between 2014 and 2015. Gummosis on sufaces of leaves, stems, petioles and fruits of cucumber was observed. The bacterial strains (45) isolated from affected cucumber were identified, based on the phenotypic, physiologic, biochemical properties and 16S ribosomal RNA sequence analysis as *P. carotovorum*. Multilocus sequence analysis confirmed that the pathogen causing soft rot of cucumber was *P. carotovorum* subsp. *brasiliense* belonging to clade II. Koch's postulates were applied to prove the isolates to be the cause of the cucumber soft rot disease. Host range tests indicated the wide host range of the pathogen (Meng et al. 2017).

A simple procedure for simultaneous detection of *Dickeya* spp. and *P. atrosepticum (Erwinia carotovoum* subsp. *atropseptica)* in tubers was developed by employing two primer sets specific for the target bacterial pathogens. Purified genomic DNAs of the type strains *Dickeya chrysanthemi* 2048$^T$ *and P. atrosepticum* 1526$^T$ were subjected to PCR analysis. The primers were specific as indicated by the tests on 61 strains belonging to various *Dickeya* and *Pectobacterium* spp. which were detected in artificially inoculated and naturally contaminated potato plants. As the procedure could detect the soft rot pathogens simultaneously in a single test, time and cost of separate test could be reduced considerably (Diallo et al. 2009). In a later study, a multiplex PCR assay was developed for simultaneous, rapid and reliable detection of principal potato soft rot and blackleg pathogens viz., *P. atrosepticum, P. carotovorum* subsp. *carotovorum, P. wasabiae* and *Dickeya* spp. employing three pairs of primers. The multiplex assay was highly specific, detecting only six strains of the target species in axenic cultures and negative results were obtained with 18 nontarget bacterial species that possibly coexisted with pectinolytic bacteria in the potato ecosystem. The multiplex assay could detect 0.01 ng/μl of *Dickeya* sp. genomic DNA and down to 0.1 ng/μl of *P. atrosepticum* and *P. carotovorum* subsp. *carotovorum* genomic DNA *in vitro*. In the presence of DNA of *Pseudomonas fluorescens* cells (competitor) the sensitivity of the multiplex PCR assay decreased ten-fold for *P. atrosepticum* and *Dickeya* sp., whereas the sensitivity of detection of *P. carotovorum* subsp. *carotovorum* and *P. wasabiae* was not affected. In spiked potato haulm and tuber samples, the threshold level for target bacteria was 10 CFU/ml of plant extract ($10^2$ CFU/g of plant tissue), $10^2$ CFU/ml of plant extract ($10^3$ CFU/g of plant tissue) and $10^3$ CFU/ml of plant extract ($10^4$ CFU/g of plant tissue), for *Dickeya* spp., *P. atrosepticum* and *P. carotovorum* subsp. *carotovorum/* and *P. Wasabiae*, respectively. The results showed that this assay could provide reliable detection and identification of soft rot and blackleg pathogens simultaneously in naturally infected symptomatic and asymptomatic potato stem and progeny

tuber samples collected from potato fields all over Poland (Potrykus et al. 2014).

*X. campestris* pv. *musacearum*, incitant of banana BW disease, infects all banana cultivars grown in east and central Africa causing heavy losses. Conventional PCR assay was developed, employing primers BXW-1 and BXW-3, based on the conserved sequences of the *hrpB* operons of the *hrp* gene cluster of the pathogen. All strains (50) of *X. campestris* pv. *musacearum* produced a single 214-bp product amplified by both primers BXW-1 and BXW-3. The LOD of the pathogen cells from culture was $10^4$ to $10^5$ CFU/ml. The presence of *X. campestris* pv. *musacearum* could be detected in bacterial ooze from infected plant tissues, and genomic DNA purified from bacterial or infected tissue used as template. The BXW primers also detected strains of *X. axonopodis* pv. *vasculorum* isolated from sugarcane and maize and strains of *X. vasicola* pv. *holcicola* isolated from sorghum. ERIC-PCR procedure generated complex fingerprinting patterns consisting of 11 or more distinct bands, ranging in size from 0.35-kb to approximately 3.0-kb, indicating all strains from Uganda, Rwanda and Tanzania were clonal. None of the *Xanthomonas* strains, including the *vasculorum* and *holcicola* pathovar strains that produced 214-bp product with BXW primers were identical to the *musacearum* strains (Ivey et al. 2010). Pathovar-specific PCR procedure was developed for the detection of *X. campestris* pv. *musacearum (Xcm)*. The *Xcm*-specific PCR assay amplified a 265-bp region of the gene encoding the general secretion pathway protein D (GspD). This amplicon was generated from the genomic DNA of all 12 *Xcm* isolates tested, but not from the DNA of other xanthomonads or plant-associated bacteria, including the two closely related species, *X. vasicola* pv. *holicicola* and *X. axonopodis* pv. *vasuclorum*. The *Xcm*-specific PCR procedure was effectively multiplexed with IC primers targeting 16S rDNA for application on DNA extracted from plant material. The PCR assay reliably differentiated healthy, *Xcm*-inoculated banana plants and naturally infected field plants with and without symptoms. The PCR assay was found to be a robust and specific diagnostic tool for sensitive and reliable detection of *X. campestris* pv. *musacearum* on symptomatic and asymptomatic banana plants (Adriko et al. 2012).

Contaminated surface water used for irrigating potato and tomato crops may be a potential source of inoculum of *R. solanacearum (Rs)*, causing BW diseases. Concentrations of *R. solanacearum* in surface water varied between $10^3$ and $10^6$ CFU/l. Rapid and reliable detection of *R. soalancearum* in water may be an effective approach to avoid dispersal of the pathogen to potato plants via irrigation water. AmpliDet RNA is a sensitive procedure, based on nucleic acid sequence-based amplification (NASBA) of RNA sequences and homogeneous real-time detection of NASBA amplicons with a molecular beacon. This assay performed in sealed tubes, reduced the risks of carryover contamination that might be associated with conventional PCR format. The RNA was extracted from 200 ml of contaminated water samples, after removing coarse particles by filtration, followed by centrifugation to concentrate the pathogen cells by about 200 times. AmpliDet RNA protocol could detect *R. solanacearum* at a concentration of 10 CFU/ml in surface water and 1 CFU/ml in demineralized water. The possibility of detecting and quantifying the viable pathogen cells is the distinct advantage of the AmpliDet RNA procedure over other PCR formats which do not differentiate live and dead cells of the pathogen (van der Wolf et al. 2004).

Among the variants of standard PCR assay, repetitive sequence-based (rep)-PCR genomic printing has been applied more frequently for the detection of bacterial pathogens. Rep-PCR format is based on PCR-mediated amplification of DNA sequences located between specific interspersed repeated sequences in prokaryotic genomes. These repeated sequences are designated REP (repetitive extragenic palindromic sequences), BOX (DNA sequences of the BOX A subunit of the BOX element of *Streptococcus pneumonia*) and enterobacterial repetitive intergenic consensus (ERIC) elements. Amplification of the DNA sequences between primers based on these repeated elements generates an array of differently sized DNA fragments from the genomes of different strains. The resolution of these fragments on agarose gels yields highly specific DNA fingerprints. The rep-PCR genomic fingerprinting protocol may be applied on whole cells (individual colonies of bacteria) bypassing the need for DNA extraction. This protocol may also be employed directly to cell suspensions obtained from symptomatic plant tissues (Louws et al. 1998).

Repetitive sequence-based PCR (rep-PCR) assay was demonstrated to be effective in detecting *R. solanacearum* in ginger, mioga and curcuma plants. Two primer sets were designed, based on sequences of polymorphic bands that were derived from repetitive sequences for application in rep-PCR fingerprinting. These primer pairs specifically detected *R. solanacearum* race 4 strains. One primer set AKIF-AKIR amplified a single band of 165-bp from genomic DNA of isolates of *R. solanacearum* infecting mioga and curcuma, whereas another primer pair set 21F-21R amplified a 125-bp band from ginger isolates of the pathogen. The rep-PCR assay along with standard PCR format was employed for detection of *R. solanacearum* in soil samples. Limits of detection of rep-PCR formats in pure culture and soil artificially infected with the pathogen were $2 \times 10^2$ cells/ml of suspension and $3 \times 10^7$ CFU/g of soil, respectively (Horita et al. 2004). In a later investigation, REP and BOX PCR analysis was applied, using selected strains of *R. solanacearum* infecting ginger, curcuma and mioga in Japan, Thailand, Indonesia, Australia and China. In addition, representative Japanese race 1 and 2 biovar 2, 3, and 4 strains from tomato, eggplant, sweet pepper and satice were also included in the analysis. The DNA profiles derived from BOX-PCR were highly reproducible. Two distinct types of DNA fingerprints based on the presence or absence of a few bands were observed among Japanese strains from plants belonging to the Zingiberaceae family. Type 1 consisted of all curcuma strains and some ginger strains. However, neither DNA patterns was similar to the strains from hosts belonging to families other than Zingiberaceae (Tsuchiya et al. 2005).

*Dickeya* spp. strains causing wilting of foliage and rotting of seed tubers of potato imported from the Netherlands to Israel were detected by using ELISA and PCR assay. Rep-PCR and biochemical assays showed that the strains from blackleg diseased plants in Israel were very similar to strains isolated from Dutch seed potatoes, suggesting that the infection of potato plants observed in Israel originated from the Dutch seed. The strains were distantly related to *Dickeya didanthicola* strains and were similar to biovar 3 *D. dadanti* or *D. zeae*. The soilborne nature of *Dickeya* spp. was demonstrated by raising potato plants in infested soil (Tsror et al. 2009). Distribution of *Dickeya* spp. and *Pectobacterium carotovorum* subsp. *Carotovorum (Pcc)* in naturally infected potato seed tubers was studied, using two potato seed lots. Different tuber tissues (peel, stolon end and peeled potato tissue at 0.5, 1.0, 2.0 and 4.0 cm from stolon end) were analyzed by enrichment PCR and plating followed by colony PCR on the resulting cavity-forming bacteria. Seed lots were contaminated with *Dickeya* spp. and *P. carotovorum* subsp. *carotovorum (Pcc)*, but not with *P. atrosepticum*. *Dickeya* spp. and *Pcc* were present at high concentrations in the stolon ends, whereas relatively low densities were present in the peel and tissues located deeper in potato tubers. Rep-PCR, 16S rDNA sequence analysis and biochemical assays, grouped all isolates of *Dickeya* spp. from two potato seed lots as biovar 3 (Czajkowski et al. 2009). Strains of *Pcc* were classified into five bvs, based on differences in physiological and biochemical properties. Further, strains of *Pcc* were also differentiated using species-specific PCR assay and 16S rDNA sequences, virulence traits, plant cell wall-degrading enzymes (CWDEs) production and phylogenetic analysis of 16S rDNA sequences. Use of four different primer sets in rep-PCR assay revealed high genetic variability, independent of the pathogenicity of isolates. Factors other than the plant host specificity appeared to correlate with genetic variability of the strains of *Pcc*. Polyphasic characterization of *P. carotovorum* susbp. carotovorum strains showed the extent of heterogeneity among them (Maisuria and Nerukar 2013). *Cms,* causing ring rot disease of potato tubers was detected, using competitive-PCR format. An internal standard template that served as control for all PCR assays was generated by the amplification of *Arabidopsis* genomic DNA under low annealing temperature with primers specific for *Cms*. The 450-bp product amplified from the internal standard DNA template was distinct from the 250-bp product characteristic of *Cms*. The ratio of PCR products amplified in the presence of a constant amount of internal DNA template increased with increase in the amount of *Cms* DNA. The PCR product ratios obtained from bacterial cells in cultures and in tissues of inoculated potato platelets were corroborated by the cell numbers estimated by the immunofluorescence antibody staining (IFAS) technique. Competitive-PCR assay was ten-fold more sensitive than IFAS procedure, as the PCR format was able to detect as few as 100 bacterial cells (Hu et al. 1995).

Nested PCR assay is effective and efficient in detecting bacterial pathogens in infected plants and soils. In nested PCR format, the sensitivity and specificity of detection of target pathogen are significantly improved by performing a second round of PCR using primers internal to the amplification product. A nested PCR assay was developed for specific and sensitive detection of *R. solanacearum (Rs)* in soil samples, since the standard PCR format was unable to detect *R. solanacearum* in inoculated soil suspensions. However, by including an overnight enrichment step prior to PCR amplification, positive results were obtained in broth that had been inoculated with soil suspension containing *R. solanacearum* at a concentration of at least $10^6$ CFU/g of soil. The primer pair OLI-1/OLI-2 was employed in the first round amplification followed by the amplification of PCR product by primer pair JE2/Y2 in the nested PCR format. The detection limit of the assay was $7.5 \times 10$ CFU/ml of soil suspension, when the suspension was incubated in the SMSA broth for 60 h. A minimum population of $10^6$ cells/ml of soil suspension was required for positive detection by nested PCR without enrichment in SMSA broth. As few as $5 \times 10^2$ CFU of *R. solanacearum*/g of soil were recovered from naturally infested soil by nested PCR assay, whereas the indirect ELISA format required a minimum concentration of $10^6$ CFU of *R. solanacearum*/g of soil (Pradhanang et al. 2000). In a later investigation, the nested PCR format was applied for the detection of *R. solanacearum*, employing the primer pair OLI-1 and OLI-2. An amplification product of 410-bp from *R. solanacearum* DNA was generated in the first round and the primer pair Y2/JE2 produced a 220-bp amplicon in the nested PCR format. These primer pairs formed a highly specific tool for detection and differentiation of *R. solanacearum* strains in soil and water samples. Availability of high quality pathogen DNA was found to be a crucial factor for sensitive and reliable detection of *R. solanacearum* in soil samples (Khakvar et al. 2008). Nested PCR assay was employed for reliable detection of *Pcc*, incitant of bacterial soft rot of crown imperial (*Fritillaria imperialis*). The primer set EXPCCF/EXPCCR amplified a single fragment of 0.55-kb from the genomic DNA of all strains of *Pcc* tested by standard PCR format. In the nested PCR procedure, the primer set INPCCR/INPCCF amplified a single fragment of 0.4-kb from the PCR product of the first round PCR. The nested PCR procedure was effective in identifying strains of *P. carotovorum* subsp. *carotovorum* in infecting crown imperial (Mahmoudi et al. 2007).

*Burkholderia (Pseudomonas) cepacia*, first reported to be the causal agent of onion slippery skin disease, emerged later as a human pathogen, causing several outbreaks especially among patients suffering from cystic fibrosis (CF) disease (Holmes et al. 1978). The species composition of *Burkholderia cepacia* complex population naturally occurring in the maize rhizosphere was examined by using culture-dependent and culture-independent methods. The DNA was extracted from root slurry and subjected to nested PCR assay. Sequences of *recA* genes were amplified with species-specific *B. cepacia* complex primers and a library of PCR-amplified *recA* genes was obtained. The culture-dependent method allowed assessment of the greater diversity of *B. cepacia* complex population naturally occurring in the rhizosphere of maize plants under field conditions. But culture-independent methods detected additional species like B. *vietnamiensis*. Culture-independent methods provided a more reliable extent of the diversity of the *B. cepacia* complex population naturally occurring on the maize root system than culture-based methods (Pirone et al. 2005). *Burkholderia* spp. constitute an important group among the soil microbial community. A DNA-based PCR denaturing gradient gel electrophoresis (DGGE) procedure was developed for detection and differentiation of *Burkholderia* species present in soil samples. Primers specific for the genus *Burkholderia* were designed based on the sequences of 16S rRNA gene. The primers showed sensitivity and specificity to the required levels for the majority of established *Burkholderia* spp. tested. Sequence analyses of amplicons generated with soil DNA exclusively revealed sequences affiliated with sequences of *Burkholderia* species, demonstrating that the PCR-DGGE procedure could be effectively employed for detecting and determining the genetic diversity of *Burkholderia* spp. in natural soils (Salles et al. 2002). Sensitivity and specificity of detection of bacterial pathogens can be enhanced by combining PCR-based methods with other detection techniques. The crown gall pathogen *A. tumefaciens* induces tumors in several crop plants. For the detection of *A. tumefaciens* infecting peach, enrichment of bacterial cells in selective liquid medium prior to PCR amplification was applied. Sample extracts were incubated for 48 h in a liquid medium selected for biovar 2 of *A. tumefaciens*. The DNA was extracted from the liquid broth. The PCR products of 172-bp for pathogen DNA and 373-bp for IC DNA were amplified specifically in most of the samples tested (Cubero et al. 2002).

*R. solanacearum* infects geranium plants, posing an important threat to their culture, as imported cuttings with latent infection might spread the disease rapidly and the pathogen might become endemic. *R. solanacearum* race 3 biovar 2 is a regulated quarantine pathogen and hence, reliable and sensitive methods have to be employed to identify infected asymptomatic plant materials. Geranium plants infected by *R. solanacearum* were found to shed high populations of bacteria in effluent water that exited from pots. A nondestructive sampling method was adopted, wherein effluent water from infected plants grown under commercial conditions was both dilution plated and filter-concentrated for real-time PCR assay. Under field conditions in Gautemala, effluent shedding of the pathogen from infested geranium plants was highly variable. Under growth chamber conditions, latently infected and mildly symptomatic geranium plants often, but not invariably shed detectable bacteria, whereas at the lowest point, 44% of plants shed detectable numbers of pathogen cells. Bacterial shedding peaked several weeks after inoculation regardless of whether plants were symptomatic or latently infected. Pathogen population sizes in stem did not correlate with either effluent population sizes or disease index rating (Swanson et al. 2007). *R. solanacearum* could be present in very low populations in asymptomatic geranium cuttings and/ or in a particular stressed physiological state that prevents isolation of the pathogen in solid media usually used. An integral procedure was developed to detect asymptomatic infection of geranium cuttings routinely. Isolation and cooperational (Co)-PCR format formed the first screening tests. Co-PCR assay based on the simultaneous and cooperational action of three primers from 16S rRNA sequences of *R. solanacearum* was highly sensitive and could detect even a single cell of *R. solanacearum/ml*, including nonculturable cells. With a failure of isolation, but a positive Co-PCR result, tomato bioassay was employed for confirmation, since stressed bacterial cells or those present in low concentration that did not develop on solid media, could be recovered from inoculated tomato plants. The integrated approach for detecting *R. solanacearum* brought out the risk of introducing the pathogen through asymptomatic geranium propagative materials (Marco-Noales et al. 2008).

A combination of immunocapture (IC) and PCR was evaluated for the efficacy in improving the sensitivity of detection of *R. solanacearum (Rs)* in soil and weeds. Anti-*Rs* PAbs were generated in rabbit and DynalR super-paramagnetic beads were coated with purified immunoglobulin G (IgG), using a IC-PCR procedure. The target DNA of the pathogen 718-bp was amplified at a detection threshold of approximately $10^4$ CFU/ml of *Rs* suspension. *R. solanacearum* in soil suspension could be detected by including the enrichment step to enhance pathogen population in nutrient broth prior to PCR amplification. IC-PCR assay detected *R. solanacearum* in tomato stems at 24 h after inoculation by stem puncture with a suspension containing approximately $10^5$ CFU/ml. IC-PCR assay detected the pathogen in 28 of 55 weed species (51%) and ten of 32 soil samples (31%). *Physalis minima*, *Amaranthus spinosus* and *Euphorbia hirta* showed high levels of infection by *R. solanacearum*. Soil samples from fallow fields did not have detectable concentration of *R. solanacearum*. The pathogen could be detected in some weed species. Asymptomatic tomato and pepper plants were found to be infected by *R. solanacearum*, indicating that the weeds and asymptomatic crop plants might play an important role in the survival of *R. solanacearum* (Dittapongpitch and Surat 2003). *R. solanacearum* was detected in planta by integrating the rapid self replication ability of bacteriophages with quantitative PCR (qPCR) assay. Six bacteriophages were tested for their ability to specifically infect and lyse 63 strains of *R. solanacearum* and 72 isolates of other bacterial species. The GW-1 strain of *R. solanacearum* infecting ginger and the phage MDS1 were selected based on the susceptibility of the pathogen strain and replication speed and reproductive burst sizes of the phage. The procedure was optimized using the primers based on the phage genome for detection of *R. solanacearum* from a number of substrates. In pure culture, the procedure could detect ~3.3 CFU/ml of *R. solanacearum*, after an incubation period of one hour with $5.3 \times 10^2$ PFU (plaque forming units)/ml of MDS1. *R. solanacearum* was detected in infected ginger plants grown in pots, using this protocol. The LOD of the pathogen was ~$10^2$ CFU/0.1 g of ginger leaf tissues (Kutin et al. 2009).

Primers and probes of real-time PCR assay and a highly sensitive BIO-PCR assay were developed for specific detection of the strains of race 3, biovar 2 of *R. solanacearum* in asymptomatic potato tubers. The biovar-specific primers and probe reacted with all 17 strains of bv 2 of *R. solanacearum*, including 12 from potato and five from geranium. None of the other 35 strains reacted with four strains of bv 1 from potato or closely related blood disease pathogen from banana. Using undiluted potato extract, inoculated with *R. solanacearum* bv 2, as few as 30 cells/ml could be detected. Two of 14 naturally infected potato tubers with no visible symptoms were positive by the real-time BIO-PCR assay, whereas none of the samples were positive with standard real-time PCR format. The real-time PCR assay procedure was found to be very simple and much less time-consuming than the standard PCR format (Ozakman and Schaad 2003). Sensitivity of detection of bacterial pathogens may be significantly enhanced by the BIO-PCR technique which is based on an enrichment of the target bacterial pathogen using appropriate nutrient medium prior to PCR amplification. *R. solanacearum* race 1 was detected in soil, weed and water samples collected from eight fields with different disease histories and cropping systems located in major tomato production areas. MSM1- broth was used to increase populations of *R. solanacearum* present in soil samples prior to PCR amplification, using the primer pair AU759/760. The sensitivity of the BIO-PCR assay was 1.9 CFU/ml and 17 CFU/g of soil for pure suspension and infested soil, respectively. The positive detection frequency of BIO-PCR assay was 66.6, 39.6, 23.1 and 31.8% for all tested samples of soil, weed and rhizosphere soil and water, respectively. The detection frequency was higher than plating on MSM-1 medium. Spatial distribution of *R. solanacearum* in the field was uneven, regardless of the presence or absence of the disease and the agro-ecosystem where the fields were located. The degree of unevenness was higher when tomato crop was not grown. Higher positive detection proportion (frequency) and population of *R. solanacearum* were present in the rhizosphere rather than the root of the weed samples tested. Pathogen populations could be detected by applying dilution plating on SMS-1 selective medium and BIO-PCR assay in irrigation water, standing water in the field planted with tomato and drainage water at the exit point of three tomato fields. The presence of *R. solanacearum* was monitored in water samples collected from the fields over several seasons. The frequency of positive detection by BIO-PCR was the highest in water samples collected from drainage water, followed by standing water and the lowest was in irrigation water. The number of positive detection among 85 water samples was 18 (21.2%) and 27 (31.8%) by plating on MSM-1 medium and BIO-PCR methods, respectively. Asymptomatic weeds and contaminated irrigation standing or drainage waters were important for the survival and dissemination of *R. solanacearum* during the absence of tomato plants (Lin et al. 2009).

Bacterial pathogens could be detected and identified by subjecting PCR amplicons to random amplified fragment length polymorphism (AFLP) analysis. *Erwinia carotovora (Ec)* subspecies could be detected by PCR-Restriction fragment length polymorphism (RFLP) analysis based on a pectate lyase-encoding gene (*pel*), since pectate lyases have an important role in the development of soft rot diseases. By including a 48-h enrichment step prior to PCR amplification, the sensitivity of the PCR-RFLP analysis was markedly enhanced. the presence of *Eca* in wash water, leaves, stem and tuber peel extract could be detected reliably (Helias et al. 1998). Strains of *Ecc* were examined for the presence of *pel* gene with RFLP, using *Sau*3AI and three patterns (1, 2 and 3) were recognized (Seo et al. 2003). RFLP pattern 3 of *pel* gene contained only type 2 strains from mulberry. Seven additional mulberry strains were subjected to RFLP analysis. Three strains of *E. carotovora* from mulberry with RFLP pattern 3 of the *pel* gene produced the 430-bp product. PCR amplification of these strains enableded them to be assigned to *E. carotovora* (Seo et al. 2004). *Erwinia carotovora* subsp. *carotovora* (*Ecc*), infecting pepper is dispersed through contaminated seeds. The pathogen was detected and identified by PCR assay, in addition to biochemical and pathogenicity test. Primers based on the *pel* gene sequence were employed for the amplification of target DNA fragment in PCR assays. RAPD-PCR analysis using nine arbitrary primers showed similarity between *Ecc* isolates obtained from pepper. *Ecc* isolates from pepper, tomato, potato and cabbage formed four distinct groups (Hadas et al. 2001). Infection of pepper plants by *Ecc* was observed in Italy. PCR amplification produced a product of 434-bp from the DNA of the pathogen isolated from infected pepper plants. RFLP analysis of amplification products showed that the isolates of *Ecc* belonged to either RFLP group 1 or RFLP group 2, indicating the limited genetic diversity of the isolates of *Ecc*, causing stem rot of pepper (Fiori and Schiaffino 2004).

A PCR-based procedure was applied for direct detection of *Streptomyces* spp., causing potato common scab disease. Primers were designed to amplify the fragment of *txtAB* (*txtA* and *txt B*) genes that have been shown to be pathogenicity determinants of pathogenic *Streptomyces* spp. Pathogenic *Streptomyces* strains in tuber lesions were detected by PCR in 70 samples. The pathogen was isolated from these lesions, confirming the pathogenic nature of the isolates. All pathogenic isolates exhibited basic general phenotypic traits of the *S. scabies* phenetic cluster. RFLP analysis of amplified rRNA sequences, together with carbon source utilization and repetitive BOX profiles, allowed most isolates to be assigned to *S. europaeiscabiei* which was the principal species causing potato common scab in Western Europe (Flores-González et al. 2008). The sequences of *atpD* gene of *Streptomyces* species were used as a molecular marker and for designing primers for the detection of the pathogen in potato tubers. PCR-RFLP assays using restriction enzymes *Hae*III and *Cfo*I were effective for differentiation of all *Streptomyces* spp. The genomospecies *S. reticuliscabiei* and *S. turgidiscabies,* differing from others, showed identical profiles between them. *Streptomyces* isolates (29) obtained from potato-growing areas in Brazil were analyzed using the PCR-RFLP of *atpD* gene technique and identified as *S. scabiei* (15 strains), *S. caviscabies/S.* setonii (12 strains) and *S. sampsonii* (12 strains). *S. europaeiscabiei* strains present in seed tubers imported from the Netherlands were detected using the PCR-RFLP procedure. Further, the genes associated with *S. turgidiscabies* PAI were detected in potato tuber slice samples, indicating the pathogenicity of the strains isolated from imported tubers (Corrêa et al. 2015).

The two primers (TP)-RAPD technique was applied for analysis of 20 strains of *Cms*, causing potato ring rot disease, from different geographic origins and other reference strains of the same or different species including potato pathogens. The TP-RAPD assay employed TP to amplify the 16S rDNA gene. At 45°C annealing, the PCR product, electrophoresed in agarose gels, produced a band pattern that was different in all species and subspecies of *Cms*. Unlike gram-negative bacteria, gram-positive bacteria with high G + C content, such as *Clavibacter*, produced low bands in TP-RAPD. By employing a different set of TP, based on the 16S rDNA sequence from *Escherichia coli*, a more adequate amplification of gram positives of high G+C including a greater number of bands was obtained. The TP-RAPD patterns using two sets of primers provided a reliable and rapid method of identifying *C. michiganensis* subsp. *sepdonicus* infecting potato tubers (Rivas et al. 2002).

A multiplex PCR assay was developed to improve the reliability of routine detection and differentiation of strains of *R. solanacearum*, using primers based on the 16S-23S rRNA gene ITS sequences. The multiplex PCR assay coamplified the pathogen-specific sequences and host plant DNA as an internal PCR control (IPC). The assay was validated by testing potato samples submitted in official surveys. Of 4,300 samples from 143 cultivars, 13 tested positive in both multiplex PCR and IF assays. The results were confirmed by bioassay in tomato seedlings and reisolation of the pathogen from inoculated plants. Additional positive results provided by IF tests were not confirmed by either multiplex PCR or tomato bioassay, indicating the greater specificity and reliability of multiplex PCR assay (Pastrik et al. 2002). The availability of rapid and cost-effective detection methods is essential for monitoring and eradication programs meant for the management of diseases caused by *R. solanacearum*, accountable for heavy economic losses in several crops like potato and tomato. Information on full genomic sequences of several strains of *R. solanacearum* enables the selection of novel specific DNA markers, using comparative genomic tools. Novel markers (17) were selected based on specific protein domains of *R. solanacearum* and were thoroughly validated for specificity and stability. PCR and hybridization-based validation assays revealed that DNA regions selected as markers were unevenly distributed among tested strains, with nine markers present throughout the species complex. The remaining eight markers were highly unevenly distributed and enabled to accomplish strain-specific dot blot hybridization patterns, particularly useful for genotype typing. PCR and hybridization-based validation assays were found to be robust and straight forward procedure for genotyping members of *R. solanacearum* species complex (Albuquerque et al. 2015).

### 4.1.2.2.3 Real-Time Polymerase Chain Reaction

The standard PCR assay has certain limitations such as the requirement of the use of hazardous chemicals,expensive equipments and longer time for obtaining results and the inability to perform tests under field conditions. Real-time PCR assay has been used to effectively overcome the limitations of standard PCR format. Real-time PCR technology has been demonstrated to be reliable, rapid and sensitive for the detection and identification of bacterial plant pathogens. Real-time PCR assay, compared with standard PCR assay, is simpler to perform, less labor-intensive and much faster, providing the results in about 60 min, if appropriate sampling procedure is followed.

*R. solanacearum* was detected, using a fluorogenic (TaqMan) PCR assay. Two fluorogenic probes were employed in a multiplex reaction: one broad range probe (Rs) detected all bvs of *R. solanacearum* and a second more specific probe (B2) detected only biovar 2A. Amplification of the target was measured by the 5' nuclease activity of TaqDNA polymerase on each probe, resulting in the emission of fluorescence. TaqMan PCR was performed with the DNA extracted from 42 genetically or serologically related strains of *R. solanacearum* to demonstrate the specificity of the assay. The LOD of the assay was $\geq 10^2$ cells/ml in pure cultures. When inoculated potato tissue extract was tested, the sensitivity was reduced. The fluorogenic probe (cox) designed with potato cytochrome oxidase gene sequence was used as an IC and it detected potato DNA in an RS-COX multiplex TaqMan PCR assay with infected potato tissues. The multiplex TaqMan procedure specific to the detection of all strains of *R. solanacearum* as well as race 3 biovar 2 allowed analysis in real-time PCR within closed tube system, reducing the risk of cross contamination between samples in the laboratory. The specificity and sensitivity of the TaqMan assay showed high speed, robustness and reliability. The presence of *R. solanacearum* in potato tubers and other plant materials could be detected routinely using the TaqMan PCR assay (Weller et al. 2000). Strains of *R. solanacearum (Rs)*, were detected by PCR assay and quantified by real-time PCR assay. PCR assay employing universal primer pair 759/760 amplified a single specific product of 280-bp from all 15 strains of *R. solanacearum* tested. However, none of the 15 strains contained the DNA fragment that was amplified by the race 3-specific primer set 630/631, indicating that none of the isolates occurring in Florida belonged to the subgroup R3B2. A quantitative real-time PCR assay, using the universal Rs primer/probe set amplified the expected *R. solanacearum* target DNA sequence from the strains tested. Real-time PCR assay using the biovar 2-specific B2 primer/probe also amplified the expected target sequence from the biovar 2 reference strains, and also potato isolate from Peru. Accurate race-specific detection of *R. solanacearum* is vital for crops like geranium susceptible to both race 1 and race 3 strains, since strict government-enforced quarantine and eradication measures have to be applied, because of the 'zero tolerance' level adopted for geranium certification (Ji et al. 2007). Diagnostic PCR formats used for regulatory purposes, have to include adequate provision of validating negative results, as well as confirming positive results. The negative results were validated through use of a reaction control plasmid, designated pRB2C2 which was designed to generate a 94-bp product, using the same amplimers, targeting the primary diagnostic 68-bp sequence of *R. solanacearum* race 3 biovar 2 DNA. SYBR Green was included in the reaction mix to facilitate the identification of post-reaction products, using a melt peak analysis. The reaction control (94-bp) and diagnostic target (68-bp) amplicons had melt peak temperatures of about 90°C and 83°C, respectively. Thus, the positive results could be easily confirmed and distinguished from control product. The modified TaqMan assay procedure developed in this investigation efficiently detected *R. solanacearum* race 3 biovar 2 in infected asymptomatic stems and leaves of tomato as well as in potato tubers and stems (Smith and De Boer 2009).

*Streptomyces* spp. causing common scab diseases of potato tubers, produce the phytotoxin thaxtomin which is the only known pathogenicity determinant. The genes encoding thaxtomin synthetase (*txtAB*) were located on the PAI characteristic of genetically diverse plant pathogenic *Streptomyces* species. The SYBR Green quantitative real-time PCR assay was developed based on the primers designed to anneal to the *txtAB* operons of *Streptomyces* and to quantify pathogenic bacterial populations in potatoes and soil. The real-time PCR format was specific for pathogenic *Streptomyces* strains. The LOD of the assay was 10 fg of target DNA, or one genome equivalent. Cycle threshold (Ct) values were linearly correlated to the concentration of the target DNA ($R^2 = 0.99$). The values were not affected by the presence of host plant DNA in extracts, indicating the reliability and specificity of the assay for quantitative analyses of pathogenic *Streptomyces* strains in plant tissues. The amount of pathogenic *Streptomyces* DNA in total DNA extracts from 1 g of asymptomatic and symptomatic tubers could be quantified by applying the assay and it ranged from 10 to $10^6$ pg. Using the standard curve, numbers of pathogenic *Streptomyces* CFU were extrapolated to range from $10^3$ to $10^6$/g of soil from potato fields where common scab disease occurred. The real-time PCR assay procedure using primers designed from the *txtAB* operons allowed rapid, accurate and cost-effective quantification of pathogenic *Streptomyces* strains in potato tubers and in the soil (Qu et al. 2008).

*Streptomyces acidiscabies* genes encoding a peptide synthase [*txtA* and *txtB* (*txtAB*)] and a cytochrome P450-type monooxigenase (*txtC*) are required for thaxtomin biosynthesis (Healy et al. 2000). Disruption mutants of the *txtA* gene were unable to synthesize thaxtomins and hence, were avirulent on potato tubers. The thaxtomin synthesis genes are conserved among all three pathogenic *Streptomyces* spp., *S. scabies*, *S. acidiscabies* and *S. turgidiscabies* associated with potato common scab disease (Healy et al. 2000). Primers were designed to amplify a fragment of the *txtAB* genes which are pathogenicity determinants of the principal pathogenic *Streptomyces* spp. Primers Stx1a and Stx1b designed to amplify by a PCR fragment of the *txtB* sequence were highly specific. The expected amplimers were obtained from DNA of relatively distant pathogenic *Streptomyces* sp. and also from potato scab lesions. Thaxtomin-producing isolates could be obtained from every sample with common scab as diagnosed by PCR assay. No amplification occurred from the DNA of nonpathogenic bacterial species. On naturally infected tubers, PCR assay with Stx1a/b primer set allowed easy determination of lesions caused by thaxtomin- producing *Streptomyces* from lesions caused by other pathogens on tubers (Flores-González et al. 2008).

Thaxtomin produced by *Streptomyces* spp. induces characteristic plant cell hypertrophy in expanding plant tissues. The *nec1* and *tomA* genes are present in a wide range of common scab-inducing *Streptomyces* strains and they are not required for pathogenicity. The *nec1* gene encodes a protein that induces necrosis in plant tissues and *tomA* encodes a virulence factor homologous to tomatinase present in pathogenic fungi. In *S. turgidiscabies*, pathogenicity and virulence genes are clustered in a PAI which can be transferred horizontally from pathogenic to saprophytic strains of *Streptomyces*, thus creating new pathogenic strains (Burkhalid and Loria 1997; Kers et al. 2005). The primer pair txtAB1/txtAB2 was used to detect and distinguish putative pathogenic isolates originated from 190 independent tubers and 130 different fields in 15 counties in Norway. All the isolates were first tested with the ScabI/ScabII primer pair. Of the 223 isolates, 152 (69%) produced amplicons with this primer set; none of those 152 isolates could be restricted with Hpv991 enzyme and hence, they were all assigned to *S. europaeiscabiei*. All 223 of the putative pathogenic isolates were also tested with primer pair TurgI/TurgII. The DNA of 71 isolates could be amplified using this primer pair and hence, they were assigned to *S. turgidiscabies*. No distinct pattern of geographical distribution of *S. europaieiscabiei* and *S. turgidiscabies* could be observed. Both pathogens could be detected in the same field and even in the same lesion. Four different PAI genotypes could be detected among the *txtAB* positive isolates; *nec1+/tomA+, nec1 –/tomA+, nec1+/tomA–* and *nec1– /tomA–* . The results indicated that there was genetic variability within species and that the species did not spread solely by clonal expansion. The pathogenicity tests showed that S. *turgidiscabies* was more aggressive causing more severe damage of the skin in general, compared with *S. europaeiscabiei* (Dees et al. 2013).

Real-time PCR assay was developed, using fluorescent quenching-based probes/primers. The procedure requires only a single dye that interacts with nuclease resulting in a decrease in fluorescence intensity. Hence, the assay is cost-effective, compared with other real-time PCR formats such as TaqMan PCR which requires two dyes (reporter and quencher) for intramolecular fluorescence resonance energy transfer (FRET) (Crockett and Wittwer 2001). *Streptomyces* spp. causing potato scab disease and their presence was detected in soil by employing reliable, a rapid and precise competitive quenching probe PCR (QCQP-PCR) procedure. The virulence gene of pathogenic *Streptomyces* spp. *nec1* gene was selected and a fluorescently labeled probe that hybridized specifically with *nec1* amplicon was designed. An internal standard DNA (ISDNA) that was identical to the *nec1* amplicon, but has a 4-base mismatch in the probe-hybridization region was included. In addition, a fluorescently labeled probe IS which specifically hybridized with IS DNA at the mutaginized region was also synthesized by PCR amplification. The target *nec1* gene was coamplified with the known copy number of IS DNA by PCR, using the same primer in the presence of the specific probes. The PCR products were monitored in

real-time by measuring the fluorescence intensity (quenching) of each probe. The initial amount of *nec1* gene was quantified based on the ratio of the PCR products of the same PCR cycle. The QCQP-PCR procedure could be employed to precisely quantify the *nec1* gene, even in the presence of PCR inhibitors in soil samples tested. The lower limit of quantification was 20 copies per tube which corresponded to 1,500 copies per g of dry soil. The results of quantification of pathogen population could be obtained within 5 h, indicating the rapidity and sensitivity of the procedure which might be useful for monitoring pathogenic *Streptomyces* species in soil (Manome et al. 2008).

*R. solanacearum* race 3 biovar 2 (R3bv2) is a designated Select Agent (SA), because of its potential to impact potato production drastically. Strains of race 3 biovar 2 (R3bv2) cause disease in cooler tropical highlands and temperate regions, although they are generally limited to the tropics and subtropics. The R3bv2 strain, a SA showed potential to devastate the potato industry in the United States and hence, it was subject to strict quarantine regulations. Quarantine screening techniques have to be rapid, sensitive and reliable for effective enforcement to prevent disease spread through seed tubers and soil. Further, the technique has to distinguish R3bv2 from other R. *solanacearum* subgroups.IMS and magnetic capture hybridization (MCH) coupled with real-time PCR could provide rapid and specific detection of *R. solanacearum* at low concentrations of R3bv2 strain present in plant tissues and soil. Beads conjugated to *R. solanacearum*-specific antibodies or an R3bv2-specific DNA capture oligonucleotides purified target cells or DNA free of PCR inhibitors, so that the subsequent RT-PCR routinely detected R3bv2 at low population levels as low as 500 CFU/ml. Both IMS and MCH effectively removed PCR inhibitors and permitted detection of R3bv2 down to 1,000 cells/ml in extracts of various plant tissues and soil that contained inhibitors of PCR amplification. Field applicability and efficacy of IMS and MCH procedures were shown by analysis of stem and rhizosphere soil samples from wilted tomato plants. Stem samples tested positive for R3bv2 using either direct, IMS- or MCH-coupled RT-PCR, whereas rhizosphere soil samples tested positive only after MCH. Generally IMS-RT-PCR assay was more sensitive than MCH-RT-PCR format, especially at lower pathogen population levels (Ha et al. 2012).

*Agrobacterium vitis* strains (50) collected from various locations in Italy and other parts of Europe, after isolation on agar plates, were screened with consensus primers from *virD2* gene. The isolates were further analyzed by PCR for the opine synthase genes and assigned to octopine, nopaline and vitopine strains. Primers designed on a octopine synthase gene did not amplify the DNA from octopine strains of *A. tumefaciens*. *virD2* fragments were sequenced and two classes of *virD2* genes were differentiated. Two primer sets were designed for the detection of octopine and nopaline strains or only vitopine strains. Multiplex real-time PCR with either primer pair and SYBR Green were performed for simultaneous identification of all opine-type strains. The combined sets of primers gave signals with DNA from any *A. vitis* strain. Several unidentified bacterial isolates from grapevines were tested to confirm the specificity of real-time PCR assay. A higher level of nonspecific background was observed, when combined primer sets were employed in multiplex PCR assays. The real-time PCR format could be used for detection of *A. vitis* directly from grapevine tumors, skipping isolation methods, resulting in enhancement of sensitivity of the assay (1 to 10 cells/assay). By using a DNA purification kit, inhibition of PCR reactions by compounds coextracted with pathogen DNA, could be avoided (Bini et al. 2008). *A. vitis*, incitant of grapevine crown gall disease, is soilborne and it can also survive systemically in vines and gets disseminated in propagation materials. An assay for indexing plants and plant materials was developed for efficient and sensitive detection of *A. vitis*. Initially, real-time PCR primers specific for diverse tumorigenic strains of *A. vitis* were designed based on a sequence of *virD2* gene. As the plant tissues contained PCR inhibitors, efficacy of DNA extraction methods that included MCH, IMS and extraction with the MoBio Powerfood Kit was compared. The assays incorporating MCH or IMS followed by real-time PCR assay were 10,000 times more sensitive than direct real-time PCR format, when tested using boiled bacterial cell suspensions, with a detection threshold of 10 CFU/ml, compared with $10^5$ CFU/ml for the direct real-time PCR format. All three assays would detect A. *vitis* in apparently healthy (asymptomatic) grapevine cuttings taken from infected vines (Johnson et al. 2013).

A. *tumefaciens* causes crown gall disease of peach, cherry and other fruit trees in China, necessitating development of diagnostic technique for disease prediction and restricting the pathogen spread. The primer pair ipt3F/ipt3R was designed based on the sequences of *ipt* gene and the primer effectively amplified a 247-bp product from the transferred DNA (T-DNA) sequences of tumorigenic *Agrobacterium*. A sensitive real-time PCR assay was developed using this primer pair for quantifying tumorigenic *Agrobacterium* in soil samples at a concentration of as low as $10^2$ CFU/g of soil. Pathogen population in 19 soil samples was quantified, using the real-time PCR assay. Population levels were < $10^2$ CFU/g in maize or uncultivated soil and $10^3$ to $10^6$ CFU/g in soil planted with

peach or cherry. Tumorigenic *Agrobacterium* density in soil was positively correlated with disease incidence in which it increased from 30–59% to > 80%, as the inoculum density increased from $10^3$ to $10^5$ CFU/g of soil. The results indicated that the density of *Agrobacterium* cells severely affected crown gall formation on peach seedlings (Li et al. 2015). *Agrobacterium* biovar1 strains containing Ri-plasmid induce extensive root proliferation (as hairy roots) on hydroponically grown cucurbitaceous and solanaceous crops, resulting in significant impact on production. SYBR Green-based quantitative real-time PCR (qPCR) assay was applied, using the *rol* locus of the Ri-plasmid for detection and quantification of biovar 1 strains in hydroponic systems. The assay was developed based on all *rolB* sequences available in GenBank and tested on a collection of both target and nontarget strains and it was specific for *Agrobacterium* biovar 1 strains. Based on a calibration with artificially contaminated water samples mimicking hydroponic conditions, unknown bacterial concentrations could be accurately quantified in water samples from surveys carried out in different greenhouses. The detection limit of the assay was one cell/ml water. The assay procedure showed potential for routine detection of rhizogenic *Agrobacterium* biovar 1 strains in aqueous samples and also for assessment of pathogen population at the presymptomatic stage of infection (Bosmans et al. 2016).

*Cms*, incitant of potato ring rot disease has the potential to cause significant economic losses. The pathogen can be readily spread through infected seed tubers that may carry latent infection. A real-time TaqMan PCR-based assay was developed by integrating an internal reaction control. The reaction control cloned into plasmid pCmsC4, consisted of a sequence unrelated to *Cms*, flanked by the primer sequences used in the TaqMan PCR, thus eliminating the need for multiplexing. The pCmsC4 insert sequence was amplified in the TaqMan assay, using the primer pair 50-2F and 133R. The 242-bp reaction control amplicon was easily distinguishable from the 152-bp *Cms* product in agarose gel electrophoresis (see Figure 4.2). Addition of SYBR Green to the reaction mix enabled the identification of both products, based on their melting profiles

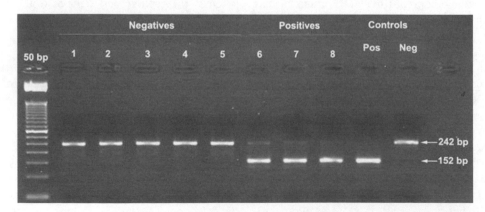

**FIGURE 4.2**   Detection of *Cms* using real-time TaqMan PCR-based assay based on the appearance of 242-bp and 152-bp bands, respectively in negative and positive samples following agarose electrophoresis Left extreme: 50-bp ladder molecular size standard. Lanes 1 to 5: negative samples generating 242-bp amplicon. Lanes 6 to 8: positive samples generating 152-bp amplicon.

[Courtesy of (Smith et al. (2008) and with kind permission of the American Phytopathological Society, MN, United States]

(85°C for *Cms* and 94°C for pCmsC4). The reaction control did not effectively compete with the primary diagnostic target in the PCR assay and had no adverse effect on the detection of *Cms*. In order to assess the effect of pCmsC4 on the specificity of TaqMan assay, known positive (28) and known negative (50) samples were analyzed for *Cms*, in the presence of 100 copies of pCmsC4/reaction. All positive samples could generate the expected PCR product, but none of the negative samples generated the pathogen-specific sequence. The results showed that the reaction control did not affect the specificity of the TaqMan assay (Smith et al. 2008).

The use of certified seed tubers is one of the key tactics for the management of bacterial ring rot disease of potato tubers caused by *Cms*. The development of methods of detection of symptomless infection by *Cms* that are rapid and reliable was considered essential. A real-time PCR assay was developed using the primers based on the sequences of the cellulase A (*Cel A*) gene to detect *Cms* in samples containing infected potato cores blended with healthy cores. The Cel A primers were able to detect two infected cores blended with 198 healthy cores. The real-time PCR assay using Cel A primers was more sensitive than the primers Cms 50/72a in detecting the pathogen. Cel A primers were specific to Cms grown *in vitro* and did not detect any other coryneform bacteria or potato pathogenic bacteria. The Cel A real-time PCR assay detected 69 strains of *Cms* and it was more sensitive than immunofluorescence assay (IFA) and *Cms* 50/72a PCR assays. The Cel A real-time PCR format was more effective in detecting the pathogen in naturally infected seed lots which were symptomatic or asymptomatic, when performed prior to planting. The real-time PCR assay using Cel A primers was found to be reliable and robust and showed potential for use as a primary screening tool for indexing certified seed tuber lots for the presence of *C. michiganensis* subsp. *sepedonicus* (Gudmestad et al. 2009). *C. michiganensis* subsp. *nebraskensis (Cmn)* causes Goss's leaf blight and wilt of maize (corn) which is reemerging as a serious disease. The conventional PCR (cPCR) and SYBR -based quantitative PCR (qPCR) assays were evaluated for their effectiveness in specific detection and quantification of *Cmn*. The primers for the gene CMN01184 and the target fragment was amplified only from the DNA of *Cmn*, but not from the genomes of a diverse collection of 129 bacterial and fungal isolates, including multiple maize bacterial and fungal pathogens, environmental organisms from agricultural fields and all known subspecies of *C. michiganensis*. Specificity of the assays for detection of *Cmn* alone was also validated with field samples of *Cmn*-infested and uninfested soil samples. The detection limits of the assays were 30 and 3 ng of pure *Cmn* DNA and 100 and 10 CFU of *Cmn* for cPCR and qPCR formats, respectively. Infection of maize leaves by *C. michiganensis* subsp. *nebaraskensis* was quantified from infected field samples and was standardized using an internal maize DNA as control. The cPCR and qPCR formats were found to be specific and sensitive for effective diagnosis of Goss's wilt disease (McNally et al. 2016).

Potato tubers are affected by *R. solanacearum* race 3 and *Cms*, causing brown rot and ring rot diseases, respectively.

These two diseases are highly destructive in temperate climates and both are listed as A2 pests in EPPO region and zero tolerance is enforced in the European Union by the quarantines. Detection tests focusing on the detection of a single pathogen were found to be ineffective for certification of seed tubers which should be free of both pathogens. A multiplex real-time PCR assay was developed for simultaneous detection of both *R. solanacearum* and *C. michiganensis* subsp. *sepedonicus* in a single assay. An IC corresponding to a potato gene was included for simultaneous amplification to avoid false negative results. The polyvalence and the specificity of each set of bacterial primers and probes were evaluated on more than 90 bacterial strains. The LOD of the triplex real-time procedure was similar to those of other molecular methods used earlier for the detection of individual pathogen. A concordance of 100% was obtained in a blind test mimicking the routing application of this assay which could be adapted to primary screening of potato tubers (Massart et al. 2014). Multiplex detection and identification procedures were developed for bacterial soft rot and black leg of potato caused by *Pectobacterium wasabiae (Pw)*, *P. atrosepticum (Pba)* and *Dickeya* spp. The methods were derived from the phylogenetic relationships of these and other Enterobacteriaceae, based on *recA* sequences. The group of *Pw* strains was highly homogenous and could be distinguished from other species. A ligation-based method for the detection of *Pw* was developed. Five padlock probes (PLPs) were designed, targeting *recA* sequences to identify the *Pw*, *Pba* or *Dickeya* spp., whereas a sixth probe recognized *recA* sequences of all soft rot coliforms, including *Pcc*. Two PLP applications viz., real-time PCR assay and universal microarrays, were developed. Sensitivity and specificity of the PLP applications were demonstrated, using 71 strains of *P. wasabiae*, *P. atrosepticum*, *P. carotovorum* subsp. *carotovorum* and *Dickeya* spp. Both multiplex procedures showed potential for seed testing and ecological investigations (Slawiak et al. 2013).

*R. solanacearum* race 3, biovar 2 (r3b2) strains are the causal agents of the highly destructive potato brown rot and BW disease of geranium. These strains are more adapted to temperate climates found at higher elevations and latitudes in the tropics and can survive and infect in relatively cooler temperatures. A simple and precise method was considered essential for the detection, identification and discrimination of the strains of *R. solanacearum* r3b2, as the methods available were time-consuming. The multiplex PCR assay was developed to detect *R. solanacearum* at the species complex level, specifically to identify whether the strain was the SA and also to exclude false negatives in a single reaction. The primers RsSC-F/RsSC-R specific to regions that were common to *R. solanacearum* species complex and also RsSA-F/RsSA-R primers specific to SA strains were designed by performing in silico genome subtraction and targeting non-phage sequences. The RsSC-F/RsSC-R primer pair amplified DNA from all 90 strains of *R. solanacearum* species complex, whereas RsSA-F/RsSA-R primer pair effectively amplified DNA from only 34 SA strains, confirming the specificity of the primer pair. The plant-specific primer pair cox1-F/cox1-R was useful for the elimination of false negatives, because of the presence of inhibitors of PCR reaction. The multiplex PCR assay amplified 641-bp band from plant DNA, a 296-bp band from DNA of *R. solanacearum* species complex strains and a 132-bp band from DNA of strains considered as SAs (see Figure 4.3). The multiplex PCR assay was effective in detecting targeted strains in tomato, potato, geranium and tobacco infected by either SA or nonSA strains of *R. solanacearum* at symptomatic or asymptomatic stage. The detection limit of the SA strain UW551 by the multiplex PCR assay was 200 CFU per PCR reaction by both RsSC and RsSA primer pairs. The multiplex PCR assay was found to be rapid, accurate and reliable for use by government officials, resulting in the development of an effective system of excluding the pathogen in desired geographical locations (Stulberg et al. 2015).

#### 4.1.2.2.4 Loop-Mediated Isothermal Amplification

The available massive amount of sequence data provides an opportunity for developing highly specific and targeted tools, such as PCR-based methods, for detection and identification of specific bacterial taxa. PCR amplification methods need temperature cycling in the range of 55–96°C in a dedicated laboratory apparatus. Loop-mediated isothermal

amplification (LAMP) has been applied for rapid detection and identification of plant pathogenic bacteria. The LAMP reaction is initiated by hybridization of inner primers to specific sites on the target DNA followed by strand extension at an elevated temperature (~ 65°C) in the presence of a strand displacement DNA polymerase (Kubota et al. 2008). Outer primers, complementary to specific priming sites just outside the inner primer binding sites, initiate synthesis of new complementary strands that displace the DNA strands initiated from the inner primers, resulting in the release of those strands. The first amplification products form stem-loop structure at each end as a result of the unique design of the inner primers, and are autoprimed, driving amplification of massive amounts of DNA. The amplified end product is a mixture of large amounts of stem-loop DNA strands with manifold inverted repeat of the target, exhibiting a cauliflower-like structure with multiple loops. LAMP-amplified DNA forms a ladder of multiple bands in gel electrophoresis and can also be visualized directly in the amplification tube by adding SYBR Green I, a fluorescent DNA-binding dye and observing under UV light (Li et al. 2009). LAMP assay may be carried out in a heat block or water bath and the need for specialized equipment as in PCR assays may be dispensed with. In addition,

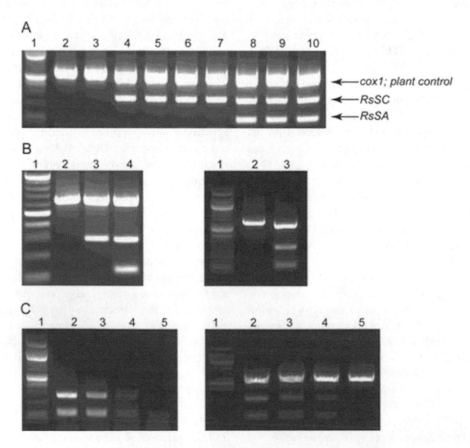

**FIGURE 4.3** Detection of *R. solanacearum* species complex and identification of SA strains of the pathogen A: amplification of potato DNA in the presence of water (lane 2), *R. pickettii* (lane 3), *R. solanacearum* non-SA strains RS5, RS124, RS126 and RS129 (lanes 4 to 7), SA strains UW224, UW276, and UW334 (lanes 8 to 10) B: (left) detection of *R. solanacearum* in extracts of tomato treated with water (lane 2), inoculated with GMI1000 (lane 3) and UE551 (lane 4); (right) geranium treated with water (lane 2) and inoculated with UW551 (lane 3) C: detection limits in the absence (left, lanes 2 to 5) and presence (right, lanes 2 to 5) of potato DNA.

[Courtesy of (Stulberg et al. (2015) and with kind permission of the American Phytopathological Society, MN, United States]

colorimetric or fluorescent dyes for visualization of positive reaction can be used, avoiding the requirement of running gel as in PCR assay. Because of the simplicity of the procedure, LAMP assay may be useful as the detection technique can be adapted for field application (Harper et al. 2010).

R. solanacearum is considered as a species complex. The fliC gene coding for the flagellar subunit protein flagellin has been used to design primers for PCR-based assay for detection of R. solanacearum. In the LAMP method developed for detection of R. solanacearum, four primers were designed to target the fliC gene sequence of the strain GM 1100. A forward inner primer (FIP) consisted of B1c (complementary sequence of F1 and F2) and a backward inner primer (BIP) consisted of B1c (complementary sequence of B1 and B2). The positive reaction could be visualized by observing the formation of magnesium pyrophosphate precipitate as a byproduct of DNA strand synthesis. The increase in the white turbidity could be assessed quantitatively using a spectrophotometer. The detection limit of LAMP reaction was determined as the minimum amount of culturable bacteria required to induce an observable LAMP reaction to be between $1.1 \times 10^4$ and $1.3 \times 10^6$ CFU/ml for R. solanacearum strain GM1100, based on visual detection of amplicons run through an agarose gel. The detection limit for other positive R. solanacearum strains varied from $6.0 \times 10^4$ to $1.5 \times 10^6$ CFU/ml. LAMP assay could be reliably employed to amplify the target gene from effluent water samples with various concentrations of R. solanacearum. A filtration step was used to preconcentrate the pathogen as well as to eliminate the inhibitor of LAMP reaction. The detection limit, based on quantitative absorbance measurements was $10^5$ CFU/ml. The most significant advantage of LAMP assay was its ability to amplify specific sequences of DNA at 65°C without the need for thermal cycling. The reaction itself occurred within 60 min and the assay could be completed within 3 h for detection from cultured bacterial cells. LAMP assay has high specificity and speed attributed to the autocycling amplification under isothermal conditions by using four primers. The visual assessment of LAMP products indicated the possibility of performing the assay under field conditions (Kubota et al. 2008). Ralstonia pseudosolanacearum race 4 causes BW of ginger. Real-time LAMP assay was developed for strain-specific detection of the pathogen in plant and soil samples. The gyrB gene sequence of R. pseudosolanacearum was used to design LAMP primers. A sigmoid amplification curve with a Ta value of $92 \pm 1°C$ was obtained only for race 4 of the pathogen. The detection limit of the assay was $10^3$ CFU/g of rhizome or soil. Detection of R. pseudosolanacearum in soil was achieved using soil supernatant as template instead of genomic DNA. The results showed that the real-time LAMP format could be employed for indexing ginger seed rhizomes and in soil infested with the BW pathogen (Prameela et al. 2017).

P. atrosepticum (earlier known as Erwinia carotovora subsp. atroseptica), causal agent of potato blackleg disease, was the first plant pathogenic bacterium whose genome was sequenced. The Cfa gene cluster has not been detected in other Pectobacterium spp. and related Erwinia spp. whose genomes have been sequenced (Bell et al. 2004). Selected genomic regions, especially the region containing a gene cluster encoding a phytotoxin coronafacic acid (Cfa)-like compound which played an important role in bacterial pathogenicity, virulence and pathogen-host interactions were used to design LAMP primer sets. The gene cluster comprising nine putative protein-encoding genes that were not present in any other Pectobacterium spp. was identified. Primers based on Cfa6 of the gene cluster (IPKS genes) were selected for the LAMP assay, because of its close association with pathogenicity of P. atrosepticum. The selected primer sets amplified DNA from P. atrosepticum at 65°C, whereas DNAs from other pectolytic bacteria were not amplified. The LAMP assay based on the Cfa6 gene of P. atrosepticum was highly specific for the species and did not cross-react with any Pseudomonas syringae strains tested, despite its homology with the Cfa6 at protein level. In addition, DNA templates from other pathogenic bacterial species, some of which had a Cfa gene cluster in their genomes, were not amplified by these primers, indicating their high level of specificity to P. atrosepticum. The sensitivity of LAMP assay, combined with the simplified DNA extraction procedure using a dilution series of suspensions of P. atrosepticum Eca1043 suspension was $2.5 \times 10^2$ CFU/ml, when amplification was evaluated by gel electrophoresis. P. atrosepticum was effectively detected in 16 field-grown blackleg-infected potato plants. Addition of SYBR Green I to the LAMP reaction tubes post-amplification clearly differentiated from the negative reactions of healthy field samples. Inclusion of the Bst DNA polymerase derived from Bacillus stearothermophilus to the reaction mix, increased the efficiency of detection, probably because of its greater resistance to inhibitors. The LAMP assays appear to be more robust than the standard PCR format, being affected less by inhibitors coextracted with pathogen DNA. The simplified DNA extraction procedure required less than ten min to be completed and DNA extracts were stable at 4°C for more than three months (Li et al. 2011). Pectobacterium spp. induce similar symptoms and share common host ranges and it was difficult to identify the pathogen precisely. Hence, a LAMP procedure was developed to detect P. carotovorum with high specificity. The isolates of P. carotovorum showed positive amplification. The assay was also tested against 15 nontarget genera of plant-associated bacteria and there were no false positives. The LAMP procedure has the potential for specific detection and identification and differentiation of P. carotovorum from other pectolytic pathogens (Yasuhara-Bell et al. 2016a, 2016b).

Clavibacter michiganensis subsp. nebraskensis (Cmn), causal agent of Goss's wilt disease of maize accounts for appreciable yield losses. Detection of the pathogen by isolation and immunoassays was found to be either time-consuming or less sensitive. Hence, the LAMP assay was developed, using the tripartite ATP-independent periplasmic (TRAP)-type C4-dicaroxylate transport system. All strains of Cmn reacted positively in the LAMP assay, whereas all the C. michiganensis subspecies failed to produce a positive reaction in the tests. LAMP assay was used to detect and identify the bacterial isolates from diseased maize plants. The 16S rDNA and *dnaA* sequence analyses were used to confirm the identity of the maize isolates of *Cmn* and validate assay specificity. The *Cmn* ImmunoStrip assay was included as a presumptive identification test for *C. michiganensis* at species level. The *Cmn*-LAMP assay was performed in a hand-held real-time monitoring device (SMART-DART) and performed equally with in-lab quantitative PCR equipment. The *Cmn*-LAMP assay precisely identified *C. michinganensis* subsp. *nebraskensis* and has potential for application as a field test (Yasuhara-Bell et al. 2016). *Dickeya* and *Pectobacterium* spp. are primarily responsible for soft rot diseases affecting potato and other plant species. As the symptoms induced by these pathogens are difficult to differentiate, it can result in the downgrading and rejection of potato seed tubers. In order to accurately diaganose the infection by these pathogens, the LAMP method was developed. The LAMP assay can be employed for detecting the pathogens in crude extracts prepared directly from symptomatic lesions. The entire procedure can be completed within 30 min, providing the results more rapidly than the currently used standard pelADE conventional PCR. Further, with LAMP assay, it was possible to detect *Dickeya* DNA in samples spiked with varying amounts of *Pectobacterium* DNA, demonstrating the highly specific and sensitive nature of the assay. The LAMP assay has the potential for application during surveys for sampling the tubers with mixed soft-rotting bacterial populations (Yasuhara-Bell et al. 2017).

### 4.1.2.2.5  Application of Nanotechnology

Nanotechnology enables the development of a new generation of biosensors and imaging techniques with higher level of sensitivity and reliability for detection of bacterial plant pathogens in soil and plants. *R. solanacearum*, causing potato BW and brown rot of tubers can survive in the soil for long periods. In order to develop sensitive on-field detection of *R. solanacearum*, a novel biosensor was developed to detect unamplified genomic DNA of the pathogen in farm soil. Gold nanoparticles functionalized with single-stranded oligonucleotides served as a probe to detect *R. solanacearum* genomic DNA. This method is simple, rapid and positive reaction could be visualized by colorimetric procedure (Khaledian et al. 2017).

## 4.2  VARIABILITY IN BACTERIAL PATHOGENS

Variations in the genetic constitution of soilborne bacterial pathogens are reflected in their phenotypic characteristics that may have an impact on the pathogenic potential (virulence). Variations in the growth and production of enzymes, toxins and antibiotics by bacterial pathogens may be recognized in nutrient media. Several techniques have been employed to assess variations in the nature of enzymes and toxins elaborated by bacterial pathogens, in addition to their immunological reactions. Various nucleic acid-based methods have been shown to be more effective in differentiating strains/races of bacterial pathogens rapidly and reliably.

### 4.2.1  VARIABILITY IN CULTURAL CHARACTERISTICS

Isolates of soilborne bacterial pathogens may exhibit variations in cultural characteristics on media. Isolates of *Xoo*, causal agent of rice bacterial leaf blight (BLB) disease, belonging to pathotypes I, II, III, IV and V existing in Japan and Indonesia showed variations in cultural and physiological characteristics. However, the variations did not have any relationship with the pathotypes tested (Hifini et al. 1975). Strains of *Xoo* were compared by computer-assisted numerical taxonomy, using 132 unit characters such as cell morphology, standard biochemical tests, acid production from carbohydrates, growth on various carbon sources, growth under different conditions and resistance to antibiotics. The races showed variations in the utilization of trehalose and sodium aconitate. Strains of race 1 (84.3%) could not utilize trehalose as a carbon source, while all strains (100%) of races 4 and 6 grew luxuriantly on the medium containing trehalose. The races 3 and 5 also could grow well on medium containing trehalose as carbon source. Sodium aconitate inhibited the growth of races 4 and 6 and high level of inhibition (89.2%) of race 1 was also seen. However, the races of *Xoo* could not be differentiated consistently based on utilization of these two compounds (Vera Cruz and Mew 1989). The cultural characteristics of *Agrobacterium* isolates from stone fruit plants on D1 medium were studied. Isolates of biovar 1, developed glistening, transparent, convex colonies with cherry red center at maturity on medium 1A, whereas biovar 2 developed typical transparent to red colonies on 2E medium. Single cell colonies of *Agrobacterium* produced red color, when streaked on yeast extract mannitol agar supplemented with Congo red. *A. radiobacter* strain K84 nonpathogenic to stone fruit plants produced agrocin. All native pathogenic isolates of *A. tumefaciens* were sensitive to agrocin produced by strain K84 (Gupta et al. 2013).

### 4.2.2  VARIABILITY IN PATHOGENICITY

Bacterial plant pathogens are differentiated basically from the saprophytes, by their ability to cause disease in one or more plant species. The isolates (populations) of pathogenic bacterial species differ in their ability to induce different grades of severity of symptoms on a crop plant species/ cultivars. The International Society of Plant Pathology has laid down the criteria for applying the term 'pathovar' for those pathogenic bacteria that do not satisfy the criteria for species designation. Pathovars are differentiated within a bacterial species primarily based on their ability or inability of isolates to infect a set of host plant species or cultivars. The basic unit of variation in classification physiologic specialization is physiologic race that is a combination of virulent and avirulent reactions induced on a standard set of differential cultivars of a crop plant species. The set of differentials that distinguish the responses induced by different pathogen isolates, is essentially required for assessing the extent of variations in the virulence of a bacterial species isolate/race/pathovar (Narayanasamy 2011).

*Streptomyces* spp., causing potato common scab, affect roots and tubers in all countries. Isolates of *Streptomyces* from scabby potato plants were evaluated for pathogenicity and virulence in radish. Scab lesions varied in appearance and severity. Some pathogenic isolates missed genes from the putative PAI. Several isolates lacked the *nec1* gene and one isolate did not have the *txtA* gene encoding thaxtomin biosynthesis, the most reliable pathogenicity determinant. Using differing inoculum density, threshold inoculum density for disease induction was determined. Disease severity increased with inoculum density over three logs, then reached the maximum, characteristic of individual *Streptomyces* strains. Lesion severity was not related to the presence of melanin, the *nec1* gene or whether an isolate reduced seedling emergence or plant survival. Differences in disease symptoms and severity, combined with absence of known pathogenicity determinants (*txtA*) or factors (*nec1*) suggested that pathogenic factors, in addition to thaxtomin, might be required for virulence of *Streptomyces* spp. (Wanner 2004).

Strains of *R. solanacearum* have been traditionally classified into five races based on differences in host range and into six bvs based on biochemical properties (Horita et al. 2014). Five races of *R. solanacearum* were differentiated based on their differential ability to produce acid from different carbohydrate substrates. Generally, no general correlation between races and bvs could be observed, except that race 3 and race 5 strains are usually placed in biovar 2 and biovar 5, respectively. Race 3 biovar 2 (R3bv2) occurring in cooler temperatures could survive in latent infections of potato tubers and this possibility was considered to be responsible for worldwide distribution of potato brown rot disease caused by *R. solanacearum*, resulting in problems in the management of the disease (Champoiseau et al. 2009). *R. solanacearum* is known to cause wilt disease in several ornamental and food plants and is carried in propagative stock imported into North America. *R. solanacearum* strains (107) were collected over a ten-year period from imported propagative stock and compared with 32 previously characterized strains of the pathogen. Most of the new biovar 1 strains of *R. solanacearum* entering the United States were genetically different from biovar 1 strains present earlier on vegetable crops. The introduced biovar 1 strains had a broader host range and could infect not only tomato, tobacco and potato, but also anthurium and pothos and induce symptoms on banana. Populations of *R. solanacearum* entering North America on ornamental crops material like geranium were analyzed by biological and genetic characteristics. The strains belonging to rep-PCR clusters B, E and F (sequevar 4) representing strains from Costa Rica, Martinique and Gudelope isolated from anthurium and pothos were virulent on tomato, potato and geranium. The race 1 strains isolated from vegetables in clusters C and D (sequevars 5 and 7) produced wilt symptoms on tomato, potato and geranium. Only 40% of the strains in clusters C and D produced wilt symptoms on tobacco. Geranium might be a universal host of biovar 1 and 2 similar to tomato and potato. Biovar 1 strains have the largest host range, although individuals within the group may have very limited

host range. Comparatively, biovar 2 strains have limited host ranges, primarily infecting solanaceous crops and weeds. All introductions of *R. solanacearum* strains of race 3 biovar 2 in the recent years were linked to geranium production and appeared to be clonal (Norman et al. 2009).

Variability in pathogenicity may arise due to horizontal gene transfer (HGT) from one strain to another strain of the same pathogenic species of bacteria. Emergence of new strains of soilborne bacterial pathogens depends on several factors, of which HGT may be a major driving force of evolution, resulting in variations in the pathogenic potential of the isolate/strain of the pathogen. The impact of natural transformation on pathogenicity of six strains, belonging to the four phylotypes of *R. solanacearum* was assessed. The genomic regions that varied between donor and recipient strains that carry genes involved in pathogenicity, such as type III effectors were examined. Strains from *R. solanacearum* species complex were naturally transformed with heterologous genomic DNA. Transferred DNA regions were then determined by comparative genomic hybridization and PCR sequencing. Three transformed strains acquired large DNA regions of up to 80-kb. The strain Psi07 (phylotype IV, tomato isolate) acquired 39.4-kb from GMI1000 (phylotype I, tomato isolate). The results showed that 24.4-kb of the acquired region contained 20 new genes. Further, an allelic exchange of 12 genes occurred, whereas 27 genes (34.4-kb) formerly present in Psi07 were lost. Virulence assessment tests showed that the three transformants of BCG20 were more virulent (aggressive) than the parent strain Psi07 on tomato. The potential importance of HGT in the pathogenic evolution of *R. solanacearum* was revealed by the results of this investigation (Coupat-Goutaland et al. 2011). Isolates of *R. solanacearum*, causing tomato BW disease were investigated to determine the relationship between the pathogen diversity and BW disease in tomato plants grown in plastic greenhouses. Pathogenic characteristics and population densities of pathogen strains were determined. Isolates of *R. solanacearum* were grouped into three pathogenic types: virulent, avirulent and interim, using the attenuation index (AI) method and plant inoculation bioassay. Populations of *R. solanacearum* ranging from 10.5 to $86.7 \times 10^5$ CFU/g were present in symptomatic tomato plants infected by virulent type isolates. On the other hand, asymptomatic plants contained interim or avirulent strains. Tomato plants displaying 1st or 2nd degree disease intensity contained interim and virulent strains, whereas plants with 3rd and 4thdegree disease intensity harbored only virulent strains. These three pathotypes of *R. solanacearum* coexisted in a competitive growth system in tomato plants or under field conditions and their distribution closely correlated with the severity of tomato BW disease (Zheng et al. 2014).

Strains of *R. solanacearum* infecting ginger and zingiberous plants varied in their pathogenic specialization as well as in their physiological properties. Isolates of *R. solanacearum* obtained from curcuma, ginger and mioga were assigned to two bvs, based on their ability or inability to produce acid from maltose, lactose, cellobiose, mannitol, sorbitol and dulcitol as carbon sources. All strains from infected curcuma, ginger and mioga in Japan belonged to biovar 4, whereas those from Thailand and Indonesia composed of both bvs 3 and 4 and those from Australia and China belonged to biovar 4 (Tsuchiya et al. 2005). *R. solanacearum* race 4 isolates induced BW disease in Zingiberaceae. Polyphasic phenotypic and genotypic analyses indicated the intraracial diversity and dominance of biovar 3 over biovar 4. Biovar 3 isolates caused severe wilting symptoms rapidly within five to seven days in ginger. 'Fast wilting' was induced in other closely and distantly related hosts such as turmeric (*Curcuma longa*), aromatic turmeric (*C. aromatica*), black turmeric (*C. caesia*), sand ginger (*Kaempferia galanga*), white turmeric (*C. zeodaria*), awapuchi (*Zingiber zerumbet*), greater galangal (*Alpinia galangal*), globba (*Globba* sp.), small cardamom (*Elettaria cardamomum*) and large cardamom (*Ammomum subulatum*) of the Zingiberaceae and in tomato. On the other hand, biovar 4 caused 'slow wilting' which progressed slowly and its distribution was restricted in some locations. Biovar 4 was genetically distinct from biovar 3 isolates (Kumar et al. 2014).

### 4.2.3 Variability in Biochemical Properties

Bacterial plant pathogens show variations in various biochemical and physiological properties that may be useful for differentiation of bvs of a pathogenic species. Variability in isolates of two strains of *A. tumefaciens* isolates from root galls of *Vicia faba* was investigated. The strain MTCC7405 and MTCC7406 were able to grow on alkaline medium as well as on 2% NaCl. They were neither able to catabolize lactose as the carbon source nor oxidize Tween 80. Based on growth on dextrose and production of lysine dihydrolase, ornithine decarboxylase and DNA G + C content, these strains could be differentiated. Both strains differed from other recognized bvs of *A. tumefaciens* (Tiwary et al. 2007). Strains of *A. vitis*, causing crown gall disease of grapevines were isolated on nonselective YMA medium. Differential physiological and biochemical tests were performed and the strains were identified as *A. vitis*. The results of multiplex PCR assay, targeting 23S rRNA gene sequences confirmed the identity of the strains. The strains were less virulent on tomato plants, compared with reference strains of *A. tumefaciens* and *A. vitis* (Kuzmanović et al. 2012). Isolates of *R. solanacearum*, causing BW disease of brinjal (eggplant) obtained from different locations in Bangladesh, were found to be pathogenic to brinjal. Gram staining and potassium hydroxide tests applied to isolates of *R. solanacearum* showed that they were gram-negative. All pathogenic isolates produced a pink or a light red or a red center with white margin on TZC medium after 24 h of incubation. All isolates oxidized disaccharides (sucrose, lactose and maltose) and sugar alcohols (mannitol, sorbitol and dulcitol) within three to five days, confirming the identity of isolates of *R. solanacearum* as biovar 3. Pathogenicity tests on tomato and chilli (pepper) indicated that the isolates of *R. solanacearum* belonged to race 1 (Rahman et al. 2010). In a later investigation, isolates of *R. solancearum*, causing BW of potato were identified as race 3 and biovar 3 based on the results

of oxidation of disaccharides and sugar alcohols (Ahamed et al. 2013). *R. solanacearum*, causal agent of banana Moko disease is widespread in Malaysia. Isolates (197) obtained from Peninsular Malaysia produced fluidal colonies, white to pink in color, after incubation for 24–48 h at 29°C on Kelman's TZC agar medium, characteristic of *R. solanacearum*. The isolates produced a positive reaction for potassium hydroxide (KOH), Kovacs oxidase, catalase and lipase activity on Tween 80 solution tests. Biovar tests showed that only 30 strains had characteristics of *R. solanacearum* biovar 1 associated with Moko disease, which was negative for utilization of disaccharides and hexose alcohols. Pathogenicity assays revealed that these 30 strains were virulent on *Musa paradisiaca* cv. Nipah explants with varying levels of virulence (Zulperi et al. 2016).

### 4.2.4 Variability in Physical Properties

Populations of *R. solanacearum* differ in the levels of virulence varying from avirulence to high virulence. A rapid method of differentiating virulent and avirulent strains of *R. solanacearum* was developed using a high performance liquid chromatography (HPLC) technique. Three chromatographic peaks P1, P2 and P3 were recognized on HPLC spectra among avirulent (68) and virulent (28) strains of *R. solanacearum*. The avirulent strains had P1 as the intense peak, while the majority of virulent strains had P3 as the peak. A chromatographic titer index (CTI) was established. The avirulent strains had high values of $CTI_1$, ranging from 63.6 to 100%, whereas the virulent strains had high values of $CTI_3$ values ranging from 90.2 to 100%. The results showed that fractionation by HPLC and their deduced CTI could be used for rapid and efficient evaluation and prediction of the avirulence/virulence nature of the strains of *R. solanacearum* (Zheng et al. 2016).

### 4.2.5 Variability in Immunological Properties

The efficacy of immunoassay for differentiation of strains of bacterial plant pathogens has been assessed. Strains (28) of *A. vitis*, including both tumorigenic and nonpathogenic phenotypes, involving 26 isolates from Japan and two strains NCPPB3554 and NCPPB2562 isolated in Australia and Greece, respectively, were characterized, by employing a slide agglutination test (SAT), using antisera raised against somatic antigens. The *A. vitis* strains separated into four serogroups A, B, C and D based on the SAT activity and one nonpathogenic strain was placed in group E. Serogroups A to C corresponded exactly to genetic groups A to C, respectively, whereas serogroups D could be further divided into two genetic groups D and E. The genetic group E was isolated in Okayama Prefecture, Japan and all strains of genetic group E coexisted with tumorigenic strains belonging to other groups within the same galled grapevine tissues (Kawaguchi et al. 2008). Among the immunoassays, ELISA has been applied more commonly for differentiation of strains of bacterial plant pathogens. MAbs were found to be efficient in differentiating strains (178) of *Xoo* into four groups (I to IV), based on their reactivity with four MAbs in ELISA tests (Benedict et al. 1989). In a later investigation, both MAbs and PAbs were employed to differentiate 63 strains of *X. oryzae* pv. *oryzae* into nine reaction types, consisting of four serovars and seven subserovars (Huang et al. 1993). PTA-ELISA format was also applied for differentiation of the strains of *R. solanacearum* which were assigned to biovar 1 (9%), biovar 2 (7%) and biovar 3 (84%). ELISA tests confirmed the widespread distribution of the extremely aggressive biovar 3 isolates causing BW of pepper (chilli) (Shahbaz et al. 2015).

## 4.2.6 Variability in Genomic Nucleic Acid Characteristics

Detection and differentiation of bacterial pathogens and their strains has been achieved with greater reliability, specificity and accuracy by employing nucleic acid-based techniques. PCR-based assays either alone or in combination have been preferentially employed, because of their greater sensitivity and reproducibility in providing the results rapidly. Universal primers/specific primers have been employed to generate an array of DNA amplified products from the target DNA and these amplicons constitute the genomic fingerprints for the bacterial species/ strains concerned. Repetitive sequence-based rep-PCR procedures have been more frequently applied to fingerprint phytopathogenic bacteria. They generate several genomic fragments via PCR amplification which are resolved as characteristic banding patterns that are useful for differentiating bacterial species/strains. The specific conserved repetitive sequences (REP sequences and ERIC sequences) and BOX elements are distributed in the genomes of bacterial species of diverse origin. REP-PCR assay are performed using primer sets that are capable of amplifying the specific conserved repetitive sequences, leading to the differentiation of bacterial strains within a species. For the identification and differentiation of *Pectobacterium* spp. (known earlier as *Erwinia* spp.), causing soft rot diseases, the 16S-23S rRNA intergenic transcribed spacer (ITS) region PCR in combination with RFLP of PCR amplification products was found to be a simple, precise and rapid method. The ITS was amplified from *Pectobacterium* and other genera using universal PCR primers. After PCR, the banding patterns generated enabled the soft rot pathogens to be differentiated from all other *Pectobacterium* or non-*Pectobacterium* species and then placed in one of three groups (I to III). Group I comprised all *P. carotovorum* subsp. *atrosepticum* and subsp. *betavasculorum* isolates. Group II contained *P. carotovorum* susbsp. *carotovorum*, subsp. *odoriferum* and subsp. *wasabiae* and *P. cacticida* isolates. In group III, all isolates of *E. chrysanthemi* were placed. For further differentiation, the ITS-PCR products were digested with one of two restriction enzymes *Cfo*I or *Rsa*I (Toth et al. 2001).

The interrelatedness, genetic diversity and geographic distribution of the common bean bacterial blight (CBB) pathogens *Xanthomonas axonopodis* pv. *phaseoli (Xap)* and *X. axonopodis* pv. *phaseoli* var. *fuscans (Xapf)* were differentiated, using RFLP analysis of PCR-amplified 16S ribosomal gene, including the 16S-23S intergenic spacer region and rep-PCR assay. The DNAs of *Xap* and *Xapf* generated six and five fragments, respectively. *Xap* isolates had 300- and 400-bp fragments which were absent in the isolates of *Xapf*, but they had a 700-bp fragment in place of these two fragments. Furthermore, due to the action of restriction enzyme *Mbo*I, polymorphic bands and fragments were generated, providing the basis for differentiation of the isolates of *X. axonopodis* pv. *phaseoli* and var. *fuscans* (Mahuku et al. 2006). Genetic diversity and pathogenicity of *X. axonopodis* pv. *phaseoli (Xap)* infecting cultivars of common bean representing two common gene pools (Andean and Middle American) were assessed, using rep-PCR and RAPD procedures. The East African strains represented distinct xanthomonads that independently evolved to be pathogenic on common bean. *Xapf* were more closely related and genetically distinct from *Xap* strains. But two distinct clusters of *Xapf* strains were identified and one of the clusters enclosed the most New World strains. Spanish strains were placed in both clusters, but regardless of the cluster grouping, all strains were highly pathogenic on bean cultivars of both gene pools. The results revealed the multiple introductions of xanthomonads associated with common bacterial blight disease (CBB) of common bean (Lopéz et al. 2006).

A phylogenetically meaningful scheme of classification of strains of *R. solanacearum* was developed, based on sequence analysis of the internal transcribed spacer (ITS) region and the *egl*, *hrpB* and *mutS* genes and this scheme divided *R. solanacearum* species complex into four phylotypes. Each phylotype may be further subdivided into a variable and additive number of sequevars which are clusters of strains whose *egl* sequences have at least 99% similarity (Fegan and Prior 2005; Wicker et al. 2009). The multilocus analyses established that phylotypes I, IIA, IIB, III and IV displayed contrasting evolutionary dynamics and identified eight clades. Phylotype I corresponded to clade 1; phylotype IIA was subdivided into clades 2 and 3; phylotype III corresponded to clade 6 and phylotype IV was subdivided into clades 7 and 8 (Wicker et al. 2012). *R. solanacearum* includes a wide range of ecotypes, sharing similar hosts and symptomatology and causing disease in similar climatic conditions (Cohan 2006). Strains of *R. solanacearum* are grouped into four phylotypes (I to IV), based on nucleotide sequences of the ITS region. Strains of *R. solanacearum*, *Pseudomonas szygii* and banana blood disease bacterium (BDB) from different countries were analyzed, using primer pair 759- and 760-bp for PCR amplification and the resultant 282-bp product was sequenced. Three groups of *R. solanacearum* strains were differentiated. Group I included strains belonging to bvs 3, 4 and 5 and biovar N2 from Japan. Most of these strains were of Asian origin, except two strains from Australia and Guyana. Group II contained strains belonging to biovar 1 and 2 and biovar N2 from Brazil. Strains of biovar N2 from Japan and the Philippines were placed in Group III. All strains of *P. szygii* and BDB clustered into Group III (Villa et al. 2003). In a further investigation, partial sequencing of 16S rDNA, *egl* and *hrpB* genes of Asian strains of *R. solanacearum* complex, including 31 strains of

*R. solanacearum* and two strains each of BDB bacterium and *P. szygii* was carried out. Various levels of polymorphisms were observed in each of these DNA regions. The highest polymorphism (ca. 25%) was seen in the *egl* gene sequence. The *hrpB* sequence showed about 22% polymorphism. Four clusters were identified based on the phylogenetic analysis. Cluster 1 included all strains of *R. solanacearum* from Asia belonging to bvs 3, 4, and 5 and N2. Biovar N2 and 1 strains forming cluster 2 were isolated from potato and clove this cluster also contained BDB and *P. szygii*. In cluster, race 3 biovar 2 strains from potato and race 2 biovar1 strains from banana and race 1 biovar 1 strains isolated from North America, Asia and other parts of the world. Strains of *R. solanacearum* from Africa were placed in a separate cluster 4 (Villa et al. 2005).

Strains of *R. solanacearum* (319) obtained from 14 different host plant species in 15 Chinese provinces were analyzed using BOX fingerprints to assess the influence of the location and host plant species on the genetic diversity of the pathogen. Phylotype, *fliC*-RFLP patterns and bvs were determined for all strains and the sequevar for 39 representative strains. The majority of strains belonged to the Asian phylotype I, shared identical *fliC*-RFLP patterns and they were assigned to four biovars, bv3, bv4, bv5 and bv 6. Phylotype II strains (20) were assigned to biovar 2 and they had distinct *fliC*-RFLP patterns. BOX-PCR fingerprints generated from the genomic DNA of each strain revealed a high diversity of the phylotype I strains where 28 types of BOX fingerprints could be differentiated. All phylotype II isolates originating from ten provinces belonged to sequevar 1 and displayed identical BOX patterns as the potato brown rot strains from various regions of the world (Xue et al. 2011). Isolates of *R. solanacearum* collected from eggplant (brinjal), tomato and pepper (chilli) were analyzed by biochemical tests and multiplex PCR assay. All isolates of *R. solanacearum* belonged to phylotype I and biovar 3. In order to assess genetic diversity of the isolates, sequences of *egl*, *pga* and *hrpB* genes of 95 isolates were determined. Indian isolates within the phylotype I did not cluster, based on the host or geographical locations. But isolates from Andaman clustered into a single group. The isolates clustered into two groups, based on the sequences of *egl* and *pga* trees, indicating the existence of two major population groups. The subgroup 1 was dominant and consisted of unknown/newer sequevars and the subgroup 2 contained sequevars based on *egl* sequences. Subgroups 1 and 2 were also distinguished based on *hrpB* gene sequences (Ramesh et al. 2014).

Strains of streptomycetes pathogenic to potato were characterized, based on phenotypic characteristics, analysis of 16S rRNA genes, production of thaxtomin A, and presence of *nec1* and ORF *tnp* gene homologs. The primary pathogenicity factor of *Streptomyces* spp. involved in potato common scab disease has been shown to be a nitrated diketopiperazine named as thaxtomin A (*ThxA*). Strains S33 and S27 possessed typical characteristics of *S. scabies* and *S. Turgidiscabies*, respectively, producing thaxtomin A and hybridizing to genes *nec1* and ORF *tnp*. Strain S71 produced thaxtomin A and had phenotypic and phylogenetic properties similar to those of

*S. acidiscabies*, except having a greater minimum growth at pH 4.5, production of a melanoid pigment on tyrosinase agar, and failure to hybridize with *nec1* and ORF *tnp* gene probes. Phylogenetic analysis of strains S63, S77 and S78 based on 16S rRNA gene sequences exhibited low homology to that of described scab pathogens (less than 97.3, 96.0 and 96.3%, respectively). Strain S78 produced thaxtomin A, but did not have sequences homologous to *nec1* and ORF *tnp* genes. Strains 63 and S77 did not produce detectable amounts of thaxtomin A and lacked gene homologs of *nec1* and ORF *tnp* genes. Strains S63, S77 and S 78 showed adaptation to acid soil conditions existing in Korea (Park et al. 2003). The inter- and intra-specific relationships of 34 strains, including 11 isolates of *Streptomyces* spp. isolated from potato scab lesions were determined, based on the sequences of 16S rRNA gene, and 15S-23S rDNA internal transcribed spacer (ITS) region sequences. Most of the isolates were classified as *S. scabiei* and *S. acidscabies*, based on phylogenetic analysis of 16S rRNA gene sequences. Isolate KJ061 was placed in an ambiguous taxonomic position between *S. reticuliscabiei* and *S. turgidiscabies*. The ITS region sequence analysis showed that tRNA genes were not present in this region of *Streptomyces* spp. The 16S-23S ITS regions of *Streptomyces* spp. showed various lengths and highly variable sequence similarities (35–100%) within strains as well as intra- and inter-species levels. *S. europaeiscabiei* could be clearly differentiated from *S. scabiei*. The results indicated that ITS region sequences were not useful in phylogenetic analyses of *Streptomyces* spp. (Song et al. 2004). In a later investigation, the phylogenetic analysis of partial *rpoB* sequences was used to differentiate *Streptomyces* spp. The *rpoB* gene-based procedure could be a means of complementing other genetic methods such as 16S rRNA gene analysis or DNA-DNA hybridization to phylogenetically differentiate potato scab-associated *Streptomyces* spp. (Mun et al. 2007).

The genetic diversity of *Streptomyces* spp. isolated from field-grown potatoes in six states of the United States was studied. The isolates were classified into species, based on the sequence of variable regions in the 16S rRNA gene and the genes associated with PAI of *S. turgidiscabies*. About 50% of the isolates belonged to *S. scabies* or *S. europaeiscabiei*, based on the 16S rDNA sequence and they had characteristics of PAI. These isolates were pathogenic to potato and radish and existed in all locations. About 10% of the isolates were pathogenic, but lacked one or more genes characteristic of PAI, although they had genes for biosynthesis of thaxtomin, the principal pathogenicity determinant. Approximately 40% of the isolates were nonpathogenic with diverse morphological characteristics. Four common scab-causing species, *S. scabies*, *S. europaeiscabiei*, *S. stelliscabies* and the new group X were identified, based on the sequences of variable regions in the 16S rRNA gene. The results of the survey suggested that differences in the incidence and severity of potato common scab in various locations and different seasons might be due to infection by different populations of pathogenic *Streptomyces* spp. (Wanner 2006). Another common scab-causing streptomycete was isolated from scabby potatoes in southeastern Idaho State, United States and this strain had hallmarks of newly characterized *Streptomyces* PAI with genes encoding the synthetase for the pathogenicity determinant thaxtominase, although it lacked a third gene *nec1*, characteristic of the Streptomyces PAI. The new strain had a unique 16S rDNA gene sequence closely related to those of other pathogenic *Streptomyces* spp. This 16S rDNA sequence was also present in isolates lacking PAI, suggesting that the new pathogenic strain might have arisen by horizontal transfer of a PAI into a saprophytic streptomycetes (see Figure 4.4). Isolates of the new strain were pathogenic to radish and potato and more virulent than *S. scabies* type strain. Further, lesions on underground stems and stolons were also induced, in addition to scab lesions on potato tubers, by the

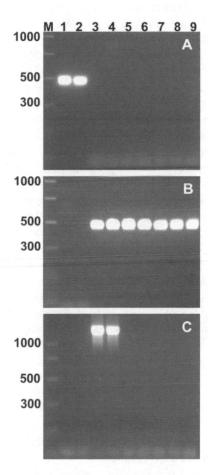

**FIGURE 4.4** Identification of a new Idaho strain of *Streptomyces* infecting potato and radish by PCR with specific primers A: DNA templates from *Streptomyces* isolates amplified with PCR primers ASE3 and Scab2m specific to *S. scabies* and *S. europaeiscabies* amplifying 474-bp fragment B: ASE3 and Aci2 primers producing 472-bp fragment from *S. acidiscabies* and Idaho isolates of *Streptomyces* C: Aci1 and Aci2 primers producing 1,278-bp fragment from *S. acidscabies* Lane M: standard size markers; lane 1- *S. scabies*ᵀ ATCC49173; lane 2- isolate ID01-16C ; lane 3- *S. acidiscabies*ᵀ ATCC49003; lane 4- isolate ME02-6987A; lane 5- isolate ID01- 62A; lane 6- isolate ID01-12Cl; lane 7- isolate ID03-1A; lane 8- isolate ID03-2A and lane 9- isolate ID03-3A.

[Courtesy of Wanner (2007) and with kind permission of the American Phytopathological Society, MN, United States]

new strain of *S. scabies*, representing additional complexity in the pathogenic populations (Wanner 2007). *Streptomyces* isolates collected from field-grown potatoes, showing scab infection between 2008 and 2009 in Germany were analyzed for morphology, pathogenicity and strain-type. The isolates could not be differentiated based on phenotypic characteristics. The isolates were characterized at genetic level by PCR-RFLP of 16S-23S ITS region with Hpy991. Based on the results of DNA fingerprinting technique, *Streptomyces* species could be differentiated genotypically. The level of diversity among scab-causing *Streptomyces* species was much higher than what was expected. Isolates belonged to various *Streptomyces* spp., associated with common scab. Among the pathogenic strains of *S. europaeiscabiei, S. stelliscabiei, S. acidiscabiei, S. turgidiscabiei* and *S. botrropensis* within Germany, *S. europaeiscabiei* was predominantly observed. Other scab-causing species showed uneven distribution. Most of the isolates possessed *txtAB* gene, indicating its requirement for pathogenicity to potato (Leiminger et al. 2013).

The *ThxA* biosynthetic cluster in *S. scabiei* (syn: *S. scabies*) is located within a mobile genomic island known as the toxicogenic region (TR). Three attachment sites (*att*) separate TR into two subregions (TR1 and TR2). TR1 contains the *ThxA* biosynthetic cluster and is conserved among several pathogenic *Streptomyces* spp. However, TR, an integrative and conjugative element is not present in most other pathogenic species. Mobilization of the whole TR element or TR2 alone between *S. scabiei* and nonpathogenic *Streptomyces* spp. was previously demonstrated. TR1 alone did not mobilize, indicating the requirement of TR2 for its mobilization. Later, self mobilization of TR1 to pathogenic *Streptomyces* spp. harboring only TR1 and integration into *att* site of TR1 were observed, resulting in the tandem accretion of resident TR1 and incoming TR2. The incoming TR2 could further mobilize resident TR1 in cis and transfer to a new recipient cell. The results showed that TR1 was a nonautonomous cis-mobilizable element and it could hijack TR2 recombination and conjugation machinery to excise, transfer and integrate, resulting in the dissemination of pathogenicity genes and the emergence of new pathogenic species. Furthermore, comparative genomic analysis of 23 pathogenic *Streptomyces* isolates from ten species showed that the composite PAI formed by TR1 and TR2 is dynamic and various compositions of the island exist within the population of newly emerged pathogenic species, indicating the structural instability of the composite PAI (Zhang and Loria 2017).

Representative strains (6) of *Dickeya* spp. isolated from potato, onion and irrigation water in Spain between 2003 and 2005 were characterized by biochemical, serological, molecular and pathogenicity assays. The strains were classified into biovar 3 and 6. Phylogenetic analysis and comparison of isolates with type strains of *Dickeya* species were performed, using concatenated partial sequences of the housekeeping genes *gapA* and *mdh*. One Spanish strain was identified as *D. diffenbachiae*, whereas other strains did not fit clearly into the *Dickeya* spp. already described. Isolation of dissimilar pathogenic strains in different rivers and irrigation water sources suggested that contaminated water could be a potential source of inoculum for the incidence of disease in different crops (Palacio-Bielsa et al. 2009). Occurrence of *Dickeya* spp. on potato increased steadily over the years in Finland. A detailed investigation of blackleg outbreaks between 2008 and 2010 indicated that *Dickeya* spp. was the major component in blackleg disease complex and the pathogen was detected and isolated from all symptomatic plants tested. Repetitive sequences PCR (rep-PCR) and PFGE analysis of strains isolated in Finland showed identical pattern with those isolated in other European countries and they were designated 'Dickeya solani'. Further, the *dnaX* gene sequence of the representative strains isolated in Finland indicated 100% similarity to the *dnaX* sequences of *D. solani*. The results indicated that strains of *D. solani* played a major role in blackleg disease incidence in Finland (Degefu et al. 2013). Multilocus sequence analysis of *recA, dnax, rpoD, gyrB* and 16S rDNA sequences of Japanese *Dickeya* spp. (*E. chrysanthemi*) strains (41) isolated in Japan, was carried out. In addition, PCR-RFLP of *recA, rpoD* and *gyrB* genes, PCR fingerprinting and biochemical tests were also performed. Based on the *recA, dnaX, rpoD, gyrB* and 16S rDNA sequences and PCR genomic fingerprinting, the strains of *Dickeya* spp. were divided into six groups (I to VI). Group I corresponded to *D. chrysanthemi*; group II corresponded to *D. dadantii*; group III to *D. dianthicola* and group IV to *D. zeae*. The group V and group VI could not be assigned to any existing *Dickeya* spp. and they were deduced to be two putative new species. The PCR-RFLP analyses of *gyrB, rpoD* and *recA* clearly differentiated the six groups of *Dickeya* strains. The results showed the usefulness of PCR-RFLP analysis of *rpoD, gyrB* and *recA* in identifying and differentiating Japanese *Dickeya* strains (Suharjo et al. 2014).

*Pcc* isolates obtained from watermelon plants showing soft rot symptoms were pathogenic causing typical water-soaked lesions. Based on phenotypic properties, the isolates (10) were identified as *Pcc*. The 16S rDNA sequences of the isolates were 99% similar to the corresponding sequences of the reference *Pcc* isolate. BOX- and ERIC-PCR analysis indicated genetic diversity among the isolates occurred and it was not related to geographical locations (Dana et al. 2015). Isolates (6) of pectinolytic bacteria isolated from soft rot diseases in potato, induced hypersensitive reaction (HR) in tobacco. Analysis of 16S-23S ITS regions of the genomic DNAs of the isolates were digested with restriction enzymes *Rsa*I and *Cfo*I and electrophoretic profiles of the digest of isolates DAPP-PG753 to DAPP-PG 757 were similar. The electrophoretic profiles of DAPP-PG752 and LMG2408 were different. Phylogenetic analysis, using partial malate dehydrogenase gene sequences, clustered DAPG-PG752 in a clade that contained the type strain of *Pcc*, whereas DAPP-PG753 to DAPP-PG-757 together with LMG2408 fell in a clade that included strains of *P. aroidearum*. Multilocus sequence analysis of concatenated partial sequences of the *atpD, gyrB, infB* and *rpoB* genes confirmed the results obtained with the *mdh* gene sequences and identified DAPP-PG752 as *P. cactorum* and LMG2408 and five related Lebanese isolates as *P. aroidearum*. These isolates showed identical BOX- and ERIC-PCR profiles that resembled profiles of known *P. aroidearum* as the causal agent of potato soft rot disease for the first time (Moretti et al. 2016).

PCR-RAPD procedure was applied for differentiation of *Agrobacterium* strains, employing one to four different 10-mer primers. The band patterns obtained with four primers for each of the 39 strains of *Agrobacterium* were distinct, facilitating differentiation of the strains from each other in a reproducible manner. Strains with similar chromosomal background, but different plasmid content as in strains C58 and A281, gave the same band pattern with all the primers. Eight *Agrobacterium* strains and isolates obtained from tumors of ten inoculated host plants were analyzed using the RAPD procedure. It was possible to rapidly identify the isolates recovered from tumors by comparing their band patterns with those of strains used as inoculum. All strains included in the investigation were differentiated by the RAPD technique. Purified bacterial cell suspensions, used for RAPD procedure, showed potential for rapid screening of *Agrobacterium*-like colonies isolated from plant tumors for epidemiological investigations (Llop et al. 2003). *Agrobacterium* spp. strains (56) isolated from grapevines in Spain were tested for biovar classification, pathogenicity on several hosts, opine utilization, 15S rRNA gene sequencing and PCR amplification, employing five primer sets targeting chromosomal and Ti-plasmid genes. Five strains belonged to *A. vitis* (biovar 3), three to *A. tumefaciens* (biovar 1), and three to *A. rhizogenes* (biovar 2). All strains induced tumors in grapevines. Most *A. vitis* strains were also pathogenic to tomato and tobacco plants, whereas three *A. tumefaciens* strains were pathogenic only to grapevines. Although most *A. vitis* strains utilized octopine, 12 strains could not utilize octopine or nopaline. The 16S rRNA gene sequencing clearly distinguished strains belonging to the three species of *Agrobacterium*. Strains of *A. vitis* could be further subdivided into three chromosomal backgrounds, according to their 16S rRNA sequences. DNA from all *A. vitis* strains was amplified with the chromosomally encoded pehA primer pair. In both *A. vitis* and *A. tumefaciens*, a correlation was seen between the amplicons obtained using the *tmr* and the *virA* Ti-plasmid-targeting primer pairs. Three types of Ti-plasmid were recognized in *A. vitis* strains, according to their PCR amplification and octopine utilization profiles. A strain with known chromosomal background harbored only one type of Ti-plasmid within the strains from each sample analyzed, showing a strong association between chromosomal backgrounds and Ti-plasmids in *A. vitis* (Palacio-Bielsa et al. 2009).

*R. solanacearum* isolates (15) from wilting plants of field-grown pepper, pot-grown hydrangea (*Hydarangea paniculata* and *H. macrophylla*) and geranium (*Pelargonium* × *hortorum*) in commercial nurseries and retention ponds in Florida were subjected to phylogenetic analysis, based on an egl gene sequence. These strains had three distinct origins. Three pepper strains belonged to phylotype I biovar 3 and were clustered with strains from diverse hosts in Asia belonging to sequevar 13. Six strains from geranium and four strains from hydrangea were closely related to strains in sequevar 5, a distinct subcluster of phylotype II biovar strains 1 isolated from French West Indies and Brazil. Two other biovar 1 strains from hydrangea and strains K60, AW and Rs5 belonged to sequevar 7 in phylotype II and probably were native to North America. None of the Florida isolates belonged to the highly regulated SA race 3 biovar 2 subgroup (Ji et al. 2007). Based on rep-PCR genomic fingerprinting and phylogenetic analysis of *egl* gene sequence, two strains from tropical region of Iran were identified as phylotype II of *R. solanacearum* species complex and they had 100% sequence similarity to a biovar 2T strain infecting potato in Peru (Nouri et al. 2009). Strains of *R. solanacearum* (239) were collected from the main Solanaceae and Cucurbitaceae crops, and 126 representative strains were selected for phylogenetic analysis. The genetic and phenotypic diversity of *R. solanacearum* strains in French Guiana was assessed using diagnostic PCR and sequence-based (*egl* and *muts*) genotyping on a 239-strain collection revealing high diversity. Strains were distributed within phylotypes I (46.9%), IIA (26.8%) and IIB (26.3%), with one new *egl* sequence type (*egl ST*) found within such group. Phylotype IIB strains consisted mostly (97%) of strains with the emerging ecotype (IIB/ sequevar 4NPB). The host range of IIB/4NPB strains from French Guiana matched the original emerging reference strain from Martinque. They were virulent on cucumber, virulent and highly aggressive on tomato, including resistant reference Hawaii 7996 and only controlled by eggplant SM6 and Surya accessions. The emerging ecotype IIB/4NPB was fully established in French Guiana in both cultivated fields and uncultivated forests, rendering the hypothesis of introduction via ornamental cuttings or banana unlikely (Deberdt et al. 2014).

The gene contents of 18 strains of *R. solancearum* were compared and about 50% of the genes were found to be conserved, including many pathogenicity-associated genes (Guidot et al. 2007). Genetic diversity in 286 strains of *R. solanacearum* from 17 plant species in 13 Chinese provinces was investigated, using biovar and phylotype classification methods. A phylotype-specific multiplex PCR assay showed that 198 isolates belonged to phylotype I (biovar 3, 4 and 5) and 68 isolates to phylotype II (biovar 2 and biovar 1). Genetic diversity of 95 representative strains was examined based on phylogenetic analysis of the partial sequence of the genes *egl* and *hrpB* of all strains. The Chinese strains were assigned to phylotype I (Asia) and II (Americas). Phylotype I strains showed considerable phylogenetic diversity, including ten different sequevars: 12–18 and three new sequevars: 34, 44 and 48. Chinese strains Z1, Z2, Z3, Z7, Pe74 and Tm82 were not genetically distinguishable from the edible ginger reference strain ACH92 (race 4 biovar 4) for sequevar 16. All Chinese biovar 2 strains in the genetically and phenotypically diverse phylotype II were placed into phylotype IIB (Xu et al. 2009). The identities of strains (75) of *R. solanacearum* isolated from wilt-infected potatoes grown in different regions in India were established by PCR assay in which a single 280-bp product was amplified from all strains tested. Phylotype-specific multiplex PCR assigned 78.7% of strains to phylotype II, 16% to phylotype I and 5.3% to phylotype IV. Phylogenetic

analysis of *egl* gene sequences clustered all 59 phylotype II (bv 2) strains with reference to strain IPO1609 (IIB-1), all four phylotype IV (bv 2T) strains with reference strain MAFF301558 (IV-8), three phylotype I (bv 3) strains with reference strain MAFF211479 (I-30) and all eight phylotype I (bv 4) and one phylotype I (bv 3) strain with strain CIP365 (I-45). The results showed that Indian potato strains of *R. solanacearum* belonged to three of four phylotypes viz., the Asian phylotype I, the American phylotype II and Indonesian phylotype IV. The occurrence of phylotype IV sequevar 8 (bv 2T) strain of *R. solanacearum*, causing potato BW in mid hills of Meghalaya State in India was also recorded (Sagar et al. 2014).

The extent of diversity in phenotypic and genetic characteristics among strains of *R. solanacearum* pathogenic to potato was assessed, based on phylotype classification. European and Mediterranean strains of *R. solanacearum* along with 57 reference strains known to cover genetic diversity in this species were included in the tests. Phylogenetic analysis was done on endoglucanase gene sequences. Pathogenicity to potato, tomato and eggplant (brinjal) was established at 24 to 30°C and 15 to 24°C, whereas the tests on banana were performed at between 24 and 30°C. The ability to induce wilt on species of Solanaceae was shared by strains in all four phylotypes. Brown rot phylotypes IIB-1 and IIB-2 and phylotype IIB-27 established latent infections in banana, and Moko disease-causing phylotypes IIA-6, IIB-3 and IIB-4 were virulent to susceptible potato and tomato. Cold-tolerance ability was shared on species of Solanaceae among brown rot phylotype IIB-1 which were in majority among European and Mediterranean strains. The results indicated that pathogenicity traits of genetically identified strains are still not clear (Collier and Prior 2010). Pathogen dissemination routes are indicated by phylogenetic investigations which are instrumental for improving import/export controls. In order to get an insight into the phylogeography of *R. solanacearum*, genomes of 17 isolates of the pathogen were analyzed. Of the 17 isolates, 13 from Europe, Africa and Asia were found to belong to a single clonal lineage. On the other hand, isolates from South America were genetically diverse and tended to carry ancestral alleles at the analyzed genomic loci consistent with a South American origin of R3bv2. These isolates shared a core repertoire of 31 type III secreted effector genes, representing excellent candidates to be targeted with resistance genes in breeding programs to develop durable disease resistance to BW. To attain this goal, 27 R3bv2 effectors were tested in eggplant, tomato, pepper, tobacco and lettuce for induction of a hypersensitive-like response, indicative of recognition by cognate resistance receptors. Fifteen effectors, eight of them being core effectors, triggered a response in one or more plant species. These could be identified and mapped, cloned and expressed in tomato or potato, for which sources of genetic resistance to R3bv2 are extremely limited (Clarke et al. 2015).

As the genome sequences of bacterial pathogens are increasingly available, it has been possible to develop genomic distance methods to determine bacterial diversity. The results of genomic methods are highly correlated with those of the standard DNA-DNA hybridization technique. The genomic methods can be performed more rapidly at lower costs and are less prone to technical and human error. The genomic comparison methods were applied to differentiate members of *R. solanacearum* species complex into three species. Three different methods were used to compare the complete genomes of 29 strains of *R. solanacearum* species complex. The proteomics of 73 strains were profiled simultaneously, using Matrix-Assisted Laser Desorption/Ionization-Time of Flight Mass Spectrometry (MALDI-TOF-MS). Proteomic profiles together with genomic sequence comparisons consistently and comprehensively described the diversity of the *R. solanacearum* complex. Furthermore, the genome-driven functional phenotypic assays supported the earlier hypothesis that closely related members of *R. solanacearum* could be identified through a simple assay of anaerobic nitrate metabolism. It was possible to clearly and easily differentiate phylotype II and IV strains from phylotype I and phylotype III strains. The assay revealed large-scale differences in energy production within the *R. solanacearum* species complex and the usefulness of proteomic and genomic approaches to delineate bacterial species (Prior et al. 2016).

The DNA microarray technology has been shown to be useful for identification of DNA fragments specifically amplified from target bacterial plant pathogens. An array of species-specific oligonucleotide probes, representing the different potato pathogens built on a solid support, such as nylon membrane or microscope slide could be readily probed with labeled PCR products amplified from a potato sample. By employing conserved primers to amplify common bacterial genome fragments from extracts of potato tubers that might contain the target bacterial pathogens, the presence of DNA sequences, indicative of the pathogen species might be detected by hybridization to species-specific oligonucleotide probes within the array. Oligonucleotides, 16 to 24 bases long, from the 3′ end of the 16S gene and the 16S-23S intergenic spacer (ITS) regions of *Cms*, *R. solanacearum*, *Pectobacterium carotovorum* subsp. *atrosepticum (Pca)*, *P. carotovorum* subsp. *carotovorum (Pcc)* and *E. chrysanthemi (Ec)* strains infecting potatoes were designed and formatted into an array by pin spotting on nylon membranes. Genomic DNA from bacterial cultures was amplified by PCR using conserved ribosomal primers and labeled simultaneously with digoxigenin (DIG)-dUTP. Thr DIG-labeled PCR amplicons generated from pure cultures of bacteria generally hybridized to the array according to specific patterns. These patterns were highly reproducible in the two to three hybridizations performed with each of the bacterial templates. Labeled amplicons generated from nine strains of *Pca* hybridized to the ten oligonucleotides designed for this subspecies. Amplicons generated from all six strains of *Cms* hybridized to five homologous oligonucleotides. Amplicons of *R. solanacearum* did not hybridize to oligonucleotides designed from the sequences of heterologous species, except for a single weak heterologous hybridization signal among all the heterologous spots probed with *R. solanacearum*. In order to determine the discrimination potential of the oligonucleotide array, several mixtures of bacterial

strains were amplified by PCR with conserved primers and hybridized to the oligonucleotide array. The mixture of amplicons did not interfere with one another and the expected pattern for each species was obtained. The hybridization patterns for *Pca* strain 31 and *E. chrysanthemi* were consistent with the expected patterns for these pathogens. Bacterial pathogens could be identified directly in potato tissue using oligonucleotide array. The PCR amplicons generated from potato tubers inoculated with *Pca* strain 31 produced hybridization pattern as the pure cultures, revealing the reliability of the results obtained from the microarray technique. Bacteria-free tissue from microplantlets did not hybridize with any of the oligonucleotide probes designed for pathogenic bacterial species tested (see Figure 4.5). The results showed that the bacterial pathogens infecting potato could be detected and identified by hybridization to the array amplicons from mixed cultures and inoculated potato tissues, based on the hybridization patterns (Fessehaie et al. 2003).

A macroarray format was developed by employing a set of 9,676 probes to detect and differentiate *Pectobacterium atrosepticum (Pa) P. carotovorum (Pc)*, *Dickeya* spp., *S. scabies*, *S. turgidiscabies* and *Cms* infecting potatoes. Gene-specific probes could be designed for all genes of *Pa*, about 50% of the genes of *S. scabies* and about 30% of the genes of *C. michiganensis* subsp. *sepedonicus*, utilizing the whole-genome sequence information available. Probes (226) designed according to the sequences of a PAI containing important virulence genes were used for the detection of *S. turgidiscabies*. In addition, probes were designed for virulence-associated necrosis-inducing protein *(nip)* genes of *P. atrosepticum*, *P. carotovorum* and *Dickeya dadantii* for intergenic spacer (IGS) sequences of the 16S-23S rRNA gene region. The probes contained on an average, 40 target-specific nucleotides and were synthesized on the array *in situ*, organized as eight sub-arrays with identical set of probes which could be used for hybridization with different samples. All bacterial species could be readily differentiated, using a single channel system for single detection. Almost all of the ca. 1,000 probes designed for *Cms*, ca. 50% and 40% of the ca. 4,000 probes designed for the genes of *S. scabies* and *P. Atrosepticum*, respectively, and over 100 probes for *S. turgidiscabies* exhibited significant signals only with the respective bacterial species. *P. atrosepticum* and *P. cactorum* and *Dickeya* strains could be detected with 110 common probes. In contrast, the strains of these pathogens were different in their signal profiles. Probes targeting the IGS region and *nip* genes could be employed to assign strains of *Dickeya* to two groups which correlated with differences in virulence. The microarray analysis could be used to evaluate thousands of new probes in a time- and cost-effective manner which would be difficult by other methods. Utilization of the whole-genome sequence information for probe design resulted in large numbers of species-specific probes that readily distinguished the bacterial species and strains infecting potatoes (Aittamaa et al. 2008). The genotypic differences between strains of *Streptomyces turgidiscabies* were determined, using microarray technique. The areas of PAI designated 'colonization region' (CR) and 'toxigenic

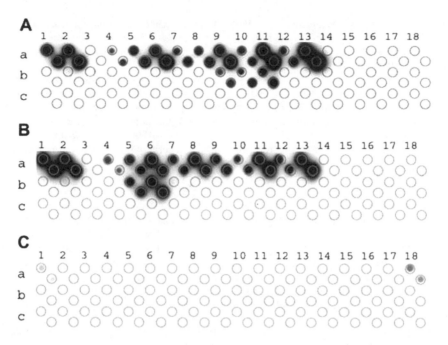

**FIGURE 4.5** Differentiation of strains of *Pca* and *E. chrysanthemi* based on hybridization patterns of digoxigenin (DIG)-labeled PCR amplicons A: mixture of *Pca* strain 31 and *Ec* strain 340 cells B: DNA from Pca-inoculated potato tuber C: DNA from potato tissue culture hybridization with oligonucleotides are revealed by development of dark spots within template circles.

[Courtesy of Fessihaie et al. (2003)) and with kind permission of the American Phytopathological Society, MN, United States]

region' (TR) contained genes required for virulence and phytotoxin production, respectively, in *Streptomyces* spp., causing common scab disease of potatoes. Microarray procedure was applied to assess genetic variability of PAI genes of strains of *S. turgidiscabies*. Four types of PAI were based on divergent CR and TR occurring in different combinations. Only one PAI type was highly similar to *S. scabies* strains 87.22 and ATTC 49173. Based on probes designed for the predicted genes of *S. scabies*, two gene clusters in S. *scabies* appeared to be similar in most strains of *S. turgidiscabies* and contained PAI genes corresponding to CR and TR. Multilocus cluster analysis indicated that two strains of *S. turgidscabies* were very closely related at the whole-genome level, but they contained distinctly different PAIs. The type strain of *S. reticuliscabies* was clustered with *S. turgidiscabies*. The results indicated a wide genetic variability of PAIs among strains of *S. turgidiscabies* and that PAI is made up of mosaic regions which might undergo independent evolution (Aittamaa et al. 2010).

# REFERENCES

Adriko J, Aritua V, Mortensen CN, Tushermereirwe WK, Kubiriba J, Lund OS (2012) Multiplex PCR for specific and robust detection of *Xanthomonas campestris* pv. *musacearum* in pure culture and infected plant material. *Plant Pathol* 61: 489–497.

Agrios GN (2005) *Plant Pathology*, Fifth edition, Elsevier/Academic Press, The Netherlands.

Ahamed NN, Islam Md R, Hossain MA, Meah MB, Hossain MM (2013) Determination of races and biovars of *Ralstonia solanacearum* causing bacterial wilt diseases of potato. *J Agrl Sci* 5

Aittamaa M, Somervuo P, Laakso I, Auvinen P, Valkonen JPT (2010) Microarray based on comparison of genetic differences between strains of *Streptomyces turgidiscabies* with focus on the pathogenicity island. *Molec Plant Pathol* 11: 733–746.

Aittamaa M, Somervuo P, Pirhonen M et al. (2008) Distinguishing bacterial pathogens of potato using a genome-wide microarray approach. *Molec Plant Pathol* 9: 705–712.

Albuquerque P, Marcal ARS, Caridade C, Costa R, Mendes MV, Travares F (2015) A quantitative hybridization approach using 17 DNA markers for identification and clustering analyses of *Ralstonia solanacearum*. *Plant Pathol* 64: 1270–1283.

Álvarez AM, Schenek S, Benedict AA (1996) Differentiation of *Xanthomonas albilineans* strains with monoclonal antibody reaction patterns and DNA fingerprints. *Plant Pathol* 45: 358–366.

Álvarez B, López MM, Bisoca EG (2007) Influence of native microbiota on survival of *Ralstonia solanacearum* phylotype II in river water microcosms. *Appl Environ Microbiol* 73: 7210–7217.

Baer D, Mitzel E, Pasche J, Gudmestad NC (2001) PCR detection of *Clavibacter michiganensis* subsp. *sepedonicus*-infected tuber samples in a plate culture assay. *Amer J Potato Res* 78: 269–277.

Bell KS, Sebaihia M, Pritchard L et al. (2004) Genome sequence of the enterobacterial phytopathogen *Erwinia carotovora* subsp. *atroseptica* and characterization of virulence factors. *Proc Natl Acad Sci USA* 101: 11105–11110.

Benedict A, Alvarez AM, Berestecky J et al. (1989) Pathovar-specific monoclonal antibodies for *Xanthomonas campestris* pv. *oryzae*. *Phytopathology* 79: 322–328.

Bini F, Geider K, Bazzi C (2008) Detection of *Agrobacterium vitis* by PCR using novel *virD2* gene-specific primers that discriminate two subgroups. *Eur J Plant Pathol* 122: 403–411.

Bosmans L, Paeleman A, Moerkens R, Wittermans L, Van Calenberge B, van Kerckhove S, De Mot R, Rediers H, Lievens B (2016) Development of a qPCR assay for detection and quantification of rhizogenic *Agrobacterium* biovar 1 strains. *Eur J Plant Pathol* 145: 719–730.

Bouri M, Chattaoui M, Gharsa HB et al. (2016) Analysis of *Agrobacterium* populations isolated from Tunisian soils: Genetic structure, avirulent-virulent ratios and characterization of tumorigenic strains. *J Phytopathol* 98

Bukhalid RA, Chung SY, Loria R (1998) *Nec1*, a gene conferring a necrogenic phenotype is conserved in plant pathogenic *Streptomyces* spp. and linked to a transposase pseudogene. *Molec Plant-Microbe Interact* 11: 960–967.

Burkhalid RA, Loria R (1997) Cloning and expression of a gene from *Streptomyces scabies* encoding putative pathogenicity factor. *J Bacteriol* 179: 7776–7783.

Caruso P, Gorris MT, Cambra M, Palomo JL, Collar J, López MM (2002) Enrichment double-antibody sandwich indirect enzyme-linked immunosorbent assay that uses a specific monoclonal antibody for sensitive detection of *Ralstonia solanacearum* in asymptomatic potato tubers. *Appl Environ Microbiol* 68: 3634–3638.

Caruso P, Palomo JL, Bertolini E, Álavarez B, López MA, Biosca EG (2005) Seasonal variation of *Ralstonia solanacearum* biovar 2 populations in a Spanish river: recovery of stressed cells at low temperatures. *Appl Environ Microbiol* 71: 140–148.

Champoiseau PG, Jones JB, Allen C (2009) *Ralstonia solanacearum* race 3 biovar 2 causes tropical losses and temperate anxieties. *Plant Health Progress*

Chandrasekhar B, Umesha S (2015) Detection, identification and phylogenetic analysis of Indian *Ralstonia solanacearum* isolates by comparison of partial *hrpB* gene sequences. *J Phytopathology* 163: 105–115.

Clarke CR, Studholme D, Hayes B et al. (2015) Genome-enabled phylogeographic investigation of the quarantine pathogen *Ralstonia solanacearum* race 3 biovar 2 and screening for sources of resistance against its core effectors. *Phytopathology* 105: 597–607.

Cohan PM (2006) Towards a conceptual and operational union of bacterial systematic, ecology and evolution. *Phil Trans Roy Soc Biol* 36: 1985–1996.

Collier G, Prior P (2010) Deciphering phenotypic diversity of *Ralstonia solanacearum* strains pathogenic to potato. *Phytopathology* 100: 1250–1261.

Corrêa DBA, Salomão D, Rodrigues-Neto J, Destefano SAL (2015) Application of PCR-RFLP technique to species identification and phylogenetic analysis of *Streptomyces* associated with potato scab in Brazil based on partial *atpD* gene sequences. *Eur J Plant Pathol* 142: 1–12.

Coupat-Goutaland B, Bernillon D, Guidot A, Prior P, Nesme X, Bertolla F (2011) *Ralstonia solanacearum* virulence increased following large interstrain gene transfers by natural transformation. *Molec Plant-Microbe Interact* 24: 497–505.

Crockett AO, Wittwer CR (2001) Fluorescein-labeled oligonucleotides for real-time PCR: Using the inherent quenching of deoxyguanosine nucleotides. *Annal Biochem* 290: 89–97.

Cubero J, Martinez MC, Llop P, Lopez M (1999) A simple and efficient PCR method for the detection of *Agrobacterium tumefaciens*. *J Appl Microbiol* 86: 591–602.

Cubero J, van der Wolf J, van Beckhoven J, López MM (2002) An internal control for the diagnosis of crown gall by PCR. *J Microbiol Meth* 51: 387–392.

Czajkowski R, Grabe GJ, van der Wolf JM (2009) Distribution of *Dickeya* spp. and *Pectobacterium carotovorum* subsp. *carotovorum* in naturally infected seed potatoes. *Eur J Plant Pathol* 125: 263–275.

Dana H, Khodakaramian G, Rouhrazi K (2015) Characterization of *Pectobacterium carotovorum* subsp. *carotovorum* causing watermelon soft rot disease in Iran. *J Phytopathol* 163: 703–710.

Deberdt P, Guyot J, Coranson-Beaudu R et al. (2014) Diversity of *Ralstonia solanacearum* in French Guiana expands knowledge of the "emerging ecotype". *Phytopathology* 104: 586–596.

Deepa J, Girija D, Mathew SK, Nazeem PA, Babu TD, Verma AS (2003) Detection of *Ralstonia solanacearum* race 3 causing bacterial wilt of solanaceous vegetables in Kerala, using random amplified polymorphic DNA (RAPD) analysis. *J Trop Agric* 41: 33–37.

Dees MW, Sletten A, Hermansen A (2013) Isolation and characterization of *Streptomyces* species from potato common scab lesions in Norway. *Plant Pathol* 62: 217–225.

Degefu Y, Potrykus M, Golanowska M, Virtanen E, Lojkowska E (2013) A new clade of *Dickeya* spp. plays a major role in potato blackleg outbreaks in North Finland. *Ann Appl Biol* 162: 231–241.

De Boer SH, Li X, Ward LJ (2012) *Pectobacterium* spp. associated with bacterial stem rot syndrome of potato in Canada. *Phytopathology* 102: 937–947.

De Boer SH, Mc Naughton ME (1987) Monoclonal antibodies to the lipopolysaccharides of *Erwinia carotovora* subsp. *atroseptica* serogroup I. *Phytopathology* 77: 828–632.

De Boer SH, Ward LJ (1995) PCR detection of *Erwinia carotovora* subsp. *atroseptica* associated with potato tissue. *Phytopathology* 85: 854–858.

De Parasis J, Roth DA (1990) Nucleic acid probes for identification of genus specific 16S rRNA sequences. *Phytopathology* 80: 618.

Diallo S, Latour X, Groboillot A et al. (2009) Simultaneous and selective detection of two major soft rot pathogens of potato: *Pectobacterium atrosepticum* (*Erwinia carotovora* subsp. *atroseptica*) and *Dickeya* spp. (*Erwinia chrysanthemi*). *Eur J Plant Pathol* 125: 349–354.

Dickey RS (1979) *Erwinia chrysanthemi*: a comparative study of phenotypic properties of strains from several hosts and other *Erwinia* species. *Phytopathology* 69: 324–329.

Dittapongpitch V, Surat S (2003) Detection of *Ralstonia solanacearum* in soil and weeds from commercial tomato fields using immunocapture and the polymerase chain reaction. *J Phytopathol* 151: 239–246.

Dong LC, Sun CW, Thies KL, Luther DS, Graves CH Jr. (1992) Use of polymerase chain reaction to detect pathogenic strains of *Agrobacterium*. *Phytopathology* 82: 434–439.

Elphinstone JG, Stanford HM, Stead DE (1998) Survival and transmission of *Ralstonia solanacearum* in aquatic plants of *Solanum dulcamara* and associated with surface water in England. *Bull OEPP* 28: 93–94.

Fegan M, Prior P (2005) How complex is the "*Ralstonia solanacearum* species complex"? In: Prior P, Allen C, Elphinstone J (eds.), *Bacterial Wilt Disease and the Ralstonia solanacearum Species Complex*, The American Phytopathological Society, Minneapolis, USA.

Fessehaie A, De Boer SH, Levesque CA (2003) An oligonucleotide array for the identification and differentiation of bacteria pathogenic on potato. *Phytopathology* 93: 262–269.

Fiori M, Schiaffino A (2004) Bacterial stem rot in greenhouse pepper (*Capsicum annuum* L.) in Sardinia (Italy): Occurrence of *Erwinia carotovora* subsp. *carotovora*. *J Phytopathol* 162: 28–33.

Flores-González R, Velasco I, Montes F (2008) Detection and characterization of *Streptomyces* causing potato common scab in Western Europe. *Plant Pathol* 57: 162–169.

Graham J, Lloyd AB (1978) An improved indicator plant method for detection of *Pseudomonas solanacearum* race 3 in soil. *Plant Dis Rept* 62: 35–37.

Griep RA, van Twisk C, van Beckhoven JRCM, van der Wolf JM, Schots A (1998) Development of specific recombinant monoclonal antibodies against lipopolysaccharide of *Ralstonia solanacearum* race 3. *Phytopathology* 88: 795–803.

Gudmestad NC, Mallik I, Pasche JS, Anderson NR, Kinzer K (2009) A real-time PCR assay for the detection of *Clavibacter michiganensis* subsp. *sepedonicus* based on the cellulase A gene sequence. *Plant Dis* 93: 649–659.

Guidot A, Prior P, Schoenfeld J, Carrere S, Genin S, Boucher C (2007) Genomic structure and phylogeny of the plant pathogen *Ralstonia solanacearum* inferred from gene distribution analysis. *J Bacteriol* 189: 377–387.

Gupta A, Gupta AK, Mahajan R et al. (2013) Protocol for isolation and identification of *Agrobacterium* isolates from stone fruit plants and sensitivity of native *A. tumefaciens* isolates against agrocin produced by *A. radiobacter* strain K84. *Natl Acad Sci Lett* 36: 79–84.

Ha Y, Kim J-S, Denny TP, Schell MA (2012) A rapid, sensitive assay for *Ralstonia solanacearum* race 3 biovar 2 in plant and soil samples using magnetic beads and real-time PCR. *Plant Dis* 96: 258–264.

Hadas R, Kritzman G, Gefen T, Manulis S (2001) Detection, quantification and characterization of *Erwinia carotovora* ssp. *carotovora* contaminating pepper seeds. *Plant Pathol* 50: 117–123.

Harper SJ, Ward LI, Clover GRG (2010) Development of LAMP and real-time PCR methods for the rapid detection of *Xylella fastidiosa* for quarantine and field applications. *Phytopathology* 100: 1282–1288.

Hartung F, Werner R, Mühlbach HP, Büttner C (1998) Highly specific PCR diagnosis to determine *Pseudomonas solanacearum* strains of different geographic origin. *Theor Appl Genet* 96: 797–782.

Haas JH, Moore LW, Ream W, Manulis S (1995) PCR primers for detection of phytopathogenic *Agrobacterium* strains. *Appl Environ Microbiol* 61: 2879–2884.

Healy FG, Wach M, Krasnoff SB, Gibson DM, Loria R (2000) The *txtAB* genes of the plant pathogen *Streptomyces acidiscabies* encode a peptide synthetase required for phytotoxin thaxtomin A production and pathogenicity. *Molec Microbiol* 38: 794–804.

Helias V, Roux AC, Ie Bertheau Y, Andrivon D, Gauthier JP, Jouan B (1998) Characterization of *Erwinia carotovora* subspecies and detection of *Erwinia carotovora* subsp. *atroseptica* in potato plants, soil, or water extracts with PCR-based methods. *Eur J Plant Pathol* 104: 689–699.

Hifini HR, Nishiyama K, Izuka A (1975) Bacteriological characteristics of some isolates of *Xanthomonas oryzae* different in their pathogenicity and locality. *Contr Cent Res Inst Agric Bogor (Indonesia)* 16: 1–18.

Hodgetts J, Karamura G, Johnson G et al. (2015) Development of a lateral flow device for diagnostic methods for *Xanthomonas campestris* pv. *musacearum*, the agent of banana Xanthomonas wilt. *Plant Pathol* 64: 559–567.

Holmes A, Govan J, Goldstein R (1978) Agricultural use of *Burkholderia* (*Pseudomonas*) *cepacia*: a threat to human health. *Emerg Infect Dis* 4: 221–227.

Hong JC, Momol MT, Jones JB et al. (2008) Detection of *Ralstonia solanacearum* in irrigation ponds and aquatic weeds associated with the ponds in North Florida. *Plant Dis* 92: 1674–1682.

Horita M, Tsuchiya K, Suga Y, Yano K, Waki T, Kurose D (2014) Current classification of *Ralstonia solanacearum* and genetic diversity of the strains in Japan. *J Gen Plant Pathol* 80: 455–465.

Horita M, Yano T, Tsuchiya K (2004) PCR-based specific detection of *Ralstonia solanacearum* race 4 strains. *J Gen Plant Pathol* 70: 278–283.

Hu X, Lai F-M, Reddy ANS, Ishimura CA (1995) Quantitative detection of *Clavibacter michiganensis* subsp. *sepedonicus* by competitive polymerase chain reaction. *Phytopathology* 85: 1468–1473.

Huang BC, Zhu H, Hu GG, Li QX, Gao JL, Goto M (1993) Serological studies on *Xanthomonas campestris* pv. *oryzae* with polyclonal and monoclonal antibodies. *Ann Phytopathol Soc Jpn* 65: 100–115.

Hyman LJ, Sullivan L, Toth IK, Perombelon MCM (2001) Modified crystal violet pectate medium (CVP)-based new pectate source (Slendid) for the detection and isolation of soft rot erwinias. *J Potato Res* 44: 265–270.

Ivey MLL, Tusiime G, Miller SA (2010) A polymerase chain reaction assay for the detection of *Xanthomonas campestris* pv. *musacearum*. *Plant Dis* 94: 109–114.

Janse JD, Schans J (1998) Experience with the diagnosis and epidemiology of bacterial brown rot (*Ralstonia solanacearum*) in the Netherlands. *Bull OEPP* 28: 65–67.

Ji P, Allen C, Sanchez-Perez A et al. (2007) New diversity of *Ralstona solanacearum* strains associated with vegetable and ornamental crops in Florida. *Plant Dis* 91: 195–203.

Johnson KL, Zheng D, Kaewnum S, Reid CL, Burr T (2013) Development of a magnetic capture hybridization real-time assay for detection of tumorigenic *Agrobacterium vitis* in grapevines. *Phytopathology* 103: 633–640.

Joshi M, Rong S, Moll S, Kers J, Franco C, Loria R (2007) *Streptomyces turgidiscabies* secretes a novel protein Nec1 which facilitates infection. *Molec Plant-Microbe Interact* 20: 599–608.

Kawaguchi A, Sawada H, Ichinose Y (2008) Phylogenetic and serological analyses reveal genetic diversity of *Agrobacterium vitis* strains in Japan. *Plant Pathol* 57: 747–753.

Kers J, Cameron K, Joshi M et al. (2005) A large mobile pathogenicity island confers plant pathogenicity on *Streptomyces* species. *Molec Microbiol* 55: 1025–1033.

Khakvar R, Sijam K, Yun WM, Radu S, Lin TK (2008) Improving a PCR-based method for identification of *Ralstonia solanacearum* in natural sources of West Malaysia. *Amer J Agric Biol Sci* 3: 490–493.

Khaledian S, Nikkah M, Shams-bakhsh M, Hoseinzadeh S (2017) A sensitive biosensor based on gold nanoparticles to detect *Ralstonia solanacearum* in soil. *J Gen Plant Pathol* 83: 231–239.

Krieg NR, Holt JG (eds.) (1984) *Bergey's Manual of Systematic Bacteriology*, Volume 1, Williams & Wilkins, Baltimore, MD.

Krimi Z, Raio A, Petit A, Neseme X, Dessaux Y (2006) *Eucalyptus occidentalis* plants are naturally infected by pathogenic *Agrobacterium tumefaciens*. *Eur J Plant Pathol* 116: 237–246.

Kubota R, Vine BG, Alvarez AM, Jenkins DM (2008) Detection of *Ralstonia solanacearum* by loop-mediated isothermal amplification. *Phytopathology* 98: 1045–1051.

Kumar A, Prameela TP, Suseelabhai R, Siljo A, Anandaraj M, Vinatzer BA (2014) Host specificity and genetic diversity of race 4 strains of *Ralstonia solanacearum*. *Plant Pathol* 63: 1138–1148.

Kuo TT, Huang TC, Wu RY, Yang CM (1967) Characterization of three bacteriophages of *Xanthomonas oryzae* pv. *oryzae* (Uyeda et Ishiyama) Dowson. *Bot Bull Acad Sin* 8: 246–254.

Kutin RK, Alvarez A, Jenkins DM (2009) Detection of *Ralstonia solanacearum* in natural substrates using phage amplification integrated with real-time PCR assay. *J Microbiol Meth* 76: 241–246.

Kuzmanović N, Gašić K, Ivanovic M, Prokić A, Obradović A (2012) Identification of *Agrobacterium vitis* as a causal agent of grapevine crown gall in Serbia. *Arch Biol Sci, Belgrade* 64: 1487–1494.

Laurila J, Ahola V, Lehtinen A et al. (2008) Characterization of *Dickeya* strains isolated from potato and river water samples in Finland. *Eur J Plant Pathol* 122: 213–225.

Lee Y, Wang C-A (2000) The design of specific primers for the detection of *Ralstonia solanacearum* in soil samples by polymerase chain reaction. *Bot Bull Acad Sin* 41: 121–128.

Lee Y-A, Yu C-P (2006) A differential medium for the isolation and rapid identification of a plant soft rot pathogen *Erwinia chrysanthemi*. *J Microbiol Meth* 64: 200–206.

Lehtonen MJ, Rantala H, Kreuze JF et al. (2004) Occurrence and survival of potato scab pathogens (*Streptomyces* species) on tuber lesions: quick diagnosis based on a PCR-based assay. *Plant Pathol* 53: 280–287.

Leiminger T, Frank M, Wenk C, Poschenrieder G, Kellermann A, Schwarzfischer A (2013) Distribution and characterization of *Streptomyces* species causing potato common scab in Germany. *Plant Pathol* 62: 611–623.

Li Q, Guo R-J, Li S-D, Li S-F, Wang H-Q (2015) Determination of tumorigenic *Agrobacterium* density in soil by real-time PCR assay and its effect on crown gall disease severity. *Eur J Plant Pathol* 142: 25–36.

Li X, De Boer SH (1995) Selection of polymerase chain reaction primers from an RNA intergenic region for specific detection of *Clavibacter michiganensis* subsp. *sepedonicus*. *Phytopathology* 85: 837–842.

Li X, Nie J, Ward L et al. (2009) Comparative genomic-guided loop-mediated isothermal amplification for characterization of *Pseudomonas syringae* pv. *phaseolicola*. *J Appl Microbiol* 107: 717–726.

Li X, Nie J, Ward LJ, Nickerson J, De Boer SH (2011) Development and evaluation of a loop-mediated isothermal amplification assay for rapid detection and identification of *Pectobacterium atrosepticum*. *Canad J Plant Pathol* 33: 447–457.

Lin C-H, Hsu S-T, Tzeng K-C, Wang J-F (2009) Detection of race 1 strains of *Ralstonia solanacearum* in field samples in Taiwan using BIO-PCR methods. *Eur J Plant Pathol* 124: 75–85.

Llop P, Lastra B, Marsal H, Murillo J, López MM (2003) Tracking *Agrobacterium* strains by a RAPD system to identify single colonies from plant tumors. *Eur J Plant Pathol* 109: 381–389.

Lopéz R, Asensio C, Gilbertson RL (2006) Phenotypic and genotypic diversity in strains of common blight bacteria (*Xanthomonas campestris* pv. *phaseoli* and *X. campestris* pv. *phaseoli* var. *fuscans*) in a secondary center of diversity of the common bean host suggests multiple introduction events. *Phytopathology* 96: 1204–1213.

Louws FJ, Beu J, Medina-Mora CM et al. (1998) rep-PCR-mediated genome fingerprinting: a rapid and effective method to identify *Clavibacter michiganensis*. *Phytopathology* 88: 862–868.

Mahmoudi E, Soleimani MJ, Taghavi M (2007) Detection of bacterial soft rot of crown imperial caused by *Pectobacterium carotovorum* subsp. *carotovorum*, using specific primers. *Phytopathol Mediterr* 46: 168–176.

Mahuku GS, Jara C, Henriquez MA, Catellanos G, Cuasquer J (2006) Genotypic characterization of the common bean bacterial blight pathogens *Xanthomonas axonopodis* pv. *phaseoli* and *Xanthomonas axonopodis* pv. *phaseoli* var. *fuscans* by rep-PCR and PCR-RFLP of the ribosomal genes. *J Phytopathol* 154: 35–44.

Maisuria VB, Nerukar AS (2013) Characterization and differentiation of soft rot causing *Pectobacterium carotovorum* of Indian origin. *Eur J Plant Pathol* 136: 87–102.

Manome A, Kageyama A, Kurata S et al. (2008) Quantification of potato common scab pathogens in soil by quantitative competitive PCR with fluorescent quenching-based probes. *Plant Pathol* 57: 887–896.

Marco-Noales E, Bertolini E, Morente C, López MM (2008) Integrated approach for detection of nonculturable cells of *Ralstonia solanacearum* in asymptomatic *Pelargonium* spp. cuttings. *Phytopathology* 98: 949–955.

Massart S, Nagy C, Jigakli MH (2014) Development of the simultaneous detection of *Ralstonia solanacearum* race 3 and *Clavibacter michiganensis* subsp. *sepedonicus* in potato tubers by a multiplex real-time PCR assay. *Eur J Plant Pathol* 138: 29–37.

Matsuyama N, Daikohara M, Yoshimura K et al. (2009) Presumptive differentiation of phytopathogenic and nonpathogenic bacteria by improved rapid-extraction TLC method. *J Fac Agric Kyushu Uinv* 54: 1–11.

Mc Garvey A, Denny TP, Schell MA (1999) Spatial-temporal and quantitative analysis of growth and EPS production by *Ralstonia solanacearum* in resistant and susceptible tomato cultivars. *Phytopathology* 89: 1233–1239.

Mc Nally RR, Ishimaru CA, Malvick DK (2016) PCR-mediated detection and quantification of the Goss's wilt pathogen *Clavibacter michiganensis* subsp. *nebraskensis* via a novel gene target. *Phytopathology* 106: 1465–1472.

Meng X, Chai A, Shi Y, Xie X, Ma Z, Li B (2017) Emergence of bacterial soft rot in cucumber by *Pectobacterium carotovorum* subsp. *brasiliense* in China. *Plant Dis* 101: 279–287.

Mills D, Russell BW, Hanus JW (1997) Specific detection of *Clavibacter michiganensis* subsp. *sepedonicus* by amplification of three unique DNA sequences isolated by subtraction hybridization. *Phytopathology* 87: 853–861.

Moline HF (1985) Differentiation of postharvest soft-rotting bacteria in two-dimensional polyacrylamide gel electrophoresis. *Phytopathology* 75: 549–553.

Moretti C, Fakhr R, Cortese C et al. (2016) *Pectobacterium carotovorum* subsp. *carotovorum* as causal agents of potato soft rot in Lebanon. *Eur J Plant Pathol* 144: 205–211.

Mun HS, Oh E-J, Kim H-J et al. (2007) Differentiation of *Streptomyces* spp. which cause potato scab disease on the basis of *rpoB* gene sequences. *Syst Appl Microbiol* 30: 401–407.

Mwangi M, Mwebaze M, Bandyopadhyay R et al. (2007) Development of a semiselective medium for isolating *Xanthomonas campestris* pv. *musacearum* from insect vectors, infected plant material and soil. *Plant Pathol* 56: 383–390.

Narayanasamy P (2001) *Plant Pathogen Detection and Disease Diagnosis*, Second edition, Marcel Dekker, New York.

Narayanasamy P (2011) *Microbial Plant Pathogens–Detection and Disease Diagnosis*, Volume 2, Bacterial and Phytoplasmal Pathogens. Springer Science + Business Media B.V., Heidelberg, Germany.

Norman DJ, Yuen JMF (1998) A distinct pathotype of *Ralstonia solanacearum* race 1, biovar 1 entering Florida in pothos (*Epipremum aureum*) cuttings. *Canad J Plant Pathol* 20: 171–175.

Norman DJ, Zapata M, Gabriel DW et al. (2009) Genetic diversity and host range variation of *Ralstonia solanacearum* strains entering North America. *Phytopathology* 99: 1070–1077.

Nouri S, Bahar M, Fegan M (2009) Diversity of *Ralstonia solanacearum* causing potato bacterial wilt in Iran and the first record of phylotype II/biovar 2T strain outside South America. *Plant Pathol* 58: 243–249.

Ou SH (1985) *Rice Diseases*, Second edition, Common Mycological Institute, Surrey, UK.

Ozakman M, Schaad NW (2003) A real-time BIO-PCR assay for detection of *Ralstonia solanacearum* race 3, biovar 2 in asymptomatic potato tubers. *Canad J Plant Pathol* 25: 232–239.

Palacio-Bielsa A, González-Abaolafio R, Álvarez B et al. (2009) Chromosomal and Ti plasmid characterization of tumorigenic strains of three *Agrobacterium* species isolated from grapevine tumors. *Plant Pathol* 58: 584–593.

Palomo JL, López MM, García-Benavides P, Valázquez E, Martinez-Molina E (2006) Evaluation of the API 50CH and API ZYM systems for rapid characterization of *Clavibacter michiganensis* subsp. *sepedonicus*, causal agent of potato ring rot. *Eur J Plant Pathol* 115: 443–451.

Park DH, Yu YM, Kim JS, Cho JM, Hur JH, Lim CK (2003) Characterization of streptomycetes causing potato common scab in Korea. *Plant Dis* 87: 1290–1296.

Pastrik K-H, Elphinstone JG, Pukall R (2002) Sequence analysis and detection of *Ralstonia solanacearum* by multiplex PCR amplification of 16S-23S ribosomal intergenic spacer region with internal positive control. *Eur J Plant Pathol* 108: 831–842.

Pirone L, Chiarini L, Dalmastric C, Bevivino A, Tabacchioni S (2005) Detection of cultured and uncultured *Burkholderia cepacia* complex bacteria naturally occurring in the maize rhizosphere. *Environ Microbiol* 7: 1734–1742.

Potrykus M, Sledz W, Golanowska M et al. (2014) Simultaneous detection of major blackleg and soft rot bacterial pathogens in potato by multiplex polymerase chain reaction. *Ann Appl Biol* 165: 474–487.

Pradhanang PM, Elphinstone JG, Fox RTV (2000) Sensitive detection of *Ralstonia solanacearum* in soil: a comparison of different detection techniques. *Plant Pathol* 49: 414–422.

Prameela TP, Suseela Bhai R, Anandaraj M, Kumar A (2017) Development of real-time loop-meidated isothermal amplification for detection of *Ralstonia pseudosolanacearum* racc4 in rhizomes and soil. *Austr Plant Pathol* 46: 547–549.

Prior P, Ailloud F, Dalsing BL, Remenant B, Sanchez B, Allen C (2016) Genomic and proteomic evidence supporting the division of the plant pathogen *Ralstonia solanacearum* into three species. *BMC Genomics* 17: 90

Priou S, Aley P, Fernandez H, Gutarra L (1999) Sensitive detection of *Ralstonia solanacearum* (race 3) in soil by post-enrichment DAS-ELISA. *Bacterial Wilt Newsletter* 16: 10–13.

Priou S, Gutarra L, Aley P (2006) An improved enrichment broth for sensitive detection of *Ralstonia solanacearum* (biovars 1 and 2A) in soil using DAS-ELISA. *Plant Pathol* 55: 36–45.

Priou S, Gutarra L, Aley P, De Mendiburu F, Llique R (2010) Detection of *Ralstonia solanacearum* (biovar 2A) in stems of symptomless plants before harvest of potato crop using post-enrichment DAS-ELISA. *Plant Pathol* 59: 59–67.

Puopolo G, Raio A, Zoina A (2007) Early detection of *Agrobacterium tumefaciens* in symptomless artificially inoculated chrysanthemum and peach plants using PCR. *J Plant Pathol* 89: 185–190.

Qu X, Wanner A, Christ BJ (2008) Using the *txtAB* operon to quantify pathogenic *Streptomyces* in potato tubers and soil. *Phytopathology* 98: 405–412.

Rahman MF, Islam MR, Rahman T, Meah MB (2010) Biochemical characterization of *Ralstonia solanacearum* causing bacterial wilt of brinjal in Bangla Desh. *Progr Agric* 21: 9–19.

Rajeshwari N, Shylaja MD, Krishnappa M, Shetty HS, Mortensen CN, Mathur SB (1998) Development of ELISA for the detection of *Ralstonia solanacearum* in tomato: its application in seed health testing. *World J Microbiol Biotechnol* 14: 697–704.

Ramesh R, Achari GA, Gaitonde S (2014) Genetic diversity of *Ralstonia solanacearum* infecting solanaceous vegetables from India reveals the existence of unknown or newer sequevars of phylotype I strains. *Eur J Plant Pathol* 140: 543–562.

Rivas R, Velazquez E, Palomo J-L, Mateos PF, Garcia-Benavides P, Martinez-Molina E (2002) Rapid identification of *Clavibacter michiganensis* subspecies *sepedonicus* using two primers random amplified polymorphic DNA (TP-RAPD) fingerprints. *Eur J Plant Pathol* 108: 179–184.

Romeiro RS, Silva HAS, Beriam LOS, Rodrigues Neto J, de Carvalho MG (1999) A bioassay for detection of *Agrobacterium tumefaciens* in soil and plant material. *Summa Phytopathol* 25: 359–362.

Sachadyn P, Kur J (1997) A new PCR system for *Agrobacterium tumefaciens* detection based on amplification of T-DNA fragment. *Acta Microbiol Polo* 46: 145–146.

Safenkova IV, Zaitsev IA, Pankratova GK (2014) Lateral flow immunoassay for rapid detection of potato ring rot caused by *Clavibacter michiganensis* subsp. *sepedonicus*. *Appl Biochem Microbiol* 50: 675–682.

Safenkova IV, Zaitsev IA, Varitsev YA, Zherdev AV, Dzantiev BB (2015) Lateral flow immunoassay for rapid diagnosis of potato blackleg caused by *Pectobacterium atrosepticum*. *Biosci Biotechnol Res Asia* 12: 1937–1945.

Sagar V, Jeevalatha A, Mian S et al. (2014) Potato bacterial wilt in India caused by strains of phylotype I, II and IV of *Ralstonia solanacearum*. *Eur J Plant Pathol* 138: 51–65.

Salles JF, de Souza FA, van Elas JD (2002) Molecular method to assess the diversity of *Burkholderia* species in environmental samples. *Appl Environ Microbiol* 68: 1595–1603.

Schaad NW, Takatsu A, Dianese JC (1978) Serological identification of strains of *Pseudomonas solanacearum* in Brazil. Proc 4th Internatl Conf Plant Pathogenic Bacteria, Angers, France, pp. 295–300.

Schumann GL (1991) *Plant Diseases: Their Biology and Social Impact*, The American Phytopathological Society, St Paul, MN, USA.

Seo ST, Furuya N, Lim CK, Takanami Y, Tsuchiya K (2003) Phenotypic and genetic characterization of *Erwinia carotovora* from mulberry (*Morus* spp.). *Plant Pathol* 52: 140–146.

Seo Tsuchiya K, Murakami R (2004) Characterization and differentiation of *Erwinina carotovora* strains from mulberry trees. *J Gen Plant Pathol* 70: 120–123.

Shahbaz MU, Mukhtar T, Ul-Haque MI, Nasem Begum (2015) Full length biochemical and serological characterization of *Ralstonia solanacearum* associated with chilli seeds in Pakistan. *Internatl J Agric Biol* ISSN Online: 1814-959614-959614-093/2015/17-1-31-40.

Shan ZH, Liao BS, Tan YL, Li D, Lei Y, Shen MZ (1997) ELISA technique used to detect latent infection of groundnut by bacterial wilt (*Pseudomonas solanacearum*). *Oilcrops, China* 19: 45–47.

Singh U, Trevors CM, De Boer SH, Janse JD (2000) Fimbria-specific monoclonal antibody-based ELISA for European potato strains of *Erwinia chrysanthemi* and comparison to PCR. *Plant Dis* 84: 443–448.

Slawiak M, van Doorn R, Szemes M et al. (2013) Multiplex detection and identification of bacterial pathogens causing potato blackleg and soft rot in Europe, using padlock probes. *Ann Appl Biol* 163: 378–393.

Smith DS, De Boer SH (2009) Implementation of an artificial reaction control in a TaqMan method PCR detection of *Ralstonia solanacearum* race 3 biovar 2. *Eur J Plant Pathol* 124: 405–412.

Smith DS, De Boer SH, Gourley J (2008) An internal reaction control for routine detection of *Clavibacter michiganensis* subsp. *sepedonicus* using a real-time TaqMan PCR-based assay. *Plant Dis* 92: 684–693.

Song J, Lee S-C, Kang J-W, Baek H-J, Suh J-W (2004) Phylogenetic analysis of *Streptomyces* spp. isolated from potato scab lesions in Korea on the basis of 16S rRNA gene and 16S-23S rDNA internally transcribed spacer sequences. *Internatl J Syst Evol Microbiol* 54: 203–209.

Stulberg MJ, Shao J, Huang Q (2015) A multiplex PCR assay to detect and differentiate select agent strains of *Ralstonia solanacearum*. *Plant Dis* 99: 333–341.

Suharjo R, Sawada H, Takikawa Y (2014) Phylogenetic study of Japanese *Dickeya* spp. and development of new rapid identification methods using PCR-RFLP. *J Gen Plant Pathol* 80: 237–254.

Swanson JK, Montes L, Mejia L, Allen C (2007) Detection of latent infections of *Ralstonia solanacearum* race 3 biovar 2 in geranium. *Plant Dis* 91: 828–834.

Tiwary BN, Prasad B, Ghosh A, Kumar S, Jain RK (2007) Characterization of two novel biovar of *Agrobacterium tumefaciens* isolated from root nodules of *Vicia faba*. *Curr Microbiol* 55: 328–333.

Tomlinson DL, Elphinstone JG, Soliman MV et al. (2009) Recovery of *Ralstonia solanacearum* from canal water in traditional potato-growing areas of Egypt, but not from designated pest-free areas. *Eur J Plant Pathol* 125: 589–601.

Toth IK, Avrova AO, Hyman LJ (2001) Rapid identification and differentiation of the soft rot erwinias by 16S-23S intergenic transcribed spacer-PCR and restriction fragment length polymorphism analyses. *Appl Environ Microbiol* 67: 4070–4076.

Toth IK, Hyman LJ, Wood JR (1999) A one-step PCR-based method for detection of economically important soft rot *Erwinia* species in micropropagated potato plants. *J Appl Microbiol* 87: 158–166.

Tran TM, Jacobs JM, Huerta A, Milling A, Weibel J, Allen C (2016) Sensitive, secure detection of race 3 biovar 2 and native US strains of *Ralstonia solanacearum*. *Plant Dis* 100: 630–639.

Tripathi L, Tripathi JN, Tushemerirwe WK, Bandyopadhyay R (2007) Development of a semiselective medium for isolation of *Xanthomonas campestris* pv. *musacearum* from banana plants. *Eur J Plant Pathol* 117: 177–186.

Tsror L, Erlich O, Hazanovsky M, Daniel BB, Zig U, Lebiush S (2012) Detection of *Dickeya* spp. latent infection in potato seed tubers using PCR or ELISA and correlation with disease incidence in commercial field crops under hot-climate conditions. *Plant Pathol* 61: 161–168.

Tsror (Lahkim) L, Erlich O, Lebiush S (2009) Assessment of recent outbreaks of *Dickeya* sp. (syn. *Erwinia chrysanthemi*) slow wilt in potato crops in Israel. *Eur J Plant Pathol* 123: 311.

Tsuchiya K, Yano K, Horita M, Morita Y, Kawada Y, d'Ursel CM (2005) Occurrence and epidemic adaptation of new species of *Ralstonia solanacearum* associated with Zingiberaceae plants under agroecosystems in Japan. In: Allen C, Prior Ph, Haywood AC (eds.), Bacterial Wilt Disease and *Ralstonia Solanacearum* Species Complex, The American Phytopathological Society, St. Paul, MN, pp. 463–469.

van der Wolf JM, Hyman LJ, Jones DAC et al. (1996) Immunomagnetic separation of *Erwinia carotovora* subsp. *atroseptica* from potato peel extracts to improve detection sensitivity on crystal violet pectate medium or by PCR. *J Appl Bacteriol* 80: 487–495.

van der Wolf JM, van Beckhoven JRCM, De Haan EG, van den Bovenkamp GW, Leone GOM (2004) Specific detection of *Ralstonia solanacearum* 16S rRNA sequences by AmpliDet RNA. *Eur J Plant Pathol* 110: 25–53.

van der Wolf JM, Vriend SGC, Kastelein P, Nijhuis EH, van Bekkum PJ, van Vuurde JWL (2000) Immunofluorescence colony-staining (IFC) for detection and quantification of *Ralstonia* (*Pseudomonas*) *solanacearum* biovar 2 (race 3) in soil and verification of positive results by PCR and dilution plating. *Eur J Plant Pathol* 106: 123–133.

Van Elsas JD, Kastelein P, de Vries PM, van Overbeck LS (2001) Effects of ecological factors in the survival and physiology of *Ralstonia solanacearum* biovar 2 in irrigation water. *Canad J Microbiol* 47: 842–854.

Vera Cruz CM, Mew TW (1989) How variable is the *Xanthomonas campestris* pv. *oryzae*? In: *Bacterial Blight of Rice*, Proc Internatl Workshop, International Rice Research Institute, Manila, Philippines, pp. 153–166.

Villa JE, Tsuchiya K, Horita M, Natural M, Opina N, Hyakumachi M (2003) DNA analysis of *Ralstonia solanacearum* and related bacteria based on 282-bp PCR-amplified fragment. *Plant Dis* 87: 1337–1343.

Villa JE, Tsuchiya K, Horita M, Naturaal M, Opina N, Hyakumachi M (2005) Phylogenetic relationships of *Ralstonia solanacearum* species complex strain from Asia and other countries based on 16S rDNA, endoglucanase and *hrpB* gene sequences. *J Gen Plant Pathol* 71: 39–46.

Wakimoto S (1957) A simple method for comparison of bacterial populations in a large number of samples by phage technique. *Ann Phytopathol Soc Jpn* 22: 159–163.

Wakimoto S (1960) Classification of strains of *Xanthomonas oryzae* on the basis of their susceptibility against bacteriophages. *Ann Phytopathol Soc Japn* 25: 193–198.

Wanner LA (2004) Field isolates of *Streptomyces* differ in pathogenicity and virulence on radish. *Plant Dis* 88: 785–796.

Wanner LA (2006) A survey of genetic variation in *Streptomyces* isolates causing common scab in the United States. *Phytopathology* 96: 1363–1371.

Wanner LA (2007) A new strain of *Streptomyces* causing common scab in potato. *Plant Dis* 91: 352–359.

Weller SA, Elphinstone JG, Smith NC, Boonham N, Stead DE (2000) Detection of *Ralstonia solanacearum* strains with a quantitative, multiplex, real-time, fluorogenic PCR (TaqMan) assay. *Appl Environ Microbiol* 66: 2853–2858.

Wicker E, Grassart L, Coranson-Beaudu R, Mian D, Prior P (2009) Epidemiological evidence for the emergence of a new pathogenic variant of *Ralstonia solanacearum* in Martinique (French West Indies). *Plant Pathol* 58: 853–861.

Wicker E, Lefeuvre P, Cambiaire JCD, Lemaire C, Poussier S, Prior P (2012) Contrasting recombination patterns and demographic histories of the plant pathogen *Ralstonia solanacearum* inferred from MLSA. *ISMEJ* 6: 961–974.

Wicks TJ, Hall BH, Morgan BA (2007) Tuber soft rot and concentrations of *Erwinia* spp. in potato washing plants in South Australia. *Austr Plant Pathol* 36: 309–312.

Wullings BA, Beuningen AR, van Janse JD, Akkermans ADL (1998) Detection of *Ralstonia solanacearum* which causes brown rot of potato by fluorescent in situ hybridization with 23S rRNA-targeted probes. *Appl Environ Microbiol* 64: 4546–4554.

Xu J, Pan C, Prior P et al. (2009) Genetic diversity of *Ralstonia solanacearum* strains from China. *Eur J Plant Pathol* 125: 641–653.

Xue Q-Y, Yin Y-N, Yang W et al. (2011) Genetic diversity of *Ralstonia solanacearum* strains from China assessed by PCR-based fingerprints to unravel host plant-and site-dependent distribution patterns. *FEMS Microbiol Ecol* 75: 507–519.

Yamada T, Kawasaki T, Nagata S, Fujiwara A, Usami S, Fujie M (2007) New bacteriophages that infect the phytopathogen *Ralstonia solanacearum*. *Microbiology* 153: 2630–2639.

Yasuhara-Bell J, de Silva A, Heuchelin SA, Chaky JL, Alvarez AM (2016a) Detection of Goss's wilt pathogen *Clavibacter michiganensis* subsp. *nebaraskensis* in maize by loop-mediated amplification. *Phytopathology* 106: 226–235.

Yasuhara-Bell J, Marrero G, Arif M, de Silva A, Alvarez AM (2017) Development of a loop-mediated isothermal amplification assay for detection of *Dickeya* spp. *Phytopathology* 107: 1339–1345.

Yasuhara-Bell J, Marrero G, De Silva A, Alvarez AM (2016b) Specific detection of *Pectobacterium carotovorum* by loop-mediated isothermal amplification. *Molec Plant Pathol* 17: 1499–1505.

Young ANL, Chichung W (2000) The design of specific primers for the detection of *Ralstonia solanacearum* in soil samples by polymerase chain reaction. *Bot Bull Acad Sin* 41: 121–128.

Young JM, Takikawa Y, Gardan L, Stead DE (1992) Changing concepts in the taxonomy of plant pathogenic bacteria. *Annu Rev Phytopathol* 30: 67–105.

Zhang Y, Loria R (2017) Emergence of novel pathogenic *Streptomyces* species by site-specific accretion and cis-mobilization of pathogenicity islands. *Molec Plant-Microbe Interact* 30: 72–82.

Zheng X, Zhu Y, Liu B, Yu Q, Lin N (2016) Rapid differentiation of *Ralstonia solanacearum* avirulent and virulent strain by cell fractioning of an isolate using high performance liquid chromatography. *Microb Pathogene* 90: 84–92.

Zheng X-f, Zhu Y-j, Liu B, Zhou Y, Che J-m, Lin N-q (2014) Relationship between *Ralstonia solanacearum* diversity and severity of bacterial wilt disease in tomato fields in China. *J Phytopathol* 162: 607–616.

Zulperi D, Sijan K, Ahmad ZAM et al. (2016) Genetic diversity of *Ralstonia solanacearum* pathotype II sequevar 4 strains associated with Moko disease of banana (*Musa* spp.) in Peninsular Malaysia. *Eur J Plant Pathol* 144: 257–270.

# 5 Biology of Soilborne Bacterial Plant Pathogens

Bacterial pathogens cause destructive diseases and remain a formidable limiting factor to crop production all over the world. The ability of bacterial pathogens to multiply and reach huge population levels rapidly under favorable conditions, poses problems in containing the incidence and spread of bacterial diseases in various crops cultivated in different ecosystems. Bacterial wilt diseases caused by *Ralstonia solanacearum* affect many economically important crops like tomato, potato, pepper and banana accounting for enormous losses, since the infected plants are ultimately killed. Generally, visible symptoms of infection appear only after the pathogen has become well established in the vascular system and at this stage, it is very difficult to save infected plants. It may be possible to protect the neighboring uninfected plants by applying preventive measures. Soft rot diseases caused by *Pectobacterium* (syn. *Erwinia)* affect several vegetable crops, inducing extensive rotting of the plant tissues, making the produce unmarketable. These bacterial diseases have worldwide distribution, resulting in variable losses, depending on the susceptibility/resistance levels of crop cultivar, the virulence of pathogen strain and the existing environmental conditions. It is essential to have knowledge on various aspects of the biology of bacterial pathogens such as biological characteristics, host-bacterial pathogen interactions, process of infection and molecular biology of disease development (pathogenesis) for the development of effective disease management systems.

## 5.1 DISEASE CYCLES OF BACTERIAL PLANT PATHOGENS

Soilborne bacterial plant pathogens differ in their virulence (aggressiveness), host range, survivability and responses to changes in the existing soil environments and to chemicals applied. Cultural practices adopted in various geographical locations have significant effects on soilborne bacterial pathogen populations and consequently on disease incidence and severity. Disease cycles of soilborne bacterial pathogens differ distinctly from that of fungal pathogens which produce different spore forms at different stages of their life cycle. On the other hand, bacterial pathogens have only one type of cells that are endowed with the ability to infect susceptible plants from seedling stage to mature plants approaching harvest. The information on sources of infection, modes of dispersal, host range, variability in virulence and survival in soil and water and sensitivity to chemicals will be useful to plan short- and long-term management strategies for containing the bacterial pathogens that leads to diseases of economically important crops.

### 5.1.1 RALSTONIA SOLANACEARUM

#### 5.1.1.1 Disease Incidence and Distribution

Among the soilborne bacterial pathogens, *Ralstonia solanacearum* has been studied more intensively, as the pathogen can infect a wide range of plants including several economically important crops. The preference of cultivation of susceptible crop cultivars, the dissemination through asymptomatically infected propagative materials, and the ability to survive in soil for a long time are some factors favoring the incidence and establishment of bacterial wilt diseases caused by *R. solanacearum* in several crops grown in tropical, subtropical and warm-temperate zones in the world. *R. solanacearum* is considered to be a species complex (containing a heterogeneous group of related strains). The species complex is subdivided into races, based on host range and into biovars, based on the ability to produce acid from a panel of carbohydrates (Hayward 1964). Biovar classification has limited capability to discern strains or predict strain origin and host specificity. With advances in molecular biology, molecular classification provided a meaningful system of classification of the isolates of *R. solanacearum*. Based on sequence analysis of the internal transcribed spacer (ITS) region and the endoglucanase (*egl*), *hrpB* and *mutS* genes, the species complex was divided into four phylotypes: phylotype I (Asia), phylotype II (Americas), phylotype III (Africa) and phylotype IV including *R. syzygii* and the blood disease of banana (BDB) pathogen (Indonesia). The phylotype II is further divided into two groups (A and B). The phylotypes have been grouped into sequevars, based on the nucleotide sequences. With the molecular classification system, it is easier to discern closely related strains and to identify genes responsible for certain characteristics unique to a subset of strains, which together may result in improved strain detection (Fegan and Prior 2005; Stulberg et al. 2015). *R. solanacearum* species complex includes a wide range of ecotypes sharing similar hosts and symptomatology and causes disease under similar climatic conditions. Among these, the 'Moko' ecotype is relatively specialized and it is limited to banana and *Musa* ornamentals, while the 'Brown rot' ecotype is specifically able to induce disease on potato and tomato in cool conditions (Milling et al. 2009).

Latent infection of the ornamental host geranium by *R. solanacearum* was studied. Following soil inoculation, latent infection by *R. solanacearum* race 3 biovar 2 (R3bv2) was detected in 12 to 26%, carrying an average population of $4.8 \times 10^8$ CFU/g of crown tissue in the absence of visible symptoms. Such latently infected plants shed an average of

1.3 x $10^5$ CFU/ml in soil run-off water, suggesting a non-destructive means of testing pools of asymptomatic plants. Similarly, symptomatic plants shed 2 x $10^6$ CFU/ml of run-off water. The introduction of a few hundred pathogen cells directly into geranium stems resulted in the death of almost all inoculated plants. However, no transmission of disease occurred after contact between wounded leaves. Exposure of latently infected plants to 28°C for two weeks did not result in symptom expression, although disease development was temperature-dependent. Holding plants at 4°C for 48 h, a routine practice during shipment of geranium cuttings, did not increase the frequency of latent infections. *R. solanacearum* cells were distributed unevenly in stems and leaves of both symptomatic and latently infected plants, indicating that random leaf sampling could be an unreliable testing method. The strain UW551 induced potato brown rot and tomato bacterial wilt diseases proving its greater level of virulence than race 1 strain K60 on tomato at relatively cool temperature of 24°C (Swanson et al. 2005). *R. solanacearum* race 1 biovar 4 (R1bv4) infects vegetable sweet potato (VSP) causing yield losses, varying from 30–80% in Taiwan. In order to identify the sources of initial inoculum of R1bv4 in VSP fields, soil and cuttings from these fields were examined during 2009 and 2010. The R1bv4 population was generally distributed throughout the natural soil of VSP fields at densities varying from 1.3 x $10^2$ to 9.5 x $10^9$ CFU/g of soil. However, the incidence of bacterial wilt was not significantly correlated with pathogen population density in soil ($R^2 = 0.084$). But asymptomatic vine tissues had population densities varying from 2.3 x $10^3$ to 5.9 x $10^5$ CFU/g of tissue in infested fields. Further, pathogen-free VSP cuttings planted in infested soils under greenhouse conditions, could harbor about 3.1 x $10^5$ R1bv4 CFU/g of tissues, suggesting the existence of a latent period for R1bv4 in VSP plants. BIO-PCR analysis revealed that VSP cuttings used for propagation in fields were infected by R1bv4 strain, to the extent of 2.0 to 98% which were positively correlated ($R^2 = 0.909$) with the incidence of bacterial wilt in the field. The results of experiments conducted under greenhouse and field conditions showed that the inoculum in the cuttings contributed more to the incidence of bacterial wilt disease than the soil inoculum (Chen et al. 2014).

Most of the strains of *R. solanacearum* are present in tropical zones, but R3bv2 strains are able to infect plants in temperate zones, as well as in tropical highlands. This distinction in ecological trait was examined by comparing the survival of tropical R3bv2 strains and warm-temperate North American strains of *R. solanacearum* under different conditions. In water at 4°C, North American strains remained culturable for the longest period (up to 90 days), whereas tropical strains remained culturable for the shortest period (ca. 40 days). However, live/dead staining indicated that pathogen cells of representative strains were viable for > 160 days. In contrast, inside the potato tubers, R3bv2 strain UW551 survived > 4 months at 4°C, whereas North American strain K60 and tropical strain GMI1000 could not be detected after < 70 days in tubers. Growth of GMI1000 and UW551 was similar in minimal medium at 20°C and 28°C, although both strains wilted

tomato plants rapidly at 28°C and the UW441 strain was much more virulent at 20°C, killing all inoculated plants under greenhouse conditions whereas GMI 1000 killed just above 50% of inoculated plants. Differences among the strains *in vitro* conditions were not predictive of their behavior in planta at cooler temperature. Interaction with plants seemed to depend on the expression of the temperate epidemiological trait of R3bv2 strain of *R. solanacearum* (Milling et al. 2009).

*R. solanacearum*, causing bacterial wilt disease in several Solanaceae crops, was spreading from lowlands to the highlands in China. *R. solanacearum* strains (97) isolated from four tobacco-growing zones over a wide range of elevations were characterized, using phylotype-specific multiplex PCR (Pmx-PCR) assay and examined their phylogenetic relationships, based on sequences of *egl* and *mutS* genes. All isolates belonged to phylotype I, which were further clustered into eight *egl*-sequence type groups (*egl*-group, sequevar): sequevars 13, 14, 15, 17, 34, 44, 54 and 55. Further, sequevar 55 detected in highlands was a new/unknown one. The basin of the Huai River, near North China, where a low level of bacterial wilt occurred, contained a single sequevar 15 group. The distribution of sequevars was associated with elevation. Sequevar 15 was overrepresented in lowland elevations, while sequevar 54 and the new one were detected only in areas of moderate to high elevations. The results suggested that phylotype I strains infecting tobacco were diverse in China (Liu et al. 2017). The optimal conditions for the development of tobacco bacterial wilt were studied. The response surface methodology, involving the use of green fluorescent gene (*gfp*) labeling, was applied to monitor the location and survival dynamics of *R. solanacearum* (*Rs::gfp*) on tobacco roots and in soil under these optimal conditions. The highest wilt incidence of 91.13% occurred, when the population reached 6.5 x $10^6$ CFU/g soil, and when the temperature and relative humidity were 30.5°C and > 81.42%, respectively. The *Rs::gfp* mutant colonized densely the root tips and root hairs, and cells of *Rs::gfp* were observed intermittently in the elongation zone or at the point of emerging lateral roots. The *Rs::gfp* number in the rhizosphere soil was 10.75-, 73.13- and 74.86-folds higher than that in the bulk soil at 10, 15 and 20 days after transplantation, respectively. Increased colonization by *Rs::gfp* was related to the population of the pathogen, the environmental temperature and the humidity in the soil. These three factors determined the pathogenic potential (virulence) of *R. solanacearum* to induce tobacco wilt disease (Li et al. 2017).

### 5.1.1.2 Host Range and Sources of Infection

Bacterial wilt diseases caused by *R. solanacearum* species complex affect over 200 plant species in 53 different botanical families and this wide host range is continuously expanding, involving new host plant species. The host range of *R. solanacearum* includes several crop plants such as tomato, potato, sweet pepper, peanut, tobacco, banana, sunflower and ornamental plants like *Anthurium* spp., *Dahlia* spp., *Lillium* spp. and marigold (Álvarez et al. 2010). Based on the host range, strains of *R. solanacearum* have been classified into six races.

Race 1 strains infect the highest number of plant species and they are well distributed in five continents, whereas race 5 infects only *Morus* spp. and is limited to China (Elphinstone 2005; EPPO and CABI 2006). Several weed species have been reported to be infected by *R. solanacearum* and they may either exhibit symptoms of infection or remain asymptomatic. Nevertheless, they are likely to be hosts of the bacterial wilt pathogen and function as potential sources of inoculum for infection of crop plants. Bitter sweet nightshade (*Solanum dulcamara*), a common perennial semi-aquatic weed inhabiting river banks and also black nightshade (*Solanum nigrum*) and stinging nettle (*Utirca dioica*) were reported to be reservoirs of *R. solanacearum* (Hayward 1991; Wenneker et al. 1999). Roots and stems of *S. dulcamara* might continuously release pathogen cells into the river water, contributing to the persistence of the pathogen in the environment (Elphinstone 2005). Common weed species growing in high hills of Nepal were examined for their ability to serve as alternative sources of inoculum by artificial inoculation with *R. solanacearum*. Pathogen populations in the roots of inoculated plants were determined at one and two months after inoculation and at various intervals after harvesting infected potato crops under natural conditions. Inoculated roots of the summer weeds *Drymaria cordata* and *Polygonum capitata* and winter weeds *Cerastium glomeratum* and *Stellaria media* yielded $10^2$ to $10^7$ CFU/g of root tissue. High populations of the pathogen could be recovered from these plants even after partial surface sterilization, indicating systemic spread of the pathogen in infected plants. Among crops tested, mustard (*Brassica juncea* cv. Fine White) was found to be susceptible under greenhouse conditions, but not under natural conditions when grown in heavily infested fields. The nonsolanaceous summer weeds might aid in the survival of *R. solanacearum*, during the absence of potato plants (Pradhanang et al. 2000a).

Outbreaks of potato brown rot disease caused by *R. solanacearum* R3bv2 caused serious losses in European countries. But incidence of this race on potatoes was not observed in the United States. In 1999, strains of *R. solanacearum* isolated from geranium plants grown in greenhouses in Wisconsin were identified as *R. solanacearum* biovar 2, which was largely synonymous with the race 3 subgroup, a classification based on host range. Based on the results of PCR assay, infection of geranium by R3bv2 was confirmed. The geranium strains were highly pathogenic to both geranium and potato. The presence of *R. solanacearum* R3bv2 raised concern that the pathogen could move from ornamental plants into potato fields, where the pathogen could cause direct economic damage and quarantine problems (Williamson et al. 2002). The host range and virulence of biovars of *R. solanacearum* may vary, when new undescribed pathotypes of *R. solanacearum* are introduced through propagative material into a new environment, the potential impact on local crops in a new environment may be unpredictable. Populations of *R. solanacearum* entering North America on ornamental crop material were compared, using repetitive (rep)-PCR and amplified fragment length polymorphism (AFLP), sequence analysis of endoglucanase (*egl*) and cytochrome b561 gene. Strains in different

clusters exhibited different host ranges. Larger host ranges were observed with strains in rep-PCR clusters B, E and F (sequevar 4), representing strains from Costa Rica, Martinque and Gudeloupe. The strains isolated from anthurium and pothos were pathogenic to tomato, potato and geranium. The race/strains showed variation in pathogenicity to tobacco, tomato, potato, banana and geranium (Norman et al. 2009).

*R. solanacearum* is a soilborne, xylem-limited pathogen, known to be responsible for bacterial wilt diseases of a wide range of crop plant species and many weed and wild species without necessarily inducing visible symptoms, contributing inoculum for infection of crops. A mutant of *R. solanacearum* harboring a transposon insertion in the gene *RSc1206* retained its virulence on tomato, but displayed reduced virulence on *Arabidopsis*. The function of *RSc1206* in bacterial pathogenesis in weed hosts was investigated. *RSc1206* encoded a putative protein homologous to *Escherichia coli* NlpD, and the organization of *RSc1206*-related gene cluster was mostly conserved among representative gram-negative bacteria analyzed and this gene was designated *nlpD*. Microscopic observations showed that the *nlpD* mutants were defective in cell separation and envelope integrity. An increased sensitivity to hydrophobic toxic chemical indicated the loss of integrity in the *nlpD* mutant. Further, the activity of the *nlpD* mutant in biofilm formation and motility was reduced. However, enzymatic activity and tobacco pathogenesis assays revealed the normal functioning of Type II and III secretion systems in the mutant. Transmutation and trans-complementation analyses confirmed that disruption of *nlpD* function resulted in observed defects in the mutant. The results indicated the contribution of nlps to *R. solanacearum* cell integrity and pathogenesis in *Arabidopsis* plant host (Hseih et al. 2012).

### 5.1.1.3 Modes of Dispersal of Bacterial Pathogens

Soilborne bacterial plant pathogens may be dispersed through infested seeds/ propagules or irrigation water from rivers and lakes/ponds, in addition to the infested soil. *R. solanacearum* can persist in the soil for a long time, once it is introduced through infected seed/propagative materials. Specific and sensitive quantitative detection in soil samples of *R. solanacearum* R3bv2 was achieved by employing a immunofluorescence colony-staining (IFC) procedure followed by confirmation of identity of fluorescent colonies by PCR-amplification or dilution plating on a semiselective SMSA medium. Rapid verification of IFC-positive results directly by PCR-amplification with primers D2/B specific to division 2 of *R. solanacearum* had a success rate of 86% and 96% with spiked and naturally contaminated soil samples, respectively. The IFC test was able to consistently detect ca. 100 CFU/g of soil, whereas tomato bioassay detected the pathogen only at a cocentratuion of $10^4$ to $10^5$ CFU/g of soil (van der Wolf et al. 2000). The sensitivities of detection of *R. solanacearum* in soil by different methods such as dilution plating on modified semiselective medium (SMSA), tomato bioassay, enrichment methods, indirect ELISA, standard PCR and nested PCR formats were compared. The efficiency of recovery of *R. solanacearum* in naturally infested soil samples varied with the techniques employed. However, dilution plating and tomato bioassay provided the estimates of live bacterial cells in the samples. Although the molecular methods had higher levels of sensitivity, they were unable to differentiate live and dead cells of the pathogen (Pradhanang et al. 2000b). The limitation of PCR assays could be overcome by employing the BIO-PCR method of detection of *R. solanacearum* in soils. Detection of *R. solanacearum* from field soils indicated that spatial distribution of the pathogen in the field was not even, regardless of the presence or absence of disease and the different ecosystems where sampled fields were located and the degree of unevenness was higher in the absence of tomato plants (Lin et al. 2009).

Dispersal of *R. solanacearum* via irrigation water, drainage and artesian wells is important, since the pathogen may be carried to distant locations. Surveys undertaken for three seasons in Egypt showed that *R. solanacearum* R3bv2 (phylotype II sequevar 1), causing potato brown rot was limited to the canals of traditional potato-growing areas in the Nile Delta region. Pathogen populations in the canals were variable (100 to 200 CFU/l) and pathogen presence was associated with potato cultivation in the reclaimed areas. Temperature and microbial activity were the principal factors affecting the survival of *R. solanacearum* in canal water, the longest and the shortest periods of survival being at 15°C and 35°C (Tomlinson et al. 2009).

### 5.1.1.4 Survival of the Pathogen

*R. solanacearum* may survive in soil and water, in addition to the plant tissues. The fate of *R. solanacearum* causing potato brown rot disease was monitored in soils of three different fields, after breakouts of the disease in the Netherlands. IFC staining supported by *R. solanacearum* division 2-specific PCR assay was employed. The *R. solanacearum* population densities were initially in the order of $10^4$ to $10^6$ CFU/g of top soil and the population decreased with lapse of time. In two fields, the pathogen persisted for 10 to 12 months. Survival rates of *R. solanacearum* biovar 2 isolate of strain 1609 in a loamy and two silt loam soils were investigated at different temperatures. At 12 or 15 and 20°C, a gradual decline of the population densities was observed in all three soils from the established $10^5$ to $10^6$ CFU/g of dry soil to significantly reduced levels between 90 and 210 days. Soil type had a significant effect on pathogen survival, with greatest decline occurring in loamy sand soil at 20°C. Occurrence of viable, but nonculturable strain 1609 cells in the loamy sandy soil was observed. Moderate soil moisture fluctuations did not affect pathogen survival, but severe drought drastically reduced the population of *R. solanacearum* strain 1609 (van Elsas et al. 2000). Survival of *R. solanacearum* (*Rs*) race 4 strains, causing bacterial wilt of edible ginger (*Zingiber officinale*) was investigated in plant-free soil and potting medium in the presence of plants inoculated by different methods viz., non-wounded, rhizome-wounded and stem-wounded and irrigated at different intervals (alternate days and daily). In the absence of ginger plants or in the presence of non-wounded plants, pathogen populations declined rapidly in the drainage

water from potting medium during the first nine days and they were undetectable after 81 days. In the presence of rhizome- and stem-wounded plants, pathogen populations increased by two or three orders of magnitude from the initial population levels for the first nine to 19 days and then gradually declined and became undetectable after 89 days. Similar results were obtained for experiments with soil except for non-wounded ginger plants, where the initial decline in pathogen populations was followed by an abrupt increase after day 11, reaching 7 log CFU/ml on day 21, then declining gradually to undetectable levels after 137 days. When irrigated on alternate days, *R. solanacearum* could be recovered from 97 to 129 days in soils and potting medium with rhizome-inoculated plants. On the other hand, when plants were watered daily, the pathogen could be recovered from soil and potting medium up to 153 days after inoculation. ELISA tests using Ps1a monoclonal antibody (MAb) detected the pathogen from > 95% of the samples from soil and potting medium, when viable populations were > 5 log CFU/ml. The immunostrip assay using the same antibody detected the pathogen from all samples (100%), when viable pathogen populations were > 3 log CFU/ml. The PCR-based assay based on the flagellin gene *fliC* detected the pathogen from 95% of the samples from soil and from 74% of samples from potting medium, when viable populations were > 4 log CFU/ml, indicating the variations in the sensitivity of the detection techniques (Paret et al. 2010).

Survival of *R. solanacearum* A1-9[Rif] race 1 phylotype I was assessed in ten different soil types in the absence of host plant as well as in infected tissues of the stem and root of bell pepper plants buried in the soil at 0, 5 and 15 cm. The period of survival of the pathogen varied from 42 to 77 days, depending on the soil type. Among the physical and chemical characteristics of the soil, clay content, residual moisture and available water were positively correlated and pH was negatively correlated with period of survival, population size at 42 days and area under the population curve. The pathogen survival differed significantly in relation to the plant tissues, but not regardless of depth of burial of plant tissues. The root tissues of bell pepper supported a larger bacterial population of $5 \times 10^4$ and $3.1 \times 10^4$ CFU/g of root tissue respectively at seven and 21 days after burial and also had a larger area under the population curve. On the other hand, the stem tissues had a greater decomposition rate and pH compared with the roots. Different soil types as well as infected pepper tissues might function as sources of *R. solanacearum* inoculum for varying periods in the absence of live pepper plants (Felix et al. 2012). Potato cultivars were evaluated under field and greenhouse conditions. Of the 28 cultivars showing low disease scores induced by races of *R. solanacearum* under greenhouse conditions, only two showed adequate resistance to *R. solanacearum*, indicating the significant impact of environment on the pathogenic activity of the isolates of pathogen and responses of potato cultivars. Percent wilt incidence and latent infection showed significant positive correlation (r = 0.9438). The results indicated that the combination of parameters of field wilt incidence and the proportion of latent infection was more reliable for evaluation of cultivars than using only the former parameter for evaluation of potato cultivars (Aliye et al. 2015).

#### 5.1.1.5 Process of Infection of Susceptible Plants

Formation of biofilms of bacteria on plant surfaces to reach the required population is essential to initiate infection. The most common sites of bacterial colonization are stomates, the base of trichomes and depressions along the veins. The biofilms, in these environments provide protection to the bacterial pathogens from external environmental conditions such as desiccation and a favorable milieu for multiplication. *Xanthomonas axonopodis* pv. *phaseoli* (Xap) can survive on leaf surfaces and endophytically (Weller and Saettler 1980). The biofilm population sizes in bean leaves remained stable throughout the growing season (around $10^5$ CFU/g of fresh weight), while solitary population sizes were more abundant (Jacques et al. 2005). Another phenomenon known as 'quorum-sensing' has been shown to occur in a wide range of microorganisms which have the ability to perceive and respond to the presence of neighboring populations and it is population-density-dependent. This system is accomplished by the extracellular accumulation of small, self-generated chemical signaling moieties that induce a concerted effort on behalf of a population to produce the desired phenotype effect (De Kievit and Iglweski 2000). Microbially derived signaling molecules may be divided into two main classes: (i) amino acids and short peptides, commonly utilized by gram-negative bacteria (Lazazzera and Grosman 1988) and (ii) fatty acid derivatives known as homoserine lactones (HSLs) frequently utilized by gram-negative bacteria (Whitehead et al. 2001).

#### 5.1.1.5.1 Initiation of Infection

Pathogenicity (ability to infect) and virulence (extent of aggressiveness) are basic traits of pathogenic bacteria and the strains of a bacterial species may be pathogenic to several or a few host species. Flagella of bacterial pathogens are considered to be helpful to seek favorable environments or to escape from adverse conditions. The flagellum has a long helical filament composed of a single protein known as flagellin encoded by *fliC* gene (Aizawa 1996). MotA and MotB are two integral membrane proteins required for flagellar rotation (Chun and Parikson 1988; Blair and Berg 1991). Flagellar motility is an important virulence factor during infection and they may also induce host defense responses (Young et al. 1999; Shimizu et al. 2003). The motility of *R. solanacearum* cell depends on culture age and high populations of bacteria might be motile in exponential phase, whereas the majority of cells might be nonmotile in the stationary phase (Clough et al. 1997). The motility of bacterial cells is required for effective invasion and colonization of host tissues. However, bacterial cells present in plant xylem vessels were nonmotile and the pathogen cells from wilted plants were nonmotile, but became motile after a few hours in fresh medium (Mao and He 1998; Tans-Kersten et al. 2001). Many soilborne plant pathogenic bacteria including *R. solanacearum* have swimming motility which makes an important quantitative contribution to bacterial wilt virulence in early stages of host invasion and colonization. The

virulence of nonmotile *fliC* mutants was significantly reduced (Tans-Kersten et al. 2001). The bacteria may use a complex behavior known as taxis to sense specific chemicals or environmental conditions and move toward attractants and/ away from repellants (Blair 1995). The chemotaxis behavior of *R. solanacearum* strain K60, which was attracted to various chemicals was studied. Qualitative and quantitative chemotaxis assays showed that *R. solanacearum* was attracted to diverse amino acids and organic acids and especially to root exudates from host plant tomato (Yao and Allen 2006). Two site-directed *R. solanacearum* mutants lacking either CheA (cytoplasmic histidine autokinase) or CheW (coupling protein) which are core chemotaxis signal transduction proteins, were entirely nonchemotactic, but retained normal swimming motility. Both nonchemotactic mutants were less virulent a nonmotile mutant, demonstrating that directed motility, but not simple random motion, was essential for full virulence. Nontactic strains could be as virulent as the wild-type strain, when they were directly introduced into the stem through wounded petiole, indicating that taxis is an important factor in the early stages for successful invasion of host tissues. The wild-type strain outcompeted the nonchemotactic mutants by 100 folds when coinoculated. The results showed that chemotaxis could be an essential trait needed for virulence and pathogenic fitness in *R. solanacearum* (Yao and Allen 2006).

*R. solanacearum* possesses diverse genes involved in different stages of disease development (pathogenesis) such as adhesion, biofilm formation, entry into the host tissues, and local and systemic spread within the infected plants. The genes involved in pathogenesis are those coding for lytic enzymes and extracellular polysaccharides (EPS), hypersensitive reaction and pathogenicity (*hrp*) genes, structural genes encoding effector proteins injected by type III secretion system (T3SS) from bacterium into the host plant cells, genes coding for factors implicated in cell adherence and other genes whose functions are yet to be clearly characterized. Many pathogenic bacteria have evolved the type III secretion system (T3SS) to successfully invade the susceptible plant species. This extracellular apparatus allows translocation of proteins known as type III effectors (T3Es), directly into the host cells. The T3Es are virulence factors that have been shown to interfere with host's immunity or to provide nutrients from the host to the bacterial pathogen. The virulence of *R. solanacearum* strongly relies on the T3SS (Lohou et al. 2014). An original association genetics approach was applied to identify type III effector (T3E) repertoires associated with virulence of *R. solanacearum*. DNA microarray and pathogenicity data on resistant eggplant, pepper and tomato accessions were combined. From the first screen, 25 T3Es were further, full-length PCR-amplified within a 35-strain field collection, to assess their distribution and allelic diversity. Six T3E repertoire groups were identified, within which 11 representative strains were chosen to challenge the bacterial wilt-resistant eggplant cvs. Dingras Multiple Purple and AG91-25 and tomato cv. Hawaii 7996. The virulence or avirulence phenotypes could not be explained by specific T3E repertoires, but rather by individual T3E genes. Seven highly avirulence-associated genes were identified. Of these genes, *ripP2* was considered to primarily confer avirulence to *Arabidopsis thaliana*. No T3E was associated with avirulence to both eggplant cultivars. The genes *ripA5_2*, *ripU* and *ripV2* were found to be highly avirulence-associated genes. The results highlighted both virulence and avirulence functions of the genes in *R. solanacearum* populations (Pensec et al. 2015).

The effects of temperature on pathogenicity of *R. solanacearum* strains were assessed. Most strains were nonpathogenic at temperatures below 20°C. R3bv2 strains were classified as quarantine, because of their ability to infect crops and survive in temperate climates. Race 1 biovar 1 phylotype IIB sequevar 4 strains present in Florida were able to produce wilt symptoms on potato and tomato at 18°C. Strains naturally nonpathogenic at 18°C were able to multiply and spread systemically inducing partial wilt symptoms, when directly inoculated into stem, suggesting that low temperatures might affect virulence of strains differently at early stages of infection. Bacterial growth *in vitro* was delayed at low temperatures. However, the strain was not attenuated. Twitching mobility observed on growing colonies was attenuated in nonpathogenic strains at 18°C, while it was not affected in the cool virulent ones. Using *pilQ* as a marker to evaluate the relative expression of the twitching activity of *R. solanacearum* strains, the ability of cool virulent strains to maintain similar level of *pilQ* expression at both temperatures could be confirmed. But in nonpathogenic strains *pilQ* was downregulated at 18°C (Bocsanczy et al. 2012). The responses of the host plants may vary depending on the site of inoculation (infection). Tomato genotypes susceptible or resistant to *R. solanacearum* were inoculated using bacterial inoculum ($10^8$ CFU/ml) by different methods viz., seed-soaking inoculation, seed-sowing followed by inoculum drenching or at two-week stage through petiole-excision inoculation, soaking of planting medium with inoculum either directly or after imparting seedling root injury. Seed-based inoculations or mere inoculum drenching at two weeks did not induce much disease in seedlings. Petiole inoculation resulted in 90 to 100% mortality in susceptible genotype, but also 50 to 60% mortality in resistant genotypes within seven to ten days. Root-injury inoculation at two-week seedling stage appeared to be the most effective for early and clear distinction between resistant and susceptible genotypes. The results suggested a role played by the root system in governing genotype resistance to *R. solanacearum*. Direct shoot inoculation was found to be a more drastic method of inoculating the tomato plants and it may be followed only for selecting highly resistant lines or to thin down segregating populations in breeding programs (Thomas et al. 2015). Resistance of tomato rootstocks to *R. solanacearum* may be due to different factors favoring or arresting the development of the disease. Resistant plants may carry high pathogen loads ($10^5$–$10^8$ CFU/g of fresh weight) without noticeable symptoms of infection. *R. solanacearum* infects the plants through the root system and early root colonization events may be critical for successful infection resulting in the development of disease symptoms. The differences in distribution and timing of bacterial invasion in the resistant and susceptible plants were assessed to determine their role in the resistance responses. Colonization of roots of soil-grown resistant and susceptible tomato cultivars was investigated at multiple time points after inoculation, employing scanning microscope and light microscope. Colonization of the root vascular cylinder was delayed in resistant cv. Hawaii 7996. Once the pathogen had entered the vascular tissue, colonization in the vasculature was spatially restricted. The results suggested that resistance might be due, in part, to the ability of the resistant cultivar to restrict root colonization by *R. solanacearum* in space and time (Caldwell et al. 2017).

In order to have an insight into the spectrum of changes in the signaling pathways during the early stages of infection, *Solanum* lines were inoculated with *R. solanacearum* phylotype I strain R008. Four tomato (*S. lycopersicum*) lines, one potato (*S. tuberosum*) line and the wild *Lycopersicon cerasiforme* and *S. commersonii* were inoculated with the bacterial pathogen. Very little activation of the jasmonic acid (JA) pathway marker genes, lipoxygenase A (*Lox A*) and protease inhibitor II (*Pin 2*) was noted. In contrast, the salicylic acid (SA) pathway marker genes, glutamate A (*GluA*) and *PR-1a*, and the ethylene pathway marker gene osmotin-like (*Osm*) and *PR-1b* were expressed significantly at higher levels in the fold change expression among the *Solanum* lines. The resistant lines *L. cerasiforme*, CRA66, Hawaii 7996 and *S. commersonii* exhibited stronger activation of SA and ethylene marker genes than the moderately resistant cultivar MST 32/1 and the susceptible lines Quatre Carreés and Spunta. The results revealed that SA and ethylene signaling pathways played an important role in defense development against *R. solanacearum* (Baichoo and Jaufeerally-Fakim 2017).

The behavior of *R. solanacearum* (R3bv2 in tomato rhizosphere and early bacterial wilt pathogenesis was studied. A taxis-based promoter-trapping strategy was employed to identify pathogen genes induced in the host plant rhizosphere. This screen identified several *rex* (root exudates expressed) genes whose promoters were upregulated in the presence of tomato root exudates. One *rex* gene encodes an assembly protein for a high affinity cbb3-type cytochrome *c* oxidase (cbb3-cco) that enables respiration in low oxygen conditions and a *cbb3-cco* mutant strain grew more slowly in a microaerobic environment (0.5% $O_2$). The *cco* mutant could wilt tomato plants, but symptom onset was significantly delayed relative to the wild-type parent strain. In addition, the *cco* mutant did not colonize host stems or adhere to roots as effectively as the wild-type strain. The results suggested that the R3bv2 strain of *R. solanacearum* may depend on this oxidase to overcome the effects of low-oxygen environments in tomato plants (Colburn-Clifford and Allen 2010). The efficiency of colonization of rhizosphere by bacterial pathogen depends directly on bacterial growth, chemotaxis, biofilm formation and interaction with host root exudates. The responses of *R. solanacearum* to the root exudates from two tobacco cv. Hongda (susceptible) and K326 (resistant) were determined. Population of *R. solanacearum* was much higher in the rhizosphere soil, resulting in higher disease index (severity) in Hongda plants (92.73%). Attraction of *R. solanacearum* to

Hongda root exudates (HRE) was stronger than the response to K326 root exudates (KRE). The amount of oxalic acid from HRE was significantly higher than from KRE. Further, oxalic acid could both significantly induce the chemotactic response and increase the biofilm biomass of *R. solanacearum*. Both malic acid and citric acid could significantly increase the chemotaxis ability *in vitro* and recruitment of *R. solanacearum* to tobacco root under gnotobiotic conditions. The results suggested that the colonization of tobacco rhizosphere by pathogenic bacterial strains was influenced by organic acid secreted from the roots (Wu et al. 2015).

*R. solanacearum* can spread through water and infect susceptible plants. The physiology and virulence of *R. solanacearum* biovar 2 strain 1609 suspended in water at 4°C and 20°C were investigated. At 20°C, total cell and plate count (CFUs) numbers were similar between log 5.03 and 5.55 CFU and log 5.03 and 5.51 cells/ml at days 0 and 132, respectively. However, CFU in the cultures at 4°C dropped from log 6.78 CFU/ml at 0 to below detection level, after 84 days. Presence of viable-but-nonculturable (VBNC) cells in the 4°C cultures was confirmed, using SYT09 viability staining. At day 84 and after 125 days log 3.70 viable cells/ml were still present. Shifts in subpopulations differing in viability were found by flow cytometric sorting of 4°C-treated cells stained with SYT09 (healthy) and propidium iodide (PI, compromised). The SYT09-stained cell fractions dropped from 99 to 39% and the PI-stained fractions increased from 0.7 to 33.3% between days 0 and 125. At 20°C, the SYT09-stained cell fraction remained stable at 99%, until day 132. Tomato plants inoculated with viable cells kept at 4°C for 100 days did not induce wilting symptoms. Cells from colonies isolated from nonwilted plants did not regain their virulence, indicating the VBNC cells of *R. solanacearum* might not pose any risk of causing disease in tomatoes (van Overbeek et al. 2004).

### 5.1.1.5.2  Colonization of Host Tissues

Bacterial plant pathogens do not produce any specialized structures for infection like fungal pathogens do. They cannot gain entry into the compact host plant tissues, but enter through natural openings such as stomata, hydathodes and lenticels and through wounds or injured plant surfaces. Studies on molecular genetics of bacterial pathogens have been useful to gaining a clear understanding of the phenomenon of pathogenesis (development of disease). Molecular genetics of *Agrobacterium tumefaciens*, *Clavibacter* spp., *Pectobacterium (Erwinia)* spp., *R. solanacearum* and *Xanthomonas* spp. have been employed to study the functions of bacterial genes involved in pathogenesis. Four major secretion pathways regulating the production of macromolecules by bacterial pathogens have been recognized. Some bacterial pathogens secrete enzymes and toxins through a type I pathway which is sec-independent. Type I secretory apparatus consisting of three accessory proteins (Prt D, E and F) governs this one-step secretion (Pugsley 1993). Proteins secreted through a type I pathway are membrane-associated and the major signal is located at the C-terminal end of enzymes. Pectinases and cellulases are secreted through a sec-dependent, two-step, type II pathway also known as the general secretory pathway (GSP). Some of these enzymes function as putative virulence factors. The genes of pectatelyase, polygalacturonase and cellulase are expressed in *E. coli*. As the first step, the respective enzymes are exported to the periplasm, but are unable to move through the outer membrane. In the second step, the periplasmic form of these enzymes moves through the outer membrane and then into the extracellular environment. Secretion of major pectatelyases, polygalacturonases and cellulases is controlled by the *out* gene cluster and the number of *out* genes may vary with different bacterial species (Lindberg and Collmer 1992).

Formation of biofilms by bacterial pathogens occurs as the initial step for increasing the pathogen population. The mechanism of colonization of intercellular spaces of tomato tissues by *R. solanacearum* strain OE 1-1 was studied. The behavior of pathogen cells in the intercellular spaces was analyzed by excising tomato leaves with lower epidermal layers after infiltration with OE 1-1 and by observing under scanning electron microscope. The OE 1-1 cells formed microcolonies on the surfaces of tomato cells adjacent to intercellular spaces and then aggregated surrounded by an extracellular matrix, forming mature biofilm structures. In addition, OE 1-1 cells produced mushroom-type biofilms when incubated in fluids of apoplasts, including intercellular spaces, but not in xylem fluids from tomato plants. Sugar application resulted in enhanced biofilm formation by OE 1-1. Mutation of *lecM* encoding a lectin, RS-ITL which might exhibit affinity for these sugars led to a significant decrease in biofilm formation. Colonization in intercellular spaces was significantly decreased in the *lecM* mutant, leading to loss in virulence on tomato plants. Complementation of the *lecM* mutant with native *lecM* resulted in recovery of mushroom-type biofilms and virulence on tomato plants. The results indicated that OE 1-1 could produce mature biofilms on surfaces of tomato cells after invasion into the intercellular spaces. RS-ITL might contribute to biofilm formation which is required for virulence of *R. solanacearum* (Mori et al. 2016). The genes of pectatelyase, polygalacturonase and cellulase are expressed in *E. coli*. As a first step, the respective enzymes are exported to the periplasm, but are unable to move through outer membrane and then into the extracellular environment. Polygalacturonase (PG) production and other virulence functions in *R. solanacearum* were under the control of regulatory locus designated *pehSR*. The endoPG activity of *pehSR* mutants were negligible, while exoPG activity was reduced by 50%. However, growth of the mutant in culture was normal, despite its reduced virulence. As the pathogen multiplied in planta, the *pehSR* expression increased (Allen et al. 1997). Involvement of the global virulence gene regulator PhcA in the negative regulation of the *pehSR* locus was revealed by reporter gene investigations. A pectin methylesterase (*pme*) gene mutant of *R. solanacearum* showed no detectable Pme activity *in vitro* and could not utilize methylated pectin (93%) as a carbon source. However, its virulence on tomato, eggplant and tobacco was not reduced, indicating its requirement only for growth, but not for virulence (Tans-Kersten et al. 1998).

Phytopathogenic bacteria produce enzymes to hydrolyze plant cell wall components to obtain nutrients and energy which are required in the early stages of the infection process, favoring the entry and progression of the pathogen in host tissues. *R. solanacearum* produces several plant cell wall degrading enzymes (CWDEs), secreted via the type II secretion system (T2SS). These enzymes include ß-1,4-cellobiohydrolase (CbhA) and ß-1,4-endoglucanase (Egl), endopolygalacturonase (PehA), two exopolygalacturonases (PehB and PehC) and pectin methylesterase (Pme). *R. solanacearum* Egl is a 43-kDa protein involved in pathogenicity (Álvarez et al. 2010). The extracellular plant CWDEs and other proteins are secreted via the highly conserved type II protein secretion system (T2SS). *R. solanacearum* with defective T2SS was weakly virulent. Mutants of wild-type strain GM100 lacking either T2SS or up to six CWDEs were generated. In soil-drench inoculation tests, virulence of mutants lacking only pectic enzymes (Peh, PehB, PehC and Pme) was not significantly different from the wild-type strain. However, all mutants lacking one or both cellulolytic enzymes (Egl or CbhA) wilted plants significantly slowly, compared with wild-type strain. The GMI-6 mutant lacking all six CWDEs was more virulent than the mutant lacking only two cellulolytic enzymes and both were significantly more virulent than the TSS mutant GMI-D, since the T2SS mutant was much less virulent than the six-fold CWDE mutant. Other secreted proteins may also contribute to the systemic colonization of tomato plants by *R. solanacearum* (Liu et al. 2005).

Production of exopolysaccharides (EPS) by many bacterial pathogens either in culture or in planta has been observed. An association of EPS production with pathogenicity has been indicated, but it is difficult to infer active participation of EPS in symptom induction or involvement indirectly in the infection process. All virulent wild-type strains of *R. solanacearum* forming mucoid colonies produce EPS, whereas EPS-deficient mutants forming non-mucoid colonies are avirulent. The EPS produced by *R. solanacearum* appeared to be highly heterologous, since it has varying composition among strains (Drigues et al. 1985). The EPS I, an acidic high molecular mass accounts for more than 90% of the total of EPS produced by *R. solanacearum* and approximately 85% appeared as a released cell-free slime, whereas 15% were present as cell wall-bound capsular form (Mc Garvey et al. 1998). EPS I was responsible for wilting and it might probably act by occluding xylem vessels, interfering directly with normal fluid movement in the plant or by breaking the vessels due to hydrostatic overpressure. In addition, EPS I mutants multiplied slowly and colonized the stem of infected plants poorly. As EPS-deficient mutants could infect and multiply to some level in planta without inducing wilting symptoms, EPS might take part mainly in the late stages of the infection process, modulating disease severity, rather than infectivity of *R. solanacearum* (Schell 2000; Hikichi et al. 2007).

The dynamics of defense responses in susceptible and resistant tomato plants infected by two biologically distinct strains of *R. solanacearum* were studied. EPS is a major virulence factor of *R. solanacearum*. The pathogen is a genetically diverse species complex, but the EPS structure is sufficiently well conserved, and an anti-EPS antibody could recognize all strains of *R. solanacearum*. Quantitative RT-PCR gene expression analysis on susceptible and resistant tomato plants infected by *R. solanacearum* showed that there was little or no activation of JAs pathway marker genes *Pin2* and *LoxA*. However, both *PR-1b*, and *Osm* which were ethylene (ET)-induced, and *GluA* and *PR-1a* which were regulated by the SA, were expressed at significantly higher levels in plants with pathogen cell densities $\geq 3 \times 10^3$ CFU/g, relative to water-inoculated control plants. Tomato plants infected with *R. solanacearum* R3bv2 strain UW551 and tropical strain GMI1000 upregulated genes in both ET and SA defense signal transduction pathways. In the tomato line Hawaii 7996 with horizontal resistance to wilt disease, activated expression of these genes was faster and to a greater degree in response to *R. solanacearum* infection than it was with the susceptible cv. Bonny Best. The role of EPS in resistant and susceptible host responses to *R. solanacearum* was studied. The interactions of wild type and eps⁻ mutant with susceptible and resistant tomato cultivars were investigated. The EPS-deficient mutant of UW551, UW551Δ*epsB* was dramatically reduced in virulence on both susceptible and resistant tomato plants (P < 0.001) and rarely killed the host plant. Tomato defense gene expression following infection by wild-type and Δ*epsB* strains of *R. solanacearum* was assessed to determine the role of EPS in cloaking *R. solanacearum* from recognition by tomato. The defense-associated *PR-1b* gene of susceptible cv. Bonny Best, was upregulated 40-fold in response to an eps⁻ mutant of UW551 strain, compared to *PR-1b* expression triggered by wild-type strain, indicating the possible role of EPS in cloaking the pathogen. This effect was observed up to a pathogen cell density of about $5 \times 10^8$ CFU/g of stem tissue. Tomato transcriptional response to purified EPS from the wild-type strain was assessed. The EPS activated the SA pathway (*GluA*) in H7996 to a significantly greater degree (seven-fold) than in Bonny Best, indicating the perception and response of the resistant tomato to *Rs* EPS. Live cells of the wild-type strain triggered much higher *PR-1b* expression in H7996 than the mutant did. But cell-free EPS alone did not significantly increase *PR-1b* expression, suggesting that EPS could activate only a subset of defense-associated responses (see Figure 5.1). The fluorescent dye dihydrorhodomine123 was employed to assess the levels of reactive oxygen species (ROS) in tomato stem, a common element of plant antimicrobial defenses. Infection by the wild-type strain triggered a strong oxidative burst in the vascular bundles of both resistant and susceptible tomato plants. However, the resistant plants infected by the mutant strain accumulated less ROS. No such response was recorded in susceptible plants. These differences in ROS accumulation triggered by wild-type and EPS-deficient *Rs* strains were also discernible in tomato leaves also, indicating that ROS accumulation was not a unique response of stem tissues to infection by *R. solanacearum* (see Figure 5.2) (Milling et al. 2011).

In order to study the metabolic activities of *R. solanacearum* the Tn5-inserted mutant of *R. solanacearum* showing various

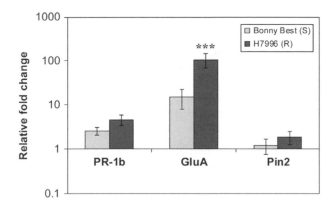

**FIGURE 5.1** Differential expression of genes (*PR-1b*, *GluA* and *Pin2*) in susceptible cv. Bonny Best (S) and horizontally resistant line H7996 (R) in response to purified EPS from *Ralstonia solanacearum*, as determined by qRT-PCR assay. Purified EPS was injected through cut petiole of the first true leaf directly into the system; significant differences in gene expression between S and R lines are indicated by asterisks above bar (P > 0.0001).

[Courtesy of Milling et al. (2011) and with kind permission of *PLoS ONE* Open Access Journal]

phenotypic changes and altered virulence, was selected for investigating glutamate dehydrogenase activity, when the *gdhA* gene encoding NAD(P)+-dependent gluatamate dehydrogenase was disrupted, the *gdhA* mutant of SL341 (SL341P2) was defective in red colony development on tetrazolium chloride-amended medium and showed less EPS production. The growth rate of the *gdhA* mutant on a rich medium did not differ from that of the wild-type strain. However, the growth of mutant on a minimal medium with glutamate as the sole carbon source was completely inhibited. SL341P2 was also defective in oxidation of several carbon sources compared with the wild-type strain. All observed defects of SL341PL

*gdhA* mutant were fully or partially restored by providing the *gdhA* gene in trans. The *gdhA* mutant showed reduced virulence after soil-soaking inoculation of tomato plants, both on susceptible tomato cv. Moneymaker and resistant cv. Hawaii 7996. Delayed disease development was due to slower multiplication of the mutant bacteria than with the wild-type strain in tomato plants. The results indicated that GdhA might be required for diverse metabolic functions in *R. solanacearum*, including normal production of EPS, the virulence factor as well as normal bacterial growth in planta and full virulence on tomato plants (Wu et al. 2015).

An interaction between lipopolysaccharides (LPS) of *Ralstonia solancearum* and plant lectins was suggested for recognition between host and pathogen. Bacterial LPS is a component of the outer membrane and has three parts: the lipid A, the oligosaccharide core and the O-specific antigen (Baker et al. 1984). The core structure of lipopolysaccharide consists of rhamnose, glucoses, heptose and 2-ketodeoxyoctonate, whereas the O-specific antigen is a chain of repeating rhamnose, N-acetylglucosamine and xylose in the ratio of 4:1:1. Presence or absence of the O-specific antigen differentiated respectively between smooth and rough LPS, which were respectively negative and positive hypersensitive response (HR)-inducers (Baker et al. 1984). However, a specific interaction between *R. solanacearum* rough LPS and a plant cell wall receptor might not be enough to initiate the HR, although many of the mutations in the LPS also affected virulence. In *R. solanacearum*, smooth LPS apparently was required to prevent agglutination by some plant lectins. *R. solanacearum* LPS and EPS appeared to be related, since a gene cluster was required for the biosynthesis of both cell surface components (Hendrick and Sequeira 1984; Kao and Sequeira 1991). The role of different forms of LPS that are critical components for fitness of most gram-negative bacteria is unclear. By screening for phage-resistant mutants of *R. solanacearum* Pss4, whose

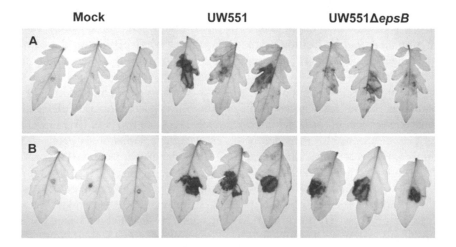

**FIGURE 5.2** Assessment of accumulation of ROS in leaves of susceptible cv. Bonny Best and horizontally resistant line H7996 following infusion with *R. solanacearum* strain UW55/or EPS1 mutant UW551ΔepsB. Staining with fluorescent diaminobenzidine (DSB) revealed ROS appearance as brown precipitate, indicating the activity of endogenous peroxidase; accumulation of ROS was less in resistant line than in susceptible cultivar and in leaves inoculated with the mutant strain than in leaves inoculated with wild-type strain.

[Courtesy of Milling et al. (2011) and with kind permission of *PLoS ONE* Open Access Journal]

genome sequence is not available, mutants with various types of structural defects in LPS were isolated. Pathogenicity tests of the mutants revealed that production of rough LPS (R-LPS) which does not contain O-polysaccharides, was sufficient to cause necrosis on *Nicotiana benthamiana* and induce the HR on *N. tabacum*. However, biosynthesis of smooth LPS (S-LPS) which contains O-polysaccharides was required for bacterial proliferation at infection sites on *N. benthamiana* leaves and for proliferation and causing wilt on tomato. Complementation tests confirmed the involvement of the cluster *Rsc 2201* to *Rsc 2204* identified earlier, in the formation of *R. solanacearum* S-LPS. Some loci involved in key steps of LPS biosynthesis in *R. solanacearum* GMI 1000 strain were identified (Li et al. 2014).

Plants produce numerous phenolic compounds to defend themselves against microbial plant pathogens. In response to root-infecting pathogens, many plants release *de novo* synthesized hydroxycinnamic acid (HCA) into the rhizosphere. Ultrastructure studies of xylem infected with vascular pathogen *R. solanacearum* revealed the release of phenolic compounds from the roots of infected plants. The bacterial pathogens have also developed the ability to degrade the phenolic compounds for their survival. The degradation pathway of HCA was shown to be genetically and functionally conserved across diverse strains of *R. solanacearum*. The feruloyl-CoA synthetase ($\Delta fcs$) mutant could not degrade HCA and it was less virulent on tomato plants. HCA degradation enabled growth of *R. solanacearum in vitro*. But HCA degradation was dispensable for growth in xylem sap and root exudates, suggesting that HCA was not an essential carbon source in planta. Acetyl bromide quantification of lignin demonstrated that *R. solanacearum* infections did not affect the gross quantity or distribution of stem lignin. However, the Δfcs mutant was significantly more susceptible to inhibition by HCA, viz., caffeate and p-coumarate. Plant colonization assays suggested that HCA degradation might facilitate early stages of infection and root colonization. Overall, the results indicated the ability to degrade HCA, which might contribute to the virulence of *R. solanacearum* by facilitating root penetration and protection of the pathogen from HCA toxicity (Lowe et al. 2015).

Pathogen-host genes are expressed at different stages of disease development (pathogenesis) in susceptible plants, whereas in resistant plants expression of pathogen genes may be silenced or the effects of products of pathogen origin may be nullified to different degrees depending on the levels of resistance of plants to the pathogen concerned. An outer membrane receptor PrhA in *R. solanacearum* senses the presence of plant cell and it transduces the signal through a complex regulatory cascade progressively integrated by PrhR, PrhI, PrhJ, HspG and HrpB regulators. Two regulatory *hrpG* and *prhJ* genes of *R. solanacearum* located at the right-hand end of *hrp* gene cluster were identified. These genes are required for full pathogenicity of *R. solanacearum*. The *hrpG* was found to be homologous to HrpG, which activates *hrp* genes in *Xanthomonas campestris* pv. *vesicatoria*. PrhJ is a novel *hrp* regulatory protein and shares homology with the

LuxR/ UhpA family of transcriptional activators; it is specifically involved in *hrp* gene expression in the presence of plant cells. Transcription of *hrpG* and *prhJ* genes is plant inducible through the Prh-dependent pathway (Brito et al. 1999). The effector protein popA of *R. solanacearum* OE 1-1 strain (pathogenic to tobacco) showed 97.6% identity to *popA* of GMI 1000 strain (not pathogenic to tobacco). RT-PCR analyses showed that *popA* in OE 1-1 was expressed at three h after inoculation (hai) in infiltrated tobacco leaves. A popABC operons-deficient mutant of OE 1-1 (ΔABC) indicated that popABC was not involved in pathogenicity of OE 1-1 directly. The results suggested that tobacco plants might be recognized by PopA, when it is expressed early in disease development and responds with an effective defense in the intercellular spaces (Kanda et al. 2003).

A 70-mer oligonucleotide-based DNA micorarray system was applied to analyze the functions of a range of genes under the control of the transcription activator Hrp that governs pathogenesis of *R. solanacearum*. Among the 5,074 genes generated, 143 *hrpB*-regulated genes and 50 *hrpB* downregulated genes were identified. The *hrpB* regulon was found to extend beyond the type III secretion system-related functions and included a number of genes governing chemotaxy, biosynthesis and catabolism of various low-molecular weight compounds and siderophore production and uptake. The results indicated that *hrpB* may function as a master regulatory gene governing physiological swing associated with the shift from saprophytic to parasitic life (Occhialini et al. 2005). Among the complex regulatory cascade components, HrpG coregulates the expression of two independent pathways. The first one regulates the expression of genes likely to be involved in adaptation to life in planta. The second pathway is dependent on HrpB which is a major regulator for pathogenicity of *R. solanacearum*, since it activates expression of *hrp* genes which encode the type III secretion system (T3SS). The translocators PopF1 and PopF2 are regulated by HrpB and also effector proteins that would be injected from the bacterial cell into the plant cell. Consequently, transcriptional activation of *hrp* genes by HrpB allows the pathogen to invade the host and proliferate in the intercellular spaces. During this stage of infection, T2SS and type III secretion system (T3SS) seem to coregulate pathogenicity factors. For example, T2SS would influence the secretion of PopB, while the *pehC* gene is positively regulated by HrpB (Álvarez et al. 2010). HrpB may also activate the operons of six genes responsible for synthesis of a HrpB-dependent factor (HDF) presumably involved in acylhomoserine lactone (AHL) receptor-mediated activity. AHLs are autoinducers, participating in bacterial quorum-sensing systems, regulatory networks that activate gene expression at high cell densities and mediate long-distance intercellular communication. The HDF produced by *R. solanacearum* might be interfering with quorum-sensing systems of other bacteria present during infection (Delaspre et al. 2009).

The *hrp* genes enclosed in a megaplasmid encoded components of type III secretory system (T3SS) and effector proteins and the conserved genes *hrc* might form the core of the T3SS in all *hrp* clusters. In some bacterial pathogens, the effectors may elicit HR due to recognition by specific plant resistance genes. Such genes of the pathogen are referred to as avirulence (Avr) genes, as the proteins coded by them may inhibit the progress of disease development in the host plant containing genes matching the pathogenicity genes. The T3SS includes extracellular appendages as the Hrp pili in bacterial pathogens considered to function either in the attachment to plant cells and / or as conduits for protein translocation, since they may penetrate the plant cell wall. Hrp pili-deficient mutants were impaired in their secretion of effectors and accessory proteins (Bogdanove et al. 1996). R. solanacearum produces Hrp-dependent pili, in addition to polar fimbriae which are dependent on the expression of the *hrp* genes. Both types of pili are located in the same pole of the bacterial cells. R. *solancearum* Hrp pili are mainly composed of the HrpY protein, essential for T3 protein secretion, but not for attachment to plant cells. Two proteins secreted via the T3SS, PopF1 and PopF2 were identified as translocators, with PopF1 playing a more important role in virulence and HR elicitation than PopF2 (Van Gijsegem ct al. 2000; Meyer et al. 2006). R. *solanacearum* T3SS secretes PopA, PopB, PopC, PopF1 and PopF2 under the control of transcriptional regulator HrpB. PopA produces a HR-like response, when infiltrated into plant tissue at high concentrations and may allow nutrient acquisition in planta and /or delivery of other effector proteins into plant cells (Arlat et al. 1994; Lee et al. 2001).

R. solanacearum is known to utilize the HR and pathogenicity (Hrp) type III secretion system (T3SS) to induce disease in plants. In order to determine the entire range of effector proteins possessed by R. *solanacearum* RS1000 strain, a transposon carrying calmodulin-dependent adenylate cyclase reporter that could be used to specifically detect Rip (*Ralstonia* protein injected into plant cells) genes by monitoring the cAMP level in plant leaves inoculated with insertion mutants, was constructed. From the new functional screen using the transposon, 38 new Rip proteins were identified and they were translocated into plant cells via the Hrp T3SS. Further, most of the 34 known effectors of RS1000 strain could be detected by the screen, except for three effectors that appear to be small in size or only weakly expressed. In the RS1000 strain, 72 Rips were identified and they include 68 effector proteins classified into over 50 families and four extracellular components of the Hrp T3SS. About one-third of effectors were specific to R. *solanacearum*. The results showed that most of the effector proteins of R. *solanacearum*, but not Hrp extracellular components, required a Hrp-associated protein, Hpa B, for their effective translocation into plant cells (Mukaihara et al. 2010). In a later investigation, the spontaneous nalixidic acid-resistant derivative of RS1000, designated RS1002 was found to induce strong HR in the nonhost wild eggplant *Solanum torvum* in a Hrp-dependent manner. The *Agrobacterium*-mediated transient expression system revealed that Rip36, a putative Zn-dependent protease effector of R. *solanacearum* induced HR in *S. torvum*. A mutation in the putative Zn-binding motif (E149A) completely abolished the ability to induce HR. Likewise, the RS1002-derived Δ*rip36* and *rip36E149*A mutants lost their ability to

induce HR in *S. torvum*. An E149A mutation had no effect on the translocation of Rip36 into plant cells. The results indicated that Rip36 was an avirulence factor that induced HR in *S. torvum* and that a putative Zn-dependent protease motif was found to be essential for HR response (Nahar et al. 2014).

The effects of infection by R. *solanacearum* GMI100 and *hrp* mutants on the root system of petunia plants were assessed. Inhibition of lateral root elongation and induction of swellings of the roots were the adverse effects caused by both wild-type and mutant strains. Further, the production of new root lateral structures (RLS) was induced by the wild-type strain, but not by the mutants. The RLS were efficient sites allowing extensive bacterial multiplication. The RLS may have a role in rhizosphere-related stages in the life cycle of R. *solanacearum*, resulting in successful infection by the wild-type strain (Zolobowska and van Gijsegan 2006). Two novel Hrp-secreted proteins, PopF1 and PopF2 from R. *solanacearum* exhibited similarity to one another and to putative translocators, HrpF and Nopx from *Xanthomonas* spp. and rhizobia, respectively. Both *popF1* and *popF2* belong to the HrpB regulon. They were not required for secretion of effector proteins, but they are needed for translocation of AvrA proteins into cells of tobacco. PopF1 and PopF2 are type III translocation belonging to the HrpF/Nopx family. The *hrpF* gene of *Xanthomonas* spp. partially restored HR-inducing ability to *popF1* and *popF2* mutants of R. *solanacearum*, suggesting that translocators of R. *solanacearum* and *Xanthomonas* are functionally conserved (Meyer et al. 2006). The type III secretion system (T3SS) and its associated type III effectors (T3Es) constitute the principal virulence determinants of R. *solanacearum*. Of the conserved T3Es among R. *solanacearum* strains, the F-box protein RipG7 is required for R. *solanacearum* pathogenesis on *Medicago truncatula*. Variability in natural *ripG7* existing in R. *solanacearum* was investigated. Eight representative *ripG7* orthologues contributed differently to pathogenicity on *M. truncatula*. Only *ripG7* from Asian or African strains could complement the absence of *ripG7* in GMI100 (Asian reference strain). However, RipG7 proteins from the American and Indonesian strains could interact with *M. truncatula* SKP1-like/MSKa protein, essential for the function of RipG7 in virulence, indicating that the absence of complementation was most likely a result of the variability in the leucine-rich repeat (LRR) domain of RipG7. Eleven sites under positive selection in the LRR domains of RipG7 were identified. Five positively selected sites contributed significantly to the functioning of RipG7CMR15 in *M. truncatula* colonization. The results revealed the genetic and functional variation of the essential core T3ERipG7 from R. *solanacearum* (Wang et al. 2016).

The *hrpB*-regulated gene *lrpE* (hpx5/lrg24) in R. *solanacearum* encoded a PopC-like LRR protein that carries 11 tandem LRR in the central region. Mutant defective in *lrpE* had slightly reduced virulence on host plants and changed morphology resulting in the aggregation of bacterial cells in a minimal medium. In the Δ*lrpE* mutant, bundled Hrp pili were observed on the cell surface even in late growth phase, in contrast to the disappearance of the Hrp pili much earlier in the wild-type strand. Such bundles were entangled and anchored the mutant cells in the aggregate preventing dispersal of cells. LrpE accumulated in bacterial cells and did not translocate into the plant cells as an effector protein. In Δ *lrpE* mutant, the expression levels of *hrp* genes attained a 3–5-fold increase in comparison to those in the wild-type strain. It was proposed that LrpE may negatively regulate production of Hrp pili on the cell surface of *R. solanacearum* to disperse the bacterial cells from aggregates leading to rapid pathogen movement in the vascular elements of host plants (Murata et al. 2006). *R. solanacearum* encodes a large number of putative T3SS effectors (up to 80) (Cunnac et al. 2004; Occhialini et al. 2005). Among these effector proteins, a group of seven genes that have homologies with plant-specific, LRR has been identified and they are designated as 'GALA' proteins after a conserved GAXALA sequence in their LRR. GALA6 was translocated into plant cells (Cunnac et al. 2004). *R. solanacearum* has effectors that contain both a LRR and an F-box domain. One *R. solanacearum* strain in which all of the seven GALA effector genes were mutated, lost pathogenicity entirely on *Arabidopsis* or became less virulent on tomato. In addition, GALA7, a host specificity factor was essential for pathogenicity on *Medicago truncatula* plants. Since the F-box domain is required for virulence function of GALA7, it was hypothesized that these effectors may act by hijacking their host SCF-type E3 ubiquitin ligases to interfere with their host ubiquitin/proteosome pathway to promote disease (Angot et al. 2006).

The type II secretory system (T2SS) contributes to colonization of xylem vessels by *R. solanacearum*. At this stage, production of extracellular pathogenicity determinants is transcriptionallly controlled by an extensive and complex regulatory network of distinct, interacting signal transduction pathways (Schell 2000). The principal transcriptional regulator PhcA (Phc, phenotype conversion) simultaneously activates diverse virulence genes such as those of EPS biosynthesis and production of Pme and Egl exoproteins and represses others such as *hrp* genes and those related to polygalacturonase synthesis and motility. In addition, the PhcA regulator controls production of second quorum-sensing molecule in *R. solanacearum*, an AHL which is involved in the additional regulatory system mediated by the SolI-SolR regulators to improve virulence gene expression (Flavier et al. 1997; Schell 2000). Mutation in *phcA* results in loss of one or more components of the EPS and strong reduction in endoglucanase activity, but an increase was observed in that of endopolygalacturonase. On the other hand, cell motility was increased in some variants, in addition to impaired wilting ability (Kelman and Hruschka 1973; Poussier et al. 2003).

Pathogenicity of *R. solanacearum* depends on the type III secretion system (T3SS) which delivers a large set of about 75 type III effectors (T3Es) into plant cells. On several plants, pathogenicity assays, based on quantification of wilting symptoms failed to detect significant contribution of the pathogen T3Es in this process, thus revealing the collective effect of T3Es in pathogenesis. A mixed infection-based method with *R. solanacearum* was applied to monitor bacterial fitness in plant tissues as a virulence assay. With this assay, it was possible to provide evidence for existence of growth defects in T3E mutants. Twelve genes contributing to bacterial fitness in eggplant leaves were identified and three of them were implicated in bacterial fitness on tomato and bean. Contribution to fitness of several T3Es appeared to be host-specific and some known avirulence determinants such as *popP2* or *avrA* might provide competitive advantages on some susceptible host plants. Furthermore, the assay revealed that the *efe* gene which directs the production of ethylene by bacteria in plant tissues and *hdfB* involved in the biosynthesis of the secondary metabolite 3-hydroxyoxindole, are also required for optimal growth in plant tissues (Macho et al. 2010). The gene family *awr* present in all sequenced strains of *R. solanacearum* and other bacterial pathogens was characterized. Five paralogues in strain GMI1000 of *R. solanacearum* encode type III secreted effectors and deletion of all *awr* genes impaired its capacity to multiply in natural host plants. Complementation studies revealed that the AWR (alanine-tryptophan-arginine triad) effectors displayed some functional redundancy, although AWR2 is the major contributor to virulence. In contrast, the strain devoid of all *awr* genes (Δ*awr 1-5*) exhibited enhanced virulence on *Arabidopsis* plants. A gain-of-function approach expressing AWR in *Pseudomonas syringae* pv. *tomato (Pst)* DC3000 proved that this might be due to effector recognition, since AWR 5 and AWR 4 restricted growth of *Pst* in *Arabidopsis*. Transient overexpression of AWR in nonhost tobacco caused macroscopic cell death to varying extents, which in the case of AWR 5, exhibited characteristics of typical HR. The results showed that AWR showing no similarity to any protein with known function, might specify either virulence or avirulence in the interaction of *R. solanacearum* with its host plant species (Solé et al. 2012).

*R. solanacearum* strains infect a wide range of plant species, but the host specificity of strains is not clearly understood. In order to identify the gene conferring host specificity of the strain SL341 (virulent to hot pepper, but avirulent to potato) and SL2029 (virulent to potato, but avirulent to hot pepper) were used as representative strains. A gene *rsa1* from SL2029 that confers avirulence to SL341 in hot pepper, was identified. The *rsa1* gene encoding an 11.8 kDa protein possessed the perfect consensus hrpII box motif upstream of the gene. Although the expression of *rsa1* was activated by HrpB, a transcriptional activator for *hrp* gene expression, Rsa1 protein was secreted in an Hrp type III secretion-independent manner. Rsa1 exhibited weak homology with an aspartic protease, cathepsin D, and possessed protease activity. Two specific aspartic protease inhibitors, pepstatin A and diazoacetyl-D,L-norleucine methyl ester, inhibited the protease activity of Rsa1. Substitution of two aspartic acid residues with alanine at positions 54 and 59 abolished protease activity. The SL2029 *rsa1* mutant was much less virulent than the wild-type strain, but did not induce disease symptoms in hot pepper. The results indicated that Rsa1 is an extracellular aspartic protease and plays an important role for virulence of SL2019 in potato (Jeong et al. 2011). HR is induced by harpins, the extracellular proteins, of bacterial pathogen origin. The harpin PopW is produced by *R. solanacearum* strain ZJ3271 and it is acidic with 380 amino acids, rich in glycine and serine and lacks cysteine. PopW induces rapid tissue collapse, when infiltrated into tobacco leaves, via a heat-stable, but protease-sensitive HR-eliciting activity. Subcellular localization and plasmolysis studies showed that PopW targeted the onion cell wall. This activity was confirmed by its ability to specifically bind to calcium pectate, a major component of the plant cell wall. However, PopW did not show detectable pectate lyase (PL) activity. Western blotting indicated that PopW was secreted by a type III secretion system in a *hrpB*-dependent manner. Gene sequencing showed that *popW* was conserved among 20 diverse strains of *R. solanacearum*. A *popW*-deficient mutant retained the ability of wild-type strain ZJ3721 to elicit HR in tobacco and to cause wilt disease in tomato, a natural host species. The results revealed that PopW was a cell wall-associated, *hrpB*-dependent, with two-domain harpin that was conserved across the *R. solanacearum* species complex (Li et al. 2010). Expression of genes involved in vascular HRs of tomato cultivars susceptible and resistant to *R. solanacearum* (Rs) was investigated. Expression of the *Phytophthora inhibitor protease 1* (*PIP1*) gene, encoding a papain-like extracellular cysteine protease, was induced in *R. solanacearum*-inoculated stem tissues of quantitatively resistant tomato cultivar LS-89, but not in susceptible cv. Ponderosa. *PIP1* was found to be closely related to *Rcr3*, which is required for the Cf-mediated HR to the leaf mold pathogen *Cladosporium fulvum* and manifestation of IIR cell death. However, upregulation of *PIP1* in *R. solanacearum*-inoculated LS-89 stems was not accompanied by visible HR cell death. Nevertheless, observations under electron microscope of inoculated stem tissues of LS-89, revealed the presence of several aggregated materials associated with HR cell death in the xylem parenchyma and pith cells surrounding xylem vessels. Further, accumulation of electron-dense substances was seen within the xylem vessel lumen of inoculated stems. In addition, in leaves of LS-89 and Ponderosa infiltrated with $10^6$ cells/ml of *R. solanacearum*, cell death appeared at 18 and 24 h after infiltration. Pathogen proliferation in infiltrated leaf tissues of LS-89 was suppressed to approximately 10–30% of that in Ponderosa, and expression of defense-related gene *PR-2* and HR marker gene *hsr203J* was induced in the infiltrated tissues. The results showed that the response of LS-89 represented true HR and induction of vascular HR in xylem parenchyma and pith cells surrounding xylem vessels and it appeared to be associated with quantitative resistance in LS-89 to *R. solanacearum* (Nakaho et al. 2017).

The role played by type III-associated regulators such as type III chaperones and T3SS control proteins in *R. solanacearum* is not clearly understood. The HpaP, a putative type III secretion substrate specificity switch (T3S4) protein, of *R. solanacearum* was characterized. HpaP was not secreted by the pathogen or translocated in the plant cells. HpaP self-interacted and interacted with Pop1T3E. HpaP modulates the secretion of early (HrpY pilin) and late (AvrA and Pop1T3Es) type III substrates. HpaP is dispensable for the translocation of T3Es into the host cells. Two regions of five amino acids in the T3S4 domain were identified and they were essential for efficient PopP1 secretion and for HpaP's role in the virulence of *R. solanacearum* on tomato and *Arabidopsis thaliana*, but not required for HpaP-HpaP and Hpap-PopP1 interactions. The results indicated that HpaP was a putative T3S4 protein in *R. solanacearum*, important for full pathogenicity on several hosts, acting as a helper for Pop1 secretion and repressing AvrA and HrpY secretion (Lohou et al. 2014).

Type IV pili have been shown to be virulence factors in different bacterial pathogens. Several subclasses of type IV pili have been described according to the characteristics of the structural prepilin subunit. A type IV pili was implicated in the virulence of *R. solanacearum*. Two distinct *tad* loci in the genome of *R. solanacearum* were characterized. The *tad* genes encode functions necessary for biogenesis of the Flp subfamily of type IV. The *tadA2* gene encoded a predicted NTPase reported to function as the energizer for Flp pilus biogenesis. In order to determine the role of the *tad* loci in *R. solanacearum* virulence, the *tadA2* gene located in the megaplasmid was mutated. Characterization of the *tadA2* mutant revealed that the mutant was not growth-impaired *in vitro* or in planta, produced wild-type levels of exopolysaccharides galactosamine and exhibited swimming and twitching motility comparable with the wild-type strain. However, the *tadA2* mutant was impaired in its ability to cause wilting of potato plants. The results showed that type IVb pili could contribute significantly during plant pathogenesis (Wairuric et al. 2012). In a later investigation, the interaction of strains of *R. solanacearum* phylotype IIB, sequevar1 (IIB-1) with different levels of virulence and potato was investigated to identify candidate virulence genes. A 33.7-kb deletion in a strain of *R. solanacearum* was identified and this mutant showed reduced virulence on potato. The fragment of 33.7-kb contained a cluster of six genes putatively involved in type IV pili (Tfp) biogenesis. Functional analysis suggested that these proteins might contribute to several Tfp-related functions such as twitching motility and biofilm formation. Further, the genetic cluster was found to contribute to early bacterial wilt pathogenesis and colonization fitness to potato roots (Siri et al. 2014). Several defense compounds like hydroxycinnamic acid (HCA) are produced by plants to combat pathogens. A HCA degradation pathway is genetically and functionally conserved across diverse strains of *R. solanacearum*. The feruloyl-CoA synthetase (Δ*fcs*) mutant could not degrade HCA and it was less virulent on tomato plants. Although HCA degradation enabled *R. solanacearum* growth on HCA *in vitro*, HCA degradation was dispensable for growth in xylem sap and root exudates, suggesting that HCAs are not a significant carbon source in planta. Acetyl-bromide quantification of lignin demonstrated that *R. solanacearum* infections did not affect the gross quantity or distribution of stem lignin. However, the Δ*fcs* mutant was significantly more susceptible to inhibition by two HCAs viz., caffeate and p-coumarate. Plant colonization assays suggested that HCA degradation may facilitate early stages of infection and root colonization. The results indicated that ability to degrade HCA could contribute to bacterial wilt pathogen virulence by facilitating root entry and by protecting the pathogen from HCA toxicity (Lowe et al. 2015).

### 5.1.2 PECTOBACTERIUM SPP. AND DICKEYA SPP.

Pectinolytic bacteria *Pectobacterium* spp. (earlier known as *Erwinia*) and *Dickeya* spp. infect potato tubers and plants, causing soft rot and wilt diseases. The members of the genus *Pectobacterium* is classified into four species: *P. atrosepticum*, *P. carotovorum*, *P. betavasculorum* and *P. wasabiae*. Another *Pectobacterium* taxon associated with monocot hosts may be assigned to a separate species or subspecies (Gardan et al. 2003; Yishay et al. 2008). *P. carotovorum* was further divided into subspecies as *carotovorum* (causing stem and soft rot of potato), *brasiliensis* (causing stem rot, tuber rot and blackleg) and *atrosepticum* (causing blackleg of potato) (Ma et al. 2007; van der Merwe et al. 2010). Other closely related phytopathogenic species include *P. wasabiae* (causing soft rot in Japanese horseradish (Goto and Matsumoto 1987), and *P. carotovorum* subsp. *odoriferum* (causing soft rot in chicory, Gallois et al. 1992). *Pectobacterium carotovorum* subsp. *atrosepticum (Pca)*, *P. carotovorum* subsp. *carotovorum (Pcc)*, *P. chrysanthemi* cause potato tuber soft rot and blackleg (stem rot). Latent infection of tubers and stems is widespread. These bacterial pathogens are gram-negative, non-sporing, facultative anaerobes, characterized by the production of large quantities of extracellular pectic enzyme. Disease development may be primarily dependent on the production of these enzymes together with a wide range of other plant CWDEs (Collmer and Keen 1986).

#### 5.1.2.1 Disease Incidence and Distribution

*Pectobacterium* spp. have a free-living or saprophytic phase in their life cycle. They are the main incitants of blackleg or stem rot under field conditions and tuber decay in storage. *Pectobacterium carotovorum* subsp. *carotovorum (Pcc)* has the broadest host range with worldwide distribution. On the other hand, *P. carotovorum* subsp. *atrosepticum (Pca)* infection restricted to potato may be due to their host specificity or lack of viability in different ecological conditions or seasonal variations. *Dickeya* (formerly known as *Erwinia*) *chrysanthemi* is pathogenic to many plant species in tropical and subtropical regions. The pathogen can also infect crops such as maize and dahlia in temperate zones (Young et al. 1992). *Pectobacterium* sp. is associated with potato early dying (PED) disease which is primarily induced by the fungal pathogen *Verticillium dahliae*. *Pectobacterium* sp. also causes aerial stem rot (ASR) of potato. The effects of coinfection by *Pectobacterium* sp. and *V. dahliae* on disease severity were investigated, using greenhouse assay and quantitative real-time PCR assay to quantify pathogen populations in planta. PED symptoms caused by *Pcc* isolate Ec 101 or *V. dahliae* isolate 653 alone were almost indistinguishable. *P. wasabiae* isolate Pw0405 caused ASR symptoms including water-soaked lesions and necrosis. Greater populations of *Pectobacterium* sp. were detected in plants inoculated with Pw0405, compared to Ec101, suggesting that ASR could

result in high *Pectobacterium* populations in potato stems. Significant additive or synergistic effects were not observed following coinoculation with these strains of *V. dahliae* and *Pectobacterium. V. dahliae* population levels were greater in basal stems of plants coinoculated with either *Pectobacterium* isolate. The results indicated that although coinfection by *Pectobacterium* and *V. dahaliae* might not always result in significant or additive interactions in potato, coinfection could increase PED severity (Dung et al. 2014).

Rapid spread of *Dickeya solani* (syn. *Erwinia chrysanthemi*), genetic clade of *Dickeya* biovar 3, causing blackleg disease in seed potato production areas in Europe was investigated. *D. solani* was more frequently detected than *D. dianthicola* existing already. Under in vitro conditions, both species were motile, had comparable siderophore production and pectinolytic activity and there was no antagonism between them when grown together. Both *D. solani* and biovar 1 and biovar 7 of *D. dianthicola* rotted tuber tissue, when inoculated at a low density of $10^3$ CFU/ml. *D. solani* was less susceptible to saprophytic bacteria than *D. dianthicola*, indicating that *D. solani* could be a stronger competitor in a potato ecosystem. In the greenhouse, at high temperatures (28°C), roots of potato plants were more rapidly colonized by *D. solani* than by biovar 1 or 7 of *D. dianthicola*. At 30 days after inoculation, higher densities of *D. solani* were found in stolons and progeny tubers. In coinoculated plants, fluorescent protein- (Green fluorescent protein (GFP) or DsRed) tagged *D. solani* outcompeted *D. dianthicola* in plants grown from vacuum-infiltrated tubers. The results indicated that *D. solani* was more efficient in colonizing potato plants at higher temperatures, but in temperate climates, this pathogen might not cause more infections on potato. However, latent infections by *D. solani* might be prevalent (Czakowski et al. 2013). A large number of *Dickeya* spp. strains cause soft rot disease of potato. In order to study regulation of virulence of *D. solani*, four strains were characterized. Significant differences in the virulence of *D. solani* strains were detected, although they were genetically similar, based on genomic fingerprinting profiles. Mutants of four *D. solani* strains were constructed by inactivating the genes coding either for one of the main negative regulators of *D. dadantii* virulence (*kdgR*, *pecS* and *pecT*) or for synthesis and perception of signaling molecules (*expI* and *expR*). Analysis of these mutants indicated that the PecS, PecT and KdgR play a similar role in both species, repressing the synthesis of virulence factors to different degrees. The thermoregulator PecT appeared to be a major regulator of *D. solani* virulence. The role of quorum sensing mediated by ExpI and ExpR in *D. solani* virulence on potato could be perceived by the results of this investigation (Potrykus et al. 2014).

#### 5.1.2.2 Host Range and Sources of Infection

*Pectobacterium carotovorum* causes wilt, soft rot and blackleg, affecting potato plant health under field conditions and storage. Tuber soft rot and ASR are often observed after plants are wounded and tuber soft rot is favored by low oxygen conditions. On the other hand, blackleg is considered as a tuber-borne disease, with the bacterial pathogen inducing an inky black decay on the lower part of the potato plant stem (Pérombelon 2002). *Erwinia (=Dickyea) chrysanthemi* could infect several plant species in the tropical and subtropical regions and maize and dahlia in temperate regions. Many strains of biovars, serovars or pathovars appear to have overlapping host ranges. Division of isolates at intraspecific level was found to be difficult (Young et al. 1992). Infection by *Pectobacterium* spp. was shown to be temperature-dependent. *P. carotovorum* subsp. *atrosepticum* tended to cause blackleg at temperatures < 26°C and *Erwinia chrysanthemi*, was favored by high temperatures, regardless of biovar. However, certain 'cold strains' could infect plants at cold temperatures causing blackleg disease (Pérombelon et al. 1987).

Various types of symptoms in potato such as non-emergence of plants, chlorosis, wilting, haulm desiccation and typical blackleg are induced by *Pectobacterium (Erwinia)*. The relationship between symptoms and the yield of tubers was investigated, under field conditions, using tubers artificially inoculated with *P. carotovorum* subsp. *atrosepticum (Pca)*. The method of inoculation and inoculum concentrations were the major factors affecting the development of symptoms. Disease severity was higher when tubers were inoculated by vacuum infiltration of the pathogen, than when inoculated by shaking tubers in contaminated sand. Disease symptoms progressed from chlorosis and/or wilting to partial or total desiccation of the plant. In both highly susceptible (Blintje) and moderately resistant (Desiree) cultivars, the yield of symptomless plants growing from inoculated seed tubers was significantly reduced, compared to uninoculated controls. Differences in symptom expression in the field between cultivars matched the level of visible infection of tubers at harvest, as Blintje tubers showed higher incidence of rot than Desiree tubers (Helias et al. 2000b). In western and central European countries, only *Pcc* could be isolated from potato plants with blackleg symptoms, but not *P. carotovorum* subsp. *atrosepticum (Pca)* or *Dickeya* spp. Vacuum infiltration of seed tubers with *Pcc* strains isolated in Europe and planted in two soil types, resulted in up to 50% infection by blackleg disease in potato plants (de Haan et al. 2008).

### 5.1.2.3 Modes of Dispersal of Pathogens

Soft rot pathogens are dispersed primarily through infected seed tubers and also through infested soil. The relationship between contamination of seed tubers with *Pca*, blackleg disease development and incidence and levels of progeny tuber contamination in field-grown crops was investigated between 1998 and 2000. Seed tubers were inoculated by vacuum infiltration at three levels (low, intermediate and high) with a streptomycin-resistant marker strain *Pca* (SCR11039Str) and planted in the field. The development of blackleg disease was directly related to the level of seed tuber contamination. High and low levels of seed tuber contamination were related to high and low incidences of progeny tuber contamination respectively, at all sampling times. The level of progeny tuber contamination, derived from seed tubers inoculated at three different levels of *Pca*, was categorized into

four contamination levels ($10^2$, $10^2 - 10^3$, $10^3 - 10^4$ and $> 10^4$ marker strain CFU/ml peel extract). As the level of seed tuber contamination increased, the proportion of contaminated progeny tubers also increased. The results suggested that progeny tuber contamination might be related to seed tuber contamination and blackleg disease incidence. The threshold level of seed tuber contamination remained an important factor for predicting seed tuber health and disease incidence in the field (Toth et al. 2003).

Dispersal of *Pectobacterium* spp. through water has been considered as a potential mode for large scale incidence and spread of soft rot diseases. Due to rainfall deficits in Florida, USA, recycled irrigation water, storm water runoff, reclaimed municipal sewage water and lakes for agricultural use were exploited, as alternative sources to meet water requirements of crop cultivation. With recycled water, the potential for both introducing and enhancing populations of plant pathogens was clearly indicated. In order to determine *Pectobacterium* spp. population levels, samples were tested at monthly intervals for one year from four hyper-eutropic lakes and eight nursery retention ponds using the direct plating method. *Pectobacterium* strains (77) were isolated by both the direct plating method and by an enrichment process. With the direct plating method, 0 to 29 CFU/ml were detected on the sodium polypectate medium. Populations of *Pectobacterium* spp. were significantly higher in retention ponds of nurseries that actively reutilized their water. Fatty acid analysis and biochemical tests were employed to identify the *Pectobacterium* strains. Rep-PCR assay was applied for differentiation of 120 strains of *Pectobacterium* spp. Most strains clustered into two heterogeneous populations of *E. chrysanthemi* and *P. carotovorum* ssp. *carotovorum* in 1:2 and 1:4 ratios for isolates from ornamentals and from water, respectively. Genetically different strains that contained high percentages of *Pectobaterium* strains in water sources could be identified. Most strains (99%) from water sources were found to be pathogenic on *Diffenbachia* (Norman et al. 2003).

Pectinolytic, pathogenic bacterial pathogen *Pectobacterium* spp. and *Dickeya* spp. form a significant threat to production of seed potato tuber stocks in Europe. *Dickeya* spp. induces various symptoms such as wilting of plants, stem rot (blackleg) and tuber soft rot. Induction of these symptoms depends on the cultivar susceptibility, pathogen virulence and environmental conditions including plant evapotranspiration. Using the flow of sap driven by plant transpiration, the pathogen could colonize the entire vascular system. To assess the consequences of colonization of the xylem vessels by *Dickeya* spp., greenhouse and field experiments were performed with plants infected via mother tuber inoculation or root inoculation. The seed- and soil-carried inocula could induce wilting and blackleg symptoms in potato plants. Further, blackleg symptoms appearing during the first three weeks after plant emergence are caused mainly by a tuber-borne inoculum. Presence of inoculum in the xylem vessels did not alter the plant transpiration process, until the appearance of wilting symptoms on rotting lesions (Ansermet et al. 2016).

#### 5.1.2.4 Process of Infection of Susceptible Plants

Pathogenesis of bacterial diseases of plant has different stages that either favor or retard disease development, based on the availability of susceptible plants, virulent strains and favorable/unfavorable environmental conditions.

##### 5.1.2.4.1 Initiation of Infection

Prior to initiation of infection, the bacterial pathogen has to undergo epiphytic multiplication and reach a certain level of population. The bacteria-to-bacteria communication is associated with a quorum-sensing (QS) process based on the production of N-AHLs (HSL). The role of HSL produced by strains of *Pectobacterium (Erwinia) atrosepticum* in plant infection was investigated and the HSL produced by a specific virulent pathogen strain on potato was identified. A derivative of this strain, expressing HSL *lactonase gene* and producing low amounts of HSL, was generated. Production of HSL and QS regulated only those traits involved in the second stage of infection, tissue maceration and HR in nonhost tobacco plants (Smadja et al. 2004). Among the proteins secreted by *P. atrosepticum*, a protein homologous to *Xanthomonas campestris* avirulence protein with unknown function was identified. This protein SVX was secreted by the type II (out) secretion apparatus which is responsible for secretion of the major virulence factors, PelC and Cel IV. Transcription of the *svx* gene was under N-AHL (HSL)-mediated QS control. The mutant of *svx* generated by transposon insertion showed decreased virulence in potato plant assays, revealing a definite role of SVX in pathogenicity of *P. atrosepticum*. The type III secretory system was found to be a conduit for virulence factors other than the main pectinases and cellulase in *P. atrosepticum* (Corbett et al. 2005). The QS signal molecule N-(3-oxohexanoyl)-1-homoserine lactone (OHHL) produced by *Pectobacterium carotovorum* subsp. *carotovorum (Pcc)* controlled seven genes of the pathogen. The mutants defective in proteins that could play a role in the interaction between *Pcc* and its host plants were enriched using TrophoA as a mutagen and Nip (*Pcc*) and its counterpart in *P. atrosepticum (Pa)* were identified. Nip (*Pcc*) induced necrosis in tobacco (nonhost), while Nip(Pa) affected potato stem rot. Both proteins affected virulence on potato tubers. They are members of a family of proteins related to Nec1 from *Fusarium oxysporum* which induced necrotic responses in many dicotyledonous plant species. In *Pcc*, *nip* was repressed weakly by LuxR regulator, PccR and it may be regulated by the negative global regulator RsmA (Pemberton et al. 2005). *Pectobacterium carotovorum* subsp. *carotovorum* strain PccS1, causing soft rot in *Zantedschia ellotiana* (colored calla) was investigated for virulence genes induced by the host plant. Using a promoter-trap transposon (mariner), 500 transposon mutants were obtained showing kanamycin resistance dependent on an extract of *Z. ellotiana*. One of these mutatants, PM86 exhibited attenuated virulence on both *Z. ellotiana* and *Brassica rapa* subsp. *pekinensis*. The growth of PM86 was reduced in minimal medium (MM), and the reduction was restored by adding plant extract to the MM. The gene containing the insertion site was identified as *rpIY*. The deletion mutant Δ*rpIY*, exhibited reduced

virulence, motility and plant cell wall-degrading enzyme production, but not biofilm formation. Analysis of gene expression and reporter fusions revealed that the *rpIY* gene I PccS1 was upregulated at both the transcriptional and translational levels in the presence of plant extract. The results suggested that *rpIY* might be induced by *Z. elliotiana* extract and might be crucial for the virulence of *Pcc* (Jiang et al. 2017).

*Pectobacterium carotovorum* uses global regulators to coordinate pathogenesis in response to different conditions. An operon required for virulence and gluconate metabolism of *P. carotovorum* was characterized. The operon contains four genes that are highly conserved among proteobacteria. A mutant with a deletion-insertion within this operon was unable to metabolize gluconate, a precursor of the pentose phosphate pathway. The mutant had a slow growth rate on leaves of potato and *Arabidopsis thaliana*. The mutant hypermacerated potato tubers and was deficient in motility. Global virulence regulators that are responsible in cell wall pectin breakdown products and other undefined environmental signals KdgR and FlhD, respectively, were misregulated in the mutant. The alteration of virulence mediating via changes in transcription of known global virulence regulators in yg6JM operons suggested a role for host-derived catabolic intermediates in *P. cartovorum* pathogenesis. The operons in *P. carotovorum* required for virulence and gluconate metabolism was renamed vguABCD (Mole et al. 2010). *Pectobacterium atrosepticum* produces plant cell wall-degrading enzymes (PCWDEs), which are key virulence determinants in pathogenesis. The impact on virulence of a transposon insertion mutation in the *metJ* gene that codes for the repressor of the methionine biosynthesis regulon was investigated. In the mutant strain defective for the small regulatory RNA rsmB, PCWDEs were not produced and virulence in potato tuber was almost totally abolished. However, when the *metJ* gene was disrupted in this background, the rsmB⁻ phenotype was suppressed and virulence and PCWDE production were restored. In addition, when the *metJ* gene was disrupted, production of the QS signal, N (3-oxohexanoyl)-homoserine lactone was increased. Genes involved in methionine biosynthesis were most highly upregulated, but many virulence-associated transcripts were also upregulated (Cubitt et al. 2013). QS regulates the secretion of PCWDEs and flagella-mediated motility via two different signaling molecules such as 3-oxohexanoyl-L-homoserine lactone and 3- oxooctanoyl-L-homoserine lactone. The phytochemical compound curcumin (from turmeric) was evaluated for its QS inhibitory activity against AHL-dependent PCWDEs production and motility in *P. wasabiae* SCC3193, *P. carotovorum* subsp. *carotovorum* Pcc21 and *P. carotovorum* subsp. *carotovorum* Pcc. Treatment with curcumin at sub-MIC (median inhibitory concentration) efficiently inhibited production of PCWDEs as well as swimming and switching motility of all tested pathogens in a non-bacterial manner. Further, curcumin treatment reduced root infection of *Arabidopsis thaliana* by pathogens by attenuating the QS-mediated virulence factors production (Sivaranjani et al. 2016).

The involvement of phenolic components in the inhibition of infection and /or progress of disease development has been indicated in various pathosystems. The effects of cinnamic acid (CA) and SA on the virulence of *Pectobacterium* spp. were studied. Both CA and SA affected QS machinery of *P. aroidearum* and *P. carotovorum* subsp. *brasiliense,* which could produce AHL QS signals, by altering the expression of bacterial virulence factors. Expression of QS-related genes increased over time, but exposure of bacteria to non-lethal concentrations of CA or SA inhibited the expression of QS genes, including *exp1, expR, PCl_1442* (luxR transcriptional regulator) and *luxS* (a component of the AI-2 system). Other virulence genes known to be regulated by the QS system, such as *pecS, pel, peh* and *yheO*, were also downregulated relative to the control. In addition, CA and SA reduced the level of AHL signal. The effects of CA and SA on AHL signaling were confirmed in combination; exogenous application of N (β-ketocaproyl)-L-homoserine lactone (eAHL) led to recovery of the reduction in virulence caused by two phenolic compounds. The results of gene expression studies, bioluminescence assays, virulence assays and compensation assays provided evidence for the potential of CA and SA in inhibiting the virulence of *Pectobacterium* spp. via QS machinery (Joshi et al. 2016).

Pectinases produced by *Pectobacterium carotovorum* subsp. *carotovorum (Pcc)* have been demonstrated to be the primary determinants of virulence. However, several ancillary factors including augmenting bacterial virulence have also been identified. Flagellum formation and bacterial movement are regulated in *Pcc* by FlhDC, the master regulator of flagellar genes and fliA, a flagellum-specific σ factor. Motility of *Pcc* was positively regulated by QS signal, N-AHL and negatively regulated by RsmA, a post-transcriptional regulator. RsmA, an RNA-binding protein induced translational repression and promoted RNA decay. The results showed that RsmA negatively regulated *flhDC* and *fliA* expression. Further, the chemical stabilities of transcripts of these genes were greater in a Rsma mutant than in Rsma+ bacteria. In the absence of AHL, the AHL receptors ExpR[1] and ExpR[2] (= AhlR) in *Pcc* negatively regulated motility and expression of *flhDC* and *fliA* by activating RsmA production. In the presence of AHL, the regulatory effects of ExpR[1] and ExpR[2] were neutralized, resulting in reduced levels of *rsmA* expression and enhanced motility (Chatterjee et al. 2010). In order to study the process of infection by *Pectobacterium caratovorum* subsp. *brasiliense (Pcb)*, causing potato soft rot disease, the mcherry-*Pcb*-tagged strain was generated. Four potato cultivars were evaluated for stem-based resistance to *Pcb*. Confocal laser-scanning microscopy and *in vitro* viable cell counts showed that *Pcb* was able to penetrate the roots of susceptible potato cultivar as early as 12 h postinoculation (hpi) and to migrate upward into aerial stem parts. In the susceptible cultivar, *Pcb* cells could colonize xylem tissue, forming 'biofilm-like' aggregates that led to occlusion of some of the vessels. In contrast, *Pcb* appeared as free-swimming planktonic cells with no specific tissue association in the tolerant cultivar. The results suggested that the resistance mechanisms operating in the tolerant cultivar might limit the aggregation of *P. carotovorum* subsp. *brasiliense* cells in planta, leading

to failure of induction of disease symptoms (Kubheka et al. 2013).

Pectinolytic *Pectobacterium (Erwinia)* spp. has been shown to be transmitted from infected mother tubers to daughter tubers produced by plants grown from artificially inoculated mother tubers. The progress of pathogen contamination in different organs through plants to progeny tubers was followed under natural conditions. *Pectobacterium atrosepticum* was detected in different symptomless plant organs, such as stolons, stems and progeny tubers and also in plant parts with symptoms on stems, collected at various stages of plant development. Infection levels in below- and above-ground organs of plants of two cultivars with different levels of resistance to P. *atrosepticum* were determined using DAS-ELISA and PCR assay. Healthy organs from symptomless plants were less frequently contaminated than symptomless organs of diseased plants and stolons were more frequently contaminated than stem and daughter tubers, regardless of the health of plant. Stem injections progressed latently and the pathogen could be recovered 10–15 cm away from visible lesions. Typical ASR symptoms could be related frequently to the upward movement of the pathogen from the infected mother tuber. Daughter tubers without symptoms were commonly contaminated usually at heal ends, suggesting internal contamination from mother tuber to progeny (Helias et al. 2000a). The type-strain of *Pectobacterium carotovorum* caused DspE/F-dependent cell death on *Nicotiana benthamiana* within 24 h postinoculation (hpi), followed by leaf maceration within 48 hpi. On the other hand, P. *carotovorum* strains with mutations in T2SS regulatory and structural genes, including the *dspE/F* operon, did not induce HR-like cell death and/leaf maceration. But a strain with a mutation in the T2SS caused HR-like plant cell death, but no maceration. P. *carotovorum* did not suppress the basal immunity function. Callose deposition along leaf vein occurred within 24 hpi and the pathogen cells were localized along the veins. Gene expression profiles in *N. benthamiana* leaves inoculated with wild-type and mutant P. *carotovorum* and *Pseudomonas syringae* were compared. The gene expression profile in *N. benthamiana* leaves infiltrated with P. *carotovorum* was similar to leaves infiltrated with P. *syringae* T3SS mutant. The results suggested that P. *carotovorum* might use T3SS to induce plant cell death, resulting in leaf maceration rather than to suppress immunity of plants (Kim et al. 2011).

GFP has been used to tag microbial plant pathogens to monitor their movement from the site of inoculation to different tissues of inoculated plants. Translocation of GGP-tagged *Dickeya* sp. from stems or from leaves to underground parts of potato plants was tracked under greenhouse conditions. The majority of potato plants inoculated with *Dickeya* sp. into the stem showed symptoms at stem base (90%) and browning of internal stem tissue (95%) at 30 days after inoculation. The GFP-tagged strain of *Dickeya* sp. was detected using dilution plating in extracts of stem interiors (100%), stem bases (90%), roots (80%), stolons (55%) and progeny tubers (24%). In the roots, the GFP-expressing *Dickeya* sp. was observed inside and between parenchyma cells, whereas in stems and stolons, xylem vessels and protoxylem cells, the pathogen could be detected. In the progeny tubers, pathogen presence was detected in the xylem end. Leaf inoculation resulted in the presence of GFP-tagged *Dickeya* sp. in extracts of 75% of the leaves, 88% of the petioles, 63% of the axils and inside 25% of stem at 30 days after inoculation, as revealed by UV-microscopy. *Dickeya* sp. did not induce blackleg or ASR symptoms and it was not translocated to underground plant parts as indicated by the absence of detectable fluorescence (Czajkowski et al. 2010). A epifluorescence stereomicroscope at low magnification (2.5 to 10 X) and a confocal laser-scanning microscope (CLSM) at a magnification of 640 X to 1,000 X were used to detect *Dickeya* sp. in different parts of potato plants. In stem-inoculated plants, GFP signal was detected in the vascular tissue of stems, stem bases and stolons. In the progeny tubers, the signal was detected in the vascular ring of the stolon end. In roots, the signal was detected in pith tissue. In plants inoculated on leaves, GFP signal was seen in the main vein of leaves and inside petioles, but not inside the stem bases. Localization of GFP-tagged *Dickeya* sp. was observed using CLSM. The pathogen cells were present mainly inside xylem vessels and between protoxylem cells of stems, stem bases and stolons of plants at 30 days after inoculation into stems. In the roots of stem-inoculated plants, GFP-tagged bacteria were detected inside parenchyma cells of pith tissue. In progeny tubers, GFP-tagged bacteria were present inside xylem and between xylem vessels of stolon ends (Czajkowski et al. 2010).

The type III secretion system (T3SS) plays a vital role, as essential virulence factor for the pathogenicity of several bacterial species. Polynucleotide phosphorylase (PNPase) is one of the major ribonucleases in bacteria and it is involved in mRNA degradation, tRNA processing and small RNA (sRNA) turnover. In the case of *Dickeya dadantii*, PNPase downregulated the transcription of T3SS structural and effector genes. This negative regulation of T3SS by PNPase occurred by representing the expression of *hrpL*, encoding a master regulator of T3SS in D. *dadantii*. The PNPase downregulated the transcription of *hrpL*, which led to a reduction in T3SS gene expression, by reducing *rpoN* mRNA stability. RsmB, a regulatory sRNA, enhances *hrpL* mRNA stability in *D. dadantii*. The results suggested that PNPase decreased the amount of functional RsmB transcripts that could result in reduction of *hrpL* mRNA stability. Furthermore, bistable gene expression (differential expression of a single gene that creates two distinct populations) of *hrpA*, *hrpN* and *dspE* was observed in *D. dadantii* under *in vitro* conditions. However, it appeared that PNPase may not be the key switch that can trigger the bistable expression patterns of T3SS genes in the bacterial pathogens (Zeng et al. 2010). In another investigation, the role of RssB, ClpXP, and RpoS in T3SS regulation in *D. dadantii* 3937 strain was studied. ClpP is a serine-type protease which associates with ClpX chaperone to form a functional Clp proteolytic complex for the degradation of proteins. ClpXP degraded the RpoS sigma factor, with the assistance of recognition factor RssB. RpoS positively regulated the expression of the *rsmA* gene, encoding an RNA-binding regulatory

protein. By interacting with the *hrpL* mRNA, RsmA reduced HrpL production and downregulated the T3SS genes in the HrpL regulon. Further, ClpXP, RssB and RpoS affected pectinolytic enzyme production in *D. dadantii* 3937 strain, possibly through RsmA. The ClpXP and RssB proteins were found to be essential for the virulence of *D. dadantii* (Li et al. 2010).

*Dickeya dianthicola* and *D. solani* induce blackleg and soft rot diseases in Europe. *D. solani* caused blackleg and soft rot symptoms in potato. *D. solani* strains were more virulent on potato plants than *D. dianthicola*, especially at higher temperatures, in addition to being capable of inducing blackleg disease from lower inoculum levels. The phenotypic characteristics of *D. solani* strains (15) from different countries (Poland, Finland and Israel), along with three strains of *D. dianthicola* were determined. The effects of temperature and geographic origin of the strains on their ability to macerate potato tissue and on the major virulence factors, such as pectolytic, cellulolytic and proteolytic activities, siderophore production and mobility were assessed. Polish *D. solani* strains possessed higher activities of CWDEs, than the Finnish and Israeli strains at all temperatures (18, 27 and 37°C) tested. The ability of Polish strains to induce more severe potato soft rot might be related to cell wall-degrading enzyme production. In addition, *D. solani* strains caused more severe tuber maceration than D. *dianthicola* strains *in vitro*. *D. solani* strains differed in their phenotypic characteristics, although they had identical pulse-field gel electrophoresis (PFGE) profiles and similar fingerprint profiles obtained via repetitive sequence-based PCR assays (Golanowska et al. 2017).

The ability to tolerate the antimicrobial compounds applied on or present naturally in host plants, is an important property required for bacterial pathogens to survive and induce disease in plants. A novel vector Tn7 was constructed for tagging and enumerating target bacteria from complex microbial communities. A cassette for inducible bioluminescence and tetracycline resistance that integrates at a defined neutral position was present in most gram-negative bacterial species. *Pectobacterium* was chromosomally tagged in such a way that it could be enumerated in mixed consortia without placing a significant bioburden on the tagged strain. Two strains of *Pectobacterium carotovorum* subsp. *carotovorum (Pcc)*, one a producer of carbapenem antibiotic and another an isogenic knockout strain, were tagged using this system. The tagged strains were used to compare the extent to which potato tuber-associated and endophytic bacteria could gain advantage and multiply in planta, utilizing the nutrients released by *Pcc* infection, when the infecting strain was either an antibiotic producer (Car⁺) or a carbapenem knockout (Car⁻) strain. The Car⁺ strain had a significant advantage in numbers of bacterial cells throughout the course of infection. While the number of other bacterial species was limited by carbapenem production, it allowed *Pectobacterium* to replicate to higher titers in the rotting tuber lesions. The results indicated that the use of Tn7 tagging vector would be helpful in studying ecological interactions in complex environments (Kovács et al. 2009). Resistance to antimicrobial peptides present in host plants was shown to be an important virulence factor in *Dickeya dadantii*, a causal agent of soft rot diseases of vegetables. A transcriptional microarray analysis was performed following treatment with sublethal concentrations of thionins, to assess the response of *D. dadantii* to treatment. Overall, 36 genes were overexpressed and were classified according to their deduced functions as transcriptional regulators, transport and modification of the bacterial membrane. The gene encoding a uricase was repressed. The majority of the genes was under the control of the PhoP/PoQ system. The results indicated that the presence of antimicrobial peptides induced a complex response that included peptide-specific elements and general stress-response elements contributing differentially to the virulence in different hosts (Rio-Alvarez et al. 2012). Plant phenolic compounds p-coumaric acid (PCA) was earlier shown to inhibit expression of the type III secretory system (T3SS) in *Dickeya dadantii* 3937 strain. Of the several derivatives of phenolic compounds tested, trans-4- hydroxycinnamohydroxamic acid (TS103) showed an eight-fold higher inhibitory level than PCA on the T3SS of *D. dadantii*. The results suggested that TS103 might inhibit HrpY phosphorylation and lead to reduced levels of *hrpS* and *hrpL* transcripts. Further, through a reduction in the RNA levels of the regulatory small RNA rsmB, TS103 also might inhibit *hrpL* at the post-transcriptional level via the *rsmB*-RsmA regulatory pathway. Finally, TS103 inhibited *hrpL* transcription and mRNA stability, which leads to reduced expression of HrpL regulon genes, such as *hrpA* and *hrpN*. The ability of TS103 to inhibit T3SS through both transcriptional and post-transcriptional pathways in the soft rot pathogen *D. dadantii* 3937 strain was demonstrated for the first time (Li et al. 2015). *Dickeya zeae*, incitant of rice root rot disease, produces a range of virulence factors, including phytotoxic zeamines and extracellular enzymes. A SlyA/MarR family transcription factor SlyA was identified in *D. zeae* strain EC1. The disruption of *slyA* gene significantly decreased zeamine production, enhanced swimming and swarming motility, reduced biofilm formation and significantly reduced pathogenicity on rice. The role of SlyA is transcriptional modulation of a range of genes associated with bacterial virulence. In trans-expression of *slyA* in *expI* mutants recovered the phenotypes of motility and biofilm formation, suggesting that SlyA is the downstream of the acylhomoserine lactone-mediated QS pathway. The results revealed a key transcriptional regulating factor production and overall pathogenicity of *D. zeae* EC1 (Zhou et al. 2016).

### 5.1.3 *Streptomyces* spp.

#### 5.1.3.1 Disease Incidence and Distribution

*Streptomyces* spp., causal agents of potato common scab disease, are common soil inhabitants and form a complex genus comprising about 500 species. However, mainly three species viz., *Streptomyces scabies, S. acidiscabies* and *S. turgidiscabies* have been well characterized (Lambert and Loria 1989a, 1989b). Other pathogenic species, S. *europaeiscabiei*, S. *stelliscabiei*, S. *niveiscabiei*, S. *luridiscabiei* and S. *puniciscabiei* have also been reported to cause common scab disease in different geographical locations (Bouchek-Mechiche 2000b;

Park et al. 2003). Among the common scab-causing species, *Streptomyces scabies* and *S. turgidscabies* have worldwide distribution, whereas *S. europaeiscabiei* has been found to be the principal causal agent of potato common scab in Europe. The incidence of common scab disease was more frequent than powdery scab disease caused by *Spongospora subterranea* in Norway (Dees et al. 2013). *Streptomyces* spp. are gram-negative, filamentous bacteria and are capable of producing different antibiotics and relatively only a few of the several hundred species in the genus are pathogenic to plants. Infection of potato cultivars by *Streptomyces* spp. usually results in superficial pitted or raised lesions on the tuber surface. *Streptomyces* spp. causing common scab of potato can infect other root crops such as radish, beet root, carrot and parsnip. Pathogenicity of *Streptomyces* spp. is primarily due to the ability to produce the toxin thaxtomin, which induces characteristic symptoms of common scab disease (Loria et al. 2006).

Soil suppressiveness is a phenomenon observed in some soils in which disease incidence shows progressive decline, after initial higher disease incidence, when monoculture of crop is practiced. Soil suppressiveness is associated with activities of microorganisms antagonistic to the pathogen(s) concerned. The potato field in East Lansing, MI, showing decline in common scab incidence was compared with an adjacent potato field to determine the effect of biological factors on the development of soil suppressiveness resulting in decline in disease incidence. The putative common scab-suppressive soil (SS) was either exposed to various temperatures or mixed with autoclaved SS at different proportions. Pathogenic isolate of *Streptomyces scabies* was incorporated into the treated soil at $10^6$ CFU/cm$^3$ of soil, followed by planting of either potato or radish. Disease severity was negatively correlated with the percentage of SS in the mixture and positively correlated with temperature above 60°C. The frequency of antagonistic bacteria in SS was higher than in common scab-conducive soil (CS) in all groups, but only pseudomonads and streptomycetes were significantly higher. The population of pathogenic *Streptomyces* spp. in the rhizosphere of CS was significantly higher than in SS. Terminal restriction fragment length polymorphism (T-RFLP) analysis could differentiate two distinct microbial communities that differ in the pathogenic/antagonistic activities, present in the suppressive and CSs (Meng et al. 2012).

#### 5.1.3.2 Pathogenicity and Host Range

The dynamics of common scab infection patterns in potato were investigated, using hydroponics and non-destructive pot culture systems that enabled inoculation of plants at specific tuber development stages, and identification of tuber physiological factors associated with susceptibility to common scab pathogen *Streptomyces scabiei*. S. *scabiei* produces a phytotoxin, thaxtomin A which is a key pathogenicity determinant in potato-*Streptomyces* pathosystem. A hydroponic system was developed to generate tubers in a soilless environment with pathogen inoculation for correlating the efficiency in a nondestructive manner. Tubers were inoculated by applying pathogen suspension as a fine spray mist with a handheld sprayer until tubers were fully wetted. Alternatively, the pathogen suspension was directly applied on tubers as a droplet (100 µl) using a micropipette. Spore suspension spray produced approximately two-fold more lesions than the droplet inoculation method. Potato cvs. Desirée and Shepody had greater (two-fold) infection rate, compared to cv. Russet Burbank. *S. scabiei* isolate #20 induced scab symptoms in tubers of cv. Desirée. The hydroponic system enabled pathogenicity testing efficiently to demonstrate the pathogenic potential of the strains of *S. scabiei*, in the absence of competition with other microorganisms. This method also provided a procedure complementary to traditional pot culture and field assessments to monitor the disease development under *in vitro* conditions (Khatri et al. 2010). At the whole-tuber level, infection percentages were greatest, when infection occurred early, at two weeks after tuberization (WAT), 68% of tubers became infected, contrasting with late inoculation (8 WAT), when only 4% infection was observed. The first-formed internodes were most susceptible to infection, whereas later-forming and slower-expanding internodes were less susceptible. Physiological examination of tubers in the second internode showed that pathogen-induced changes, including increased phellem (periderm) thickness, cell layers and phellem suberization (the key physiological factor believed to be critical to *S. scabiei* infection) were promoted through *S. scabiei* infection. Differences in cultivar response, had a bearing to their levels of resistance to common disease. Greater phellem suberization was observed at ten days after tuberization (DAT) in tolerant cv. Russet Burbank than in susceptible cv. Desiree. Likewise, Russet Burbank had thicker and more numerous cell layers in the phellem (up to eight cell layers) during early tuber growth (20–30 DAT) than Desiree (up to six cell layers) (Khatri et al. 2011).

The pathogenic potential of *Streptomyces* spp. viz., *S. scabies*, *S. europaeiscabiei*, *S. stelliscabiei* and *S. reticuliscabiei* infecting potatoes was studied. Three pathogenicity groups were differentiated. Group 1 included isolates of *S. scabies*, *S. europaeiscabiei* and S. *stelliscabiei* from common scab lesions of potato and other susceptible crops. These isolates induced similar symptoms and they were pathogenic to potato, carrot and radish. Group 2 included all isolates from *S. reticuloscabiei* inducing netted scab lesions and they were pathogenic to both tubers and roots of only a few potato cultivars. These isolates could not infect carrot or radish. In group 3, three isolates of *S. europaeiscabiei* induced netted scab lesions on cv. Bintje. But they caused either common or netted scab symptoms on other cultivars or plant species tested. Soil temperature showed significant influence on the type of symptoms induced by the three groups of *Streptomyces* spp. Three isolates of group 1 were highly pathogenic at higher temperatures (20°C or 20/30°C), whereas two isolates of group 2 were most pathogenic at lower temperature (17°C). Group 3 isolates caused netted scab symptoms at 20°C and deep-pitted lesions at higher temperature. Variation in ecological conditions favoring development of *Streptomyces* spp. should be taken into account, while disease management strategies are

planned against potato scab disease (Bouchex-Mechiche et al. 2000a). The identity of a new strain DS3024 was established, based on morphology, biochemistry and genetic analysis. Analyses of 16S rRNA gene sequence indicated that the strain DS3024 was most similar to an isolate of *Streptomyces* sp. ME02-6979.3a which was not pathogenic to potato tubers, but distinct from other known pathogenic strains of *Streptomyces* spp. Strain DS3024 had genes that encoded thaxtomin synthetase (*txtAB*) which is required for pathogenicity and virulence, and tomatinase (*tomA*) used as a marker for many pathogenic *Streptomyces* spp. But this strain lacked *nec1* gene, associated with virulence in most pathogenic *Streptomyces* spp. The strain DS3024 could grow at pH 4.3, induce scab symptoms on potato tubers like *S. scabies* on tubers of susceptible cv. Atlantic potato. The strain DS3024 was identified as a new strain of *Streptomyces* capable of causing common scab of potato tubers in Michigan State (Hao et al. 2009).

The genes governing pathogenicity of *Streptomyces turgidiscabies*, causing potato common scab disease were studied. Pathogenicity island (PAI) containing pathogenicity gene was found to be mobile and appeared to transfer and disseminate like an integrative and conjugative element (ICE). The pathogenicity genes of *S. scabiei* were clustered into two regions designated toxigenic region (TR) and colonization region (CR). A composite TR structure was not present in all *S. scabiei* and *S. acidiscabiei* strains tested. The majority of strains lacked TR2 and TR excision was not observed in any of the strains tested, suggesting the involvement of TR2 in the mobilization of *S. scabiei* TR (Chapleau et al. 2016). The genome of *S. scabiei* 87.22 was analyzed to determine the potential mobility of TR. Attachment sites (*att*), short homologous sequences that delineate ICEs were identified at both extremities of the TR. Among ten *Streptomyces* spp. infecting underground plant parts, *S. scabies* (syn. *S. scabiei*) has been studied more frequently. The principal pathogenicity determinant of *Streptomyces* spp. is a nitrated diketopiperazine known as thaxtomin A (Thxa). The ThxA is a biosynthetic cluster located within a mobile genomic island designated toxicogenic region (TR) and in S. *turgidiscabies*, it resides on a mobile PAI. Three attachment (*att*) sites further divide TR into two subregions (TR1 and TR2). TR1 contains the Thx A biosynthetic cluster and it is conserved among several pathogenic *Streptomyces* species. However, most pathogenic *Streptomyces* spp. lack TR2, an ICE. Mobilization of the whole TR element or TR2 alone between *S. scabies* and nonpathogenic *Streptomyces* spp. was previously demonstrated. TR1 alone did not mobilize and the data suggested that TR2 was required for mobilization of TR1. On the other hand, TR2 could self-mobilize to pathogenic *Streptomyces* species, harboring only TR1 and integrate into the *att* site of TR1, resulting in tandem accretion of resident TR1 and incoming TR2 which might further mobilize resident TR1 in *cis* and transfer it to a new recipient cell. Further, TR1 was a nonautonomus *cis*-mobilizable element and it could hijack TR2 recombination and conjugation machinery to excise, transfer and integrate, leading to the dissemination of pathogenicity genes and the emergence of new pathogenic species. Comparative genomic analyses of 23 pathogenic *Streptomyces* isolates from ten species revealed that the composite PAI formed by TR1 and TR2 was dynamic. Various compositions of the island could exist within the population of newly emerged pathogenic species, indicating the structural instability of the composite PAI. PAI of *S. scabies* 96-12 strain contained a PAI that was almost identical to the PAI of *S. scabies* 87-22, despite significant differences in their genome sequences. Further, the strain 96-12 was phylogenetically grouped with nonpathogenic species. The results suggested that direct or indirect *in vivo* mobilization of PAI might occur between *S. scabies* and nonpathogenic *Streptomyces* spp. *S. scabies* 87-22 strain deletion mutants containing antibiotic resistance markers in the PAI were mated with *S. diastatochromogenes*, a nonpathogenic species. The PAI of *S. scabies* was site-specifically inserted into the *avrXI* gene of *S. diastatochromogenes* and conferred pathogenicity in radish seedling assays, indicating that *S. scabies* could be the source of a PAI responsible for the emergence of new pathogenic species capable of infecting different crops (Chapleau et a. 2016; Zhang et al. 2016; Zhang and Loria 2017).

*Streptomyces scabiei*, causal agent of common scab disease of potato tubers, produces the toxin compound thaxtomin A as well as indole-3-acetic acid (IAA). Tryptophan is a biosynthetic precursor of both IAA and thaxtomin A. The effect of tryptophan on production of thaxtomin A and IAA and also its effect on the transcription of the corresponding biosynthetic genes in *S. scabiei* were analyzed. The availability of tryptophan determined *in vitro* production of IAA. However, culture medium amended with tryptophan inhibited the biosynthesis of thaxtomin A. The expression of biosynthetic genes of thaxtomin A *nos* and *txtA* were strongly repressed in the presence of tryptophan. However, modulation of the expression was not observed for the IAA biosynthetic genes *iaaM* and *iaaH*. The effects of the exogenous application of tryptophan on the virulence of *S. scabiei* on radish seedlings roots were assessed. Reduction in disease severity in the roots of radish seedling was due to the inhibition of the production of thaxtomin A and an increase in IAA biosynthesis (Legault et al. 2011). *Streptomyces scabies* produces the phytotoxins concanamycin and thaxtomin. The contribution of concamycin to lesion development was studied using potato tuber slice assay. Concanamycin A showed weak necrosis-inducing activities. An amount of > ten-fold of thaxtomin A was required for concanamycin to produce equivalent severity of lesion. Concanamycins were detected in tubers inoculated with *S. scabies*, which induced deep-pitted lesions, but not in those inoculated with *S.acidscabies*, which caused corky, raised lesions. In field-grown, diseased potatoes, concanamycin content tended to be higher in tubers with deep-pitted lesions than in those with cork, raised lesions (Natsum et al. 2017).

Due to the influence of environmental factors on symptom development and the erratic distribution of scab pathogens (*Streptomyces* spp.) in the field, it was difficult to select resistant breeding lines under field conditions. The toxin thaxtomin produced by *Streptomyces* spp. could be employed to screen large numbers of potato seedlings for tolerance *in vitro*. However, the relationship between tolerance to thaxtomin and expression of resistance to scab disease under field conditions was not examined precisely. Hence, potato F1 progenies (120) from a single cross were screened *in vitro* by treating the seedlings to thaxtomin A added to the culture medium. Eighteen genotypes were selected, based on high sensitivity or tolerance based on shoot growth and they were multiplied *in vitro* and tested under greenhouse and field conditions. Evaluation of tubers (about 6,500) showed that 18 potato genotypes differed in scab indices and disease severity (P < 0.0001). The relative shoot height *in vitro* (treated with thaxtomin A at 0.5 µg/ml) and the scab index in the field showed significant correlation. The results of *in vitro* tests were corroborated by the assessment of scab severity under field conditions, indicating that *in vitro* tests might be useful to discard scab-susceptible genotypes at the initial stages of screening for resistance to potato scab disease (Hiltunen et al. 2011). *Streptomyces ipomoeae*, the causative agent of sweet potato *Streptomyces* soil rot disease, produced the phytotoxin, a less-modified thaxtomin derivative thaxtomin C whose role was unclear. The thaxtomin gene cluster (*txt*) of *S. ipomoeae* was cloned and sequenced. The *txt* mutants, lacking ability to produce thaxtomin C, were constructed and they were unable to penetrate intact adventitious roots, but they could induce necrosis on storage root tissue. The results suggested that thaxtomin C might have an essential role in inter- and intra-cellular penetration of adventitious roots of sweet potato by *S. ipomoeae* (Guan et al. 2012). Functioning of the type VII protein secretion system (T7SS) in *Streptomyces scabies* was investigated. The T7SS was essential for the virulence of gram-negative bacteria. The hallmarks of a functional T7SS are an Ecc C protein forming a crucial component of secretion apparatus and two small, sequence-related substrate proteins EsxA and EsxB. Strains of *S. scabies* were constructed carrying marked mutations in genes coding for EccC, EccD, EsxA and EsxB. All four mutants retained full virulence toward seven plant species. However, disruption of the *esxA* or *esxB*, but not *eccC* or *eccD* genes affected *S. scabies* development, including delayed sporulation, abnormal spore chains and resistance to lysis by *Streptomyces*-specific phage ØC31. Further, three phenotypes were specific to the loss of the T7SS substrate EsxA and EsxB. But these phenotypes were not observed, when components of the T7SS secretion machine were absent, implying an intracellular role for EsxA and EsxB (Fyans et al. 2015).

The genome of *Streptomyces scabies* harbors the virulence-associated biosynthetic gene cluster called coronafacic acid (CFA)-like gene cluster which was previously predicted to produce metabolites that resembled coronatine (COR), a phytotoxin produced by *Pseduomonas syringae*. COR consists of CFA linked to an ethylcyclopropyl amino acid, coronamic acid, derived from L-allo-isoleucine. Combination of genetic and chemical analyses showed that the *S. scabies*-like gene cluster was responsible for producing CFA-L-isoleucine as the major product as well as other minor COR-like metabolites. The *cfl* gene, governing production of toxic metabolites, was present within the CFA-like gene cluster and encodes an enzyme involved in ligating CFA to its amino acid partner. CFA-L-isoleucine purified from *S. scabies* had similar bioactivities, but less toxic, compared with COR. The results revealed the involvement of coronafacoyl phytotoxins by *S. scabies* in the induction of common scab disease in potato (Fyans et al. 2015).

Key virulence determinants produced by *Streptomyces* spp. include the cellulose synthesis inhibitor, thaxtomin A and the secreted Nec1protein, required for colonization of susceptible plants. A biosynthetic cluster in the genome of *S. scabies* 87-22 strain was identified and it was predicted to synthesize a compound similar to CFA, a component of the virulence-associated coronatine phytotoxin produced by *Pseudomonas syringae*. Southern analysis indicated that the *cfa*-like cluster in *S. scabies* 87-22 strain was likely conserved in other strains of *S. scabies*, but it was absent in two other species viz., *S. turgidiscabies* and S. *acidiscabies*. Transcriptional analysis showed that the cluster was expressed during host-pathogen interaction and the expression required a transcriptional regulator embedded in the cluster as well as the *bldA* tRNA. A knockout strain of the biosynthetic cluster displayed reduced virulence phenotype on tobacco seedlings compared with the wild-type strain (Bignell et al. 2010). Additional regulatory genes required for thaxtomin A production were studied in a later investigation. Thaxtomin A, the main virulence factor of *S. scabies* functions as a cellulose synthesis inhibitor. Thaxtomin A production is controlled by the cluster-situated regulator TxtR which activates the expression of the biosynthetic genes in response to cellooligosaccharides. At least five additional regulatory genes were shown to be required for wild-type levels of thaxtomin A production and pathogenicity of *S. scabies* to plants. These regulatory genes belonged to the *bld* gene family of global regulators that controlled secondary metabolism or morphological differentiation in *Streptomyces* spp. Quantitative reverse transcriptase polymerase chain reaction (Q RT-PCR) showed that expression of the thaxtomin biosynthetic genes was significantly downregulated in all five *bld* mutants and in four of these mutants, downregulation was attributed to the reduction in expression of *txtR*. In addition, all mutants displayed reduced expression of other known or predicted virulence genes, suggesting that the *bld* genes might function as global regulators of virulence gene expression in *S. scabies* (Bignell et al. 2014).

### 5.1.4 *Agrobacterium* spp.

#### 5.1.4.1 Disease Incidence and Distribution

*Agrobacterium* spp. are ubiquitous in soil and form one of the major components of soil microflora, the vast majority of which are saprophytes, surviving primarily on decaying organic matter. But several species of agrobacteria induce neoplastic diseases in plants, such as *Agrobacterium rhizogenes* (causing hairy root disease), *A. rubi* (causing sugarcane gall disease), *A. tumefaciens* (causing crown gall disease in fruit trees) and *A. vitis* (causing crown gall disease in grapevines). *Agrobacterium* pathogenesis in susceptible host plant species has been studied intensively, since it is a unique and

highly specialized process involving bacterium-plant interkingdom gene transfer. Crown gall and hairy root symptoms were considered as a form of genetic colonization in which the transfer and expression of a suite of *Agrobacterium* genes in a plant cell causes uncontrolled cell proliferation and synthesis of nutritive compounds that can be metabolized specifically by the pathogen. Infection effectively creates a new niche specifically suited to *Agrobacterium* survival (Escobar and Dandekar 2003). Crown gall diseases severely damage the nursery industry, since infected seedlings with galls become unmarketable. *A. tumefaciens* has been reported to infect over 90 dicotyledonous and three monocotyledonous families (De Cleene and De Ley 1996). Symptomless (latent) infection of grape, rose and weeping fig by *A. tumefaciens* was observed and this has important implications in nurseries, since transmission of the pathogen via vegetative propagation and a micropropagation system may lead to disastrous consequences (Cooke et al. 1992; Peppenberger et al. 2002). *A. vitis* causes crown gall resulting in poor growth and death of vines ultimately. In addition to the formation of tumors, *A. vitis* also induces tissue-specific necrosis on grapevines and hypersensitivity-like response on nonhosts such as tobacco (Herlache et al. 2001). The mechanisms associated with necrosis and HR were studied. *A. vitis* strains F2/5 Tn5 mutants were selected based on altered necrosis and HR phenotypes. The mutant M1154 entirely lacked the ability to induce necrosis and HR and it had the *Tn5* insertion in the *luxR* homolog that was named *aviR*. These responses of necrosis and HR were regulated by QS (Zhang et al. 2002). *Agrobacterium vitis* occurs in severe proportions in several countries in the world. Grapevine (*Vitis vinfera*) cultivars are highly susceptible to freeze injury, providing the wounds necessary for infection by *A. vitis*. The wound position in relation to the uppermost bud of cuttings was found to be important for the development of tumors. Inoculated wounds below buds developed tumors, whereas wounds opposite to bud did not, implying that indole-3-acetic acid (IAA) flow contributed to tumor formation. The application of auxin to wounds prior to inoculation with a tumorigenic strain of *A. vitis*, resulted in formation of tumors at all sites, accompanied by an increased amount of callus in the cambium. On the other hand, the biological control strain F2/5 of *A. vitis*, applied prior to pathogen inoculation provided protection to the plants, as revealed by the absence of galls. The results suggested that the strain F2/5 might inhibit normal wound healing by inducing necrosis in the cambium (Creasap et al. 2005).

Walnut production in the United States depends on walnut rootstock *Juglans hindsii* x J. *regia* 'Paradox' which was highly susceptible to infection by *Agrobacterium tumefaciens*. When seeds were germinated and grown in the presence of *A. tumefaciens* and in the absence of wounding, 94% of seedlings exhibited tumors, while 89% contained systemic pathogen populations. Wound-inoculation with *A. tumefaciens* resulted in stem infection and the pathogen often migrated from the site of inoculation. Distribution of *A. tumefaciens* in the stem was random and may exhibit seasonal variation. *A. tumefaciens* populations in root tissue were more readily detected than in stem tissue and may serve as a reservoir for subsequent infection of aerial portions of the tree. Asymptomatic seedlings (7%) contained endophytic pathogen populations. Seedlings (17%) growing from inoculated seeds developed galls at secondary stem-wound sites. The results indicated the need for modification of existing tree-handling practices in both nursery and orchard production environments for effectively reducing the incidence of crown gall disease (Yakabe et al. 2012).

### 5.1.4.2  Process of Infection

The interaction between *A. tumefaciens* and the host plant species involves a complex series of chemical signals communicated between the pathogen and the host. As the first step in the process of induction of tumor, the pathogen cells get attached to the plant surface. The requirement of attachment of bacterium to the host surface was revealed in grapevine-*A. vitis* pathosystem. The non-tumorigenic *A. vitis* E26 strain could prevent crown gall infection, when applied to wounds on grapevine prior to or simultaneously with tumorigenic strain. ME19, a *Tn5* mutant of strain E26, was impaired in its ability to be chemo-attracted by grapevine root tissue extracts and its attachment to grapevine roots. The mutant ME19 did not differ significantly from the wild-type strain of E26 in phenotypes of agrocin production, growth in minimum medium or swarming activity. Complementation of ME19 with a cosmid clone of CP1543 from an E26 DNA library restored the chemotaxis, attachment and biocontrol phenotypes. Complements of the *mcp* gene [involved in the biosynthesis of methyl-accepting chemotaxis proteins (MCPs)] restored the affected phenotypes to the level of the wild-type E26 strain (Yang et al. 2009). Pathogenic as well as nonpathogenic strains of *A. tumefaciens*, *A. rhizogenes* and *A. vitis* growing in MM adhered to different abiotic surfaces, forming biofilms at initial stages of development. *A. tumefaciens* and *A. vitis* strains attached to both polystyrene and polypropylene materials, whereas *A. rhizogenes* strains only bound to polystyrene surfaces. Tumorigenic *A. tumefaciens* and *A. vitis* strains and the biological control agent *A. rhizogenes* strain K84, bound tightly to and formed complex biofilms on the surface of tomato root tips ex planta. All three *Agrobacterium* spp. strains efficiently colonized tomato seedlings and also formed biofilms on roots. These complex structures, as revealed by scanning electron microscopy, were composed of numerous bacterial cells arranged in different ways, either dense and continuous carpets, large aggregates embedded in extracellular material or globular mushrooms traversed internally by channels. Confocal laser-scanning microscopy, using GFP-marked derivative strains, revealed the presence of live, three-dimensional and thick fluorescent structures attached to plant materials. The results showed that all *Agrobacterium* spp. tested, were capable of forming biofilms both on abiotic and plant root surfaces (Abarca-Grau et al. 2011).

*A. tumefaciens* infects susceptible plant and transfers a piece of its tumor-inducing (Ti) plasmid, the transferred DNA (T-DNA), to several dicotyledonous plants, thereby modifying the host plant genome and inducing hyperparasitic response, leading to crown gall symptom development. After attachment of pathogen cells to plant surface, the cDNA transfer process is initiated, when *Agrobacterium* perceives the presence of certain phenolic and sugar compounds released from the wounded plant cells, acting as chemoattractants which bind the bacteria to plant cells in a polar orientation upon reaching the wound site. Weak attachment to the plant cell is first established through synthesis of acetylated polysaccharides, followed by strong binding through extrusion of cellulose fibrils (Gelvin 2000). A highly motile strain of *A. tumefaciens* showed significantly greater level of chemotaxis toward phenolic compounds that strongly induce *vir* genes compared to poorly motile strain. This pTi-dependent chemotaxis depends on *virA* and *virG* genes (Shaw et al. 1988). The genes *chvA*, *chvB* and *pscA* (or *exoC*) are located in the pathogen chromosome and are needed for the attachment of bacterial cells. Mutations in these loci may lead to loss of virulence to many plant species. The *att* (*attA* to *attH*) genes involved in the bacterial attachment to the plant surface, appear to be essential for signaling between the bacteria and host plant (Matthysee 1994). *AttA1-H* deletion mutants may be defective in sensing the plant signal or in responding to it, whereas att⁻ mutants may lack bacterial adhesion or blocked steps prior to adhesion (Reuhs et al. 1997). The *cel* genes located on the bacterial chromosome encode the cellulose fibrils required for bacterial cell attachment to plant surface. The *cel* mutants exhibit reduced virulence, indicating the requirement of *cel* genes for the full virulence of *A. tumefaciens* (Matthysee and McMahan 1998). Mutations in *chvA* and *chvB* genes result in reduced motility and failure to bind plant cells (Cangelosi et al. 1990). The *exoC⁻* mutants are defective in synthesis of EPS, whereas defective cellulose synthesis and aggregate formation are observed in *cel* mutants which bind loosely to plant wound sites (Uttaro et al. 1990). The phenolic compounds may serve as inducers (or coinducers) of bacterial *vir* genes, although these compounds may be normally involved in the production of defense-related substances, thus subverting part of the host defense mechanism. In addition, the pathogen can use these compounds to signal the availability of a potentially susceptible host plant (Bolton et al. 1986). Acetosyringone and other related compounds are perceived by *Agrobacterium* via the Vir protein (Stachel et al. 1986; Dye et al. 1997). Simultaneously the vir regulon, a set of operons required for the transfer of virulent DNA, is activated by the VirA/VirG, two-component regulatory system (Stachel and Zambryski 1986). The presence of acidic extracellular conditions (pH 5.5–5.5), phenolic compounds and monosaccharide at the plant wound site directly or indirectly induce autophosphorylation of the transmembrane receptor kinase VirA (Winans 1992).

Autophosphorylation of VirA protein and the subsequent transphosphorylation of VirG protein leads to the activation of *vir* gene transcription (Jin et al. 1990). The *vir* genes present in one-half of Ti-plasmid (*vir* region) control the virulence and tumor formation, whereas the genes for replication of opine catabolism and conjugation are located in the other half of Ti- plasmid. The octopine Ti-plasmid contains eight operons

(VirA to VirH) in the *vir* region. Mutations in the *virA*, *virB*, *virO* and *virG* operons result in the loss of tumor formation function. On the other hand, the host range of *Agrobacterium* is affected, following mutations in *virC*, *virE* and *virH*. The functions of Ti-plasmid genes encoding virulence proteins in the pathogen and the host plant have been determined to some extent. The Vir-regions of octopine and nopaline Ti-plasmids direct the transfer of oncogenic T-DNA to nuclei of host plant cells. The *vir* genes form a regulon consisting of a set of operons coregulated by the same regulatory proteins (Duban et al. 1993). A two-component regulatory system consisting of VirA and VirG protein mediates the expression of the *vir* genes. The *virA* gene, which is constitutively expressed, produces a protein located in the inner membrane that responds to plant wound metabolites. VirA is a membrane-spanning protein with an N-terminal periplasmic 'sensor' domain (capable of sensing acetosyringone and related phenolics), a 'linker' domain (responding to pH changes and interacting with ChvE –a sugar-binding protein), a 'kinase' domain and a 'receiver' domain. Acetophosphorylation of VirA protein results in the activation of the intracellular VirG which in turn is phosphorylated to become the transcriptional activator for all *vir* genes (Hooykaas and Beijersbergen 1994). After induction of the *vir* genes, the bacterial cells generate a linear single-stranded DNA designated T-DNA or T-strand. The T-strand is the coding (bottom) strand of the T-DNA region of the Ti-plasmid. The *virD2* gene encodes an enzyme VirD2 which liberates the linear T-DNA (Stachel et al. 1986). The enzyme VirD1 is a cleave-joining enzyme that cuts the lower strand of the border sequences in a site-specific manner. The excised T-DNA is removed and the resulting single-stranded gap is repaired probably by replacement DNA-strand synthesis (Yanofsky et al. 1986). The VirD2 effector may be covalently bound to the T-DNA and pilots the transfer of T-DNA into plant cell. The VirD4 protein functions as the coupling protein located at the inner membrane which recognizes and directs the substrates to the translocation machinery (Atmakuri et al. 2004).

The transfer of T-DNA and the effectors seems to occur due to the function of the type IV secretion system of *A. tumefaciens*. But it is unclear how they cross the host plasma membrane. After induction of transfer machinery, *A. tumefaciens* forms a pilus which is composed of the cyclized VirB2 protein (Fullner et al. 1996; Lai et al. 2002). The gene products of the complex *virB* locus of the Ti-plasmid regulate the transfer of the T-complex. An operon of the *virB* locus encodes 11 proteins (VirB1–VirB11). The VirB membrane proteins are considered to form a pore for the transport of the T-DNA across the bacterial membranes. The nature of the transport pore complex was investigated, using immunofluorescence (IF) and immunoelectron microscopy (ISEM) techniques. The VirB8, VirB9 and VirB10 proteins localized primarily to the inner membrane, outer membrane and periplasm, respectively, may form the components of the transport pore. In a virB8 deletion mutant, VirB9 and VirB10 proteins were randomly distributed on the cell membrane, indicating the requirement of VirB8 protein for assembling the transport complex (Kumar et al. 2000). Since the VirB2

protein is required for virulence, the pilus was proposed to either pierce the plasma membrane to deliver the T-DNA and the effectors into the host cell cytoplasm or to mediate intimate contact with the plant cell (Christie 2004). In addition, the ability of the effector VirE2 protein to integrate in black lipid membranes (BLM) and to form large anion-selective channels that transfer ssDNA across membranes *in vitro* also suggested a role for VirE2 in assisting T-DNA passage across the eukaryotic plasma membrane (Dumas et al. 2001; Duckey and Hohn 2003).

The VirE2 has multiple functions, such as the ability to bind to ssDNA (Gietl et al. 1987), to protect the T-DNA from nuclease digestion (Citovsky et al. 1988) and to assist in the nuclear import (Citovsky et al. 2004; Lacroix et al. 2005). The T-DNA, after insertion into the host plant cell, is transcribed, resulting in the synthesis of opines required for bacterial growth and the growth regulators that accelerate host cell enlargement and division leading to formation of characteristic tumors (Burr et al. 1998). The genes of T-DNA are involved in the synthesis of low-molecular weight amino acid and sugar phosphate derivatives known as opines. About 20 different opines have been identified in crown galls and hairy roots, but only a small subset of opines are encoded for opine uptake and catabolism genes corresponding to the particular opine, whose synthesis is directed by the resident T-DNA. Thus, opine production in transformed host plant cells creates a distinct ecological niche for the infecting *Agrobacterium* strain. The nature of opines such as ocotopine, nopaline, succinamopine and leucinopine present in the tumors, forms the basis of classifying the strains of *A. tumefaciens*. The chromosomal gene *acvB* isolated from *A. tumefaciens* strain A208 with nopaline Ti-plasmid was also shown to be essential for virulence (Wirawan et al. 1993). Another gene, *virJ*, homologous to *acvB* was also detected in an octopine Ti-plasmid (Pan et al. 1995). A transposon (*Tn5*) insertion in the *acvB* gene abolished virulence, indicating the requirement of this gene for virulence. The *acvB* gene was found to be involved in the transfer of T-DNA to the tobacco cell nuclei (Fujiwara et al. 1998). The *virJ* mutants and *acvB* mutants were avirulent, indicating their requirement for the virulence of *A. tumefaciens*. VirF may also be exported to plant cells. Nopaline-type strains lacking a functional *virF* gene were either avirulent or could induce very small tumors on *Nicotiana glauca* (Melchers et al. 1990). Octopine-type T-DNAs contain four opine synthesis genes catalyzing the production of octopine (*ocs*) and agropine (*ags*) and mannopine (mas1, mas2). Correspondingly, octopine *Agrobacterium* strains possess about 40 Ti-plasmid-encoded genes related to octopine, agropine and mannopine and their use. Opines provide growth susbtrate(s) for the development of the specific strain of *Agrobacterium*. In addition, they favor conjugal Ti-plasmid exchange and chemotaxis (Escobar and Dandekar 2003).

The nuclear targeting of the T-DNA in plant cells is likely to be mediated by the agrobacterial proteins. Active import of proteins and nucleoprotein complexes in eukaryotic cells needs specific nuclear localization signals (NLS) that are recognized by nuclear import cytosolic factors such as importins. The VirD2 and VirE2 proteins have plant-active NLS. When fused to ß-galactosidase, VirD2 tightly associated with the 5′ end of the T-DNA, a peptide containing NLS could target the chimeric protein to plant nuclei (Herrera-Estrella et al. 1990). On the other hand, VirE2 protein containing two separate bipartite NLS regions may target linked reporter proteins to plant cell nuclei (Citovsky et al. 1997). Following entry into the host plant cell, the bacterial T-DNA is integrated into the plant genome in the cell nucleus by illegitimate recombination, a mechanism that joins two DNA molecules which do not share extensive homology. The *chv* genes located on other replicons are also required, in addition to *vir* genes on Ti-plasmid, for successful interaction of *Agrobacterium* with its hosts. The *chv* genes have dual functions, whereas *vir* genes seem to be dedicated entirely to the interaction of *Agrobacterium* with its host plants. The *chv* genes govern the physiological functions of *Agrobacterium* developing in the absence of host plants as well as in the interaction of this pathogen with its host plants. The gene *chvE* is involved in the transport of specific sugars required for bacterial cells as a carbon source, in addition to its role in activating *vir* genes (Cangelosi et al. 1990). Likewise, the gene *katA* assists the pathogen in overcoming the host plant defenses and may play a role in the stress response of the cell (Zupan et al. 2000). With insertion mutagenesis using Tn5 or derivative such as TnPhoA, new *chv* genes were identified. Citrate synthase (CS) governs the entry of carbon into the TCA cycle. A CS deletion mutant of *A. tumefaciens* C58 was found to be highly attenuated in virulence. The mutation in CS led to a ten-fold decrease in *vir* gene expression, possibly resulting in reduced virulence. When the plasmid containing a constitutive *virG* [virG (con)] locus was introduced into this mutant, induction of *vir* gene occurred to the level comparable to that of wild-type strain. In addition, the size and number of tumors induced by the *virG* (con)-complemented CS mutant strain were similar to the wild-type strain (Suksomtip et al. 2006). The neoplastic growth at sites of infection by *A. tumefaciens*, is caused by transferring, integrating and expressing transfer DNA (T-DNA) from the pathogen into susceptible plant cells. A trans-zeatin synthesizing (*tzs*) gene is located in the nopaline-type tumor-inducing plasmid and causes trans-zeatin production in *A. tumefaciens*. Known virulence (Vir) proteins are induced by the *vir* gene inducer acetosyringone (AS) at acidic pH 5.5. Likewise, Tzs protein was highly induced by AS under this growth condition, but also constitutively expressed and moderately upregulated by AS at neutral pH 7.0. The promoter activities and protein levels of several AS-induced *vir* genes increased in the *tzs* deletion mutant, a mutant with decreased tumorigenesis and transient transformation efficiencies, in *Arabidopsis* roots. During AS induction and infection of roots of *Arabidopsis*, the *tzs* deletion mutant conferred impaired growth, which could be rescued by genetic complementation and supplementing exogenous cytokinin. Exogenous cytokinin also repressed *vir* promoter activities and Vir protein accumulation in both wild-type and *tzs* mutant strains with AS induction. The results indicated

that the *tzs* gene or its product, cytokinin, might be involved in regulating AS-induced *vir* gene expression and therefore, alter bacterial growth and virulence during *A. tumefaciens* infection (Hwang et al. 2013).

The proteins VirE2 and VirF are secreted directly into the plant cells via the same VirB/VirD4 transport system. Both proteins assist in the transformation of normal cells into tumor cells. Deletion of *virF* led to reduced virulence of *A. tumefaciens*. It could be complemented by expression of *virF* in the host plant, indicating the intracellular function of the VirF protein which may be involved in the targeted proteolysis of specific host proteins during early stages of transformation of plant cells (Schrammeijer et al. 2001). The VirE2 protein has an important role in the transformation process. The *virE2* mutants exhibited drastically reduced virulence, although they had low level of virulence on some host plant species (Dombek and Ream 1997). Among the Vir proteins, VirE2 was the most abundant in acetosyringone-induced *Agrobacterium* cells (Engstrom et al. 1987). The VirE2 protein and T-DNA appeared to be transferred to plant cells separately or VirE2 protein can form a complex with incoming T-DNA in the plant cytoplasm. The critical role of VirE2 protein in the transfer of ssDNA from *A. tumefaciens* to the nucleus of the plant host cell was studied, because of its ssDNA binding activity and assistance in nuclear import. The VirE2 is associated with its specific chaperone, the VirE1 protein which is a small, acidic, 7-kDa protein and is capable of preventing the VirE2 protein from forming aggregates and tripling its half life (Zhao et al. 2005). The native form of VirE2 in the bacterial cytoplasm is complex with its chaperone, VirE1. Electron microscopic observations revealed that upon binding of VirE1-VirE2 to ssDNA, helical structures similar to those reported for the VirE2-ssDNA complex were formed. The VirE1-VirE2 complex is associated with different kinds of lipids. The black lipid membrane tests showed the ability of the VirE1-VirE2 complex to form channels (Duckely et al. 2005).

Studies on the biological functions of VirE3 showed that VirE3 was transferred from the pathogen *A. tumefaciens* to plant cell and then imported into the nucleus via the karyopherin α-dependent pathway. In addition, VirE3 interacted with VirE2, a major bacterial protein that was directly associated with T-DNA and facilitated its nuclear import. The nuclear import of VirE2, in turn, was mediated by a plant protein VIP1 and acted as an 'adapter' molecule between VirE2 and karyopherin α and 'piggy-backing' VirE2 into host cell nucleus (Lacroix et al. 2005). In a later investigation, double mutation of *virF* and *virE3* strongly reduced the ability of A. tumefaciens to form tumors on tobacco, tomato and sunflower. The VirE3 protein was translated from a polycistronic mRNA containing the *virE1*, *virE2 and virE3* genes in *A. tumefaciens*. The VirE3 protein had nuclear localization sequences, suggesting that it might be transported into the plant cell nucleus upon translocation. VirE3 protein interacted *in vitro* with importin-α. The VirE3-GFP fusion protein was localized in the nucleus. VirE3 interacted with two other proteins viz., pcsn5, a component of the CoP9 signalosome and pBrp, a plant cell specific general transcription factor belonging to the TFIIB family. VirE3 protein could induce transcription in yeast, when bound to DNA. The translocated effector protein VirE3 was transported into the nucleus, where it might interact with the transcription factor pBrP to induce the expression of genes required for tumor formation (García-Rodriguez et al. 2006).

T-DNA of *Agrobacterium* contains two classes of genes: oncogenes and opine-related genes. The oncogenes alter the phytohormone synthesis and sensitivity in infected cells, thus generating the tumor phenotype. Several genes of *A. tumefaciens* transferred into T-DNA encode proteins that are involved in developmental alterations, resulting in the formation of tumors in infected plants. *Atu6002* gene expression occurred in *Agrobacterium*-induced tumors and it was also activated on activation of plant cell division by growth hormones. Within the expressing host cells, the Atu6002 protein was targeted to the plasma membrane. Constitutive ectopic expression of *Atu6002* in transgenic tobacco plants led to a severe developmental phenotype characterized by stunted growth, shorter internodes, lanceolate leaves, increased branching and modified flower morphology. Transgenic plants expressing *Atu6002* displayed impaired response to auxin. However, auxin cellular uptake and polar transport were not significantly inhibited in these plants. The results suggested that *Atu6002* might interfere with auxin perception or signaling pathway (Lacroix et al. 2014).

A QS process is involved in the regulation of many important physiological functions of the bacteria, such as symbiosis, conjugation and virulence. *Agrobacterium vitis* causes tissue-specific necrosis, in addition to induction of tumors in grapes and HR-like symptoms on nonhosts like tobacco (Herlache et al. 2001). The *luxR* homolog *aviR* in *A. vitis* strain F2/5 was associated with induction of a HR on tobacco and necrosis on grapevine plants, indicating that the responses were regulated by QS. The M1320 with disruption in a second *luxR* homolog, *avhR* exhibited HR-negative and reduced grape necrosis characteristics. The expression of *aviR*, *avrh* and the *clpA* (encoding one of two ATPase subunits of the ClpA protease of *E. coli*) in F2/5 and mutants M1154 and M1320 and their complemented derivatives was determined by RT-PCR assay. The gene *avhR* was expressed in F2/5 and in complemented M1320 and also in the *aviR* mutant M1154. Further, *aviR* expression in F2/5 and M1320 could be detected, indicating that *aviR* and *avhR* were likely to be expressed independently and not in a hierarchical manner. The expression of *clpA* was observed in F2/3 and in all mutants (Hao et al. 2005). *Agrobacterium vitis* produces polygalacturonase which is an important virulence factor for this pathogen. Polygalacturonase (PG), encoded by a single gene, degrades the pectin component of the xylem cell wall of susceptible grapevine plant. Disruption of PG gene resulted in a mutant with reduced virulence, leading to the production of significantly fewer root lesions on grapevines. In order to identify the peptides or proteins that could inhibit the activity of PG, a phage-displayed combinatorial peptide library was employed to isolate peptides with a high binding affinity to *A. vitis* PG. These peptides showed sequence similarity to regions of PG-inhibiting proteins (PGIPs) of *Oryza sativa* (Japonica) and *Triticum urartu* (wild wheat). A peptide capable of inhibiting the activity of *A. vitis* PG by 35% *in vitro* was identified (Warren et al. 2016).

Recognition of the pathogen by the host plants results in the activation of host defense systems. As the first step in the perception of pathogens by plants, recognition of pathogen-associated molecular patterns (PAMPs) or general elicitors by the host plant results in rapid activation of defense mechanisms such as cell wall reinforcement by callose deposition, production of ROS and induction of several defense-related genes. On the other hand, virulence factors produced by pathogens may inhibit the PAMP-elicited basal defenses (Kim et al. 2005; Nomura et al. 2005). The structures characteristic of bacterial pathogens such as LPS, bacterial cold-shock protein (CSP), flagellin and EF-Tu may be perceived by plants (Nurnberger et al. 2004; Zipfel and Felix 2005). The pathogens are recognized by the plants through the activity of an array of pattern recognition receptors (PRRs) which are capable of recognizing characteristic molecular structures, the PAMPs. *Arabidopsis* plants can detect a variety of PAMPs including conserved domains of bacterial flagellin and EF-Tu. *A. tumefaciens* has an EF-Tu that is fully active as an elicitor in *Arabidopsis* (Kunze et al. 2004). Flagellin and EF-Tu activate a common set of signaling events and defense responses. A targeted reverse-genetic approach was applied to identify a receptor kinase essential for EF-Tu perception named as EFR. Following transient expression of EFR, *Nicotiana benthamina* (with inability to perceive EF-Tu), acquired EF-Tu binding sites and responsiveness. Susceptibility to *A. tumefaciens* was increased in *efr* mutants of *Arabidopsis* into which T-DNA transformation occurred with higher efficiency. The results indicated that EFR was the EF-Tu receptor and that plant defense responses induced by PAMPs-like EF-Tu reduced transformation by *A. tumefaciens* (Zipfel et al. 2006).

Horizontal gene transfer (HGT) is recognized as a natural phenomenon, especially between bacteria, but it is also being increasingly detected in eukaryotic genomes also (Dunning Hotopp 2011). Horizontally transferred genes have been shown to be correlated with the occurrence of a specific phenotype. Transfer of carotenoid biosynthetic genes from fungi to aphids, resulting in the red or green coloration of aphids was observed (Moran and Jarvik 2010). Evidence was obtained for HGT between *Agrobacterium* spp. and an ancestor of the sweet potato (*Ipomoea batatus*). The presence of *Ib* T-DNA is a general feature of the domesticated sweet potato gene pool and *Ib* T-DNA could be detected in all hexaploid cultigens examined and the lack of segregation in the progeny of the analyzed cross suggested that this T-DNA fragment of *Agrobacterium* was fixed in the cultivated sweet potato genome, in contrast to its close wild relatives. The gene sequences identified in *Ib* T-DNA1 and *Ib* T-DNA2 indicated that the transforming *Agrobacterium* most likely was *A. rhizogenes*, or a species closely related to *A. rhizogenes*. The data showed that T-DNA integration and the subsequent fixation of foreign T-DNA into the sweet potato genome occurred during evolution and domestication of this crop which is one of the world's most consumed foods (FAO 2013). This fact could influence the current perception of the public that transgenic crops are unnatural (Kyndt et al. 2015).

### 5.1.5 XANTHOMONAS SPP.

#### 5.1.5.1 Disease Incidence and Distribution

Among the *Xanthomonas* spp. infecting crops, *X. oryzae* pv. *oryzae*, causing rice bacterial leaf blight disease, has multiple modes of dispersal through infected stubbles left in the field after harvest, contaminated irrigation water and to some extent through infested seed. *X. campestris* pv. *musacearum* (*Xcm*), causing a destructive disease in banana, survives in the soil and propagules which remain asymptomatic. Infected propagules form the important means of pathogen dissemination. Symptoms of banana Xanthomonas wilt (BXW), also known as banana bacterial wilt (BBW) include shriveling of the male flower bud, followed by yellowing and wilting of leaves and premature fruit ripening and discoloration. Bacterial ooze from vascular tissue and fruit is characteristic of BXW disease. The disease occurs in severe proportions in many African countries, including Ethiopia, Democratic Republic of Congo, Uganda and Rwanda. The annual loss due to BXW disease was estimated to be over 500 million dollars across East and Central Africa (Biruma et al. 2007). Bananas are the most important food crop in Uganda and the extent of loss was determined to be about 70% in one year (Mwebaze et al. 2006). Many symptoms of BXW are similar to those of other bacterial and fungal wilts of banana, including Fusarium wilt caused by *F. oxysporum* f.sp. *cubense* and Moko/bugtok disease caused by *R. solanacearum* (Thwaites et al. 2000). At preflowering stage, banana plants may be infected directly from diseased mother plants or from the inoculum in the contaminated tools. The first visible symptom of inflorescence infection is wilting of the male buds, followed by the decay of the rachis and finally rotting of the bunches. The pathogen in the ooze is spread by visiting insects to healthy inflorescence (Tinzaara et al. 2006). *Xcm* in infected inflorescence of cv. Pisang Awak, causing wilting of male bud was restricted to upper parts of the stem. On the other hand, *Xcm* could move down further in cv. Matooke (Ssekiwoko et al. 2010).

Banana plants were inoculated with *Xanthomonas campestris* pv. *musacearum (Xcm)* through abscission wounds of female bracts and male bud bracts with contaminated machete to study the movement of *Xcm* from the point of inoculation to other plant tissues/organs. Inoculation through all floral entry points resulted in infection, with the highest infection occurring in the combined male bract and male flower abscission wound inoculations. The systemicity of *Xcm* could be observed by the ability of the pathogen to live within the mat for long periods (5–16 months) asymptomatically (Ocimati et al. 2013). In a subsequent study, investigations on *X. campestris* pv. *musacearum (Xcm)* survival and latent infections through symptomless plants in mats having diseased plants in subsequent generations, were carried out. Bacteria at low levels were detected in up to 20% of the third generation suckers with a significant (P < 0.05) reduction (43–20%) in

subsequent generations. Only 3–6% of latently infected suckers succumbed to Xanthomonas wilt disease. Incidence of *Xcm* in symptomless suckers from farmers' fields (with up to 70% incidence) was low (3%), whereas it increased (8–25%) with disease severity in mats in controlled experiments. The results indicated that if new infections could be prevented, it would be possible to rejuvenate field with high Xanthomonas wilt disease incidence. Incomplete systemic movement of *Xcm* in mats coupled with gradual decline of pathogen load occurred in subsequent generations to levels that might not initiate disease. Not all suckers from infected mats were found to be infected and some suckers remained free of *Xcm*, indicating the possibility of selecting pathogen-free suckers from such mats in the subsequent years (Ocimati et al. 2015).

The comparative efficiency of infection of *X. campestris* pv. *musacearum* (*Xcm*) through the corm- and pseudostem-inoculation at bunch at harvest, leaf and female and male bud bracts was compared. The male and female bud bract inoculations led to the highest Xanthomonas wilt (XW) disease incidence (81% and 93%, respectively), compared with 0–44% infection for harvest and corm inoculations. The nature of insect-mediated transmission resulted in up to 99% disease incidence in cv. Pisang Awak. Floral inoculations and insect-mediated infections only caused floral symptoms. Male bud bract wounds form the principal entry points for insect-mediated transmission. Leaf and harvest inoculations induced leaf symptoms only, whereas harvest inoculations caused only leaf symptoms. Corm inoculations resulted in late floral symptoms. Floral inoculations formed the main mode of infection by *Xcm*. A significant difference in the populations of *Xcm* (CFU/g) was observed between symptomatic and asymptomatic leaves. The results suggested that cultural practices capable of reducing pathogen population below the threshold levels could be applied to reduce the disease incidence (Nakato et al. 2014). Several insects, including honeybees visit the male flowers of both symptomatic and asymptomatic plants. *Xanthomonas campestris* pv. *musacearum (Xcm)* could be isolated from honey bees and other insects such as stingless bees, fruit flies and grass flies visiting male flowers. Transmission of the pathogen is likely to occur through the male flower or the neuter part of the floral raceme (Tinzaara et al. 2006). The spatial and temporal distribution of insect vectors of *Xcm* and their activity across banana cultivars grown in Rwanda were investigated. Variations in vector populations due to elevation and banana varieties were observed. Higher populations of insects were recorded in low altitude areas and during rainy seasons. A decline in vector populations with an increase in altitude correlated with incidence of Xanthomonas wilt disease of banana. Incidence of floral infections was higher in low altitudes. The activity of insects on banana male buds varied among banana cultivars. The banana cultivars with persistent male bracts and neuter flowers were less attractive to insects visiting flowers. Hence, such cultivars could be preferred for cultivars to reduce Xanthomonas wilt disease (Rutikanga et al. 2015).

*Xanthomonas axonopodis* pv. *phaseoli (Xap)* and *X. axonopodis* pv. *phaseoli* var. *fuscans (Xapf)* are associated with common bacterial blight (CBB) disease. All aerial plant parts are affected by these pathogens, but symptoms are generally more severe on leaves and pods. Infected seeds form the primary sources of inoculum. But the pathogen *Xap* could survive for months in leaves and other plant debris on the soil surface (Saettler 1989). Both *Xap* and *Xapf* cause similar symptoms on bean, but *Xapf* appeared to be more virulent (Opio et al. 1996). Further, *Xapf* produces a brown diffusible pigment in various culture media and the pigment is not associated with pathogenicity (Fouri 2002). Molecular analysis indicated that *X. axonopodis* pv. *phaseoli* and *X. axonopodis* pv. *phaseoli* var. *fuscans* were genetically distinct, although there was no geographical differentiation between these bacterial pathogens, causing common bean bacterial blight disease of common bean (Mahuku et al. 2006). Another pathovar, *X. campestris* pv. *vignicola (Xcv)* causes cowpea bacterial blight disease which was considered as one of the constraints of cowpea production in African countries. Survival of *X. campestris* pv. *vignicola* in infested soil, cowpea seeds and plant debris on the soil during three cropping seasons (1996–1998) was observed. Populations of the pathogen varied in soil and plant debris depending on the season. Plant debris contained higher populations of *Xcv*, compared to soil. During the off-season in the absence of cowpea plants, much lower populations of *Xcv* were detected in plant debris and no detectable levels of *Xcv* were present in soil in the absence of cowpea plants. The seed and plant debris infested by *X. campestris* pv. *vignicola* formed major sources of pathogen survival (Okechukwu and Ekpo 2008).

### 5.1.5.2 Process of Infection

Bacterial pathogens causing bacterial leaf blight (BLB) and bacterial leaf streak (BLS) diseases affecting rice were elevated to the current status as new species and named *Xanthomonas oryzae* pv. *oryzae (Xoo)* and *X. oryzae* pv. *oryzicola (Xoc)*, respectively (Swings et al. 1990). *X. oryzae* produced copious capsular EPS which is required for the formation of droplets or strands of bacterial exudates from infected leaves, providing protection from desiccation and aiding in wind- and rain-borne dispersal (Swings et al. 1990). Pathovars *oryzae* and *oryzicola* may be differentiated by (i) acetoin production (*Xoo*⁻ and *Xoc*⁺), (ii) growth on L-alanine as sole carbon source (*Xoo*⁻ and *Xoc*⁺), (iii) growth on 2% vitamin-free casamino acids (*Xoo*⁻ and *Xoc*⁺) and (iv) resistance to 0.001% copper nitrate (*Xoo*⁺ and *Xoc*⁻) (Gossele et al. 1985). *Xoo* induces leaf blight and vascular wilt, resulting in scorching of foliage and the eventual death of infected plants. 'Kresek' is the destructive manifestation of the disease rapidly killing the infected plants. Opaque and turbid drops of bacterial ooze may fall into irrigation water or dry up into yellowish beads on infected leaves during dry weather. *Xoo* enters the rice leaves typically through hydathodes at the leaf tip and leaf margin (Ou 1985). Within the xylem, *Xoo* presumably interacts with xylem parenchyma cells. The pathogen moves vertically through commissural veins. Within a few days bacterial cells and EPS fill the xylem vessels and ooze out from hydathodes, forming beads or strands of exudates on the leaf surface and source of secondary inoculum. The exudates may fall into irrigation water or be dispersed by wind, rain, insects or other means that contribute to the pathogen spread (Mew et al. 1993; Nyvall 1999). *Xoo* enters through cut leaf tips and root injury made while lifting and planting seedlings (cutting leaf tips being a common practice). The pathogen spreads through the vascular system to the growing point of the plant, infecting the base of other leaves and killing entire plants in two to three weeks (Ou 1985; Nyvall 1999). Wind and rain disseminate *Xoo* from infected rice plants and other hosts, as well as infected rice stubbles that form important sources of primary inoculum. *Xoo* may also be disseminated in irrigation water (Nyvall 1999), as well as by humans, insects and birds (Ou 1985; Nyvall 1999). *Xanthomonas oryzae* pv. *oryzae (Xoo)* can infect wild species such as *Oryza rufipogon*, O. *australiensis* and graminaceous weeds like *Leersia oryzoids* and *Zizania latifolia* in temperate regions and *Leptochloa* spp. and *Cyperus* spp. in the tropics. *Xoo* can survive the winter in the rhizosphere of *Leersia* and *Zizania* as well as in the stem bases and roots of rice stubble (Mizukami and Wakimoto 1969). Furthermore, *Xoo* can survive in the soil for one to three months in the temperate regions, depending on the soil moisture and acidity. *Xoo* may also overwinter in piled straw, which may become important in the absence of other alternative weed hosts. High temperature, humidity and the availability of several alternative host plant species in the tropics provide conditions favorable for persistence of *X. oryzae* pv. *oryzae* throughout the year (Ou 1985).

*Xanthomonas oryzae* pv. *oryzae (Xoo)* is considered to be a model organism for analysis of host-pathogen interactions. More than 30 races differing in virulence and 25 resistance genes in rice have been identified (Zhu et al. 2000; Ochiai et al. 2005). Development of strains of *Xoo* showing insensitivity/resistance to antibiotics and fungicides has become problematic to growers as well as to researchers (Zhang et al. 2013; Zhu et al. 2013). *Xoo* was able to multiply on the leaf surface of susceptible cultivar TN1. The number of cells of virulent strain PX061 increased significantly at 24 h after inoculation (*hai*). They were densely distributed on or around the water pores and some had gained entry into the leaves through water pores at *72 hai*, as revealed by scanning electron microscopy (Mew et al. 1984). In contrast, in incompatible host-pathogen interaction, the bacterial cells might be immobilized and their multiplication might be inhibited at the water pores (Mew et al. 1984). The hydathodes, but not stomata, appeared to serve as portals of entry for *Xoo*. The pathogen could survive epiphytically on nonhosts like maize (corn) as shown by scanning electron microscopy (SEM) (Huang and De Cleene 1989). Many gram-negative bacteria have type IV fimbriae (pili) for adhesion to plant surface. The presence of type IV fimbriae genes distributed in several loci on the genome of *Xanthomonas oryzae* pv. *oryzae (Xoo)* was observed. In addition, a non-fimbrial adhesion designated XadA detected in *Xoo* might also have a role in virulence. Two *xadA* homologs at different loci have been identified in *Xoo* MAFF311018 strain (Ray et al. 2002; Ochiai et al. 2005). *Xoo* has homologs of two sets of characteristic two-component regulators involved in microbial interactions with plants. One set is the *virA/virG* two-component system of *A. tumefaciens* and the other one is the *nodV* and *nodW* two-component system of *Bradyrhizobium japonicum* (Stachel and Zambryski 1986; Göttfert et al. 1990). The similarity to *vir* and *nod* gene regulators suggested that they might play a role in regulation of pathogenicity in response to plant environmental signals. Comparisons of the *hrp* cluster in *X. oryaze* pv. *oryzae* revealed that the gene orders were similar, but the clusters were located in different regions in the genomes (Ochiai et al. 2005). *X. oryzae* pv. *oryzae (Xoo)* produces AvrXa7 protein, the product of the pathogen gene *avrXa7* which is a virulence factor. The gene *avrXa7* is included in the *avrBs* avirulence gene family which encodes proteins targeted to plant cells by type III secretion system (T3SS) (Yang et al. 2000). The pathogenicity of *Xoo* depends on a functional T3SS, while members of the *avrBs/pthA* gene family of T3SS substrate effectors are required for its ability to induce BLB disease in rice (Yang and White 2004). The whole-genome-based microarray of rice genes was employed to identify genes that were expressed in rice plants challenged with *Xoo*. The expression of rice gene *Os8N3*, was enhanced upon infection by *Xoo* strain gene *PthX01*. The gene *Os8N3* resides near R gene *Xa13*. The strain PXO99A did not induce *Os8N3* in rice lines with *Xa13*. Rice plants with *Os8N3* silenced by inhibiting RNA synthesis were resistant to infection by strain PXO99A only, but not to other strains. The results indicated that *Os8N3* was a host susceptibility gene for BLB targeted by the transcription activator-like (TAL) effectors *PthX01* (Yang et al. 2006).

Several virulence factors, in addition to *hrp*-controlled pathogenic factors (secreted via T3SS), are required for inducing different types of symptoms in susceptible plants. The induction of frequently observed water-soaked lesions in infected plants in the early stages of pathogenesis, is attributed to the action of EPS produced by bacterial pathogens. Large amounts of EPS are secreted by many bacterial pathogens including *Xanthomonas oryzae* pv. *orzae (Xoo)*. The quantities of EPS appeared to be related to the levels of virulence of *Xoo* strains. Further, the compositional variation of EPS may have some effects on the virulence of *Xoo* (Singh et al. 2006). Rice BLB pathogen *Xoo* secretes effector proteins (T3Es) directly into rice cell through a *hrp*-encoded specialized type III secretion system to induce blight. The function of XopF, one of the conserved effectors of *Xoo*, was investigated, using a null mutant developed through a PCR-based homologous recombination strategy. Analysis of a *hrp*-dependent expression pattern of *xopF* showed that XopF was translocated in rice cytosol through type III secretion system (T3SS). XopF regulated the in planta *Xoo* growth and suppressed PAMP-triggered immune (PTI) response in rice. *Xoo* wild-type strain, but not mutant, produced intense blight lesions following inoculation using the leaf clipping method (Mondal et al. 2016). Many CWDEs secreted by *X. oryzae* pv. *oryzae* have been investigated to elucidate their role in induction of defense responses that are suppressed by the pathogen using type III secretion system (T3SS) effectors. A type III secretion mutant (T3SS⁻) of *Xoo* induced defense responses in inoculated plants. The role of individual CWDEs in induction of rice defense responses during infection was studied by mutating five genes in the genetic background of a T3SS⁻ mutant. Mutants defective in two cellulases (*clsA* and *cbsA*), one xylanase (*xyn*), one pectinase (*pglA*) and an esterase (*lipA*) were employed to assess their role in inducing defense responses in rice. The results indicated that collective action of secreted proteins ClsA, CbsA and Xyn proteins was essential for inducing defense responses in rice during pathogenesis (Tayi et al. 2016).

*Xanthomonas oryzae* pv. *oryzae (Xoo)* has intracellular signaling systems that are designated two-component regulatory systems (TCSs) which regulate pathogenesis and other biological processes. In order to identify the TCSs involved in *Xoo* pathogenesis, knockout strains, lacking response regulators (RRs, a cytoplasmic signaling component of TCS) were generated. The knockout strain *detR* lacking the PXO_04659 gene showed drastic reduction in virulence, relative to the wild-type strain. Analysis of *detR* function in *Xoo* pathogenesis revealed reduction in EPS production, intolerance to reactive oxygen (ROS) and deregulation of iron homeostasis in the *detR* ⁻ strain. Further, gene expression of regulatory factors, including other RRs and transcription factors (TFs), was altered in the absence of DetR protein, as determined by RT-PCR assay and/real-time quantitative RT-PCR analyses. The results indicated that DetR was required for virulence of *Xoo* through regulation of the *Xoo* defense system, including EPS synthesis, ROS detoxification and iron homeostasis, either independently or in combination with other regulatory factors (Nguyen et al. 2016).

During infection, *Xanthomonas oryzae* pv. *oryzae (Xoo)* produces virulence factors and EPS and enzymes, iron chelating siderophores and effectors regulated by type III secretory system. A biochemical approach was applied to study the rice-*Xoo* interactions in rice genotypes with differing levels of susceptibility/resistance and virulence of pathogen strains. *Xoo* strains Mai1, PX088, Dak1 and Dak16 selected for biochemical analysis of virulence factors were virulent on rice genotypes IRBB4 and FKR14. PX088 was the most virulent among the pathogen strains tested. Virulence levels of PX088 on IRBB4 and FKR14 showed variations at 14 days postinoculation (dpi), but at 21 dpi, the virulence was at almost the same level on both rice genotypes, with a lesion length of 26.7 cm. The West African strain Mai 1 was the least virulent. PX088 strain induced lesions that progressed faster, covering the entire leaf length in FKR14 after 14 dpi. The East African strains Dak1 and Dak16 were more virulent than the West African strain Mai 1. The *Xoo* culture filtrate (CF), heated CF and proteinase K-treated CF, induced typical BLB symptoms on IRBB4 and FKR14, with a maximum lesion length of about 23.1 cm for CF (see Figure 5.3). Lesions induced by heated CF were generally smaller at 14 days postinoculation (dpi) than in unheated CF treatment. Significant reduction in lesion length was observed due to treatment with proteinase K. The proteinase K-treated fraction induced lesions 13.4 cm in length. The ethylacetate fraction of CF was used to investigate the role of small molecules as virulence factors of *Xoo*.

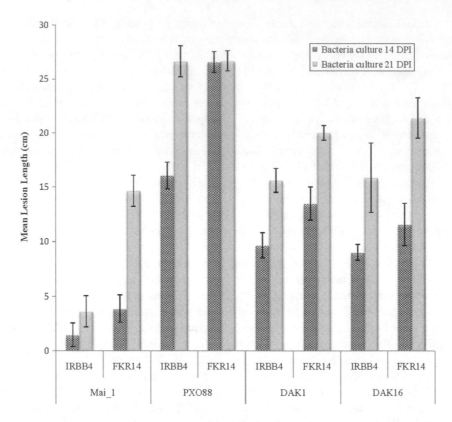

**FIGURE 5.3** Assessment of virulence levels of strains of *Xanthomonas oryzae* pv. *oryzae (Xoo)* strains Mai_1, PX088, DAK1 and DAL16, based on lesion length induced on rice genotypes IRBB4 and FK14 at four and 21 days postinoculation.

**[Courtesy of Dossa et al. (2014) and with kind permission of *Journal of Plant Pathology and Microbiology* Open Access Journal]**

Inoculation of CF extracted with ethylacetate using the leaf clipping method resulted in significant reduction in phytotoxicity of the CF. The results suggested that low-molecular compounds in the ethylacetate fraction of the CF might not have a major role in the induction of phytotoxicity by *X. oryzae* pv. *oryzae* (Dossa et al. 2014).

### 5.1.6 *CLAVIBACTER MICHIGANENSIS*

*Clavibacter michiganensis* subsp. *sepedonicus (Cms)*, causing potato ring rot disease, is a gram-positive phytopathogenic bacterium. Potato bacterial ring rot (BRR) caused by *Cms* has worldwide distribution. Occurrence of BRR has been observed in Asia, Europe and North America and South America (EPPO and CABI 1996). All known *C. michiganensis* subspecies impact agriculturally important crops such as tomato, potato and alfalfa and they have a limited host range and most of them are vascular pathogens, causing wilt as the principal symptom (Eichenlaub et al. 2006). Use of certified seed tubers has been followed as one of the key tactics for management of BRR disease, which is a highly regulated 'zero tolerance' disease in the production of certified seed tubers. However, zero tolerance for BRR alone may not be enough for long-term disease management, since ring rot infections can remain symptomless or latent (Nelson 1982). Several factors such as potato cultivar, pathogen strain, initial inoculum dose and environmental factors, influence BRR

symptom expression. Further, *Cms* could persist for extended periods of time on potato tuber surfaces, that allow low concentration of pathogen to exist, thereby increasing the chances of spreading inoculum to healthy seed tubers (Nelson 1982). The pathogen can survive and remain infectious on potato bags, barn wall and in self-sown plants from infected crop. The bacterium remains infectious at and above freezing temperature for at least 18 months on burlap and for 63 months in infected potato stems (Nelson 1984). Ring rot pathogen *C. michiganensis* subsp. *sepedonicus (Cms)* can infect potato, eggplants and tomato and it is unlikely to spread from one plant to another under field conditions, but it can spread from one tuber to another through direct contact or through contact with contaminated containers, equipments (cutting knife) and agricultural machinery contaminated during handling, grading and processing. *Cms* can survive for different periods of time in soil. It can survive in moist soil for a few months and under dry conditions for more than one year. As dry bacterial ooze on the surface of the equipment, *Cms* may survive for many years (Robert 2013).

The rotation crops such as maize (corn), wheat, barley, oat, bush bean, broad bean, rape seed (canola), pea and onion and five cultivars of sugar beet were tested for their ability to provide inoculum for potato crops for infection by *Clavibacter michiganensis* subsp. *sepedonicus (Cms)*. The pathogen could be detected, at 6–8 weeks after inoculation, in large numbers ($10^5$ to $10^6$ cells/g of tissue) using the immunofluorescence (IF) cell staining method. Of the 12 solanaceous weeds and 13 other weeds commonly present in potato fields in Europe, *Cms* persisted only in *Solanum rostratum* in high numbers ($10^8$ cells/g of tissue) in stems and leaves, where symptoms appeared. In *Cms*-contaminated potato field plots, the pathogen was detected in roots, but not in stems of *Elymus repens* plants growing through rotten potato tubers left in the field. *Clavibacter michiganensis* subsp. *sepedonicus(Cms)* could be detected in both stems and roots of *Viola arvensis* and *Stellaria media* plants using the AmpliDet RNA technique (van der Wolf et al. 2005). The risk of dissemination of *Cms* through potato residues from potato processing industries was estimated, since the residues were used as amendments in agriculture. The effect of composting and pasteurization of residues under varying conditions was assessed using samples (ready-made compost mold) contaminated with *Cms*. Viable bacteria could be extracted using eggplant bioassay and plating on semiselective media on which characteristic colonies were formed. Viable pathogen could be extracted after composting for six days at maximum temperatures at 70°C, 13 days at 55°C and 90-min pasteurization at 70°C. The results indicated that these sanitation measures were ineffective in inactivating *Cms* and the risk of applying the potato residues from processing industries as manures to potato crops could not be ignored (Steinmöller et al. 2013).

The infection process of *Clavibacter michiganensis* subsp. *sepedonicus (Cms)* was studied using eggplant, an alternative host and strains that differed in phenotype. Two parameters affecting virulence viz., the ability to induce HR and production of cellulase by the strains were considered in this investigation. A plasmid-free isolate of *Cms* causing HR on tobacco and unable to produce cellulase, multiplied efficiently in plants, but induced only weak symptoms. In contrast, another strain with an inability to induce HR on tobacco and was capable of producing cellulase, was impaired in ability to multiply in tobacco and caused no symptoms. Coinoculation of these strains on eggplant resulted in production of typical symptoms of the disease. The results suggested that both cellulase production and the ability to induce HR were required for successful completion of the infection process, leading to induction of symptoms by *Cms* (Nissinen et al. 2001). EPS produced by bacterial pathogens play an important role in the infection process, leading to development of disease symptoms in susceptible plants. The interaction of EPS of potato ring rot pathogen *C.michiganensis* subsp. *sepedonicus (Cms)* with protoplasts isolated both from leaf cells of plants grown *in vitro* and membrane fractions obtained from cell suspension cultures of two populations with different resistance levels was investigated. The EPS was firmly bound to protoplast surfaces and microsomal membranes of susceptible potato cultivars, but not to those of resistant cultivar. Treatment with protease, excess of unlabeled EPS and with dextran, did not lead to the binding of fluorochrome-labeled EPS to protoplasts and microsomal membranes from both resistant and susceptible cultivars. The results suggested that the plasma membranes of cells of susceptible and resistant cultivars with variable numbers of proteinaceous sites might have a role in

determining the nature of interactions between the host and the bacterial pathogen (Romanenko et al. 2003).

## REFERENCES

Abarca-Garu AM, Penyalver R, Lopez MM, Marco-Noales E (2011) Pathogenic and nonpathogenic *Agrobacterium tumefaciens*, *A. rhizogenes* and *A. vitis* strains form biofilms on abiotic as well as root surfaces. *Plant Pathol* 60: 416–425.

Aizawa SI (1996) Flagellar assembly in *Salmonella typhimurium*. *Molec Microbiol* 19: 1–5.

Aliye N, Dilbo C, Pillay M (2015) Understanding reaction of potato (*Solanum tuberosum*) to *Ralstonia solanacearum* and relationship of wilt incidence to latent infection. *J Phytopathol* 153: 444–455.

Allen C, Gray J, Simon-Buela L (1997) A regulatory locus *pehSR* controls polygalacturonase production and other virulence functions in *Ralstonia solanacearum*. *Molec Plant-Microbe Interact* 10: 1054–1064.

Álvarez B, Biosca EG, López MM (2010) On the life of *Ralstonia solanacearum*, a destructive bacterial plant pathogen. In: Mendez-Vilas A (ed.), *Current Research, Technology and Education Topics in Applied Microbiology and Microbiology Technology*, Formatex Research Center, Badajoz, Spain, pp. 267–279.

Angot A, Peters N, Lechner E et al. (2006) *Ralstonia solanacearum* requires F-box like domain containing type III effectors to promote disease on several host plants. *Proc Natl Acad Sci USA* 103: 14620–14625.

Ansermet M, Schaerer S, Kellenberger I, Tallant M, Dupuis B (2016) Influence of seed-borne and soil-carried inocula of *Dickeya* spp. on potato plant transpiration and symptom expression. *Eur J Plant Pathol* 145: 459–467.

Arlat M, van Gijsegem F, Huet JC, Permollet JC, Boucher CA (1994) PopA1, a protein which induces a hypersensitive-like response on specific *Petunia* genotypes, is secreted via the Hrp pathway of *Pseudomonas solanacearum*. *EMBO J* 13: 543–553.

Atmakuri K, Cascales E, Christie PJ (2004) Energetic components VirD4, VirB11 and VirB4 mediate early DNA transfer reactions required for bacterial type IV secretion. *Appl Environ Microbiol* 64: 1199–1211.

Baichoo Z, Jaufeerally-Fakim Y (2017) *Ralstonia solanacearum* upregulates marker genes of the salicylic acid and ethylene signaling pathways, but not those of the jasmonic acid pathway in leaflets of *Solanum* lines during early stage of infection. *Eur J Plant Pathol* 147: 615–625.

Baker CJ, Neilson MJ, Sequeira L, Keegstra KG (1984) Chemical characterization of the lipopolysaccharides of *Pseudomonas solanacearum*. *Appl Environ Microbiol* 47: 1096–1100.

Bignell DRD, Francis IM, Fyans JK, Loria R (2014) Thaxtomin A production and virulence are controlled by several *bld* gene global regulators in *Streptomyces scabies*. *Molec Plant-Microbe Interact* 27: 875–885.

Bignell DRD, Seipke RF, Huguet-Tapia JC, Chambers AH, Parry RJ, Loria R (2010) *Streptomyces scabies* 87–22 contains a coronafacic acid-like biosynthetic cluster that contributes to plant-microbe interactions. *Molec Plant-Microbe Interact* 23: 161–175.

Biruma M, Pillay M, Tripathi L et al. (2007) Banana Xanthomonas wilt: a review of the disease management strategies and future research directions. *Afr J Biotechnol* 6: 953–962.

Blair DF (1995) How bacteria sense and swim? *Annu Rev Microbiol* 49: 489–520.

Blair DF, Berg HC (1991) Mutations in the MotA protein of *Escherichia coli* reveal domains critical for proton conduction. *J Molec Biol* 221: 1433–1442.

Bocsanczy AM, Achenbach UCM, Mangravita-Novo A, Yuen JMF, Norman DJ (2012) Comparative effect of low temperature on virulence and twitching motility of *Ralstonia solanacearum* strains present in Florida. *Phytopathology* 102: 185–194.

Bogdanove AJ, Beer SV, Bonas U et al. (1996) Unified nomemclature for broadly conserved *hrp* genes of phytopathogenic bacteria. *Molec Microbiol* 20: 681–683.

Bolton GW, Nester GW, Gordon MP (1986) Plant phenolic compounds induce expression of the *Agrobacterium tumefaciens* loci needed for virulence. *Science* 232: 983–985.

Bouchek-Mechiche K, Gardan L, Normand P, Jouan B (2000a) DNA relatedness among strains of *Streptomyces* pathogenic to potato in France: description of three new species, *Streptomyces europaeiscabiei* sp. nov. and *S. stelliscabiei* sp. nov. associated with common scab and *S. reticuliscabiei* sp. nov. associated with netted scab. *Internatl J Syst Evol Microbiol* 50: 91–99.

Bouchek-Meichiche K, Pasco, Andrivon, Jouan (2000b) Differences in host range, pathogenicity to potato cultivars and response to soil temperature among *Streptomyces* species causing common scab and netted scab in France. *Plant Pathol* 49: 3–10.

Brito B, Marenda M, Berberis P, Boucher C, Genin S (1999) *PrhJ* and *hrpG*, two new components of the plant signal-dependent regulatory cascade controlled by *PhhA* in *Ralstonia solanacearum*. *Molec Microbiol* 31: 237–251.

Burr TJ, Bazzi C, Süle S, Ottent L (1998) Crown gall of grape: biology of *Agrobacterium vitis* and the development of disease control strategies. *Plant Dis* 82: 1288–1297.

Caldwell D, Kim B-S, Iyer-Pascuzzi AS (2017) *Ralstonia solanacearum* differentially colonizes roots of resistant and susceptible tomato plants. *Phytopathology* 107: 528–536.

Cangelosi GA, Ankenbauer RG, Nester EW (1990) Sugars induce the *Agrobacterium* virulence genes through a periplasmic binding protein and a transmembrane signal protein. *Proc Natl Acad Sci USA* 87: 6708–6712.

Chapleau M, Guertin JF, Farrokhi A, Lerat S, Burrus V, Beaulieu C (2016) Identification of genetic and environmental factors stimulating excision from *Streptomyces scabiei* chromosome of the toxigenic region responsible for pathogenicity. *Molec Plant Pathol* 17: 501–509.

Chatterjee A, Cui Y, Chakrabarty P, Chatterjee AK (2010) Regulation of motility in *Erwinia carotovora* subsp. *carotovora*: Quorum-sensing signal controls FlhDC, the global regulator of flagellar and exoproteins genes by modulating the production of RsmA, an RNA-binding protein. *Molec Plant-Microbe Interact* 23: 1316–1323.

Chen Y-J, Lin Y-S, Tseng K-J, Chung W-H (2014) Vine cuttings as possible initial inoculum sources of *Ralstonia solanacearum* race 1 biovar 4 on vegetable sweet potato in fields. *Eur J Plant Pathol* 140: 83–95.

Christie PJ (2004) Type IV secretion: the *Agrobacterium* VirB1D4 and related conjugation systems. *Biochem Biophys Acta* 1694: 219–234.

Chun SY, Parikson JS (1988) Bacterial motility: membrane topology of the *Escherichia coli* motB protein. *Science* 239: 276–278.

Citovsky V, De Vos G, Zambryski P (1988) Single-stranded DNA-binding protein encoded by the *virE* locus of *Agrobacterium tumefaciens*. *Science* 240: 501–504.

Citovsky V, Guralmick B, Simon MN, Wall JS (1997) Nuclear import of *Agrobacterium* VirD2 and VirE2 proteins in maize and tobacco. *Proc Natl Acad Sci USA* 91: 3210–3214.

Citovsky V, Kapelnikov A, Oliel S, Zakai N, Rojas MR, Gilbertson RL (2004) Protein interactions involved in nuclear import of *Agrobacterium* VirE2 protein in vitro and in vivo. *J Biol Chem* 279: 29528–29533.

Clough SJ, Flavier AB, Schell MA, Denny TP (1997) Differential expression of virulence genes and motility in *Ralstonia* (*Pseudomonas*) *solanacearum*. *Appl Environ Microbiol* 63: 844–850.

Colburn-Clifford J, Allen C (2010) A cbb3-type cytochrome c oxidase contributes to *Ralstonia solanacearum* R3bv2 growth in microaerobic environments and to bacterial wilt disease development in tomato. *Molec Plant-Microbe Interact* 23: 1042–1052.

Collmer A, Keen NT (1986) The role of pectic enzymes in plant pathogenesis. *Annu Rev Phytopathol* 24: 383–409.

Cooke DL, Waites WM, Leifert C (1992) Effects of *Agrobacterium tumefaciens*, *Erwinia carotovora*, *Pseudomonas syringae* and *Xanthomonas campestris* on plant tissue cultures of *Aster*, *Cheiranthus*, *Delphinuum*, Iris and Rosa: disease development in vivo as a result of latent infection in vitro. *J Plant Dis Protect* 99: 459–481.

Corbett M, Virtue S, Bell K et al. (2005) Identification of a new quorum-sensing controlled virulence factor in *Erwinia carotovora* subsp. *atroseptica* secreted via type II targeting pathway. *Molec Plant-Microbe Interact* 18: 334–342.

Creasap JE, Reid CL, Goffinet MC, Aloni R, Ullrich C, Burr TJ (2005) Effect of wound position, auxin and *Agrobacterium vitis* strain F2/5 on wound healing and crown gall in grapevine. *Phytopathology* 95: 362–367.

Cubitt MF, Hedley PE, Williamson NR et al. (2013) A metabolic regulator modulates virulence and quorum-sensing signal production in *Pectobacterium atrosepticum*. *Molec Plant-Microbe Interact* 26: 356–366.

Cunnac S, Occhialini A, Berberis P, Boucher C, Genin S (2004) Inventory and functional analysis of the large Hrp regulon in *Ralstonia solanacearum*: Identification of novel effector proteins translocated to plant cells through type III secretion system. *Molec Microbiol* 53: 115–128.

Czajkowski R, de Boer WJ, van Veen JA, van der Wolf JM (2010) Downward vascular translocation of a green fluorescent protein-tagged strain of *Dickey* sp. (biovar 3) from stem and leaf inoculation sites on potato. *Phytopathology* 100: 1128–1137.

De Cleene M, De Ley J (1996) The host range of crown gall. *Botanical Rev* 42: 389–466.

de Haan EG, Dekker-Nooren TCEM, van den Bovenkamp GW, Speksmjder AGCL, van der Zouwen PS, van der Wolf JM (2008) *Pectobacterium carotovorum* subsp. *carotovorum* can cause potato blackleg in temperate climates. *Eur J Plant Pathol* 122: 561.

De Kievit TR, Iglweski BH (2000) Bacterial quorum-sensing in pathogenic relationships. *Infection Immunol* 68: 4839–4849.

Dees MW, Sletten A, Hermansen A (2013) Isolation and characterization of *Streptomyces* species from potato common scab lesions in Norway. *Plant Pathol* 62: 217–225.

Delaspre F, Peñalver CGN, Saurel O et al. (2009) The *Ralstonia solanacearum* pathogenicity regulator HrpB induces 3-hydroxyoxindole synthesis. *Proc Natl Acad Sci USA* 104: 15870–15875.

Dombek P, Ream W (1997) Functional domains of *Agrobacterium tumefaciens* single-stranded DNA-binding protein VirE2. *J Bacteriol* 179: 1165–1173.

Dossa SG, Karlovsky P, Wydra K (2014) Biochemical approach for virulence factors' identification in *Xanthomonas oryzae* pv. *oryzae*. *J Plant Pathol Microbiol* 5: 222

Drigues P, Demery-Lafforgue D, Trigalet A, Dupin P, Samain D, Asselineau J (1985) Comparative studies of lipopolysaccharides and exopolysaccharides from a virulent strain of *Pseudomonas solanacearum* and from 3 avirulent mutants. *J Bacteriol* 162: 504–509.

Duban ME, Lee K, Lynn DG (1993) Strategies in pathogenesis: mechanistic specificity in the detection of genic signals. *Molec Microbiol* 7: 637–645.

Duckely M, Oomen C, Axthelm F, Van Gelder P, Waksman G, Engel A (2005) The VirE1-VirE2 complex of *Agrobacterium tumefaciens* interacts with single-stranded DNA and forms channels. *Molec Microbiol* 58: 1130–1142.

Duckey M, Hohn B (2003) The VirE2 protein of *Agrobacterium tumefaciens*: the Yin and Yang of T-DNA transfer. *FEMS Microbiol Lett* 223: 1–6.

Dumas F, Duckey M, Pelczar P, Van Gelder P, Hohn B (2001) An *Agrobacterium* VirE2 channel for transferred-DNA transport into plant cells. *Proc Natl Acad Sci USA* 98: 485–490.

Dung JKS, Johnson DA, Schroeder BK (2014) Role of coinfection by *Pectobacterium* and *Verticillium dahliae* in the development of early dying and aerial stem rot of Russet Burbank potato. *Plant Pathol* 63: 299–307.

Dunning Hotopp JC (2011) Horizontal gene transfer between bacteria and animals. *Trends Genet* 27: 157–153.

Dye F, Berthelot K, Griffon B, Delay D, Delmotte FM (1997) Alkylsyringamides new inducers of *Agrobacterium tumefaciens* virulence genes. *Biochimie* 79: 3–6.

Eichenlaub R, Grateman K-H, Burger A (2006) *Clavibacter michiganensis*, a group of gram-positive phytopathogenic bacteria. In: Gnanamanickam SS (ed.), *Plant-associated Bacteria*, Springer, Netherlands, pp. 385–421.

Elphinstone JG (2005) The current bacterial wilt situation: a global overview. In: Allen C, Prior P, Hayward AC (eds.), *Bacterial Wilt Disease and the Ralstonia solanacearum Species Complex*, The American Phytopathological Society, St Paul, MN, USA, pp. 9–28.

Engstrom P, Zambryski P, Van Montagu M, Stachel S (1987) Characterization of *Agrobacterium tumefaciens* virulence proteins induced by the plant factor acetosyringone. *J Molec Biol* 197: 635–645.

EPPO/CABI (1996) *Data Sheets on Quarantine Pests: Clavibacter michiganensis subsp. sepedonicus*, pp. 1–5. EPPO Publications: https/www.eppo.int/RESOURCES/eppo_publications

EPPO/CABI (2006) *Distribution maps of plant diseases: Ralstonia solanacearum (2003–2006)*. www. Cabi.org/DMPD 2006.

Escobar MA, Dandekar AM (2003) *Agrobacterium tumefaciens* as an agent of disease. *Trends Plant Sci* 8: 380–386.

FAO (2013) FAOSTAT vol. 2014 faostat3.fao.org/browse/rankings/commodities-by-regions/E.

Fegan M, Prior P (2005) How complex is the "*Ralstonia solanacearum* species complex"? In: Allen C, Prior P, Hayward AC (eds.), *Bacterial Wilt Disease and the Ralstonia solanacearum species complex*, The American Phytopathological Society, St Paul, MN, USA. pp. 449–452.

Felix KCS, Souza EB, Michereff SJ, Mariano RLR (2012) Survival of *Ralstonia solanacearum* in infected tissues of *Capsicum annuum* and in soils of the state of Pernambuco, Brazil. *Phytopathology* 40: 53–62.

Flavier AB, Ganova-Raeva LM, Schell MA, Denny TP (1997) Hierarchical autoinduction in *Ralstonia solanacearum*: Control of acylhomoserine lactone production by a novel autoregulatory system responsive to 3-hydroxypalmitic acid methyl ester. *J Bacteriol* 179: 7089–7097.

Fouri D (2002) Distribution and severity of bacterial diseases of dry beans (*Phaseolus vulgaris* L.) in South Africa. *J Phytopathol* 150: 220–226.

Fujiwara A, Takamura H, Majumder P, Yoshida H, Kojima (1998) Functional analysis of the protein encoded by the chromosomal virulence gene (*acvB*) of *Agrobacterium tumefaciens*. *Ann Phytopathol Soc Jpn* 64: 191–193.

Fullner KJ, Lara JC, Nester EW (1996) Pilus assembly of *Agrobacterium* T-DNA transfer genes. *Science* 273: 1107–1109.

Fyans JK, Altowairish MS, Li Y, Bignell DRD (2015) Characterization of the coronatine-like phytotoxins produced by the common scab pathogen *Streptomyces scabies*. *Molec Plant-Microbe Interact* 28: 443–454.

Gallois A, Samson R, Ageron E, Grimont PAD (1992) *Erwinia carotovora* subsp. *odorifera* subsp. nov. associated with odorous soft rot of chicory (*Cichorium intybus* L.). *Internatl J Syst Bacteriol* 42: 582–588.

García-Rodriguez FM, Schrammeijer B, Hooykaas PJ (2006) The *Agrobacterium* VirE3 effector protein: A potential plant transcription activator. *Nucleic Acids Res* 34: 6496–6504.

Gardan L, Gouy C, Christen R, Samson R (2003) Elevation of three subspecies of *Pectobacterium carotovorum* to species level: *Pectobacterium atrosepticum* sp. nov., *Pectobacterium betavasculorum* sp. nov. and *Pecotbacterium wasabiae* sp. nov. *Internatl J Syst Evol Microbiol* 53: 381–391.

Gelvin SB (2000) *Agrobacterium* and plant genes involved in T-DNA transfer and integration. *Annu Rev Plant Physiol Plant Molec Biol* 51: 223–256.

Gietl C, Koukolikova-Nicola Z, Hohn B (1987) Mobilization of T-DNA from *Agrobacterium* to plant cells involves a protein that binds single-stranded DNA. *Proc Natl Acad Sci USA* 84: 9006–9010.

Golanowska M, Kielar J, Lojkowska E (2017) The effects of temperatures on the phenotypic features and the maceration ability of *Dickeya solani* strains isolated in Finland, Israel and Poland. *Eur J Plant Pathol* 147: 803–817.

Gossele F, Cruz CMV, Outryve MFV, Swings J, Ley JD (1985) Differentiation between the bacteria causing bacterial blight (BB), bacterial leaf streak (BLS) and bacterial brown blotch on rice. *Internatl Rice Res Newslett* 10: 23–24.

Goto M, Matsumoto K (1987) *Erwinia carotovora* subsp. *wasbiae* subsp. nov isolated from diseased rhizomes and fibrous roots of Japanese horseradish (*Eutrema wasabi* Maxim). *Internatl J Syst Evol Microbiol* 37: 130–135.

Göttfert M, Grob P, Hennecke H (1990) Proposed regulatory pathway encoded by the *nodV* and *nodW* genes, determinants of host specificity in *Bradyrhizobium japonicum*. *Proc Natl Acad Sci USA* 87: 2680–2684.

Guan D, Grau BL, Clark CA, Taylor CM, Loria R, Pettis GS (2012) Evidence that thaxtomin C is a pathogenicity determinant of *Strepotmyces ipomoeae*, the causative agent of Streptomyces soil rot disease of sweet potato. *Molec Plant-Microbe Interact* 25: 393–401.

Hao G, Zhang H, Zheng D, Burr TJ (2005) *LuxR* homolog *avrR* in *Agrobacterium vitis* affects the development of a grape-specific necrosis and a tobacco hypersensitive response. *J Bacteriol* 187: 185–192.

Hao JJ, Meng QX, Yin JF, Kirk WW (2009) Characterization of a new *Streptomyces* strain DS3024 that causes potato common scab. *Plant Dis* 93: 1329–1334.

Hayward AC (1964) Characteristics of *Pseudomonas solanacearum*. *J Appl Bacteriol* 27: 265–277.

Hayward AC (1991) Biology and epidemiology of bacterial wilt caused by *Pseudomonas solanacearum. Annu Rev Phytopathol* 29: 65–87.

Helias V, Andrivon D, Jouan B (2000a) Internal colonization pathways of potato plants by *Erwinia carotovora* subsp. *carotovora. Plant Pathol* 49: 33–42.

Helias V, Andrivon D, Jouan B (2000b) Development of symptoms caused by *Erwinia carotovora* ssp. *atroseptica* under field conditions and their effects on yield of individual potato plants. *Plant Pathol* 49: 23–32.

Hendrick CA, Sequeira L (1984) Lipopolysaccharide-defective mutants of the wilt pathogen *Pseudomonas solanacearum. Appl Environ Microbiol* 48: 94–101.

Herlache TC, Zhang HS, Reid CL et al. (2001) Mutations that affect *Agrobacterium vitis*-induced grape necrosis also alter its ability to cause a hypersensitive response on tobacco. *Phytopathology* 91: 966–972.

Herrera-Estrella A, Van Montagu M, Wang K (1990) A bacterial peptide acting as a plant nuclear targeting signal: the amino-terminal portion of *Agrobacterium* VirD2 directs a β-galactosidase fusion protein into tobacco nuclei. *Proc Natl Acad Sci USA* 87: 9534–9537.

Hikichi Y, Yoshimochi T, Tsujimoto S et al. (2007) Global regulation of pathogenicity mechanism of *Ralstonia solanacearum. Plant Biotechnol* 24: 149–154.

Hiltunen LH, Alanen M, Laakso I, Kangas A, Virtanen E, Valkonen JPT (2011) Elimination of common scab sensitive progeny from a potato breeding population using thaxtomin A as a selective agent. *Plant Pathol* 60: 426–435.

Hooykaas PJJ, Beijersbergen AGM (1994) The virulence system of *Agrobacterium tumefaciens. Annu Rev Phytopathol* 32: 157–179.

Hseih C-Y, Wang J-F, Huang P-C et al. (2012) *Ralstonia solanacearum nlpD (RSc 1206)* contributes to host adaptation. *Eur J Plant Pathol* 133: 645–656.

Huang JS, De Cleene M (1989) How rice plants are infected by *Xanthomonas campestris* pv. *oryzae*? In: *Bacterial Blight of Rice*, The International Rice Research Institute, Philippines, pp. 31–42.

Hwang H-H, Yang F-J, Cheng T-F et al. (2013) The Tzs protein and exogenous cytokinin affect virulence gene expression and bacterial growth of *Agrobacterium tumefaciens. Phytopathology* 103: 888–899.

Jacques MA, Josi K, Darrasse A, Samson R (2005) *Xanthomonas axonopodis* pv. *phaseoli* var. *fuscans* is aggregated in stable biofilm population size in the phyllosphere of field-grown beans. *Appl Environ Microbiol* 71: 2008–2015.

Jeong Y, Cheong H, Choi O et al. (2011) An HrpB-dependent but type III-independent extracellular aspartic protease is a virulence factor of *Ralstonia solanacearum. Molec Plant Pathol* 12: 373–380.

Jiang H, Jiang M, Yang L et al. (2017) The ribosomoal protein RPlY is required for *Pectobacterium carotovorum* virulence and is induced by *Zantedeschia ellotiana* extract. *Phytopathology* 107: 1322–1330.

Jin S, Prusti RK, Roitsch T, Ankenbauer RG, Nester EW (1990) Phosphorylation of the VirG protein of *Agrobacterium tumefaciens* by the autophosphorylated VirA protein: essential role in biological activity of VirG. *J Bacteriol* 172: 4945–4950.

Joshi JR, Burdman S, Lipsky A, Yariv S, Yedidia I (2016) Plant phenolic acids affect the virulence of *Pectobacterium aroidearum* and *P. carotovorum* ssp. *brasiliense* via quorum-sensing regulation. *Molec Plant Pathol* 17: 487–500.

Kanda A, Yasukohchim M, Ohnishi K, Kiba A, Okuno T, Hikichi Y (2003) Ectopic expression of *Ralstonia solanacearum* effector protein *popA* early in invasion results in loss of virulence. *Molec Plant-Microbe Interact* 16: 447–455.

Kao CC, Sequeira L (1991) A gene cluster required for coordinated biosynthesis of lipopolysaccharide of extracellular polysaccharide also affects virulence of *Pseudomonas solanacearum. J Bacteriol* 173: 7841–7847.

Kelman A, Hruschka J (1973) Role of motility and aerotaxis in selective increase of avirulent bacteria in still broth cultures of *Pseudomonas solanacearum. J Gen Microbiol* 76: 177–188.

Khatri BB, Tegg RS, Brown PH, Wilson CR (2010) Infection of potato tubers with the common scab pathogen *Streptomyces scabiei. J Phytopathol* 158: 453–455.

Khatri BB, Tegg RS, Brown PH, Wilson CR (2011) Temporal association of potato tuber development with susceptibility to common scab and *Streptomyces scabiei*-induced responses in the potato periderm. *Plant Pathol* 60: 776–786.

Kim H-S, Thammarat P, Lommel SA, Hogan CS, Charkowski AO (2011) *Pectobacterium carotovorum* elicits plant cell death with Dsp E/F but the *P. carotovorum* DspE does not suppress callose or induce expression of plant genes early in plant-microbe interactions. *Molec Plant-Microbe Interact* 24: 773–786.

Kim MG, Cunha L, da McFall AJ et al. (2005) Two *Pseudomonas syringae* type III effectors inhibit RIN4-regulated basal defense in *Arabidopsis. Cell* 121: 749–759.

Kubheka GC, Coutinho TA, Moleleki N, Moleleki LN (2013) Colonization patterns of an mcherry-tagged *Pectobacterium carotovorum* subsp. *brasiliense* strain in potato plants. *Phytopathology* 103: 1268–1279.

Kumar RB, Xie YH, Ananth Das (2000) Subcellular localization of the *Agrobacterium tumefaciens* T-DNA transport pore proteins: VirB8 is essential for the assembly of the transport pore. *Molec Microbiol* 36: 608–617.

Kunze G, Zipfel C, Robatzek S, Nienhaus K, Boller T, Felix G (2004) The N-terminus of bacterial elongation factor Tu elicits innate immunity in *Arabidopsis* plants. *Plant Cell* 16: 3496–3507.

Kyndt T, Quispe D, Zhai H et al. (2015) The genome of cultivated sweet potato contains *Agrobacterium* T-DNAs with expressed genes: An example of a naturally transgenic food crop. *Proc Natl Acad Sci USA* 112: 5844–5849.

Lacroix B, Gizatullina D, Babst BA, Gifford AN, Citovsky V (2014) *Agrobacterium* T-DNA encoded protein Atu 6002 interferes with the host auxin response. *Molec Plant Pathol* 15: 275–283.

Lacroix B, Vaidya M, Tzfira T, Citovsky V (2005) The VirE3 of *Agrobacterium* mimics a host cell function required for plant genetic transformation. *EMBO J* 24: 428–437.

Lai EM, Eisenbrandt R, Kalkum M, Lamka E, Kado CI (2002) Biogenesis of T pili in *Agrobacterium tumefaciens* requires precise cleavage and cyclization. *J Bacteriol* 184: 327–330.

Lambert DH, Loria R (1989a) *Streptomyces scabies* sp. nov nom. rev. *Internatl J Syst Bacteriol* 39: 393–392.

Lambert DH, Loria R (1989b) *Streptomyces acidiscabies* sp. nov. *Internatl J Syst Bacteriol* 39: 393–396.

Lazazzera BA, Grossman AD (1988) The ins and outs of peptide signaling. *Trends Microbiol* 7: 288–294.

Lee YA, Fan SC, Chiu LY, Hsia KC (2001) Isolation of an insertion sequence from *Ralstonia solanacerum* race 1 and its potential use for strain characterization and detection. *Appl Environ Microbiol* 67: 3943–3950.

Legault GS, Lerat S, Nicolas P, Beaulieu C (2011) Tryptophan regulates thaxtomin A and indole-3-acetic acid and production in *Streptomyces scabiei* and modifies its interactions with radish seedlings. *Phytopathology* 101: 1045–1051.

Li C-H, Wang K-C, Hong Y-H et al. (2014) Roles of different forms of lipopolysaccharides in *Ralstonia solanacearum* pathogenesis. *Molec Plant-Microbe Interact* 27: 471–478.

Li J-G, Liu H-X, Cao J et al. (2010) PopW of *Ralstonia solanacearum*, a new two-domain harpin targeting the plant cell wall. *Molec Plant Pathol* 11: 371–381.

Li X, Liu Y, Cai L, Zhang H, Shi J, Yuan Y (2017) Factors affecting the virulence of *Ralstonia solanacearum* and its colonization of tobacco roots. *Plant Pathology* 66: 1345–1356.

Li Y, Hutchins W, Wu X et al. (2015) Derivative of plant phenolic compound inhibits the type III secretion system of *Dickeya dadantii* via HrpX/HrpY two-component signal transduction and Rsm systems. *Molec Plant Pathol* 16: 150–163.

Li Y, Yamazaki A, Zou L, Biddle E, Zeng Q, Wang Y, Lin H, Wang Q, Yang C-H (2010) ClpXP protease regulates the type III secretion system of *Dickeya dadantii* 3937 and is essential for the bacterial virulence. *Molec Plant-Microbe Interact* 23: 871–878.

Lin C-H, Hsu S-T, Tzeng K-C, Wang J-F (2009) Detection of race 1 strains of *Ralstonia solanacearum* in field samples in Taiwan using BIO-PCR method. *Eur J Plant Pathol* 124: 75–85.

Lindberg M, Collmer A (1992) Analysis of eight *out* genes in a cluster required for pectic enzyme secretion by *Erwinia chrysanthemi*: sequence comparison with secretion genes from other Gram-negative bacteria. *J Bacteriol* 174: 7385–7397.

Liu H, Zhang S, Schell MA, Denny TP (2005) Pyramiding unmarked deletions in *Ralstonia solanacearum* shows that secreted proteins in addition to plant cell wall-degrading enzymes contribute to virulence. *Molec Plant-Microbe Interact* 18: 1296–1305.

Liu Y, Wu D, Liu Q, Zhang S, Tang Y, Jiang G, Li S, Ding W (2017) The sequevar distribution of *Ralstonia solanacearum* in tobacco growing zones of China is structured by elevation. *Eur J Plant Pathol* 147: 541–551.

Lohou D, Turner M, Lonjon F, Cazale A-C, Peeters N, Genin S, Vailleau F (2014) HpaP modulates type III effector secretion in *Ralstonia solanacearum* and harbours a substrate specificity switch domain essential for virulence. *Molec Plant Pathol* 15: 601–614.

Loria R, Kers J, Joshi M (2006) Evolution of plant pathogenicity in *Streptomyces*. *Annu Rev Phytopathol* 44: 469–487.

Lowe TM, Ailloud F, Allen C (2015) Hydrocinnamic acid degradation, a broadly conserved trait, protects *Ralstonia solanacearum* from chemical plant defences and contributes to root colonization and virulence. *Molec Plant-Microbe Interact* 28: 286–297.

Ma B, Hibbing ME, Kim H-S et al. (2007) Host range and molecular phylogenies of the soft rot enterobacterial genera *Pectobacterium* and *Dickeya*. *Phytopathology* 97: 1150–1163.

Macho AP, Guidot A, Barberis P, Beuzon CR, Genin S (2010) A competitive index assay identifies several *Ralstonia solanacearum* type III effector mutant strains with reduced fitness in host plants. *Molec Plant-Microbe Interact* 23: 1197–1205.

Mahuku GS, Jara C, Henriquez MA, Castellanos G, Cuasquer J (2006) Genotype characterization of the common bean bacterial blight pathogens *Xanthomonas axonopodis* pv. *phaseoli* and *Xanthomonas axonopodis* pv. *phaseoli* var. *fuscans* by rep-PCR and PCR-RFLP of the ribosomal genes. *J Phytopathol* 154: 35–44.

Mao GZ, He LY (1998) Relationship of wild type strain motility and interaction with host plants in *Ralstonia solanacearum*. In: Allen C, Prior P, Elphinstone J (eds.), Bacterial Wilt Disease–Molecular and Ecological Aspects. Springer-Verlag, Heidelberg, Germany, p. 184.

Matthysee AG (1994) Conditioned medium promotes the attachment of *Agrobacterium tumefaciens* NT1 to carrot cells. *Protoplasma* 183: 131–136.

Matthysee AG, McMahan S (1998) Root colonization by *Agrobacterium tumefaciens* is reduced by *cel*, *attB*, *attD* and *attR* mutants. *Appl Environ Microbiol* 64: 2341–2345.

Mc Garvey JA, Bell CJ, Denny TP, Schell MA (1998) Analysis of extracellular polysaccharide in culture and in planta using immunological methods: new insights and implications. In: Allen C, Prior P, Elphinstone J (eds), *Bacterial Wilt Disease: Molecular and Ecological Aspects*, Springer-Verlag, Berlin, Gerp. 157.

Melchers LS, Maroney MJ, de Dulk Ras A, Thompson DV, van Vuuren HAJ (1990) Octopine and nopaline strains of *Agrobacterium tumefaciens* differ in virulence: molecular characterization of the *virF* locus. *Plant Molec Biol* 14: 249–259.

Meng Q, Yin J, Rosenzweig N, Douches D, Hao JJ (2012) Culture based assessment of microbial communities in soil suppressive to potato common scab. *Plant Dis* 96: 712–717.

Mew TW, Alvarez AM, Leach JE, Swings J (1993) Focus on bacterial blight of rice. *Plant Dis* 77: 5–12.

Mew TW, Mew IC, Huang JS (1984) Scanning electron microscopy of virulent and avirulent strains of *Xanthomonas campestris* pv. *oryzae* on rice leaves. *Phytopathology* 74: 635–641.

Meyer D, Cunnac S, Guéneron M et al. (2006) PopF1 and PopF2, two proteins secreted by the type III protein secretion system of *Ralstonia solanacearum* are translocators belonging to the HrpF/NopX family. *J Bacteriol* 188: 4903–4917.

Milling A, Babujee L, Allen C (2011) *Ralstonia solanacearum* extracellular polysaccharide is a specific elicitor of defense responses in wilt-resistant tomato plant. *PLoS ONE* 6(1): e15853

Milling A, Meng F, Denny TP, Allen C (2009) Interactions with hosts at cool temperatures, not cold tolerance, explain the unique epidemiology of *Ralstonia solanacearum* race 3 biovar 2. *Phytopathology* 99: 1127–1134.

Mizukami T, Wakimoto S (1969) Epidemiology and control of bacterial leaf blight of rice. *Annu Rev Phytopathol* 7: 51–72.

Molc B, Habibi S, Dangl, Grant SR (2010) Gluconate metabolism is required for virulence of the soft rot pathogen *Pectobacterium carotovorum*. *Molec Plant-Microbe Interact* 23: 1335–1344.

Mondal KK, Verma G, Junaid A, Mani (2016) Rice pathogen *Xanthomonas oryzae* pv. *oryzae* employs inducible *hrp*-dependent XopF type III effector protein for its growth, pathogenicity and for suppression of PTI response to induce blight disease. *Eur J Plant Pathol* 144: 311–322.

Moran NA, Jarvik T (2010) Lateral transfer of genes from fungi underlies carotenoid production in aphids. *Science* 328: 624–627.

Mori Y, Inoue K, Ikeda K et al. (2016) The vascular plant pathogenic bacterium *Ralstonia solanacearum* produces biofilms required for its virulence on the surfaces of tomato cells adjacent to intracellular spaces. *Molec Plant Pathol* 17: 890–902.

Mukaihara T, Tamura N, Iwabuchi M (2010) Genome-wide identification of a large repetitive of *Ralstonia solanacearum* type III effector proteins by a new functional screen. *Molec Plant-Microbe Interact* 23: 251–262.

Murata Y, Yamura N, Nakaho K, Mukaihara T (2006) Mutations in the *lrpE* gene of *Ralstonia solanacearum* affects Hrp pili production and virulence. *Molec Plant-Microbe Interact* 19: 884–895.

Mwebaze JM, Tusiime G, Tushemereirwe WK, Maina M (2006) Development of a semi-selective medium for *Xanthomonas campestris* pv. *musacearum*. *Afr Crop Sci J* 14: 129–135.

Nahar K, Matsumoto I, Taguchi F et al. (2014) *Ralstonia solanacearum* type III secretion system effector Rip36 induces a hypersensitive response in the nonhost wild eggplant *Solanum torvum*. *Molec Plant Pathol* 15: 297–303.

Nakaho K, Seo S, Ookawa K et al. (2017) Involvement of a vascular hypersensitive response in quantitative resistance to *Ralstonia solanacearum* on tomato rootstock cultivar LS-89. *Plant Pathol* 66: 150–158.

Nakato GV, Ocimati W, Blomme G, Fiaboe KKM, Beed F (2014) Comparative importance of infection routes for banana Xanthomonas wilt and implications on disease epidemiology and management. *Canad J Plant Pathol* 36: 418–427.

Natsum M, Tashiro N, Doi A, Nishi Y, Kawaide H (2017) Effects of concanamycins produced by *Streptomyces scabies* on lesion type of common scab of potato. *J Gen Plant Pathol* 83: 78–82.

Nelson GA (1982) *Corynebacterium sepedonicum* in potato: Effect of inoculum concentration on ring rot symptoms and latent infection. *Canad J Plant Pathol* 4: 129–133.

Nelson GA (1984) Survival of *Corynebacterium sepedonicum* in potato stems and on surfaces held at freezing and above-freezing temperatures. *Amer Potato J* 62: 23–28.

Nguyen M-P, Park J, Cho M-H, Lee S-W (2016) Role of DetR in defence is critical for virulence of *Xanthomonas oryzae* pv. *oryzae*. *Molec Plant Pathol* 17: 601–613.

Nissinen R, Kassuwi S, Peltola R, Melzer MC (2001) In plant-complementation of *Clavibacter michiganensis* subsp. *sepedonicus* strains deficient in cellulase production or HR induction restores virulence. *Eur J Plant Pathol* 107: 175–182.

Nomura K, Melotto M, He SY (2005) Suppression of host defense compatible plant-*Psuedomonas syringae* interactions. *Curr Opin Plant Biol* 8: 361–368.

Norman DJ, Yuen JMF, Resendiz R, Boswell J (2003) Characterization of *Erwinia* populations from nursery retention ponds and lakes infecting ornamental plants in Florida. *Plant Dis* 87: 193–196.

Norman DJ, Zapata M, Gabriel DW et al. (2009) Genetic diversity and host range variation of *Ralstonia solanacearum* strains entering North America. *Phytopathology* 99: 1070–1077.

Nurnberger T, Brunner F, Kemmerling B, Piater L (2004) Innate immunity in plants and animals: striking similarities and obvious differences. *Immunol Rev* 198: 249–266.

Nyvall RF (1999) *Field Crop Diseases*, Ames, Iowa State University Press, USA.

Occhialini A, Cunnac S, Reymond N, Genin S, Boucher C (2005) Genome-wide analysis of gene expression in *Ralstonia solanacearum* reveals that the *hrpB* gene acts as a regulatory switch controlling multiple virulence pathways. *Molec Plant-Microbe Interact* 18: 938–949.

Ochiai H, Inoue Y, Takeya M, Saski A, Kaku H (2005) Genome sequence of *Xanthomonas oryzae* pv. *oryzae* suggests contribution of large number of effector genes and insertion sequences to its race diversity. *J Agric Res Quart* 39: 275–287.

Ocimati W, Nakato GV, Fiaboe KM, Beed F, Blomme G (2015) Incomplete systemic movement of *Xanthomonas campestris* pv. *musacearum* and the occurrence of latent infections in Xanthomonas wilt-infected banana mats. *Plant Pathol* 64: 81–90.

Ocimati W, Ssekiwoko F, Karamura E, Tinzaara W, Eden-Green S, Blomme G (2013) Systemicity of *Xanthomonas campestris* pv. *musacearum* and time to disease expression after inflorescence infection in East African highland and Pisang Awak bananas in Uganda. *Plant Pathol* 62: 775–785.

Okechukwu RV, Ekpo EJA (2008) Survival of *Xanthomonas campestris* pv. *vignicola* in infested soil, cowpea seed and cowpea debris. *Tropic Agric Res Extn* 11: 45–48.

Opio AF, Allen DJ, Teri JM (1996) Pathogenic variation in *Xanthomonas campestris* pv. *phaseoli*, the causal agent of common bacterial blight in *Phaseolus* beans. *Plant Pathol* 45: 1126–1133.

Ou SH (1985) *Rice Diseases*, Commonwealth Agricultural Bureau International, Kew, Surrey, United Kingdom.

Pan SQ, Jin S, Boulton MI, Hawes M, Gordon MP, Nester EW (1995) An *Agrobacterium* virulence factor encoded by a Ti-plasmid gene or a chromosomal gene is required for T-DNA transfer into plants. *Molec Microbiol* 17: 259–269.

Paret ML, Kubota R, Jenkins DM, Alvarez AM (2010) Survival of *Ralstonia solanacearum* race 4 in drainage water and soil with immunodiagnostic and DNA-based assays. *HortTechnology* 20: 539–548.

Park DH, Kim JS, Kwon SW et al. (2003) *Streptomyces luridiscabiei* sp. nov., *Streptomyces puniscabiei* sp.nov., *Streptomyces niveiscabiei* sp. nov which cause potato common scab disease in Korea. *Internatl J Syst Evol Microbiol* 62: 3489–3493.

Pemberton CL, Whitehead NA, Sebaihia M et al. (2005) Novel quorum-sensing-controlled genes in *Erwinia carotovora* subsp. *carotovora*: identification of a fungal elicitor homologue in a soft-rotting bacterium. *Molec Plant-Microbe Interact* 18: 343–353.

Pensec F, Lebeau A, Daunay MC, Chiroleu F, Guidot A, Wicker E (2015) Towards the identification of type III effectors associated with *Ralstonia solanacearum* virulence on tomato and eggplant. *Phytopathology* 105: 1529–1544.

Peppenberger B, Leonhardt W, Reddi H (2002) Latent persistence of *Agrobacterium vitis* in micropropagated *Vitis vinifera*. *Vitis* 41: 113–114.

Pérombelon MCM (2002) Potato diseases caused by soft rot erwinias: An overview of pathogenesis. *Plant Pathol* 51: 1–12.

Pérombelon MCM, Lumb VM, Zutra D (1987) Pathogenicity of soft rot erwinias to potato plants in Scotland and Israel. *J Appl Bacteriol* 63: 73–84.

Potrykus M, Golanowska M, Hugouvieux-Cotte-Pattat N, Lojkowska E (2014) Regulators involved in *Dickeya solani* virulence, genetic conservation and functional variability. *Molec Plant-Microbe Interact* 27: 700–711.

Poussier S, Thoquet P, Trigalet-Demery D et al. (2003) Host plant-dependent phenotypic reversion of *Ralstonia solanacearum* from nonpathogenic to pathogenic forms via alterations in the *phcA* gene. *Molec Microbiol* 49: 991–1003.

Pradhanang PM, Elphinstone JG, Fox TRV (2000a) Identification of crop and weed hosts of *Ralstonia solanacearum* in the hills of Nepal. *Plant Pathol* 49: 403–413.

Pradhanang PM, Elphinstone JG, Fox RTV (2000b) Sensitive detection of *Ralstonia solanacearum* in soil: A comparison of different detection techniques. *Plant Pathol* 49: 414–422.

Pugsley AP (1993) The complete general protein secretory pathway in Gram negative bacteria. *Microbiol Rev* 57: 50–108.

Ray SK, Rajeswari R, Sharma Y, Sonti RV (2002) A high-molecular weight membrane protein of *Xanthomonas oryzae* pv. *oryzae* exhibits similarity to nonfimbrial adhesions of animal pathogenic bacteria and is required for optimum virulence. *Molec Microbiol* 46: 637–647.

Reuhs BL, Kim JS, Matthysee AG (1997) Attachment of *Agrobacterium tumefaciens* to carrot cells and *Arabidopsis* wound sites is correlated with the presence of cell-associated acid polysaccharide. *J Bacteriol* 179: 5372–5379.

Rio-Alvarez I, Rodriguez-Herva JJ, Cuartas-Lanza R et al. (2012) Genome-wide analysis of the response of *Dickeya dadantii* 3937 to plant antimicrobial peptides. *Molec Plant-Microbe Interact* 25: 523–533.

Robert M (2013) General aspects of the prevention and control of the potato ring rot disease (*Clavibacter michiganensis* subsp. *sepedonicus*). *J Hortic Forestry Biotechnol* 17: 122–124.

Romanenko AS, Lomovatskaya LA, Shafikova TN, Borovskii GB, Krivolapova NV (2003) Potato cell plasma membrane receptors to ring rot pathogen extracellular polysaccharides. *J Phytopathol* 151: 1–6.

Rutikanga A, Night G, Tusiime G, Ocimati W, Blomme G (2015) Spatial and temporal distribution of insect vectors of *Xanthomonas campestris* pv. *musacearum* and their activity across banana cultivars grown in Rwanda. In: Marčić D, Galvendekić M, Nicot P (eds.), *Proc 7th Congress on Plant Protection*, Plant Protection Society, Serbia, Belgrade, pp. 139–153.

Saettler AW (1989) Common bacterial blights. In: Schwartz HF, Pastor-Corrales MA (eds.) *Bean Production Problems in the Tropics*, Second edition, Centro Internacional de Agricultura Tropical (CIAT), Cali, Colombia, pp. 261–301.

Schell MA (2000) Control of virulence and pathogenicity genes of *Ralstonia solanacearum* by an elaborate sensory network. *Annu Rev Phytopathol* 38: 263–292.

Schrammeijer B, Risseauw E, Pansegru W, Resensburg-Tuink TJG, Crosby WL, Hooykas PJJ (2001) Interaction of the virulence protein VirF of with plant homologs of the yeast Skp1 protein. *Curr Biol* 11: 258–262.

Shaw CH, Ashby AM, Brown A, Royal C, Loake GJ, Shaw C (1988) *virA* and *virG* are Ti-plasmid function required for chemotaxis of *Agrobacterium tumefaciens* toward acetosyringone. *Molec Microbiol* 2: 413–417.

Shimizu R, Taguchi F, Marutani M et al. (2003) The *ΔfliD* mutant of *Pseudomonas syringae* pv. *tabaci* which secretes flagellin monomers induces a strong hypersensitive reaction (HR) in nonhost tomato cells. *Molec Genet Genom* 269: 218–225.

Singh VB, Kumar A, Kirubakaran SI, Ayyadurai N, Samish Kumar R, Sakthivel N (2006) Comparison of exopolysaccharides produced by *Xanthomonas oryzae* pv. *oryzae* strains BXO1 and BXO8 that vary in degrees of virulence in rice (*Oryza sativa* L.). *J Phytopathol* 154: 410–413.

Siri MI, Sanabria A, Boucher C, Pianzzola MJ (2014) New type IV pili-related genes involved in early stages of *Ralstonia solanacearum* potato infection. *Molec Plant-Microbe Interact* 27: 712–724.

Sivaranjani M, Krishnan SR, Kannappan A, Ramesh M, Ravi AV (2016) Curcumin from *Curcuma longa* affects the virulence of *Pectobacterium wasabiae* and *P. carotovorum* subsp. *carotovorum* via quorum-sensing regulation. *Eur J Plant Pathol* 146: 793–806.

Smadja B, Latour X, Faure D, Chevalier S, Dessaux Y, Orange N (2004) Involvement of N-acylhomoserine lactones throughout plant infection by *Erwinia carotovora* subsp. *atroseptica* (*Pectobacterium atrosepticum*). *Molec Plant-Microbe Interact* 17: 1269–1278.

Solé M, Popa C, Mith O et al. (2012) The *awr* gene family encodes a novel class of *Ralstonia solanacearum* type III effectors displaying virulence and avirulence activities. *Molec Plant-Microbe Interact* 25: 941–953.

Ssekiwoko F, Turyagyenda LF, Mukasas H, Eden-Green S, Blomme G (2010) Spread of *Xanthomonas campestris* pv. *musacearum* in banana (*Musa* spp.) plants following infection of male inflorescence. *Acta Hortic* 879: 349–356.

Stachel SE, Nester EW, Zambryski PC (1986) A plant cell factor induces *Agrobacterium tumefaciens vir* gene expression. *Proc Natl Acad Sci USA* 83: 379–383.

Stachel SE, Zambryski PC (1986) *virA* and *virG* control the plant induced activation of the T-DNA transfer process of *Agrobacterium tumefaciens*. *Cell* 46: 325–333.

Steinmöller S, Müller P, Bandte M, Büttner C (2013) Risk of dissemination of *Clavibacter michiganensis* subsp. *sepedonicus* with potato waste. *Eur J Plant Pathol* 137: 573–584.

Stulberg MJ, Shao J, Huang Q (2015) A multiplex PCR assay to detect and differentiate select agent strains of *Ralstonia solanacearum*. *Plant Dis* 99: 333–341.

Suksomtip M, Liu P, Anderson T, Tungpradabkul S, Wood DW, Nester EW (2006) Citrate synthase mutants of *Agrobacterium* are attenuated in virulence and display reduced *vir* gene induction. *J Bacteriol* 187: 4844–4852.

Swanson JK, Yao J, Tans-Kersten J, Allen C (2005) Behavior of *Ralstonia solanacearum* race 3 biovar 2 during latent and active infection of geranium. *Phytopathology* 95: 136–143.

Swings J, Van Den Mooter M, Vauterin L, Hoste B, Gills M, Mew TW, Kersters K (1990) Reclassification of the causal agents of bacterial blight *Xanthomonas campestris* pathovar *oryzae* and bacterial leaf streak *Xanthomonas campestris* pathovar *oryzicola* of rice as pathovars of *Xanthomonas oryzae* new species Ex Ishiyama 1922 Revised Name. *Internatl J Syst Bacteriol* 40: 309–311.

Tans-Kersten J, Guam Y, Allen C (1998) *Ralstonia solanacearum* pectin methyl esterase is required for growth on methylated pectin but not for bacterial virulence. *Appl Environ Microbiol* 64: 4918–4923.

Tans-Kersten J, Guan Y, Allen C (2001) *Ralstonia solanacearum* needs motility for invasive virulence on tomato. *J Bacteriol* 183: 3597–3605.

Tayi L, Maku R, Patel HK, Sonti RV (2016) Action of multiple cell wall degrading enzymes is required for elicitation of innate immune responses during *Xanthomonas oryzae* pv. *oryzae* infection in rice. *Molec Plant-Microbe Interact* 29: 599–608.

Thomas P, Sadashiva AT, Upreti R, Mujawar MM (2015) Direct delivery of inoculum to shoot tissue interferes with genotypic resistance to *Ralstonia solanacearum* in tomato seedlings. *J Phytopathol* 163: 320–323.

Thwaites R, Eden-Green SJ, Black R (2000) Diseases caused by bacteria. In: Jones DR (ed.), *Diseases of Banana: Abaca and Enset*, CAB International, Wallingford, UK, pp. 213–240.

Tinzaara W, Gold CS, Ssekiwoko F et al. (2006) Role of insects in the transmission of banana bacterial wilt. *Afr Crop Sci J* 14: 105–110.

Tomlinson DL, Elphinstone JG, Soliman MY et al. (2009) Recovery of *Ralstonia solanacearum* in traditional potato-growing areas of Egypt, but not from designated pest-free areas (PFAs). *Eur J Plant Pathol* 125: 589.

Toth IK, Sullivan L, Brierley JL et al. (2003) Relationship between potato seed tuber contamination by *Erwinia carotovora* ssp. *atroseptica*, blackleg disease development and progeny tuber contamination. *Plant Pathol* 52: 119–126.

Uttaro AD, Cangelosi GA, Geremia PA, Nester EW, Ugalde RA (1990) Characterization of avirulent *exoC* mutants of *Agrobacterium tumefaciens*. *J Bacteriol* 172: 1640–1646.

van der Merwe JJ, Coutinho TA, Korsten L, van der Waals JE (2010) *Pectobacterium carotovorum* subsp. *brasiliensis* causing blackleg on potatoes in South Africa. *Eur J Plant Pathol* 126: 175–185.

van der Wolf JM, van Beckhoven JRCM, Hukkanen A, Karjalainen R, Müller P (2005) Fate of *Clavibacter michiganensis* ssp. *sepedonicus*, the causal organism of bacterial ring rot of potato, in weeds and field crops. *J Phytopathol* 153: 358–365.

van der Wolf JM, Vriend SGG, Kastelein P, Nijhuis EH, van Bekkum PJ, van Vuurde JWL (2000) Immunofluorescence colony-staining (IFC) for detection and quantification of *Ralstonia* (*Pseudomonas*) *solanacearum* biovar 2 (race 3) in soil and verification of positive results by PCR and dilution plating. *Eur J Plant Pathol* 106: 123–133.

van Elsas JP, Kastelein P, van Bekkum P et al. (2000) Survival of *Ralstonia solanacearum* biovar 2, the causative agent of potato brown rot in field and microcosm soils in temperate climates. *Phytopathology* 90: 1358–1366.

van Gijsegem V, Vasse J, Camus JC, Marenda M, Boucher C (2000) *Ralstonia solanacearum* produces Hrp-dependent pili that are required for PopA secretion, but not for attachment of bacteria to plant cells. *Molec Microbiol* 36: 249–260.

van Overbeek S, Bergervoet JHW, Jacobs FHH, van Elsas JD (2004) The low-temperature-induced viable-but-nonculturable state affects virulence of *Ralstonia solanacearum*. *Phytopathology* 94: 463–465.

Wairuric K, van der Waals JE, van Schalkwyk A, Theron J (2012) *Ralstonia solanacearum* needs Flp pili for virulence on potato. *Molec Plant-Microbe Interact* 25: 546–556.

Wang K, Remigi P, Anisimova M et al. (2016) Functional assignment to positively selected sites in the core type III effector RipG7 from *Ralstonia solanacearum*. *Molec Plant Pathol* 17: 553–564.

Warren JG, Kasun GW, Leonard T, Kirkpatrick BC (2016) A phage display-selected peptide inhibitor of *Agrobacterium vitis* polygalacturonase. *Molec Plant Pathol* 17: 480–486.

Weller DM, Saettler AW (1980) Colonization and distribution of *Xanthomonas phaseoli* and *Xanthomonas phaseoli* var *fuscans* in field-grown navy beans. *Phytopathology* 70: 500–506.

Wenneker M, Verdel MSW, Groeneveld RMW, Kempenaar C, van Beuningen AR, Janse JD (1999) *Ralstonia (Pseudomonas) solanacearum* race 3 (biovar 2) in surface water and natural weed hosts: first report on stinging nettle (*Urtica dioica*). *Eur J Plant Pathol* 105: 307–315.

Whitehead NA, Barnard AML, Slater H, Simpson NJL, Salmond GPC (2001) Quorum-sensing in Gram-negative bacteria. *FEMS Microbiol Rev* 25: 365–404.

Williamson L, Nekaho K, Hudelson B, Allen C (2002) *Ralstonia solanacearum* race 3 biovar 2 strains isolated from geranium are pathogenic to potato. *Plant Dis* 86: 987–991.

Winans SC (1992) Two-way chemical signaling in *Agrobacterium*-plant interactions. *Microbiol Rev* 56: 12–31.

Wirawan IGP, Kang HW, Kojima M (1993) Isolation and characterization of a new chromosomal virulence gene of *Agrobacterium tumefaciens*. *J Bacteriol* 175: 3208–3212.

Wu J, Kong HG, Jung EJ et al. (2015a) Loss of glutamate dehydrogenase in *Ralstonia solanacearum* alters dehydrogenase activity, extracellular polysaccharide production and bacterial virulence. *Physiol Molec Plant Pathol* 90: 57–64.

Wu K, Yuan S, Xun G et al. (2015b) Root exudates from two tobacco cultivars after colonization of *Ralstonia solanacearum* and the disease index. *Eur J Plant Pathol* 141: 667–677.

Yakabe LE, Parker SR, Kluepfel DA (2012) Role of systemic *Agrobacterium tumefaciens* populations in crown gall incidence on the walnut hybrid rootstock 'paradox'. *Plant Dis* 96: 1415–1421.

Yang B, Sugio A, White FF (2006) *Os8N3* is a host disease susceptibility gene for bacterial blight of rice. *Proc Natl Acad Sci USA* 103: 10503–10508.

Yang B, White FF (2004) Diverse members of the AvrBs3/PthA family of type III effectors are major virulence determinants in bacterial blight disease of rice. *Molec Plant-Microbe Interact* 17: 1192–1200.

Yang B, Zhu W, Johnson LB, White FF (2000) The virulence factor AvrXa7 of *Xanthomonas oryzae* pv. *oryzae* in a type III secretion pathway-dependent nuclear-localized double-stranded DNA-binding protein. *Proc Natl Acad Sci USA* 97: 9807–9812.

Yang YL, Li JY, Wang JH, Wang HM (2009) Mutations affecting chemotaxis of *Agrobacterium vitis* strain E26 also alter attachment to grapevine roots and biocontrol of crown gall disease. *Plant Pathol* 58: 594–605.

Yanofsky MF, Porter SG, Young C, Albright LM, Gordon MP, Nester EW (1986) The *virD* operon from *Agrobacterium tumefaciens* encodes site-specific endonuclease. *Cell* 7: 471–477.

Yao J, Allen C (2006) Chemotaxis is required for virulence and competitive fitness of the bacterial wilt pathogen *Ralstonia solanacearum*. *J Bacteriol* 188: 3697–3708.

Yishay M, Burdman S, Valverde A, Luzzatto T, Ophir R, Yedidia I (2008) Differential pathogenicity and genetic diversity among *Pectobacterium carotovorum* ssp. *carotovorum* isolates from monocot and dicot hosts support early genomic divergence within this taxon. *Environ Microbiol* 10: 2749–2759.

Young GM, Schmiel DM, Miller VL (1999) A new pathway for the secretion of virulence factors by bacteria: the flagellar export apparatus functions as a protein-secretion system. *Proc Natl Acad Sci USA* 96: 6456–6461.

Young JM, Takikawa G, Gardan I, Stead DE (1992) Changing concepts in the taxonomy of plant pathogenic bacteria. *Annu Rev Phytopathol* 30: 67–195.

Zeng Q, Ibekwe AM, Biddle E, Yang C-H (2010) Regulatory mechanisms of exoribonuclease PNPase and regulatory small RNA on T3SS of *Dickeya dadantii*. *Molec Plant-Microbe Interact* 23: 1345–1355.

Zhang C, Wu H, Li X, Shi H, Wei F, Zhu G (2013) Baseline sensitivity of natural populations and resistance mutants of *Xanthomonas oryzae* pv. *oryzae* to a novel bactericide zinc thiazole. *Plant Pathol* 62: 1378–1383.

Zhang R-G, Pappas T, Brace JL et al. (2002) Structure of bacterial quorum-sensing transcription factor complexed with pheromone and DNA. *Nature* 417: 971–974.

Zhang Y, Bignell DRD, Zuo R et al. (2016) Promiscuous pathogenicity islands and phylogeny of pathogenic *Streptomyces* spp. *Molec Plant-Microbe Interact* 29: 640–650.

Zhang Y, Loria R (2017) Emergence of novel pathogenic *Streptomyces* species by site-specific accretion and *cis*-mobilization of pathogenicity islands. *Molec Plant-Microbe Interact* 30: 72–82.

Zhao Z, Sagulenko E, Ding Z, Christie PJ (2005) Activities of VirE1 and Vir E11 secretion chaperone in export of the multifunctional VirE2 effector via an *Agrobacterium* type IV secretion system pathway. *J Bacteriol* 183: 3855–3865.

Zhou J-N, Zhang H-B, Lv M-F et al. (2016) Sly A regulates phytotoxin production and virulence in *Dickeya zeae* EC1. *Molec Plant Pathol* 17: 1398–1408.

Zhu W, Ma Gabanua MM, White FF (2000) Identification of two novel *hrp*-assoicated genes in the *hrp* gene cluster of *Xanthomonas oryzae* pv. *oryzae*. *J Bacteriol* 182: 1844–1853.

Zhu XF, Xu Y, Peng D et al. (2013) Detection and characterization of bimerthiazol-resistance of *Xanthomonas oryzae* pv. *oryzae*. *Crop Protect* 47: 24–29.

Zipfel C, Felix G (2005) Plants and animals: a different taste for microbes? *Curr Opin Plant Biol* 8: 353–360.

Zipfel C, Kunze G, Chinchilla D et al. (2006) Perception of the bacterial PAMP EF-Tu by the receptor EFR restricts *Agrobacterium*-mediated transformation. *Cell* 125: 749–760.

Zolobowska L, van Gijsegem F (2006) Induction of lateral root structure formation on petunia roots: a novel effect of GMI1000 *Ralstonia solanacearum* infection impaired in *Hrp* mutants. *Molec Plant-Microbe Interact* 19: 597–606.

Zupan J, Muth TR, Draper O, Zambryski P (2000) The transfer of DNA from *Agrobacterium tumefaciens* into plants: A feast of fundamental insights. *Plant J* 23: 11–28.

# 6 Detection and Biology of Soilborne Viral Plant Pathogens

All organisms from unicellular prokaryotic bacteria to highly evolved human beings are affected by diseases caused by viruses. The viruses are obligate, intracellular molecular parasites, requiring the presence of living cells for their development. They were proved to be causal agents of plant diseases only in the late nineteenth century. Plant viruses remain to be elusive and enigmatic causes of several destructive crop diseases defying human efforts to restrict the incidence and spread of diseases induced by them. Although rapid advancements have been made to obtain information on various aspects of host-virus interactions, management of virus diseases is still restricted primarily to the elimination of infected seeds, propagules, and infected plants and to reducing the population of vectors spreading the viruses from infected plants to healthy plants. This situation brings into focus the imperative need for the development of techniques for rapid, reliable, specific and sensitive detection of viruses in planta, soil and water to assess the quantum of inoculum available for infection of different crops susceptible to them (Narayanasamy 2002, 2017).

Plant viruses are ultramicroscopic, molecular pathogens much smaller than bacteria and fungi, with limited variability in particle morphology that can be used as basis for their identification and differentiation. They have a simple structural constitution, primarily composed of a protein coat which provides a protective covering for viral genomic nucleic acid. The viral nucleic acid may be either ribonucleic acid (RNA) or deoxyribonucleic acid (DNA) molecules which carry the information for their pathogenicity and replication in susceptible host cells. Replication of viruses follows a distinctly different process which is not observed in any other organisms. The viral components – coat protein and nucleic acid – are synthesized separately in different sites in susceptible host cells. When these components attain the required concentrations, they are assembled together to form mature progeny virus particles (virions). Plant viruses can be purified from the infected plant sap and stored as a chemical substance under refrigerated conditions. The purified virus preparation, when inoculated on appropriate susceptible plant host, infection occurs, producing symptoms characteristic of the viral nucleic acid, indicating the unique nature of the plant viruses which differ from all other microbial plant pathogens. All plant viruses, except a few, depend on different kinds of vectors, such as insects, mites and nematodes belonging to the animal kingdom and a few viruses are transmitted by fungus-like/fungal vectors from infected plants to healthy plants. Under natural conditions, parasitic dodders have been shown to transmit the viruses from infected plants to healthy plants to some extent. Under experimental conditions, mechanical (sap) inoculation and grafting and budding have been employed for transmitting the viruses. Horticultural propagation methods have been responsible for dissemination of viruses from asymptomatic propagules that are carried from one location to another. Of the large number of viruses infecting plants, some of the viruses are transmitted by soil- and waterborne fungi or nematodes. They infect the plants through roots and the symptoms of infection are expressed much later in the aerial plant parts. It is essential to detect and identify these viruses rapidly and reliably based on their biological, physical, biochemical, immunological and genetic characteristics. The usefulness of various diagnostic techniques employed for detection, differentiation and quantification and biology of soil- and waterborne plant viruses is discussed hereunder.

As the knowledge on physical, biochemical, immunological and genetic characteristics of plant viruses available was previousy limited, they were named based on the host plant and the type of external symptoms produced by the virus e.g., *Tobacco mosaic virus* (TMV). Beijerinck (1898) named the agent causing tobacco mosaic disease as *contagium vivum fluidum* (contagious living fluid). Since then, enormous and fascinating data have been collected by various researchers on these ultramicroscopic, diminutive pathogenic agents, capable of infecting all organisms existing on the Earth. As the number of viruses reported increased rapidly, this system of naming the viruses based on criteria used previously became inadequate, indicating the need for developing a comprehensive system of nomenclature and classification of viruses. The International Committee on Nomenclature of Viruses (ICNV) was established in 1966 to develop a universal taxonomy of viruses. In 1973, the committee was renamed The International Committee on Taxonomy of Viruses (ICTV) to develop systems for naming and the classification of viruses. The ICTV presented a proposal for classifying plant viruses into 20 families and 71 genera with another 17 genera yet to be assigned to appropriate families in due course (Mayo and Brunt 2007). Plant viruses are divided into two groups, based on the nature of viral genome as RNA viruses and DNA viruses. The RNA viruses are further subdivided into single-stranded (ss) + sense RNA viruses, double-stranded (ds) + sense RNA viruses and single-stranded (ss) – sense RNA viruses. The DNA viruses are much less in number and they are divided into ss-DNA viruses and reverse-transcribing ds-DNA viruses. Particle morphology and presence of membranous envelope, immunological properties and sequences of nucleotides of genome, and amino acid sequences of capsid (coat protein) and presence of virus-associated protein have also been included as criteria for the identification and classification of plant viruses and their strains (Mayo and Brunt 2007).

## 6.1   DETECTION OF SOILBORNE VIRUSES

Plant viruses have been detected in the natural ecosystems, including soil, water bodies and rivers, providing water for irrigating crops, in addition to their respective host plants. Some highly stable viruses like TMV, may remain in soil in a free state and infect the susceptible plants, when available. They do not appear to have any biological vectors for their dissemination. Most of the soilborne viruses have either the nematodes or fungal vectors which by themselves cause diseases in addition to being vectors of viruses. These viruses may be grouped into two classes: (i) viruses with abiotic transmission and (ii) viruses with biotic transmission.

### 6.1.1   DETECTION OF VIRUSES WITH ABIOTIC TRANSMISSION

Many viruses, infecting forest trees and some infecting crop plant species lack biotic vectors and are soilborne. Viruses belonging to the genera *Tobamovirus*, *Potexvirus*, and *Tombusvirus* are quite stable in soil. Abiotic soil transmission of *Tomato mosaic virus* (ToMV) and TMV in agricultural and greenhouse settings was demonstrated. But such transmission under natural conditions, is yet to be proven. Tomato plants were infected, when the seedlings were raised in soil infested with ToMV-infected plant debris from a previous crop. Root infection of tomato by ToMV allowed pathogen migration to young shoots where it could be detected. No causal relationship could be established between the microflora and microfauna of glasshouse soils and persistence of ToMV in the soil (Broadbent 1965; Lanter et al. 1982). Similar abiotic transmission of ToMV in tomato plants grown on soil containing infected plant debris to high proportions was observed. Almost all tomato plants were infected by ToMV through roots (Pares et al. 1996).

ToMV is one of the several soilborne plant viruses that spread under natural conditions without any known biotic vector. ToMV is soilborne and waterborne and has a wide host range. ToMV is extremely stable and infects plants through roots. Infection of red spruce by ToMV was observed, when inoculated with purified virus preparation (Hollings and Huttinga 1976; Jacobi and Castello 1992). Büttner and Nienhaus (1989) successfully transmitted members of the genera *Potexvirus*, *Tobamovirus*, *Necrovirus* and *Potyvirus* from German forest soils to herbaceous and/ or woody hosts. About one-third of 284 soil samples tested were positive for one or more of the viruses. It is possible that many European forests have developed on sites where agricultural crops were raised earlier and the viruses detected on forest trees might have originated from infected crop residues (Filhart et al. 1998).

#### 6.1.1.1   Biological Methods

Infectious tobamoviruses present in forest soils of New York State were detected, using elution and bait plant methods. Soils from two forest sites, White Face Mountains (WF) and Heiberg Forest (HF) were tested. Purified ToMV preparation was used to amend the forest soil samples. More ToMV was bound to WF mineral than to organic soil, because water alone

eluted more than 100 ng ToMV from virus-amended organic soil and approximately 50 ng from virus-amended mineral soil. The detection sensitivity of elution method was approximately ten to 20 ng of ToMV/g of soil. *Chenopodium quinoa* was used as bait plant to detect ToMV in soil samples. Abiotic transmission of ToMV to forest trees may result in localized spread and persistence of the tobamoviruses may be possible in forest ecosystems. The bait plant method appeared to be more sensitive than the elution method for the detection of tobamoviruses in forest soils (Filhart et al. 1998). *Cucumber soilborne virus* (CSBV), occurring in Lebanon on cucumber had a number of properties in common with viruses included in the genera *Tombusvirus* and *Dianthovirus*. CSBV was transmitted through infested soil. The biological properties, physical properties of the purified virus particles, presence of isometric virus particles and immunological properties were useful to differentiating the virus from other viruses used for comparison. CSBV produced local lesions on several plant species, but did not infect any of those tested systematically. The virus could be included in the *Carmovirus* group (Koenig et al. 1983). The soilborne nature of the *Lettuce necrotic stunt virus* (LNSV) associated with necrosis and dieback of lettuce was demonstrated by growing lettuce plants in soil samples from dieback-affected fields. The pathogenicity tests, using soil inoculation of lettuce plants with LNSV-infected *N. clevelandii* plant sap resulted in systemic infection of lettuce plants and typical dieback symptoms, including necrotic ringpots on the leaves were formed, suggesting that LNSV might be the cause of the dieback in lettuce (Obermeier et al. 2001). *Pepper mild mottle virus* (PMMoV) infecting pepper (chilli) was shown to have abiotic soil transmission. The extracts of soils infested with PMMoV produced local lesions on *Chenopodium* spp., the numbers of which could be related to the virus concentrations of the soil extracts. The results of infectivity tests were corroborated by the absorbance values of indirect-ELISA tests performed with soil extracts (Takeuchi et al. 2000). Incidence of a virus causing severe mosaic and leaf rugose symptoms on soybean was recorded in Japan. The virus was not transmissible via seeds or aphids species tested. But soybean plants became infected, when grown in virus-infested soil, indicating abiotic transmission of the causal virus. Reverse transcription (RT)-polymerase chain reaction (PCR) assay and phylogenetic analyses showed that the virus was not related to any of the previously described soybean viruses. The virus causing soybean leaf rugose mosaic disease was tentatively named as *Soybean leaf rugose mosaic virus* (SLRMV) (Kuroda et al. 2010).

#### 6.1.1.2   Immunoassays

An Enzyme-linked immunosorbent assay (ELISA) test was applied to detect ToMV in tomato plants grown on soils amended with infected plant debris. The infection level, as determined by visual examination was only 0.1%. But the presence of ToMV was detected in up to 70% of tomato plants by ELISA tests. Immunoelectron microscopy (IEM) revealed the presence of ToMV particles in roots and leaves of tomato plants grown on soils contaminated with infected plant debris

(Pares et al. 1996). *C. quinoa* plants were used as bait plants growing on soils amended with ToMV. The virus was also eluted from organic mineral fractions of soil samples. Necrotic local lesions developed on *C. quinoa* leaves inoculated with the concentrated eluate from HF soil. ELISA was employed to quantify ToMV in the eluates from the forest soils. Roots collected from plants grown in all virus-amended soils tested positive by ELISA. The composite root sample from plants grown in non-virus-amended organic soil, but not mineral soil, also tested positive for ToMV by double antibody sandwich (DAS)-ELISA format. The mean virus concentration within infected roots was approximately 10 ng/g. Transmission electron microscopic (TEM) observations showed the presence of many rigid rods (250–300 nm × 18 nm). A DAS-ELISA test was employed to detect the *Tobamovirus* in the roots of *C. quinoa* plants grown on forest soil samples from WF and HF. No relationship was seen between virus detection and the time of collection of soil samples. Virus was detected in soils collected in June and August (WF organic), December (WF mineral) and February and September (HF). As the red spruce trees have a long life span (about 400 years), they are likely to get infected by the ToMV even if the virus concentration is low, because of the long period of exposure of roots to the virus present in the soil, making them vulnerable to abiotic transmission of virus to healthy trees (Filhart et al. 1998).

PMMoV was detected in extracts of field soils using indirect-ELISA format. Absorbance values were relatively low, when PMMoV-infested soil extracts were tested directly. But heat treatment of the soil extract prior to tests improved the sensitivity of tests as indicated by absorbance values. Heat treatment was found to be essential for efficient detection of PMMoV in the infested soil samples, although the efficiency of virus recovery varied, depending on the soil properties. Infectivity of soil extracts was assessed by inoculating *Chenopodium* spp. which reacted by producing local lesions. The results of immunoassay were similar to those of infectivity tests conducted on indicator host plants. Heat treatment combined with indirect-ELISA format showed potential for use for detection of viruses present in soil samples (Takeuchi et al. 2000). The effectiveness of ELISA formats for detection of viruses in planta as well as in soil and other environmental samples has been demonstrated. DAS-ELISA format was effective in checking the presence of PMMoV in the extracts of soil from fields cropped to green pepper (*Capsicum annuum*) and this format was optimized for the detection of PMMoV in soil samples which are likely to contain inhibitory materials. Positive results obtained with DAS-ELISA were confirmed by inhibition testing, using specific anti-PMMoV antibody, IEM, RT-PCR assay and infectivity tests on indicator host plant species. Levels of soil infestation with PMMoV could be determined by employing a DAS-ELISA test. Based on the results, fields with high infestation levels could be avoided for the cultivation of green pepper (Ikegashira et al. 2004). Rhizomania disease causes serious losses in sugar beet crops. *Beet necrotic yellow vein virus* (BNYVV) associated with rhizomania disease could be detected by planting sugar beet bait plants and an ELISA test was applied to confirm the presence of BNYVV in bait plants. Breakdown of resistance of sugar beet cultivars was recognized by quantifying the virus concentration in sugar beet cultivars by ELISA test. Sugar beet cultivars showing resistance, when grown in BNYVV-infested soil from Salinas, California, became infected. Infection by BNYVV could be confirmed by ELISA test. Variations in the reactions of sugar beet cultivars might be due to the presence of BNYVV isolates with different virulence levels (Liu et al. 2005).

### 6.1.1.3 Polymerase Chain Reaction-Based Assays

ToMV was detected in water and clouds by employing Real-time-PCR assay for amplification of virus-specific product. ToMV was detected in glacial ice subcores < 500 to approximately 140,000 years old from drill sites in Greenland. Subcores that contained multiple ToMV genotypes suggested diverse atmospheric origins of the virus, whereas those containing ToMV sequences nearly identical to contemporary ones suggested that the existing *Tobamovirus* populations may have an extended age structure. Detection of ToMV in ice indicated the possibility that stable human and other viruses might be preserved there, and that entrapped ancient viable viruses might be continually or intermittently released into the present environment (Castello et al 1999). A quantitative RT real-time PCR assay was developed to monitor the health status of environmental irrigation waters. Convective Interaction Media® (CIM) Chromatographic Columns were used to concentrate ToMV present in water samples from several rivers in Slovenia. The sensitivities of DAS-ELISA test and RT real-time PCR in detecting ToMV were compared. The presence of ToMV in the samples from the river Krka and Vipava could be detected after concentrating the virus populations by CIM monolithic chromatographic columns containing positively charged membranes. ELISA and infectivity tests on *Nicotiana glutinosa* and *N. clevelandii* were performed. *N. glutinosa* developed local lesions when mechanically inoculated with water samples. *N. clevelandii* developed systemic symptoms following watering with original water samples (without concentration). Real-time PCR assay was more sensitive than ELISA tests. Real-time PCR assay could detect ToMV in 60 of 78 samples (77%), as against 38 of 78 samples (49%) by ELISA test. The real-time PCR assay was found to be a reliable and rapid method of screening environmental water samples for the presence of plant viruses (Boben et al. 2007). TMV, the type member of the genus *Tobamovirus*, causes heavy losses in tomato crops in Parana State, Brazil. The Sapopema strain of TMV differed from the other strains of TMV in the type of symptoms induced and also it could not infect the Petunia hybrid, described as a host for *TMV* and ToMV. RT-PCR assay was applied, using primers for an internal region within the movement protein (MP) gene of TMV and ToMV. An amplicon of 409-bp cDNA fragment was generated only by the primers for TMV. Deduced amino acids showed 100% identity, when compared to TMV MP and 94% with ToMV. The RT-PCR assay was rapid and provided conclusive results for detecting and differentiating TMV and ToMV infecting tomatoes (da Silva et al. 2008).

Lettuce dieback disease was found to be soilborne, as the disease symptoms developed in lettuce plants grown in soil from infested fields. Further, typical dieback symptoms appeared in lettuce seedlings grown in sterile soil amended with the sap of plants infected by tombusvirus-L2 isolate. The virus from lettuce plants with dieback symptoms inoculated to indicator plants *C. quinoa*, *Nicotiana benthamiana* and *N. clevelandii* caused symptoms indistinguishable from those of the virus isolate L2, indicating the causal nature of the virus. Virus particles measuring 30 nm in diameter were detected under electron microscope in the sap of symptomatic indicator plants that were inoculated with inoculum from greenhouse-grown tomato and field-grown eggplants with malformed leaves . Isolates from lettuce, tomato, sugar beet and eggplant reacted positively with antiserum raised against *Tomato bushy stunt virus* (TBSV)-cherry strain in western blot analysis. Antiserum prepared against the *Tombusvirus* isolate L2 from dieback-affected lettuce did not react in western blots with standard TBSV strains, but was highly specific for the L1, L2, L4, L9 and To3 isolates. The major ds-RNA bands from *N. clevelandii* plants inoculated with sap from infected lettuce leaves or necrotic tomato fruits were similar to those obtained from TBSV- and *Cucumber necrosis virus* (CNV)-infected *N. cleavelandii*. The ds-RNA bands from *N. clevelandii* plants infected with *Tobacco necrosis virus* (TNV) were distinct from those of TBSV, CNV and lettuce and tomato isolates. Isolate To1, which induced attenuated symptoms had one small ds-RNA band, in addition to the three major genomic and subgenomic ds-RNA bands (see Figure 6.1). Based on the genomic and serological properties, the new *Tombusvirus* was named as *Lettuce necrotic stunt virus* (LNSV) (Obermeier et al. 2001).

## 6.1.2 Detection of Viruses with Biotic Transmission

A large number of soilborne viruses cannot exist in a free state in the soil environments. Hence, they need protection offered by plant hosts and vectors living in the soil. Vectors of soilborne viruses are primarily root parasites on different plant species which express the symptoms induced by viruses as well as by vectors. Various species of nematodes are involved in the transmission of more number of viruses than those transmitted by fungus-like and fungal vectors. Generally, nematodes have been reported to be vectors of viruses infecting horticultural crops such as grapevine, strawberry and tomato, whereas viruses infecting agricultural crops such as wheat, barley and peanut are transmitted by fungal vectors. Nematode vectors are all free-living ectoparasites and belong to three genera viz., *Xiphinema*, *Longidorus* and *Trichodorus*. Nematode-transmitted viruses are divided into two groups based on virus particle morphology. Nepoviruses with isometric particles are transmitted by different species of *Xiphinema* and *Longidorus*, whereas tobraviruses with rod-shaped particles are transmitted by *Trichodorus* spp. Fungus-like and fungal vectors of viruses belong to three genera *Olpidium*, *Polymyxa* and *Spongospora*. Isometric viruses are transmitted by *Olpidium* spp., whereas rod-shaped viruses are transmitted

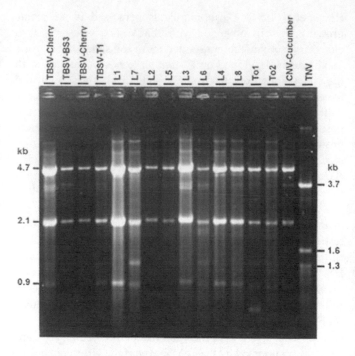

**FIGURE 6.1** Identification and differentiation of tombusviruses based on ds-RNA analysis *Tombusvirus* isolates from lettuce (L1 to L8) and tomato (To1 and To2) in comparison with TBSV-Cherry, TBSV, BS3 and TBSV-T1), CNV-cucumber, and TNV isolates Sizes of genomic and subgenomic ds-RNAs for standard tombusvirus isolates are indicated on the left and for TNV on the right.

**[Courtesy of Obermeier et al. (2001) and with kind permission of the American Phytopathological Society, MN, United States]**

by *Polymyxa* spp. and *Spongospora* sp. (Narayanasamy 2001). Various techniques, based on biological, biochemical, physical, immunological and genetic properties of plant viruses with biotic transmission have been shown to be effective for the detection, identification and quantification of soilborne viruses.

### 6.1.2.1 Viruses Transmitted by Fungal Vectors

#### 6.1.2.1.1 Biological Methods

##### 6.1.2.1.1.1 Infectivity Assays

Bait plants are grown on virus-infested soil for a certain period of time required for the development of visible symptoms characteristic of the virus under investigation. Alternatively, the diagnostic/assay host plants are inoculated with the inoculum prepared from plants/soil/vectors and the development of symptoms is observed. *Indian peanut clump virus* (IPCV) was detected by inoculating leaves of *Phaseolus vulgaris* with extracts of plants showing disease symptoms. IPCV-H (Hyderabad isolate) caused veinal necrosis on inoculated and systemically infected leaves of *P. vulgaris* (Reddy et al. 1998). The diagnostic method, involving use of bait plant assay for trapping zoospores of *Spongospora subterranea* f.sp. *subterranea (Sss)*, the vector of *Potato mop-top virus* (PMTV), was employed. Sss is the causal agent of powdery scab disease affecting potato tubers. PMTV causes spraing disease characterized by brownish arcs and rings on the tuber surface and flesh. As *Sss* is an obligate root parasite, the vector has to be trapped using susceptible host plant species, tomato, as bait plant to detect PMTV in soil suspension. *Sss* and PMTV could be detected by incubating tomato plants for nine days in hydroponic culture. PMTV was detected in soil samples from 137 of 224 fields (61.2%) in Japan. Although PMTV was detected in a high percentage of the fields tested, spraing incidence was limited to a single field and other PMTV-positive fields were still spraing-free, in spite of planting susceptible potato cv. Sayaka in some of these fields included in the investigation. The results indicated that soil infestation with PMTV might not lead to spraing disease incidence (Nakayama et al. 2008). In another investigation, *Nicotiana debneyii* plants were grown as bait plants on putative infested soils from 20–26 locations to detect PMTV. The bait plants showed characteristic V-shaped yellow leaf markings induced by PMTV. The sap from roots and leaves of plants showing symptoms was back-inoculated onto indicator hosts *N. debneyii* and *N. benthamiana*. The root samples of bait plants grown on soils from 25 locations and leaves from 24 locations induced systemic symptoms on indicator plants on back-inoculation (Arif et al. 2014).

*Lettuce big-vein associated virus* (LBVaV, genus *Varicosavirus*) was demonstrated to be the cause of necrotic symptoms observed in combination with big-vein symptoms in lettuce plants by employing the nutrient film technique (NFT). Viruliferous vector *Olpidium virulentus* spores were used to inoculate lettuce seedlings. Lettuce plants showing symptoms of lettuce big-vein disease (BVD) are commonly infected by two viruses viz., *Mirafiori lettuce big-vein virus* (MiLBVV, genus *Ophiovirus*) and LBVaV. Inoculation methods were developed to separate these two viruses and to transfer them to indicator plants and lettuce. Following mechanical inoculation of lettuce plants, vein-band chlorosis, characteristic of MiLBVV and necrotic spots and rings, characteristic of LBVaV developed. The symptoms induced by LBVaV were similar to those observed previously in the NFT system. It was believed previously that LBVaV caused latent infection in lettuce. De Novo next generation sequencing showed that LBVaV was able to induce visible symptoms – necrotic spots and rings – on lettuce and the virus which was named as LBVaV-*associated necrosis virus* (LAN) (Verbeek et al. 2013).

### 6.1.2.1.1.2 Grow-Out Test

Some of the soilborne viruses have been reported to be transmitted through seeds. For the detection of these viruses, seeds from infected plants are sown in the greenhouse and the percentage of plants growing from the seeds showing infection is determined by visual observation of disease symptoms. Transmission of IPCV by fungal vector *Polymyxa* sp. to roots of finger millet (*Eleusine corocana*), fox tail millet (*Setaria italica*) and pearl millet (*Pennisetum glaucum*) resulted in infection by IPCV-H isolate. Frequency of seed transmission of ICPV-H in these graminaceous hosts varied from 0.9% to 9.7%, depending on the host plant species. Transmission of ICPV in the seeds of millets assumes importance, since these crops are grown in rotation or as intercrops with peanut. The infected millet plants might provide inoculum for infection of peanut crops (Reddy et al. 1998).

### 6.1.2.1.2 Physical Methods

Hyper-spectral leaf reflectance and multispectral canopy reflectance techniques were used to differentiate the physical differences between healthy and BNYVV-infected sugar beet plants. Total leaf nitrogen, chlorophyll and carotenoid levels were significantly less in symptomatic plants, compared with healthy plants. Betacyanin levels estimated from leaf spectra showed reduction at the end of season in 2000, but not in 2001. The ratio of betacyanins to chlorophyll, estimated from canopy spectra, was increased in symptomatic sugar beet plants on four of seven sampling dates. Differences in betacyanin and carotenoid levels seemed to be related to disease and not nitrogen content. Vegetative indices calculated from leaf spectra showed reductions in chlorophyll and carotenoids in symptomatic sugar beet plants. Regression models that incorporated vegetative indices and reflectance correctly predicted 88.8% of the observations for canopy reflectance into healthy or symptomatic plants. Differentiation of healthy and BNYVV-infected sugar beet plants was best in August with a gradual decrease in accuracy until harvest. The results indicated the possibility of using a remote sensing technique for detection of rhizomania-infected sugar beet crops (Steddom et al. 2003).

### 6.1.2.1.3 Immunoassays

Different types of immunoassays have been applied for the detection, identification and quantification of plant viruses in plants, seeds, propagules, alternative sources of infection such as weeds, soil, water and other environmental samples. Because of the structural simplicity, antisera specific to a large number of plant viruses has been prepared. The exquisite specificity and sensitivity of mammalian and avian immune systems have been effectively and extensively employed for the development of antisera containing antibodies specific to target viruses. Labeled antibody techniques have been developed by attaching labels or markers such as enzymes, fluorescent dyes or radioactive materials to either antigens or antibodies and these methods are currently used extensively. Use of monoclonal antibodies (MAbs) targeting specific antigenic reactive sites has resulted in the enhancement of levels of specificity, sensitivity, reliability and rapidity, compared to polyclonal antibodies (PAbs) which may cross-react with related viruses. Among the labeled antibody techniques, ELISA and its variants have been shown to be effective in detecting viruses in soil, water and other environmental samples. Immunoassays have been employed for the sensitive detection of soilborne viruses and their fungal vectors (Narayanasamy 2011).

*Wheat soilborne mosaic virus* (WSBMV), infecting winter wheat causes serious losses (up to 50%) in the United States, Italy and Japan. The plasmodiophoraceous fungus-like *Polymyxa graminis* acts as the vector WSBMV, which persists in the vector in infested field (Rao and Brakke 1969).

MAbs generated against WSBMV reacted positively with four isolates of the virus, but not with 13 other viruses tested, indicating the specificity of MAbs in protein A sandwich (PAS)-ELISA format. MAbs were more sensitive than PAbs. Satisfactory level of detection of WSBMV by dot immuno-binding assay could be possible, only if MAbs were employed in the PAS-ELISA tests (Bahrani et al. 1988). *Polymyxa graminis* and *P. betae* have been reported to be the vectors of several viruses belonging to the genera *Benyvirus*, *Bymovirus*, *Furovirus*, and *Pecluvirus*. *P. graminis*, the vector of *Barley mild mosaic virus* (BMMV) was tested for the presence of the virus in the fungal vector cells. Labeled bundles of virus-like particles were detected in the zoospores of P. *graminis* released from barley roots infected by BMMV (Jianping et al. 1991). Presence of labeled virions of BNTVV was observed inside the zoosporangia and zoospores of *P. betae* (Peng et al. 1998).

Incidence of *Soilborne wheat mosaic virus* (SBWMV) was observed on nine wheat cultivars in Zambia, Africa. DAS-ELISA format was employed to identify SBWMV in symptomatic wheat plants using virus-specific antisera. Occurrence of SBWMV along field edges and in poorly drained areas was detected. Waterlogged conditions were conducive for the multiplication and spread of *P. graminis*, the vector of SBWMV (Kapooria et al. 2000). The serological relationships between four members of the genus *Furovirus* (fungus-transmitted rod-shaped virus), *Chinese wheat mosaic virus* (CWMV), SBWMV, *Oat golden stripe virus* (OGSV) and *European wheat mosaic virus* (EWMV) were investigated. MAbs raised against CWMV could be employed to detect and differentiate these viruses. OGSV appeared to be more similar to CWMV than others, since in Western blotting tests, five CWMV MAbs reacted to OGSV and three of them were reactive in ELISA, indicating the presence of at least one common epitope that was sequential and conformational. ELISA tests with MAbs also showed that the conformational antigenic structure of CWMV was partially similar to that of three other furoviruses. Thus, both ELISA and Western blotting showed that the dominant epitopes of SBWMV were totally shared with CWMV, but only partially with OGSV and EWMV (Ye et al. 2000).

*Indian peanut clump virus*-Hyderabad (IPCV-H), causal agent of peanut clump disease, is transmitted by *P. graminis*. *Peanut clump virus* (PCV), occurring in West Africa was found to be serologically and genomically distinct from IPCV-H. ELISA tests efficiently detected IPCV-H isolates from peanut, finger millet, fox tail millet and pearl millet plants growing from seeds collected from infected plants (Reddy et al. 1998). Direct antigen coating (DAC)-ELISA format was effective in detecting the vector of IPCV in the roots of sorghum. The assay could detect even one sporosorus per well of the ELISA plate. The presence of resting spores of *P. graminis* was detected by conjugating specific antibodies to fluorescein-5-isothiocyanate (FITC) in the root sections (Delfosse et al. 2000). In a later investigation, development of PCV in the roots of sorghum was studied. After transfer of *P. graminis* viruliferous zoospores in the root hair or the

epidermis of root cells, the distribution of the virus was traced in the roots, using immunoreactions and *in situ* hybridization assay. The roots showed viral infection of the epidermis, exodermis, endodermis, vascular parenchyma pith and phloem vessels and heavy infection of cortical cells, but the xylem was hardly infected. The coat protein (CP) was present in phloem vessels and companion cells. PCV could infect the vascular parenchyma and pith, when it was present in the endodermis. The results obtained from immunoreactions and *in situ* hybridization were similar and corroborated each other (Dieryck et al. 2011).

Barley yellow mosaic disease, one of the most economically important diseases in northwestern Europe, is caused by BaMMV and *Barley yellow mosaic virus* (BaTMV). Both viruses belonging to *Bymovirus* (family Potyviridae) are transmitted by *P. graminis* (Adams et al. 2009). But these two taxonomically related viruses differ in immunological and genetic characteristics. Strains of BaMMV and BaYMV have been differentiated, based on their pathogenicity on barley cultivars which respond differently. DAS-ELISA format was effective in detecting BaMMV in barley plants growing on infested soil. In contrast, BaYMV could not be detected by DAS-ELISA test. Immunoelectron microscopic observations revealed the presence of virus particles decorated with BaMMV-antibodies only. On the other hand, in preparations with BaYMV antibodies, the virus particles remained undecorated, indicating the effectiveness of immunoassays for detecting BaMMV, but not BaYMV in barley plants (Habekuss et al. 2008). In a later investigation, the effectiveness of DAS-ELISA format for the detection of BaYMV in field samples collected in Iran, was demonstrated. The coat protein (CP) gene region (971-bp long) of eight isolates of BaYMV was sequenced. Based on the phylogenetic analysis of CP gene of BaYMV isolates, the Asian and European isolates could be separated into two main groups. But Iranian isolates formed a minor cluster within the European group. Iranian isolates of BaYMV were found to be more similar to European than to Asian isolates, following alignment of CP amino acid sequences (Hosseini et al. 2014).

The effects of cultivar mixtures on the incidence of *Soilborne cereal mosaic virus* (SBCMV) were investigated. One susceptible (Soissons) and one resistant (Tremie) wheat cultivar were grown in mixtures of 1:1 and 1:3 and in pure stands in a field infested by SBCMV. The presence of the virus was detected, using DAS-ELISA format from January to May. The resistant cultivar did not develop recognizable foliar symptoms and the virus could not be detected in the resistant plants or in roots by DAS-ELISA test. In contrast, about 88% of plants of susceptible cultivar grown in pure stands were infected. The disease reduction relative to pure stands was 32.2% in 1:1 mixture and 39.8% in the 1:3 mixture. Absorbance values in ELISA tests of infected plants in two mixtures were consistently lower than that of the infected plants of susceptible cultivar in pure stands. The results indicated the ability of resistant cultivar to reduce the infection levels in susceptible cultivar when grown as a mixture with resistant cultivar (Hariri et al. 2001). Responses of 21 UK winter wheat cultivars to natural infection conditions were assessed using ELISA tests, in addition to visual assessment of virus disease symptoms. At most heavily infected sites, all UK cultivars except Aarvark, Charger, Claire, Cockpit, Hereward and Xi 19 were infected by SBCMV and these cultivars showed mild symptoms. Novak cultivar tested positive for SBCMV. The results indicated the usefulness of ELISA tests for assessing the levels of resistance of wheat cultivars to the viruses (Budge et al. 2008b).

Sugar beet crops suffer heavily in Turkey due to infection by BNYVV and *Beet soilborne virus* (BSBV) which are transmitted by *Polymyxa betae*. To determine the extent of soil infestation, soil samples collected from 26 fields, where incidence of rhizomania disease was observed, were tested by DAS-ELISA and TAS-ELISA formats. Of the 26 samples, 85.5% and 92.3% were infected by BNYVV and BSBV, respectively. Root samples from beet plants grown in these fields showed the presence of these viruses. Infection rate by BNYVV (67%) was greater than that of BSBV (8%). The vector *P. betae* was detected by using sugar beet bait plants (Yilmaz et al. 2004). In the later investigation, in addition to DAS-ELISA and TAS-ELISA tests, root samples were examined, after staining with lactophenol containing 0.1% acid fuchsin under light microscope. By employing ELISA tests, BNYVV and BSBV were detected in 54 (37.5%) and 53 (36.8%) fields. Single infections with BMYVV (17.4%) or BSBV (16.7%) were detected in 49 fields (34.0%), whereas infection by both viruses was detected in 29 fields (20.1%). Presence of cytosori (resting spores) in root tissues of infected sugar beet plants could be visualized under the light microscope (Yilmaz et al. 2010). In another investigation, a DAS-ELISA test was applied to verify the presence of BNYVV, the primary causal agent of rhizomania disease of sugar beet in 38 of 40 fields surveyed in Greece. BNYVV, the type species of *Benyvirus* genus, is transmitted by the plasmodiophorid vector *P. betae* present in the soil (Pavli et al. 2010).

*Lettuce big-vein virus* (LBVV) transmitted by the fungal vector *Olipidium brassicae* was considered as the cause of the lettuce big-vein (LBV) disease. As a second soilborne virus, *Mirafiori lettuce virus* (MiLV) was also shown to be associated with LBV disease, the need for investigating the causative agent and the vector was realized. Evidence indicated that both MiLV and LBVV were transmitted by *O. brassicae* and that MiLV, not LBVV, was the real cause of LBV disease. These two viruses contained coat proteins of similar size, but had different morphologies and were serologically unrelated. Plants infected by MiLV and LBVV individually in LBV-prone fields and plants from protected crops in France and Italy were tested, using antisera specific to MiLV and LBVV in DAS-ELISA tests. Both MiLV and LBVV were found at high levels, often together, but sometimes separately. Symptoms were frequently found to be associated with MiLV alone or both viruses, rarely with LBVV alone. However, no absolute correlations emerged, because sometimes MiLV was present in the absence of symptoms and vice-versa. During the field surveys in France and Italy, all combinations of MiLV and LBVV were observed: the presence of both MiLV

and LBVV, or the presence of each one alone in plants that developed BVD symptoms. At the end of surveys, nearly all plants had BVD and they contained both viruses. The results indicated the involvement of both viruses in BVD (Roggero et al. 2003). But later, elucidation of the etiology of lettuce big-vein disease was considered essential, since the results obtained earlier remained inconclusive. Specific antibodies to MiLBVV and LBVaV were employed in Western blotting, in addition to host reaction of cucumber and temperature dependence in lettuce as criteria for differentiation of MiLBVV and LBVaV. Virus-free isolates of *Olpidium brassicae* were allowed to acquire these two viruses individually or together. After acquisition, the zoospores were used for serial inoculations of lettuce seedlings 12 times successively. Lettuce seedlings were infected at each transfer either with MiLBVV alone, LBVaV alone or with both viruses together, depending on the virus carried by the vector. Lettuce seedlings inoculated with MiLBVV alone developed big-vein symptoms, while those infected with LBVaV alone did not develop any visible symptoms. During field surveys, MiLBVV was consistently detected in big-vein affected lettuce plants, whereas LBVaV was detected in lettuce plants, not only in big-vein affected fields, but also from big-vein-free fields. The results provided evidence to conclude that MiLBVV, but not LBvaV, was the causal agent of BVD in Japan (Sasaya et al. 2008).

MiLBVV causing BVD is transmitted by *Olpidium virulentus* which is an obligate pathogen and hence not culturable. The resting spores of the vector persist in the soil and retain the virus for several years, making the management of the disease difficult. Viable resting spores of *O. virulentus* were purified from infected roots, using an enzymatic treatment and two-step density gradient centrifugation. Polyhedral antiserum was prepared using the resting spores as immunogen. The antibody was species-specific in a direct immunostaining assay in detecting the presence of resting spores. In Western blotting assay, the antibody-antigen reaction resulted in the formation of two specific bands (ca. 30.5, and 29.0 kDa) (Nomiyama et al. 2013). In a further study, resting spores of *O. virulentus* present in lettuce roots were quantified, using DAS-ELISA with a PAb. The relationship between the resting spore density in soil and disease severity was investigated. The soil was amended with the powder made from diseased roots of lettuce plants with known number of resting spores. Development of disease symptoms was observed, when the number of resting spores was ca. $10^2$ spores/g of soil or more. This procedure showed potential for soil diagnosis and risk assessment of BVD (Nomiyama et al. 2015).

PMTV, with fungus-like vector *Spongospora subterranea*, could be detected by ELISA test. Storing potato tubers at 20°C for three weeks facilitated the detection of PMTV by ELISA test (Sokmen et al. 1998). The PAbs raised against the nonstructural triple gene block protein 1 (TGBp1) of PMTV were employed for the detection of the virus in diagnostic hosts *Nicotiana benathamiana* and potato (*Solanum tuberosum*) by ELISA as well as by Western blotting tests. The antiserum was more suitable for detection of PMTV by Western blots than by ELISA test (Cerovska et al. 2006). DAS-ELISA test was employed for detection of PMTV in 17 weed species commonly present on potato fields in Denmark grown in a hydroponic system infested with viruliferous zoospores of *Sss*, carrying PMTV. Two weed species *Chenopodium album* and *Solanum nigrum* were infected by PMTV, as revealed by ELISA tests and development of disease symptoms, whereas 13 weed species were infected by *Sss*. *C. album* showed systemic infection, the leaves exhibiting necrotic lesions, while *S. nigrum* developed local infection due to infection by PMTV, when sap-inoculated. When *S. nigrum* plants were inoculated using viruliferous zoospores of *Sss*, PMTV was confined to the roots of *S. nigrum* and the virus did not spread to aerial plant parts. *N. benthamiana* was susceptible to both PMTV and the vector which causes potato powdery scab disease (Birgitte et al. 2002). In a later investigation, DAS-ELISA format was employed to identify PMTV present in the bait plants, *N. debneyi* grown in soils from different locations. PMTV was detected in ten potato cultivars commercially grown, using DAS-ELISA tests. Triple antibody sandwich (TAS)-ELISA format was applied to detect zoospores of *S. subterranea* derived from peels of selected scabby tubers. The results indicated the effectiveness of ELISA formats in detecting PMTV in roots and leaves of bait plants, field samples and zoospores of the vector and tubers of commercial potato cultivars (Arif et al. 2014).

### 6.1.2.1.4    Nucleic Acid-Based Assays

Development of effective techniques like nucleic acid-based assays is required to detect the viruses and their soil-inhabiting vectors for obtaining information about various aspects of virus-host interaction, epidemiology and assessment of disease risk prior to planting the crops. *Polymyxa* spp. are involved in the transmission of WSBMV, BaYMV and BNYVV which are responsible for considerable yield losses in various countries around the world. Two protocols of RT-PCR assay were developed for the detection of all SBWMV strains and for specific detection of European SBWMV strain. The RT-PCR procedure detected effectively 21 isolates of SBWMV from different European countries. Both procedures were efficient in detecting the virus in purified total RNA in one- or two- step RT-PCR or immunocapture (IC) RT-PCR formats. IC-RT-PCR assay was more sensitive (> 100 folds) than ELISA test. No amplification occurred with either *Wheat spindle streak mosaic virus* or WYMV which were associated with SBWMV, or with related viruses, IPCV, PMTV and BNYVV. The RT-PCR protocols had the potential for precise identification of the causal virus affecting wheat crops (Clover et al 2001). In a later investigation, among the three buffers tested, EDTA lysis buffer gave the highest recovery of DNA consistently from all the six soils tested. Hence, DNA extracted from three soils infested with *P. betae* and three soils infested with *P. graminis* was analyzed, using the EDTA lysis buffer in combination with a Magnesil™ DNA extraction kit and Kingfisher™ magnetic particle processor. By employing an automated approach to nucleic acid extraction and purification, clear DNA extracts could be obtained. Primers and probes corresponding to sequences within ITS region 2

(ITS2) of ribosomal DNA were designed. The primers and probes enabled recovery and amplification of *P. betae* and *P. graminis* DNA, using real-time PCR and TaqMan chemistry. For *P. graminis*-infested soils, the purity of DNA obtained was sufficient to allow *Polymyxa* DNA to be amplified without dilution to remove inhibitors. On the other hand, for *P. betae*-infested soils, the DNA extracts had to be diluted (1:10) for effective amplification of fungal DNA. Using TaqMan PCR format, a standard curve was constructed from uninfected soil spiked with known numbers of cystosori of *P. betae*. Then the quantity of *P. betae* populations from naturally infested soils was extrapolated from the standard curve. The real-time PCR was rapid and reliable for direct detection and quantification of *Polymyxa* DNA in less than one day. Quantification may be required for epidemiological investigations to determine the number of viruliferous *Polymyxa* cysts in the soil. The procedure developed in this investigation has the potential for providing reliable data for studying cultivar resistance to soilborne virus diseases (Ward et al. 2004).

The SBWMV species was divided into three related new species differing in biological, serological and molecular properties. SBWMV contained American isolates; European isolates were named as SBCMV and virus isolates present in China as CWMV (Koenig and Huth 2000; Adams et al. 2009). SBCMV is transmitted by *P. graminis*. The possibility of SBCMV RNA2 being persistent in seed of winter wheat was examined. Over 7,000 seedlings were raised from seed collected from two cultivars of SBCMV-infected winter wheat in glasshouse. The seedlings were tested by real-time RT-PCR assay. The majority of batches of seedlings tested positive for SBCMV, indicating an RNA2 transmission rate of 1.8–9.4% in wheat. The presence of the virus was confirmed by amplifying and sequencing a larger 400-bp fragment of viral RNA 2 in a subset of the seedlings testing positive by real-time RT-PCR assay. Root extracts from this subset were negative for *P. graminis* using real-time PCR assay. The results indicated that the seedlings grown from seeds carrying viral RNA-2 could be sources of inoculum for transmission through *P. graminis* present in the soil (Budge et al. 2008a). A broad spectrum RT-PCR assay was developed for detection of SBCMV isolates inducing mosaic diseases in European countries. Primers were designed based on the sequence of conserved 3'-untranslated region (UTR) of RNA-1 and RNA-2 of SBCMV. The 3'-end region was a privileged target for the detection of a wide range of isolates, because of the sequence conservation of the tRNA-like structure. In addition, primers were also designed for virus quantification using real-time RT-PCR with SYBR chemistry. No cross-reaction could be seen with *Wheat spindle streak mosaic virus* that frequently occurred along with SBCMV. The procedure adopted was more sensitive than the ELISA test in detecting and quantifying SBCMV isolates from different countries in Europe and the United Kingdom. The real-time RT-PCR assay could be useful for reliable comparison of soil inoculum potential as well as for screening cultivars for resistance to SBCMV (Vaïanopoulos et al. 2009).

PCV is transmitted by *P. graminis* which was classified into five *formae speciales*, based on ecological properties, geographical distribution and rDNA sequence relationships: *P. graminis* f.sp. *colombiana* infecting rice in Colombia; *P. graminis* f.sp. *subtropicalis* infecting barley, maize, sorghum, pearl millet or wheat in the Indian subcontinent and Africa; *P. graminis* f.sp. *temperata* infecting barley or wheat in Belgium, Canada, China, France, Germany and the United Kingdom; *P. graminis* f.sp. *tepida* infecting barley, oat or wheat in Belgium, Canada and the United Kingdom and *P. graminis* f.sp. *tropicalis* infecting maize, sorghum or pearl millet in India and West Africa (Legréve et al. 2002; Vaïanopoulos et al. 2007). *P. graminis* can transmit 14 different plant viruses belonging to the genera *Benyvirus, Bymovirus, Furovirus* and *Pecluvirus* (Torrance and May 1997). Characteristics of transmission of PCV by *P. graminis* formae speciales were studied, using RT-PCR assay and sugarcane as the host for virus and the vector. PCV was detected by RT-PCR in the leaves and root samples of sugarcane inoculated using *P. graminis*. After three weeks of coculture, PCV and *P. graminis* were detected by specific RT-PCR in zoospore suspensions released from the roots of PCV-infected sugarcane plants. In contrast, PCV was not detected in the solution of root exudates from PCV-infected sugarcane plants that had not been infected by *P. graminis*, used as negative control. RT-PCR assay was applied to quantify the vector population and virus titer to assess the effect of virus on vector multiplication and viceversa. The results showed that there was no significant favorable or unfavorable effect of either virus on vector or vector on virus multiplication in sugarcane plants. PCV and *P. graminis* developed independently (Dieryck et al. 2011).

BNYVV, belonging to the genus *Benyvirus*, and transmitted by *P. betae*, causes rhizomania disease in sugar beet. Resting spores of *P. betae* (sporosori) protects BNYVV in soil for more than ten years in the absence of sugar beet plants (Asher 1999). BSBV and *Beet virus Q* (BVQ), belonging to the genus *Pomovirus* were also found to be associated with rhizomania disease and the viruses are transmitted by the same vector *P. betae*. The multiplex (m) RT-PCR procedure was developed for the simultaneous detection of BNYVV, BVQ and *P. betae*, employing combination of four pairs of primers targeting each virus and vector. Detection of *P. betae* repetitive *Eco*RI-like fragment directly in the RNA extract was easier than by the PCR-based procedure employed earlier. The sensitivity of mRT-PCR assay was compared with DAS)-ELISAformat. A commercial kit was used for the dection of BNYVV by DAS-ELISA test. Soil samples from Bulgaria, France, Germany, Hungary, Italy, the Netherlands, Sweden and Turkey were tested. The mRT-PCR amplification profiles showed a systematic association between BNYVV and BSBV or BSBV and BVQ. The most frequently detected virus was BSBV, followed by BVQ and then by BNYVV. BVQ was not present alone in any of the samples tested (see Figure 6.4). The presence of BSBV was detected predominantly in more than 80% of the samples analyzed, with a frequency much higher than that which was observed for BNYVV. The mRT-PCR assay was effective for simultaneous detection of BNYVV, BSBV, BVQ and *P. betae*. The detection threshold of mRT-PCR assay was greater than 128 times that of the DAS-ELISA

test. The mRT-PCR assay provided the advantage of detecting three major soilborne viruses infecting sugar beet and their vector, facilitating estimation of the extent of infestation by the vector and virus disease incidence required for epidemiological and resistance breeding programs (Meunier et al. 2003). In a later investigation, the extent of occurrence of BNYVV, BSBV and their vector *P. betae* was assessed by employing bait plant tests, ELISA and root staining techniques. The percentages of BNYVV and BSBV infested soil samples were 38.0 and 49.8%, respectively. *P. betae* was present commonly (87.7%) in the sugar beet fields sampled. In addition, RT-PCR assay using primers specific for p26 coding region was applied to detect the RNA-5 component of BNYVV. Of the 85 BNYVV-infected samples, BNYVV RNA-5 segment was detected in 62.4% of the samples. BNYVV populations containing RNA-5 was highly widespread in sugar beet production areas in Turkey (Yilmaz et al. 2016).

BNYVV, BSBV and BVQ occurred in mixed infections often, making it difficult to identify the individual virus. RT-PCR assay was performed, using primers for amplifying part of the coat protein (CP) gene and obtaining products of the size expected for each virus. The CP gene sequences were determined for three isolates each of BNYVV, BSBV and BVQ. BNYVV isolates occurring in Poland belonged to the A and B type groups. The CP gene of three isolates of BNYVV had 98–100% identity with comparable sequences available in GenBank. The CP genes of three BVQ isolates were almost identical to each other (98–99%) and contained three additional nucleotides, resulting in an additional amino acid residue (arginine) at position 86 and a serine residue at 115, instead of glycine (Borodynko et al. 2009). Multiplex (m) RT-PCR assay was employed for identification of BNYVV, BSBV and BVQ in sugar beet plant, showing rhizomania symptoms in Greece. The presence of BNYVV, the primary causal agent of rhizomania disease was first ascertained by employing a DAS-ELISA test prior to mRT-PCR assay. The presence of BSBV and BVQ was confirmed by mRT-PCR assay in nine and 23 rhizomania-infected fields, respectively. BVQ was present invariably in dual infections with BNYVV, whereas BSBV was present only in two sugar beet cultivation areas in Greece. Nine of the samples were infected with all three viruses. BSBV was detected in all triple infections (Pavli et al. 2010).

MiLBVV, causing BVD and its vector *O. virulentus* have to be detected directly in the soil. DNA and RNA were extracted from 5-g soil samples, using bead beating procedure, followed by purification through adsorption to a column. Real-time and a TaqMan probe were applied to detect and quantify pathogen and vector genomic nucleic acids. TaqMan probes were prepared based on the CP region of MiLBVV and rDNA-ITS region of *O. Virulentus*, respectively. In addition, using a visual assessment of the incidence rate of BVD of lettuce in agricultural fields, the Ct values of MiLBVV and *O. virulentus* from soil were also determined, using real-time PCR assay. MiLBVV concentrations in the soil were high in the field as determined by real-time PCR assay. Visual assessment of the disease incidence rate showed similar results.

However, the amount of *O. virulentus* in the soil was not correlatable to the incidence of MiLBVV. The results suggested that the amount of MiLBVV content in the soil might be an index of the risk of lettuce crops for contracting BVD (Momonoi et al. 2015). Detection and identification of *O. virulentus*, is essential, since the vector is also pathogenic to lettuce. When the roots of infected plants were tested by PCR assay, using primers specific to the rDNA of ITS region of *O. brassicae* and *O. virulentus*, only *O. virulentus* was detected in the LBV-infected plants. The nucleotide sequences of the complete rDNA-ITS regions of isolates from five of the root samples and ten isolates of *O. virulentus* from Europe and Japan showed 97.9 to 100% identities. But six isolates of *O. brassicae* had identities ranging from 76.9 to 79.4%. LBVaV and MiLBVV were both detected, when symptomatic lettuce tissue samples corresponding to the root samples from southwest Australia were tested, using virus-specific primers in RT-PCR assay. Both viruses were associated with *O. virulentus* (Maccarone et al. 2010). *Melon necrotic spot virus* (MNSV), causing necrotic spot disease of melon occurs in several Latin American and European countries accounting for considerable losses. Of the 112 cucurbit samples, 69 were DAS-ELISA-positive for MNSV. *Olpidium boronovanus* and *O. virulentus* infections and MNSV infection mixed with *Olpidium* spp. were observed in the samples from Gautemala, Honduras, Panama and Spain. MNSV isolates (29) from different locations were detected by RT-PCR assay. Three groups of MNSV isolates were recognized according to geographical origins: (i) EU-LA genotype group (European and Latin American isolates), (ii) JP melon type group (Japanese melon isolates) and (iii) JP watermelon genotype (Japanese watermelon isolates) (Herrera-Vásquez et al. 2010).

PMTV, the type member of the genus *Pomovirus*, is transmitted through spores of *Sss*, the incitant of powdery scab disease of potato tubers. PMTV causes spraing disease (necrotic arcs in tuber flesh) on sensitive potato cultivars and the symptoms are similar to those caused by *Tobacco rattle virus* (TRV), type member of the genus *Tobravirus* transmitted by the nematode, *Longidorus* sp. and potato tuber necrotic strain of *Potato virus Y* (PVY[NTN], *Potyvirus* genus). RT-PCR assay, using primers targeting the coat protein (CP) gene in RNA-3, was applied for detection of PMTV. Of the 3,221 lots of seed and ware potatoes tested, 4.3% were positive for PMTV. The results were confirmed by additional RT-PCR tests, using two primer sets targeting gene segments in RNA-2 and RNA-3. Amplicons generated from RT-PCR assays were subjected to restriction fragment length polymorphism (RFLP) analysis, after digestion with restriction enzymes *Bam*HI and *Hind*III, respectively. The results of ELISA test, bioassay on *N. debneyii* (an indicator host) and transmission electron microscopy (TEM) corroborated those from RT-PCR-RFLP analysis (Xu et al. 2004). Isolates of PMTV (37) detected in tubers from fields in Finland and screenhouse in Latvia, had two distinguishable types of RNA-2 and RNA-3, each of which showed only little genetic variability. Sequencing and RFLP analysis of PCR amplicons indicated that the majority of PMTV isolates infecting tubers comprised restrictotypes RNA2-II

and RNA3-B. Asymptomatic infection of potato tubers varied depending on cultivars and years (2006/2007). Relying on visual inspection of seed tubers alone could lead to the escape of large percentages of infected tubers and the risk of spreading PMTV into new locations. No specific combination of the types of RNA2 and RNA3 was associated with spraing-expressing or symptomless tubers. Sprout sap of tubers might be more suitable for testing by ELISA for PMTV detection than tuber sap inoculation procedure (Latvia-Kilby et al. 2009). A soil diagnostic technique consisting of a bioassay using tomato as bait plant to trap the vector of PMTV, *S. subterranea*, followed by RT-PCR-microplate hybridization (MPH) was developed to detect PMTV in root hairs of bait plants. After incubation of tomato seedlings grown with their roots immersed in soil suspension at 18°C for nine days, the total RNA extracted from bait roots, was analyzed by RT-PCR-MPH, employing PMTV-specific primers and a digoxigenin (DIG)-labeled probe. Soil diagnosis to map PMTV-infested field showed that 137 of 224 of the fields (61.2%) were infested by PMTV, although tubers harvested from only one of these fields showed incidence of spraing symptoms in Japan (Nakayama et al. 2008, 2010).

PMTV isolates occurring in China and Canada, were characterized after sequencing the complete genome comprising three genomic nucleic acids, RNA1, RNA2 and RNA3. Two open reading frames (ORFs) were present in RNA1 of 6.1 kb encoding a readthrough RNA-dependent RNA polymerase (RdRp). A coat protein (CP)-readthrough protein was encoded by RNA2 of 3.1 kb. Four ORFs encoding the triple gene block proteins (TGBps) and a cysteine-rich protein (CRP) were detected in RNA3 of 2.9 kb of Chinese isolate 'Yunnan', whereas only three ORFs encoding TGBps were present in RNA3 of three Canadian isolates and one Chinese isolate. Based on phylogenetic and sequence similarity analysis of these isolates, each of RNA1, RNA2 and RNA3 could be divided into at least two groups A and B. A duplex RT-PCR was developed and evaluated for differentiating A and B groups of RNA3. All PMTV samples collected in Guangdong, China and New Burnswick, Canada contained RNA3 belonging to group A, whereas the samples collected in Yunnan, China possessed RNA3 belonging to group B (Hu et al. 2010). PMTV was detected in infested soil using bait plants such as *N. benthamiana* which showed diagnostic systemic symptoms and also by immunoassays and PCR assay in the roots of bait plants. The transmission of PMTV from infected seed tubers to foliage and daughter tubers was investigated under field conditions by employing real-time RT-PCR soil assay. PMTV was detected in potato soil samples. The relative importance of infected seed tubers derived from PMTV-infected tubers was assessed. The rate of transmission of PMTV from infected seed tubers to daughter tubers ranged from 18 to 54%. Transmission was influenced by cultivar and/ or origin of seed tubers for a cultivar, but not by the sensitivity of the cultivar to PMTV infection. The incidence of PMTV in daughter tubers was not correlated with that in the seed tubers, but appeared to be strongly associated with soil inoculum. Detection of PMTV, using RNA2 primer set in real-time RT-PCR assay was successful in all soil samples, whereas RNA3 primer set did not detect PMTV in 23.6% of soil samples. In 2004 and 2006, PMTV was detected in all fields which produced crops with > 10% infection by PMTV. The presence of PMTV could be detected also in some soils producing crops in which daughter tubers showed no infection. The incidence of PMTV was related to infection of crops infected by both powdery scab and potato mop-top pathogens. The results indicated that planting PMTV-infected seed potatoes could increase the risk of introducing the virus in PMTV-free soils (Davey et al. 2014).

Incidence of severe mosaic and leaf rugose symptoms in soybean plants was observed in Japan in 2005. Flexuous and filamentous virus-like particles with a nodal length of 500 nm were present in infected soybean plants. Double-stranded (ss)-RNA analysis revealed a single band of ≈7,000-bp. RT-PCR assay for Potyviridae diagnosis generated a 1.5-kbp cDNA fragment, which contained a partial NIb gene, a coat protein (CP) gene, a 3′-noncoding region and a poly (A) tract. The CP gene coded for 248 amino acids and had less than 40% amino acid sequence identity with CPs of other member of Potyviridae. The C-terminal region of the NIb gene had ≈ 60% amino acid sequence identity with BMMV and other *Bymovirus* isolates. The CP of the virus had no motifs associated with aphid transmission. Phylogenetic analyses of amino acid sequences of the CP region showed that the virus differed from other members of Potyvirus. The virus was not transmitted via seeds or aphids. But soybean plant grown in virus-infested soil became infected. As this virus was not related to any of the viruses in the genus Potyvirus, it was tentatively named as Soybean leaf rugose mosaic (SLRMV) (Kuroda et al. 2010).

*Japanese soilborne wheat mosaic virus* (JSBWMV) and CWMV are rod-shaped viruses that are transmitted by *P. graminis* in soil. Later JSBWMV was classified as a distinct species and included under the genus *Furovirus*, based on the genome sequence analysis. WYMV is another important virus infecting wheat in Japan. In order to detect and differentiate those three wheat-infecting viruses simultaneously, a differential detection (DD)-RT loop-mediated isothermal amplification (LAMP) procedure was developed. This method involves monitoring fluorescence during DNA amplification and annealing, making it possible to detect three viruses simultaneously from wheat or barley leaf samples in one tube within one hour. All three primer sets, which were designed, based on genome sequences of WYMV, JSBWMV and CWMV, respectively, were most effective at 65°C and could detect each virus RNA within ten min by fluorescence monitoring, using an isothermal DNA amplification and fluorescence detection device. Further, these primer sets showed unique annealing curves. The peak denaturing temperatures of WYMV, JSBWMV and CWMV primer sets were 87.6°C, 84.8°C and 86.4°C, respectively and these viruses could be clearly distinguished by isothermal DNA amplification and fluorescence detection device. The RT-LAMP assay developed in this investigation including three primer sets was 100 times more sensitive than RT-PCR assay for WYMV and JSBWMV and as sensitive as RT-PCR for CWMV.

The RT-LAMP procedure was validated for the simultaneous detection of the viruses in wheat and barley leaves. The RT-LAMP was superior because of its high sensitivity along with the advantage of DD of three viruses infecting wheat and barely (Fukuta et al. 2013).

### 6.1.2.2 Detection of Viruses Transmitted by Nnematodes

Nematodes are involved in the transmission of plant viruses, in addition to their ability to induce different diseases. Nematodes belonging to the genera *Xiphinema*, *Longidorus* and *Trichodorus* have been demonstrated to be vectors of several plant viruses. *Grapevine fan leaf virus* (GFLV) was the first plant virus demonstrated to be transmitted by the ectoparasitic nematode *Xiphinema index* (Hewitt et al. 1958). The nematode-transmitted viruses have several properties in common and they are divided into two genera *Nepovirus* and *Tobravirus*, based on the virus particle morphology (Mayo and Brunt 2007). Nepoviruses have spherical (isometric) particles, while tobraviruses have rod-shaped particles. Nepoviruses are vectored by *Xiphinema* spp. and *Longidorus* spp., whereas tobraviruses are transmitted by *Trichodorus* spp. and *Paratrichodorus* spp. Many nematode-transmitted viruses are readily transmitted by mechanical (sap) inoculation and some are seed- and pollen-borne in many host plant species, especially weed plants which are of epidemiological significance (Narayanasamy 2011).

#### 6.1.2.2.1 Biological Methods

Infectivity of nematode-transmitted viruses is tested by inoculating the primary hosts and diagnostic host plant species. GFLV was transmitted from infected grapevine to *Chenopodium amaranticolor* and back to grapevine, using *X. index* as the vector, fulfilling Koch's postulates to prove the causal nature of the virus and the ability of *X. index* to transmit GFLV (Hewitt et al. 1962). Typical symptoms of fan leaf were induced by graft transmission from GFLV-infected *C. amaranticolor* and also GFLV was sap-inoculated successfully on *C. amaranticolor*, using the inoculum from grapevine (Dias 1963). GFLV was readily transmissible by sap inoculation to herbaceous species in the families, Amaranthaceae, Solanaceae and Fabaceae (Horvath et al. 1994). The ability of GFLV to infect graminaceous plant species was reported. Infection of Bermuda grass (BG) by GFLV was detected by employing ELISA test, using antiserum specific for a North American isolate of the virus and the RT-PCR assay, using GFLV-specific primers. The virus induced few or no symptoms in BG. However, the virus could be transmitted from BG to *C. quinoa* by mechanical inoculation. The diagnostic tests confirmed the presence of GFLV in *C. quinoa* plants (Izadpanah et al. 2003). Nepoviruses *Tobacco ringspot virus* (TRSV) and *Tomato black ring virus* (TBRV) were isolated from tomato plants and inoculated onto additional plant species, in addition to tomato. Infected tomato plants developed systemic symptoms such as yellow discoloration of leaves with brownish veins and downward rolling of leaves, whereas *N. glutinosa*, *N. rustica*, *Chenopodium ambrosioides* and

*Gomphrena globosa* developed necrotic local lesions with bright center and brown rings around them. Similar necrotic local lesions were formed on four other assay plant species. Another nepovirus TBRV induced local lesions on *G. globosa*, *C. amaranticolor*, *C. murale* and *C. quinoa*. *N. glutinosa* showed systemic mottling, while *N. rustica* and tobacco cv. Samsun developed necrotic ring spots. TRSV and TBRV could be differentiated, based on the symptoms induced by them on these host plant species (Šneideris et al. 2012).

The genus *Tobravirus* includes three species viz., Tobacco rattle virus (TRV), *Pepper ringspot virus* (PepRSV) and *Pea early-browning virus* (PEBV) which have wide host ranges including economically important crops such as *Beta vulgaris, C. annuum, Hyacinthus* sp. *Narcissus pseudonarcissus, Nicotiana tabaccum, Pisum sativum, Phaseolus vulgaris, S. tuberosum, Spinacia oleracia, Tulipa* sp. and *Viola arvensis* (Brunt et al. 1996a, b). TRV, is capable of infecting many crops, including ornamental plants as well as weed species. TRV is the type member of the genus *Tobravirus*, family Virgaviridae. TRV causes oak-leaf patterns and ring spots on leaves of bleeding heart (*Dicentra spectabilis*). TRV was transmitted from bleeding heart to *N. benthamiana*, *C. amaranticolor* and *C. quinoa* by sap inoculation (Robertson 2013). Presence of TRV in the nematode vector was detected by homogenizing viruliferous nematodes and inoculating the extract onto indicator plants (Sanger et al. 1962). PCR assay was employed later to identify TRV in viruliferous nematodes (Van der Wilk et al. 1994).

### 6.1.2.2.2  Immunoassays

Immunoassays have been employed to detect and visualize the viruses in their nematode vectors. Thin-section electron microscopy was used to locate the virus particles in the nematode tissues and to determine the sites of virus retention for viruses transmitted by nematodes (McGuire et al. 1970; Raski et al. 1973). Immunosorbent electron microscopy was effective in detecting six nepoviruses in the homogenized suspension of their nematode vectors (Roberts and Brown 1980). Indirect immunofluorescence technique was developed to localize TRSV in the vector *Xiphinema americanum*. Known populations of *X. americanum*, capable of transmitting TRSV efficiently were given acquisition access for ten days on TRSV-infected cucumber plants. Treatment of fragments of viruliferous nematodes with TRSV-specific antiserum, followed by fluorescein isothiocyanate (FITC)-conjugated goat anti-rabbit immunoglobulin G, resulted in virus-specific bright fluorescence only in the lumen of the stylet extension and esophagus. Virus-specific fluorescent signals were detected and the virus-specific fluorescence increased as the acquisition access period (AAP) increased from 0 to 22 days. The increase in AAP and transmission efficiency of nematode population showed positive correlation. The indirect immunofluorescence procedure provided the possibility of visualizing the entire virus-retention region in individual nematode within a population functioning as vector/nonvector of TRSV (Wang and Gergerich 1998). TRV was detected in the nematode vector *Paratrichodorus*

*anemones* by employing immunogold labeling technique. TRV particles were found attached to the cuticle lining of the posterior tract of pharyngeal lumen of *P. anemones* (Karanastasi et al. 2001). TRSV and *Tomato ringspot virus* (TomRSV) transmitted by *X. americanum* were detected in the nematodes by employing immunofluorescent labeling and electron microscopy techniques. TRSV and TomRSV were localized in different regions of the food canal of *X. americanum*. TRSV was primarily localized to the lining of the lumen of the stylet extension and esophagus, but only rarely in the triadiate lumen, whereas TomRSV was localized only in the triadiate lumen. The results indicated the differential tissue preference for virus retention, probably determined by the nematode-virus combination (Wang et al. 2002).

GFLV particles have been shown to be relatively good immunogens, as indicated by its ability to generate PAbs in antisera with titers of 1/20,000 (Etienne et al. 1990). GFLV is a serologically homogeneous virus, as revealed by MAbs (Huss et al. 1987). GFLV was distantly related serologically to *Arabis mosaic virus* (ArMV) (Cadman et al. 1960). GFLV could be detected in a graminaceous host plant species, BG and also in the diagnostic host *C. quinoa*, employing antiserum specific for a North American isolate of GFLV in ELISA tests (Izadpanah et al. 2003). GFLV could be routinely detected by ELISA tests in extracts of leaves, rootlets, cortical scrapings from mature canes and petioles of infected grapevines (Andret-Link et al. 2004a). DAS-ELISA format and RT-PCR assay were compared for their efficacy in detecting GFLV in grapevine plant samples collected during a survey conducted in Iran. GFLV was detected in 31 of 134 samples by DAS-ELISA tests. Infection of *C. quinoa* by GFLV was confirmed by DAS-ELISA. RT-PCR assay using two primer sets targeting CP gene of GFLV did not detect the virus in 22 ELISA-positive samples. Further, no ELISA-negative sample tested positive by RT-PCR assay. The lower levels of virus detection by RT-PCR assay might be due to the presence of PCR inhibitors in the plant samples (Bashir and Hajizadeh 2007).

### 6.1.2.2.3  Nucleic Acid-Based Techniques

GFLV is transmitted by *X. index* from infected grapevine plants to healthy plants under natural conditions. The nematodes remain viruliferous for a long time and they can acquire GFLV not only from infected plants, but also from root debris scattered in the soil after the removal of infected grapevine plants (Raski et al. 1973). RT-PCR assay, employing two different pairs of primers specific to GFLV, detected the virus in both fresh and dried bermuda grass (BG) tissues and also in virus preparations purified from infected plants. Cloning and sequencing of the RT-PCR amplicons confirmed that the amplified sequences were sections of GFLV coat protein (CP) gene. RT-PCR assay efficiently detected GFLV in the diagnostic host *C. quinoa* plants which were inoculated with the inoculum from BG (Izadpanah et al. 2003). *Chenopoidium quinoa* plants were mechanically inoculated with GFLV and the virus purified from the extracts of *C. quinoa* was used to raise PAbs against GFLV. The antiserum was used to detect

GFLV in grapevine samples by employing a DAS-ELISA test. ELISA-positive samples were subjected to oligoprobe-RT-PCR to amplify a 606-bp product from viral CP gene. PCR products were analyzed by restriction RFLP procedure, after digestion with endonuclease *Alu*I. Three restriction profiles were differentiated among the GFLV strains tested. RFLP data enabled the clear distinction of two GFLV strains from Tunisia. The nucleotide sequences of strains showed 93.4% similarity at amino acid sequence level, compared with that of the French strain GFLV-F13 of GFLV (Fattouch et al. 2005).

In order to assess the nematode populations (*X. index*) that might serve as sources of inoculum for GFLV, real-time nested RT-PCR assay was performed on *X. index* collected from the rhizosphere of GFLV-infected grapevine at Palagiano, Italy. The 1,157-bp fragment of GFLV RNA-2 coat protein (CP) gene was amplified and sequenced. A fluorescent Scorpion probe was designed to detect a highly conserved CP region. A second region with isolate-specific multiple nucleotide polymorphisms was used to detect GFLV isolates, using molecular beacons (MB). The Scorpion probe permitted quantitative estimation of GFLV RNA-2 in single nematode, using a dilution series of a 692-nucleotide transcript of the CP gene. The real-time nested RT-PCR format allowed detection of GFLV RNA-2 in individual *X. index* with a minimum template threshold of 800 fg or $2.8 \times 10^6$ RNA-2 molecules/nematode. Real-time nested RT-PCR assay showed higher sensitivity in detecting the target virus, when compared to RT-PCR alone, as suggested by lower Ct values. However, detection efficiency was affected, since in both formats, positive signals were obtained in the presence of GFLV. This technique has the potential for investigating the molecular biology of virus transmission by nematode vectors (Finetti-Sialer and Ciancio 2005).

The *in vitro* dual culture system was developed for GFLV transmission via *X. index* for assessing levels of resistance of grapevine lines to GFLV, since field testing over several years was found to be laborious, expensive and time-consuming. The dual culture system required less time and space than inoculation experiments carried out in the greenhouse. GFLV infection of *in vitro* grapevines was assessed using IC-RT-PCR assay. The development of root galls in grapevines induced by feeding nematodes *in vitro* was determined by inoculating 20 rootstock plantlets with 20 nematodes per jar after three weeks. Virus infection in grapevines in the dual culture inoculation, using viruliferous nematodes was detected at six weeks postinoculation. Roots of 14 plants tested positive for GFLV using IC-RT-PCR assay. Six plants with infected roots also revealed systemic symptoms in the shoot. Root galls were always absent from parasitized *in vitro* grapevine plants with detectable virus titer, whereas root galls developed on some parasitized, but virus-negative tested grapevine. The results indicated that presence of root galls might not be a reliable indicator for parasitism and virus transmission. The dual culture system might be effective in assessing resistance of grapevine lines to GFLV using the natural vector of the virus (Winterhagen et al. 2007).

Nucleic acid-based amplification assay (NASBA) amplification of RNA and simultaneous detection of the amplified products by means of a molecular beacon probe are combined in AmpliDet RNA procedure. Isothermal AmpliDet RNA system was applied for the detection of nematode-transmitted nepoviruses infecting fruit crops in *in vitro* generated plantlets. Detection systems for the detection of ArMV, *Raspberry ringspot virus* (RpRSV), *Strawberry latent ringspot virus* (SLRSV), TBRV and ToRSV were formulated. Based on sequence homology of RNA-2, five different virus-specific primers were designed. A large number of virus isolates from the plant virus collection of Plant Research International, were detected by NASBA and MBs. Using probes with different fluorescent labels, it was possible to simultaneously detect ArMV/SLRSV or RpRSV/TBRV. By simultaneous amplification and detection of RNA in a single reaction tube, carryover contamination introduced by gel analysis and northern blot methods, could be prevented. The results indicated the usefulness of AmpliDet RNA procedure for routine diagnosis of viruses in propagation materials. However, the high costs of the technique might preclude its wider application (Klerks et al. 2001). In a later investigation, SLRSV transmitted by the nematode *Xiphinema diversicaudatum*, infecting olive trees was found to be widespread reducing the yield considerably in the European countries. The one-step RT-PCR assay using primers 5D and 3D amplified a 293-bp product from plant samples infected by SLRSV. The primers showed high specificity to the virus, since no amplification occurred in negative or water control samples. The one step RT-PCR assay was very rapid and sensitive and cross-contamination among samples could be avoided by performing the assay in a single tube (Faggioli et al. 2002). Incidence of SLRSV and ArMV on rose and lily crops was observed in Himachal Pradesh, India. Both SLRSV and ArMV were detected in three rose cultivars, whereas only SLRSV could be detected in three lily cultivars. In order to detect the presence of viruliferous *X. diversicaudatum* and *Longidorus microsoma*, they were collected from the rhizosphere of rose and lily cultivars and the total RNA from the nematodes was extracted. RT-PCR assay was performed on the total RNA extract, using SLRSV- and ArMV-specific primers. Both viruses could be detected in the total RNA extracts. *Cucumis sativus* was used as bait plant to confirm the nematode transmission of SLRSV and ArMV (Kulshreshta et al. 2005). A one-step RT-PCR procedure was employed for detection of *Cherry leafroll virus* (CLRV) and SLRSV in apricot and peach plants maintained in the screenhouse. The RT-PCR assay was also effective in detecting *X. diversicaudatum* carrying SLRSV in vineyard soils in Italy. The one-step RT-PCR was rapid and sensitive for reliable diagnosis of CLRV and SLRSV infections (Kumar 2009).

*X. americanum* is involved in the transmission of nepoviruses ToRSV and TRSV, whereas *Paratrichodorus allius* is the vector of TRV, the type member of the genus *Tobravirus*. A procedure to extract viral RNA from nematodes present in the soil was developed using collagenase, a commercially available enzyme. A sensitive nested RT-PCR assay was developed for the detection of ToRSV and TRSV in *X. americanum* and TRV in *L. allius*. Detection of TRSV using nested RT-PCR assay in groups of ten and 50 nematodes was achieved. The intensity of bands (257-bp) was comparable with that obtained from TRSV-infected cucumber leaves. Likewise, the amplicon 707-bp from viruliferous *P. allius* carrying TRV was comparable to the one present in TRV-infected tobacco leaves. Viruliferous nematodes carrying ToRSV produced an amplicon of 435-bp, after nested RT-PCR amplification. The nested RT-PCR was much faster than the bioassay method, although the nested RT-PCR required intensive labor for handpicking the nematodes. The amount of virus retained by nematodes appeared to be very different for various nematode-virus combinations, since *X. americanum* retained TRSV in the stylet and esophagus, whereas ToRSV was present in the posterior part of the esophagus. ToRSV was detected in the nematodes throughout the season with similar frequencies by RT-PCR assay and transmission bioassay (Martin et al. 2009). Total RNA extracted from infected tomato fruits and infected assay hosts, *N. glutinosa* and *Datura stramonium* was subjected to one-step RT-PCR assay. TBRV was detected based on the amplification of specific product of 348-bp from infected plants. Likewise, one-step RT-PCR assay was effective in detecting TBRV. The virus-specific product of 221-bp was amplified from RNA extracts of tomato and test plants *D. stramonium*, *C. quinoa* and *Nicotiana debney* (Šneideris et al. 2012).

TBRV isolates from tomato, cucumber, potato and black locust plants, showing differences in the type of symptoms induced and host range, were characterized, using RT-PCR assay combined with restriction analysis of almost full-length cDNA derived from both viral genomic RNA-1 and RNA-2. RT-PCR amplification produced four products, representing RNA-1 and RNA-2 of each isolate. Restriction analysis of RT-PCR products showed variations among TBRV isolates. A higher diversity in the restriction patterns was observed for RNA-1 and the restriction patterns were more diverse for RT-PCR products corresponding to the genes in the 5' end proximity of both genomic RNAs. RNA-1 of TBRV encodes all proteins considered to be highly conserved (like RNA-dependent RNA polymerase or protease), while RNA-2 encodes capsid and MPs. Most of the TBRV isolates occurring in Poland showed the presence of additional RNA molecules associated and encapsidated with viral RNAs. The relationships between isolates of TBRV were established, based on the phylogenetic analysis (Jończyk et al. 2004).

Nepoviruses are transmitted by different species of *Xiphinema* or *Longidorus*. The identity of the vector species has to be established precisely for studying various aspects of nematode vector-virus relationship. Species-specific diagnostic primers from ribosomal genes for *X. diersicaudatum* and *X. index*, vectors of ArMV and GFLV and SLRSV, were employed. Specific PCR products were amplified with respective *Xiphinema* spp. No PCR product was observed, when the primers were tested against six *Longidorus*, one *Paralongidorus* and one *Xiphinema* nontarget species. Further, when challenged with a range of nontarget nematode species, comprising the nematode community typical of viticulture soil, no PCR product was amplified, indicating the specificity of the PCR primers employed (Hübeschen

et al. 2004). *Longidorus* spp. are involved in the transmission of the nepoviruses, RpRSV and TBRV, infecting grapevines. Species-specific primers were designed from rDNA for all species of *Longidorus*. Multiplex PCR assay employing a common forward primer combined with species-specific reverse primers, enabled detection of *L. attenuatus*, *L. elongatus* and *L. macrosoma* in the same PCR reaction. All primers were highly specific. Detection of all nematode developmental forms from disparate populations was carried out. The multiplex PCR assay was significantly sensitive to detect a single target nematode within a whole nematode community typical of vineyard soil comprising of a range of nontarget species (Hübschen et al. 2004). Real-time PCR assay was developed for specific and reliable detection of *X. index*, and *X. diversicaudatum*, vectors of GFLV and the nonvectors present in the vineyard soil. Specific primers and TaqMan probes were designed from rDNA internal transcribed spacer 1 (ITS1), enabling specific detection of a single individual of each of *X. index*, *X. diversicaudatum* and nonvectors *X. italicae* and *X. vuittenezi*, whatever might be the nematode population. The specificity of detection and absence of false positive reaction were confirmed in samples of each species mixed with the three other *Xiphinema* species or mixed with nematodes representative from other genera. The real-time PCR assay was found to be an effective alternative to the conventional time-consuming microscopic identification and counting nematodes (Van Ghelder et al. 2015).

In order to establish the involvement of *Paratrichodorus pachydermus* and *Trichodorus similis* in the transmission of TRV, the real-time fluorogenic 5′ nuclease PCR (TaqMan) technique was applied. Two independent primer probe sets were designed targeting the 18S gene of the ribosomal cistron of the vector species. Relative quantification of target DNA present in unknown samples to those obtained from plasmid standard dilutions was carried out. In addition, three more primer/probe sets were also employed to target TRV: one set for RNA-1 and two other sets for RNA-2 of specific isolates. By employing the protocol developed in this investigation, both vector species and TRV RNA-1 and RNA-2 could be detected in the field soil samples. This assay provided accurate and sensitive molecular data rapidly for disease risk assessments under field conditions (Holeva et al. 2006). *Raspberry bushy dwarf virus* (RBDV), transmitted by *Longidorus juvenilis*, infects *Rubus* species worldwide and many infected *Rubus* spp. and cultivars remain asymptomatic. RBDV infection of grapevine was first observed in Slovenia in 2003. Hence, the grapevine isolates of RBDV were characterized biologically, serologically and genetically for comparison with *Rubus* isolates. The nematodes were extracted from fresh soil samples by the whirling-motion method and the total RNA was extracted using RNeasy Plant Mini Kit (Qiagen). Nested RT-PCR procedure was applied for detection of RBDV in the nematode extracts. Specific amplification products representing RBDV were detected in *L. juvenilis*. The virus could be detected in the fresh soil samples and also after storage for four to eight months in the refrigerator. The PCR products were sequenced for confirming the nested RT-PCR results.

In raspberry, seed transmission of RBDV was up to 77% of the seedlings growing from red raspberry seeds, when both parents were infected by the virus. But RBDV was not transmitted via seeds of grapevine plants infected by the virus. Grapevine-infecting isolates of RBDV were found to be widespread in Slovenia (Pleško et al. 2009).

RT-PCR assay was employed for the detection of TRV in potato tubers. Primer A complementary to residues 6555 to 6575 and primer B, identical to residues 6113 to 6132 of TRV-SYM RNA-1 were used. After amplification, the TRV-specific amplicon of 463-bp was cloned, sequenced and compared to sequences of TRV isolates from Europe and Canada. Nucleotide sequence homology of the 16 kDa ORF between the Florida TRV isolate and the isolates from other countries varied between 90% and 94%. A nonradioactive biotin-labeled probe detected TRV by tuber tissue blotting assay. TRV was also detected by using tobacco cv. Samsun, *Nicotiana clevelandii* and Petunia hybrida plants, as well as by ELISA test (Pérez et al. 2000). The TRV isolate from the Netherlands capable of causing spraing disease in tubers of potato cv. Bintje was characterized. The isolate contained two phenotypically distinguishable RNA-1 variants, one of which carried the determinant for the ability to cause spraing in cv. Bintje. The nucleotide sequence of the coding region of this RNA-1 was determined and found to differ at 5.2–5.4% of positions from other TRV RNA-1 sequences in the database. The amino acid sequences of the predicted translation products between 92 and 99% were identical to those of a TRV RNA-1 that did not cause spraing in cv. Bintje (Robinson 2004). Commercially processed potato chips showed dark brown arcs and rings typical of corky ringspot disease induced by TRV in tubers. RT-PCR assay was applied to confirm the presence of TRV in the processed potato chips. A portion of TRV RNA-1 was amplified by RT-PCR in each of the eight discolored chips from three different bags purchased at three locations. Sequence analysis of the 463-bp amplicon showed that the products were of TRV in origin and were 97% identical to the TRV sequences of isolates originating from Washington, Florida and Wisconsin. However, extracts from symptomatic chips failed to infect tobacco plants, indicating the loss of infectivity of TRV after processing at about 160°C temperature (Crosslin 2009). Multiplex RT (mRT)-PCR assay was evaluated for simultaneous detection of TRV, along with *Potato yellow vein virus* (PYVV) and *Tomato infectious chlorosis virus* (TICV) infecting potatoes. A plant internal amplification control (PIAC) was also included to enhance the reliability of the results of the assay. Primers specific to individual viruses and PIAC were evaluated for amplification of the targets and then optimized by adding bovine serum albumin (BSA) during cDNA synthesis, increasing the PCR extension time, reducing the PCR extension temperature and optimizing concentration of each pair of primers. The multiplex RT-PCR format showed the same detection sensitivity of the same order as standard RT-PCR format. The multiplex RT-PCR assay was found to be a reliable and sensitive procedure for simultaneous detection of three potato viruses that could be spread through tubers to new geographical locations (Wei et al. 2009).

Three virus species viz., TRV, PepRSV and PEBV are included in the genus *Tobravirus*. Tobraviruses have a bipartite genome consisting of RNA-1 and RNA-2 encapsidated separately in two rod-shaped long (L) and short (S) particles. RNA-1 is the larger strand that contains replicating capabilities and may occur as M-type isolate inducing severe symptoms without the presence of RNA-2. The shorter RNA-2 contains a coat protein (CP) gene required for particle integrity and nematode transmission of the virus, but it requires RNA-1 for infectivity. Isolates containing both RNA-1 and RNA-2 are designated NM-type. PCR detection of tobraviruses is complicated, because of the presence of genomic RNA in two different rod-shaped particles. Only one of the RNAs can be detected by standard RT-PCR format. Hence, a generic PCR assay was developed to detect all known members of the genus *Tobravirus*, using two forward and three reverse PCR primers with little or no degeneracy. Only the 194 K RNA polymerase gene in the RNA-1 section of the bipartite genome from plant tissue infected with TRV, PEBV and PepRSV was amplified. All primer combinations produced a band of the expected size, when used with cDNA derived from plant tissue infected with each of the three viruses, following reverse transcription using either a poly-T or specific primer. A single primer pair was selected after optimization of annealing temperature and $MgCl_2$ concentration. The primer set was sensitive and amplified expected PCR product in both virus-infected plant tissue and a purified plasmid containing the 194 K RNA polymerase gene from TRV. This primer pair detected all three *Tobravirus* species using a single PCR procedure (Jones et al. 2008). Total RNA was extracted from bleeding heart (*D. spectabilis*), infected by TRV having two single-stranded RNA-1 and RNA-2 encapsidated in two different rod-shaped particles. . Using RNeasy Kit (Qiagen) RT-PCR assay detected TRV in all four symptomatic *Dicentra* plants tested. But two plants tested negative in Western blot tests and the isolate could not be transmitted to indicator plants *C. amaranticolor* and *C. quinoa*, suggesting that these plants were infected by TRV-M-type isolate. The results indicated that RT-PCR assay was more effective in detecting both M- and NM-type isolates of TRV which could readily spread by vegetative propagation of new roots to new geographic locations (Robertson 2013).

## 6.2  BIOLOGY OF SOILBORNE VIRUSES

Plant viruses, ultramicroscopic in size and simple in construction, cause several economically important diseases in various crops grown in different agroecosystems. Except for a few plant viruses, they are generally transmitted in nature by biological vectors with which they have different types of relationships. The vectors of plant viruses may be insects, mites, nematodes and fungi. Insects/mites form the major groups of vectors involved in the transmission of viruses infecting aerial plant organs/tissues. On the other hand, fungus-like, true fungi and nematodes, which are present in the soil, have been demonstrated to be vectors of several viruses that have worldwide distribution. Such viruses with soilborne vectors have distinct biological, immunological and genetic characteristics, based on which they are detected, identified and differentiated by employing one or more techniques. The soilborne viruses may be transmitted via multiple modes, in addition to vector transmission. The soilborne viruses may be transmitted via seeds, vegetative propagules and irrigation water from rivers, lakes and ponds.

### 6.2.1 VIRUSES WITH ABIOTIC TRANSMISSION

Soilborne viruses may be grouped into two classes as (i) those lacking biotic vectors (abiotic transmission) and (ii) those with one or more species of fungi/nematodes present in the soil, capable of transmitting them from infected plants to healthy plants (biotic transmission). Viruses belonging to the genera *Tobamovirus, Potexvirus* and *Tombusvirus* exist in the soil in a free state and involvement of any biological vector for these viruses is not known so far. Soil transmission of TMV and ToMV in agricultural and greenhouse conditions has been observed (Jones 1993). ToMV is one of several plant viruses spreading under natural conditions without a biological vector and it is soilborne and waterborne with a wide host range. ToMV is extremely stable and infects plants through roots. The presence of ToMV in river water and clouds in ancient glacial ice and high elevations has been detected (Filhart et al. 1998; Castello et al. 1999; Boben et al. 2007). Tobamoviruses are readily transmitted through sap inoculation under experimental conditions. ToMV spreads through infected seeds also. Soilborne viruses like ToMV with multiple modes of transmission have the potential to spread rapidly and cause epidemics.

*Pepino mosaic virus* (PepMV) belonging to the genus *Potexvirus*, is typically highly infectious with a narrow host range that is confined to Solanaceae. PepMV, first recorded on pepino (*Solanum muricatum*), has no known arthropod or soilborne vector and does not appear to be seed-transmitted. The importance of PepMV was recognized, when tomatoes were infected by the virus in the United Kingdom, the Netherlands, Germany, Spain, France and Italy (Roggero et al. 2001). The experimental host range of an Italian isolate of PepMV was determined by mechanically inoculating 21 plant species, included in Solanaceae, Cucurbitaceae, Amaranthaceae, Chenopodiaceae and Labiatae. PepMV infected only solanaceous plants systematically such as tomato, tobacco, eggplant, *N. benthamiana, N. clevelandii, N. rustica* and *S. nigrum* and *S. muricatum*. Presence of PepMV in the inoculated plants was confirmed by DAS-ELISA and lateral flow tests. The virus was not transmitted via seeds of any of the plant species tested (Salomone and Roggero 2002). In a later study, the possibility of transmission of PepMV by *O. virulentus* was investigated. Two characterized cultures of *O. virulentus* from lettuce (A) and tomato (B) roots were maintained in their original host and irrigated with sterile water. The drainage water collected after irrigating these stock cultures, was used for watering PepMV-infected and non-infected tomato plants to constitute the acquisition-source plants of the assay. Healthy plants (36) grouped into six pots, which constituted the virus acquisition-transmission plants of the assay, were irrigated

with different drainage waters obtained by watering the different pots of the acquisition-source plants. PepMV was transmitted only to plants irrigated with drainage water collected from PepMV-infected plants whose roots contained the fungal culture B from tomato with a transmission rate of 8%. No infection occurred in plants irrigated with the drainage water collected from pots with only *O. virulentus* or PepMV. Both PepMV and *O. virulentus* were detected in water samples collected from drainage water of acquisition-source plants of the assay. The results indicated the possibility of transmission of PepMV by *O. virulentus* culture maintained in tomato roots (Alfaro-Ferenández et al. 2010).

Cucumber soilborne virus (CSBV), transmitted via soil, causes local lesions on several plant species including sugar beet, tomato, tobacco, French bean, cowpea and other plant species commonly used as indicator plants, such as *C. amaranticolor, C. quinoa, D. stramonium, G. globosa, N. clevelandii, N. glutinosa* and *Petunia hybrida*. None of the plant species tested, showed systemic infection. Cowpea cv. California Blackeye No. 5 was found to be the best source, allowing the virus to reach high concentrations required for purification of CSBV. The virus has isometric particles with a diameter of about 31 nm, detectable in crude sap from lesions produced on infected plants as well as in purified preparations (see Figure 6.2). Virus-like particles (~ 28 nm) were found scattered or in small aggregates in non-necrotic epidermal, parenchyma and vascular parenchyma cells of local lesions). CSBV particles were detected in the cytoplasm of living cells as well as in necrotic cells (see Figure 6.3). No changes in nuclei of infected cells were observable, except for membranous structure present sometimes at the periphery (Koenig et al. 1983). TBSV, is soilborne and considered to be transmitted via water also. Involvement of any biological vector in the transmission of TBSV is yet to be demonstrated. TBSV is readily transmitted by mechanical inoculation of sap from infected plants to a wide range of experimental plant species. TBSV, the type member of the genus *Tombusvirus*

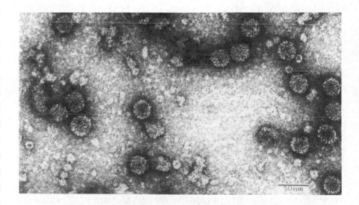

**FIGURE 6.2** Identification of CSBV in partially purified preparation Negatively stained with vinyl acetate using transmission electron microscope (TEM) Presence of spherical virus-like particles with ~ 28 nm in diameter.

**[Courtesy of Koenig et al. (1983) and with kind permission of the American Phytopathological Society, MN, United States]**

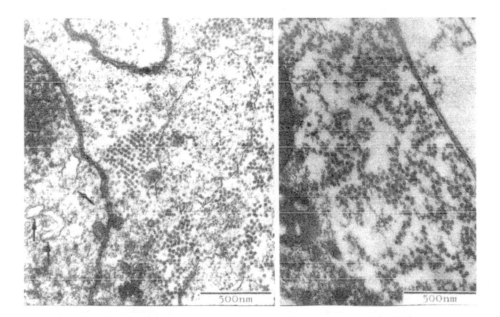

**FIGURE 6.3** Visualization of CSBV in ultrathin sections of palisade cells of infected cowpea plants under TEM Spherical virus-like particles in groups are seen in infected cells of cowpea plants.

[Courtesy of Koenig et al. (1983) and with kind permission of the American Phytopathological Society, MN, United States]

in the family Tombusviridae, has icosahedral, single-stranded positive sense RNA (Yamamura and Scholthof 2005). The genomic RNA (gRNA) is about 4.8-kb with five main open reading frames (ORFs). The internal p41 ORF encodes the 41-kDa coat protein (CP) that is translated from subgenomic RNA1. Two nested genes on different ORFs, p22, and p19, are located near the 3′ terminus of the gRNA and these are translated from subgenomic RNA2. The p22 protein is required for cell-to-cell movement of viral RNA and the p19 product (P19) suppresses the host defense RNA silencing process to permit systemic invasion (Turina et al. 2003; Scholthof 2006). TBSV CP was strictly required for systemic spread, but accelerated the process in older plants. In addition, P19 was shown to be dispensable to initiate systemic infection upon *N. benthamiana* leaf inoculation, but is needed to maintain viral RNA load, because in the absence of P19, plants exhibited recovery associated with RNA silencing-mediated degradation of viral RNA (Qu and Morris 2002; Lakatos et al. 2004; Hsieh et al. 2009).

### 6.2.2 Viruses with Biotic Transmission

Soilborne plant viruses that have biotic transmission mode, may be transmitted by either fungus-like/ true fungi or nematodes from infected plants to healthy plants under natural conditions.

#### 6.2.2.1 Viruses Transmitted by Fungal Vectors

SBWMV vectored by *P. graminis* was the first *Furovirus* recorded by McKinney (1923) in the United States. All rod-shaped viruses transmitted by plasmodiophorid fungus-like vectors were included in the genus *Furovirus*, but they were later shown to have very diverse genome organization. Only

the viruses that have two genome segments which are not polyadenylated and lack a triple gene block were included in the genus *Furovirus*. SBWMV, the type member, OGSV and *Sorghum chlorotic spot virus* (SCSV) were included in the genus *Furovirus* (Pringle 1999). Wheat-infecting CWMV and EWMV were also included in the genus *Furovirus* (Ye et al. 1999). WYMV, JSBWMV and CWMV were found to be economically important in Japan. In the 1990s, WYMV occurred in more than 666,700 ha and the yield loss was estimated to be 20–40% (Zhang et al. 2011). During a survey conducted in Zambia, the incidence of SBWMV was estimated to be 72% in four provinces. The identity of the virus was established by inoculating indicator plants and a DAS-ELISA test (Kapooria et al. 2000). Infection of wheat plants by a New York strain of SBWMV was detected using a sequence analysis approach. In addition, the occurrence of SBWMV was also recorded. The results indicated at least two separate introductions of SBWMV into Germany from other locations (Ziegler et al. 2013).

Sugar beet rhizomania caused by BNYVV is transmitted by the soilborne obligate root parasite *P. betae*, a protist initially considered as a fungus. The resting spores of *P. betae*, carrying BNYVV, are extremely resistant to biodegradation, drought and fungicide treatment that allow viruliferous spores to rest for more than two decades in the infested soils. Symptoms of rhizomania (root madness) include dwarfing of plants, formation of wine-glass-like tap root and proliferation of rootlets that are abundant, fragile and become necrotic later. Vein yellowing, necrosis and local lesions in leaves may be exhibited when the virus becomes systemic. BNYVV has a multipartite, linear, positive sense ss-RNA genome consisting of four to five RNAs possessing 5′ cap and polyadenylated 3′ ends (Quillet et al. 1989; Tamada 1999). RNA-1 is involved in the replication, cell-to-cell movement and RNA silencing suppression. RNA-1 and RNA-2 are required and sufficient for infection through mechanical inoculation for infection through mechanical inoculation on leaves. However, for natural infection, other smaller RNAs are essential. RNA-3 allows viral amplification in sugar beet roots and its expression influenced symptom expression. RNA-4 is required for viral transmission and RNA-4-encoded p31 functions as a root-specific silencing suppressor. Immunosorbent electron microscopic examination revealed the morphology of virus particles. BNYVV particles were not enveloped and they had a diameter of 20 nm in width with lengths proportional to the sizes of the encapsidated RNAs viz., 390, 265, 105, 89 and 80 nm. BNYVV behaves as a bipartite virus, when mechanically inoculated or as a tetra- or penta-partite virus in natural infection (Peltier et al. 2008).

BaYMV causes a destructive disease of winter barley in northwestern Europe and East Asia (Plumb 2002). Another virus, BaMMV also infected barley crops. *P. graminis* can transmit both viruses, although they are genotypically and serologically distinct. In addition, the responses of barley genotypes showed significant variations. Based on the pathogenicity tests on barley cultivars, two strains of BaMMV and three strains of BaYMV were recognized (Kashiwazaki et al. 1989). Multiple infections by different soilborne viruses have been observed in sugar beet. BNYVV is the principal causal agent of rhizomania disease affecting sugar beet crops in several countries around the world. In addition, incidence of other viruses such as *Beet soilborne mosaic virus* (BSBMV), BSBV, BVQ and *Beet black scorch virus* (BBSV) was also observed. Transmission of BNYVV, BSBV and BVQ by *Poymyxa betae* was demonstrated in the Netherlands (Prillwitz and Schlosser 1992). *O. brassicae* transmits BBSV in a nonpersistent manner. The other three viruses are also transmitted by the same vector and they are often found in mixed infections in sugar beet roots (Pavli et al. 2010).

*Polymyxa* spp., belonging to the order Plasmodiophorales are soilborne, obligate, intracellular parasites of roots of monocot and dicot hosts. They are involved in the transmission of several plant viruses included in the genera *Benyvirus*, *Furovirus* and *Pecluvirus* (Adams 1990; Mayo and Pringle 1998). PCV and IPCV, belonging to the genus *Pecluvirus* are responsible for significant yield losses in peanut and millet crops in West Africa and India. PCV and IPCV are transmitted also via seeds at a high frequency, in addition to transmission by *P. graminis* and they are considered as high-risk pathogens for germplasm exchange. IPCV infects grains such as sorghum and millets that are commonly cultivated as intercrops or rotation crops with peanut without expressing recognizable symptoms (Legréve et al. 1996). The dynamics oIPCV and its vector *P. graminis* were investigated in the rainy seasons of 1994, 1995 and 1996 in a naturally infested field in India. Young seedlings of test crops were exposed for short periods to the conditions existing in the field. Wheat followed by barley showed the highest virus incidence, although *P. graminis* was rarely observed in the roots of wheat and it was not detected in roots of barley. Roots of maize, pearl

millet and sorghum plants infected by *P. graminis* showed intense colonization by sporosori. IPCV accumulated in systemically infected maize plants, whereas sorghum and millet cultivars showed a transient presence of IPCV-H (Hyderabad isolate). Rice was never infected by IPCV-H and *P. graminis* was not detected in its roots. Groundnut (peanut) was infected systemically, although the vector was not detected in the roots. Groundnut appeared to be susceptible to infection by IPCV, mostly in the early stages of crop development and the rate of IPCV-H transmission via groundnut seed was highest (13%) when young plants were infected. The results indicated that the quantity and distribution of rainfall influenced incidence of IPCV-H and *P. graminis*. The temperatures (23 to 30°C) that existed during the period of experimentation were conducive to natural virus transmission by the vector (Delfosse et al. 2002).

PMTV, a member of the genus *Pomovirus* is transmitted via spores of *Sss*, causing powdery scab disease of potato tubers. PMTV persists in the soil in resting spores of *Sss*. PMTV causes the spraing disease characterized by necrotic areas in tuber flesh of sensitive potato cultivars with symptoms similar to those caused by nematode-transmitted TRV and potato tuber necrotic strain of *Potato virus Y* (PVY[NTN]). But PMTV causes a wide range of foliar symptoms, such as stunting, mottling and yellow blotches or rings, in addition to spraing symptoms on tubers. Incidence of PMTV has been observed in most of the European countries, South and North Americas, Canada, China and Japan growing potatoes. Of the 224 fields located within about 7 Km radius from the field with spraing disease incidence, 137 fields (61.2%) were found to be already infested by PMTV (Nakayama et al. 2008). During the survey conducted in Scotland, incidence of PMTV varied (37.5–47.2%) and the virus incidence was greater when the potato tubers were also infected by powdery scab disease. Significant differences in crop infection were recorded among the four major seed-producing regions and the countries within these regions. The incidence of crop and tuber infection by PMTV was least for class 'Prebasic' seed potatoes and greatest for class 'Super Elite 3' and 'Elite' seed potatoes (Carnegie et al. 2012). Widespread occurrence of PMTV was observed in Scandinavia. In order to assess the incidence of PMTV in other countries of the Baltic Sea region, an international research program determined the distribution of PMTV in the Nordic and Baltic countries, Germany and Poland. The geographic disease distribution was used as the basis for applying measures to restrict the spread of the disease (Latvia-Kilby et al. 2009). Responses of potato cultivars to infection by PMTV were determined by growing in the same field cultivars Kardal, Saturna and Nicola which showed 55%, 33% and 62% infection, respectively in 2006 and 60%, 39% and 68%, respectively in 2007. Incidence of asymptomatic infection in tubers was significantly higher in 2006 than in 2007, indicating the use of other reliable diagnostic tests, in addition to visual inspection to prevent the escape of diseased seed tubers. Under field conditions, PMTV consisted of heterogeneous RNA populations. Analysis of PMTV from 37 potato tubers naturally infected in different parts of Finland, revealed two distinguishable types of RNA-2 and RNA-3. The combination of RNA-2II and RNA-3B was more common than any other combinations and comprised typical genomic composition of PMTV strains infecting potatoes (Latvia-Kilby et al. 2009).

The relative importance of transmission of PMTV, via seed tubers and vector *S. subterranea*, was investigated to assess the transmission of PMTV from seed tubers to daughter plants, using six potato cultivars for two years. Development of foliar symptoms varied with year and cultivar. Infection of daughter tubers derived from PMTV-infected seed tubers was more prevalent on plants affected by foliar symptoms than those without symptoms. The rate of transmission of PMTV from infected seed tubers to daughter tubers ranged from 18 to 54%. Incidence of PMTV in daughter tubers in cv. Cara grown from seed tubers was not correlated with that in the seed tubers, but appeared to be strongly associated with soil inoculum. The incidence of PMTV was correlated with powdery scab in those crops in which both seed tuber and soil inocula were available. The results of tomato bait plant assay and real-time RT-PCR assay indicated that planting PMTV-infected seed tubers could increase the risk of introducing the virus into the soil that was not infested by PMTV (Davey et al. 2014). Variations in symptoms induced by PMTV have been observed in different ecological zones in the world. In temperate countries, PMTV causes necrotic flecks inside and on the surface of potato tubers, whereas in South America, these symptoms have not been seen, although the presence of the virus has been confirmed in the Andes and in Central America. In order to characterize PMTV isolates occurring in the Andes, soil samples were collected from the main potato-growing regions in Colombia and PMTV was recovered by planting *N. benthamiana* as bait plants. Complete genome sequence analyses of five isolates were performed and three of the isolates were inoculated to four different indicator plants. Three types of RNA-CP (RNA-2) and RNA-TGB (RNA-3) were differentiated based on RNA sequence comparison. The isolates CO3 and CO4 from the central areas of Colombia were similar to the isolates from Europe. The genomes of CO1, CO2 and CO5 differed from other PMTV isolates and they were placed in a separate clade in phylogenetic trees. These isolates induced only slightly different symptoms on indicator plants. But the isolate CO1 from the northwest region induced stronger symptoms like severe stunting on *N. benthamiana* (Gil et al. 2016).

The effect of temperature during plant growth on the transmission of PMTV from infected seed tubers and from infested growing media was assessed under glasshouse conditions. Symptoms induced by PMTV developed on foliage of plants derived from infected seed tubers. On the other hand, foliar symptoms were not produced following PMTV transmission by *S. subterranea* in soil. The development of foliar symptoms was on the greatest number of plants grown at 12°C, less at 16°C, few at 20°C and absent at 24°C. When PMTV was transmitted by the vector, minimal tuber infection was observed at 24°C and no influence could be seen on symptom expression on plants grown at temperatures between 12 and 20°C. Incidence of powdery scab on potato tubers was greatest at 12°C and 16°C and very low at 20°C and 24°C. Incidence of powdery scab was greater on tubers of plants derived from infected seed tubers grown in a fluctuating temperature regime of 12 h at 20°C, followed by 24 h at 12°C, than on those grown at a constant temperature of 20°C, whereas the incidence of tuber infection by PMTV and spraing was similar for both regimes. The results showed that infection of roots might occur at a higher temperature than that for powdery scab on tubers and root infection by powdery scab pathogen might enable the transmission of PMTV onto the potato plant (Carnegie et al. 2010). The influence of soil moisture regimes on the development of tuber necrosis due to PMTV was assessed, using two commercial potato cultivars with differing levels of susceptibility. Potato cv. Dakota Crisp and Ivory Crisp were grown in soil collected from a PMTV-infested field, under the moisture regimes of wet throughout (90% field capacity, WT), wet early/dry late (WEDL), dry early/wet late (DEWL) and dry throughout (60% field capacity, DT). Early and late referred to the first and last 50 days after planting, respectively. Visual assessment was made at three months after storage showed significant differences in root gall formation, powdery scab on tubers, and PMTV tuber necrosis was determined among soil moisture regimes. The incidence of PMTV tuber necrosis was significantly lower in the DT regime than other regimes. Severity of PMTV tuber necrosis was significantly higher in the WT than in other regimes, which did not differ significantly among themselves. A significant interaction was observed between cultivar and moisture regime on root gall formation, with the highest number of galls found on Ivory Crisp grown in the WT moisture regime. A significant correlation was observed between powdery scab incidence on tubers and PMTV-induced tuber necrosis incidence (Domfeh and Gudmestad 2016).

The causal nature of MiLV and LBVV in inducing BVD was investigated. The role of LBVV in the etiology of LBV became doubtful, because another soilborne virus, MiLV belonging to the genus *Ophiovirus* associated with LBVV was isolated and differentiated using ELISA tests. Both MiLV and LBVV(belonging to the genus *Varicosavirus*) could be transmitted by *O. brassicae* and mechanical inoculation from infected lettuce to healthy lettuce plants. LBVV was transmitted by *O. brassicae* zoospores, but lettuce infected with this virus alone, never showed symptoms of infection. MiLV was also transmitted by the zoospores in the same manner and lettuce infected with this virus alone, developed big-vein symptoms consistently, regardless of the presence or absence of LBVV. With repeated mechanical transmission, isolates both MiLV and LBVV appeared to lose the ability to be transmitted by the fungal vector and MiLV seemed to lose the ability to induce big-vein symptoms (Lot et al. 2002). The potential role of common weed species present in Spanish lettuce crops, as host plants for LBVaV and MLBVV was investigated. Among ten weed species commonly present in Spanish lettuce crops, MLBVV and LBVaV could infect the naturally growing *Sonchus oleraceus* (common sowthistle) plants. *S. oleraceous* could function as a source of lettuce infection, since the weed species supported the development of *O.brassicae*, the vector of both viruses (Navarro et al. 2005). Development of BVD caused by MiLBVV and LBVaV was suppressed when the field soil became acidic. Hence, the effect of soil pH on the activities of the vector *O. virulentus* was assessed. In acidic soils with less than pH 6.0, a significant reduction in infection of roots by *O. virulentus* was observed and the lower pH levels influenced the rate of release of zoospores in the surrounding water. The results indicated that acidic soils reduced the rate of release of zoospores, resulting in the reduction in the incidence of BVD (Iwamoto et al. 2017).

### 6.2.2.2   Viruses Transmitted by Nematode Vectors

Plant viruses transmitted by nematodes form another group of soilborne viruses with biotic transmission, causing heavy losses in high value crops like grapevine, tomato and strawberry. Nematodes belonging to the genera *Xiphinema*, *Longidorus* and *Trichodorus* are the principal vectors of plant viruses. GFLV, belonging to the genus *Nepovirus* is the major cause of grapevine degeneration and is associated with yield reductions of as much as 80%, reduced fruit quality and shortened life spans of vines (Martelli and Savino 1988). *X. index* is the natural biological vector of GFLV which can be experimentally transmitted by inoculation with sap from infected grapevine plants. The GFLV naturally infects *Vitis* spp. and some weed species in the vineyards. The experimental host range of GFLV is limited to species in the families, Amaranthaceae, Chenopodiaceae, Cucurbitaceae, Leguminaceae, Solanaceae and Fabaceae (Andret-Link et al. 2004). Crop losses due to grapevine fan leaf disease, may vary from a moderate (10%) to a very high level (>80%), depending on the cultivar susceptibility/resistance, extent of selection of virus-free planting material and the population of viruliferous nematode vectors. *X. index* may remain viruliferous for long periods and is capable of acquiring GFLV particles even by feeding on root debris scattered in soil after the removal of infected plants. Virus-vector specificity was observed between GFLV and *X. index*. The virus particles acquired, during feeding, remained adsorbed on the inner lining of the stylet and esophagus, where they might be retained for long periods of time. The virus was released through salivation, when the nematode molted (Tylor and Robertson 1970). Grapevines are normally propagated by cuttings, a process that perpetuates GFLV infections. Planting vine stock with latent infection is an important mode of dispersal of GFLV to long distances, while the nematodes may be responsible for virus spread within or to adjacent fields (Narayanasamy 2017).

The location of transmissible and non-transmissible plant viruses in the vector nematode species *X. americanum* was investigated, using three nepoviruses, TRSV, TomRSV and CLRV and one non-transmissible virus, *Squash mosaic virus* (SqMV). Of the three nepoviruses, TRSV and TomRSV are transmitted by *X. americanum*, while CLRB is transmitted by *X. diversicaudatum*. SqMV is not transmitted by *X. americanum*. The transmission efficiency was highest for TomRSV (38% with 1 and 100% with ten nematodes, respectively) and intermediate efficiency for TRSV (27% with 1 and 65%

with ten nematodes, respectively). CLRV and SqMV were not transmitted by *X. americanum*. Electron microscopy and immunofluorescent labeling procedures showed that TRSV was primarily localized to the lining of the lumen of the stylet extension and the anterior esophagus, but only rarely on the triadiate lumen. It appeared that there was a gradual and random loss of TRSV from these locations and it could not be detected in these locations at ten weeks after acquisition. Likewise, loss of virus particles occurred when the viruliferous nematodes were fed on nonhost of the virus. The loss of immunofluorescent labeling paralleled the decrease in the ability of the nematodes to transmit the virus. TomRSV was localized only in the triadiate lumens, as revealed by thin-section electron microscopy. No particle of CLRV was present in any part of the food canal of nematodes fed on CLRV-infected plants. Virus-like particles were observed only in the triadiate lumen of nematodes that had fed on SqMV-infected plants. The nematode transmissible viruses TRSV and TomRSV were localized in different regions of the food canal of *X. americanum* (Wang et al. 2002).

The nepoviruses included in the genus *Nepovirus* (family Comoviridae) have polyhedral particles and consist of three serologically distinguishable density components called top (T), middle (M) and bottom (B). The T component particles are empty shells, M component particles contain the genomic nucleic acid RNA-2 and B component particles contain both genomic nucleic acids RNA-1 and RNA-2 (Quacquarelli et al. 1976). All nepoviruses have a bipartite genome consisting of two linear single-stranded positive sense RNA species, both of which are transcribed as polyproteins. RNA-1 is ~ 800 nucleotides in length and encodes the helicase, polymerase and proteinase. RNA-2 is ~ 4,000–7,000 nucleotides in length and encodes the capsid protein (CP) and MP. Both RNA molecules have their 3' end polyadenylated and virus-specific peptides linked to their 5' ends. RNA-1 and RNA-2 are encapsidated in separate particles. Nepoviruses have isometric, polyhedral particles which are ~ 30 nm in diameter. They are composed of 60 copies of single coat protein species and sediment as three components. The nepoviruses are transmitted by soil-inhabiting free-living nematodes belonging to the genera *Xiphinema* and *Longidorus* that feed on roots of susceptible plant species. Members of *Nepovirus* are classified into three subgroups: A (8 species), B (9 species) and C (15 species), based on their serological relationships, sequence similarities and particle length and arrangements of RNA-2 (Fauquet et al. 2005).

TRSV is the type species of the genus *Nepovirus* in the family Secoviridae and it has a wide range of host plant species, including several herbaceous crop plants and woody plants. Bud blight disease of soybean, the most severe disease caused by TRSV, resulting in total yield loss (100%) (Demski and Kuhn 1989). TRSV has bipartite genome of positive sense single-stranded, polyadenylated RNA molecules, RNA-1 and RNA-2, encapsidated in separate particles of similar size. The RNA molecules possess a genome-linked protein (Vpg) covalently bound to their 5' ends. RNA-1 encodes a large polyprotein precursor that is proteolytically processed into protease cofactor (PIA), putative ATP-dependent helicase (Hel), picornain 3C-like protease (Pro), and RNA-directed RNA polymerase (Pol). RNA-2 encodes the capsid protein (CP), a putative MP and an N-terminal domain involved in RNA-2 replication (P2A). Proteins encoded by RNA-1 are required for RNA replication, whereas protein encoded by RNA-2 is required for cell-to-cell movement and viral RNA encapsidation. RNA-1 can replicate independently, both RNA-1 and RNA-3 are required for systemic infection (CMI and AAB 1985). In addition to transmission by nematodes, transmission of TRSV via seeds, which is important for long-distance dispersal, also has been demonstrated (Scarborough and Smith 1977).

Transmission of TRSV through honeybees, *Apis mellifera* was studied critically. Honeybees transmit TRSV, when the contaminated pollen from infected plants is transferred to healthy flowers (Card et al. 2007). Honeybees carry strong electrostatic charge that ensures the adherence of pollen to their bodies and also actively store pollen in specialized pollen baskets on their hind legs. In addition, honeybees exposed to TRSV-contaminated pollen could also be infected and the infection could become systemic and spread through the entire body of honeybees. Electron microscopic examination of the TRSV preparation purified from A. mellifera showed the presence of TRSV particles, the identity of which was confirmed by RT-PCR assay. Negatively-stained virus particles with a diameter of 25 to 30 nm and icosahedral shape were observed. Distribution and replication of TRSV in infected honeybees were studied. TRSV particles were widespread in different tissues of infected honeybees. Presence of TRSV could be visualized in all tissues, except compound eyes, including hemolymph, wings, legs, antennae, brain, fat bodies, salivary glands, guts, nerves, tracheae and hypopharangeal gland. The intensity of PCR signals representing virus titer varied between samples. Tissues of the gut and muscle had weaker PCR bands than other tissues, indicating relatively lower concentrations of TRSV infection. Further, the presence of TRSV was detected also in the ectoparasitic mite *Varroa destructor*, that fed on honeybee hemolymph collected from TRSV-infected colonies, using *in situ* hybridization technique, suggesting the possibility of *Varroa* mites facilitating the spread of TRSV via honeybees. The results revealed that honey bees exposed to TRSV-contaminated pollen could also be infected and the virus might multiply in the bees that might serve as inoculum source for TRSV (Li et al. 2014).

The knowledge of spatial distribution of *X. index*, the vector of GFLV, both horizontally and vertically, may be useful for efficient management of the disease. In two infested blocks of a French vineyard, vertical distribution data showed that the highest numbers of individuals occurred at 4 cm to 110 cm depth, corresponding to the two layers where the highest densities of free roots were present. Horizontal distribution analysis of 10 × 15 cm grids indicated a significant aggregative pattern, but no significant neighborhood structure of nematode densities. At finer scale (~ 2 × 2 m) confirmed a significant aggregative pattern, with patches of 6 to 8 m diameter, together with a significant neighborhood structure of nematode densities, thus identifying the relevant sampling scale to determine nematode population distribution. Nematode patches correlated significantly with those of GFLV-infected grapevine plants. Further, nematode and virus spread extended preferentially parallel to the vine rows, probably due to tillage during mechanical weeding (Villate et al. 2008).

TBRV, transmitted by the nematodes *Longidorus elongatus* and *L. attenuatus*, infects a wide range of economically important crops such as grapevine, cherry, apricot, peach, berry-fruits (raspberry, strawberry and blueberry and currant), solanaceous crops (potato, tomato, pepper and tobacco) and a number of ornamental and weed species (Brunt et al. 1996a; Edwardson and Christie 1997). TBRV causes asymptomatic to severe symptoms on various crops and it has been difficult to estimate yield losses. Strawberry showed mild to asymptomatic infections, whereas artichoke (*Cynara cardunculus*) suffered losses of up to 40%. TBRV is transmitted via seeds which carry the virus to long distances, as well as to subsequent generations. TBRV is regarded as a regulated pathogen in the North American, Australia, New Zealand and European Union countries (Brunt et al. 1996a; Harper et al. 2011). TRSV is transmitted by *Xiphinema* and also via seeds. TRSV also has a wide host range, including grapevine, soybean, tobacco, blue berry and members of Cucurbitaceae (melon, cucumber, squash and pumpkin) which are seriously affected. In addition, natural infection by TRSV has also been observed on apple, eggplant, black berry, pepper, cherry, papaya and weed species (Sneideris et al. 2012). Occurrence of TRSV has been reported from Europe, North America, Australia, Africa, India, Japan and New Zealand (Brunt et al. 1996b).

RBDV, transmitted by *L. juvenilis*, infects *Rubus* spp. all over the world. Asymptomatic infection of many *Rubus* spp. is common. In sensitive *Rubus* sp., leaf curling, necrosis, premature defoliation and death of lateral roots appear as major symptoms. RBDV is naturally transmitted via pollen from infected plants and also through sap inoculation experimentally. Infection of grapevines by RBDV was first reported in Slovenia, and grapevine was shown to be a natural host of RBDV outside the genus *Rubus* (Mavrič et al. 2003). Grapevine isolates of RBDV infected *Chenopodium murale* systematically, which was not infected by raspberry isolates of RBDV. Grapevine RBDV isolates could be transmitted by grafting, using scions infected by type-strain of RBDV to *Cydonia oblonga* and *Pyronia veitchi* which developed symptoms, but *Fragaria vesca* and *Prunus mahaleb* remained symptomless (Jones and Mayo 1998). RBDV has a bipartite genome with the bicistronic RNA-2 producing the MP and a subgenomic RNA that codes for the coat protein (CP) (Jones 2001). RBDV strains could be differentiated, based on serological and genomic characteristics. RBDV was detected by nested RT-PCR assay in *L. juvenilis* from the rhizosphere soil of RBDV-infected grapevine plants. Further, RBDV could be detected in the nematodes after storage of soils for four to eight months in a refrigerator. Grapevine-infecting RBDV isolates were widespread in Slovenia where the virus could be detected in all grapevine-growing areas and in many grapevine cultivars (Pleško et al. 2009).

TRV, the type member of the genus *Tobravirus*, is transmitted by soil-inhabiting nematodes, belonging to the genera *Trichodorus* and *Paratrichodorus* (Harrison and Robinson 1986). *Tobravirus* includes three virus species, TRV, PepRSV and *Pea early-browning virus* (PEBV). These viruses infect a large number of crops, including sugar beet, pepper, tobacco, pea, bean, tomato, potato, spinach and the ornamental plant, tulip (Brunt et al. 1996b). TRV incidence is more common in soils with sandy texture which favors development of the nematode vectors. In the northwestern United States, the primary vector of TRV was found to be *P. allius*. Infection of developing potato tubers results in production of internal symptoms, consisting of rings, arcs and various dark patches with corky texture earning the name 'corky ringspot' disease (Crosslin et al. 2007). Tobraviruses have a bipartite genome, consisting of two single-stranded positive sense RNA strands that are encapsulated separately within two rod-shaped particles. The longer RNA-1 (L particle) is ~ 7-kb in length and contains four ORFs, enclosing genes responsible for replication, intercellular transport (*P1a*) and pathogenicity (*P1b*). The shorter strand RNA-2 (S particle) may vary from ~ 2 to 4-kb in length and encodes viral coat protein (CP) and genes (*P2b*, *P2c*) involved in nematode transmission (Fauquet et al. 2005). TRV RNA-1 is infectious in the absence of RNA-2, but produces infections that lack CP, since this gene is located in RNA-2 (MacFarlane 1999). The NM-type isolates (without RNA-2) cause severe symptoms. The isolates containing both RNA-1 and RNA-2 are known as M-type isolates which can be transmitted by using infective sap or purified virus preparations to tobacco and *N. benthamiana*, *C. amaranticolor* and *C. quinoa*. Infection by TRV M-type isolates of potato cultivars may lead to systemic infection and consequent persistence through generations by vegetative propagation, although spraing symptoms rarely develop in tubers. Under field conditions, nine potato cultivars were grown and TRV infection delayed plant emergence and retarded further growth of plants, reduced tuber yield and mean tuber weight, increased tuber numbers and adversely affected tuber quality. Concentrations of TRV in leaf extracts from individual plants were highly variable. As the selected potato cultivars were grown on large areas in the UK and other European countries, significant adverse impact on potato industry might be expected (Dale et al. 2004). In another investigation, the effects of TRV infection of 15 potato cultivars, using nematode vectors for inoculation, were assessed. Failure to produce daughter tubers, in addition to general mottling of leaves, was observed. The incidence and severity of spraing symptoms differed at three locations included in the study. Spraing symptoms were induced in all cultivars except cv. Record. The incidence of spraing symptoms were greater in larger than in small tubers and increased when harvest was delayed. Further, TRV was detected in the roots of several weed species growing at one of the sites and in the roots of a few plants of the subsequent barley crop (Robinson 2004). Natural infection by TRV occurred on bleeding heart (*D. spectabilis*) and both M and NM-type isolates of TRV could spread readily by vegetative propagation of infected roots to new locations (Robertson 2013).

### 6.2.3 PROCESS OF INFECTION

Plant viruses constitute a distinct group of obligate, intracellular, molecular pathogens, in contrast to the cellular and multicellular bacterial and fungal pathogens. Viruses have a simple structural constitution composed of a coat protein and genomic nucleic acid which may be either RNA or DNA, requiring only nucleotides and amino acids from the host plant for their replication. The viruses have no known physiological functions and enzymatic activities aiding infection/pathogenesis. Entry of viruses into the host cells depends on the presence of wounds/injuries created by the natural biological vectors or abrasives used for artificial inoculation of plants. Although viruses have a simple constitution, infection of susceptible host plants is a complex process. The process of infection of plants by viruses passes through four distinct phases: (i) entry into plant cells, (ii) replication in the initially infected cells, (iii) cell-to-cell movement through plasmodesmata (PD) and (iv) long-distance movement through vascular system. Plant cell wall is a formidable barrier to the entry of virus and there is no receptor to mediate the initial transfer of plant virus into a susceptible cell. The infectious virus particle has to be introduced physically by their vector(s), breaching the cell wall (Narayanasamy 2008).

SBWMV, causing significant economic losses in wheat, is transmitted by *P. graminis*. Successful inoculation of wheat plants, using *P. graminis* is hampered because of the poor infectivity of SBWMV. A procedure was developed for eliminating wound-induced necrosis normally associated with rub-inoculation of virus to wheat leaves and this procedure was useful for uniformly inoculating wheat plants *in vitro*. Wheat plants were grown hydroponically in seed germination pouches and their roots were inoculated with SBWMV either by placing *P. graminis*-infested root material in the pouch or by mechanically inoculating the roots with purified virus. The procedure was applied for inoculating one susceptible and three field resistant wheat cultivars to determine the responses of the test cultivars and compared with the conventional inoculation method of growing plants in *P. graminis*-infested soil. The results suggested that susceptibility/resistance of wheat cultivars might function in the roots to allow or block virus infection (Driskel et al. 2002).

Inoculation of plants with viruses employing an agroinoculation method is a rapid and easy technique that can be used to establish a high percentage of infection of susceptible plants. This method involves infiltration of *Agrobacterium tumefaciens* carrying binary plasmids harboring full-length cDNA copies of viral genome components. When transferred into host cells, transcription of the cDNA produces RNA copies of the viral genome that initiate infection. Full-length cDNA corresponding to BNYVV) RNAs and derived replicon vectors expressing viral and fluorescent proteins in pLJ89 binary plasmid under the control of the *Cauliflower mosaic virus* 35S promoter were produced. *N. benthamiana* and *Beta macrocarpa* plants were successfully infected by leaf agroinfiltration of combinations of agrobacteria carrying full-length cDNA clones of BNYVV RNAs. Root agroinfection of *B. vulgaris* by BNYVV was also accomplished and this technique has the potential for use as a tool for assessing the resistance of beet cultivars and genotypes to rhizomania disease caused by BNYVV (Delbianco et al. 2013).

### 6.2.4 Mechanisms of Transmission of Viruses by Vectors

Mechanisms of acquisition of viruses by *Polymyxa* spp. from plants infected by soilborne viruses have been investigated. These viruses have been internalized by the vector and hence, they can remain in the soils in the body of the vectors for several years. As the developmental cycle of *Polymyxa* spp. and development of viral pathogens vectored by them are intimately connected, basic knowledge of the vector life cycle would be useful. In the case of *Polymyxa* spp. infection begins with penetration of the plant cell wall by the swimming zoospores. Zoospores transfer their cytoplasm into the plant cell and a multinucleate sporangial plasmodium develops. The plasmodium matures into a zoosporangium containing numerous secondary zoospores. Mature sporangia have several lobes divided by cross-walls. Exit tubes are formed from the zoosporangium into the plant extracellular space. Secondary zoospores are released into the plant extracellular space through exit tubes extending from zoosporangium.

The secondary spores penetrate new cells and sporogenic plasmodia develop. These mature into sporosori containing several resting spores. Thick-walled resting spores often remain in root debris in the soil after harvest. With heavy rain or irrigation, resting spores germinate, releasing primary zoospores to infect new roots or new plants and commences subsequent rounds of infection (Keskin and Fuchs 1969; Campbell 1996). Accumulation of viral RNAs and MPs of SBWMV and capsid proteins of *Wheat spindle streak virus* (WSSV) within the resting spores of the vector *P. graminis* possibly indicated the presence of ribonucleoprotein (RNP) complexes (Driskel et al. 2004).

Presence of BNYVV capsid protein was detected in both sporospores and the plasmodia of *P. betae*, using virus-specific labeled antibodies (Doucet 2006). Immunofluorescence and immunogold labeling techniques revealed the presence of all BNYVV proteins in *P. betae* resting spores and zoospores. The specificity of immunolabeling of BNYVV protein was indicated by the negative results obtained with buffer or heterologous antiserum. Comparison of the immunogold label in BNYVV and healthy *P. betae* zoospores provided further evidence for the presence of each BNYVV protein inside the viruliferous *P. betae* zoosporangia (see

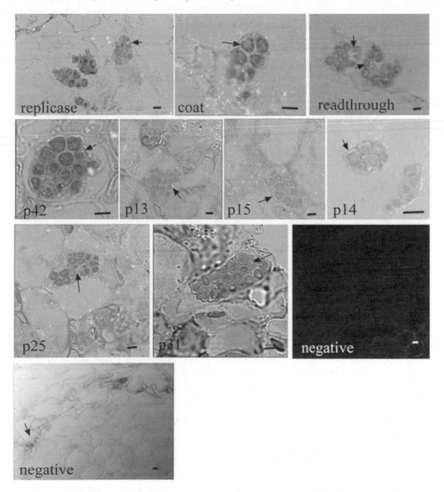

**FIGURE 6.4** Detection of BNYVV proteins associated with sporosori of *Plasmodiophora betae* using confocal microscope Presence in the root cells of BNYVV proteins replicase, coat protein, readthrough proteins, p42, p13, p15, p14 and p25 is indicated by arrows; none of the proteins were seen in the negative samples.

[Courtesy of Lubicz et al. (2007) and with kind permission of *Virology Journal*, BioMed Central Open Access]

Figure 6.4). In resting spores, the concentrations of viral replicase were greatest in the cell wall, but significant in most structures. Each of the BNYVV protein showed distinct subcellular accumulation patterns in resting spores, making it difficult to identify specific centers for viral activities. In zoospores, concentrations of most of the BNYVV proteins were greatest in the cytoplasm, vesicles and areas between spores (see Figure 6.5). Detection of viral proteins in the vector might be due to the mix up of cytoplasms of *Polymyxa* and host plant and the virus might be freely exchanged (Lubicz et al. 2007). Campbell (1996) suggested that this might happen before membranes are laid down to form the sporangial plasmodium. Another alternative suggestion was that *P. betae* might be a host as well as a vector of BNYVV. Presence of viral replicase inside *P. betae* resting spores and zoospores might indicate the possibility of BNYVV replication in the vector (Lubicz et al. 2007).

Nematodes belonging to the families *Longidoridae* and *Trichodoridae* are involved in the transmission of plant viruses. Transmission of viruses by nematodes has been shown to be semipersistent and nonpropagative in nature. The virus particles are retained by binding to specific sites on the surface of esophagus of the nematode vector. The specificity of transmission of virus by a nematode species is considered to be due to the ability or inability of the

virus to be retained within the mouth parts of the nematode. The nematode-transmitted viruses belonging to the genera *Nepovirus* and *Tobravirus* have two genomic RNAs: RNA-1 and RNA-2. The smaller RNA-2 encodes viral coat protein (CP) that determines the transmission characteristics of both nepoviruses and tobraviruses (Harrison and Murrant 1977; Ploeg et al. 1993). GFLV is acquired and transmitted by both juvenile and adult forms of *X. index*, but not after molting. No transovarial passage of GFLV in the nematode vector is known (Andret-Link et al. 2004b). Electron microscopic observations revealed the presence of GFLV particles adhering to the cuticular lining of the esophageal tract from the most anterior part of the esophageal bulb (Taylor and Robertson 1970). GFLV could be acquired by *X. index* from infected rooted grapevine cuttings, after as little as five-minute feeding and a similar short period of transmission feeding on the roots of healthy cuttings (Das and Raski 1968).

In GFLV, RNA-2-encoded polyprotein is cleaved into three proteins, 2A (28 kDa), 2B (38 kDa) and 2C (56 kDa). The 2C protein forms the viral CP (Gaire et al. 1999). Although the GFLV and ArMV are serologically related, they are transmitted by two different nematode species, *X. index* and *X. Diversicaudatum,* respectively. The CP is the sole determinant of specificity of transmission by the nematode species

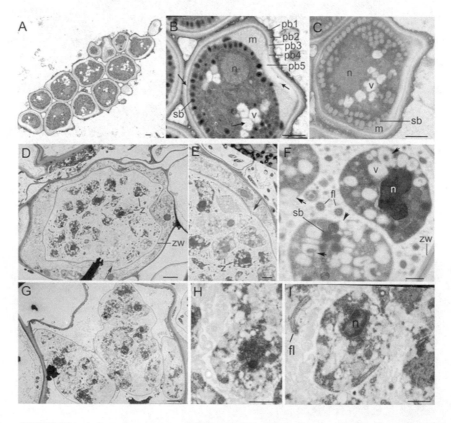

**FIGURE 6.5**  Detection of BNYVV in infected zoospores of *P. betae* A and B: presence of immunogold labels in the cytoplasm vesicles and extracellular space is indicated by arrows; C and D: zoospores and vesicles labeled with coat protein antiserum; E and F: zoospores labeled with P14 and P31 antiserum show labels in zoospore vacuoles/zoosporangial wall; G and H: presence of BNYVV in host plant cell I: gold particles labeling P31 protein on zoosporangial wall.

[Courtesy of Lubicz et al. (2007) and with kind permission of *Virology Journal*, BioMed Central Open Access]

(Belin et al 1999, 2001). The viral determinants involved in the specific transmission of GFLV by *X. index* are located within the 513 C-terminal residues of the RNA-2 encoded polyproteins, that is, the 9C-terminal amino acids of the MP (2B$^{MP}$) and contiguous 504 amino acids of the coat protein (2C$^{CP}$). To further delineate the viral determinants responsible for the specific spread, the four amino acids that are different within the 9C-terminal 2B$^{MP}$ residues between GFLV and ArMV, another nepovirus which is transmitted by *X. diversicaudatum*, but not by *X. index*, were subjected to mutational analysis. Of the recombinant viruses derived from transcripts of GFLV RNA-1 and RNA-2 mutants that systemically infected herbaceous host plants, all with the 2 C$^{CP}$ of GFLV were transmitted by *X. index*, unlike none with the 2C$^{CP}$ of ArMV, regardless of the mutation with the 2B$^{MP}$ terminus. The results revealed that the coat protein of GFLV is the sole viral determinant for the spread of the virus by *X. index* (Andret-Link et al. 2004b).

In the nepoviruses, the coat protein (CP) that assembles into icosahedral particles is the sole viral determinant involved in the specificity of transmission by nematode vectors. GFLV is transmitted by *X. index*, whereas ArMV is vectored by *X. diversicaudatum*, as determined by the differences in the amino acid sequences of respective coat proteins. The icosahedral capsid of GFLV, consisting of 60 identical coat protein (CP) subunits, carries the determinants of specificity of transmission by *X. index*. GFLV has a bipartite, linear, single stranded, positive sense RNA. The RNA1 has an essential role in viral replication and RNA2 is required for movement and encapsidation of viral nucleic acid. The strain FGLV F13 obtained after four decades of successive passage by mechanical inoculation onto *C. quinoa*, was poorly transmitted by *X. index* and a variant of this strain was named as GFLV-TD which showed variations in nematode transmission (see Figure 6.6) No differences in symptoms induced on *C. quinoa* and reactivity to GFLV-specific antibodies in DAS-ELISA could

be recognized. However, the ability of *X. index* to ingest GFLV-F13 and GFLV-TD was not affected, as shown by RT-PCR assay, after a monthly AAP (see Figure 6.7, top panel). At the end of the inoculation access period (IAP), GFLV was not detectable by RT-PCR in *X. index* (see Figure 6.7, bottom panel), indicating that GFLV-TD was poorly or not retained by *X. index*. The results were consistent with the transmission deficiency of GFLV-TD. The GFLV-TD CP coding sequence was characterized by IC-RT-PCR assay and sequencing to identify potential amino acid mutations. A single Gly to Asp mutation at position 297 was detected. The results of site-directed mutagenesis showed that the mutant was not retained by the vector after IAP, confirming the critical role of Gly$^{297}$ in the efficiency of GFLV transmission by *X. index*. The crystal structures of the wild-type and mutant strains were compared. The Gly297Asp mutation mapped to an exposed loop at the outer surface of the capsid and did not affect the conformation of the assembled capsid nor of the individual CP molecules. The loop is part of a positively charged pocket that includes a previously identified determinant of nematode transmission. The data suggested that perturbation of the electrostatic landscape of the pocket might affect the interaction of the virion with specific receptors of the feeding apparatus of the nematode. Consequently, it might severely diminish the transmission efficiency of the nematode vector (Schellenberger et al. 2011).

Tobraviruses are transmitted by root-feeding ectoparasitic nematodes, belonging to the genera *Trichodorus* and *Paratrichodorus* in a noncirculative manner. TRV, PEBV and PepRSV included in the genus *Tobravirus*, have a bipartite, single-stranded (ss)-RNA genome of positive polarity which is encapsidated in two rod-shaped particles. RNA-1 which is highly similar between isolates from the same group, encodes the replicase and MPs and small nonstructural protein with a putative function in seed transmission. On the other hand, RNA-2 is highly variable in size and nucleotide sequence among different isolates, even within the same subgroup. In addition to the coat protein (CP), RNA-2 may encode one to three nonstructural proteins (Harrison and Robinson 1986;

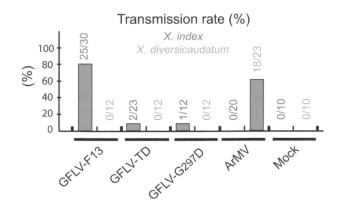

**FIGURE 6.6** Comparative rates of transmission of *Grapevine fan leave virus* (GFLV) strains and ArMV by *X. index* and *X. diversicaudatum* GFLV-TD strain was indistinguishable from wild-type strain GFLV-F13, except for its poor transmission rate by *X. Index*.

[Schellenberger et al. (2011) and with kind permission of *PLoS Pathogens* Open Access Journal]

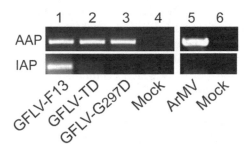

**FIGURE 6.7** Assessment of ability of *X. index* to ingest wild-type strain GFLV-F13 and variant GFLV-TD, using RT-PCR assay after a monthly AAP wild-strain GFLV-F13 was detectable in *X. index* (top panel); GFLV-TD was in the vector after IAP (bottom panel).

[Schellenberger et al. (2011) and with kind permission of *PLoS Pathogens* Open Access Journal]

Visser et al. 1999). Nematode transmission investigations with pseudorecombinant TRV isolates produced from nontransmissible isolate PLB and the transmissible isolate PpK20 showed that vector transmissibility segregated with RNA-2 (Ploeg et al. 1993). The mutational analysis of genes in RNA-2 of the isolates of TRV-PpK20 and PEBV-TpA56 showed that CP and nonstructural proteins are both involved in the transmission process (Hernández et al. 1995; Visser et al. 1999). RNA-2 of TRV isolate PpK20 encodes the viral CP and two nonstructural proteins of 40 kDa (40-k protein) and 32.8-kDA (32.8-k). The 40-k protein is required for transmission of TRV by *P. pachydermus*, whereas the 32.8-k protein may be involved in transmission by other nematode species. The expression and subcellular localization of 40-k protein in *N. benthamiana* was studied by raising an antiserum against 40-k protein expressed in *Escherichia coli*. The time-course of the expression of the 40-k protein in leaves and roots was similar to that of CP and both proteins were distributed in leaf and root in a similar fashion. Isolates of TRV may fall into two types based on the nature of RNA contained in the virus particles. NM-type isolates containing RNA-1 cause severe symptoms in the absence of RNA-2. But these isolates may not be transmitted by the nematode vectors. The M-type isolates contain both RNA-1 and RNA-2 which contains coat protein gene and provides particle integrity and facilitates nematode transmission of TRV. Both M- and NM- type isolates may be spread by vegetable propagation to new locations (Robertson 2013).

## 6.2.5    Movement of Viruses

The general pattern of virus replication involves the release of viral genome from the protein coat (CP) shortly after the entry of the virion (virus particles) into the susceptible host plant cell, either through wounds or injuries made by feeding vectors. The viral genome (+ sense)-single-stranded (ss)-RNA of the infecting virus is either transcribed directly or mRNAs are formed and translated to produce virus replicase, as an early product as well as virus-specific proteins. The viral replicase or replication associated proteins are involved in the synthesis of viral genomes. The viral coat protein (CP) and MP required to encapsidate the viral genome and for cell-to-cell movement of the virus are synthesized as the next step in viral replication. The coat protein and the viral genome are assembled to form new progeny viral particles which accumulate within the cell, usually in the cytoplasm. The infectious virus particles have to move from the initially infected cells to adjacent cells and later to other tissues as the virus spreads systemically. The steps in virus replication may vary depending on host-virus combinations. Plant viruses replicate in different patterns based on the nature of the genomes, regardless of their mode of transmission to healthy plants to initiate infection (Narayanasamy 2008). Movement of plant viruses, after replication in initially infected host plant cells, occurs in two phases viz., cell-to-cell or short-distance movement and long-distance movement to other tissues of infected plants

following systemic infection. If the virus is able to replicate in the initially infected cell, but unable to move to neighboring cells, the infection of the plant species is termed as subliminal infection. Such plant species show a high form of resistance (extreme sensitivity) (Cooper and Jones 1983). The extent of virus movement may be restricted to some chlorotic areas which may later turn necrotic and the virus is not able to become systemic in such host plant species that are used as assay host or indicator plant for rapid detection of the virus(es) under investigation. The primary hosts (crops) permit the long-distance movement of viruses and various kinds of symptoms are induced, when the virus infects different tissues of the plant species. The levels of susceptibility/resistance of the genotype are reflected by the severity of symptoms caused by the virus concerned.

### 6.2.5.1    Movement of Viruses with Abiotic Transmission

Plant viruses require plant proteins, normally involved in host-specific activities, during replication, accumulation of virions, and intracellular-, local- and systemic movement (Nelson and Citovsky 2005). They achieve cell-to-cell movement via plasmodesmata (PD) which may be modified to different extent and reach distant organs through vascular system. The PD function as cytoplasmic bridges that establish membrane [endoplasmic reticulum (ER) and plasma membrane], as well as cytoplasmic continuity between adjacent plant cells. The PDs form an important pathway for communication between plant cells, regulating cell-to-cell communication. As such, PDs are too small to allow free passage of virions (mature virus particles) or viral genomes which are far larger than the size exclusion limit (SEL) of PDs. The viruses have to move first from replication sites to PD at the cell periphery and then traverse the intercellular channels to enter the neighboring cells. The cell-to-cell movement of TMV, the type member of the genus *Tobamovirus* was investigated. The involvement of MP was indicated by the studies, using transgenic plants expressing TMV MP which allowed larger molecules (9.4 kDa) to pass through PD, whereas the SEL of PDs of nontransgenic plants was between 0.75 and 1.0 kDa (Wolf et al. 1989).

Regulation of plasmodesmatal transport by TMV MP (P30) is dependent on phosphorylation in a host-dependent manner. P30 was shown to be associated with the cytoskeletal structures, as revealed by experiment using the antibody to P30 and infectious TMV clones with green fluorescent protein (GFP) fused to P30. Colocalization of P30 with microtubules was observed in protoplasts and whole plants (Heinlein et al. 1995). Participation of microfilaments in the cell-to-cell movement of TMV and involvement of microtubules and microtubule-associated proteins in the degradation of MP have been demonstrated (Gillespie et al. 2002; Kragler et al. 2003). Disruption of microfilaments by virus-induced gene silencing, significantly reduced the spread of TMV from cell to cell. However, accumulation of virus within the infected cells was not affected. In addition, another TMV factor, the 126-kDa protein was also involved in viral transport with microfilaments. This protein was associated with viral replication complexes, modulates their size and potentially mediates their interaction with and movement along with microfilament network (Liu et al. 2005). The intracellular and intercellular movement of virus replication complexes (VRCs) of TMV was monitored by using confocal microscopy in intact leaf tissue from 12 to 36 h postinoculation (*hpi*). The VRCs in primary infected cells were associated with ER from 12 hpi. High intracellular motility of VRCs was observed at 14 hpi ($\approx$ 160 nm/sec), followed by production immobility between 14 and 16 hpi, becoming stationary at 18 hpi, adjacent to PD. The VRCs traversed the PD between 18 and 20 hpi. The process of formation and movement of VRCs in adjacent cells needed only three to four h, as against 20 h taken in primary infected cells. The application of inhibitors of filamentous actin and myosin blocked the rapid movement of VRCs and the spread to adjacent cells. Inhibitors of microtubules did not have any effect on this process. The results suggested that cell-to-cell movement of TMV infection might be accomplished by subviral replication complexes that initiate TMV replication immediately after entry to adjacent cells (Kawakami et al. 2004).

### 6.2.5.2 Movement of Viruses Transmitted by Fungal Vectors

The cell-to-cell movement of certain groups of viruses like potexviruses and pomoviruses is mediated by proteins translated from a set of three overlapping ORFs, known as triple gene block (TGB). These proteins (TGBs) encoded by TGB of potexviruses are required for their cell-to-cell movement (Angell et al. 1996). Two classes of triple gene blocks have been distinguished: class 1 TGB includes the hordeiviruses having a TGB1 encoding a 42–63 kDa protein and class 2 TGB contains potexviruses and carlaviruses with a TGB encoding a product of 25 kDa (Solovyev et al. 1996). Viral coat protein (CP) is also required for the movement of viruses with class 2 TGB, whereas class 1 TGB in hordeiviruses mediates their movement independently of viral CP (Donald and Jackson 1996). The accumulation levels of TGBp1 of BNYVV was adversely affected by mutations in TGBp2 and TGBp3, indicating the existence of a highly coordinated control of the interactions among elements of the movement machineries (Lauber et al. 1998).

Based on sequence analysis, the protein region carrying the NTPase/helicase domain in potex-like TGBp1 was predicted to possess NTPase activity *in vitro* (Kalinina et al. 2002). The ATPase activity may be required for cell-to-cell movement, particularly for PD dilation by the potex-like TGBp1 (Howard et al. 2004). The TGBp1 helicase activity was essential for protein/RNA translocation through PD to adjacent cell for both types of viruses with TGB (Morozov and Solovyev 2003). The suppressing activity of RNA silencing was shown to be another function of PVX-TGBp1 protein. Further, it was suggested that silencing suppression was essential for TGBp1 protein to mediate cell-to-cell movement of viruses (Bayne et al. 2005). The transport component of RNPs or virions involved in the cell-to-cell movement of PVX genome include TGBp1 and CP. Both proteins are able to move through PD (Batten et al. 2003; Howard et al. 2004). The PVX TGBp1 is able to modify PD and move between cells. In rod-shaped viruses, TGBp2 and TGBp3 move to PD and act in concert to transport TGBp1 to and through PD. In PVX, the 8 kDa TGBp3 is a membrane-embedded protein that has an N-terminal hydrophobic sequence segment and hydrophilic C-terminus. The TGBp3 mutants with deletion in the C-terminal hydrophilic region retain the ability to be targeted to cell peripheral structures and to support limited PVX cell-to-cell movement (Schepetilnikov et al. 2005). For BNYVV, three proteins are encoded by subgenomic RNA-2 sub-a for TGBp1 (42 kDa) and RNA-2 sub-b for TGBp2 (13 kDa) and TGBp3 (15 kDa) (Gillmer et al. 1992). TGBp1 protein contains a helicase domain and is able to bind nucleic acids *in vitro* (Bleykasten et al. 1996). Complementation studies of TGBp1 protein with a virally expressed GFP::TGBp1 fusion protein permitted to localize TGBp1 in PD only in the presence of the TGBp2 and TGBp3 proteins. Furthermore, TGBp2 and TGBp3 were detected within the PD only, when all the TGB proteins were expressed in the same cell (Erhardt et al. 2005).

The major difference between two groups of viruses containing TGB is the requirement of coat protein (CP) for Group 2 viruses and dispensability of CP for Group 1 viruses for cell-to-cell movement. This factor suggests that the RNP of Group 1 viruses cannot move in the form of intact virions. The investigations on subcellular localization of TGB proteins of PMTV, when expressed as N-terminal fusions to GFP from a TMV-based vector showed that both the PMTV GFP-TGB2 and GFP-TGB3 fusions were associated with cellular endomembranes, particularly the ER network and membranes surrounding the nucleus. The GFP-TGB3-labeled opposing pairs of fluorescent spots, across neighboring cell walls that suggested localization at or near PD, were observed (Cowan et al. 2002). The TGB1 protein interaction with RNA and its transport as part of the RNP complex out of the cell and long-distance throughout the plant has been reported. But the details of the process in which TGB2 and TGB3 proteins facilitate RNP in intercellular transport are unclear. The interaction of TGB2 and TGB3 proteins of PMTV with components of the secretory and endocytic pathways was investigated. These proteins were expressed as N-terminal fusions to GFP or monomeric red protein (mRFP). The fluorophore-labeled TGB2 and TGB3 showed an early association with ER and colocalized in mobile granules that used the ER-actin network for intracellular movement. Both proteins were able to increase the SEL of PD and TGB3 accumulated at PD in the absence of TGB2. Both fusion proteins were incorporated into vesicular structures during expression cycle. TGB2 was associated with its structures independently, while TGB3 required TGB2 for its incorporation into the vesicles. In addition to colocalization to the ER and motile granules, mRFP-TGB3 was incorporated into vesicles, when expressed in PMTV-infected epidermal cells, indicating recruitment by virus-expressed TGB2. The experiments involving labeling of the TGB fusion proteins-containing vesicles with FM4-64 indicated that the endocytic pathway was involved in viral intracellular movement (Haupt et al. 2005).

The process of increasing the SEL of PD is known as gating and this is a characteristic of the MPs of plant viruses. Several movement complexes (MCs) are more complex than originally visualized and different functions required for cell-to-cell movement may be performed by separate proteins. Gating appears to be a basic requirement for cell-to-cell movement of non-tubules-forming viruses. Actin is involved in depolymerizing the actin leading to increase in SEL. In contrast, callose deposition has been shown to close PD during defense and wound responses (Roberts and Oparka 2003). The interaction between PVX TGB2 and proteins that interact with ß-1,3-glucanase, a callose-degrading enzyme has been suggested to be a strategy of PVX employed to gate PD to accelerate callose degradation (Fridborg et al. 2003). In the case of TMV, the interaction of MP with pectin methylesterase (PME) may regulate the activity of PME, loosening the cell wall around PD. This may result in permitting the PD to open more easily. Alternatively, the MP may recruit additional PME to PD to assist gating rather than relying on PME for targeting (Chen et al. 2000). A nonisotopic molecular dot blot hybridization technique and multiplex transcription-polymerase chain reaction (mRT-PCR) assay were employed to study the distribution of MiLV and LBVV) in lettuce plants inoculated using viruliferous O. brassicae . The highest concentrations of LBVV and MiLV were found in root and old leaves, respectively. LBVV infection progress in lettuce crops was faster than that of MiLV. However, at about 100 days after transplantation, most of the plants were infected by both viruses. The appearance of both viruses in lettuce crops was preceded by a peak in the population of resting spores and zoosporangia of the vector O. brassicae in lettuce roots, indicating the relationship between vector population and disease incidence (Navarro et al. 2004).

The ability of a virus to counter host defenses requires an active mechanism to either bypass or disarm the host defense machinery. Silencing suppressors encoded by viruses reduce degradation of viral RNAs by the RNA silencing machinery. Among plant viruses, some silencing suppressor proteins also affect symptom development and increase virus titer. Small cysteine-rich proteins (CRPs) near the 3' ends of viral genomes have been identified as both silencing suppressor proteins and pathogenicity factors of TRV. The 16 K CRP was found to be a pathogenicity factor and also a suppressor of RNA silencing (Reavy et al. 2004). The 19 K CRP of SBWMV RNA-2 encodes for the viral replicase and putative viral MP. SBWMV RNA-2 encodes four proteins – 25 K protein, coat protein (84 K), the CP readthrough domain (RTD) required for vector transmission and 19 K CRP encoded by the 3' proximal ORF of RNA-2. In order to determine the potential of SBWMV 19 K CRP functioning as suppressor of RNA silencing in plants, the reversal of silencing assay procedure was applied. The SBWMV 10 K CRP prevented RNA silencing only in emerging leaves where RNA silencing had not developed prior to virus infection. The SBWMV 19 K ORF was inserted into *Potato virus X* (PVX) genome and PVX 19 K infectious transcripts were inoculated onto *N. benthamiana, N. clevelandii, C. quinoa* and *C. amaranticolor*. As control plants were also inoculated with PVX-GFP which

had GFP gene inserted into PVX genome. The SBWMV 19 K CRP, when expressed from PVX genome induced systemic necrosis on all test plant species. These symptoms were distinct from symptoms associated with PVX infection in those hosts and from symptoms induced by SBWMV in its natural hosts. The presence of 19 K CRP of SBWMV aggravated the symptoms induced by PVX and restored GFP expression to GFP silenced tissues. The results indicated that SBWMV 19 K CRP was a pathogenicity determinant, as well as a suppressor of RNA silencing in plants (Te et al. 2005). The use of an immunogold labeling technique revealed that the replicase and P42 antisera labeled the zoospore cytoplasm, storage bodies, vacuoles. The highest concentration of coat protein and RTD of the coat proteins was in the vacuoles/vesicles. RTD was also prevalent in the zoospore cytoplasm. P13 was prevalent in the cytoplasm and vacuoles. P15 and P14 were scattered and did not localize in any specific site. Significant levels of P25 and P31 proteins which are required for transmission and virus accumulation in roots, were detected in most locations. Further, P25 was detected in the nucleus as observed in plant cells (Lubicz et al. 2007; Figure 6.10).

BSBMV and BNYVV are transmitted by the same vector, *P. betae* and have similar host ranges, particle number and morphology. BNYVV has a worldwide distribution, whereas incidence of BSBMV has been reported only in the United States. BNYVV and BSBMV are not serologically related, but they have similar genomic organization. Field isolates usually consist of four RNA species, but some BNYVV isolates contain a fifth RNA. RNAs 1 and 2 are essential for infection and replication, while RNAs 3 and 4 have important roles in plant and vector interactions, respectively. Nucleotide and amino acid sequence analysis showed that BSBMV and BNYVV had significant differences required for considering them as two different species. In *B. macrocarpa*, long-distance movement of BNYVV was dependent on the presence of RNA-3 and involved a nucleotide sequence named 'core;. Sequence comparisons of BNYVV and BSBMV RNA-3 led to the identification of a stretch of 22 conserved nucleotides present within the core sequences. The possibility of using BSBMV RNA-3 to replace BNYVV for long-distance movement was investigated. Complementary base changes found within the BSBMV RNA-3 5' UTR had resemblance to BNYVV 5' RNA-3 structure, whereas the 3' UTRs of both species were more conserved. The cDNA clones were allowed complete copies of BSBMV RNA-3 to be trans-replicated, transencapsidated by the BNYVV viral machinery. Long-distance movement was observed, indicating that BSBMV RNA-3 could substitute BNYVV RNA-3 for systemic spread, even though the P29 encoded by BSBMV RNA-3 was much closer to the RNA-5-encoded P26 than BNYVV RNA-3-encoded P25. Competition was observed when BSBMV RNA-3-derived replicons were used together with BNYVV-derived RNA-3, but not when the RNA-5-derived component was employed (Rattin et al. 2009).

Plant viruses have to move over long distances to become systemic and to infect plant organs away from the site of inoculation/infection. Virus-specific proteins may have a role in invasion of different host tissues. The molecular basis of long-distance movement of plant viruses via vascular elements has been investigated only in a few pathosystems. Using a modified TMV, expressing GFP, the systemic vascular invasion routes of TMV in *N. benthamiana* were investigated. TMV could enter minor, major or transport veins directly from non-vascular cells to cause systemic infection (Cheng et al. 2000). The gene expression profiles of *Arabidopsis thaliana* ecotype Shahdara which is as susceptible as tomato and tobacco to TMV and permits rapid accumulation of the virus in both inoculated and systemically infected host plant tissues (Dardick et al. 2000). Gene expression levels in Shahdara were monitored in inoculated leaves at four days postinoculation (*dpi*) and in systemically infected leaves at 14 *dpi*, using cDNA microarrays developed by the Arabidopsis Functional Genomics Consortium (AFGC). The cDNA microarray technique permitted the simultaneous screening of approximately 8,000–10,000 *Arabidopsis* genes for their response to TMV infection. Based on the expression analysis, 68 genes that displayed significant and consistent changes in expression levels, either up or down, in either TMV-inoculated or systemically infected tissues or both were identified. At 4 *dpi*, all of the identified TMV-responsive genes were induced in transcription. In contrast, 35 of the 53 genes regulated at 14 dpi showed repression. This increase in the number of repressed genes at 14 dpi may reflect the general repression of host genes. The identified TMV-responsive genes encode a diverse array of functional proteins that include transcription factors, antioxidants, metabolic enzymes and transporters. The TMV-responsive genes may alter a wide range of host cellular functions that may reflect the biochemical and physiological changes involved in the development of this syndrome (Golen and Culver 2003).

### 6.2.5.3 Movement of Viruses Transmitted by Nematode Vectors

Plant viruses replicate in the initially infected susceptible plant cells to reach adequate concentration before the commencement of their movement to neighboring cells/tissues in plants. Plant viral replication proceeds in association with membranes originating from different sources such as endomembrane system, chloroplasts and mitochondria. GFLV-infected cells exhibit proliferation of membranes that generally accumulate in the nuclear periphery to form a viral compartment. The membranes of the viral compartment originated primarily from an ER, as revealed by cells expressing the GFP targeting different cell components as a marker. The perinuclear compartment was found to be the site of GFLV-replication, as indicated by observations under immunoconfocal microscope. Double-stranded replicative forms (RFs), newly synthesized viral RNA and the RNA-1-encoded VPg (virus-associated genomic protein) were generated (Ritzenthaler et al. 2002). The polyprotein P1 encoded by RNA-1 of GFLV was required for viral RNA replication and it was responsible for the formation of viral compartment, although the nature of maturation product(s) responsible for the recruitment of vesicles from the ER was not clearly determined. RNA-2 was replicated in trans by RNA-1 encoded replication machinery and protein 2A might be necessary for RNA-2 replication. Protein 2A accumulated in the viral compartment and its subcellular distribution was affected by the replication process (Andret-Link et al. 2004).

After replication of GFLV possibly in the perinuclear area where all replication proteins accumulate, GFLV virions have to move to the cell periphery and probably to PD through which they move further to adjacent cells. Such movement of GFLV may be in two distinct steps: (i) intracellular movement from perinuclear site of RNA synthesis and (ii) virus assembly to the cell periphery and then intercellular movement across the cell wall through PD. The highly stable MP did not accumulate in the viral compartment, but formed virion-filled tubules that either protruded from the surface of infected protoplasts or were embedded within highly modified PD in infected cells (Ritzenthaler et al. 1995a, b). GFLV movement was dependent on both MP and CP, as indicated by the RNA-2 deletion mutants and chimeric GFLV/ArMV RNA-2 constructs (Belin et al. 1999; Gaire et al. 1999). The requirement of only MP for tubule formation was demonstrated using transgenic tobacco BY-2 cell line, expressing GFP:MP (Laporte et al. 2003). This transgenic cell line allowed the spatio-temporal analysis of GFLP MP intracellular transport and assembly into tubules. By employing various inhibitors, the requirement of a functional secretory pathway and intact microtubules for the intracellular MP transport to cross cell walls through PD was demonstrated. On the other hand, tubule formation *per se* was independent of cytoskeleton. Further, MP:GFP accumulated within the cell plate in dividing cells. The MP self-assembled into tubules in the proximity of PD at the cell periphery (Laporte et al. 2003). Transport of GFLV particles for long-distance to reach other tissues of infected plants has not been studied and it is presumed that long-distance transport of GFLV may occur in a pattern similar to that of the viruses transmitted by fungal vectors.

TRV, belonging to the genus *Tobravirus*, family Virgaviridae, has a wide host range and worldwide distribution. In addition to transmission by *Trichodorus* sp. and *Paratrichodorus* sp., TRV is also transmitted via seed and pollen. The distribution of TRV in tobacco and pepper in vegetative tissues and pollen and seed was studied. The presence of TRV in pollen grains and inside ovaries was linked with seed transmission and it could have effects on transport of virus particles during pollination and fertilization process. The impact of TRV on pepper and tobacco anthers and ultrastructural changes in ovaries were assessed. The presence of two types of TRV particles in ovary wall parenchyma and vascular tissues as well as in placenta cells was revealed by ultrastructural analysis. Regular inclusion of virus particles in both ovules integuments and nucellus parenchyma cells was observed. Immunolocalization of TRV capsid proteins indicated the deposition of TRV-CP epitope in ovary vascular bundles and in placenta cells. Further, the presence of TRV particles was visualized inside pepper seeds in endothelium and integument parenchyma layers, as well as on the embryo cell wall. TRV particles were present not only on the surface of pollen grains, but also inside pepper pollen protoplasts in mature anthers. The presence of TRV was observed in both differentiated and endothelium cells and the remaining tapetum cells. In addition, detection of TRV CP epitope in tobacco and pepper vascular anther tissues as well as in tapetum and endothelium cells was correlated with TRV distribution in infected anthers. The results indicated that pollen grains and ovaries with ovules could function as natural sources of TRV inoculum (Katarzyna et al. 2016).

## 6.2.6 ROLE OF RNA SILENCING IN VIRUS DISEASE DEVELOPMENT

Most of the soilborne viruses infect the roots of susceptible host plants. Roots of plants, compared to shoots, have distinct metabolism and physiological characteristics, due to differences in development, cell composition, gene expression patterns and surrounding environmental conditions to which the root system is continuously exposed. Hence, the viruses have to adapt to the root environments to establish successful infection. RNA silencing involves downregulation of gene expression, mediated by small RNAs in eukaryotes. The host plant initiates RNA silencing as an antiviral defense strategy soon after the infection of roots. To counteract antiviral silencing, most of the plant viruses encode silencing suppressor proteins (Li and Ding 2000; Csorba et al. 2015). In plants, RNA silencing is initiated when imperfect or true ds-RNA derived from cellular sequences or viral genomes are processed by ribonucleases III-like protein in the Dicer family called 'Dicer-like (DCL) proteins' to generate 21–22 nt microRNAs (mRNAs) or 21–26 nt short interfering RNAs (siRNAs) duplexes. Each strand of small RNA is then incorporated into effector complexes designated 'RNA-induced silencing complexes' (RISCs) which contain ARGONAUTE (AGO) proteins, to guide the sequence specificity in the downregulation processes (Axtell 2013; Martinez de Alba et al. 2013). Plant-encoded RNA-dependent RNA polymerases (RDRsRdRp) could contribute to the generation of ds-RNA substrates for DCL processing, resulting in either the initiation of RNA silencing or the production of secondary small RNAs that further intensify the potency of RNA silencing (Dalmay et al. 2000; Wang et al. 2010). Plants encode multiple DCL, AGO and RDR proteins to meet the demands of diverse endogenous RNA silencing pathways (Zhang et al. 2015).

Several plant-infecting single-stranded (ss)-RNA viruses belonging to about 17 genera in eight virus families, but not the DNA or ds-RNA viruses, are known to be transmitted by soil-inhabiting organisms. It is possible that some of the viruses with abiotic transmission may be shown to have natural biological vectors present in the soil. Several studies have indicated the occurrence and mechanism of RNA silencing in the roots. By analyzing the gene regulation, involving RNA silencing in roots, these studies revealed some unique characteristics of RNA silencing in roots, relative to those observed in leaves or other aerial plant organs. The accumulation of small interfering (si) RNAs derived from various ss-RNA viruses was detected in the roots of infected plants, including

tomato, cucumber, melon and *N. benthamiana*, demonstrating that viruses could induce antiviral RNA silencing responses in roots (Andika et al. 2005, 2013; Herranz et al. 2015). BNYVV siRNA accumulation was lower in roots than in leaves of *N. benthamiana* and inversely related with RNA genome accumulation, suggesting that BNYVV might be able to suppress RNA silencing in roots more effectively than in leaves (Andika et al. 2005).

In virus-infected plants, RNA silencing involves a mobile silencing signal that can move cell-to-cell and also systemically through plants. The silencing signal may be linked to the defense activity, if it moves through the plant either with or ahead of an invading virus. Plant viral genomes (as in *Tombusvirus*) encode suppressors of RNA silencing that are also effectors of long-distance movement of the virus. The MP of *Potato virus X* (PVX) was assessed for its ability as a suppressor of silencing. This protein is one of the three TGB proteins of potexviruses required for cell-to-cell movement of the virus (Yang et al. 2000). Through random mutation analysis of PVX silencing suppressor p25, evidence was obtained to show that suppression of silencing is necessary, but not sufficient, for cell-to-cell movement through PD. All mutants that were defective for silencing suppression, were also nonfunctional in viral cell-to-cell movement. It appeared that a second P25 functioning independent of silencing, might also be required for cell-to-cell movement. Two classes of suppressor inactive P25 mutants were identified, one of them being functional for the accessory function. Based on the analyses of short interfering RNA accumulation, it was concluded that P25 may suppress silencing by interfering with either assembly or function of the effector complexes of RNA silencing (Bayne et al. 2005).

If silencing signal moves through the plant either with or ahead of invading virus, it may have a role in virus silencing. The silencing machinery in cells receiving the signal could then be primed to target the viral RNA as it enters the cell, so that virus movement through the plant would be impeded. Systemic movement of *Potato virus X* (PVX) into the growing point of infected plants was shown to be dependent on a host-encoded RNA-dependent RNA polymerase (RDR) which formed an integral part of the RNA silencing machinery (Belkahadir et al. 2005). The RDR6 in *N. benthamiana* was involved in the defense against PVX at the level of systemic spread and in exclusion of the virus from the apical growing point. The RDR6 did not affect primary replication and cell-to-cell movement of the virus. In addition, it had no significant influence on the formation of virus-derived small interfering (siRNA) in a fully established PVX infection. The results suggested that RDR6 might use the incoming silencing signal to generate ds-RNA precursors of secondary siRNA. The secondary siRNAs may mediate RNA silencing as an immediate response that slows down the systemic spread of the virus into the growing point and newly emerging leaves (Schwach et al. 2005). Analyses using next-generation sequencing indicated that siRNAs derived from *Potato virus X* (PVX) and CWMV, genus *Furovirus*, are predominantly 21-nt in both leaves and roots, indicating that DCL4 protein is also the major DCL component for the biosynthesis of viral siRNAs in roots (Andika et al. 2013; Herranz et al. 2015). In roots, DCL proteins might preferentially target the sense strand of the genome of PVX and CWMV through cleaving of the secondary structures within viral RNA to generate sense siRNAs (Herranz et al. 2015). RNA silencing strongly inhibits PVX replication in roots of susceptible plants of *A. thaliana*. *A. thaliana* is not a susceptible host of PVX, but inactivation of DCL4 enables high accumulation of PVX in inoculated leaves, while inactivation of both DCL4 is critical for intracellular antiviral silencing against PVX replication in shoots. There are strong functional redundancies among DCL proteins, in which other DCLs (most probably DCL2) functionally complement DCL4 in roots. These redundancies might result in potent inhibition of PVX replication in roots possibly by providing multiple layers of antiviral defense (Andika et al. 2015).

Posttranscriptional gene silencing (PTGS) functions as an endogenous defense mechanism against plant virus infection by directly targeting viral genome integrity and consequently lowering the titer of the invading virus. It is possible to redirect this mechanism to target indigenous host mRNAs by introducing corresponding plant cDNA fragments within the viral genomes, thus providing a means to downregulate host genome expression. This approach known as virus-induced gene silencing (VIGS) permits rapid transient knockdown of host gene expression. VIGS takes advantage of the PTGS defense system to silence endogenous RNA sequences that are homologous to a sequence engineered into a viral genome which generated ds-RNA that mediates silencing, whereas VIGS offers more flexibility. PTGS-based approaches need stable transformation of antisense or inverted repeat sequences (Wesley et al. 2001). The RNA viruses TMV, *Potato virus X* (PVX) and TRV have been used as VIGS vectors. Many host plant genes have been successfully silenced in foliar tissue, meristematic tissue and potato tubers (Ratcliff et al. 2001; Faivre-Rampant et al. 2004). The efficiency of VIGS is dependent mainly on the ability of the virus to invade the host and replicate to the required level in the targeted tissue. Tobraviruses such as TRV are transmitted by nematodes and need a viral helper protein encoded by the RNA-2-2b gene (Vassilakos et al. 2001; Vellios et al. 2005). A modified TRV vector retaining the helper protein 2b required for transmission of TRV was constructed. This TRV vector replicated extensively in whole plants, including meristems and also triggered a pervasive systemic VIGs response in the roots of *N. benthamiana*, *Arabidopsis* and tomato. Roots and silenced plants exhibited reduced levels of target mRNA. The TRV-2b vector exhibited enhanced infectivity and meristem invasion, both key requirements for efficient VIGS-based functional characterization of genes in root tissues. The results suggested that the TRV-2b helper protein might have a vital role in the host regulatory mechanism capable of controlling TRV invasion. The p25 protein of potexviruses was considered to be involved in the inhibition of PTGS and the inhibition of enhancement of systemic silencing signal of PTGS (Voinnet et al. 2000). The product of ORF2 of *Potato virus X* (PVX),

p25 was thought to interfere with the assembly of siRNAs in RISC (Voinnet et al. 2000). On the other hand, *Tombusvirus* p19 was able to bind with siRNAs, thus making them unavailable for processing by RISC (Lakatos et al. 2004). These proteins have been shown to interfere with the steady levels of microRNAs (Chapman et al. 2005). The results of these studies suggest that the suppression of RNA silencing or other antiviral defense mechanisms is one of the factors that critically influence the efficiency of virus transmission through roots of susceptible host plant species.

SBWMV, a bipartite RNA virus, is the type member of the genus *Furovirus*. RNA encodes the viral replicase and putative viral MP. The viral replicase is encoded by a single large ORF and is phylogenetically related to the TMV replicase. SBWMV RNA encodes four proteins. The 5′ proximal ORF of RNA2 encodes a 25 K protein. The coat protein (CP) ORF has an opal translational termination codon that produces a large 84 K protein. The CP readthrough (RT) domain is needed for plasmodiophorid transmission of the virus. The 3′ proximal ORF of RNA2 encodes a 19 K CRP (Shirako and Wilson 1993; Shirako 1998; An et al. 2003). SBWMV has a CRP similar to those in the members of the genera *Tobamovirus* and *Hordeivirus*, for which CRPs were described as pathogenicity determinants that regulated symptom severity in infected plants. The 19 K protein of SBWMV with nine Cys residues was characterized to study its role in suppressing RNA silencing in infected plants. The reversal of silencing assay was employed to determine, if SBWMV 19 K CRP was a suppressor of RNA silencing in plants. The role of SBWMV 19 K protein on symptom development was investigated. The SBWMV 19 K ORF was inserted into the *Potato virus X* (PVX) genome and PVX-19 K infectious transcripts were used to inoculate *N. benthamiana*, *N. clevelandii*, *C. quinoa* and *C. amaranticolor* leaves. Control plants were inoculated with GFP gene to monitor viral movement. The spread of PVX-GFP expression was monitored, using a hand-held UV lamp and its systemic accumulation may be verified by this technique. The SBWMV 19 K CRP, when expressed from PVX genome, functioned as a suppressor of RNA silencing. The GFP-expression in the 16C transgenic *N. benthamiana* plants was silenced by infiltrating young leaves with a suspension of *Agrobacterium*-expressing GFP. Symptoms began to appear in plants inoculated with PVX-GFP and PVX.19K between ten and 12 days postinoculation (*dpi*). Systemic necrosis was observed in N. *benthamiana* and *N. clevelandii* plants inoculated with PVX.19K by 21 *dpi*, whereas systemic mosaic symptoms were induced in plants infected with PVX-GFP. Stunting of *N. benthamiana* plants infected with PVX-GFP could be recognized. N. *clevelandii* leaves collapsed, following infection with PVX.19K by 21 *dpi* (see Figure 6.8). The progression of GFP silencing could be visualized first locally and then systemically. Within two weeks, the spread of GFP silencing was observed systemically and by three weeks, the only visible fluorescence was red fluorescence from chloroplasts. At this stage, the silenced plants were inoculated with PVX.19K. As the PVX.19K virus spread locally and then systemically, there was no change

**FIGURE 6.8** Effects of inoculating assay plants with PVX.19K infectious transcripts or PVX.GFP on symptom development at 21 days postinoculation (dpi) A: *N. benthamiana* plants infected with PVX.GFP (left) and PVX.19K (right); B and D: development of systemic necrosis in *N. benthamiana* and *N. clevelandii* plants, respectively at 21 dpi; C: *N. clevelandii* plants infected with PVX.GFP; E and F: *C. quinoa* and *C. amaranticolor* leaves infected with PVX.10K (left in both panels) and PVX.GFP (right in both panels).

**[Courtesy of Te et al. (2005) and with kind permission of *Virology Journal* Med-Central, Open Access Journal]**

in GFP expression in the inoculated leaves or in the upper leaves. However, GFP expression was observed in the emerging leaves. The SBWMV 19K CRP prevented RNA silencing only in emerging leaves, where RNA silencing had not developed prior to virus infection. Immunoblot analysis revealed high levels of PVX.CP in plants systemically infected with PVX-GFP (see Figure 6.9A) and PVX.19K (see Figure 6.9B). The SBWMV 19 K CRP did not induce a clear effect on PVX accumulation in upper inoculated leaves. Viral RNA accumulation was analyzed by Northern blot which showed high levels of genomic RNA in the upper leaves inoculated with PVX.GFP (see Figure 6.10A) and PVX.19K (see Figure 6.10B) in inoculated plants. The results indicated that the SBWMV 19 K did not appear to induce an adverse effect on PVX accumulation (Te et al. 2005).

TBSV often infects roots of plants under natural conditions. But TBSV can readily infect plants through mechanical inoculation on leaves. A method was developed to compare leaf versus root inoculation of *N. benthamiana* and tomato with transcripts of the wild-type TBSV (wtTBSV), capsid (Tcp)

replacement construct, expressing GFP ((T-GFP), or mutants not expressing the silencing suppressor P19 (TBSVΔp19). Two week-old *N. benthamiana* seedlings (plantlets), after inoculation with transcripts of wtTBSV (expressing Tcp or T-GFP) were transferred into water-filled Faclcon tubes covered with aluminum foil. Upon mechanical inoculation of *N. benthamiana* leaves with T-GFP transcripts, little or no systemic GFP expression occurred. However, when Tcp was provided via expression from a heterologous PVX vector, systemic spread of T-GFP was readily observed, indicating a beneficial role of Tcp in accumulation of T-GFP. The results suggested that Tcp expression both stabilized and enhanced the spread of T-GFP. On the other hand, in inoculated aerial tissues, unlike roots, the Tcp might be required for entry into or exit from internal phloem. The presence of Tcp in xylem vessels in aerial tissue above the inoculated leaves indicated the potential of the Tcp to influence entry or exit from vascular tissue when establishing systemic infection. The Tcp-mediated accelerated infection after aerial inoculation could be related to entry or exit into both xylem and internal phloem. Leaf inoculation with

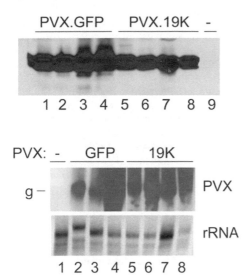

**FIGURE 6.9** Accumulation of PVX in the leaves of *N. benthamiana* plants inoculated with PVX.GFP and PVX-19K as determined by immunoblot and northern analyses Top: Similar levels of PVX. GFP virus (lanes 1 to 4) and PVX-19K virus (lanes 5 to 8) detected by immunoblot assay; Bottom: RNA contents of healthy plant (lane 1) upper noninoculated leaves of PVX.19K-infected plant (lanes 2 to 4) and upper noninoculated leaves of PVX.GFP-infected plant (lanes 5 to 8) quantified by northern blot technique using GFP sequence as probes. Bottom image obtained after staining the gel with ethidium bromide for detection of ribosomal RNAs.

[Courtesy of Te et al. (2005) and with kind permission of *Virology Journal* Med-Central, Open Access Journal]

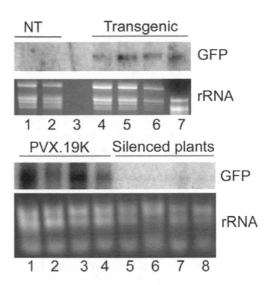

**FIGURE 6.10** Suppression of RNA silencing by SBWMV 19K CRP. Top: Northern blot analyses of total RNAs from tissues of non-transgenic plants (lanes 1 and 2); tissues of GFP transgenic plants (lanes 4 to 7) probed with a labeled GFP sequence probe; and lane (3) blank; presence of ribosomal RNAs was detected on ethidium bromide-stained gel. Bottom: Northern analysis of total RNAs from 16 plants infiltrated with *Agrobacterium* containing constructs and probed with a labeled GFP sequence probe; RNA samples from plants inoculated with PVX.19K (lanes 1 to 4); RNA samples from plants inoculated with PVX.GVS (lanes 5 to 8); ribosomal RNA detectable by staining the gel.

[Courtesy of Te et al. (2005) and with kind permission of *Virology Journal* Med-Central, Open Access Journal]

wtTBSV and TBSΔp19 transcripts resulted in rapid systemic invasion of *N. benthamiana* prior to the onset of apical necrosis for wtTBSV or silencing-mediated viral RNA clearance and recovery for TBSVΔp19 infections. Similar results were obtained with tomato plants inoculated with wtTBSV and TBSVΔp19. Root inoculations with T-GFP transcripts showed green fluorescence in upper stem parts, starting at between five and seven days postinoculation (*dpi*), as confirmed by GFP signal in Western blots and at 10–12 *dpi*, fluorescence was observed in the leaf sections. Immunoblot analysis verified the visual observations on accumulation of P19. Similar results were obtained upon root inoculations of pepper and tomato plants that also clearly revealed the presence of P19 monomer. In order to determine the contribution of P19 to systemic infection upon root inoculation, tissues of root-inoculated plants were harvested approximately one week after inoculation and Tcp accumulation was monitored to determine the spread of wtTBSV and TBSVΔp19. Western blot analysis of extracts of leaf-inoculated plants with wtTBSV transcripts showed that Tcp (41-kDa CP) accumulated in upper leaves, stem tissues and roots. Extracts from root-inoculated plants also showed similar results. However, systemic spread of TBSVΔ-p19 only occurred upon leaf inoculation, whereas no CP accumulated in root-inoculated plants. Similar results were obtained also from tomato and pepper plants. Upon root inoculation with transcripts, systemic infection occurred only for wtTBSV, but not for TBSVΔp19. The contribution of Tcp or p19 in establishing systemic infection depended on the point of entry by TBSV in plants (Manabayeva et al. 2013).

## REFERENCES

Adams MJ (1990) Epidemiology of fungally-transmitted viruses. *Soil Use Manag* 6: 184–189.

Adams MJ, Antoniw JF, Kreuze J (2009) Virgaviridae: A new family of rod-shaped plant viruses. *Arch Virol* 154: 1967–1972.

Alfaro-Fernández A, Córdoba-Sellés MDC, Herrera-Vasquez JA, Cebrián MDC, Jordá C (2010) Transmission of *Pepino mosaic virus* by the fungal vector *Olpidium virulentus*. *J Phytopathol* 158: 217–226.

An H, Melcher U, Doss P et al. (2003) Evidence that the 37 kDa protein of *Soilborne wheat mosaic virus* is a virus movement protein. *J Gen Virol* 84: 3153–3163.

Andika IB, Kondo H, Tamada T (2005) Evidence that RNA silencing-mediated resistance to *Beet necrotic yellow vein virus* is less effective in roots than in leaves. *Molec Plant-Microbe Interact* 18: 194–201.

Andika IB, Maruyama K, Sun I, Kondo H, Tamada T, Suzuki N (2015) Differential contributions of plant Dicer-like proteins to antiviral defences against *Potato virus X* in leaves and roots. *Plant J* 81: 781–793.

Andika IB, Sun L, Xiang R, Li J, Chen J (2013) Root-specific role for *Nicotiana benthamiana* RDR6 in the inhibition of *Chinese wheat mosaic virus* accumulation at higher temperatures. *Molec Plant-Microbe Interact* 26: 1165–1175.

Andret-Link C, Laporte C, Valet L et al. (2004a) *Grapevine fan leaf virus*: still a major threat to grapevine industry. *J Plant Pathol* 86: 183–195.

Andret-Link P, Schmitt-Keichinger C, Demangeat G, Komar V, Fuchs M (2004b) The specific transmission of *Grapevine fan leaf virus* by its nematode vector *Xiphinema index* is solely determined by the viral coat protein. *Virology* 320: 12–22.

Angell SM, Davies C, Baulcombe DC (1996) Cell-to-cell movement of *Potato virus X* is associated with a change in size exclusion limit of plasmodesmata in trichome cells of *Nicotiana clevelandii*. *Virology* 216: 197–201.

Arif M, Ali M, Rehman A, Fahim M (2014) Detection of *Potato mop-top virus* in soils and potato tubers using bait-plant assay, ELISA and RT-PCR. *J Virol Meth* 195: 221–227.

Asher M (1999) Sugar beet rhizomania: the spread of a soilborne disease. *Microbiol Today* 26: 120–122.

Axtell MJ (2013) Classification and comparison of small RNAs from plants. *Annu Rev Plant Biol* 64: 137–159.

Bahrani Z, Sherwood JL, Sanborn MR, Keyser GC (1988) The use of monoclonal antibodies to detect *Wheat soilborne mosaic virus*. *J Gen Virol* 69: 1317–1322.

Bashir NS, Hajizadeh M (2007) Survey for *Grapevine fan leaf virus* in vineyards of north-west Iran and genetic diversity of isolates. *Austr Plant Pathol* 36: 46–52.

Batten JS, Yoshinari S, Hemenway C (2003) *Potato virus X*: a model system for virus replication, movement and gene expression. *Molec Plant Pathol* 4: 125–131.

Bayne EM, Rakitina DV, Morozova SY, Baulcombe DC (2005) Cell-to-cell movement of *Potato potexvirus X* is dependent on suppression of RNA silencing. *Plant J* 44: 471–482.

Beijerinck MW (1898) Over bean contagium vivum fluidum also orzaak van de vlekziekte der tabaksbladen. *Versl.gewone Vergad Wisen naturrk Afd K Akad Wet Amst* 7: 229–235.

Belin C, Schmitt C, Demangeat G, Komar V, Pinck L, Fuchs M (2001) Involvement of RNA-2 encoded proteins in the specific transmission of *Grapevine fan leaf virus* by its nematode vector *Xiphinema index*. *Virology* 291: 161–171.

Belin C, Schmitt C, Gaire F, Walter B, Demangeat G, Pinck L (1999) The nine C-terminal residues of the grapevine fan leaf nepovirus movement protein are critical for systemic virus spread. *J Gen Virol* 80: 1347–1356.

Belkahadir Y, Nimchuk Z, Hubert DA et al. (2005) Cell-to-cell movement of *Potato potexvirus X* is dependent on suppression of RNA silencing. *Plant J* 44: 471–482.

Birgitte A, Andersen M, Steen N, Nielsen L (2002) Alternative hosts for *Potato mop-top virus*, genus *Pomovirus* and its vector *Spongospora subterranea* f.sp *subterranea*. *Potato Res* 45: 37–43.

Bleykasten C, Gillmer D, Gulley H, Richards KE, Jonard G (1996) *Beet necrotic yellow vein virus* 42 kDa triple gene block protein binds nucleic acid in vitro. *J Gen Virol* 77: 889–897.

Boben J, Kramberger P, Petrović N et al. (2007) Detection and quantification of *Tomato mosaic virus* in irrigation waters. *Eur J Plant Pathol* 118: 59–71.

Borodynko N, Rymelska N, Hasiow-Jaroszewska B, Pospieszny H (2009) Molecular characterization of three soil-borne sugar beet-infecting viruses based on the coat protein gene. *J Plant Pathol* 91: 191–193.

Broadbent L (1965) The epidemiology of tomato mosaic. VIII. Virus infection through tomato. *Ann Appl Biol* 55: 57–66.

Brunt AA, Crabtree K, Dallwitz M, Gibbs A, Watson L (1996a) *Viruses of Plants, Descriptions and Lists from the VIDE Database*. CAB International, Wallingford, UK.

Brunt AA, Crabtree K, Dallwitz MJ, Gibbs J, Watson I (1996b) *Arabis mosaic nepovirus*, *Cucumber mosaic cucumovirus*, *Tomato black ring nepovirus*, and *Tomato ringspot nepovirus* of plants. Descriptions and lists from VIDE database-Cambridge, UK. www.agls.uidaho.edu/ebi/vide/refs.htm.

Budge GE, Loram J, Donovan G, Boonham N (2008a) RNA2 of *Soil-borne cereal mosaic virus* is detectable in plants of winter wheat grown from infected seeds. *Eur J Plant Pathol* 120: 97–102.

Budge GE, Ratti C, Rubies-Autonell C et al. (2008) Response of UK winter wheat cultivars to *Soilborne cereal mosaic virus* (SBCMV) and *Wheat spindle streak mosaic virus* across Europe. *Eur J Plant Pathol* 120: 259–272.

Büttner C, Nienhaus F (1989) Virus contamination of soils in forest ecosystems of the Federal Republic of Germany. *Eur J For Pathol* 19: 47–53.

Cadman CH, Dias HF, Harrison BD (1960) Sap transmissible viruses associated with disease of grapevines in Europe and North America. *Nature* 187: 577–579.

Campbell RN (1996) Fungal transmission of plant viruses. *Annu Rev Phytopathol* 34: 87–108.

Card SD, Pearson MN, Clover GRG (2007) Plant pathogens transmitted by pollen. *Austr Plant Pathol* 36: 455–461.

Carnegie SF, Davey T, Saddler GS (2010) Effect of temperature on the transmission of *Potato mop-top virus* from seed tuber and by its vector, *Spongospora subterranea*. *Plant Pathol* 59: 22–30.

Carnegie SF, Davey T, Saddler GS (2012) Prevalence and distribution of *Potato mop-top virus* in Scotland. *Plant Pathol* 61: 623–631.

Castello JD, Rogers SO, Starmer WT et al. (1999) Detection of tomato mosaic tobamovirus RNA in ancient glacial ice. *Polar Biol* 22: 207–212.

Cerovska N, Filigarova M, Pečenková T (2006) Production of polyclonal antibodies to a recombinant *Potato mop-top virus* nonstructural triple gene block protein1. *J Phytopathol* 154: 422–427.

Chapman EJ, Prokhnevsky AI, Gopinath K, Dolja VV, Carrington JC (2004) Viral RNA silencing suppressors inhibit the microRNA pathway at an intermediate step. *Genes Develop* 18: 1179–1186.

Chen MH, Sheng J, Hind G, Handa A, Citovsky V (2000) Interaction in between the *Tobacco mosaic virus* movement protein and host cell pectin methylesterases is required for viral cell-to-cell movement. *EMBO J* 19: 913–920.

Cheng NH, Su CL, Carter SA, Nelson RS (2000) Vascular invasion routes and systemic accumulation patterns of *Tobacco mosaic virus* in *Nicotiana benthmiana*. *Plant J* 23: 349–362.

Clover GRG, Ratti C, Henry M (2001) Molecular characterization and detection of European isolates of *Soil-borne wheat mosaic virus*. *Plant Pathol* 50: 761–767.

CMI/AAB (1985) *Tobacco ringspot virus*. CMI/AAB Descriptions of plant viruses No. 309. Association of Applied Biologists, Wellesbourne, UK.

Cooper JI, Jones AT (1983) Responses of plants to viruses: proposals for the use of terms. *Phytopathology* 73: 127–128.

Cowan GH, Loliopoulou F, Ziegler A, Torrance L (2002) Subcellular localization protein interactions and RNA binding of *Potato mop-top virus* triple gene block proteins. *Virology* 298: 106–115.

Crosslin JM (2009) Detection of *Tobacco rattle virus* RNA in processed potato chips displaying symptoms of corky ringspot disease. *HortScience* 44: 1790–1791.

Crosslin JM, Hamm PB, Pike KS, Mowry TM, Nolite P, Mojltahedi H (2007) Managing diseases caused by viruses, viroids and phytoplasmas. In: Johnson DA (ed.), *Potato Health Management*, Second edition, The American Phytopathological Society, St Paul, MN, 161–169.

Csorba T, Kontra L, Burgyán J (2015) Viral silencing suppressors: Tools forged to fine-tune host-pathogen coexistence. *Virology* 17: 85–103.

da Silva R, de Souto ER, Pedroso JC et al. (2008) Detection and identification of TMV infecting tomato under protected cultivation in Paraná State. *Braz Arch Biol Technol* 51: 903–909.

Dale MFB, Robinson DJ, Todd D (2004) Effects of systemic infections with *Tobacco rattle virus* on agronomic and quality traits of a range of potato cultivars. *Plant Pathol* 53: 788–793.

Dalmay T, Hamilton A, Rudd S, Angell S, Baulcombe DC (2000) An RNA-dependent RNA polymerase gene in *Arabidopsis* is required for post-transcriptional gene silencing mediated by a transgene but not by a virus. *Cell* 101: 543–553.

Dardick CD, Golen S, Culver JN (2000) Susceptibility and symptom development in *Arabidopsis thaliana* to *Tobacco mosaic virus* is influenced by virus cell-to-cell movement. *Molec Plant-Microbe Interact* 13: 1139–1144.

Das S, Raski DJ (1968) Vector efficiency of *Xiphinema index* in the transmission of *Grapevine fan leaf virus*. *Nematologica* 14: 55–62.

Davey T, Carnegie SF, Saddler GS, Mitchell WJ (2014) The importance of the infected seed tuber and soil inoculum in transmitting *Potato mop-top virus* to potato plants. *Plant Pathol* 63: 88–97.

Delbianco A, Lanzoni C, Klein E, Autonell CR, Gilmer D, Ratti C (2013) Agroinoculation of *Beet necrotic yellow vein virus* cDNA clones results in plant systemic infection and efficient *Polymyxa betae* transmission. *Molec Plant Pathol* 14: 422–428.

Delfosse P, Reddy AS, Legréve A et al. (2000) Serological methods for detection of *Polymyxa graminis*, an obligate root parasite and vector of plant viruses. *Phytopathology* 90: 537–545.

Delfosse P, Reddy AS, Thirumala Devi K et al. (2002) Dynamics of *Polymyxa graminis* and *Indian peanut clump virus* (IPCV) infection on various monocotyledonous crops and groundnut during the rainy season. *Plant Pathol* 51: 546–550.

Demski TW, Kuhn CW (1989) *Tobacco ringspot virus*. In: *Compendium of Soybean Diseases*, Third edition, The American Phytopathological Society, MN, USA, pp. 57–59.

Dias HF (1963) Host range and properties of grapevine fan leaf and grapevine yellow mosaic viruses. *Ann Appl Biol* 51: 85–95.

Dieryck B, Weynes J, Doucert D, Bragard C, Legréve A (2011) Acquisition and transmission of *Peanunt clump virus* by *Polymyxa graminis* on cereal species. *Phytopathology* 101: 1149–1158.

Domfeh O, Gudmestad NC (2016) Effect of soil moisture management on the development of *Potato mop-top virus* induced tuber necrosis. *Plant Dis* 100: 418–423.

Donald RGK, Jackson AO (1996) RNA-binding activities of *Barley stripe mosaic virus* gamma b fusion proteins. *J Gen Virol* 77: 879–888.

Doucet DC (2006) Characterization of the *Pecluvirus* movement in plant roots (development of a model and comparison with a benyvirus). Ph. D. Thesis, Universite' Catholique de Louvain, pp. 226.

Driskel BA, Doss P, Littlefield IJ, Walker NR, Verchot-Lubicz J (2004) *Soilborne wheat mosaic virus* movement protein and RNA and *Wheat spindle streak mosaic virus* coat protein accumulate inside resting spores of their vector *Polymyxa graminis*. *Molec Plant-Microbe Interact* 17: 739–748.

Driskel BA, Hunger RM, Payton ME, Verchot-Lubicz J (2002) Response of hard red winter wheat to *Soilborne wheat mosaic virus*, using novel inoculation methods. *Phytopathology* 92: 347–354.

Edwardson JR, Christie RG (1997) *Viruses Infecting Peppers and Other Solanaceae Crops*. Gainville, USA, vol 2, pp. 337–390.

Etienne L, Clauzel JM, Fuchs M (1990) Simultaneous detection of several nepoviruses infecting grapevines in a single DAS-ELISA test using mixed antibodies. *J Phytopathol* 131: 89–100.

Faggioli F, Ferretti L, Pasquini G, Barba M (2002) Detection of *Strawberry latent ringspot virus* in leaves of olive trees in Italy using a one-step RT-PCR. *J Phytopathol* 150: 636–639.

Faivre-Rampant O, Gilony EM, Hrubikova K et al. (2004) *Potato virus X*-induced gene silencing in leaves and tubers in potato. *Plant Physiol* 134: 1308–1316.

Fattouch S, Acheche H, M'hirsi S, Marrakchi M, Marzouki N (2005) Detection and characterization of two strains of Grapevine fan leaf nepovirus in Tunisia. *OEPP/EPPO Bull* 35: 265–270.

Fauquet CM, Mayo MA, Maniloff J, Desselberger U, Ball LA (2005) *Virus Taxonomy*. 8th Report of International Committee on Taxonomy of Viruses, San Diego, USA, pp. 813–818.

Filhart RC, Bachand GD, Castello JD (1998) Detection of infectious tobamoviruses in forest soils. *Appl Environ Microbiol* 64: 1430–1435.

Finetti-Sialer MM, Ciancio A (2005) Isolate-specific detection of *Grapevine fan leaf virus* from *Xiphinema index* through DNA-based molecular probes. *Phytopathology* 95: 262–268.

Fridborg I, Grainger J, Page A, Coleman M, Finlay K, Angell S (2003) TIP, a novel host factor linking callose degradation with cell-to-cell movement of *Potato virus X*. *Molec Plant-Microbe Interact* 16: 132–140.

Fukuta S, Tamura M, Maejima H et al. (2013) Differential detection of *Wheat yellow mosaic virus*, *Japanese soil-borne wheat mosaic virus* and *Chinese wheat mosaic virus* by reverse transcription loop-mediated isothermal amplification reaction. *J Virol Meth*.

Gaire F, Schmitt C, Stussi-Garaud C, Pinck L, Ritzenthaler C (1999) Protein 2A of *Grapevine fan leaf nepovirus* is implicated in RNA-2 replication and colocalises to the replication sites. *Virology* 264: 25–36.

Gil JF, Adams N, Boonham N, Nielsen SL, Nicolaisen M (2016) Molecular and biological characterization of *Potato mop-top virus* (PMTV, *Pomovirus*) isolates from the potato-growing regions of Colombia. *Plant Pathol* 65: 1210–1220.

Gillespie T, Boevink P, Haupt S et al. (2002) Movement protein reveals that microtubules are dispensable for the cell-to-cell movement of *Tobacco mosaic virus*. *Plant Cell* 14: 1207–1222.

Gillmer D, Bouzoubaa S, Hehn A, Guilley H, Richards K, Jonard G (1992) Efficient cell-to-cell movement of *Beet necrotic yellow vein virus* requires 3′ proximal genes located on RNA-2. *Virology* 189 40–47.

Golen S, Culver JN (2003) *Tobacco mosaic virus* induced alterations in the gene expression profile of *Arabidopsis thaliana*. *Molec Plant-Microbe Interact* 16: 681–688.

Habekuss A, Kühne T, Kramer I, Rabenstein F, Ehrig F, Ruge-Wehling B, Huth W, Ordon F (2008) Identification of *Barley mild mosaic virus* isolates in Germany breaking *rym5* resistance. *J Phytopathol* 156: 36–41.

Hariri D, Fouchard M, Prud'homme H (2001) Incidence of *Soilborne wheat mosaic virus* in mixtures of susceptible and resistant wheat cultivars. *Eur J Plant Pathol* 107: 625–631.

Harper SJ, Delmiglio C, Ward LI, Clover GRG (2011) Detection of *Tomato black ring virus* by real-time one-step RT-PCR. *J Virol Meth* 171: 190–194.

Harrison BD, Murant AE (1977) Nematode transmissibility of pseudo-recombinant isolates of *Tomato black ring virus*. *Ann Appl Biol* 86: 209–212.

Harrison BD, Robinson DJ (1986) Tobraviruses. In: Van Regenmortel MHV, Fraenkel-Conrat H (eds.), *The Plant Viruses*, Volume 2, The rod-shaped plant viruses, Plenum Press, New York, pp. 339–369.

Haupt S, Cowan GH, Ziegler A, Roberts AG, Oparka KJ, Torrance L (2005) Two plant viral movement proteins traffic in endocytic recycling pathway. *Plant Cell* 17: 164–181.

Heinlein M, Epel BL, Padgett HS, Beachy RN (1995) Interaction of tobamovirus movement proteins with the plant cytoskeleton. *Science* 270: 1983–1985.

Hernández C, Mathis A, Brown DJF, Bol JFC (1995) Sequence of RNA-2 of a nematode-transmissible isolate of *Tobacco rattle virus*. *J Gen Virol* 76: 2847–2851.

Herranz MC, Navarro JA, Sommen E, Pallas V (2015) Comparative analysis among the small RNA populations of source, sink and conductive tissues in two different plant-virus pathosystems. *BMC Genomics* 16: 116

Herrera-Vásquez JA, Córdoba-Sellés MC, Cebrián MC, Rossello JA, Alfaro-Fernández A, Jord C (2010) Genetic diversity of *Melon necrotic spot virus* and *Olpidium* isolates from different origins. *Plant Pathol* 59: 240–251.

Hewitt WB, Goheen AC, Raski DJ, Gooding GV (1962) Studies on virus diseases of grapevine in California. *Vitis* 3: 57–83.

Hewitt WB, Raski DJ, Goheen AC (1958) Nematode vector of soilborne fan leaf virus of grapevines. *Phytopathology* 48: 586–595.

Holeva R, Phillips MS, Neilson R et al. (2006) Real-time PCR detection and quantification of vector trichodorid nematodes and *Tobacco rattle virus*. *Molec Cell Probes* 20: 203–211.

Hollings M, Huttinga H (1976) *Tomato mosaic virus*: Descriptions of plant viruses, number 156. CMI/Association for Applied Biologists, Kew, England.

Horvath J, Tobias I, Hunyadi K (1994) New natural herbaceous hosts of grapevine leaf nepovirus. *Hortic Sci* 26: 31–32.

Hosseini A, Jafarpour B, Mehrvar M et al. (2014) Occurrence of soil-borne cereal viruses and molecular characterization of the coat protein gene of *Barley yellow mosaic virus* isolates from Iran. *J Plant Pathol* 96: 391–396.

Howard AR, Heppler ML, Ju HJ, Krishnamurthy K, Payton ME, Verchot-Lubicz J (2004) *Potato virus X* TGBp1 induces plasmodesmata gating and moves between cells in several host species whereas CP moves only in *Nicotiana benthamiana* leaves. *Virology* 328: 185–197.

Hsieh YC, Omaro RT, Scholthof HB (2009) Diverse and newly recognized effects associated with short interfering RNA binding site modification on the *Tomato bushy stunt virus* P19 silencing suppressor. *J Virol* 83: 2188–2200.

Hu X, Dickinson V, Lei Y et al. (2010) Molecular characterization of *Potato mop-top virus* isolates from China and Canada and development of RT-PCR differentiation of two sequence variant groups. *Canad J Plant Pathol* 38: 231–242.

Hübschen J, Kling L, Ipach U et al. (2004a) Validation of the specificity and sensitivity of species-specific primers that provide a reliable molecular diagnostic for *Xiphinema diversicaudatum*, X. *index* and X. *vuittenezi*. *Eur J Plant Pathol* 110: 779–788.

Hübschen J, Kling L, Ipach U, Zinkernagel V, Brown DJF, Neilson R (2004b) Development and validation of species-specific primers that provide a molecular diagnostic for virus-vector longidorid nematodes and related species in German viticulture. *Eur J Plant Pathol* 110: 883–891.

Huss B, Muller S, Sommermeyer G, Walter B, van Regenmortel MHV (1987) *Grapevine fan leaf virus* monoclonal antibodies: their use to distinguish different isolates. *J Phytopathol* 119: 358–370.

Ikegashira Y, Ohki T, Ichiki UT et al. (2004) An immunological system for detection of *Pepper mild mottle virus* in soil from green pepper fields. *Plant Dis* 88: 650–656.

Iwamoto Y, Inoue K, Nishiguchi S et al. (2017) Acidic soil conditions suppress zoospore release from zoosporangia in *Olpidium virulentus*. *J Gen Plant Pathol* 83: 240–243.

Izadpanah K, Zaki-Aghl M, Zhang YP, Daubert SD, Rowhani A (2003) Bermuda grass as a potential host for *Grapevine fan leaf virus*. *Plant Dis* 87: 1179–1182.

Jacobi V, Castello JD (1992) Infection of red spruce, black spruce and balsam fir seedlings with *Tomato mosaic virus*. *Canad J For Res* 22: 919–924.

Jianping C, Swaby AG, Adams MJ, Yili R (1991) *Barley mild mosaic virus* inside its fungal vector *Polymyxa graminis*. *Ann Appl Biol* 118: 615–621.

Jończyk M, Borodynko N, Pospieszny H (2004) Restriction analysis of genetic variability of Polish isolates of *Tomato black ring virus*. *Acta Biochim Polon* 51: 673–681.

Jones AT (1993) Virus transmission through soil and by soil-inhabiting organisms. In: Matthews REF (ed.), Diagnosis of Plant Virus Diseases, CRC Press Inc., Boca Raton, FL, 73–100.

Jones AT (2001) Genus *Idaeovirus*. In: Fauquet CM, Mayo MA, Maniloff J, Desselberger U, Ball LA (eds.), *Virus Taxonomy, Classification and Nomenclature of Viruses*. Eighth Report of International Committee on Taxonomy of Viruses, New York, pp. 1063–1065.

Jones AT, Mayo MA (1998) *Raspberry bushy dwarf virus*. CMI/AAB Description of Plant Viruses No. 360.

Jones D, Farreyrol K, Clover GRG, Pearson MN (2008) Development of a generic PCR detection method for tobraviruses. *Austr Plant Pathol* 37: 132–136.

Kalinina NO, Rakitna DV, Solovyev AG, Schiemann J, Morozov SY (2002) RNA helicase activity of the plant virus movement proteins encoded by the first gene of the triple gene block. *Virology* 296: 321–329.

Kapooria RG, Ndunguru J, Clover GRG (2000) First reports of *Soilborne wheat mosaic virus* and *Wheat spindle streak mosaic virus* in Africa. *Plant Dis* 84: 921.

Karanastasi E, Vassilokos N, Roberts IM, McFarlane SA, Brown DJF (2001) Immunogold localization of *Tobacco rattle virus* particles with *Paratrichodorus anemones*. *J Nematol* 32: 5–12.

Kashiwazaki S, Ogawa K, Usugi T, Tsuchizaki T (1989) Characterization of several strains of *Barley yellow mosaic virus*. *Ann Phytopathol Soc Jpn* 55: 16–25.

Katarzyna O, Koziel E, Garbaczewska G (2016) Ultrastructural impact of *Tobacco rattle virus* on tobacco and pepper ovary and anther tissues. *J Phytopathol* 164: 226–241.

Kawakami S, Watanabe Y, Beachy RN (2004) *Tobacco mosaic virus* infection spreads cell-to-cell as intact replication complexes. *Proc Natl Acad Sci USA* 101: 6291–6296.

Keskin B, Fuchs WH (1969) The process of infection by *Polymyxa betae*. *Arch Mikrobiol* 68: 218–226.

Klerks MM, van den Heuvel JFJM, Schoen CD (2001) Detection of nematode-transmitted nepoviruses by the novel one-tube AmpliDet RNA assay. In: Clark MF (ed.), *Proc Internatl Symp on Fruit Tree Virus Diseases*, Acta Hortic 550: 53–58.

Koenig R, Huth W (2000) Nucleotide sequence analyses indicate a furo-like virus from cereal, formerly considered to be strain of *Soilborne wheat mosaic virus* should be regarded as a new virus species: *Soilborne cereal mosaic virus*. *Zeits Pflanzenk Pflanzenschut* 107: 445–446.

Koenig R, Lesemann D-E, Huth W, Makkouk KM (1983) Comparison of a new soilborne virus from cucumber with Tombus-, Diantho-, and other similar viruses. *Phytopathology* 73: 515–520.

Kragler F, Curin M, Trutneyva K, Gansch A, Waigmann E (2003) MPB2C, a microtubule associated plant protein binds to and interfaces with cell-to-cell transport of *Tobacco mosaic virus* movement protein. *Plant Physiol* 132: 1870–1883.

Kulshrestha S, Hallan V, Raikhy G et al. (2005) Reverse transcription polymerase chain reaction-based detection of *Arabis mosaic virus* and *Strawberry latent ringspot virus* in vector nematodes. *Curr Sci* 89: 1759–1762.

Kumar S (2009) Detection of *Cherry leaf roll virus* and *Strawberry latent ringspot virus* by one-step RT-PCR. *Plant Protect Sci* 45: 140–143.

Kuroda T, Nabata K, Hori T, Ishikawa K, Natsuaki T (2010) *Soybean leaf rugose mosaic virus*, a new soilborne virus in the family Potyviridae, isolated from soybean in Japan. *J Gen Plant Pathol* 76: 382–388.

Lakatos L, Szittya G, Silhavy D, Burgyan J (2004) Molecular mechanism of RNA silencing suppression mediated by the P19 protein of tombusviruses. *EMBO J* 23: 876–884.

Lanter JM, McGuire JM, Goode MJ (1982) Persistence of *Tomato mosaic virus* in tomato debris and soil under field conditions. *Plant Dis* 66: 552–555.

Laporte C, Vetter G, Loudes AM et al. (2003) Involvement of the secondary pathway and the cytoskeleton in intracellular targeting and tubule assembly of *Grapevine fan leaf virus* movement protein in tobacco BY-2 cells. *The Plant Cell* 15: 2058–2075.

Latvia-Kilby S, Aura JM, Pupola N, Hannukkala A, Valkonen JPT (2009) Detection of *Potato mop-top virus* in potato tubers and sprouts: combinations of RNA2 and RNA3 variants and incidence of symptomless infections. *Phytopathology* 99: 519–531.

Lauber B, Bleykasten-Grosshans C, Erhardt M et al. (1998) Cell-to-cell movement of *Beet necrotic yellow vein virus*. I. Heterologous complementation experiments provide evidence for specific interaction among triple gene block proteins. *Molec Plant-Microbe Interact* 11: 618–625.

Legréve A, Delfosse P, Maraite H (2002) Phylogenetic analysis of *Polymyxa* species based on nuclear 5.8S and internal transcribed spacers ribosomal DNA sequences. *Mycol Res* 106: 138–147.

Legréve A, Vanpee B, Risopoulos J, Ward E, Maraite H (1996) Characterization of *Polymyxa* spp. associated with the transmission of *Indian peanut clump virus*. In: Sherwood JL, Rush CM (eds.), *Proc 3rd Symp Internatl Working Group on Plant Viruses with Fungal Vectors*, Dundee, Scotland, pp. 157–160.

Li F, Ding S (2006) Virus counterdefense: diverse strategies for evading the RNA-silencing immunity. *Annu Rev Microbiol* 60: 503–531.

Li JL, Cornman RS, Evans JD et al. (2014) Systemic spread and propagation of a plant pathogenic virus in European honeybees, *Apis mellifera*. *mBio* 5 (1): e00898–13

Liu HY, Sears JL, Lewellen RT (2005) Occurrence of resistance-breaking *Beet necrotic yellow vein virus* of sugar beet. *Plant Dis* 89: 464–468.

Liu JZ, Blancaflor EB, Nelson RS (2005) The *Tobacco mosaic virus* 126-kilodalton protein, a constituent of the virus replication complex, alone or within the complex aligns with and traffics microfilaments. *Plant Physiol* 138: 1853–1865.

Lot H, Campbell RN, Souche S, Milne RG, Roggero P (2002) Transmission of *Olpidium brassicae* of *Mirafiori lettuce mosaic virus* and *Lettuce big-vein virus* and their roles in lettuce big-vein etiology. *Phytopathology* 92: 288–293.

Lubicz JV, Rush CM, Payton M, Colberg T (2007) *Beet necrotic yellow vein virus* accumulates inside resting spores and zoosporangia of its vector *Polymyxa betae*, BNYVV infects *P. betae*. *Virology J* 4: 37

Maccarone LD, Barbetti MJ, Sivasithamparam K, Jones RAC (2010) Molecular genetic characterization of *Olpidium virulentus* isolates associated with big-vein diseased lettuce plants. *Plant Dis* 94: 563–569.

MacFarlane SA (1999) Molecular biology of the tobraviruses. *J Gen Virol* 80: 2799–2807.

Manabayeva SA, Shamekova M, Park J-W et al. (2013) Differential requirements of *Tombusvirus* coat protein and P19 in plants following leaf versus root inoculation. *Virology* 439: 89–96.

Martelli GP, Savino C (1988) Fan leaf degeneration. In: Pearson RC, Goheen AC (eds.), *Compendium of Grape Diseases*, The American Phytopathology Press, St Paul MN, USA, pp. 48–89.

Martin RR, Pinnkerton JN, Kraus J (2009) The use of collagenase to improve the detection of plant viruses in vector nematodes by RT-PCR. *J Virol Meth* 155: 91–95.

Martinez de Alba AEM, Elvira-Matelot E, Vaucheret H (2013) Gene silencing in plants: A diversity of pathways. *Biochim Biophys Acta* 1829: 1300–1308.

Mavrič I, Marn V, Koron D, Žežlina I (2003) First report of *Raspberry bushy dwarf virus* on red raspberry and grapevine in Slovenia. *Plant Dis* 87: 1148.

Mayo MA, Pringle CR (1998) Virus taxonomy–1997. *J Gen Virol* 79: 649–759.

McGuire JM, Kim KS, Douthi LB (1970) *Tobacco ringspot virus* in the nematode *Xiphinema americanum*. *Virology* 42: 212–216.

McKinney HH (1923) Investigations on the rosette disease of wheat and its control. *J Agric Res* 23: 771–800.

Meunier A, Schmit J, Stas A, Kutluk N, Bragard (2003) Multiplex reverse transcription-PCR for simultaneous detection of *Beet necrotic yellow vein virus*, *Beet soilborne virus* and *Beet virus Q* and their vector *Polymyxa betae* Keskin on sugar beet. *Appl Environ Microbiol* 69: 2356–2360.

Momonoi K, Mori M, Matsuura K, Moriwaki J, Morikawa T (2015) Quantification of *Mirafiori lettuce big-vein virus* and its vector *Olpidium virulentus*, from soil using real-time PCR. *Plant Pathol* 64: 825–830.

Morozov SY, Solovyev AG (2003) Triple gene block: modular design of multifunctional machine for plant virus movement. *J Gen Virol* 84: 1351–1366.

Nakayama T, Hataya T, Tsuda S et al. (2008) Soil diagnosis by detection of *Potato mop-top virus* using bait plant bioassay and RT-PCR-microplate hybridization. *Proc 7th Symp Internatl Working Group on Plant Viruses with Fungal Vectors*, Quedlinburg, Germany, pp. 128–132.

Nakayama T, Maoka T, Hataya T et al. (2010) Diagnosis of *Potato mop-top virus* in soil using bait plant bioassay and RT-PCR microplate hybridization. *Amer J Potato Res* 87: 218–225.

Narayanasamy P (2001) *Plant Pathogen Detection and Disease Diagnosis*, Second edition, Marcel Dekker, Inc., New York.

Narayanasamy P (2002) *Microbial Plant Pathogens and Crop Disease Management*, Science Publishers, Enfield, USA.

Narayanasamy P (2008) *Molecular Biology of Plant Pathogenesis and Disease Management: Disease Development*, Volume 2, Springer Science + Business Media B. V., Heidelberg, Germany.

Narayanasamy P (2011) *Microbial Plant Pathogens–Detection and Disease Diagnosis*, Volume 3, Viral and Viroid Pathogens, Springer Science + Business Media, B. V., Heidelberg, Germany.

Narayanasamy P (2017) *Microbial Plant Pathogens: Detection and Management in Seeds and Propagules*, Volume 1, Wiley-Blackwell, United Kingdom.

Navarro JA, Botella F, Marhuenda A, Sastre P, Sánchez-Pina M, Pallas V (2005) Identification and partial characterisation of *Lettuce big-vein associated virus* (LBVaV) and *Mirafiori lettuce big-vein virus* in common weeds found amongst Spanish lettuce crops and their role in lettuce big-vein disease transmission. *Eur J Plant Pathol* 113: 25–34.

Navarro JA, Botella F, Marhuenda A, Sastre P, Sánchez-Pina M, Pallas V (2004) Comparative infection progress analysis of *Lettuce big-vein virus* and *Mirafiori lettuce virus* in lettuce crops by developed molecular diagnosis techniques. *Phytopathology* 94: 470–477.

Nelson RS, Citovsky V (2005) Plant viruses: invaders of cells and pirates of cellular pathways. *Plant Physiol* 138: 1809–1814.

Nomiyama K, Osaki H, Sasaya T, Ishikawa (2013) Preparation and characterization of polyclonal antibody against resting spores of *Olpidium virulentus*, fungal vector of lettuce big-vein disease. *J Gen Plant Pathol* 79: 64–68.

Nomiyama K, Sasaya T, Sekiguchi H et al. (2015) DAS-ELISA quantification of resting spores of *Olpidium virulentus* in roots and correlations between resting spore density in soil and severity of lettuce big-vein disease. *J Gen Plant Pathol* 81: 243–248.

Obermeier C, Sears JL, Liu HY et al. (2001) Characterization of distinct tombusviruses that cause diseases of lettuce and tomato in the western United States. *Phytopathology* 91: 797–806.

Pares RD, Gunn LV, Keskula EN (1996) The role of infective plant debris and its concentration in soil in the ecology of tomato mosaic tobamovirus, a non-vectored plant virus. *J Phytopathol* 144: 147–150.

Pavli OI, Prins M, Skaracis GN (2010) Detection of *Beet soil-borne virus* and *Beet virus Q* in sugar beet in Greece. *J Plant Pathol* 92: 793–796.

Peltier C, Hleibieh K, Thiel H, Klein E, Bragard C, Gilmer D (2008) Molecular biology of the *Beet necrotic yellow vein virus*. *Plant Viruses* 2: 14–24.

Peng R, Han CG, Yan L, Yu J-L, Liu Y (1998) Cytological localization of *Beet necrotic yellow vein virus* transmitted by *Polymyxa betae*. *Acta Phytopathol Sinica* 28: 257–261.

Pérez EE, Weingartner DP, Hiebert E, McSorley R (2000) *Tobacco rattle virus* detection in potato tubers from northwest Florida by PCR and tissue blotting. *Amer J Potato Res* 77: 363–368.

Pleško IM, Marn MV, Širca S, Urek G (2009) Biological, serological and molecular characterization of *Raspberry bushy dwarf virus* from grapevine and its detection in the nematode *Longidorus juvenilis*. *Eur J Plant Pathol* 123: 261–268.

Ploeg AT, Robinson DJ, Brown DJF (1993) RNA-2 of *Tobacco rattle virus* encodes the determinants of transmissibility by trichorid vector nematodes. *J Gen Virol* 74: 1463–1466.

Plumb RT (2002) Viruses of Poaceae: a case history in plant pathology. *Plant Pathol* 51: 673–682.

Prillwitz H, Schlösser E (1992) *Beet soilborne virus*: occurrence, symptoms and effect on development. *Meded Fac Landbouwwet Univ Gent* 57/2a: 295–302.

Pringle CR (1999) Virus taxonomy–1999. *Arch Virol* 144: 421–429.

Qu F, Morris TJ (2002) Efficient infection of *Nicotiana benthamiana* by *Tomato bushy stunt virus* is facilitated by the coat protein and maintained by p19 through suppression of gene silencing. *Molec Plant-Microbe Interact* 15: 193–202.

Quacquarelli A, Gallitelli D, Savino V, Martelli GP (1976) Properties of *Grapevine fan leaf virus*. *J Gen Virol* 32: 349–360.

Quillet I, Guilley H, Jonard G, Richards K (1989) In vitro synthesis of biologically active *Beet necrotic yellow vein virus* RNA. *Virology* 172: 293–301.

Rao AA, Brakke MK (1969) Relation of *Soilborne wheat mosaic virus* and its fungal vector *Polymyxa graminis*. *Phytopathology* 59: 581–587.

Raski DJ, Maggenti AR, Jones NO (1973) Location of grapevine fan leaf and yellow mosaic virus particles in *Xiphinema index*. *J Nematol* 5: 208–211.

Ratcliff F, Martin-Hernandez AM, Baulcombe DC (2001) *Tobacco rattle virus* as a vector for analysis of gene function by silencing. *Plant J* 25: 237–245.

Rattin C, Hleibieh K, Bianchi L, Schirmer A, Autonell CR, Gilmer D (2009) *Beet soil-borne mosaic virus* RNA-3 is replicated and encapsidated in the presence of BNYVV RNA-1 and -2 and allows long-distance movement in *Beta macrocarpa*. *Virology* 385: 392–399.

Reavy B, Dawson S, Canto T, MacFarlane SA (2004) Heterologous expression of plant virus genes that suppress post-transcriptional gene silencing results in suppression of RNA interference in *Drosophila* cells. *BMC Biotechnol* 4: 18.

Reddy AS, Hobbs HA, Delfosse P, Murthy AK, Reddy DVR (1998) Seed transmission of *Indian peanut clump virus* (ICPV) in peanut and millets. *Plant Dis* 82: 343–346.

Ritzenthaler C, Laporte C, Gaire F et al. (2002) *Grapevine fan leaf virus* replication occurs on endoplasmic reticulum-derived membranes. *J Virol* 76: 8808–8819.

Ritzenthaler C, Pinck M, Pinck L (1995a) Grapevine fan leaf nepovirus P38 putative movement protein is not transiently expressed and is a stable final maturation product in vivo. *J Gen Virol* 76: 907–915.

Ritzenthaler C, Schmitt AC, Michler P, Stussi-Garaud C, Pinck L (1995b) Grapevine fan leaf nepovirus P38 putative movement protein is located on tubules in vivo. *Molec Plant-Microbe Interact* 8: 379–387.

Roberts AG, Oparka KJ (2003) Plasmodesmata and the control of symplastic transport. *Plant Cell Environ* 26: 103–124.

Roberts IM, Brown DJF (1980) Detection of nepoviruses in their nematode vectors by immunosorbent electron microscopy. *Ann Appl Biol* 96: 187–192.

Robertson NL (2013) Molecular detection of *Tobacco rattle virus* in bleeding heart [*Dicentra spectabilis* (L.) Lem.] growing in Alaska. On line. *Plant Health Progress*. 0227–01-BR

Robinson DJ (2004) Identification and nucleotide sequence of a *Tobacco rattle virus* RNA-1 variant that causes spraing disease in potato cv. Bintje. *J Phytopathol* 152: 286–290.

Roggero P, Lot H, Souche S, Lenzi R, Milne RG (2003) Occurrence of *Mirafiori lettuce virus* and *Lettuce big-vein virus* in relation to development of big-vein symptoms in lettuce crops. *Eur J Plant Pathol* 109: 261–267.

Roggero P, Masenga V, Lenzi R, Coghe F, Ena S, Winter S (2001) First report of *Pepino mosaic virus* in tomato in Italy. *Plant Pathol* 50: 798.

Salomone A, Roggero P (2002) Host range, seed transmission and detection by ELISA and lateral flow of an Italian isolate of *Pepino mosaic virus*. *J Plant Pathol* 84: 65–68.

Sanger HL, Allen MW, Gold AH (1962) Direct recovery of *Tobacco rattle virus* from its nematode vector. *Phytopathology* 52: 750 (Abs).

Sasaya T, Fujii H, Ishikawa K, Koganezawa H (2008) Further evidence of *Mirafiori lettuce big-vein virus*, but not of *Lettuce big-vein virus* associated with big-vein disease in lettuce. *Phytopathology* 98: 464–468.

Scarborough BA, Smith (1977) Effects of tobacco- and tomato ringspot viruses on reproduction tissues of *Pelargonium xhortorum*. *Phytopathology* 67: 292–297.

Schellenberger P, Sauter C, Lorber B et al. (2011) Structural insights into viral determinants of nematode mediated *Grapevine fan leaf virus* transmission. *PLoS Pathog* 7 (5): e1002034

Schepetilnikov MV, Manske U, Solovyev AG, Zamayatnin AA Jr, Schiemann J, Morozov SY (2005) The hydrophobic segment of *Potato virus X* TGBNp3 is a major determinant of the protein intracellular trafficking. *J Gen Virol* 86: 2379–2391.

Scholthof HB (2006) The tombusvirus-encoded P19 from irrelevance to elegance. *Nat Rev Microbiol* 4: 405–411.

Schwach F, Vaistij FE, Jones L, Baulcombe DC (2005) An RNA dependent RNA polymerase prevents meristem invasion by *Potato virus X* and is required for the activity, but not the production of a systemic silencing signal. *Plant Physiol* 138: 1842–1845.

Shirako Y (1998) Non-AUG translation initiation in a plant RNA virus a forty-amino acid-extension is added to the N terminus of the *Soilborne wheat mosaic virus* capsid protein. *J Virol* 72: 1677–1682.

Shirako Y, Wilson TM (1993) Complete nucleotide sequence and organization of the bipartite RNA genome of *Soilborne wheat mosaic virus*. *Virology* 195: 16–32.

Šneideris S, Zitikaite I, Žižyte M, Girgaliunaite B, Staniulis J (2012) Identification of nepoviruses in tomato (*Lycopersicon esculentum* Mill.). *Agriculture* 99: 173–178.

Sokmen MA, Barker H, Torrance L (1998) Factors affecting the detection of *Potato mop-top virus* in potato tubers and improvement of test procedures for more reliable assays. *Ann Appl Biol* 133: 55–63.

Solovyev AG, Savenkov EJ, Agranovsky AA, Morozov SY (1966) Comparison of the genomic cis elements and coding regions in RNA components of hordeivirus *Barley stripe mosaic virus*, *Lychnis ringspot virus* and *Poa semilatent virus*. *Virology* 219: 9–18.

Steddom K, Heidel G, Jones D, Rush CM (2003) Remote detection of rhizomania in sugar beets. *Phytopathology* 93: 720–726.

Takeuchi S, Hikichi Y, Kawada Y, Okuno T (2000) Detection of tobamoviruses from soils by non– precoated indirect ELISA. *J Gen Plant Pathol* 66: 153–158.

Tamada T, Uchino H, Kusume T, Saito M (1999) RNA3 deletion mutants of *Beet necrotic yellow vein virus* do not cause rhizomania disease in sugar beets. *Phytopathology* 89: 1000–1006.

Taylor CE, Robertson WM (1970) Sites of virus retention in the alimentary tract of the nematode *Xiphinema diversicaudatum* (Micol.) and *X. index* (Thorne and Allen). *Ann Appl Biol* 65: 375–380.

Te J, Melcher U, Howard AR, Verchot-Lubicz J (2005) *Soilborne wheat mosaic virus* (SBWMV) 19 K protein belongs to a class of cysteine-rich proteins that suppress RNA silencing. *Virology J* 2: 18

Torrance L, Mayo MA (1997) Proposed reclassification of furoviruses. *Arch Virol* 142: 435–439.

Turina M, Omarov R, Murphy IF, Bazaldua-Hernandez C, Desoves B, Schothof B (2003) A newly identified role for the *Tomato bushy stunt virus* P19 in short-distance spread. *Molec Plant Pathol* 4: 67–72.

Vaïanopoulos C, Bragard C, Moreau V, Maraite H, Legréve A (2007) Identification and quantification of *Polymyxa graminis* f.sp. *tepida* in barley and wheat. *Plant Dis* 91: 857–864.

Vaïanopoulos C, Legréve A, Moreau V, Bragard C (2009) Broad-spectrum detection and quantification methods of *Soil-borne cereal mosaic virus* isolates. *J Virol Meth* 159: 227–232.

Van der Wilk F, Korsman M, Zoom F (1994) Detection of *Tobacco rattle virus* in nematodes by reverse transcription and polymerase chain reaction. *Eur J Plant Pathol* 100: 109–112.

Van Ghelder C, Reid A, Kenyon D, Esmenjaud D (2015) Development of a real-time PCR method for the detection of the dagger nematodes *Xiphinema index*, *X. diversicaudatum*, *X. vuittenezi* and *X. italicae* and for the quantification of *X. index* numbers. *Plant Pathol* 64: 489–500.

Vassilakos N, Vellios EK, Brown DJF, MacFarlane SA (2001) *Tobravirus* 2b protein acts in trans to facilitate transmission by nematodes. *Virology* 279: 478–487.

Vellios EK, Brown DJF, MacFarlane SA (2002) Substitution of a single amino acid in the 2b protein of *Pea early browning virus* affects nematode transmission. *J Gen Virol* 83: 1771–1775.

Verbeek M, Dullemans AM, van Bekkum PJ, van der Vlugt RAA (2013) Evidence for *Lettuce big-vein associated virus* as the causal agent of a syndrome of necrotic rings and spots in lettuce. *Plant Pathol* 62: 444–451.

Villate L, Fievet V, Hanse B et al. (2008) Spatial distribution of the dagger nematode *Xiphinema index* and its associated *Grapevine fan leaf virus* in French vineyard. *Phytopathology* 98: 942–948.

Visser PB, Mathis A, Linthorst HJM (1999) Tobraviruses. In: Granoff A, Webster RG (eds.), *Encyclopedia of Virology*, Second edition, Academic Press, New York, 1784–1789.

Voinnet O, Lederer CC, Baulcombe DC (2000) A viral movement protein prevents spread of the gene silencing signal in *Nicotiana benthamiana*. *Cell* 103: 157–167.

Wang S, Gergerich RC, Wickizer SL, Kim KS (2002) Localization of transmissible and nontransmissible viruses in the vector nematode *Xiphinema americanum*. *Phytopathology* 92: 646–653.

Wang SH, Gregerich RC (1998) Immunofluorescence localization of *Tobacco ringspot virus* in the vector nematode *Xiphinema americanum*. *Phytopathology* 88: 885–889.

Wang SH, Gregerich RC, Wickizer SL, Kim KS (2002) Localization of transmissible and nontransmissible viruses in the vector nematode *Xiphinema americanum*. *Phytopathology* 92: 646–653.

Wang XB, Wu Q, Ito T, Cillo F Li WX, Chen X et al. (2010) RNA-mediated viral immunity requires amplification of virus-derived siRNAs in *Arabidopsis thaliana*. *Proc Natl Acad Sci USA* 107: 484–489.

Ward LI, Fenn MGC, Henry CM (2004) A rapid method for direct detection of *Polymyxa* DNA in soil. *Plant Pathol* 53: 485–490.

Wei T, Lu G, Clover GRG (2009) A multiplex RT-PCR for the detection of *Potato yellow vein virus*, *Tobacco rattle virus* and *Tomato infectious chlorosis virus* in potato with a plant internal amplification control. *Plant Pathol* 58: 203–209.

Wesley SV, Helliwell CA, Smith NA et al. (2001) Construct design efficient effective and high-throughput gene silencing in plants. *Plant J* 27: 581–590.

Winterhagen P, Brendell G, Krczal G, Reustle GM (2007) Development of an in vitro dual culture system for grapevine and *Xiphinema index* as a tool for virus transmission. *S Afr J Enol Vitic* 28: 2–5.

Wolf S, Deom CM, Beachy RN, Lucas WJ (1989) Movement protein of *Tobacco mosaic virus* modified plasmodesmatal size exclusion limit. *Science* 246: 377–379.

Xu H, De Haan T-L, De Boer SH (2004) Detection and confirmation of *Potato mop-top virus* in potatoes produced in the United States and Canada. *Plant Dis* 88: 363–367.

Yamamura Y, Scholthof HB (2005) Pathogen profile: *Tomato bushy stunt virus*, a resilient model system for studying virus-plant interactions. *Molec Plant Pathol* 6: 491–502.

Yang Y, Ding B, Baulcombe DC, Verchot-Lubicz J (2000) Cell-to-cell movement of the 25 K protein *Potato virus X* is regulated by three other viral proteins. *Molec Plant Microbe-Interact* 13: 599–605.

Ye R, Xu L, Gao ZZ, Yang JP, Chen J, Chen JP, Adams MJ, Yu SQ (2000) Use of monoclonal antibodies for serological differentiation of wheat and oat furoviruses. *J Phytopathol* 148: 257–262.

Ye R, Zheng T, Chen J, Diao A, Adams MJ, Yu S, Antoniw JF (1999) Characterization and partial sequence of a new *Furovirus* of wheat in China. *Plant Pathol* 48: 379–387.

Yilmaz NDK, Sokmen M, Gulser C, Saracoglu S, Yilmaz D (2010) Relationships between soil properties and soilborne viruses transmitted by *Polymyxa betae* Keskin in sugar beet field. *Spanish J Agric Res* 8: 766–769.

Yilmaz NDK, Sokmen MA, Kaya R, Sevik MA, Tunali B, Demirtas S (2016) The widespread occurrences of *Beet soilborne virus* and RNA-5 containing *Beet necrotic yellow vein virus* isolates in sugar beet production areas in Turkey. *Eur J Plant Pathol* 144: 443–455.

Yilmaz NDK, Yanar Y, Günal H, Erkan S (2004) Effects of soil properties on the occurrence of *Beet necrotic yellow vein virus* and *Beet soilborne virus* on sugar beet in Tokat, Turkey. *Plant Pathol J* 3: 56–60.

Zhang C, Wu Z, Li Y, Wu J (2015) Biogenesis, function and applications of virus-derived small RNAs in plants. *Front Microbiol* 6: 1237

Zhang ZY, Liu XJ, Li DW, Yu JL, Han CG (2011) Rapid detection of *Wheat yellow mosaic virus* by reverse transcription loop-mediated isothermal amplification. *Virol J* 8: 550.

Ziegler A, Golecki B, Kastirr U (2013) Occurrence of the New York strain of *Soil-borne wheat mosaic virus* in Northern Germany. *J Phytopathol* 161: 290–292.

# 7 Ecology and Epidemiology of Soilborne Microbial Plant Pathogens

The ecological conditions required for the development of various microbial pathogens – oomycetes, fungi, bacteria and viruses differ widely. Conditions favoring the development of the microbial plant pathogens, may result in epidemics, when susceptible cultivar is available in large areas. The susceptibility/resistance levels of the cultivars and the environmental conditions existing in the various agroecosystems are known to exert significant influence on both the pathogens and their host plants. The soil provides water and nutrients not only to the plants and microbial pathogens, but also to a wide range of organisms that interact with the plant hosts and the microbial pathogens.

## 7.1 NATURE AND FUNCTIONS OF SOILS

Soil is the resultant of weathering of parent rock materials and it functions as a habitat for various kinds of microorganisms. Soils originate from rocks via a complex process of physical, chemical and biological factors that reduce the rocks first to regolith (rock rubble) and then to the soil. The principal factors involved in soil formation include parent material, climate, topography, biological activity and adequate time. Diverse microbial communities present in the soils may either provide support to or adversely affect the growth of plants. Several microorganisms that contribute significantly to soil fertility and plant growth have been identified. The plants, in turn, have a major influence on the microbial communities (Atlas and Bartha 1998). The soil is a natural body of minerals and organic materials differentiated into horizons which differ among themselves, as well as from the underlying material in morphology, physical makeup, chemical composition and biological characteristics. All soils originate from weathered rock, volcanic ash deposits or accumulated plant debris. Soils have three principal functions, as a natural medium for plant growth, providing mechanical support as well as supplying nutrients and water to plants. Soils have four major components: (i) mineral matter, (ii) organic matter, (iii) soil air and (iv) soil water. Differences in soil properties and communities of microorganisms were evaluated, using both univariate and multivariate statistical analyses and canonical discriminant analysis (CDA) was performed to assess the strength of the association of soil variables within communities from 83 locations. In all, 21 species of *Pythium* were identified, but only six were recovered from > 40 locations. Five communities were formed using cluster analysis, and significant differences were observed in disease incidence, as well as soil pH, calcium, magnesium and cation exchange capacity between communities. Stepwise, multiple discriminant analysis and CDA identified pH, calcium, magnesium and field capacity as the highest contributing factors to the separation of five *Pythium* communities. A strong association was observed between abiotic components and the structure of *Pythium* communities, as well as diversity of *Pythium* spp. collected from agronomic production fields in Ohio (Broders et al. 2009).

The microorganisms play an important role in degrading complex organic compounds from which they obtain nutrients required for growth and reproduction. Dark colloidal mass known as humus is left on the soil, after the microbial degradation, and it is associated with high soil fertility and soil structure. Humus promotes formation of soil aggregates, allowing movement of plant roots, air and water. Soil air is required for all organisms, as exchange of gas (air) is essential for their survival. A well-aerated soil may have rapid oxygen exchange between soil air and atmospheric air. Soil temperature significantly influences seed germination and plant growth in the early stages. Water is required for the development of plants and associated microorganisms. The soil water potential is defined as the work which water can carry out, while it moves from its present state to a pool of water (at a specified elevation, temperature, soluble salt content, etc.). The moisture is held with soil water potential less than $-1/3$ or 0.3 bars or $-33$ kPa. Soil water potential indicates the amount of irrigation water required and the amount of reserve soil water available to plants.

### 7.1.1 INTERACTIONS OF MICROORGANISMS WITH COMPONENTS OF SOILS

#### 7.1.1.1 Interactions among Microorganisms

The ecosphere or biosphere constitutes the various kinds of organisms living on Earth in their entirety and the abiotic surroundings they inhabit. A habitat may be defined as the physical location where an organism is present. Numerous habitats exist in each of the major divisions of the ecosphere. The habitat is a component of the ecological niche which indicates where an organism lives and performs its activities as well. The niche is the functional role of an organism within an ecosystem. Plant roots provide suitable habitats for growth of microorganisms attaining high population levels. Interactions between soil microorganisms and plant roots may be beneficial for both or harmful to one partner of the association. The rhizosphere includes plant roots and the surrounding soil adhering to the root system, after shaking to remove the loose soil. The rhizoplane is the actual root surface and within the rhizosphere and this region of soil exerts greater influence on the associated microorganism (Bouwen and Rovira 1976). Within the rhizosphere, plant roots exert direct influence on the composition and density of the soil microbial community known as the 'rhizosphere effect'. The plant cover of the soil is an important factor in determining the types and numbers of microorganisms in the soil. Plant root exudates and senescent plant parts form important sources of nutrients for soil microorganisms. Rhizosphere and mycorrhizal interactions, as well as plant susceptibility to pathogens exert selective influences on the soil microbial community which include microorganisms antagonistic to soilborne microbial plant pathogens (Metting Jr. 1993).

Microbial communities in roots, rhizoplane, rhizosphere soil and nonrhizosphere soil in potato present in organic and integrated production systems were compared between 2005 and 2007. In the rhizoplane, rhizosphere and nonrhizosphere soil, the total density of pathogens was greater in organic systems and of antagonists in the integrated system. Dominant pathogens *Colletotrichum coccodes, Fusarium culmorum, Haematonectria haematococca* and *Gibellulopsis nigrescens* and dominant antagonists, *Clonostachys + Gliocladium* and *Trichoderma* were in greater populations in the organic system. Subdominant pathogens, *Alternaria + Ulocladium, Pythium* and *Thanetophorus cucumeris* and subdominant antagonists, *Mortierella* and *Umbelopsis vinacea* were in greater density in the integrated system. Incidence of sprout rot was more frequently observed in an organic system, whereas Fusarium dry rot and black scurf diseases occurred at greater levels in an integrated system. The organic system provided a less disease-suppressive environment than the integrated system and resulted in lower potato yield (Lenc et al. 2012). *Aspergillus flavus*, causing aflatoxin contamination of cotton seed, is a natural soil inhabitant. The isolates of *A. flavus* have been differentiated into small (S) and large (L) strains. S-strain isolates produce greater concentrations of aflatoxin than L strain isolates. The community structure of *A. flavus* in soils of South Texas was investigated, using 326 samples collected from 152 fields between 2001 and 2003. Significant differences in the incidence of isolates of *A. flavus* S-strain were observed among regions. The CFU/g of soil was not significantly different among regions. S-strain incidence was positively correlated with clay content and negatively correlated with sand content. Fields cropped to cotton in the previous year had a higher incidence of S-strain, whereas fields cropped to corn had greater total quantities of *A. flavus* propagules. Maps of S-strain patterns showed that the S-strain constituted > 30% of the overall *A. flavus* community in a particular area (Central Coastal Bend to Central Upper Coast). The West RioGrande Valley had the lowest S-strain incidence (< 10%). Geographic variations in populations of S-strain isolates might influence the distribution of aflatoxin contamination in South Texas (Jaime-Garcia and Cotty 2006). The community structure in soils of *A. flavus* and *A. parasiticus*, infecting nut crops, including almond, pistachio and walnut was studied, since the nuts of these crops are contaminated with mycotoxins produced by these pathogens. The distribution of two sclerotial size morphotypes of *A. flavus* [small (S) and large (L) strains] and *A. parasiticus* was monitored in the soil of almond orchards in California over a five-year period (2007 to 2011, excluding 2009). Among the 4349

isolates collected from 28 orchards, overall *A. flavus* L strain was the most frequently occurring in the southern region (79.9 to 95.1%), compared with the northern region (21.4 to 47%). In the northern region, *A. parasiticus* was recovered more frequently (28.5 to 61.0%) in different years. Frequency of aflatoxin-producing isolates among L strains fluctuated from year to year. Aflatoxin-producing L strain isolates were significantly more prevalent than atoxigenic isolates in each region, during the survey period, except in 2011, in the northern region where prevalence of atoxigenic isolates was more common (56%) than aflatoxin-producing isolates. The results showed that the structure of *A. flavus* and *A. parasiticus* communities in the soil and the proportion of toxigenic isolates differed across regions and years. Such assessment of potential of *Aspergillus* isolates would be useful for managing aflatoxin levels efficiently (Donner et al. 2015).

### 7.1.2 Pathogen Suppression in Soils

Development of microbial pathogens may be favored or hampered by soils to different degrees. The soils allowing pathogen development, resulting in incidence of disease in a crop expressing characteristic symptoms in the infected plant are known as 'conducive soils'. In contrast, the soils that suppress pathogen development and consequently cause failure of disease incidence are designated 'suppressive soils'. The term suppressive soil was applied by Menzies (1959) to describe the phenomenon of disease suppression observed against Streptomyces scab disease of potato in certain soils. In all natural soil, expression of disease symptoms occurs, leading to various extents of incidence and degrees of severity. The disease suppression occurs along a continuum from highly suppressive to highly conducive stages, rather than being either suppressive or conducive (Alabouvette et al. 1996). Pathogen suppressive soil, as defined by Baker and Cook (1974) is

> 'one in which the pathogen does not establish or persist, or establishes, but causes little or no damage, or establishes and cause disease for a while, but thereafter the disease is less important, although the pathogen may persist in the soil'.

Soil suppression of pathogen development may be either general or specific. General suppression of pathogen may be directly related to the quantum of microbial activity at a time critical to the pathogen concerned, such as propagule germination and prepenetration growth in the host rhizosphere. General suppression of a pathogen in soil may be due to high degree of soil fungistasis, and it may not be brought out by any one or group of microorganisms. On the other hand, specific suppression may be due to more specific effects of individual or group(s) of microorganisms with antagonistic activity against the pathogen at some stage in its life cycle. Often both general and specific suppression of pathogens may operate in combination (Cook and Baker 1983).

Existence of soils suppressive to specific pathogens, such as *Gaeumannomyces graminis* var. *tritici (Ggt), Fusarium oxysporum, Phytophthora cinnamomi, Rhizoctonia solani* and *Phymatotrichum omnivorum* has been reported (Cook and Baker 1983). Attempts have been made to identify the functional biological entities that contribute to the observed disease suppression. Microbial resources with potential have been recognized so that they can be utilized as agents for managing the soilborne plant pathogens. Disease suppression in a soil may require the interactions among multiple biotic and abiotic factors. Functions of specific agent(s), at sites external to the specific system, could be effectively diminished. Two kinds of biological suppression of soilborne diseases may be differentiated. Specific suppression occurs, when there are exclusive interactions involving one or more agents that suppress the development of a soilborne pathogen. On the other hand, in soils that are suppressive, disease suppression is due to cumulative effects of complex interactions between the pathogen and a multiplicity of biotic and abiotic factors (van Bruggen and Semenov 2000). The soil suppressive to wheat take-all disease caused *by Ggt* forms a model for suppressive soil. Continuous monoculture of wheat during the initial years, increased the incidence of wheat take-all disease. But a spontaneous decline occurred, after an indeterminate period of time. Then the soil remained suppressive, as long as wheat monoculture was continued. This suppression apparently is the resultant of a qualitative change in the soil microbial population, following monoculture of wheat (Gerlagh 1968; Shipton et al. 1973).

The soil microbial community is primarily responsible for soil suppressiveness of plant diseases, as indicated by the evidences: (i) soil pasteurization or sterilization abolishes the disease-suppressive capacity of soils and (ii) transfer of suppressiveness could be accomplished by mixing a small quantity of suppressive soil with the sterilized soil (Stutz et al. 1986). The rhizosphere microbial community is highly diverse and they may act by protecting the plant directly through the release of pathogen inhibitors or indirectly by promoting plant growth or enhancing rhizosphere functioning of antagonistic pseudomonads (Lemanceau and Alabouvette 1991; Bally and Elmerich 2005; Raaijmakers et al. 2008). Community-level assessment of taxa more prevalent in suppressive soil has to be carried out, but such assessment in naturally suppressive soils is rarely performed (Hjort et al. 2007). Soil suppressiveness has been observed across geographical locations and elicitation of suppressiveness by the same biological factor has been demonstrated across soils. These evidences indicated the potential of resident microbial resource management to be exploited as an effective strategy for management of soilborne diseases (Weller et al. 2002). Resource manipulation has been used to alter microbial densities and community structure in such a way that may limit pathogen activity. Production of antimicrobial secondary metabolites has been shown to be an important factor in many soils that have pathogen suppressive activities (Weller et al. 2002; Hjort et al. 2007). Streptomycetes are a ubiquitous component of soil microbial communities and they are known to be efficient producers of antimicrobial secondary metabolites that are effective in inhibiting the development of several soilborne pathogens. Among the different factors influencing the soil microbial communities,

host plants function as a distinct selective force shaping the streptomycetes community structure and activities. In addition, land-use changes altering plant diversity or community composition might indirectly affect the structure and function of microbial communities. Inhibitory activity against the fungal pathogens viz., *Fusarium graminearum*, *R. solani*, and *Verticillium dahliae* by streptomycetes was greater among monoculture, whereas praire streptomycetes showed greater inhibitory activity against pathogenic *Streptomyces scabies*, causing potato scab disease. Inhibition of *V. dahliae* was more intense by *Streptomyces* isolates than other fungal pathogens. The range of pathogens that each isolate of *Streptomyces* inhibited varied widely, probably as a function of the quantity and variety of secondary metabolites produced. The isolates of *Streptomyces* were characterized, based on inhibition of plant pathogens as a measure of functional activity and 16S rDNA sequence to determine community structure. High and low diversity in plant communities supported streptomycetes communities with similar diversity, phylogenetic composition and pathogen suppressive activity. The results indicated the impacts of plant diversity on a narrow range of organisms present in soil microbial communities (Bakker et al. 2010).

Soils suppressive and conducive to *R. solani* AG2-1 were cropped to cauliflower for five successive cropping cycles or allowed to remain fallow in the greenhouse experiment. Soils were inoculated with the pathogen only once or before every crop. Disease decline occurred in all treatments cropped with cauliflower, either because of a decreased pathogen population or increased suppressiveness of the soil. Suppressiveness was determined, using a seed germination test for assessing preemergence damping-off and also by measuring the spread of disease symptoms in young seedlings. Conducive soil became suppressive, after five subsequent cauliflower crops, inoculated in each cycle with *R. solani*. Suppressiveness was significantly stimulated by successive pathogen inoculations and the presence of cauliflower had less effect. Suppressiveness was of biological origin, as sterilization of soil abolished this property. Further, suppressiveness was transferable to sterilized soil by adding 10% of the suppressive soil. Suppressiveness could be correlated to populations of actinomycetes or pseudomonads or parasitic fungi present in the suppressive soil. A potential role of *Lysobacter* in soil suppressiveness could be confirmed by quantitative TaqMan polymerase chain reaction (PCR) detection which indicated the presence of large population of *Lysobacter* in the suppressive soil, compared to conducive soil. The results showed that successive cauliflower plantings could cause significant decline in disease incidence and *Lysobacter* spp. could be the potential key organism inducing suppressiveness in soils against *R. solani* (Postma et al. 2010).

The influence of biotic changes on the local decrease in soil conduciveness in disease patches toward the disease caused by *R. solani* AG2-2 in a sugar beet field in France was studied. Soil samples from healthy and diseased areas were analyzed for bacterial and fungal densities, molecular and physiological microbial community structures and antagonistic activity of *Trichoderma* isolates collected from the healthy and diseased areas. The inoculum density was higher inside the diseased patches, but the respective soil was less conducive toward disease incited by *R. solani* AG2-2. The pathogen, although present in healthy areas, was unable to induce disease in sugar beet under field conditions. The response of the microflora to previous development of *R. solani* in diseased areas prevented further pathogenic activity. The genetic and physiological structures of fungal communities and physiological structures of bacterial communities were modified in diseased patches, compared to healthy areas. The terminal restriction fragment length polymorphism (T-RFLP) analysis revealed that peaks corresponding to *Trichoderma* isolates were higher in diseased patches than in healthy areas. *Trichoderma* isolates from diseased patches were more antagonistic than those from healthy areas. The results suggested that the disease caused by *R. solani* AG2-2 induced changes in genetic and physiological structures of microbial populations and development of antagonists. The decrease in conduciveness inside patches might result in patch mobility in the next season (Anees et al. 2010).

In Australia, biologically based disease suppression has been reported in long-term experimental plots and farmers' fields. Suppression of disease in soil has been attributed to diverse microbial communities, including bacteria, fungi and protozoa and it is reported to affect pathogen survival, growth in bulk soil, rhizosphere and infection of plants. Soil microorganisms and microbial pathogens may interact, prior to crop sowing and/ or in the rhizosphere, subsequently influencing both plant growth and productivity. Fungal populations were analyzed in co-located soils 'suppressive' or 'nonsuppressive' for bare patch disease caused by *R. solani* AG-8 at two sites in South Australia, using 454 pyrosequencing, targeting the fungal 28S LSU rRNA gene. The DNA was extracted from samples of not less than 125 g soil per replicate to reduce the microsite community variability, and from soil samples taken at sowing and from the rhizosphere at seven weeks to cover the peak *Rhizoctonia* infection period. A total of 994,000 reads were classified into 917 genera covering 54% of the RDP Fungal Classifier database, a high diversity for an alkaline, low organic matter soil. Significant differences in fungal community composition between suppressive and nonsuppressive soils and between soil type/location were observed. The majority of differences associated with suppressive soils were attributed to less than 40 genera including a number of endophytic species with pathogen suppression potential and mycoparasites such as *Xylaria* spp. Nonsuppressive soils contained primarily fungi belonging to the genera, *Alternaria*, *Gibberella* and *Penicillium*, which include many species known to be plant pathogens (Penton et al. 2014).

The effects of fungal infestation levels and soil moisture on both root necrosis and foliar sudden death syndrome (SDS) caused by *Fusarium virguliforme* (*Fv*) (syn. *F. oxysporum* f.sp. *glycines*) along with *Heterodera glycines* (Hg), soybean cyst nematode (SCN) were studied in fumigated and nonfumigated soils. Soybean cv. Spencer was grown in nonfumigated and bromide-fumigated soil infested with increasing levels of *F. virguliforme*, either under rainfall or irrigated

conditions in 2003. Interactions between *F. virguliforme* and *H. glycines* were investigated in a functional inoculation design in fumigated and nonfumigated soil, planted with Williams 82 or Cyst-X-20-18 in 2004. In both years, higher levels of foliar SDS severity and root necrosis were recorded in *Fv*-infested soils with Hg, than in soils without Hg on the soybean cultivars susceptible to both pathogens. Both natural infestations of Hg in 2003, and an artificially amended population of Hg in 2004, contributed to higher foliar SDS severity. More severe SDS symptoms were always associated with more root necrosis, but elevated levels of root necrosis did not predict severe leaf symptoms. In contrast to the critical role of Hg, increasing fungal infestation levels had no significant effects on increasing either foliar SDS symptoms or root necrosis. High soil moisture resulted in higher levels of SDS root necrosis. In the greenhouse, root necrosis increased at a higher rate in low soil moisture than the rate in high soil moisture conditions. The fungal pathogen and the nematode functioned together to form complex disease, which was strongly dependent on high soil moisture (Xing and Westphal 2006).

Disease complex in soybean due to infection by *Fv*, causing SDS and HG, the SCN was further investigated in fields where a monoculture system was adopted between 2003 and 2007. Susceptible soybean cv. Spencer was planted, while inoculating *F. virguliforme* into nonfumigated or preseason-fumigated plots with methyl bromide (MB) at 450 kg/ha. Incidence of SDS and SCN were monitored. In one field, SCN population densities declined in nonfumigated plots, but increased in fumigated plots. SDS appeared later, after limited incidences in 2003 and 2004, in nonfumigated rather than in fumigated plots. In the greenhouse, nondisturbed or disturbed soil cores (10 cm diameter, 30-cm depth) from field plots received two-level factors: (i) nonfumigated or fumigated (1,070 kg/ha MB) and (ii) noninoculated or inoculated with 9,000 second-stage juveniles of SCN. At harvest, nonfumigated plots had fewer nematodes and less SDS, regardless of disturbance or inoculation than the corresponding fumigated cores and any cores from fumigated plots. In the second field, SCN could be detected after 2003, during the monoculture in nonfumigated plots and lagged in fumigated plots. Both treatments had low incidence of SDS. Suppressiveness of SDS in one of the fields could be exploited as a disease management strategy (Westphal and Xing 2011).

Monoculture of peanut has been practiced for a long time (over 20 years) in the hilly red soil regions in southern China. In such areas high incidence of Fusarium wilt, caused by *F. oxysporum* and *F. solani* was observed. The responses of these pathogens to root exudates of susceptible cultivar Ganhua-5 (GH) and moderately resistant cultivar Quanuya-7 (QH) were investigated. The components and contents of the amino acids, sugars and phenolic acids in the peanut root exudates were determined. The root exudates from both susceptible and moderately resistant cultivars significantly promoted the spore germination, sporulation and mycelial growth of *F. oxysporum* and *F. solani*, compared with control. The degree of stimulation depended on the pathogen strain and the stimulation of pathogen development increased with increase in proportion to the concentration of root exudates applied. Analysis of the root exudates by high performance liquid chromatography (HPLC) technique revealed that the contents of sugar, alanine, total amino acids in the root exudates of cv. GH were significantly higher than in cv. QH. On the other hand, the contents of *p*-hydroxybenzoic acid, benzoic acid *p*-coumaric acid and total phenolic acids were significantly lower than in cv. QH. The results suggested that the differences in the root exudates from different peanut cultivars might have a role in the operation of wilt-resistance mechanism functioning in the peanut rhizosphere (Li et al. 2013).

## 7.2 INFLUENCE OF AGRICULTURAL PRACTICES ON SOIL MICROORGANISMS

Various agricultural practices are applied to improve and preserve soil quality which is essential for plant and animal health, maintenance of desirable environment and sustained crop productivity. Soil microorganisms play a central role in maintaining soil quality. Microbial diversity and interactions between plants and microorganisms in the rhizosphere are either useful or harmful for crop cultivation. Agricultural practices, in turn, alter the structure and composition of soil microbial communities. The effects of tillage, nutrient application, irrigation methods and cropping systems affect the soil microbial communities in different agroecosystems and significantly influence crop production.

### 7.2.1 CROPPING SYSTEMS

A cropping system to be adopted in a field or location may involve cultivation of sequence or combination of crops with a view to limiting the incidence and severity of disease(s). The effects of rotation crops/ crop sequences, as well as green manures on the incidence of soilborne diseases have been investigated to choose the crop(s) suitable for the field/location. Rotation crops may reduce the soilborne pathogen by (i) interrupting or breaking the pathogen cycle of inoculum production, growth or survival; (ii) altering the soil physical, chemical or biological characteristics and stimulating biological activity, diversity of plant-beneficial microorganisms, making the soil environment less conducive for pathogen development, and (iii) direct inhibition of pathogens either through production of inhibitory or toxic compounds in the roots or plant residues or by stimulating specific microbial antagonists which may directly suppress pathogen development. Several studies have indicated significant differences among microbial communities from different long-term cropping systems (Larkin and Honeycutt 2002; Larkin 2003).

The root zone of different crops, comprising of the root system, rhizoplane and rhizosphere, forms a habitat where specific microbial activity may be observed and root pathogens and other beneficial microorganisms interact, during plant growth and incorporation of plant debris in the soil after harvest. Crop rotation aims primarily to slow down the buildup of populations of harmful organisms and prevent the development of diseases to epidemic level. Growing soybean as a cover crop and green manure effectively prevented buildup of potato common scab disease caused by *Spongospora subterranea* (Weinhold et al. 1964). Several two-year rotations, such as wheat-potato, wheat-corn, wheat-beans or wheat-sugar beet were found to be effective in reducing the incidence of wheat take-all disease caused by *Ggt* in the irrigated Columbia Basin and Snake River Plains districts of Pacific Northwest in the United States (Cook and Baker 1983).

Agricultural management practices have been reported to have differential effects on soil microbial communities. Effects of rotational crops and crop sequences, as well as green manures on the incidence of Verticillium wilt disease affecting several economically important crops have been investigated. Reduction in incidence of Verticillium wilt disease caused by *V. dahliae* on potato crop grown in two- or three- year rotations that included monocotyledonous crops and green manure crops was observed. Green manures generally included sudangrass (*Sorghum vulgare* var. *sudanense*), sweet corn (as either an annual rotation or a summer rotation, depending on the location) and brassica crops. A closer examination of the soil microbial community after two or three seasons, in corn green manure-applied plots showed that although the disease incidence was reduced in subsequent potato crop, microsclerotia densities in soil remained the same as before incorporation of corn to soil and an increase in *Fusarium* spp. was also seen (Davis et al. 2010; Larkin et al. 2010).

Green manures of various crops, including sudangrass, rapeseed, broccoli, winter pea, corn, oat and rye have been reported to reduce the incidence of Verticillium wilt disease of potato (Davis et al. 2010). The influence of previous cropping history on the incidence of *V. dahliae* on potato was investigated. The significant influence of rotations could be inferred on soil microbial communities and their characteristics, based on fatty acid profiles. The single season green manures grown prior to potato crops, mustard blend (mixture of white mustard (*Sinapsis alba*) and oriental mustard (*Brassica juncea*) and a sorghum sudangrass hybrid) were evaluated for their effect on the incidence of Verticillium wilt disease of potato. All rotation treatments in the first potato crop, following each single-season rotation treatment in 2007 and 2008 reduced Verticillium disease development, relative to the standard barley rotation. The mustard blend and sudangrass manure rotations reduced the incidence of wilt disease by 15 to 19% and 20 to 31% in 2007 and 2008, respectively. Averaged over both seasons, sudangrass and mustard blend reduced the disease incidence by 17 and 24%, respectively (see Figure 7.1). In 2009 potato crop, representing the second cycle disease incidence was much higher in all plots. The extent of reduction in wilt disease incidence in the second cycle was less effective, compared to the first cycle with an average of only 8 to 10%. Fumigation with metam sodium was the most effective in reducing Verticillium wilt disease of potato. The barley (as rotation crop)-fumigated soils had higher bacterial populations and microbial activity and lower fungal populations (including *Verticillium* and *Trichoderma* subpopulations), reflecting the fumigant effect of eliminating

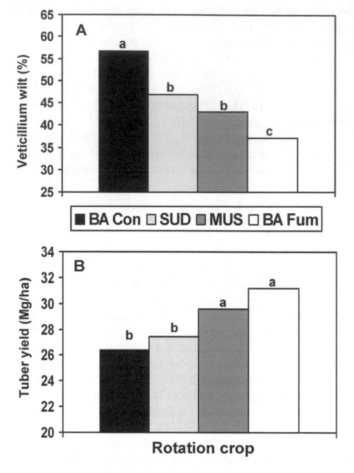

**FIGURE 7.1** Suppressive activities of rotation crops on Verticillium wilt disease of potato A: extent of suppression of disease development by sudangrass (SUD), and mustard blend (MUS) applied as green manures in comparison with barley (BA Cont) and barley (BA Fum) fumigated with metam sodium B: effect of rotation crops on tuber yield averaged over two potato seasons (2007 and 2008).

**[Courtesy of Larkin et al.(2011) and with kind permission of the American Phytopathological Society, MN, United States]**

fungal propagules and stimulating bacterial growth. This effect was observed at a much lesser degree with mustard rotation, although there was no corresponding reduction in *Verticillium* populations (Larkin et al. 2011).

The effects of ten successive monocultural cycles involving different legume species/cultivars on the inoculum potential (IP) of soils naturally infested with *Aphanomyces euteiches* were assessed under greenhouse conditions. The IP of naturally infested soil could be significantly modified by nonhost/host status of crop plant species as well as by the level of resistance/susceptibility of the cultivar. Continuous cultivation of susceptible cultivars of pea, lentil or faba bean was favorable for pathogen multiplication and increased the IP from 1.9 to 3.5 after ten cycles. In contrast, nonhost species/resistant cultivars of vetch or faba bean resulted in reduction of IP values of soils, regardless of the initial values from 1.9 to 0.5 and from 4 to 2, respectively, after ten cycles. The disease severity in resistant legume species/cultivars was not affected by the successive cultural cycles. The results indicated the need for

planting appropriate legume cultivars to contain the development of *A. euteiches* under greenhouse conditions (Moussart et al. 2013). The use of forage and cover crops in rotations, has an important role in increasing the organic matter contents of soils, thereby contributing to carbon sequestration and soil productivity. Leguminous forage crops such as red clover (*Trifolium pretense*) and white clover (*Trifolium repens*) offer beneficial effects on plant growth by producing nitrogen for the following crops and by breaking the life cycles of soilborne pathogens like Ggt, causal agent of wheat take-all disease. The legacy effects of cover crops on soil microbial communities and structure were studied, using crops viz., perennial ryegrass (*Lolium perenne*), chicory (*Cichorium intybus*), red clover and white clover between 2010 and 2012. Cultivation of chicory, red clover or white clover resulted in significantly divergent soil fungal communities, with a notably lower diversity of fungal populations under clover, suggesting a link to soil nitrogen dynamics. Spring wheat was sown on the same plots, followed by winter barley. A legacy effect of the preceding forage crop on fungal community was detected after both cereal crops, with plots previously planted with ryegrass being most divergent. Arbuscular mycorrhizal fungi (AMF) were more abundant on direct-drilled plots and pathogenic fungi were more abundant on plowed plots after the sowing of winter barley (Detheridge et al. 2016).

### 7.2.2 TILLAGE

Tillage systems influence physical, chemical and biological properties of soils and they have a major impact on soil productivity and sustainability. Tillage may affect the activity of a soilborne pathogen and its interaction with antagonists and other microbes, as well as with hosts due to change in the physical environment, the position of crop residue infested by pathogens. Loosening the soil by tillage up to a depth of about 30 cm facilitated bean roots to penetrate the layer of soil, where *Fusarium solani* f.sp. *phaseoli* inoculum was present. This condition resulted due to infection by the wilt pathogen and yield loss (Cook and Baker 1983).

### 7.2.3 APPLICATION OF NUTRIENTS

Nutrients applied to the soils, either as inorganic or organic, should be in available form. Soil nutrients may affect disease development directly, due to their effects on the pathogen or indirectly by their effects on plant growth or through their influence on microbial antagonists. Effects of nutrients, applied as a disease management strategy, are discussed in Chapter 8.

### 7.2.4 IRRIGATION PRACTICES

Soil moisture levels influence the development of both plants and microbial plant pathogens with neutral and positive effects on plant health. Irrigation significantly affects the biological balance existing prior to irrigation, particularly in the soil, where microbial balance is the result of long-time adjustments to arid and semi-arid conditions. Different types of irrigation systems, such as flood irrigation, furrow irrigation, drip irrigation and sprinkler irrigation, exert different effects on pathogen development and plant growth. The effects of soil type, soil moisture and depth of planting tubers on infection of potato tubers by *Phytophthora infestans*, causative agent of late blight disease, were studied under greenhouse conditions. Healthy tubers were hand-buried in soil at specific depths or naturally produced tubers from potato plants growing in the soil were also examined. A spore suspension of *P. infestans* was chilled to induce zoospore formation and a suspension of resulting zoospores and sporangia were applied to the soil. The soil depth at which tubers became infected was used as a measure for spore movement in the soil. Tuber infection decreased significantly with increasing soil depth. Tubers at the soil surface showed highest infection level, whereas infection was rare on tubers at 5 cm or deeper in soil. Amount of infection varied among soil types. Tuber infection in the Shano silt loam was significantly less than in medium and fine sands. Only tubers on the surface of Shano silt loam showed infection, but not other tubers below soil surface. Increased soil moisture did not significantly increase infection of tubers at deeper soil, regardless of soil type (Porter et al. 2005).

The effects of irrigation types on incidence of charcoal rot of melon caused by *Macrophomina phaseolina* were assessed, since the disease incidence and severity increased in Arizona over the years in fields where the subsurface drip irrigation system was adopted. But incidence of the disease was rarely seen in fields with furrow irrigation. Investigation was taken up to determine the influence of edaphic factors on inoculum density. Soil samples were collected once from the fields irrigated by subsurface drip, with and without plastic mulch and by furrow at 10, 20 and 30 cm depths. Percentage of soil moisture, pH, salinity and inoculum density were determined. Percentage of soil moisture was significantly higher at 20 and 30 cm depths in furrow-irrigated field, compared with drip-irrigated field with plastic mulch, but not in the field without plastic mulch. Average minimum and maximum temperature and inoculum density were significantly lower at all depths in the furrow-irrigated field, compared with both types of drip irrigation. The pH was significantly higher in the furrow-irrigated field, compared with both types of drip irrigation at 20 and 30 cm depth, but not at 10 cm depth. The results suggested that furrow irrigation might reduce disease by creating conditions unfavorable for pathogen development (Nischwitz et al. 2004).

Soil water potential is an important factor, influencing the development of soilborne pathogens. Soil moisture requirements differ based on the crop plant species and the nature of microbial pathogens – fungi or bacteria. Soil water at – 0.4 bar or wetter was found to arrest development of potato scab disease caused by *S. scabies*. Susceptible lenticels on potato tubers were protected in wet soil probably due to abundance of bacteria in the infection court of *S. scabies* which might exclude the pathogen through competition for space, nutrients and oxygen (Lapwood and Hering 1970). The wet soil might facilitate swarming of bacterial cells (at – 0.4 bars) which

might colonize the lenticels before the slower-growing actinomycete, *S. scabies* (Lewis 1970). The effects of soil matric water potential on the interaction among melon (*Cucumis melo*), *Monosporascus cannonballus* and *Olpidium bornovanus* (root pathogens) were studied. Colonization of cantaloupe roots by zoospores of *O. bornovanus* and germination of ascospores of *M. cannonballus* were highest at a soil matric potential of 0.001 MPa, but significantly inhibited at a matric potential of only – 0.01 MPa. Matric potential of – 0.01 MPa or drier were decidedly inhibitory to the motility of zoosporic pathogens, but not hyphal growth of filamentous fungi like *M. cannonballus*. The germination of ascospores of *M. cannonballus* could be mediated by *O. bornovanus*, an obligate zoosporic destructive root pathogen of cantaloupe (Stanghellini et al. 2014).

Productivity of crops and the outcome of host-pathogen interactions in natural plant populations are dependent on several factors, including quantity of water applied and its distribution during crop growth. Tissue culture plants of the East African highland banana cultivar Mbwazijume were established in a screenhouse to mimic drought conditions in the field, so that the effect of water stress on the development of banana Xanthomonas wilt (BXW), caused by *Xanthomonas campestris* pv. *musacearum (Xcm)* could be assessed. All inoculated banana plants were killed after expressing symptoms at ca. 14 days postinoculation (dpi). Water stress effects were significant ($P < 0.05$) for incubation period and significant ($P < 0.01$) for disease incidence and severity. BXW development was hastened by the combined effect of water stress before and after inoculation or plants under stress-free (SF) conditions. A two-fold increase in evaluated disease parameters suggested that both timing of infection and duration of exposure to water stress were critical in determining the shortened resident phase of *Xcm* to multiply and rapidly spread in the vascular tissues of banana. The results indicated that water-stressed banana plants might be physiologically weak resulting in increased vulnerability to infection by *X. campestris* pv. *musacearum* (Ochola et al. 2015). The effects of three systems of irrigation viz., furrow, sprinkler and subsurface drip irrigation, on spatial distribution patterns of lettuce big-vein disease were assessed. Since the causal virus, *Lettuce big-vein virus* (LBVV) is vectored by the chytrid *Olpidium brassicae*, different irrigation systems might influence the movement of the vector. Lettuce plants were mapped in arbitrarily selected plots with varying sizes, and the disease incidence was recorded in each quadrat. LBV incidence was aggregated in all furrow-irrigated fields, four of five surface drip-irrigated fields and two of three sprinkler-irrigated fields. Other fields had random distribution of infected plants. As the quadrat size increased, the aggregation index increased and vice-versa. Under furrow or subsurface drip irrigation, aggregation of infected plants occurred mostly across rows. The differential effects of irrigation type on *O. brassicae* were not clearly indicated by the results (Hao and Subbarao 2014).

Common bean (*Phaseolus vulgaris*) is subjected to stresses like waterlogging (abiotic) and *Pythium irregulare* (biotic), incitant of damping-off and root rot disease. The effects of timing (one, three, five, seven and nine days after sowing) and duration (three, six, 12 and 24 h) of soil saturation (waterlogging) on damping-off, as well as hypocotyl, and root diseases of common bean caused by P. *irregulare* were assessed. In the presence of P. *irregulare*, waterlogging one day after sowing resulted in the least emergence (55.2 ± 5.6%), although the plants that survived after five weeks, had less hypocotyl disease index (% HDI) and percent root disease index (% RDI) than non-waterlogged plants. The most severe disease was observed on plants exposed to waterlogging for nine days after sowing. In general, both hypocotyl and root disease severity increased as the duration of waterlogging increased from one to 24 h. The common bean cvs. Pioneer, Brown Beauty and Gourmet Delight responded differently to waterlogging, as reflected by disease development. Despite being susceptible to hypocotyl and root disease, Pioneer had the greatest emergence, suggesting its greater degree of tolerance to waterlogging and infection by P. *irregulare*. Gourmet Delight showed less hypocotyl and root disease, when waterlogging occurred. These two cultivars appear to have the potential for tolerance to waterlogging and infection by P. *irregulare*. Gourmet Delight showed less hypocotyl and root disease, when waterlogging occurred. These two cultivars appear to have tolerance to abiotic and biotic stresses, which could be exploited to breed cultivars with tolerance to these stresses (Li et al. 2015).

## 7.3 EPIDEMIOLOGY OF SOILBORNE CROP DISEASES

Epidemiology deals with one of the several aspects of host-pathogen interactions and it indicates the relative contributions of the pathogen, host plant and the environments, leading to varying levels of disease incidence and spread and consequent losses in quantity and quality of produce. The term epidemiology refers to the production and spread of pathogen inoculum in individual, as well as in a population of plants. Variability in the pathogenic potential of isolates/strains of soilborne microbial pathogens and levels of resistance of host plant genotypes are critical factors determining that outcome of host-pathogen interactions – incidence or absence of disease in a geographical location. Availability of susceptible host plant species or cultivar, virulent strains/isolates of the pathogen and environmental conditions favorable for the development and spread of the disease(s) to larger areas are primarily required for the development of epidemics. These parameters may be individually or collectively influenced by the duration for which it remains favorable. The term 'disease triangle' applied previously, has been modified to 'disease tetrahedron' to include the effect of duration of time for which each of the three components – host, pathogen and environment – on disease development is available (Fry 1982; Narayanasamy 2017).

An efficient and reliable diagnostic system for early detection and precise identification of strains/races/varieties of the pathogen and information on the pathogen genomics are the basic need for the epidemiological investigations. Generally, infection of plants by soilborne pathogens can be recognized only by the symptoms induced on aerial plant parts. Hence, the detection of pathogen presence in the asymptomatic plants whose roots may have been infected assumes great importance for the elimination of primary sources of inoculum to prevent further spread of the disease. Molecular diagnostic techniques have been developed for rapid, reliable and sensitive detection of microbial plant pathogens in plants, soil and water from lakes, ponds and rivers which form irrigation sources.

Infectious propagules of the pathogen residing in the soil, the inability to visualize initial symptoms induced in the roots hidden beneath the soil and a complex soil environment, are some of the problems associated with soilborne diseases, which are not encountered with diseases affecting aerial plant parts. Researchers investigating soilborne diseases have to face the challenges for quantifying the propagules present in the soil and for characterizing the components of the epidemic without destructive sampling of the affected crops. Basic information about the pathogens and the diseases caused by them are required to study pathogen development and disease incidence. The fungal pathogens may produce different spore forms which have different sensitivities to environmental conditions, chemicals and the changes occurring in the microbial community structure in the soil. Bacterial pathogens possess different mechanisms such as production of enzymes/toxins and formation of spores, capable of resisting adverse environmental conditions. The information on pathogen life cycle may be useful in determining the weak link in the life cycle, facilitating the application of suitable disease management strategies.

Epidemiology encloses several subprocesses. Infection cycle, infection chain or cycle of pathogenesis, the major subprocess represents the duration of time required for one dispersal unit (or infecting unit) to develop and produce the next generation of dispersal units. Soilborne pathogens like *F. oxysporum*, causing wilt diseases, have monocyclic life cycles. During one crop season, they have only one cycle. The pathogen infects a plant, develops systemically and then pathogen propagules are released when the plant tissues decompose in the soil to initiate infection of plants in the next crop season. Such pathogens are called monocyclic pathogens. In contrast, several pathogens such as *Aphanomyces* spp., *Phytophthora* spp., *Rhizoctonia* spp. and *Sclerotium rolfsii* are able to produce two or more cycles during one crop season, initiating new infections repeatedly in the same crops, leading to epidemics. These pathogens are designated polycyclic pathogens (Van der Plank 1963). *Gaeumannomyces graminis* var *tritici* (*Ggt*), causative agent of wheat take-all disease, forms perithecia on wheat stem bases at about the harvest time and asci and ascospores are formed later. As the fungal biomass increases, the active parasitic phase of *Ggt* alternates with a declining resting phase outside the living host, generally on colonized host tissues. The life cycle of *Ggt* is characteristic of a highly specialized root-infecting soilborne fungus, confined to the existence upon its host, because of microbial competition in soil. The ascospores of *Ggt* do not appear to be important for the initiation of take-all epidemics in the United

Kingdom. The take-all pathogen has to be placed in a different category, since direct initial infection occurs primarily by mycelial growth from a resting or survival spore or from the plant residues colonized by mycelium and dispersal by growth of mycelium from infected plant residues (Hornby 1998). New infection of plants may be initiated by spores produced from the same plants (autoinfection) or spores produced from other plants (alloinfection). In the case of polycyclic pathogens, both types of infection may be possible, facilitating occurrence of epidemics. Take-all disease may be considered as oligocyclic disease, because alloinfection may be relatively less frequent within a season. Take-all epidemic in the UK is generally polycyclic (increasing incidence due to accumulation of inoculum annually), requiring a succession of annual host crops (Zadoks and Schein 1979; Gilligan 1985).

Some of the soilborne pathogens have multiple modes of transmission from infected plant to healthy plants, enhancing the chances of epidemic development. *Fusarium oxysporum* f.sp. *radicis-lycopersici* (FORL), causing crown and root rot disease in tomato and *F. oxysporum* f.sp. *basilici* (FOB), inducing wilt disease in basil, are capable of producing conspicuous masses of macroconidia along the stems of infected plants. The role of the airborne propagules in the epidemics of the disease in tomato crops was investigated. Airborne propagules of FORL were trapped in the field using a selective medium. Plants grown in both covered and uncovered pots, obtained from field soil, and exposed to natural aerial inoculum, developed typical symptoms in 82 to 87% of plants. The distribution of inoculum in the growth medium in the pots also indicated the incidence of foliar infection by FORL. In the greenhouse, foliage and root inoculations were carried out with both tomato and basil and their respective pathogens. Temperature and a duration of high relative humidity (RH) affected the rate of colonization of tomato, but not of basil, by the pathogen concerned. Disease incidence in foliage-inoculated plants reached 75 to 100%. Wounding enhanced pathogen invasion and establishment in the foliage-inoculated plants. The results indicated the epidemiological importance of airborne propagules of *Fusarium* spp. infecting tomato and basil (Rekah et al. 2000).

The severity/intensity of disease symptoms varies with the nature and aggressiveness (virulence) of the pathogen races/varieties/strains, levels of resistance of cultivars and environmental conditions. The disease intensity is frequently analyzed more easily by visual observations, using a disease rating scale (0–9), than to estimate pathogen populations at a specific time. The amount of diseased tissue is expressed as a proportion of total plant tissue or percentage of plants infected and the values are plotted over time. The shapes of disease progress curves may vary with location and time. The disease progress curve for disease induced by monocyclic pathogen usually resembles a saturation curve. In the case of polycyclic pathogens, the disease progress curve is likely to be sigmoid. The dynamics of polyetic pathogen-induced disease may result in an exponential curve. Quantification of pathogens present in the soil may be accomplished by employing plate count and bait plant methods which are time-consuming, but molecular quantitative techniques provide more reliable results rapidly, facilitating the application of disease management strategies to effectively prevent development of epidemics.

## 7.3.1 Components of Epidemics of Soilborne Diseases

Epidemics of soilborne diseases may occur, following the interactions among three major components viz., the pathogen, host plant species/cultivars and environmental factors. Production of inocula may be affected, when the susceptible host plants are available in restricted areas and the environmental conditions favorable for pathogen development exist only for a short period. The contributions of various factors that influence the major components of epidemics are discussed hereunder:

### 7.3.1.1 Dynamics of Fungal Pathogen-Host Plant Interactions

#### 7.3.1.1.1 Pathogen Factors

Various factors are known to influence the vegetative growth, sexual reproduction and formation of resistant structures required for overwintering/survival of the fungal pathogens, during adverse environmental conditions. The inoculum of soilborne pathogens is the potentially infective material capable of initiating infection in susceptible plants. In the soil, the inocula may be in the form of mycelium, spores or other propagules and infected plant residues (Hornby 1981). Soil inoculum densities, (varying from 0 to $10^6$ conidia/g of potting mix) of *Fusarium oxysporum* f.sp. *vasinfectum (Fov)* race 4, were evaluated for their effects on plant growth, severity of Fusarium wilt symptoms, vascular discoloration and number of CFU/g of stem tissue for five cotton cultivars under greenhouse conditions. Symptoms of wilt and reduced plant growth were observed in susceptible cv. DP744 at inoculum levels of $10^3$ conidia/g of soil. On the other hand, the resistant cv. Pima Ph800 did not show any adverse effect on growth at all soil inoculum densities. All other cvs. DP340, Ph72 and UltEF responded to inoculation with *Fov* with a relatively moderate negative growth response to different inoculum densities at above $10^4$ conidia/g of soil. *Fov* race 4 could be recovered from stems of all cultivars. The stems of resistant Pima Ph800 yielded the pathogen less frequently at any inoculum density of conidia used for inoculation. The results indicated that *F. oxysporum* f.sp. *vasinfectum* race 4 was an inoculum density-dependent pathogen, inducing mild symptoms at levels less than $10^4$ conidia/g of soil (Hao et al. 2009).

The effect of horizontal/vertical distribution of *Ggt* inoculum on take-all disease severity was assessed under greenhouse and field conditions. Oat kernel inoculum of *Ggt* was placed at 0 (seed level), 5, 10, or 15 cm between the wheat seeds or 5, 10 or 15 cm to the side of wheat seed at a depth of 5 cm. Inoculum spatial location and distance greatly influenced disease incidence. Placement of inoculum below the seed induced more severe losses than when the inoculum was placed to the side of the seed. Within the same direction, take-all decreased as the inoculum was placed at greater distances from the seed, often to insignificant levels at 10 to 15 cm. The results indicated that significant reduction (≥ 50%) in take-all incidence might be possible by plowing under the infested residues (crowns) to depths greater than 15 cm (Kabbage and Bockus 2002). The spatial pattern of distribution of take-all disease incidence on a second consecutive crop of winter wheat was studied. The spatial pattern of distribution of take-all disease was aggregated in 48% of the data sets, when disease incidence was assessed at plant level and in 83%, when it was assessed at the root level. Clusters of diseased roots were in general less than 1 m diameter for crown roots and 1–1.5 m for seminal roots. When present in clusters of diseased plants, they were 2–5 m in diameter. Clusters did not increase in size over the cropping season, but increased spatial heterogeneity of the disease level was observed (Gosme et al. 2007). In a later investigation, within season gradient of wheat take-all disease was determined under field conditions, using two isolates G1i and G2i of *G. graminis var. tritici (Ggt)*, representing the G1 and G2 genotype groups in terms of virulence. Root disease incidence and severity were assessed at six intervals from early March to late June on plants located at regular distances from inoculum sources. A strong disease aggregation was observed both within and among plants. Disease severity on source plants placed near the inoculum source increased over time, ranging from 5 to 46% at the first assessment, and from 55 to 98% at the last assessment, the increase being larger for G2i than for G1i. In line-sown plots, the disease progressed steadily along the line, but did not extend beyond 20 cm, seldom reaching the neighbor line. Disease rarely reached 25 cm in direct-seeded crop stands. The results showed that *Ggt* infection intensified, but did not spread to a large extent during cropping season. The disease severity was significantly influenced by distance from the source, pathogen genotype and assessment date. No significant effect of host spatial distribution was evident (Willocquet et al. 2008).

The importance of inoculum source for the development of stem canker and black scurf disease of potato caused by *R. solani* was evaluated. The pathogen survived as sclerotia on tubers, in soils and crop residues. Disease-free minitubers and seed tubers contaminated with low levels of *R. solani* inoculum were planted in fumigated or artificially inoculated growth mixture in the greenhouse. Black scurf incidence and severity on tubers were significantly higher when inoculum was present in both seed tubers and soil, compared with either of them separately. Disease severity on underground plant parts of the plants was also significantly higher in plots where both seed tubers and soil were contaminated. Thus, infested tubers and soil were found to be major sources of inoculum for *R. solani*. Under field conditions, disease incidence and severity on daughter tubers were positively correlated with levels of contamination in seed tubers and soil. Disease incidence was reduced by in furrow fungicide treatment, which was ineffective when the seed tuber and soil inoculum levels were high, indicating the importance of selecting disease-free tubers for planting (Tsror and Peretz-Alon 2005). The effect of soilborne inoculum density and survival of sclerotia of *R. solani* AG3PT on the development of stem canker and scurf disease on potato were studied. Sclerotia produced by *R. solani* AG3PT were

buried in small plots. After 18 months on average across all experiments, 20% of retrieved sclerotia were viable. Burial of sclerotia in field soil at 20 cm depth was significantly reduced to 50%, compared to 60% at 5 cm. In pot experiments, amending the growing medium and soil with increasing number of stem infected, stem canker and black scurf severity increased, regardless of whether this soilborne inoculum was derived from mycelia or sclerotia. Black scurf incidence and severity depended on the range of soilborne inoculum densities. Both parameters of disease increased with an increase in inoculum density (Ritchie and Mcquiken 2013).

Pathogen movement in the soil environment is one of the important factors influencing the incidence and spread of soilborne pathogens, affecting crops in different ecosystems. The modes of pathogen spread (movement) can be expected to influence development of epidemics. The relative contribution of migration of *R. solani* anastomosis group 3 (AG-3) on potato seed tubers originating from production areas in Canada, Maine and Wisconsin (source population) to genetic diversity and population structure of *R. solani* AG-3 in North Carolina (NC) soil (recipient population) was investigated. Analyses of molecular variations, using Multilocus PCR-RFLP, heterozygosity at individual loci and gametic phase disequilibrium between all pairs of loci were performed. Little variation between seed source and NC recipient soil populations or between subpopulations within each region could be detected. The one-way (unidirectional) migration of genotypes of *R. solani* AG-3 into NC on infected potato seed tubers from different locations, was possibly responsible for the lack of genetic differentiation between populations in NC soils (Ceresini et al. 2003). Real-time PCR assay was employed to determine the spatial distribution of *R. solani* AG 2-1 in field soils. Soil from shallow depths of a field planted with *Brassica oleracea* tested positive for *R. solani* AG 2-1 more frequently than soil collected from greater depths. Quantification of *R. solani* inoculum in field samples was difficult, due to low levels of inoculum in natural soils (Budge et al. 2009).

The ability of *R. solani* to access and exploit its host radish (*Raphanus sativus*) was investigated. The spread of *R. solani* is driven by two types of resources: (i) organic matter, when the fungus acts a saprotroph, and (ii) infected host tissues, when it functions as a pathogen. The pathogen spread depends on host/pathogen interaction, which is significantly affected by the ontogenic resistance (varying ability of host to resist or tolerate disease, as it develops). The link between host development and its infection by *R. solani* was assessed by monitoring disease incidence and symptoms on radish inoculated at different stages of development (ages). Then the host exploitation in terms of pathogenic spread was quantified through the ability of *R. solani* to grow into soil and colonize baits located at different distances. Disease incidence at harvest (day 30) was negatively linked to host phenological age at inoculation and it decreased from 88± 6% for inoculation at seedling stage to 69 ± 4%, when tuberization occurred. Disease incidence finally dropped to 13%, when inoculation was done at 24 days after sowing. The type of symptoms changed along with the phenological development and appeared markedly

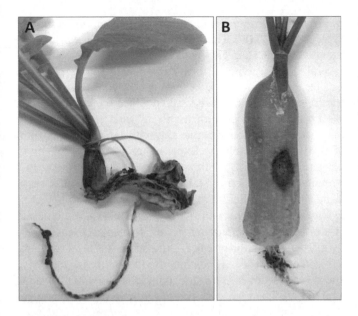

**FIGURE 7.2** Relationship between type of symptoms induced by *R. solani* and host development at inoculation A: damping-off symptoms on radish seedling inoculated at six days after sowing B: necrosis symptoms on root of radish plant inoculated at tuber initiation stage (18 days after sowing).

**[Courtesy of Simon et al. (2014) and with kind permission of *PLoS ONE* Open Access Journal]**

linked to host development at inoculation (see Figure 7.2). When the host root system was well-developed, the proportion of damping-off symptoms was drastically reduced to 3% in plants inoculated at 18 days after sowing. In the absence of host plant, the fungus was able to spread from a mycelial disk and colonize the surrounding soil surface relatively rapidly. The baits (74%) placed at 1 cm away from mycelial disk were colonized at two days after inoculation. However, saprophytic spread was limited to about 10% of baits at 2.5 cm at 14 days after soil inoculation and no colonization was observed at 5 cm and beyond. Further, without any host plant to exploit, *R. solani* could not sustain its spread. Colonization at 1 cm started to decline after ten days after inoculation. Pathogenic spread was essentially slower, but more sustainable and more extensive than saprophytic spread. Host development impacted the infection process, decreasing incidence and switching the type of symptoms caused by *R. solani* from damping-off to necrosis of roots. Disease characteristics changed markedly with host age at the time of inoculation. It was suggested that the host tuberization process might lead to remarkable change in the interaction between radish and *R. solani* (Simon et al. 2014).

Contributions of seed tuber- and soilborne inocula of *R. solani* AG3-PT for the development of potato black scurf disease epidemics was investigated, using an integrated experimental approach combining field trials and molecular techniques. Two distinct sets of genetically marked isolates of *R. solani* were used as seed tuberborne and soilborne inocula in a mark-release-recapture experiment. Disease assessments were made at tuber initiation and harvest stages. Both

inoculum sources were found to be equally important for incidence of black scurf disease, whereas soilborne inocula appeared to be more important for root and stolon infection, and tuberborne inocula contributed more to the stem canker. However, seed tuber-transmitted genotypes accounted for 60% of the total recovered isolates, when genotyped, using three PCR-RFLP markers. The changes in population structure of the experimental *R. solani* population over the course of the growing season and across two growing seasons, were investigated, using eight microsatellite markers. The populations at different sampling times were somewhat genetically differentiated. The proportion of isolates with genotypes that differed from the inoculants ranged from 13 to 16%, suggesting the possible emergence of new genotypes in the field. As both soilborne and tuberborne inocula are vital for infection and disease incidence, use of pathogen-free seed tubers should be ensured to reduce disease incidence and introduction of new genotypes of *R. solani* for sustainable potato production (Muzhinji et al. 2018).

*Gibberella zeae* (anamorph *Fusarium graminearum*), causal agent of wheat Fusarium head blight (FHB) disease, produces macroconidia, which are splash-dispersed by rain and are involved in the secondary infection of wheat earheads. Sexual fruiting bodies, perithecia, are formed in the spring on the infected crop residue left on the soil after harvest with rainfall greater than 5 mm. Ascospores formed in the perithecia are forcibly discharged and dispersed by wind, leading to initiation of primary infections. With increase in no-till or conservation tillage practices, large amounts of crop residues accumulate on the soil surface and form an important source of inoculum of FHB pathogen. Macroconidia and perithecia could be produced from the residues on or above soil surface for up to three years (Konga and Sutton 1988). Survival of *G. zeae* in *Fusarium*-damaged kernel (FDK) was investigated under field conditions. FDKs were either left on soil surface or buried 5 or 10 cm deep and monitored for 24 months. Perithecia developed on FDKs from all locations and treatments, but ascospores developed only in perithecia present on kernels located on the soil surface. Ascospores were formed in perithecia produced on FDKs incubated at 20°C, but not at lower temperatures (Inch and Gilbert 2003). In most of the soilborne diseases, the infested soil is the source of inoculum for primary infections, whereas diseased plants, in turn provide inoculum for secondary infections (both auto- and allo-infections). But in the case of carrot cavity spot (CCS) caused by *Pythium violae*, fragments of CCS lesions were used to infest soil and it was shown that CCS lesions on carrot residues could cause primary infection of healthy roots. A novel soil infestation method was developed by placing an artificially infested carrot root (the donor plant) close to healthy roots (receptor plants) and formation of typical CCS lesions was induced more efficiently than the use of soil inoculum. CCS could spread from root to root by alloinfection from transplanted diseased roots. The procedure revealed the polycyclic nature of a CCS epidemic caused by *P. violae* under controlled conditions. Secondary infections developed disease symptoms and reduced root weight as early as two weeks after transplantation of diseased carrot (Suffert and Montfort 2007).

*Leptosphaeria maculans* and *L. biglobosa*, causal agents of oilseed rape blackleg disease, coexist in the United Kingdom survive on naturally infected oilseed rape debris. Pseudothecial maturation on naturally infected plant debris were studied both in controlled environments (5, 10, 15 or 20°C), under continuous wetness and in natural conditions (debris exposed in September and November 2002). Four maturation classes of pseudothecia viz., A (asci undifferentiated), B (asci differentiated), C (ascospores differentiated) and D (ascospores matured). Progress in pseudothecial maturation (assessed by time until 50% pseudothecia reached each class) was similar for both *L. maculans* and L. *biglobosa* at 15–20°C, but *L. biglobosa* matured more slowly at < 10°C. A negative correlation between maturation time and temperature increase from 5 to 20°C under continuous wetness was observed, but maturation time was longer under natural conditions, especially when periods of dry weather existed. Differences in the requirements of pseudothecial maturation in *L. maculans* and *L. biglobosa* might be of epidemiological significance in the United Kingdom (Toscano-Underwood et al. 2003). The role of soilborne ascospores and pycnidiospores of *L. maculans* in the development of epidemics of blackleg disease of oilseed rape (*Brassica napus*) was investigated. Both spore forms caused seedling death, even after the spores had remained in the growth medium for up to 21 days before sowing. Highest severity of seedling death was observed, when seeds were sown at seven days before infestation (59 and 40% seedlings dead for pycnidiospores and ascospores, respectively). In the case of pycnidiospores in moist sand, a significantly greater percentage of seedlings was killed for sowings seven, 14 or 21 days after infestation, compared with concurrent sowing and infestation. Allowing the sand to dry between infestation and sowing, mortality due to infection by pycnidiospores was significantly reduced, but not when ascospores were used. Buried slide assay showed that both ascospores and pycnidiospores were able to germinate in the sand with lengths of germ tubes similar to those observed in controls incubated in a sterile moist chamber. Under conditions prevailing in Australia, blackleg disease was considered as a polycyclic disease, since soilborne pycnidiospores might be produced in more than one cycle during a single crop season and they may contribute significantly to occurrence of epidemics in oilseed rape crops (Li et al. 2007).

*L. maculans*, incitant of root rot disease of *B. napus*, infects aerial plant organs also. A survey was carried out in Australia to assess the incidence and severity of root rot and to study the pathway of root infection under field conditions. Root rot infection was up to 95% in 127 crops surveyed. The severity and incidence was significantly correlated with that of crown canker. Root rot symptoms appeared before flowering and disease severity increased at maturity. The infection of the root appeared to be an extension of the canker phase of the disease cycle. All isolates were pathogenic to both *B. napus* and *B. juncea*. Under field conditions, the main pathway of root infection was via invasion of cotyledons or leaves

by airborne ascospores, rather than from inoculum in the soil (Sprague et al. 2009). The development of *L. maculans*, causal agent of Phoma stem canker of oilseed rape, is influenced by several factors such as competitive effects between pathogen isolates, number of infection sites on cotyledons and levels of host resistance. Two oilseed rape doubled haploid lines, one susceptible and other with high polygenic resistance, were inoculated via wounded cotyledons with conidial suspensions of two pathogen isolates. Exposure of inoculated plants to low temperature (6°C), followed by warm temperature (20°C) stimulated expression of stem canker symptoms. PCR assay using three minisatellite markers was employed to detect *L. maculans* in all stems with visible stem canker symptoms and also in stems of asymptomatic (14 of 50) plants. Disease severity increased with the number of infection sites on cotyledons. Polygenic resistance significantly reduced disease incidence and severity from 74% (in susceptible line) to 16% (Travadon et al. 2009). The relationship between severity of blackleg or Phoma stem canker induced by *L. maculans*/*L. biglobosa* and subsequent primary inoculum production on stubbles of oilseed rape (*B. napus*) was studied at two sites over three years in France. The quantity of primary inoculum produced in the following year increased with canker severity index to 1.9 and 10.8 pseudothecia/cm² on stubble with the least and most severe cankers, respectively. Planting cv. Darmor with good level of quantitative resistance, reduced the severity of canker in the field, but not the subsequent inoculum production from stubble of the same canker severity class. At both sites, maturation of pseudothecia occurred after 63–75 days of incubation and increased with canker severity, with a mean of 0.5 and 3% mature pseudothecia appearing per favorable day, on stubble with the least and most severe cankers, respectively. The total area occupied by pseudothecia was correlated with the number of pseudothecia. The results showed that the quantity of available primary inoculum for a given severity might form the basis of forecasting the quantity of available inoculum for a given disease severity (Lô-Pelzer et al. 2009).

*V. dahliae* causes wilt diseases in a wide range of agricultural and horticultural crops in different agroecosystems. The relationship between inoculum density (number of microsclerotia per gram of air-dried soil) of *V. dahliae* at the time of planting and the severity and incidence of root discoloration of horseradish at harvest was investigated in a two-year study conducted in the greenhouse, microplots and commercial production fields. Significant correlations were noted between inoculum density and severity and incidence of root discoloration in the greenhouse and microplots assessments. However, in commercial production fields such correlations were not evident. Fields with low inoculum densities had high levels of disease ratings and root discoloration even in partially resistant horseradish cultivar 769A, whereas with susceptible cv. 647A, low disease ratings and root discoloration were recorded in fields with high inoculum densities. The results suggested the unreliability of disease forecast system based only on the inoculum densities of *V. dahliae*. The data on inoculum densities may be probably useful to avoid fields that are highly infested with microsclerotia of *V. dahliae* for planting susceptible crops (Khan et al. 2000). Isolates of *V. dahliae* associated with epidemics of Verticillium wilt in pepper fields of the central coast of California were characterized. Inoculum density of *V. dahliae* ranged from 2.77 to 66.66 microsclerotia/g of dry soils surveyed. A high correlation between disease incidence and density of microsclerotia (r = 0.81, P> 0.01) was observed. Distribution of wilt was aggregated in a majority of pepper fields, but the degree of aggregation varied. Of the 67 isolates of *V. dahliae*, 67% belonged to VCG2, 22% to VCG4 and 11% to a new group named as VCG6. Pathogenicity tests showed that bell pepper seedlings (one month old) were susceptible to pepper isolates only, whereas tomato seedlings were susceptible to both pepper and tomato isolates of *V. dahliae*. Pepper isolates belonging to VCG2, VCG4 and VCG6 were highly pathogenic to bell pepper and chilli pepper. PCR-based random amplified polymorphic DNA (RAPD) revealed unique polymorphic banding patterns in VCG isolates, but only minor variation could be detected in the isolates of VCG2 and VCG4. Intensive cropping practices in the region might have contributed to the selection of more aggressive isolates of *V. dahliae*, resulting in epidemics in pepper fields in California (Bhat et al. 2003).

*V. dahliae* induces the destructive Verticillium wilt disease of olive in southern Spain. The dominant cultivars, Picual and Arbequina were found to be highly susceptible to the disease caused by the defoliating pathotype of *V. dahliae*. Differential reactions of olive cultivars were observed during the surveys of naturally infected orchards with different inoculum densities of the pathogen. The relationship between inoculum density and disease incidence fit a logarithmic function for cvs. Picual and Arbequina. The percentage of affected trees of cv. Arbequina per year increased linearly with inoculum density in the soil, whereas no such relationship could be observed for the cv. Picual. Further, planting density of olive did not have any effect on disease incidence in both olive cultivars (Roca et al. 2016). Verticillium wilt disease caused by *V. dahliae* was widely distributed in the Mediterranean region, accounting for heavy losses and high tree mortality in olive orchards. The relationship between inoculum density of microsclerotia (MS) of *V. dahliae* in soil and disease incidence was investigated. The results showed that a high frequency of *V. dahliae* in soils in Lebanon (75.3%), coupled with a mean soil inoculum density of 17.0 MS/g of soil greatly impacted the production in susceptible olive cultivars. Molecular methods to determine microsclerotia inoculum density showed greater frequency of pathogen presence in infested soil, than the traditional plating method. The overall Verticillium wilt pathogen prevalence in the surveyed orchards was 46.2% and the frequency of *V. dahliae*-infected trees was 25.7%. The widespread distribution of *V. dahliae* in all olive orchards demanded the application of strategies directed toward reduction of soil inoculum density before new plantations of olive are raised and strict enforcement of phytosanitary regulations for certification programs (Habib et al. 2017). Persistence of *V. dahliae* in soil as melanized microsclerotia (MS) for up to 14 years is a strong hurdle to be overcome for effective management of Verticillium wilt disease, affecting members of Brassicaceae.

MB has been applied as a preplant soil fumigant for a long time to reduce MS of *V. dahliae*. The dynamics of MS persistence in soil before and after MB + chloropicrin (CP) fumigation were studied for over three years in six 8 × 8 m sites in two fields. In separate fields, the dynamics of MS in the 60-cm deep vertical soil profile at pre- and postfumigation with MB + CP, followed by various cropping patterns were monitored over four years. In addition, MS densities were determined in six 8 × 8 m sites in separate field prior to and following a natural six-week flood. MB + CP significantly reduced, but did not eliminate the MS of *V. dahliae* in either the vertical or horizontal soil profiles. In the field trials, increases in MS were highly dependent upon the crop rotation pattern, followed by postfumigation. In the vertical profile, densities of MS were highest in the top 5 to 20 cm of soil, but were consistently detected at 60 cm depths. Natural flooding for six weeks reduced significantly, but did not eliminate viable MS of *V. dahliae* (Short et al. 2015b).

Soil samples from sugar beet fields were evaluated for studying spatial distribution of *Aphanomyces cochlioides*, causal agent of sugar beet root rot disease. Sugar beet seedling assay was performed in the greenhouse to determine root rot index value (0–100-point scale) which was employed to obtain indirect estimates of relative activity and density of pathogen inoculum. Field assessments of Aphanomyces root rot of sugar beet (0–7 scale) were made at harvest at each on-site soil sample collection. Greenhouse root rot index values were positively correlated with disease severity values determined in the field. Variance-to-mean ratios of greenhouse root rot index values and of field disease severity ratings among samples within each plot were calculated to compare the spatial distribution of midseason inoculum with root rot at harvest. Ratios of greenhouse root rot indices revealed that the inoculum of *A. cochlioides* was aggregated in the field midseason, but root rot was uniform within plots at harvest. Wet weather in July to August was favorable for disease incidence and development. Combination of factors such as root growth reaching inoculum foci, redistribution of inoculum and inoculum densities available above minimum threshold levels might have contributed to uniform distribution of the disease at harvest (Beale et al. 2002). Dispersal and movement mechanisms of sporangia of *Phytophthora capsici* were studied. Direct laboratory observations showed that dispersal of sporangia of *P. capsici* occurred in water with capillary force, but did not occur in response to wind or a reduction in RH. On the other hand, dispersal of sporangia by wind was infrequent (0.7% of total hours monitoring) during sampling, as determined by a volumetric spore sampler in a commercial cucurbit field and also in an experimental setting, where large numbers of sporangia were continuously available in close proximity to the spore trap. The results of both direct laboratory observations and volumetric spore sampling showed that dispersal of sporangia via wind currents was infrequent and sporangia were unlikely to be spread by wind alone under natural field conditions (Granke et al. 2009).

Spatial distribution of *Aphanomyces euteichus*, causal agent of pea root rot disease, was assessed in naturally infested pea field. Estimates of IP revealed a horizontal distribution of inoculum among several foci in the field, these foci differing in size and disease intensity. Disease severity was highly correlated with soil IP. Assessment of vertical distribution of inoculum in the soil showed that the maximal depth at which the pathogen could be detected was 10 to 40 cm, but the inoculum was detectable up to a depth of 60 cm. In general, IP was the lowest for soil layers at depths of 50 to 60 cm (Moussart et al. 2009). Spatial patterns of mint plants with symptoms induced by *V. dahliae* were studied in 10 commercial fields in Washington. The disease incidence was recorded in 0.76 × 0.76 quadrat with a width of mint rows of 0.76 m in size from 5 to 76 m × 57 to 396 m. In general, there was more clustering within than across rows, as per both doublets and runs analyses. Total number of wilt foci ranged from five to 170 per field and mean size of foci ranged from one to 20 quadrats. In one field, total foci increased from 24 to 104 and the mean size of foci increased from 1.0 to 1.3 quadrats in the same section of the field from one year to the next. Size of foci increased to 2.7 quadrats in the third year of sampling in the same field. Verticillium wilt spread during the life of the perennial mint crop. Inoculum for much of the secondary increase apparently did not originate from MS present in the soil before planting the crop or from infected rhizomes that were originally planted (Johnson et al. 2006).

The possibility of host specificity being responsible for outbreaks of Verticillium wilt disease of chrysanthemum in the greenhouse in the Netherlands was investigated. Following artificial inoculation, five isolates of *V. dahliae* were found to be pathogenic on chrysanthemum. On the other hand, five isolates of *V. dahliae* from potato were nonpathogenic to chrysanthemum. On potato, all isolates from potato and chrysanthemum induced early senescence with significant difference among the isolates. Potato isolates formed a distinct cluster from all other isolates, as indicated by amplified fragments length polymorphism (AFLP) analysis. As a group, the chrysanthemum isolates were no more diverse than potato isolates, but did not form a distinct cluster from 12 other isolates tested. The MS produced by chrysanthemum isolates had significantly lower average lethal temperature tolerance than those produced by potato isolates (Ispahani et al. 2008).

Quantification of MS of *V. dahliae* in soil is an important requirement for disease risk predictions for Verticillium wilt disease, affecting several crop plants. Dry or wet-sieving and plating of soil samples on selective medium is a time-consuming and resource-intensive procedure. The molecular quantification method based on Synergy Brands (SYBR) Green real-time quantitative PCR of wet-sieving samples (wet-sieving qPCR) was developed. This method could detect *V. dahaliae* MS at concentrations as low as 0.5 CFU/g of soil. Wilt inoculum was estimated, using > 400 soil samples taken from the rhizosphere of individual cotton plants with or without visual symptoms in experimental and commercial cotton fields at the boll-forming stage. The estimated inoculum threshold varied with cultivar, ranging from 4.0 and 7.0 CFU/g of soil for susceptible and resistant cultivar, respectively. The wilt inoculum as estimated by wet-sieving qPCR method was positively related to wilt development. Further, an overall relationship of wilt incidence was observed with inoculum density across 31 commercial fields, where a single composite soil sample was taken in each field, with an estimated inoculum threshold of 11 CFU/g of soil. Based on the results, pathogen populations of 4.0 and 7.0 CFU/g of soil could be recommended as the inoculum threshold for susceptible and resistant cotton cultivars, respectively in practical risk prediction investigations (Wei et al 2015).

*V. dahliae* has been reported to be transmitted via seeds of spinach in the United States and Europe. Planting infected seeds increased the soil inoculum density, in addition to the possibility of introducing exotic strains that contribute to epidemics of Verticillium wilt disease of lettuce and other crops grown in rotation with spinach. Further, other crop residues that are returned to the soil harbor the pathogen inoculum for infection of lettuce or other crops (Duressa et al. 2012). *V. dahliae* may be transmitted via seed and/or soil. Infected seeds of spinach carry the pathogen to long-distance and also introduce the pathogen into noninfested field soil. It was considered important to investigate the relationship between *V. dahliae* soil inoculum and infection of harvested spinach seed. The threshold level for *V. dahliae* in soil was determined as the pathogen DNA content of 0.003 ng/g of soil for seed infection to occur by semifield experiment where the spinach was grown in soil with different inoculum levels. Soils from spinach production fields were sampled during and before planting in 2013 and 2014 and also the harvested seed were tested for the extent of infection by *V. dahliae*. Seed from plants grown in infested field soil were infected with *V. dahliae* in samples from both semifield and open field experiments. Pathogen populations were low in seed from spinach grown in soils with a scattered distribution of *V. dahliae*, than in soils with uniform pathogen distribution. The results showed that infection of *V. dahliae* in harvested spinach seed strongly depended on the quantity of pathogen inoculum in the soil, indicating the epidemiological significance of soil inoculum and the possibility of long-distance dispersal of *V. dahliae* via spinach seed (Sapkota et al. 2016).

*Fusarium solani* f.sp. *piperis* (*Fsp*) (teleomorph, *Nectria haematococca* f.sp. *piperis*) induces two symptom types: root rot (RR) type and stem rot (SR) type in pepper. The temporal and spatial association between perithecial formation of *Fsp* and the development of SR were investigated in naturally infested fields. The locations in two neighboring fields of pepper vines with perithecia and all vines showing SR type symptoms were mapped. The frequency of vines with perithecia increased during April and May (in the late rainy season). In June, the early dry season, the number of vines with SR type symptoms greatly increased. The vertical range of perithecial formation on the vines was restricted to 30 cm in dry season in both fields. The spatial association between vines with perithecia and vines with SR type symptoms was variable between the fields, only one of them showing spatial association. The ascospores from perithecia of *F. solani* f.sp. *piperis* on pepper vines might be an inoculum source for SR type disease on adjacent vines (Ikeda 2010). *F. oxysporum* f.sp. *cucumerinum (Foc)*, causing Fusarium vascular wilt of cucumber, is responsible for heavy losses in soilless greenhouse cucumber crops. The risk associated infection of pruning wounds in cucumber plants by airborne *Foc* propagules was investigated. Both macroconidia and microconidia were identified as sources of airborne inoculum, using quantitative real-time PCR assay. A potential relationship between fluctuation in RH and spore release was observed. In addition, crop disturbance during handling might also influence liberation of pathogen propagules. However, the experimental inoculation of stem with conidia failed to establish infections. The results suggested that infection of cucumber plants by *Foc* occurred through roots, and the use of resistant rootstock might provide protection against this pathogen (Scarlett et al. 2015).

Soil temperature is one of the important factors affecting pathogen development and disease incidence/severity. Effects of different temperatures on formation of MS by six isolates of *V. dahliae in vitro* and on *Arabidopsis thaliana* were investigated. Mycelia growth was at optimal rate at 25°C, but MS production was maximum for two isolates at 20°C and at 15–20°C for another isolate. On seedlings of *A. thaliana*, the optimum for MS formation was 15–20°C. Two isolates were more efficient by producing ten times more MS than other isolates. Exposure of plants to a constant temperature of 20°C, after the onset of senescence resulted in the production of moderate amount of MS, compared with plants maintained at other temperatures. The results suggested that the rate of vascular colonization and MS formation might be influenced by temperature in a similar manner (Soesanto and Termorshuizen 2001). The temperature tolerance of *F. oxysporum* f.sp. *lactucae*, causing Fusarium wilt of lettuce, was investigated under controlled conditions. Nine cultivars of lettuce planted in infested soil (500 or 5,000 CFU/g of soil) were maintained under high/low diurnal temperature regimes of 26/18°C, 28/20°C or 33/26°C. Three resistant cultivars were not affected by differences in inoculum levels or temperature regimes tested. Five susceptible cultivars were more severely affected at higher inoculum levels and higher temperatures. The effect of temperature on the pathogen growth was reflected on the severity of symptoms. The radial growth rates for six isolates of *F. oxysporum* f.sp. *lactucae* increased with increase in temperature from 10°C to near 25°C. The apparent thermotolerance of the pathogen has to be considered, by avoiding planting susceptible cultivars during warm periods (Scott et al. 2010).

*Sclerotinia sclerotiorum* and *S. minor* cause lettuce drop disease, which was found to be an important limiting factor in California between 1995 and 1998. Incidence of *S. minor* predominantly in 14 of 25 commercial fields, *S. sclerotiorum* predominantly in nine of 25 fields and at varying levels of both pathogens in two fields, was observed. Infection by sclerotia directly (type I) and by airborne ascospores (type II) was differentiated, based on the symptoms. The precise locations of diseased and healthy lettuce plants were mapped and disease progress in the marked plants was monitored. Disease incidence with type I infection showed an aggregated pattern in the majority of the fields and random pattern, typical of soilborne diseases, was also observed. In fields with aggregated distribution, spatial dependence was observed up to 10 m and was either isotropic or random in direction, suggesting the potential influence of tillage operations on inoculum distribution and disease incidence. Lettuce drop incidence in fields with type II infection was erratic in time and peaked within a very short time. However, disease incidence assumed an aggregated pattern in all fields investigated. Spatial dependence of quadrats was generally detected in two adjacent directions. Increasing quadrat sizes generally increased the degree of aggregation of lettuce drop-infected plants, but not the distribution pattern itself. The results showed that the source of inoculum and the types of infection caused, are most likely to determine spatial patterns of lettuce drop disease in the fields (Hao and Subbarao 2005).

Survival of sclerotia of *Sclerotinia minor* and *S. sclerotiorum* was assessed in irrigated fields during summer in California. After burial of sclerotia at 15 cm depth, > 75% of sclerotia of *S. sclerotiorum* remained viable, whereas only 4 and 5% of sclerotia of *S. minor* buried at 15 and 5 cm depth, respectively were viable in the San Joaquin Valley. On the other hand, > 80% of sclerotia of both pathogens were viable in Salinas Valley, indicating the differential effects of location. More than 90% of sclerotia of both pathogens survived for at least three months in sterilized dry soils at temperatures between 15°C and 40°C. Soil moisture did not affect sclerotial survival at 15°C and 25°C, with less than 2% sclerotia surviving over four weeks, compared with about 45% of sclerotia surviving at ambient oxygen level. The combined effects of high temperature, high soil moisture and reduced oxygen in irrigated fields adversely affected the survival of *S. minor* and *S. sclerotiorum* (Wu et al. 2008). In another investigation, the effect of moisture contents on carpogenic germination (CG) of *S. sclerotiorum* sclerotia and the dynamics of sclerotial water imbibitions were studied under controlled conditions, using silty clay, sandy loam and sandy soils, maintained at 100, 75, 50 and 20% soil saturation. Smaller sclerotia imbibed water at a significantly faster rate than larger sclerotia (P = 0.05) in water and in soil at saturation percentages. When buried in soil, small, medium and large sclerotia were fully saturated for 5, 15 and 25 h, respectively in all types of soils and moisture percentages. The effect of sclerotial moisture content on CG was assessed, using sclerotia maintained at 95 to 100, 70–80, 40–50 and 20–30% of their water saturation capacity. Sclerotial moisture content significantly influenced CG, reaching the peak by fully saturated sclerotia, whereas no germination occurred below 70–80% of saturation. The results revealed the ability of sclerotia of *S. sclerotiorum* to produce apothecia (sexual spore formers) in soils with relatively low moisture levels (Nepal and del Rio Mendoza 2012).

The impact of inoculum densities of *Plasmodiophora brassicae*, incitant of clubroot disease of crucifers on disease risk was investigated. In order to quantify the risk potential of P. *brassicae*, a semicontrolled experiment was performed, using artificially infested soil at inoculum densities between $10^6$ and $10^{10}$ spores/l of soil. Another greenhouse experiment was conducted, using the soil of the semicontrolled experiment after cropping two cultivars to quantify the influence of cultivar resistance on soil inoculum. The results indicated that disease rating was positively correlated with the amount of inoculum added to the soil. Linear regression analyses indicated a negative correlation between seed yield and inoculum density. Even at the lowest inoculum density, yield was reduced by 60%. The seed yield of resistant cultivar was not affected significantly, although clubroot symptoms were observed. The greenhouse experiment showed that clubroot disease severity could be reduced, by first cropping resistant cultivar, followed subsequently by growing the susceptible cultivar (Strehlow et al. 2015). The ability of *P. brassicae* to spread via another mode, in addition to infested soil, was investigated. Clubroot pathogen was suspected to spread via infested soil sticking onto farm machinery. The potential for the spread of *P. brassicae* in wind-blown dust from infested fields was investigated. Dust samples were collected at four locations in Alberta. Soil particles in suspension moving through saltation were collected in samples at heights ranging from 0 to 1 m. Quantitative PCR assay was applied to quantify the pathogen DNA in the dust samples. The DNA of *P. brassicae* could be detected in samples collected at all locations in 2011 and 2012. No clear association could be established between sampling height and the frequency of pathogen DNA-positive samples or between sampling height and pathogen DNA contents of the samples. The results indicated the possibility of movement of *P. brassicae* through windborne dust and wind-mediated dispersal of *P. brassicae* could be a potential factor contributing to development of clubroot epidemics in canola crops (Rennie et al. 2015).

*Phytophthora* spp. can infect a wide range of plant species, including several economically important crop plants which suffer seriously, resulting in huge losses, because of epidemics occurring in many countries. Mechanisms of dispersal of primary inoculum of *P. capsici*, incitant of Phytophthora blight of bell pepper were studied under field and growth chamber conditions. *P. capsici* may be dispersed by several mechanisms in soil, including inoculum movement to roots, root growth to inoculum and root-to-root spread. The inoculum was applied in field plots by introducing (i) sporangia and mycelia directly in soil, so that all three mechanisms of dispersal are possible; (ii) a plant with sporulating lesions on soil surface in plastic polyvinyl chloride (PVC) tube, so that inoculum movement to roots was possible; (iii) a wax-coated peat pot containing sporangia and mycelia in soil, so that root growth to inoculum was possible; (iv) a wax-coated peat pot containing infected roots in soil for root-to-root spread; (v) noninfested V8 vermiculite media into soil directly as a control and (vi) wax-coated noninfested soil as a control. The final disease incidence was highest in pots containing sporangia and mycelia buried directly in the soil, where all mechanisms of dispersal were operative (60 and 32% in 1995 and 1996, respectively). In pots where infected plants were placed in PVC tubes on soil surface and inoculum movement to roots occurred with rainfall (89 and 23%). Disease onset was delayed in 1995 and 1996. The final disease incidence was lower in plants grown in wax-coated pots containing

sporangia (6 and 22%) or infected roots (22%) buried in soil. Root growth toward inoculum or root-to-root spread occurred. Incidence of root infections was higher over time in pots, where inoculum moved to roots or all mechanisms of dispersal were possible. In growth chamber, all plants were infected, regardless of dispersal mechanisms of primary inoculum. But disease onset was delayed when plant roots had to grow through a wax layer to inoculum or infected roots in tension funnels containing a small amount of soil. The results showed that *P. capsici* could infect bell pepper roots through different mechanisms (Sujkowski et al. 2000). Dispersal and movement mechanisms of sporangia of *P. capsici* were studied. Direct laboratory observations showed that dispersal of sporangia of *P. capsici* occurred in water with capillary force, but did not occur in response to wind or a reduction in RH. On the other hand, dispersal of sporangia by wind was infrequent (0.7% of total hours monitoring) during sampling, as determined by a volumetric spore sampler in a commercial cucurbit field and also in an experimental setting, where large numbers of sporangia were continuously available in close proximity to the spore trap. The results of both direct laboratory observations and volumetric spore sampling showed that dispersal of sporangia via wind currents was infrequent and sporangia were unlikely to be spread by wind alone under natural field conditions (Granke et al. 2009).

*P. infestans* causes the highly destructive late blight disease of potato of historic importance, since it is considered to be the major factor for the infamous Irish potato famine. P. *infestans* also seriously affects tomato production in all countries. *P. infestans* has multiple modes of spread viz., through air, soil, seed tubers, tomato seeds and irrigation water. The pathogen is polycyclic, capable of producing several cycles of asexual spores in a single crop season and the asexual sporangia are well adapted for aerial dispersal. The sporangia may be carried to long distances (40–60 km) by air (Fry and Goodwin 1997; Aylor et al. 2001). Aerial dispersal of sporangia is the important factor for development of potato late blight disease epidemics. Different decision support systems (DSS) have been developed in various countries to predict the risk of infection by *P. infestans* (Aylor et al. 2001). It has been found to be difficult to determine when, where and how abundant airborne inoculum would be available for initiation of epidemic in a geographic location. As *P. infestans* is able to produce multiple cycles of infective propagules, new infections may increase proportionally to both initial amount of inoculum present and amount of new inoculum produced *in situ* or carried by wind from external sources. The usefulness of real-time monitoring of *P. infestans* airborne inoculum as a complement to DSS was assessed between 2010 and 2012 potato production seasons at two locations in Canada. Airborne sporangia concentrations (ASCs) were monitored, using 16 rotatingram spore samplers at 3 m above ground level. In general, no correlations could be established between ASCs and weather conditions, such as temperature, duration of high humidity and rainfall, except for a weak significant correlation between the ASCs and a high humidity period between 2010 and 2012. The results suggested that the two

DSS underestimated the late blight risk in 2010 and overestimated it in 2012 (Fall et al. 2015). Changes in the composition of pathogen populations may lead to epidemic outbreaks of late blight disease. The phenotypic and genotypic characteristics of 63 potato isolates and 94 tomato isolates of *P. infestans* were compared, based on the mating types, *in vitro* metalaxyl sensitivity, mitochondrial DNA haplotype, DNA fingerprint patterns, simple sequence repeat (SSR) markers and aggressiveness (virulence) on potato and tomato leaves. The results showed that the 13 A2 lineage causing severe late blight outbreaks in South Indian States had replaced the US-1 and other genotypes of *P. infestans* that were present earlier in this region (Chowdappa et al. 2015).

*Tilletia horrida* and *T. indica*, causal agents of kernel smut of rice and Karnal bunt of wheat, produce teliospores that can persist in the soil for several years in field soils. Sporidia (basidiospores) produced from germinating teliospores present on the soil surface are thin-walled and filiform, and they produce hyphae or allantoids secondary sporidia, which are forcibly discharged. The secondary sporidia from agar cultures of *T. horrida* and *T. indica* naturally discharged onto petridish lids were air-dried and maintained in the laboratory at 10 to 20% RH at 20–22°C and 40 to 50 RH at 18°C. The sporidia dried on lids regenerated secondary sporidia for 31 to 34 days at 10 to 20% RH, and after 56 to 88 days at 40 to 50% RH. Sporidia regenerated new sporidia within 18 h after inverting dried sporidia over fresh agar. The regenerated sporidia germinated and produced long hyphae. No difference in viability of sporidia that were initially dried rapidly or dried slowly over ten h (see Figure 7.3). Sporidia of *T. horrida* or *T. indica* dried on petridish lids placed in Idaho, and Arizona, United States during early flag leaf stage to soft dough stages.

Sporidia held in the field for several days to a few weeks had the same appearance as the originally dried sporidia. The results suggested that the secondary sporidia produced prior to the susceptible growth stage of host plant might lay dormant in dry field conditions and they might rapidly regenerate under humid rainy conditions favorable for disease development (Goates 2010). The relationship of initial soil inoculum and composition of *S. subterranea* f.sp. *subterranea (Sss)* with incidence and severity of powdery scab on potato tubers, was studied. Soil inoculum level of *Sss* (sporeballs/ g of soil) was determined, using quantitative real-time PCR assay. Of the 113 commercial potato fields across the United Kingdom, soil inoculum was detected in 75%, ranging from 0 to 148 *Sss* sporeballs/g of soil. When arbitrary soil inoculum threshold levels of 0, < 10 and > 10 sporeballs/g of soil, were used, the number of progeny crops developing powdery scab showed positive relationship with the level of inoculum quantified in the field soil at preplanting stage. In the four field trials, disease incidence and severity of powdery scab on progeny tuber showed significant correlation (P < 0.01) with levels of inoculum incorporated. A cultivar effect was observed in all years, with disease incidence and severity scores being significantly greater in cv. Agria and Estima than on Nicola (P < 0.01) (Brierly et al. 2013).

### 7.3.1.1.2 Host Factors

Several host factors such as plant structure, plant age at the time of infection, genetic resistance of cultivars, alternative host plant species and susceptibility/resistance levels of rotational crops and cover crops have been reported to influence the development of microbial pathogens and incidence of diseases caused by them. Effects of factors, affecting development

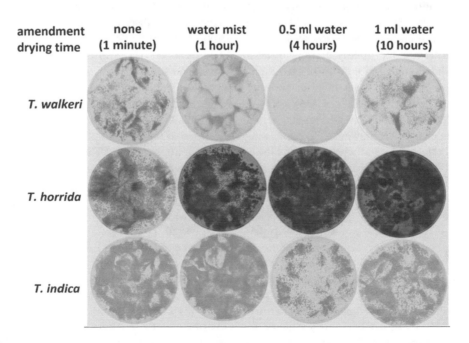

**FIGURE 7.3** Effect of drying sporidia of *Tilletia walkeri*, *T. horrida* and *T. indica* regenerated from petridish lids dried rapidly or slowly over a period of ten hours No significant differences in the viability of sporidia was recorded following rapid or slow drying.

of collar rot of tobacco caused by *S. sclerotiorum*, were investigated under greenhouse conditions. Test plants were maintained in a phytotron growth chamber with a 28/18°C day/night regime. Seedlings were inoculated with ascospores at different growth stages. Seedlings at 35–45 days of age were more susceptible to collar rot than younger or older seedlings. Inoculum efficiency was highest and disease was most severe, when an external source of nutrients was present on leaf surfaces. Clipping leaf tips did not increase disease severity. But leaving leaf pieces after clipping led to more severe collar rot symptoms, compared to removal of clipped leaf pieces. Heat injury resulting in necrosis of tissues provided favorable conditions for more infections by ascospores of *S. sclerotiorum*. However, exposure to low temperature did not increase infections. The results indicated that avoiding factors capable of inducing necrosis and removal of infected plant debris would be effective in reducing the severity of tobacco collar rot disease (Gutierrez and Shaw 2000). The contribution of disease-induced root production to the spread of take-all disease caused by *Ggt* on two contrasting winter wheat cultivars Genghis and Savannah was assessed. The parameters considered, were primary infection, secondary infection, inoculum decay and disease-induced root growth for determining changes in the numbers of infected and susceptible roots over time at a low or high inoculum density. Disease progress curves were characterized by consecutive phases of primary and secondary infections. In the absence of disease, no difference in root growth in both cultivars was discernible. But significant differences in disease-induced root production were detected among cvs. Savannah and Genghis. At low inoculum density, the transition from stimulation to inhibition occurred, when an average of 4.5 and 6.7 roots were infected in cvs. Genghis and Savannah, respectively. At high inoculum density, differences in the rates of primary and secondary infection between the wheat cultivars were observed. At low inoculum density, Genghis was marginally more resistant to secondary infection, whereas at high density of inoculums, Savannah was marginally more resistant to primary infection. Further, the period of stimulated root production was extended by seven and 15 days for Savannah, respectively at low and high inoculum densities (Bailey et al. 2006).

Yield potential of single plants and whole plot of oilseed rape (OSR) may be differentially impacted by *Verticillium longisporum* under natural and artificial soil infestation conditions. Disease incidence (DI) and disease severity (DS) correlated with the amount of inoculum provided at four different levels by addition of infested rapeseed straw to soil. Maximum DI and DS levels achieved with artificial infestation were 54.3 and 0.57 (in a rating scale of 0–2), respectively which was insufficient to reduce yield levels significantly in whole plots. On the other hand, a significant reduction in yield was induced on single naturally infected plants at field growth stage (GS) 83 to 85. Yield loss in single plants accounted for 20 to more than 80% DS levels above 5 (on a 0–9 scale). The systemic spread of *V. longisporum* was significantly delayed in plants in the field with substantial colonization of the shoots, not occurring before maturity stages. The pathogen

systemically spread after 28 days in susceptible cv. Falcon plants in the greenhouse. Root dip inoculation with conidial suspension induced earlier and more severe symptoms than microsclerotial inoculum added to the soil. Under greenhouse conditions, cv. Falcon showed high DS, stronger reduction in root and shoot length and a faster pathogen spread in plant tissue than the moderately susceptible cv. Talent (Dunker et al. 2008).

The impact of different plant residues and cultivation of wheat (nonhost) and different maize (corn) varieties on the soil inoculum density of *R. solani*, causal agent of late crown and RR of sugar beet was investigated. Sugar beet residues were either tilled or removed from the field and maize was grown during the two following years and also tilled or removed. The soil IP of *R. solani* was determined, using three different on- and off-site monitoring systems. Root damage indices of maize and sugar beet and broad bean as an indicator plant were determined. Further, an indirect real-time PCR assay using quinoa seed baits was also employed to assess the populations of *R. solani* AG2-2 in soil samples at the end of each year. Wheat, as a precrop to sugar beet, reduced *R. solani* IP in the soil significantly. Incorporation of host plant debris (sugar beet/maize) into soil increased the IP in the soil as well as incidence of sugar beet RR disease. Maize genotypes with varying degrees of susceptibility to *R. solani* did not significantly influence growth and survival of *R. solani* in the soil (Boine et al. 2014).

*Cercospora beticola*, incitant of Cercospora leaf spot (CLS) of sugar beet can infect predominantly aerial plant parts, as well as root of susceptible plants. Sugar beet root infection by *C. beticola* was investigated in a climate chamber and also in the field. Root incubation of susceptible seedlings with conidial suspension in the climate chamber, resulted in differential incidences of disease in two sugar beet cultivars, Auris and A00170. In the field trial with susceptible Savannah with soil-incorporated CLS-infested leaf material, disease symptoms developed four weeks earlier in infested plots, compared to control plots. Symptomless plants from infested field plots were used to induce leaf symptoms in the glasshouse and a significant higher probability to induce symptom development (0.4 ± 0.08) than plants from control plots (0.02 ± 0.02), at 14 days after transfer. The results showed *C. beticola* could infect sugar beet seedlings through roots and the latent CLS infections in seedlings might lead to symptom development at high temperatures (> 20°C) and high RH (> 95%) in the climate chamber or after canopy closure in the field (Vereijssen et al. 2005). Host plant organs differ in their ability to serve as sources of inoculum for the fungal pathogen *Didymella rabiei*, causing chickpea Ascochyta blight which survives in infected seeds and seedlings. Diseased seedlings developing from infected seeds, occasionally serve as the source of primary infection in chickpea crops. The temporal and spatial distribution of Ascochyta blight from initial infections and the relationship between the amount of initial infection and intensity of subsequent epidemics on cultivars with varying susceptibility were investigated in Australia and Israel. Disease spread over short distances (< 10 m) from individual primary infections was governed by rain and wind, and it was up to five times greater down-wind than up-wind direction. Cultivar response to *D. rabiei* significantly influenced distance and area over which disease spread and the intensity of the disease on infected plants. At the onset of epiphytotic, the relationship between spread and time was exponential ($P < 0.05$, $R^2 > 0.95$) and the area of the resulting foci was over ten times greater in susceptible cultivars than in resistant cultivars (Kimber et al. 2007). *V. dahliae* was considered to be introduced into lettuce fields through infested spinach seed. Field experiments were conducted to obtain evidence for linking spinach seedborne inoculum of *V. dahliae* to wilt epidemics in lettuce in four coastal counties of California, United States. *V. isaachi* was more frequently isolated from spinach seed than *V. dahliae*. The frequency of recovery of *Verticillium* spp. was unrelated to the Verticillium wilt incidence, but was related to the area under spinach production in individual counties. Verticillium wilt incidence on lettuce was recorded in microplots. The disease developed on lettuce, when *Verticillium*-infested spinach seeds were planted in the same soil two or three times. The identity of the pathogen recovered from lettuce was confirmed by PCR assays. Furthermore, green fluorescent protein (GFP)-tagged mutant strain of *V. dahliae* was employed to demonstrate transmission of *V. dahliae* from infested spinach seed to lettuce roots under greenhouse conditions, after two cycles of incorporation of infected residue into the soil. The results demonstrated that *V. dahliae* could be transmitted from infested spinach seed into soil and then to the lettuce crop planted subsequently (Short et al. 2015a).

*Phytophthora ramorum* can infect many understory forest species, and becomes a serious threat to ornamentals and forest trees. Different plant species (21 genera in 14 families) constituting major forest plant community in the Eastern United States were evaluated for their susceptibility to *P. ramorum*. Seedlings of the selected species were spray-inoculated with a mixture of four isolates of the pathogen (at 4,000 sporangia/ml), followed by incubation in dew chamber at 20°C in darkness for five days. Estimates of symptom development on individual leaves/leaflets were made. Mean percentage leaf area infected ranged from 0.7% for *Smilax rotundifolia* to 93.8% for *Kalmia latifolia*. Larger lesions were formed on eight species, compared to susceptible control *Rhododendron* cv. Cunningham White. Sporangial production by *P. ramorum* differed considerably among plant species tested, ranging from 36 per cm² of lesion area on *Myrica pennsylvanica* to 2,001per cm² of lesion area on *Robina pseudoacacia*. The number of chlamydospores (survival structures) produced per 6-mm diameter leaf disk incubated in a *P. ramorum* sporangial suspension, ranged from 25 on *Ilex verticillata* to 493 on *Rhus typhina*. The results indicated the possibility of several understory plant species in the forests functioning as potential sources of inoculum of *P. ramorum* for development of epidemics on *Rhododendron* (Tooley and Browning 2009). Sudden oak death (SOD) caused by *P. ramorum* is a serious problem in tree nurseries, particularly *Rhododendron* which has a potential role as a vector. The effects of *Rhododendron* host factors on susceptibility to *P. ramorum* and sporulation

were studied. Different inoculation methods were employed, using wounded or nonwounded detached leaves to inoculate 59 *Rhododendron* cultivars and 22 botanical species, replicated in three years. All *Rhododendron* species and cultivars were susceptible when inoculated on wounded leaves, but not on unwounded leaves, suggesting the operation of a resistance mechanism at leaf penetration. Young leaves were more susceptible than mature leaves when wounded, but less susceptible, when nonwounded. The type of rootstock did not affect the cultivar susceptibility level. Production of sporangia and chlamydospore in the leaf lesions varied widely among cultivars and it was not correlated with level of susceptibility. The levels of susceptibility of *Rhododendron* to *P. ramorum* correlated well with susceptibility to *Phytophthora citricola* and *P. hedraiandra* x *cactorum*, suggesting that the resistance mechanisms against *Phytophthora* spp. are nonspecific (De Deobbelaere et al. 2010). The effect of growing resistant and susceptible canola genotypes on soil resting spore populations of *P. brassicae*, incitant of clubroot disease, was assessed under greenhouse, mini-plot and field conditions. One crop of susceptible canola contributed to an increase of $1.4 \times 10^8$ spores/g soil in mini-plot experiment, and $1 \times 10^{10}$ spores/g gall under field conditions. Under repeated cropping of susceptible canola resulted in greater gall mass, compared to resistant canola lines. Growing susceptible canola genotypes also led to reduced plant height, increased clubroot severity in susceptible canola, and increased numbers of resting spores in the soil mix (Hwang et al. 2013).

Canopy structure may be a host factor that may influence incidence and spread of soilborne diseases, as in the case of Sclerotinia rot of carrot caused by *S. sclerotiorum*. The effects of four canopy management treatments, viz., unclipped control, unclipped canopy with manual removal of collapsed senescing leaves at two-week intervals, lateral clipping of the canopy at the initial emergence of apothecia, leaving the debris in the furrow and lateral clipping of the canopy with manual removal of the debris from the furrow were determined. Clipping reduced the canopy by about 20% on both sides of the carrot bed. Maximum air and soil temperatures were up to 9.2°C and 3.1°C lower, respectively, whereas RH was higher up to 30% in unclipped control compared with clipped canopies. Total number of apothecia in clipped plots was reduced by 74% and 76%, compared with unclipped plots in 2001 and 2002, respectively. Presence of clipped foliar debris on the furrow influenced the apothecia formed in 2001, but not in 2002. The apothecia under the debris did not appear to contribute to build up of inoculum. The results indicated the influence of canopy structure on the microclimate and development of apothecia and subsequent incidence of Sclerotinia rot in carrot crops (Kora et al. 2005a, 2005b). Disease development is either favored or limited by environmental conditions prevailing at different stages of plant development from sowing (planting) to harvest. Soybean SDS induced by *Fv* is favored by planting in cool soil, but epidemics can be severe even when planted later in the season into warmer soil. The effects of soil temperature on susceptibility of soybean plants exposed to *F. virguliforme* at different stages were assessed.

Soybean plants were grown in rhizotrons in water baths at 17, 23 and 29°C. Subsets of plants were inoculated at 0, 3, 7 and 13 days after planting (DAP) by drenching the soil with conidial suspension. RR developed in all inoculated plants, but DS decreased with increasing temperature and plant age at inoculation. Severity of foliage symptoms also decreased with increasing plant age. Plants inoculated at three and four DAP, developed symptoms only at 17 and 23°C and those inoculated at 13 DAP did not develop foliar symptoms at any temperature. Plants were highly susceptible, when inoculated at 0 DAP, and symptoms developed at all temperatures. Root length at inoculation was negatively correlated with DS. The results suggested that roots were most susceptible to infection during the first days after seed germination and accelerated root growth in warmer temperatures reduced susceptibility to root infection (Gongora-Canul and Leandro 2011).

### 7.3.1.1.3 Environmental Factors

Environmental factors, existing around the plants and more importantly in the soil, including microbial populations, chemical composition, soil temperature, soil moisture and crop cultivation inputs, have marked influence on microbial pathogen populations and disease incidence. Production of sporangia and zoospores by *P. capsici* in extracts of soils from different soil environments was assessed: (i) agricultural environments with a long history of pepper cultivation and incidence of the pathogen (CP), (ii) agricultural environments with no earlier pepper cropping (non-CP) and nonagricultural environments (forests and range highlands, Non-Ag). Significant differences were observed in production of *P. capsici* asexual propagules among the three environments. Production of propagules was 9 to 13% greater in non-Ag than in CP or Non-CP environments. The results showed that soils from agricultural and nonagricultural environments differentially influence production of sporangia and zoospores of *P. capsici* (Sanogo 2007). The role of *P. ramorum* inoculum present above ground level in the soils at infested nurseries in disease incidence was investigated. Soil cores were collected and sampled from three Washington State retain nurseries at which presence of *Phytophthora* spp. had been confirmed. The pathogen was recovered from the soil, using rhododendron leaves as bait, and pure cultures were isolated and maintained on V8 juice agar. Recovery frequencies were compared by species of *Phytophthora* at organic layer, 0 to 5 cm, 5 to 10 cm, and 10 to 15 cm depth classes. The identity of the isolates was confirmed by a combination of DNA sequencing of the ITS region of rDNA, real-time PCR assay and cultural morphology. *P. citricola* (32%), *P. drechsleri* (32%) and *P. ramorum* were recovered from the soil cores. *P. citricola* and *P. drechsleri* were more evenly distributed throughout the soil profile, whereas *P. ramorum* was primarily detected in the organic and 0 to 5 cm depth class, the frequency of recovery being 86% and the pathogen species was not detected in the soil below 10 cm depth (Dart et al. 2007).

The acidic soils of the maritime Pacific Northwest were found to be highly conducive for the development of Fusarium wilt disease of spinach caused by *Fusarium oxysporum* f.sp.

*spinaciae*, although this region was suitable for spinach seed production. A greenhouse soil bioassay method was developed to assess the relative risk of Fusarium wilt in fields intended for spinach seed production. Test soils with a range of *Fusarium* IPs and three spinach inbred parent lines that were highly susceptible, moderately susceptible and moderately resistant to Fusarium wilt, were used to evaluate the sensitivity of the bioassay to different levels of risk of Fusarium wilt disease. Soil samples from 147 fields (submitted by seed growers between 2010 and 2013) were evaluated, using the bioassay method. Differences in susceptibility to Fusarium wilt of three inbred lines were vital for detecting differences in the levels of risk among soils. Visual examination of spinach crops planted in the fields evaluated by the bioassay, as well as test plots of the three inbred lines planted in growers' seed crops, confirmed the predictive value of the bioassay for Fusarium wilt risk. Multiple regression analyses identified different statistical models for prediction of Fusarium wilt risk, depending on the spinach inbred lines. But even the best fitting model could explain only < 34% of the variability in Fusarium wilt risk among 121 fields evaluated in the soil bioassay. The results showed that no model was robust enough to replace the bioassay for the purpose of predicting Fusarium wilt risk (Gatch and du Toit 2015).

Plants vary in their responses to infection by soilborne plant pathogens, depending on the prevailing seasonal conditions in different geographical locations. The influence of seasonal conditions on the susceptibility of citrus scions to infection by *Phytophthora citrophthora* and *P. nicotianae* was assessed. Clementine mandarin cv. Hernandina, hybrid Fortune mandarin and the sweet orange cv. Lane-Late, were branch-inoculated with the pathogens under field and laboratory conditions. The cultivars inoculated with *P. citrophthora* in the field, developed the highest lesion area during spring (March–June) and autumn (September–October) and with *P. nicotianae* during summer (June–August). However, the lesion areas varied significantly on detached citrus branches and no pattern of disease incidence could be recognized. The lesion area induced by *P. nicotianae* in different citrus scions correlated significantly with monthly mean values of temperature, RH and the percentage of relative water content in the 24-month period of inoculations. In contrast, no such relationship between environmental parameters and the extent of colonization of host tissues by *P. citrophthora* was evident. However, a significant relationship was seen between weather variables and *P. citrophthora* during a specific period i.e., between October and May of each year (Álvarez et al. 2009). *Tilletia controversa* causes dwarf bunt of winter wheat which has a limited geographic distribution due to specific winter climatic requirements. The risk of disease introduction into new areas was evaluated under field conditions in Kansas State, United States. The soil surface of plots (2.8 × 9.75 m) planted with highly susceptible wheat cultivar, was inoculated with six teliospore concentrations ranging from 0.88 to 88,400 teliospores/cm². A single inoculation during separate seasons was carried out, followed by visual examination for disease incidence for four to six years afterward. The diseased spikes produced were crushed and returned to the plots where they were produced. One nursery did not show any infection at lower inoculation rates, and the disease was induced at trace levels at the highest inoculation rate, but no disease incidence was seen in the following three seasons. A duplicate nursery planted in a disease conducive area in Utah showed that the highest rate of inoculum used, was enough to induce 100% infection. The results showed that in an area with marginal climatic conditions, only transient trace levels of dwarf bunt incidence occurred and the disease was not increased even on highly susceptible wheat cultivar and by providing high doses of inoculum (Goates et al. 2011).

The effects of environmental factors and seasonal inoculum levels of *P. ramorum* in stream water used for irrigation on SOD disease incidence on nursery stock were assessed. A pear bait monitoring system was employed to detect and quantify *P. ramorum* propagules in streams that flowed through woodland areas with SOD in California between 2001 and 2007. Propagules of *P. ramorum* were detected in stream water, most frequently or occurred at highest concentrations of propagules, in samplings was preceded by about two months at low maximum daily temperatures and by four days with high rainfall. Presence of pathogen propagules in streams in the summer was most associated with infected leaves from the native host *Umbellaria californica* that prematurely abscized and fell into the water. The disease occurred only three times in 2005 and 2006, when the stream was used, via sprinkler irrigation system. The concentration of infective propagules were significantly reduced after water was pumped from stream and applied through sprinklers (Tjosvold et al. 2008). Survival of *P. ramorum* in soil infested at densities ranging from 0.2 to 42 chlamydospores /cm$^3$ of soil was studied. Recovery was assessed by baiting with *Rhodendron* leaf disks and dilution plating at 0 and 30 days of storage at 4°C. The baiting method was slightly more sensitive than dilution plating in recovering *P. ramorum* immediately following infestation of soil and it was possible to detect as little as 0.2 chlamydospore /cm$^3$, compared with 1 chlamydospore/cm$^3$ for dilution plating method. At 30 days after storage of infested soil at 4°C, the pathogen could be detected significantly at higher levels than at 0 day with both recovery methods. The results indicated that storage of *P. ramorum*-infested soil at 4°C could facilitate pathogen development such as production of sporangia, resulting in enhancement of recovery from soil at higher levels (Tooley and Carras 2011).

The effects of different soil factors on incidence and severity of root infection and root galling caused by *S. subterranea* in potato cv. Estima plants were studied under controlled environmental conditions. Molecular methods and visual examinations were performed, at two potato growth stages. Root galling severity was scored at harvest, followed by extraction of DNA from roots and its quantification, using real-time PCR assay specific for *S. subterranea*. At 17°C, root galling was severe with DS rating of 3.1 on a 0–4 scale. Root galling was reduced (0.6) at 12°C and failed to appear at 9°C, indicating the negative relationship between temperature and root galling. The level of inoculum in soil, added as sporosori,

had no effect on the disease incidence and severity of visual symptoms. Incidence of infection, as revealed by real-time PCR assay was greater with increasing soil inoculum levels, ranging from 48% at 5 sporosori/g to 59% at 15 sporosori/g and 73% at 50 sporosori/g infections at maturity. However, the differences were not statistically significant. Further, no relationship could be observed between occurrence of galls on roots and powdery scab on tubers of the same plants (Van de Graaf et al. 2007).

Damping-off of sugar beet seedlings caused by *Pythium* spp. is a serious limiting factor affecting yield levels. Survival and dry weight of plants in soils infested with *Pythium* spp. decreased with increasing soil temperature and matric potential, indicating an increase in disease pressure. However, there was no significant interaction between these two factors. At –330 × 10$^3$ Pascals (Pa), soil dryness was the limiting factor for plant emergence, but not for *Pythium* infection. At temperatures from 7 to 25°C and matric potentials of –7 × 10$^3$ to 120 × 10$^3$ Pa, treatment with the biological control agent *Pseudomonas fluorescens* B5, increased plant survival and dry weight. In regimes with different day and night temperatures, the maximum temperature was decisive for disease development and antagonistic activity (Schmidt et al. 2004). The effects of soil moisture levels varying from 15% to 42% (v/v) and sowing depth from 1.5 to 6.0 cm on the development of bean plants grown on sterile soil infested with the pathogen *R. solani* and its antagonist *Trichoderma harzianum* were investigated under greenhouse conditions. DS, percentage of plants emerged, plant height and dry weight at three weeks after sowing formed the criteria to determine the effects of soil moisture and depth of planting seeds. Soil moisture did not affect the emergence rate and plant growth inoculated only with *R. solani*. But in the presence of both pathogen and antagonist, plant emergence, plant height and dry weight decreased, when soil moisture diminished. Deep sowing significantly reduced emergence rate and growth of plants that were inoculated with *R. solani* alone. The planting depth did not affect the parameters in the presence of both fungi in the soil. At a sowing depth of 6.0 cm, plant emergence was 50% in the presence of *T. harzianum*, but it was reduced to 6.7% in the absence of the antagonist. The antagonist provided protection to bean seedlings against preemergence damping-off, especially in moist soil infested with *R. solani* (Paula Júnior et al. 2007).

The effects of management practices such as tillage, herbicide, manure and fertilizer application and seed treatment with fungicides and summer weather variables such as mean monthly air temperature and precipitation for the months of June to August as inputs were assessed to find their relationship with regional prevalence of soybean Sclerotinia stem rot (SSR) caused by *S. sclerotiorum*. Logistic regression analysis indicated the relationship of management practices and weather variables with soybean yield. Variables significant to SSR incidence, such as average air temperature during July and August, precipitation during July, seed treatment, liquid manure, fertilizer and herbicide applications, were also associated with high attainable yield. The results suggested that

SSR occurrence in the north central regions of the United States (Illinois, Iowa, Minnesota and Ohio) was associated with environments of high potential yield. The results suggested that management practices could be adopted to enhance attainable yield, in spite of their association with high disease risk (Mila et al. 2003). Influence of environmental factors on infection of lettuce by ascospores of *S. sclerotiorum* and subsequent disease development was studied under controlled and natural field conditions. Lettuce plants were inoculated with a suspension of ascospores in water or with dry ascospores and exposed to a range of wetness durations or RH at different temperatures. All inoculated plants were infected, but no relationship was observed between leaf wetness or RH and percentage of diseased plants. Ascospores began to germinate on lettuce leaves after two to four h of continuous leaf wetness at optimum temperatures of 15 to 20°C. The rate of development of Sclerotinia rot disease and the final percentage of infected plants after 50 days were greater at 16–27°C, with disease symptom appearing at seven to nine days after inoculation, and reached final disease incidence level of 96%. At lower temperatures, 8 to 11°C, symptoms of infection appeared first at 20–26 days after inoculation, reaching the final disease level of 10%. Under field conditions, with artificial inoculation with ascospores, disease occurred only on lettuce planted in May and June with disease incidence of 20–49% at eight weeks after inoculation. Lettuce crops planted after May had little (< 4%) or no disease. The results showed that effects of temperature prevailing at different months after planting lettuce on the incidence of *S. sclerotiorum* had to be considered while planting lettuce crops (Young et al. 2004).

Sclerotinia rot disease of carrot caused by *S. sclerotiorum* occurred in epidemic proportions, causing significant crop losses. Apothecia were first observed in the crop in early August to mid-September, after carrot canopy closed and after soil matric potentials between –0.1 and – 0.4 bars and soil temperatures between 14 and 23°C. On the other hand, ascospores were first detected in mid-July to mid-August, usually before production of apothecia in the crop and after 7–12 days with matric potentials between – 0.1 and – 0.3 bars and air temperatures between 15 and 21°C. The numbers of apothecia and ascospores were positively correlated with soil matric potential. Preharvest epidemics commenced in mid-August to mid-September, after the closure and lodging of the canopy, after the appearance of senescing leaves on the soil and ascospores in the crop along with leaf surface wetness for 12–24 h/day, after initiation of rain. Disease incidence was negatively correlated with air and soil temperatures. The results suggested that severe epidemics of Sclerotinia rot might occur when disease in the field progressed rapidly in association with soil matric potentials of ≥ 0.2 bars and leaf wetness of ≥ 14 h/day, especially close to harvest (Kora et al. 2005b). *S. sclerotiorum* produces apothecia from germinating sclerotia (carpogenic germination, CG). But apothecia formation in *S. minor* seldom occurs under natural conditions. CG of these two pathogens was compared under different temperature, soil moisture, burial depths and short periods of high temperature and low soil moisture. The optimal temperatures for

rapid germination and for maximum germination rates were lower for *S. minor* than for *S. sclerotiorum*. The temperature range for CG was narrower for *S. minor* than for *S. sclerotiorum*. The percentages of CG of sclerotia showed increases in both pathogens as soil water potential increased from – 0.3 to 0.01 M Pascals (Pa). In naturally infested fields, the number of sclerotia in 100 cc of soil decreased as the depth increased from 0 to 10 cm before tillage, but became uniform between 0 and 10 cm after conventional tillage for both species. Most apothecia of *S. minor* were produced from sclerotia located at depths shallower than 0.5 cm, whereas some apothecia of *S. sclerotiorum* were produced from sclerotia located as deep as 4 to 5 cm. The results indicated that variations in the CG of *S. sclerotiorum* and *S. minor* could be used to assess the epidemiological roles of inoculum for sexual reproduction by these pathogens in different geographical locations (Wu and Subbarao 2008).

Effects of different combinations of soil moisture and temperature on sclerotial germination of three isolates each of *S. minor* and *S. sclerotiorum* were studied, using soils from two locations in California. Sclerotia from the pathogen isolates in soil disks equilibrated at 0, 1, – 0.03, – 0.07, – 0.1, – 0.15 and – 0.3 MPa, were transferred into petridishes and incubated at 5, 10, 15, 25 and 30°C. Petridishes with apothecia formation were transferred into growth chamber at 15°C with a 12-h light/dark regime. Soil type did not affect either the type [direct/indirect (apothecial)] of germination and levels of sclerotia. Mycelial germination was the predominant mode of germination in *S. minor* and it occurred between – 0.03 and 0.3 MPa and 5 and 25°C, with an optimum at – 0.1 MPa and 15°C. No germination occurred at 30°C or 0.0 MPa. Soil temperature, moisture or soil type did not affect the variability of sclerotia of either species. CG of *S. sclerotiorum* sclerotia (measured as the number of sclerotia producing stipes and apothecia), was the predominant mode that was significantly affected by soil conditions. Optimal conditions for CG were 15°C and – 0.03 or – 0.07 MPa. Experiments to determine the influence of sclerotial size on CG at 15°C and – 0.03 MPa showed that solitary *S. minor* did not form apothecia, but aggregates of attached sclerotia readily formed apothecia. The number of stipes formed by both pathogens was highly correlated with sclerotial size. The results suggested the requirement of sclerotial size threshold for successful CG, explaining the absence of formation of apothecia by *S. minor* in nature (Hao et al. 2003). In a later investigation, the influence of fluctuating soil temperature and water potential on sclerotia germination and apothecial production by *S. sclerotiorum* were studied under growth chamber conditions. Fluctuating temperatures of 4, 8, 12 and 16°C around a median of 20°C and a constant of 20°C were tested. Daily temperature fluctuations of 8°C resulted in highest levels of sclerotial germination and apothecial production at 24 days after incubation. Soil water potential levels tested were constant saturation (approx – 0.001 MPa) and three levels of fluctuating soil water potential (0.09 to 0.1 MPa). Constant saturation favored maximum number of sclerotial germination and apothecial formation. All soil water potential fluctuations adversely affected sclerotial germination and apothecial production, requiring more than twice the duration taken by fluctuations of water potential (76 days), compared with constant saturation treatment (35 days) (Mila and Yang 2008).

The temporal patterns of *S. sclerotiorum* ascospore dispersal during flowering of canola (OSR) were studied at two locations in North Dakota between 2005 and 2007. Airborne ascospores populations were monitored, using volumetric spore samplers and electronic data loggers that were employed to record hourly changes in air temperature, RH, soil moisture and soil temperature under canola canopy. Ascospore dispersal occurred during single periods lasting fur to six hours. Most ascospores were collected between 10 AM and 1 PM in 2005 and 2007. But in 2006 with a drier environment, the most ascospores were dispersed early in the morning between 5 AM and 7 AM, revealing the influence of environment on ascospore dispersal. The first sharp increase in ascospore dispersal was preceded by a ten-unit drop in RH from close to saturation and an increase in air temperature of 5°C in 2005 and 2007. On the other hand, no significant changes were recorded in RH remaining around 90% or in air temperature around 15°C prior to commencement of spore discharges. Peak days of ascospore discharges were related to the preceding periods of seven consecutive days with a mean RH > 85% in the canola canopy (Qandah and del Rio Mendoza 2011).

*Monosporascus cannonballus* causes melon collapse resulting in considerable yield losses. Effects of soil temperature in pathogen development were assessed under laboratory and field conditions. Ascospore germination and hyphal penetration into melon roots were increased by increasing the temperature from 20 to 32°C under *in vitro*, the optimum temperature for mycelia growth being 30°C. Under field conditions between 1995 and 1998, disease progressed at a faster rate in the autumn than in winter crop seasons, disease incidence reaching the maximum (100%) in the three consecutive autumn seasons. In the winter season, early planting at the beginning of January resulted in low disease incidence (6–26%, at 125 DAP), whereas planting at the end of January contracted higher disease incidence (72–88%, at 95–119 DAP). In plots artificially heated to 35°C during the winter season, disease incidence reached 85% as in autumn season, indicating the influence of date of planting and prevailing temperature. Plants grown in unheated soil or in heated soil treated with MB, did not show collapse symptoms. Root colonization by *M. cannonballus* was higher in the autumn and in heated soil than in winter season in nontreated soil. A high correlation was observed between soil temperatures above 20°C during the first 30 DAP and DS. The results indicated the critical role of soil temperature during early stages of plant development on disease symptom expression of melon collapse caused by *M. cannonballus* (Pivonia et al. 2002). In another investigation, the effects of temperature, wetness and darkness on the formation of pseudothecia and the effect of temperature on the release of ascospores of *L. maculans* on OSR stubble were investigated *in vitro* in South Australia. The optimum temperature for maturation of pseudothecia was between 15°C and 20°C. Exposure to a 12-h photoperiod enhanced pseudothecium formation on the stubble, compared with continuous

darkness. Wetting the stubbles twice in a day favored pseudothecium maturation. A higher number of ascospores was released from the pseudothecia for a longer period at 20°C than at 5–15°C. The infected stubbles of OSR formed an important source of inoculum of *L. maculans* (Naseri et al. 2009).

The effects of extremes of high and low temperatures on the recovery of *P. ramorum*, both as free chlamydospores and within infected rhododendron tissue were assessed over a seven-day period. Chlamydospores held in moistened sand were incubated at 30, 35, 40, 0, – 10 and – 20°C for up to seven days. Infected rhododendron cv. Cunningham White leaf disks held in sandy loam, loam or sand at two different soil moisture levels also were exposed to these temperatures for up to seven days and to a variable temperature regimen for 12 weeks. Growth of *P. ramorum* on selective agar medium represented the quantum of pathogen recovery. Chlamydospores held in moistened sand showed a high recovery rate at 30°C, steadily at 40°C over the seven-day period. Chlamydospores could be recovered from treatments at 0°C after seven days, with little or no recovery at – 10°C or – 20°C. From infected rhododendron tissue, *P. ramorum* was recovered at 20 and 30°C after seven days with a decline at 35°C within two days and no recovery from infected tissue after two days. *P. ramorum* could be recovered from infected leaf disks exposed to 0 and – 10°C after seven days. Rapid decline in recovery occurred at – 20°C after one to three days and complete loss of recovery after four days. The pathogen could be recovered from all infected leaf disks subjected to a 12-week variable temperature treatment based on ambient summer temperatures in Lewisburg. The results indicated that *P. ramorum* present as free chlamydospores or within infected host plant tissue might survive and be potential sources of inoculum during summer and winter seasons in many eastern states of the United States (Tooley et al. 2008).

The effects of temperature on infection and development of *P. brassicae*, causing clubroot disease of *Brassica rapa* subsp. *chinensis* were investigated. Ten-day old seedlings were grown individually, inoculated with resting spores and maintained in growth chamber at 10, 15, 20, 25 and 30°C. Roots of seedlings were examined at four-day intervals for incidence of cortical infection and stage of infection (presence of young plasmodia, mature plasmodia and resting spores), at two-day intervals for symptom development and clubroot severity and at eight-day intervals for the number of spores/g of gall tissues. Temperature affected every stage of clubroot development. Cortical infection was highest and symptoms were observed earlier at 25°C, intermediately at 20 and 30°C and at the lowest and least at 15°C. Cortical infection or symptoms were not observed at 42 days after inoculation (*dai*) in plants grown at 10°C. Delay in the development of the pathogen was substantial at 15°C. Resting spores were first observed at 38 *dai* in plants at 15°C, 26 *dai* at 20 and 30°C and 22 *dai* at 25°C. The quantity of resting spores formed in galls was higher at 20 to 30°C than those that developed at 15°C over a period of 42 days. The results indicated that cool temperatures might result in slower development of clubroot symptoms in *Brassica* crops and the temperature could induce a consistent pattern of effect throughout the life cycle of *P. brassicae* (Kalpana Sharma et al. 2011). In a later investigation, the effects of temperature and soil type, pH and micronutrients on incidence clubroot on canola caused by *P. brassicae* were studied to assess the risk of spread of the disease in western Canada. Temperatures below 17°C reduced or inhibited the development of *P. brassicae* at all stages of its life cycle. Alkaline pH also decreased infection and symptom development under both *in vitro* and field conditions. But alkaline pH did not eliminate clubroot disease, when other favorable conditions existed for infection. Soil moisture, especially in the rhizosphere during primary and secondary infection had significant impact on clubroot development. However, manipulation of soil moisture was found to be difficult by the grower. Soil type had a small effect on clubroot severity under *in vitro* conditions, but a strong interaction of soil type with soil moisture was considered to be likely under field conditions. Assessment of the interaction of environmental factors affecting infection and mechanism of pathogen dispersal indicated that clubroot on canola had the potential to spread across large areas of Canadian prairies (Gossen et al. 2014).

Among the soil factors, the pH is a major limiting factor affecting pathogen development at different stages in the life cycles of microbial plant pathogens. The pH sensitivity of 82 strains of wheat take-all pathogen, *Ggt* was investigated under controlled conditions. Two types of populations, G1 and G2 were identified as the main genetic groups among the strains of *Ggt*. Assessments were made in petridishes on Fahraeus solid medium buffered at pH 4.6, 6.0 or 7.0 with citrate disodium phosphate solutions. All strains had different hyphal growth rates at the three pH levels tested. G2 group strains grew faster than G1 strains on slightly acidic pH (6.0) and neutral pH (7.0) buffered media. Three selected strains were able to alkalinize the acid medium (adjusted to pH 5.6). The results indicated the intraspecific variability in pH sensitivity with the soilborne pathogen, *Ggt* (Leberton et al. 2014). Suppression of the development of lettuce big-vein disease caused by *Mirafiori lettuce big-vein virus* and *Lettuce big-vein associated virus* by acid pH condition of soil is known. The effect of soil pH on the activities of *Olpidium virulentus*, the vector of the causal viruses was studied. The soil pH of less than 6.0 significantly reduced infection of roots of lettuce by *O. virulentus* and influenced the detection rate of zoospores released from zoosporangia and consequent reduction in development of lettuce big-vein disease in acidic soils (Iwamoto et al. 2017).

Irrigation water recycling has become an essential modification in water use pattern, due to diminishing water availability. But this practically resulted in potential accumulation and dissemination of microbial plant pathogens, including *Phytophthora* spp. which cause several economically important diseases in different agriculture and forest ecosystems. The distribution and diversity of *Phytophthora* spp. in an irrigation reservoir of a commercial nursery in east Virginia were investigated over two consecutive winters. Multiple baits were employed at surface water at run-off entrance, 20, 40, 60 and 80 m from the entrance and near the pump inlet and at various depths at a 20-m station. *P. citrophthora*, *P. gonapodyides*,

*P. hydropathica*, *P. inundata*, *P. irrigata*, *P. megasperma*, *P. pini*, *P. polonica*, *P. syringae* and *P. tropicalis* were recovered from the irrigation water. Recovery of *Phytophthora* declined through winters from November to March and also it declined with distance from the run-off entrance. The results suggested the need for decontaminating water during winter irrigation practices in the nursery (Ghimire et al. 2011). Soil salinity and water stress are known to influence plant growth and ion composition in soil. The effects of these factors on the development of Verticillium wilt disease of pistachio caused by *V. dahliae* were assessed under greenhouse conditions at 18–22°C. Soil salinity levels, 0, 1, 200 and 2,400 mg NaCl/kg of soil, and water stress levels, three-, seven- and 14-day irrigation regimes and two pistachio cultivars, Sarakhs and Qazvini rootstocks formed the treatments. Infested soil containing 50 MS/g of pistachio isolate of *V. dahliae* was applied for all treatments, using noninfested soil as a control and planting eight-week old pistachio seedlings. Shoot dry weights of both rootstocks were significantly decreased with increasing NaCl levels. Increasing irrigation frequency reduced salt injury. Salt stress significantly increased shoot and root colonization by the pathogen in both cultivars. Concentrations of $Na^+$, Ca and Cl showed positive and negative correlations, respectively with salinity and irrigation frequencies. The rootstocks Sarakhs and Qazvini were sensitive and tolerant, respectively to different irrigation regimes, salinity and *V. dahliae* infection and disease development (Saadatmand et al. 2008). Effects of salinity on development of cucumber damping-off disease and the pathogen *Pythium aphanidermatum* were investigated. Isolates of *P. aphanidermatum* from different geographical origins and genetic background were evaluated for their tolerance to salinity. Increasing irrigation water salinity from 0.01 to 5 $dSm^{-1}$ significantly increased mortality in cucumber seedlings inoculated with *P. aphanidermatum* and reduced dry weight of noninoculated seedlings. Growth of *P. aphanidermatum*, *P. spinosum* and *P. splendons* isolates was stimulated or unaffected at salinity levels stressful for cucumber [electrical conductivity (EC) = 5 $dSm^{-1}$]. Significant differences were observed in tolerance to salinity among 47 isolates of *P. aphanidermatum*. Oospore formation was more sensitive to salinity than hyphal growth and no oospores were produced above 20 $dSm^{-1}$. No difference in the tolerance to salinity of isolates of different geographical origin could be observed. Isolates of *P. aphanidermatum* from greenhouses with no salinity problems were equally tolerant to salinity. Increased mortality in cucumber seedlings at higher salinity levels might imply a synergistic interaction between salinity stress and salinity-tolerant *Pythium* spp. on cucumber seedlings, resulting in greater seedling losses (Al-Sadi et al. 2010).

### 7.3.1.2 Dynamics of Host-Bacterial Pathogen Interactions

Bacterial cells are surrounded by mucilage which protects them against desiccation, loss of viability and dry weather. The bacterial pathogens rapidly multiply in the soil on the available organic matter and infected plant residues added to the soil after harvest. During rainfall, conditions favor splash dispersal and free water required for activation of bacterial cells and initiation of new infections coexist. Wind-driven rain is responsible for dissemination of the majority of bacterial pathogens, including soilborne pathogens. In addition to soil transmission, soilborne bacterial pathogens are disseminated through infected seeds also. The rate of spread is likely to be more rapid in the case of bacterial pathogens with multiple modes of dissemination, compared with bacterial pathogens that spread only through soil. The number of infective propagules (pathogen population) and their virulence constitute IP of bacterial species pathogenic to different host plant species. The disease potential of bacterial pathogens depends on the levels of host susceptibility/resistance which may be positively or negatively influenced by environment and availability of nutrition. Furthermore, host susceptibility/resistance is governed by its genetic constitution. The environmental conditions prevailing in the soil are affected by various biotic and abiotic components of the soil which is more complex, compared with environmental conditions existing above ground level. In addition, agricultural practices applied primarily to enhance yield levels, can also markedly influence soil environments.

### 7.3.1.2.1    Pathogen Factors

*Ralstonia solanacearum*, causal agent of bacterial wilt diseases of several crops and brown rot of potato tubers, has a wide host range including weeds. Strains of *R. solanacearum* have been classified into five races, based on host range. Apart from races, *R. solanacearum* isolates have been grouped into biovars, based on their ability to utilize and / or oxidize several hexose alcohols and disaccharides (Hayward 1991). The pathogen attaches itself to the plant roots, infects the cortex, colonizes the xylem and spreads systemically to other plant organs and tubers of potato plants. After the death of infected plants, *R. solanacearum* comes back to the soil/water and/ or infects other susceptible plants adjacent to the dead plant and survives as a saprophyte or parasite. *R. solanacearum* could survive up to one year in the agricultural soil, even after treatment with herbicide to eliminate alternative host plant species. *R. solanacearum* has a remarkable ability to survive in the soil environment in different forms such as viable but nonculturable (VBNC) forms, starved cells, physiological characteristics (PC)-type and biofilms on host tissues for variable periods (Álvarez et al. 2010). The effects of tuberborne *Streptomyces* spp., causing potato common scab disease, on disease incidence in daughter tubers was investigated under microplot conditions. Visually healthy tubers, surface-sterilized healthy tubers and tubers that were 25% covered by scab infection were planted in pasteurized soil. Total populations of actinomycetes were determined by plating on a semiselective medium and colonies of pathogenic *Streptomyces* spp. were identified using PCR assay, targeting the thaxtomin A synthesizing *txtA* gene. The populations of the pathogen were below detectable levels at 39 DAP. After proliferation of the pathogen, the population attained measurable levels at 93 DAP in below-ground plant parts and in soil adjacent to scabby mother tubers ($10^4$–$10^5$ CFU/g of soil). Incidence of scab in progeny tubers was 89% at harvest. Progeny tubers produced by visually healthy seed tubers had an incidence of 60% scab and high populations of pathogenic *Streptomyces* were detected in the zone near the mother tuber. On the other hand, plants growing from surface-disinfested seed tubers had a very low pathogen population in the tuber zone and yielded 100% marketable tubers. Significant correlation (r = 1.00) between population densities of the pathogen in the root zone and daughter tuber disease incidence was observed. After harvest, high populations of *Streptomyces* ($10^6$–$10^7$ CFU/lesion in tissue) on mother tubers were observed. The results demonstrated that measurement of pathogen populations in the field could be a reliable predictor of scab disease incidence in potato (Wang and Lazarovits 2005).

Factors affecting the survival of *Clavibacter michiganensis* subsp. *sepedonicus (Cms)*, causing bacterial ring rot disease of potato tubers, were studied to assess the risk of dissemination of *Cms* via surface water and infection of potato through irrigation. *Cms* survived for a maximum period of seven days in non-sterile surface water at 10°C and during this period the pathogen could be disseminated to long distances, although the pathogen population might be highly diluted. Survival of a fluidal and non-mucoid strain was investigated, using sterile drainage water and simulated 'drainage water', in sterile MilliQ water, in tap water, in physiological salt and in artificial xylem fluid. A maximum survival period of 35 days in sterile tap water at 20°C was observed for the strains of *Cms* tested. The survival period ranging from 0 to 21 days in other diluents were recorded. Relatively poor survival in MilliQ water and artificial xylem fluid was seen. Low temperature had no beneficial effect on the survival of *Cms*, whereas oxygen level adversely affected pathogen survival, when it was depleted. The results indicated that C. *michiganensis* subsp. *sepedonicus* did not form cells in a viable but nonculturable state (van der Wolf and van Beckhoven 2004). Bacterial pathogens may resort to diverse mechanisms when they are challenged by stress conditions like nutrient deprivation. Different starvation-survival responses have been reported, with starved bacterial populations maintaining their numbers over time in a non-growing, but culturable state at different levels, depending on the bacterial species. The starvation-induced state of survival proved to be distinctly different from that of active growth. Survival strategies over four years by *Ralstonia solanacearum* phylotype (PH) II biovar (bv) 2 in environmental water microcosms were investigated. Outbreaks of the bacterial wilt disease caused by *R. solanacearum (Rs)* might originate from dissemination of the pathogen in watercourses, where nutrient limitation conditions might exist. The effects of long-term starvation on survival and pathogenicity of *Rs* in natural water microcosm were assessed. Microcosms were prepared from different sterile river water samples, inoculated separately with two European strains of PH2 at 1–6 CFU/ml and maintained at 24°C for four years. In all water samples, starved *Rs* remained in a non-growing but culturable state during the first year, maintaining approximately the initial pathogen populations. After that, a portion of the pathogen population progressively lost the ability to form colonies and became nonculturable, but metabolically active cells were present. During the whole period, *Rs* remained pathogenic on host plants and underwent a transition from typical bacilli to small cocci which tended to aggregate. Some starved *Rs* cells formed filaments and buds. The existence of long-starved pathogenic cells of *R. solanacearum* in environmental waters appeared to be a potential source of inoculum with epidemiological significance (Álvarez et al. 2008).

Bacterial pathogens require soil environments that are different from those needed for fungal pathogens to build up the IP for infection and disease development. Two model systems were constructed to assess the horizontal and vertical movement of *Ralstonia solanacearum* race 1, biovar 3, causal agent of bacterial wilt disease of tomato in andosol and sand at 28°C. The horizontal movement of the pathogen in the soil packed in a narrow horizontal frame was measured by applying the bacterial suspension to soil at one end of the frame and movement of the bacteria was measured over distance from the inoculation point, after four days. The bacterial front advanced at 2.2 cm/day in andosol and at 8.1 cm/day in sand. Vertical movement of *R. solanacearum* was measured, using a cylinder of soil packed in short tube soaked with water and soil at the top of the tube was inoculated with bacterial suspension and the soil cylinder were turned upside down and the vertical movement of bacteria was determined at seven days after inoculation. In andosol, the capillary water rose to 32.5 cm over seven days after inoculation and the pathogen reached 18.8 cm height. In sand, capillary water rose to 20.0 cm and the bacteria reached a height of 16.3 cm. The results indicated the influence of soil type in permitting the horizontal and vertical movement of *R. solanacearum*, which is an important factor affecting pathogen inoculum buildup in these two types of soil (Satou et al. 2006). The distribution patterns observed in commercial fields may be useful to determine the factors that affect the pattern and spread of disease. Aggregated incidence of tomato bacterial wilt caused by *R. solanacearum* within rows of a commercial field could indicate that irrigation might be a factor contributing to the spread of the disease. Irrigation with drip or trickle tape might deliver a large volume of water to the root zone, increasing the soil moisture around the roots, where the pathogen is present. A more random distribution would indicate that initial soilborne inoculum is the greatest contributing factor to epidemics, signifying that secondary spread is not a significant contributor to the epidemic. Four field trials were conducted over three tomato growing seasons in the Easter Shore of Virginia (ESV), to determine the temporal and spatial distribution of bacterial wilt throughout the commercial fields. Individual plants were examined for bacterial infection at one week intervals throughout the growing seasons. The data were analyzed both within rows and across rows to determine if there were significant patterns of clustering. Rows exhibiting a clustered distribution were divided by the total number of rows in that particular replication and assessment date to determine the percentage of rows showing clustered distribution. The ordinary runs analysis showed that tomato bacterial wilt became more clustered within rows, as the growing season progressed. Results of all four trials indicated a strong positive correlation

between disease incidence and the percentage of rows showing a clustered distribution of the disease. Spread of the disease within the rows was more important than across rows in the development of bacterial wilt epidemics. Constant increase in disease incidence and rows with a clustered distribution within rows, indicated that secondary spread between adjacent plants occurred. One of the major differences between spread within rows and across rows was the plant-to-plant proximity. Plant spacing was ~ 2 ft within rows, whereas across the rows, the plant spacing was about three times greater (~ 6 ft). Spread of *R. solanacearum* via irrigation water and root exudates might enhance the chances of the bacterial wilt disease spreading within rows (Wimer et al. 2011).

*Pectobacterium* spp. and *Dickeya* spp. cause blackleg and soft rot of potato which accounts for heavy losses in all countries around the world. *P. carotovorum* susbsp. *carotovorum (Pcc)* has a wide host range, whereas *P. atrosepticum* is restricted only to potato, mainly in the temperate regions. In contrast, *Dickeya* spp. have a restricted range of host plant species in temperate, subtropical and tropical regions (Ma et al. 2007; Toth et al. 2011). *Dickeya dianthica*, causing blackleg in potato was found to be widespread in Western Europe and introduced into several countries via seed potatoes. The possibility of potato getting inoculum of *Dickeya* spp. from infected hyacinth to potato via irrigation water was indicated (Slawiak et al. 2009). *P carotovorum* subsp. *brasiliensis* was reported to be highly virulent and responsible for the majority of blackleg incidences in Brazil (Duarte et al. 2004). *P. carotovorum* susbsp. *wasabiae* was reported to cause blackleg of potato in New Zealand (Pitman et al. 2010). Bacterial pathogens causing soft rot disease do not overwinter in soil, as they could survive in soil only for a period of one week to six months, depending on environmental conditions such as soil temperature, moisture and pH. However, the pathogen could survive for longer periods in plant debris and volunteer plants (Pérombelon and Hyman 1988). Latently infected seed (mother) tubers form the major source of inoculum by introducing the pathogen into the field at the earliest stage of crop growth. As rotting sets in, the pathogen in the mother tuber, is released into the soil and spread by soil water to contaminate neighboring progeny tubers. The bacteria in soil could also colonize potato roots and subsequently move via the vascular system into the progeny tubers. The pathogen in the stem could survive in the latent form without causing SR symptoms (Czajkowski et al. 2010). Soft rot bacteria may be carried from diseased plants by airborne insects over long distances to contaminate other potato crops. In addition, rain-splash dispersal of soft rot bacteria and haul pulverization prior to harvest may also aid in pathogen dispersal. Presence of *P. carotovorum* subsp. *carotovorum* and *P. atrosepticum* was detected in the air in large number on rainy rather than on dry days in Scotland (Czajkowski et al. 2011). Predominance of *Pectobacterium* spp. as reflected by frequency of recovery from soils, may vary depending on the soil type, soil environment and season. Frequencies of recovery of *P. carotovorum* subsp. *carotovorum* (Pcc), *P. carotovorum* subsp. *atrosepticum* (Pca) and *Erwinia chrysanthemi* causing blackleg and soft rot were variable and seed potatoes were the most important source of inoculum and the bacterial populations increased throughout the observational period of three months (Ali et al. 2012). *Dickeya* sp. was the predominant pathogen among the bacteria associated with potato blackleg in Switzerland. Among the factors influencing the disease incidence and spread, aggressiveness (virulence) of isolates of *Dickeya* spp. and levels of susceptibility/resistance of potato cultivars were more important (Gill et al. 2014). The soft rot disease of potatoes, caused by *Pectobacterium atrosepticum (Pa)* and Pcc, is widely distributed in all countries around the world. The influence of temperature (10, 15 and 20°C), RH (86, 96 and 100%) and initial concentration of bacterial inoculum ($10^5$, $10^7$ and $10^9$ CFU/ml) on the pathogen population and development of soft rot symptoms at the surface of wounded potato tubers were assessed *in vitro*. With both pathogen species, significant effects of temperature, RH and initial pathogen inoculum were observed on population dynamics and disease symptom development at the tuber surface. Multiple regression analyses showed that temperature is the most important factor, followed by initial pathogen inoculum and RH. Presence of *Pa* and *Pcc* at the level of wounded potato tubers was considered to be responsible for more than 64% variability of soft rot symptoms, when the effects of combined factors were considered (Moh et al. 2012).

*Erwinia rhapontici*, a facultative anaerobic bacterial species, causes pink seed and crown rot or soft rot on different plant species. It produces a characteristic diffusible pink pigment in sucrose-peptone agar and the pigment was identified as proferrosamine A which chelates iron, converting ferrosamine. Proferrosamine A was suggested to be a virulence factor of *E. rhapontici* (Feistner et al. 1997). *E. rhapontici* causes rotting of onion bulbs or cucumber slices. All strains of *E. rhapontici* were pathogenic to pea and bean, inducing pink or pinkish-brown lesions on pods and discoloration of seeds. A strain of *E. rhapontici* from pea could infect kernels of durum wheat,

resulting in pink wheat grains. *E. rhapontici* could be isolated from water, soil and plant leaf surfaces. The pathogen is an opportunistic parasite, capable of infecting host plant species through wounds (Mc Mullen et al. 1984). *E. rhapontici* was readily isolated from tap roots and basal stems. The pathogen could survive Canadian prairie winters on infected seeds and stems of pea, regardless of depth of burial of infected seeds. Further, *E. rhapontici* could also survive on the weed species *Amaranthus hybridus* (Gonzalez-Mendoza and Rodriguez 1990). The host range of *E. rhapontici* included dry pea, dry bean, common wheat (*Triticum aestivum*) and durum wheat (*T. durum*) (Huang et al. 2003).

*Agrobacterium tumefaciens*, causing crown gall in several economically important fruit and nut crops affects the production levels considerbly. *A. tumefaciens* is a soil inhabitant and it can bind to soil particles and survive for many years in the absence of susceptible host plant species. When dried, the *A. tumefaciens*-colonized soil particles, in the form of wind-borne dust, could potentially transport the pathogen to long distances. *A. tumefaciens* is responsible for high incidence of crown gall on walnut rootstock Paradox (*Juglans hindsii* x *J. regia*) which was resistant to *Phytophthora*, resulting in reduction in orchard yield and tree longevity. Paradox seed, which was traditionally collected from orchard floor, was able to acquire *A. tumefaciens* from the soil, as a function of the harvest method used, prior to planting. Paradox seeds were collected over a period of two years at two commercial nurseries, directly from the mother tree without contacting the soil, or gathered after the fall of seed on the floor of up to 28 days. Virulent and avirulent strains of *A. tumefaciens* were detected on the exterior of seed placed on the orchard floor under mother trees. The proportion of seed groups with detectable virulent and avirulent strains was positively correlated with the duration of seed contact with soil ($P < 0.000\,1$ and $P < 0.000\,1$, respectively) (see Figure 7.4 and Figure 7.5). On an average, virulent and avirulent *A. tumefaciens* strains

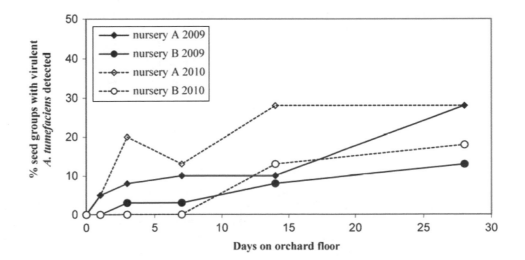

**FIGURE 7.4** Proportion of seed groups with viable virulent strains of *A. tumefaciens* in walnut seeds harvested directly from mother tree (time 0) or picked off the orchard floor at different periods (1–28 days).

**[Courtesy of Yakabe et al. (2014) and with kind permission of the American Phytopathological Society, MN, United States]**

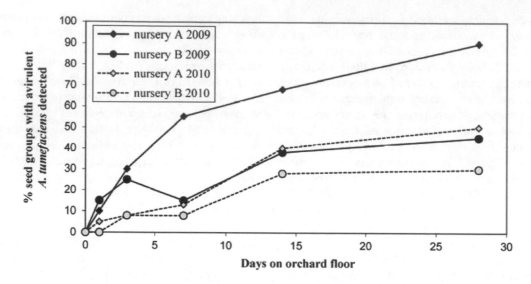

**FIGURE 7.5**   Proportions of seed groups with viable avirulent strains of *A. tumefaciens* in walnut seeds harvested directly from mother tree (time 0) or picked off the orchard floor at different periods (1–28 days).

**[Courtesy of Yakabe et al. (2014) and with kind permission of the American Phytopathological Society, MN, United States]**

were detected on $22 \pm 7.2$ and $54 \pm 26\%$ of the groups, respectively, in contact with orchard floor for a maximum of 28 days. On the other hand, *A. tumefaciens* could not be detected in or on the 2,650 seeds collected directly from the mother plants. Both virulent and avirulent strains were detected in and on the husk of nuts incubated on the orchard floor. Presence of *A. tumefaciens* was not detected in the interior seed tissues. Soilborne populations of *A. tumefaciens* in the orchard floor appeared to be a potential inoculum source of contamination of seeds of Paradox hybrid by *A. tumefaciens*. Avoiding seed contact with soil could be a simple, cost-effective strategy to reduce crown gall incidence in walnut (Yakabe et al. 2014).

### 7.3.1.2.2   Host Factors

The bacterial pathogens may be able to infect only a few or large numbers of plant species, in addition to the primary host (crop plant species), depending on their pathogenic potential (virulence). The ability of the pathogen to adapt to various environmental conditions determines the extent of disease incidence and distribution in different agroecosystems. *R. solanacearum*, causing bacterial wilt diseases, infects over 200 plant species, including tomato, potato, banana and tobacco and several weed species that remain asymptomatic (Hayward 1991). *R. solanacearum* finds suitable conditions for supply of required inoculum for infecting susceptible crop plant, when they are planted in the infested soil. *R. solanacearum* race 1 is able to infect the largest number of plant species and is the most widespread. Epidemics of bacterial wilt disease in tomato, potato, banana, groundnut (peanut) and ginger crops resulted in the breakdown of production system in several countries (Elphinstone 2005). Infected alternative plants (acting as a reservoir of inoculum) and infected plant debris/residue on the soil could support the survival of *R. solanacearum* in soil or water, favoring overwintering of the pathogen in the temperate regions. Some of the alternative

host plant species might be tolerant and allow the pathogen to invade xylem in the roots and to colonize at stem level (Álvarez et al. 2008). Further, movement of the pathogen was curtailed onto the stem in tolerant plants, whereas it progressively colonized other tissues in susceptible plants (Prior et al. 1994). Among the weed species functioning as reservoirs of inoculum, *Solanum dulcamara* (bitter sweet nightshade), *S. nigrum* (black nightshade) and *Urtica dioica* (stinging nettle) remain asymptomatic and they might be potential sources of inoculum for *R. solanacearum* (Hayward 1991; Pradhanang et al. 2000).

In Europe, *Cms* causes ring rot disease in potato which is rotated with other crops. In order to assess the ability of *Cms* to infect rotation crops, suspensions of the bacterial pathogen cells were used to inoculate them busing the stem inoculation method. The presence of *Cms* in inoculated plants was detected using theimmunofluorescence cell-staining (IFC) method. *Cms* could be detected in maize, wheat, barley, oat, bush bean, broad bean, OSR and pea, but not in onion or five cultivars of sugar beet tested. *Cms* was present in higher concentrations in stems ($10^5 - 10^6$ cells/g of tissue) than in roots ($< 10^3$ cells/g of tissues). *Cms* could be isolated only from IFC-positive stem samples of maize, bush bean, rape and pea, but not from roots. Of the 12 solanaceous weed plant species tested, only *Solanum rostratum* was susceptible to *Cms* which persisted in large numbers ($10^8$ cells/g of tissues) in stems and leaves which exhibited symptoms of infection. Of the 14 weed species tested, using IFC and AmpliDet RNA techniques, the pathogen was detected in the roots, but not in the stems of *Elymus* repens growing through rotten potato tubers. In addition, *Viola arvensis* and *Stellaria media* also showed the presence of *Cms* in stems and roots. The AmpliDet RNA technique was more sensitive than the IFC procedure in detecting *C. michiganensis* subsp. *sepedonicus* in plant tissues (van der Wolf et al. 2005).

### 7.3.1.2.3 Environmental Factors

*R. solanacearum*, identified as a quarantine pathogen in Portugal was recognized as an endemic pathogen in the main irrigated agricultural ecosystems, because of several outbreaks of bacterial wilt in potato crops in Portugal. All isolates of *R. solanacearum* obtained between 1999 and 2006 belonged to biovar 2A. But in 2007, biovar 1 strains were detected in potato fields under a confined area. Following a polyphasic approach, 101 Portugese isolates obtained from biotic and environmental samples were analyzed. Occurrence of an adaptive microevolution phenomenon for surface water strains of *R. solanacearum* was indicated. Coexistence of different populations of the pathogen in the same habitats was revealed and the presence of *R. solanacearum* phylotype II strains in Western Europe was observed for the first time (Cruz et al. 2012). The effects of environment and other factors on the severity of bacterial ring rot (BRR) symptoms induced by *Clavibacter michiganensis* susbp. *sepedonicus* were studied. Analysis of field data from 154 cultivated genotypes over 15 years indicated that moisture, temperature and cultivar were major factors influencing BRR symptom expression. Sensitivity analysis showed that late season temperatures were more important than midseason moisture. Unexpectedly cultivar susceptibility was of less importance than weather parameters in determining BRR symptoms. A neural network model successfully predicted severity of BRR symptom expression based on late season temperature, precipitation and cultivar susceptibility (Hill et al. 2011).

### 7.3.1.3 Dynamics of Host-Viral Pathogen Interactions

An epidemic is the resultant of positive interactions of components, such as the pathogen, host and environment in the case of the majority of fungal and bacterial diseases. On the other hand, most virus diseases have an additional component, vector(s) involved in the transmission of the virus from infected plant to healthy plant, making the interaction more complex. Most of the plant viruses have one or more species of vectors which show varying degrees of efficiency and specificity of transmission. The type of relationship between the virus with its vector viz., nonpersistence (styletborne), semipersistence (circulative) or persistence (propagative), is determined only by the virus. The extent of incidence of soilborne virus diseases and their distribution in a geographic location are variable, depending on the virus characteristics, nature of relationship with fungal or nematode vectors, geographic location and climatic conditions. Soilborne viruses causing diseases fall into three categories: (i) viruses lacking vector transmission, (ii) viruses transmitted by fungal vectors and (iii) viruses transmitted by nematode vectors. Virus disease epidemics commence with the introduction of the causal virus into the field or a geographical location via different modes of transmission such as seed/propagules, pollen, in addition to soil transmission. Further, information on the nature and disposition of the primary infection source(s), mode of spread of the virus from infected plant to healthy plant within a crop, survival of the virus in the absence of primary host (crop) plant, type of vector and virus-vector relationship, ability to survive in the soil in a free state or in plant debris/volunteer plants and the presence of additional sources of inoculum has to be gathered. In addition, viruses may infect annuals, perennials, wild relatives and weeds and crops grown in rotation in the same farm. Latently infected plants may be more dangerous than those that express symptoms of infection. Virus epidemics have been reported under different crop-growing conditions, ranging from monocultures to mixed cultivation, rainfed to irrigated, intensive high input to extensive low input, under controlled conditions and hydroponics. Sources of virus infection may be internal (within the fields) or external (existing outside the field) (Jones et al. 2010).

### 7.3.1.3.1 Viruses with Abiotic Transmission

Viruses belonging to the genera *Tobamovirus, Potexvirus, Necrovirus* and *Tombusvirus* spread under natural conditions without a known biotic vector. *Tomato mosaic virus* (ToMV) is soilborne and waterborne and has a wide host range. ToMV is extremely stable and infects plants through roots. *Tobacco mosaic virus* (TMV) the type member of *Tobamovirus* and ToMV could be detected in infested forest soils, using *Chenopodium quinoa* as the bait plants, the roots of which are infected by these viruses. Potex-, tobamo-, necro- and poty-virus from German soils could be transmitted to herbaceous and/ woody hosts, indicating the extended host range of these viruses (Bütner and Nienhaus 1989). As the forests seem to have developed on soils where agricultural crops had been grown previously, viruses detected in the forest trees might have originated from infected crop residues. Roots of *C. quinoa* grown on ToMV-infested soils tested positive by ELISA tests for the virus, indicating the stable nature of ToMV in forest soils in the New York State (Fillhart et al. 1998). *Pepino mosaic virus* (PepMV), a potexvirus, originally infecting pepino *(Solanum muricatum)* had a narrow host range, primarily limited to Solanaceae. The importance of PepMV increased, when tomato was infected in North America (French et al. 2001) and also PepMV was detected in some asymptomatic weed species present around affected tomato glasshouses in Spain (Jordá et al. 2001). Host range of an Italian isolate of PepMV-To was investigated, using ELISA tests. The experimental host range of the Italian isolate from pepino and to less extent from that of the Spanish isolate from tomato was determined. Among the crops, eggplant was susceptible to PepMV-To, but not to pepper or potato isolates. Further, the virus was detected in tomato seed by immunoassay, but it was not sap transmissible. The results indicated that strains of PepMV differed in their host range and seed transmissibility which are important in epidemiological investigations to assess virus disease incidence and spread (Salomone and Rogerro 2002). Presence of soilborne viruses in water resources (river, streams and ponds) and in hydroponic systems has been detected. Viruses in water could infect plants through root systems and induce symptoms characteristic of the disease later. The viruses might be released from infected plants into drainage water and then spread to other plants (Koenig 1986). Several plant viruses (26) have been detected in 47 environmental water samples in Hungary (Horvath et al. 1999). The occurrence of *Melon necrotic spot virus* (MNSV) in hydroponic cultures was reported to have significant adverse effects on crops (Gosalves et al. 2003). ToMV was detected and quantified in irrigation water using highly sensitive quantitative RT-real-time PCR assay at a concentration as low as 12 viral particles/PCR reaction. Seven of nine water sources from different locations in Slovenia tested positive for ToMV (Boben et al. 2007). These investigations indicated the need for monitoring irrigation waters for the presence of viruses like ToMV which are quite stable in a free state to prevent the spread of the viruses through irrigation water.

*Tomato bushy stunt virus* (TBSV) causes epidemics in tomato, eggplant and pepper. Distinct strains of TBSV were responsible for disease outbreaks in tomato and eggplant. Artichoke mottle crinkle (AMCV) and *Carnation Italian ringspot virus* (CIRV) were previously considered as distinct species. But based on the high degree of sequence similarity with TBSV-type strain, they were considered as strains of TBSV (Martelli et al. 1988), indicating the need for establishing the precise identity of the viruses to facilitate determination of the extent of disease incidence and losses caused by different strains/species of virus responsible for the epidemics. The characteristics of a virus causing chlorosis, necrosis and dieback in lettuce and tomato were studied and compared with other strains of TBSV, *Petunia asteroid mosaic virus* (PAMV), AMCV and CIRV. Based on genomic and serological properties, the virus infecting lettuce and tomato was named as *Lettuce necrotic stunt virus* (LNSV). The virus infected 32 species in nine families that developed local/systemic symptoms. Crops such as *B. oleracea*, sugar beet, spinach, lettuce, cucumber, melon, pea, cotton, pepper, tomato and petunia were susceptible to the virus. The soilborne nature of LNSV was demonstrated by growing lettuce plants in soil collected from infested fields. The results showed the potential of LNSV to cause epidemics in lettuce and tomato, as large numbers of plant species might serve as sources of inoculum for LNSV (Obermeier et al. 2001).

### 7.3.1.3.2  *Viruses Transmitted by Fungal Vectors*

Fungal vectors of soilborne plant viruses are themselves pathogens infecting roots causing specific symptoms. In addition, they function as vectors of several viruses causing economically important diseases of various crops. *Polymyxa graminis* and *P. betae* are soilborne, obligate, intracellular parasites of roots and they are placed under the order Plasmodiophorales. *Polymyxa* spp. are able to transmit at least 12 different plant viruses, belonging to the genera *Benyvirus*, *Bymovirus*, *Furovirus* and *Pecluvirus* (Mayo and Pringle 1998). It is essential to detect and quantify the populations of *P. graminis*, since the spread of the viruses depends on the populations and activities of the vectors. Direct antigen coating (DAC)-ELISA format was effective in detecting various stages in the life cycle of *P. graminis* and infection occurring under natural and controlled conditions indicating the potential for application of the immunoassays for epidemiological investigations on peanut clump disease (Delfosse et al. 2000).

The virus genus *Furovirus* includes *Soilborne wheat mosaic virus* (SBWMV), the type member, *Oat golden stripe virus* (OGSV), *Sorghum chlorotic spot virus* (SCSV), *Chinese wheat mosaic virus* (CWMV) and *European wheat mosaic virus* (EWMV), based on the presence of two genome segments which are not polyadenylated and lack of triple gene block in the viral genome (Pringle 1999). Incidences of SBWMV-infected plants were found to be irregularly distributed and circular patches of severely stunted sparse plants inspired the name 'crater disease'. Infection was more common in light to medium sandy-loam clay soils. Affected plants were especially common along field edges and in poorly drained areas and such conditions were conducive for the multiplication and spread of *P. graminis*, the vector of SBWMV. *Wheat spindle streak mosaic virus* (WSSMV) infection was observed along with SBWMV commonly in wheat grown in African countries. Both viruses were sap transmissible. Several indicator plants, including both dicots (*Chenopodium amaranticolor*) and monocots (*T. aestivum*) could be used for the detection and differentiation of these two viruses. DAC-ELISA format was efficient in confirming the identity of these viruses, indicating the usefulness of the immunoassay for epidemiological studies (Kapooria et al. 2000).

Occurrence of *Japanese soilborne wheat mosaic virus* (JSWMV), in addition to CWMV, transmitted by *P. graminis* was reported on winter wheat and barley in Japan. JSWMV was earlier considered as a strain of SBWMV. Later, as the Japanese SBWMV strains were shown to be genomically different, JSWMV was named as a distinct species under the genus *Furovirus* (Adams and Carstens 2012). JSWMV and CWMV were serologically related and they induced similar symptoms. The reverse transcription loop-mediated isothermal amplification (RT-LAMP) reaction technique was effective in differentiating JSWMV and CWMV and it was 100 times more sensitive than the standard RT-PCR assay. The results showed that appropriate sensitive techniques have to be employed to apportion the adverse effects of multiple infections by two or more viruses and to develop effective management systems for virus disease complexes (Fukuta et al. 2013). Barley yellow mosaic disease is caused by *Barley mild mosaic virus* (BaMMV) and *Barley yellow mosaic virus* (BaYMV) which are transmitted by *P. graminis*. The viruses belong to the genus *Bymovirus* in the family Potyviridae (Adams et al. 2005). These two viruses, taxonomically related, differed in their immunological properties, in nucleotide sequences of the two genomic RNAs. They showed differences in pathogenicity levels to different barley cultivars. Based on the differential responses of barley cultivars, two strains of BaMMV (BaMMV-Ka1 and BaMMV-Na1) and three strains of BaYMV (BaYMV-I, -II and -III) were differentiated in Japan (Kashiwazaki et al. 1989; Nomura et al. 1996). Likewise, different strains of both viruses were identified in Europe and China. The existence of different strains of the viruses with varying virulence has to be recognized in different locations to assess the possible losses that could be induced by the variants of the soilborne viruses. *Indian peanut clump virus* (IPCV) causes peanut clump disease in India and the virus is transmitted via peanut seed and *P. graminis* (Ratna et al. 1991). *Peanut clump virus* (PCV) occurring in West Africa and IPCV are serologically and genomically distinct. Both viruses are considered as high-risk pathogens, because of the high frequency of seed transmission. In addition, IPCV can infect sorghum and millets that are either grown as intercrops or rotated with peanuts, without expressing any visible symptoms of infection by the virus. Seed transmission in the field-infected peanut plants ranged from 3.5 to 17%, depending on the genotype. The transmission frequency was markedly increased (up to 55%) in seed collected from plants growing from infected seed. Further, ICPV-H (Hyderabad) strain was also transmitted via seeds of finger millet (*Eleusine corocana*), foxtail millet (*Setaria italica*) and pearl millet (*Pennisetum glaucum*). Multiple modes of transmission of ICPV is an important factor, influencing the extent of incidence and rapidity of spread of the disease (Reddy et al. 1998).

The effects of matric potentials of $-1, -5, -20$ and $-40$ kPa in cores of field soil infested with *Wheat soilborne mosaic virus* (WSBMV) and *P. graminis* on transmission to wheat seedlings were assessed under controlled conditions. Air pressure cells controlling soil matric potential precisely were utilized in this investigation to maintain different levels. Equilibrated soil cores were planted to wheat (*T. aestivum*) and virus transmission was determined by immunoassay at different intervals of plant growth. WSBMV was transmitted to all, except the driest soil matric potential tested, $-40$ kPa, in which only pores with a diameter of 7.4 μm or less were water-filled, possibly obstructing movement of zoospores of *P. graminis*. By placing plants at $-40$ kPa for 10.5 days and then watering them to conducive matric potential, WSBMV transmission occurred between 12 to 24 h at 15°C and within 36 h at 20°C. No significant transmission was observed within 96 h at 6.5°C. On the other hand, WSSMV was not transmitted at 15°C, suggesting either that WSSMV was unable to infect at 25°C or a different vector was involved in its transmission (Cadle-Davidson et al. 2003).

Rhizomania is one of the major diseases with worldwide distribution affecting sugar beet production in all countries. *Beet necrotic yellow vein virus* (BNYVV, genus *Benyvirus*), transmitted by *P. graminis* was considered to be the primary cause of the disease. Resting spores (sporosori) could protect BNYVV in soil for more than ten years in the absence of susceptible host plants, as the virus is internalized in the resting spores (Asher 1999; Rush 2003). The involvement of two more viruses viz., *Beet soilborne virus* (BSBV) and *Beet virus Q* (BVQ, genus *Pomovirus*) was later demonstrated. Both viruses were also transmitted by *P. betae* (Henry et al. 1986; Koenig et al. 1998). A multiplex reverse transcription-PCR technique was shown to be effective in simultaneously detecting BNYVV, BSBV and BVQ together in *P. betae*. This assay was found to be 128 times more sensitive than an ELISA test (Meunier et al. 2003). Generally BNYVV was associated with one or two pomoviruses in the vector. *Beet soilborne mosaic virus* (BSBMV), a member of the genus *Benyvirus*, was also

reported to infect sugar beet extensively in the United States (Rush 2003). Plant species belonging to Amaranthaceae, Chenopodiaceae and Tetragoniaceae were susceptible to BNYVV. Systemic infection on sugar beet and *Beta macrocarpa* occurred under natural conditions. *Nicotiana benthamiana* and spinach (*Spinacia oleracea*) could be infected under experimental conditions. BNYVV was maintained in *C. quinoa* or *Tetragonia expansa* that responded with local necrotic or chlorotic local lesions to inoculation with BNYVV (Tamada et al. 1989). As the virus could infect *Beta* spp. systemically only under natural conditions, BNYVV might have limited sources of inoculum and the spread of the disease might be virtually dependent on the activities of *P. betae* (Rush 2003).

The causal nature of LBVV, vectored by *O. brassicae* and *Mirafiori lettuce virus* (MiLV) transmitted by *O. virulentus* in inducing lettuce big-vein disease was investigated. The results indicated that MiLV, but not LBVV, was the cause of LBV disease. Individual lettuce plants were tested by ELISA test for the presence of the viruses in BV-prone fields in France and Italy. Both MiLV and LBVV occurred often together, but sometimes separately. Symptoms of LBV were found to be frequently associated with MiLV alone or both viruses, but rarely with LBVV alone. Presence or absence of LBVV did not appear to have any effect on symptom development under field conditions (Roggero et al. 2003). In a later investigation, *Mirafiori lettuce big-vein virus* (MiLBVV), causing lettuce big-vein disease, was shown to be transmitted by the soilborne chytrid, *O. brassicae*. The zoospores of *O. brassicae* carried the virus internally and they attach to roots of lettuce plants to initiate infection by both vector and virus after entry into the host cell. The virus survived within the resting sporangia of *O. brassicae* and persisted under adverse conditions for more than ten years, facilitating the spread of the MiLBVV between crops and survival between seasons (Campbell 1985). It is possible that fields could be infested with MiLBVV due to movement of equipments and workers in the production fields. As elimination of the virus from the infested soils was found to be very difficult, the influence of irrigation systems on spatial patterns of distribution of LBV disease was assessed. LBV disease incidence was aggregated in all furrow-irrigated fields, four of five surface drip-irrigated fields and two of three sprinkler-irrigated fields. Other fields had random distribution of infected lettuce plants. The sprinkler irrigation applied in lettuce production until thinning might influence the vector distribution and the subsequent irrigation methods used for the remainder of the crop season in individual plots, could significantly influence disease incidence (Hao and Subbarao 2014). The dynamics of MiLBV) and its vector *O. virulentus* were studied by quantifying them directly in soil and visual assessment of big-vein disease incidence on lettuce in fields. MiLBVV concentrations in the field soil were high and also the LBV disease incidence rate was also high. However, the population of *O. virulentus* in the soil was not correlated with the incidence of MiLBVV. The risk of incidence of LBVV might be estimated, using the MiLBVV concentration as an index (Momonoi et al. 2015).

### 7.3.1.3.3  Viruses Transmitted by Nematode Vectors

Nematode-transmitted polyhedral (NEPO) viruses induce diseases of economic importance in a wide range of cultivated annual, perennial and woody plants. The wide host range combined with multiple modes of transmission via seed and/or pollen, in addition to nematode transmission make these viruses more dangerous and difficult to apply eradication and other methods of management of the diseases. All nepoviruses have a bipartite genome, consisting of two linear single-stranded, positive sense RNA-1 and RNA-2. Nepoviruses have isometric, polyhedral particles with 30 nm diameter. All nepoviruses are transmitted by free-living nematodes belonging to the genera *Xiphinema*, *Longidorus* and *Trichodorus*, feeding on roots of susceptible plant species. *Grapevine fan leaf virus* (GFLV), included in the genus *Nepovirus*, family Comoviridae, is the major cause of grapevine degeneration, accounting for losses of as much as 80% due to reduction in yield and life span of vines (Andret-Link et al. 2004a; Brunt et al. 2006). GFLV is transmitted by *Xiphinema index* in nature and by mechanical inoculation experimentally. The experimental host range of GFLV includes species belonging to families Amaranthaceae, Chenopodiaceae, Cucurbitaceae, Leguminosaceae, Solanaceae and Fabaceae (Andret-Link 2004a). Under natural conditions, GFLV infects *Vitis* spp. and some weed species present in the vineyards occasionally (Horvath et al. 1994). The pathogen spreads through asymptomatic vinestock commonly. Hence, use of virus-free certified seed material is emphasized to prevent spread of the disease to other locations. Infested fields need a long-term quarantine period (four to five years) and repeated tests to quantify nematode densities. *X. index* has a life span of about one year. It may remain viruliferous for long periods and is able to acquire GFLV even by feeding on root debris scattered in soil after the removal of infected plant (Taylor and Raski 1964; Raski et al. 1965). The relationship between GFLV and *X. index* was shown to be specific and virus coat protein (CP) was the sole determinant of the specific transmission of the virus by its vector (Andret-Link et al. 2004b).

*Tobacco ringspot virus* (TRSV) and *Tomato black ring virus* (TBRV) are also transmitted by nematode vectors. *Xiphinema americanum*, transmits TRSV, while *Longidorus elongatus* and *L. attenuatus* are involved in the transmission of TBRV. Visualization of TRSV in the viruliferous nematodes, using immunofluorescence-labeling method provided the possibility of determining the nematode population carrying TRSV and restricting their movement to other locations, as a disease management strategy (Wang and Gergerich 1998). TRSV can infect a wide range of herbaceous and woody hosts. Grapevine, tobacco, blueberry, melon, cucumber, squash and pumpkin are seriously affected by diseases induced by TRSV infection. Other crops infected by TRSV include apple, eggplant, blackberry, pepper, cherry, papaya and weed species (Jossey and Babadoost 2006). The virus could be transmitted through seeds also, resulting in long-distance dissemination of TRSV via seeds. Occurrence of TRSV in several regions such as Europe, North America, Australia, Africa, India, Japan and New Zealand indicated the lack of effectiveness of quarantine activities (Brunt et al. 1996; Ward et al. 2009). TBRV infects several crop plant species, including grapevine, cherry, apricot, peach, raspberry, strawberry, blueberry, potato, tomato, pepper and tobacco, as well as several ornamental plant and weed species (Harrison and Murrant 1977; Brunt et al. 1996; Edwardson and Christie 1997). Dynamics of *Tomato ringspot virus* (ToRSV) transmission and population of its vector *X. americanum* were studied in a red raspberry field in Washington State. Population densities of *X. americanum* reached the maximum in the winter and remained at the lowest level in the summer. All nematode life cycle stages were present throughout the year. Cucumber seedlings were planted in soil collected every month from the field and the transmission of ToRSV was confirmed by ELISA tests. ToRSV was detected by ELISA in fine roots of raspberry plants at five months after planting in the field soil infested with viruliferous nematodes, in all subterranean plant parts at 12 months after planting and in all aerial plant parts in the second year. The rate of spread of ToRSV in a raspberry field was 70 cm per year (Pinkerton et al. 2008).

*Tobacco rattle virus* (TRV), the type member of the genus *Tobravirus*, has rod-shaped particles, containing a bipartite RNA genome. The genus *Tobravirus* includes three species *Pepper ringspot virus* (PepRSV) and *Pea early browning virus* (PEBV), in addition to TRV. Tobraviruses are transmitted by soilborne trichorid nematodes included in the genera *Trichodorus* and *Paratrichodorus* and they are seedborne also (Robinson 2003). TRV is transmitted by *P. allius* which retained the virus particles in esophageal lining. The virus was retained by quiescent, nonfeeding nematodes for many months. Nematode transmission of tobraviruses appeared to show low specificity between viruses or virus strains and the nematode species (Mojtahedi et al. 2003). Tobraviruses infect several economically important plant species, including sugar beet, pepper (chilli), tobacco, pea, bean, tomato, potato, spinach and tulips which may provide inoculum to other susceptible crop plants (Brunt et al. 1996). In the later investigation, weed species serving as hosts for virus and the vector were found to be kochia, prickly lettuce, henbit, nightshade species (black, hairy and cutleaf), common chickweed and annual sowthistle. Virus-free *Paratrichodorus allius* acquired TRV from three nightshade species, volunteer potato grown from infected tubers and prickly lettuce and subsequently transmitted TRV to tobacco cv. Samsun (indicator plant). The results showed that the weed species could play a role in supplying virus inoculum as well as viruliferous vectors for infection of newly planted potato by TRV, causing corky ringspot disease (CRS) of potato (Mojtahedi et al. 2003).

## 7.4  FORECASTING SYSTEMS FOR DISEASES CAUSED BY SOILBORNE MICROBIAL PATHOGENS

The contributions of various pathogen, host and environmental factors to the development of epidemics of soilborne diseases in different ecosystems are considered for developing models for predicting disease incidence. The interactions among pathogen, host and environments in relation to spatial and temporal distribution of pathogens in soil have been investigated to provide required parameters for models and risk analysis primarily for diseases caused by soilborne fungal pathogens.

## 7.4.1 Pathogen-Based Forecasting

Production and germination of pathogen propagules have been used as the basis for the development of predictive models. A forecasting system was developed for carpogenic germination of sclerotia of *S. sclerotiorum*, causing Sclerotinia disease of lettuce. Relationships among temperature, soil water potential and CG of sclerotia were determined for two isolates of *S. sclerotiorum* under laboratory conditions. Germination of multiple burials of sclerotia to produce apothecia under field conditions was also investigated. CG of sclerotia occurred between 5 and 25°C, but only for soil water potentials at ≥ 100 kilopascals (kPa) for both pathogen isolates. At optimum temperatures of 15 to 25°C, sclerotia buried in soil and placed in illuminated growth cabinets produced stipes after 20 to 27 days and apothecia after 27 to 34 days, indicating significant influence of temperature on sclerotial germination. Thermal time analysis of field data with constraints for temperature and water potential showed that the main degree of days to 10% germination of sclerotia was 285 and 279, respectively in 2000 and 2001 and it was generally a good predictor of formation of apothecia. Neither thermal time nor relationship determined *in vitro* could account for a decline in the final percentage of germination for sclerotia buried from mid-May, compared with earlier burials (Clarkson et al. 2007).

A predictive model for production of apothecia by CG of *S. sclerotiorum* was developed. This model was based on the assumption that a conditioning phase had to be completed before the germination phase could occur. Sclerotia were transferred from one temperature regime to another, enabled temperature-dependent rates to be derived for conditioning and germination for two isolates of *S. sclerotiorum*. The sclerotia were fully conditioned after two to six days at 5°C in soil, but it required up to 80 days at 15°C. Subsequent germination needed more than 200 days at 5°C and 33 to 52 days at 20°C. Upper temperature thresholds for conditioning and germination were 20 and 25°C, respectively. The model for production of apothecia was effective in simulating the germination of multiple burials of sclerotia in the field, when the soil water potential between – 4.0 and – 12.25 kPa was imposed (Clarkson et al. 2009). Ascospore dispersal and its relationship with incidence of Sclerotinia stem rot (SSR) disease of canola, caused by *S. sclerotiorum* were investigated in commercial fields in two locations in North Dakota betweeb 2005 and 2007. Ascospores released from apothecia were collected on dishes containing a semiselective medium placed at different positions. SSR incidence and inoculum gradients declined with distance from the area source in 2005 and 2007. In 2006, neither ascospores nor SSR were detected. The number of pathogen colonies per dish (COL) declined between 30% and 75%, within 40 m from the source, whereas SSR incidence declined by 50% within 12–17 m from the source. Every unit increase in the weekly mean COL represented an increase in SSR incidence that ranged between 0.5 and 0.75% (P ≤ 0.001) in 2005 and between 0.13 and 0.17 (P ≤ 0.001) in 2007. The results indicated the existence of short distance dispersal, disease gradients and close association between inoculum concentration and disease incidence (Qandah and Del Rio Mendoza 2012).

A system for predicting incidence of Ascochyta blight disease of field peas, caused by *Didymella pinodes*, was developed by assessing the availability of airborne primary inoculum for inducing disease. Field peas were sown in eight dates between mid-September and mid-December in two consecutive years. The seasonal dynamics of airborne inoculum were studied, using trap plants. The weekly availability of airborne primary inoculum was extremely low during autumn and winter and was partially influenced by mesoclimatic conditions. Disease onset occurred between mid-October and early March, depending on sowing date. Disease onset was observed at 14–35 days after emergence. A disease onset model, based on the calculation of weather-dependent daily infection values (DIVs) was established, assuming that the disease might occur once the temperature and moisture requirements of incubation were available. Cumulative daily infection values (CDIVs) were determined by sowing date and experiment through the addition of consecutive DIVs between emergence and disease onset. A frequency analysis of CDIVs was performed to determine the 10th and 90th percentiles of distribution. Analysis of the observed and predicted values showed that observed disease onset dates were almost always included in the forecast window defined by the two percentiles. The results showed that the onset of Ascochyta blight on field peas could be predicted, based on the airborne primary inoculum availability at different dates after sowing (Schoeny et al. 2007). The impact of pea canopy architecture and development on microclimate and infection by *Mycosphaerella pinodes*, causal agent of Ascochyta blight was investigated under field conditions between 2009 and 2010 in France. Leaf wetness duration (LWD) either due to rainfall or dew, was positively correlated with leaf area index (LAI) until canopy closure during these periods. During dry periods, when dew was the only source of leaf wetness, average daily LWD was short, decreasing as the canopy developed. Shorter LWDs were observed at the base than at the middle level of canopies and longer LWDs were observed outside canopy and inside the less dense canopies, regardless of the cultivar. LWD was negatively correlated with canopy height and LAI during these periods. Slow wind speeds were recorded inside the canopies (less than 0.5 km/h) and no significant canopy effect was observed on air temperature. The infection model showed that only rainfall periods, which induced long LWDs inside canopy, were favorable for infection by *M. pinodes* under the climatic conditions existing in the experimental location in France and more favorable environmental conditions were available inside dense canopies (Richard et al. 2013).

A model was developed to predict the spread of ascospores of *D. pinodes* from Ascochyta blight-infected field pea stubble of previous season's crops to the subsequent season crop. Field experimental data consisting of release events of ascospores of *D. pinodes* from known sources for 21 consecutive weeks, under natural environmental conditions, in a 400 × 400 m area collected earlier formed the basis for the model. The model was applied to the crop raised in a 30.9 km × 36.8 km area in a major field pea growing region of Western Australia to show the magnitude and spatial diversity of the dispersal of ascospores, generated in the previous season's field pea stubble, could differ between growing seasons. The validated simulation of the model has the potential to reveal the value of physical separation of the current season field pea crop from previous season's stubble and visualize the scale and diversity of ascospore dispersal in a large area of field pea cultivation (Salam et al. 2011). In a later investigation, the relationship between the number of ascospores of *D. pinodes* infecting field pea at sowing and DS at crop maturity was studied. Field pea stubble infested with the pathogen from one site was exposed to ambient conditions at two sites, repeated in two years. Three batches of stubble with different DS were exposed at one site, repeated in three years. At two-week intervals, stubble samples were retrieved, wetted and placed in a wind tunnel and ascospores were released up to 2,500 ascospores/g/h. Secondary inoculum, monitored, using field pea seedlings as trap plants in canopies arising from three sowing dates and external to field pea canopies, was at peak level in early sown crops. Maximum disease intensity was predicted, based on the calculated number of effective ascospore, soilborne inoculum and spring rainfall over two seasons. The threshold amount of ascospores of *D. pinodes* was determined to be 294/g of stubble/hour. The DS did not increase above the threshold level, whereas a linear relationship was noted between ascospore number and maximum disease intensity, when the inoculum was below threshold level (Davidson et al. 2013).

## 7.4.2 PATHOGEN DISTRIBUTION-BASED FORECASTING

The spatial and temporal distribution of soilborne fungal pathogens has significant influence on the incidence of diseases and subsequent development of epidemics, when other factors are favorable. Spatiotemporal analysis of spread of infections by *V. dahliae* pathotypes within a high tree density olive orchard in southern Spain was carried out. The development of Verticillium wilt epidemics in olive cv. Arbequina in a drip-irrigated, nontillage orchard in the soil without a history of the disease was studied between 1999 and 2003. Disease incidence, determined at monthly intervals, increased from 0.2 to 7.8% during the experimental period. Of the symptomatic trees, 87.2% and 12.8% were infected by D (defoliating) and ND (nondefoliating) pathotypes, respectively. Spatial analysis by distance indices showed a nonrandom arrangement of quadrats, containing infected trees. Spatial pattern was characterized by the presence of several clusters of infected trees. Increasing clustering over time was generally suggested by stronger values of clustering index over time and by the increase in size of patch clusters. Significant spatial association

was observed in clustering of diseased trees over time across cropping seasons. However, clustering was significant only for infections by D pathotype isolates of *V. dahliae*, indicating that infections by the D pathotype isolates were aggregated around initial infections. The number and size of clusters of D pathotype isolates increased over time. Microsatellite-primed PCR assays indicated that the majority of infecting D isolates shared the fingerprinting profile with the D pathotype of *V. dahliae* from a naturally infested cotton field in close proximity to the orchard, suggesting that a short-distance dispersal of the pathogen from cotton field soil to the olive orchard might have occurred (Navas-Cortés et al. 2008). Coffee wilt disease (CWD) caused by *Fusarium xylarioides* is an endemic disease in several African countries. Development of CWD in naturally infested field was studied from 2001 to 2006. Temporal and spatial spread of the wilt disease was monitored by mapping all infected trees. Host influence on the spatiotemporal structure was deduced from the distribution pattern of dead and healthy trees and analysis of variance. Temporal disease spread was at a slow rate. The disease spread from the initial infections to healthy neighboring trees was at a slow rate, resulting in an aggregated pattern. An infected tree could infect up to three healthy trees away, in all directions. Disease foci formed and expanded with time, coalescing, but punctuated in spots planted with resistant hosts. Host genotypes possessed varying degrees of susceptibility/resistance to wilt disease, which determined the rates and levels of disease incidence and spread (Musoli et al. 2008).

*S. sclerotiorum*, causing Sclerotinia rot of carrot (SRC) is one of the factors adversely impacting production in several countries. A disease risk forecasting model was developed by detecting and quantifying airborne ascospores that formed a component of the forecast model. Sclerotinia semiselective medium (SSM) in the blue plate test (BPT), used to trap the ascospores, had a threshold of 5 ascospores/plate. Data from nine years of air sampling were analyzed to determine the number of sampling sites required to accurately estimate the airborne ascospore populations in a region (ca. 40 km²) of intense carrot production. In six years, ascospore counts were correlated between sites in 20–100% of the possible pairwise comparisons among air sampling sites. The most consistent relationships among the sites were observed during periods of no detection or detection below BPT threshold. In most years, periods when ascospore counts surpassed the BPT threshold were consistent among sites. The results showed that one air sampling site would be sufficient to detect ascospores, when counts might be low, and increase the number to two or more sites during periods, when ascospores were present near or above threshold levels and cropping and environmental conditions were conducive to disease development (Parker et al. 2014).

## 7.4.3 ENVIRONMENT-BASED FORECASTING

The environmental conditions influence host plant development, as well as infection, disease progress in infected plant and spread of disease from infected plant to healthy plant(s). The effects of temperature and moisture contents on infection of *Rhododendron* cv. Cunningham's White by *P. ramorum* were determined, using a single isolate in most experiments. The optimum temperature was 20.5°C for infection of the greatest proportion of leaves, when whole plants were incubated in dew chambers at a range of temperatures from 10 to 31°C. For whole plants exposed to varying dew periods at 20°C and then incubated at 20°C for seven days, a dew period of as short as 1 h resulted in a small amount of disease. However, the least dew for 4-h period was required for >10% of the leaves to become diseased. The number of diseased leaves reached the maximum when moisture periods of 24 and 48 h were maintained. The results showed that *P. ramorum* could infect *Rhododendron* cv. Cunningham's White over a wide range of temperatures and moisture levels and it might become established in new areas in the United States (Tooley et al. 2009). In a further study, the effect of temperature and moist period on the sporangial production by *P. ramorum* on *Rhododendron* cv. Cunningham's White was investigated, using misted detached leaves held in a humidity chamber. Leaves, after wound inoculation with sporangia, were pre-incubated at 20°C for either 24 or 72 h prior to placing them at 4, 10, 15, 20, 25 and 30°C. The overall mean moist period required for first occurrence of sporulation over all temperatures was 3.24 days with the 24-h preincubation time, compared with 1.49 days for the 72-h preincubation time. Sporulation of *P. ramorum* reached its peak at 20°C. After 72-h preincubation at 20°C, sporulation generally commenced within a day, even at temperatures of 4 and 30°C that were suboptimal for sporulation. The time taken for sporangial production by *P. ramorum* on host tissue under moist conditions was similar to the time required for *P. phaseoli*, *P. palmivora* and *P. nicotianae*, but significantly longer time was required for *P. infestans*. The results provide information to understand conditions needed for epidemics of *P. ramorum* that can infect a large number of crop plant species (Tooley and Browning 2015).

Rhizoctonia web blight of container-grown azalea (*Rhododendron* spp.) is caused predominantly by binucleate *Rhizoctonia* AG-U (teleomorph *Ceratobasidium*). Infected plants may be killed due to the severe stem blight phase of disease. A predictive model was considered to be necessary to determine the timing of preventive application of fungicide in advance of increase in leaf blight intensity (LBI) significantly. Conducive conditions were temperature (18 or more hours between 20 and 30°C, when maximum temperature was less than 35°C), leaf wetness (16 or more hours), and /or rainfall (greater than 6.7 mm above the maximum daily irrigation). Weather measurements taken over 30 min from 2006 to 2011, and weekly or biweekly LBI assessments were used for analysis. The weather-based web blight predictive model was developed using the data collected during 11 site-years that represented weather conditions ranging from wet (2006), to moderate (2007 and 2008), to dry (2009 to 2011). The temperature, leaf wetness and rainfall provided a measurable daily risk that accumulated over time to hypothetically predict the need for fungicide application and to prevent significant LBI development under diverse weather conditions. The forecast model developed in this investigation established an accumulated risk prediction for Rhizoctonia web blight development on container-grown azalea produced under a daily irrigation regime (Copes 2015).

Races of *Fusarium oxysporum* f.sp. *ciceris* (Foc) differing in virulence have been identified. Races 0 (Foc-0) and 5 (Foc-5) induced yellowing and wilting syndromes, respectively in chickpea. The combined effects of soil temperature and inoculum density of Foc-0 and Foc-5 on disease development in chickpea cvs. P-2245 and PV-61 differing in their susceptibility were assessed. A temperature range of 10 to 30°C and inoculum densities between six and 8,000 chlamydospores/g of soil formed the factors to be tested. The response variables were the reciprocal incubation period, the final disease intensity, the standardized area under disease progress curve and the intrinsic rate of disease development. The favorable temperature was estimated to be 22 to 26°C for infection of the cvs. P-2245 and PV-61 by Foc-5 and 24 to 28°C for infection of cv. P-2245 by Foc-0. Disease symptoms did not develop at 10°C, except in cv. P-2245 inoculated with Foc-5. At optimum soil temperature, maximum DS developed with Foc-5 and Foc-0 at 6 and 50 chlamydospores/g of soil, respectively in cv P-2245 and with Foc-5 at 1,000 chlamydospores/g of soil in cv. PV-61. Risk threshold charts were constructed to estimate the potential risk of Fusarium wilt epidemics in a geographic location, based on soil temperature, the pathogen race and inoculum density in soil and the level of susceptibility of the chickpea cultivar (Navas-Cortés et al. 2007).

A weather-based forecasting system was formulated for *S. sclerotiorum*, causal agent of Sclerotinia rot of Indian mustard (*B. juncea*), based on the data collected from 2004 to 2012. CG of sclerotia and infection by ascospores was initiated and continued during the first three weeks of January. Disease symptoms were observed after the closure of the crop canopy and the commencement of flowering. The mean daily maximum and minimum air temperatures were respectively, 19.4 and 5.1°C, RH of 95% and 62% in the morning and afternoon, respectively, with bright sunshine hours of 4.9 and rainfall of 1.4 mm. These conditions were conducive for the development of Sclerotinia rot disease. The forecasting system was validated between 2012 and 2013 under field conditions (Sharma et al. 2015). The weather-based model for assessing the risk of formation of apothecia by *S. sclerotiorum*, incitant of Slerotinia stem rot disease in soybean fields was developed, since accurate prediction of the *S. scleriotiorum* apothecial risk, during susceptible soybean growth stages might improve in-season management of the disease. The presence or absence of apothecia was monitored from 2014 to 2016, in three irrigated (n = 1,505 plot-level observations), and six non-irrigated (n = 2,361 plot-level observations) field trials located in Iowa (n = 156), Michigan (n = 1,400) and Wisconsin (n = 2,310), for a total of 3,866 plot-level observations. Hourly air temperature, RH, dew point, wind speed, leaf wetness and rainfall were also recorded continuously throughout the season, at each location, using high resolution gridded weather data. Logistic regression models were developed for irrigated and nonirrigated conditions, using apothecial presence as a binary response variable. In irrigated soybean fields, apothecia formation was best explained by row width (r = − 0.41, P < 0.000 1), 30-day moving averages of daily maximum air temperature (r = − 0.30, P < 0.000 1), and wind speed (r = − 0.27, P < 0.000 1). The models could be used for predicting apothecial presence (overall accuracy of 67 to 70%) during soybean flowering period for four independent datasets (n = 1,102 plot level) observations or daily mean observations (Willbur et al. 2018).

### 7.4.4 MANAGEMENT-BASED FORECASTING

Crop management strategies are applied primarily to enhance production levels. Sclerotinia stem rot (SSR), caused by *S. sclerotiorum*, became serious threat to winter OSR in Germany and other European countries. The weather variables, air temperature, RH, rainfall and sunshine duration were used to calculate the microclimate in the plant canopy. Temperature of 7–11°C and 80–86% RH were indicated by the climate chambers study as the essential conditions for stem infection with ascospores and expressed as an index to discriminate infection hours (*Inh*). Disease intensities (DIs) showed significant relationship with *Inh* at post-growth stage (GS) 58 (late bud stage). Using the sum of *Inh* from continuous infection periods exceeding 23 h significantly improved the correlation with DI. The forecasting system SkleroPro, involved a two-tiered approach, the first providing a regional assessment of the disease risk, which was assumed, when 23 *Inh* had accumulated, after the crop had passed GS58. The next tier provided the field-site-specific economy-based recommendation. Based on spray cost, yield and price of rapeseed, the number of *Inh* corresponding to DI at the economic damage threshold $Inh_1$ was calculated. Decision to spray was considered, when *Inh* was < $Inh_1$. The impact of crop management strategies on SSR incidence was assessed. The two-year crop rotation enhanced disease risk and consequently lowered infection threshold, whereas four-year rotations elevated the threshold by a factor 1.3. Other agronomic factors viz., number of plants/sq.m, nitrogen fertilization and soil management method, did not have significant effects on DI. The effectiveness of SkleroPro was evaluated with 76 historical (1994 to 2004) and 32 actual field trials conducted in 2005. The percentages of economically correct decisions were 70 and 81%, respectively. The decision support for the fungicide application against Sclerotinia stem rot (SSR) of OSR offered a significant saving of fungicide cost, compared with routine application of fungicides. The forecasting model could be useful, particularly in areas with high inoculum density, to predict SSR from conditions of stem infection during late bud or flowering stage with sufficient accuracy and stimulation of apothecial development and ascospore dispersal might not be required. SkleroPro model was the first crop-loss-related forecasting model developed for Sclerotinia stem rot disease of OSR (Koch et al. 2007).

Fusarium root rot (FRR) disease is one of the limiting factors affecting bean production in several countries. In order to reduce chemical use for controlling FRR, the possibility of identifying planting strategies associated with FRR development and seed production in bean crops was explored. In 122 commercial bean farms in Iran, the association of farming indicators with FRR was examined. Low disease and high productivity were linked to herbicide and manure application, fungicidal treatment of seeds, manual sowing and sprinkler irrigation. Furrow irrigation, mechanical sowing, planting on raised seedbeds and lack of fertilizer were associated with high disease incidence and low seed production. The results indicated that the management strategies, favoring high disease incidence leading to low seed production, should be avoided by the growers (Naseri et al. 2016).

## REFERENCES

Adams MJ, Antoniw JF, Fauquet CM (2005) Molecular criteria for genus and species discrimination within the family Potyviridae. *Arch Virol* 150: 459–479.

Adams MJ, Carstens EB (2012) Ratification vote on taxonomic proposals to the International Committee on Taxonomy of Viruses (2012). *Arch Virol* 157: 1411–1422.

Alabouvette C, Höper H, Lemanceau P, Steinberg C (1996) Soil suppressiveness to diseases induced by soilborne pathogens. *Soil Biochem* 9: 371–413.

Ali HF, Ahmad M, Junaid M et al. (2012) Inoculum sources, disease incidence and severity of bacterial blackleg and soft rot of potato. *Pak J Bot* 44: 825–830.

Al-Sadi AM, Al-Masoudi RS, Al-Habsi N et al. (2010). Effect of salinity on Pythium damping-off of cucumber and on the tolerance of *Pythium aphanidermatum*. *Plant Pathol* 59: 112–120.

Álvarez B, Biosca E, López MM (2010) On the life of *Ralstonia solanacearum*, a destructive bacterial pathogen. In: Mendez-Vilas A (ed.), *Current Research, Technology and Education Topics in Applied Microbiology and Microbial Technology*, Formatex Research Center, Badajoz, Spain, pp. 267–279.

Álvarez B, López MA, Biosca EG (2008) Survival strategies and pathogenicity of *Ralstonia solanacearum* phylotype II subjected to prolonged starvation in environmental water microcosms. *Microbiology* 154: 3590–3598.

Alvarez LA, Gramaje D, Abad-Campos P, Garcia-Jiménez J (2009) Seasonal susceptibility of citrus scions to *Phytophthora citrophthora* and *P. nicotianae* and the influence of environmental and host-linked factors on infection development. *Eur J Plant Pathol* 124: 621–635.

Andret-Link P, Laporte C, Valet L et al. (2004a) *Grapevine fan leaf virus*: still a major threat to grapevine industry. *J Plant Pathol* 86: 183–195.

Andret-Link P, Schmitt-Keichinger C, Demangeat G, Komar V, Fuchs M (2004b) The specific transmission of *Grapevine fan leaf virus* by its nematode vector *Xiphinema index* is solely determined by the viral coat protein. *Virology* 320: 12–22.

Anees M, Tronsmo A, Edel-Hermann V, Gautheron N, Faloya V, Steinberg C (2010) Biotic changes in relation to local decrease in soil conduciveness to disease caused by *Rhizoctonia solani*. *Eur J Plant Pathol* 126: 29–41.

Asher M (1999) Sugar beet rhizomania: the spread of a soilborne disease. *Microbiol Today* 26: 120–122.

Atlas RM, Bartha R (1998) *Microbial Ecology–Fundamentals and Applications*, Benjamin/Cummings Publishing Company, Inc., USA.

Aylor DE, Fry WE, Mayton H, Andrade-Piedra JL (2001) Quantifying the rate of release and escape of *Phytophthora infestans* sporangia from a potato canopy. *Phytopatholoy* 91: 1189–1196.

Bailey DJ, Kleczkowski A, Gilligan CA (2006) An epidemiological analysis of the role of disease-induced root growth in the differential response of two cultivars of winter wheat to infection by *Gaeumannomyces graminis* var. *tritici*. *Phytopathology* 96: 510–516.

Bally R, Elmerich C (2005) Biocontrol of plant diseases by associative and endophytic nitrogen-fixing bacteria. Elmerich C, Newton WE (eds.), *Associative and Endophytic Nitrogen-fixing Bacteria and Cyanobacterial Associations*, Kluwer, Dordrecht, 171–190.

Baker KF, Cook RJ (1974) *Biological Control of Plant Pathogens*, The American Phytopathological Society, MN, USA.

Bakker MG, Glover JD, Mai JG, Kinkel LL (2010) Plant community effects on the diversity and pathogen suppressive activity of soil streptomycetes. *Appl Soil Ecol* 46: 35–42.

Beale JW, Windels CE, Kinkel LL (2002) Spatial distribution of *Aphanomyces cochlioides* and root rot in sugar beet fields. *Plant Dis* 86: 547–551.

Bhat RG, Smith RF, Koike ST, Wu BM, Subbarao KV (2003) Characterization of *Verticillium dahliae* isolates and wilt epidemics of pepper. *Plant Dis* 87: 789–797.

Boben J, Kramberger P, Petrovič N et al. (2007) Detection and quantification of *Tomato mosaic virus* in irrigation waters. *Eur J Plant Pathol* 118: 59–71

Boine B, Renner A-C, Zellner M, Nechwatal J (2014) Quantitative methods for assessment of the impact of different crops on the inoculum density of *Rhizoctonia solani* AG2-2 IIIB in soil. *Eur J Plant Pathol* 140: 745–756.

Bouwen GD, Rovira AD (1976) Microbial colonization of plant roots. *Annu Rev Phytopathol* 14: 121–144.

Brierly JL, Sullivan L, Wale SJ, Hilton AJ, Kiezebrink DT, Lees AK (2013) Relationship between *Spongospora subterranea* f.sp. *subterranea* soil inoculum level, host resistance and powdery scab on potato tubers in the field. *Plant Pathol* 62: 413–420.

Broders KD, Wallhead MW, Austin GD et al. (2009) Association of soil chemical and physical properties with *Pythium* species diversity community composition and disease incidence. *Phytopathology* 99: 957–967.

Brunt AA, Crabtree K, Dallwitz MJ, Gibbs AJ, Watson L (1996) Arabis mosaic nepovirus, Cucumber mosaic cucumovirus, Tobacco ringspot nepovirus // Viruses of plants. Descriptions and lists for VIDE database, Cambridge, UK. www.agls.uidaho.edu/ebi/vide/refs.htm

Budge GE, Shaw MW, Colyer A, Pietravalle S, Boonham N (2009) Molecular tools to investigate *Rhizoctonia solani* distribution in soil. *Plant Pathol* 58: 1071–1080.

Büttner C, Nienhaus F (1989) Virus contamination of soils in forest ecosystems of the Federal Republic, Germany. *Eur J For Pathol* 19: 47–53.

Cadle-Davidson L, Schindelbeck RR, van Es HM, Gray SM, Bergstrom GC (2003) Using air pressure cells to evaluate the effect of soil environment on the transmission of soilborne viruses of wheat. *Phytopathology* 93: 1131–1136.

Campbell RN (1985) Longevity of *Olpidium brassicae* in air-dry soil and persistence of the lettuce big-vein agent. *Canad J Bot* 63: 2288–2289.

Ceresini PC, Shew HD, Vilgalys RJ, Gale LR, Cubeta MA (2003) Detecting migrants in populations of *Rhizoctonia solani* anastomosis group 3 from potato in North Carolina using multilocus genotype probabilities. *Phytopathology* 93: 601–615.

Chowdappa P, Nirmal Kumar BJ, Madhura S et al. (2015) Severe outbreaks of late blight on potato and tomato in South India caused by *Phytophthora infestans* population. *Plant Pathol* 64: 191–199.

Clarkson JP, Phelps K, Whipps JM, Young CS, Smith JA, Watling M (2007) Forecasting Sclerotinia disease on lettuce: A predictive model for carpogenic germination of *Sclerotinia sclerotiorum* sclerotia. *Phytopathology* 97: 621–631.

Clarkson JP, Phelps K, Whipps JM, Young CS, Smith JA, Watling M (2009) Forecasting Sclerotinia disease on lettuce: Toward developing a prediction model for carpogenic germination of sclerotia. *Phytopathology* 99: 268–279.

Cook RJ, Baker KF (1983) *The Nature and Practice of Biological Control of Plant Pathogens*. The American Phytopathological Society, MN, USA.

Copes WE (2015) Weather-based forecasting of Rhizoctonia web blight development on container-grown azalea. *Plant Dis* 99: 100–105.

Cruz L, Eloy M, Quirino F, Oliveira H, Tenreiro R (2012) Molecular epidemiology of *Ralstonia solanacearum* strains from plants and environmental sources in Portugal. *Eur J Plant Pathol* 133: 687–706.

Czajkowski R, de Boer WJ, Velvis H, van der Wolf J (2010) Systemic colonization of potato plants by a soilborne, green fluorescent protein-tagged strain of *Dickeya* sp. Biovar 3. *Phytopathology* 100: 134–142.

Czajkowski R, Pérombelon MCM, van Veen JA, van der Wolf JM (2011) Control of blackleg and tuber soft rot of potato caused by *Pectobacterium* and *Dickeya* species: a review. *Plant Pathol* 60: 990–1013.

Dart NL, Chastagner GA, Rugarber EF, Riley KL (2007) Recovery frequency of *Phytophthora ramorum* and other *Phytophthora* spp. in the soil profile of ornamental retail nurseries. *Plant Dis* 91: 1419–1422.

Davidson JA, Wilmshurst CJ, Scott ES, Salam MU (2013) Relationship between Ascochyta blight on field pea (*Pisum sativum*) and spore release patterns of *Didymella pinodes* and other causal agents of Ascochyta blight. *Plant Pathol* 62: 1258–1270.

Davis JR, Huisman OC, Everson DO, Nolte P, Sorenson LH, Schneider AT (2010) The suppression of Verticillium wilt of potato using corn as a green manure crop. *Amer J Potato Res* 17: 195–208.

De Dobbelaere I, Vercauteren A, Speybroeck N et al. (2010) Effect of host factors on the susceptibility of *Rhododendron* to *Phytophthora ramorum*. *Plant Pathol* 59: 301–312.

Delfosse P, Reddy AS, Legréve A et al. (2000) Serological methods for detection of *Polymyxa graminis*, an obligate parasite and vector of plant virus. *Phytopathology* 90: 537–545.

Detheridge AP, Brand G, Fychan R et al. (2016) The legacy effect of cover crops on soil fungal populations in a cereal rotation. *Agric Ecosyst Environ* 228: 49–61.

Donner M, Lichtemberg PSF, Doster M et al. (2015) Community structure of *Aspergillus flavus* and *A. parasiticus* in major almond-producing areas in California, United States. *Plant Dis* 99: 1161–1169.

Duarte V, De Boer SH, Ward LJ, de Oliveira AM (2004) Characterization of a typical *Erwinia carotovora* strains causing blackleg of potato in Brazil. *J Appl Microbiol* 96: 535–545.

Dunker S, Keunecke H, Steinbach P, von Tiedmann A (2008) Impact of *Verticillium longisporum* on yield and morphology of winter oilseed rape (*Brassica napus*) in relation to systemic spread in the plant. *J Phytopathol* 156: 698–707.

Duressa D, Rauscher G, Koike ST et al. (2012) A real-time PCR assay for detection and quantification of *Verticillium dahliae* in spinach seed. *Phytopathology* 102: 443–451.

Edwardson JR, Christie RG (1997) *Viruses Infecting Peppers and Other Solanaceae Crops*. Gainesville, USA, Vol 2, pp. 337–390.

Elphinstone JG (2005) The current bacterial wilt situation: A global review. In: Allen C, Prior P, Hayward A (eds.), Bacterial Wilt Disease and the *Ralstonia Solanacearum* Species Complex, The American Phytopathological Society, St Paul, MN.

Fall ML, van der Heyden H, Bordeur L, Leclerc Y, Moreau G, Carrise D (2015) Spatio-temporal variation in airborne sporangia of *Phytophthora infestans*: Characterization and initiatives towards improving potato late blight risk estimation. *Plant Pathol* 64: 178–190.

Feistner G, Mavridis A, Rudolph K (1997) Proferrorosamines and phytopathogenicity in *Erwinia* spp. *Biometals* 10: 1–10.

Fillhart RC, Bachand GD, Castello JD (1998) Detection of infectious tobamoviruses in forest soils. *Appl Environ Microbiol* 64: 1430–1435.

French CJ, Bouthillier M, Bernardy M et al. (2001) First report of *Pepino mosaic virus* in Canada and the United States. *Plant Dis* 85: 1121.

Fry WE (1982) *Principles of Plant Disease Management*, Academic Press Inc., London, UK.

Fry WE, Goodwin SB (1997) Resurgence of the Irish potato famine fungus. *Bioscience* 47: 363–371.

Fukuta S, Tamura M, Maejima H et al. (2013) Differential detection of *Wheat yellow mosaic virus*, *Japanese soilborne wheat mosaic virus* and *Chinese wheat mosaic virus* by reverse transcription loop-mediated isothermal amplification reaction. J Virol Meth.

Gatch EW, du Toit LJ (2015) A soil bioassay for predicting the risk of spinach Fusarium wilt. *Plant Dis* 99: 512–526.

Gerlagh M (1968) Introduction of *Ophiobolus graminis* into new polders and its decline. *Neth J Plant Pathol* 74: 1–97.

Ghimire SR, Richardson PA, Kong P et al. (2011) Distribution and diversity of *Phytophthora* species in nursery irrigation reservoir adopting water recycling system during winter months. *J Phytopathol* 159: 713–719.

Gill ED, Schaerer S, Dupuis B (2014) Factors impacting blackleg development caused by *Dickeya* spp. in the field. *Eur J Plant Pathol* 140: 317–327.

Gilligan CA (1985) Construction of temporal models II. Disease progress of soilborne pathogens. In: Gilligan CA (ed.), *Advances in Plant Pathology*, *Volume 3*, *Mathematical Modeling of Crop Disease*, Academic Press, London, pp. 67–105.

Goates BJ (2010) Survival of secondary sporidia of floret-infecting *Tilletia* species: Implications for epidemiology. *Phytopathology* 100: 655–662.

Goates BJ, Peterson GL, Bowden RL, Maddux LD (2011) Analysis of introduction and establishment of dwarf bunt of wheat under marginal climatic conditions. *Plant Dis* 95: 478–484.

Gongora-Canul CC, Leandro LFS (2011) Effect of soil temperature and plant age at time of inoculation on progress of root rot and foliar symptoms of soybean sudden death syndrome. *Plant Dis* 95: 436–440.

Gonzalez-Mendoza L, Rodriguez MM, de L (1990) Aislmiento, identificaccion y patogenicidad de bacterias en quelite *Amaranthus hybridus* L, y su possibilidad en el control biologic. *Revista Chapingo* 15: 66–69.

Gosalves B, Navarro JA, Lorca A, Botella F, Sánchez-Pina MA, Pallas V (2003) Detection of *Melon necrotic spot virus* in water samples and melon plants by molecular methods. *J Virol Meth* 113: 87–93.

Gosme M, Willocquet L, Lucas P (2007) Size, shape and intensity of aggregation of take-all during natural epidemics in second wheat crops. *Plant Pathol* 56: 87–96.

Gossen BD, Deora A, Peng G, Hwang S-F, McDonald MR (2014) Effect of environmental parameters on clubroot development and the risk of pathogen spread. *Canad J Plant Pathol* 36: 37–48.

Granke LL, Windstam ST, Hoch HC, Smart CD, Hausbeck MK (2009) Dispersal and movement mechanisms of *Phytophthora capsici* sporangia. *Phytopathology* 99: 1258–1264.

Gutierrez, WA, Shaw HD, (2000) Factors that affect development of collar rot on tobacco seedlings grown in greenhouse. *Plant Dis* 54: 1076–1080.

Habib W, Choueiri E, Baroudy F et al. (2017) Soil inoculum density of *Verticillium dahliae* and Verticillium wilt of olive in Lebanon. *Ann Appl Biol* 170: 150–159.

Hao JJ, Subbarao KV (2005) Comparative analyses of lettuce drop epidemics caused by *Sclerotinia minor* and *S. sclerotiorum*. *Plant Dis* 89: 717–725.

Hao JJ, Subbarao KV (2014) Distribution of lettuce big-vein incidence under three irrigation systems. *Plant Dis* 98: 206–212.

Hao JJ, Subbarao KV, Duniway JM (2003) Germination of *Sclerotinia minor* and *S. sclerotiorum* sclerotia under various soil moisture and temperature combinations. *Phytopathology* 93: 443–450.

Hao JJ, Yang ME, Davis RM (2009) Effect of soil inoculum density of *Fusarium oxysporum* f.sp. *vasinfectum* race 4 on disease development in cotton. *Plant Dis* 93: 1324–1328.

Harrison BD, Murrant AF (1977) *Nepovirus* group: CMI/AAB Descriptions of plant viruses, number 185. www.dpvweb.net

Hayward AC (1991) Biology and epidemiology of bacterial wilt caused by *Pseudomonas solanacearum*. *Annu Rev Phytopathol* 29: 65–87.

Henry CM, Barker I, Jones RAC, Coutts RRA (1986) Occurrence of a soilborne virus of sugar beet in England. *Plant Pathol* 35: 585–591.

Hill BD, Kalischuk M, Waterer DR, Bizimungu B, Howard R, Kawchuk LM (2011) An environmental model predicting bacterial ring rot symptom expression. *Amer J Potato Res* 88: 294–301.

Hjort K, Lembke A, Speksnijder A, Smalla K, Jansson JK (2007) Community structure of actively growing bacterial populations in plant pathogen suppressive soil. *Microbial Ecol* 53: 393–413.

Hornby D (1981) Inoculum. In: Asher MJC, Shipton PJ (eds.), *Biology and Control of Take-all*, Academic Press, London, pp. 271–293.

Hornby D (1998) Diseases caused by soilborne pathogens. In: Jones DG (ed.), *The Epidemiology of Plant Diseases*, Kluwer Publishers, Dordrecht, Netherlands, pp. 308–322.

Horvath J, Boscai E, Kazinczi G (1999) Plant virus contamination in natural waters in Hungary. In: Maček J (ed.), 4th Slovenian *Conf* Plant Protect Soc, Slovenian Plant Protection Society, Slovenia, pp. 353–356.

Horvath J, Tobias I, Hunyadi K (1994) New natural herbaceous hosts of grapevine fan leaf nepovirus. *Hortic Sci* 26: 31–32.

Huang HC, Hsieh TF, Erickson RS (2003) Biology and epidemiology of *Erwinia rhapontici*, causal agent of pink seed and crown rot of plants. *Plant Pathol Bull* 12: 69–76.

Hwang SF, Ahmed HU, Zhou Q et al. (2013) Effect of susceptible and resistant canola plants on *Plasmodiophora brassicae* resting spore populations in the soil. *Plant Pathol* 62: 404–412.

Ikeda K (2010) Role of perithecia as an inoculum source for stem rot type of pepper root rot caused by *Fusarium solani* f.sp. *piperis* (teleomorph: *Nectria haematococca* f.sp. *piperis*). *J Gen Plant Pathol* 76: 241–246.

Inch SA, Gilbert (2003) Survival of *Gibberella zeae* in Fusarium-damaged wheat kernels. *Plant Dis* 87: 282–287.

Ispahani SK, Goud JC, Termorshuizen AJ, Morton A, Barbara DJ (2008) Host specificity, but not high temperature-tolerance, is associated with recent outbreaks of *Verticillium dahliae* in chrysanthemum in the Netherlands. *Eur J Plant Pathol* 122: 437–442.

Iwamoto Y, Inoue K, Nishiguchi S et al. (2017) Acidic soil conditions suppress zoospore release from zoosporangia in *Olpidium virulentus*. *J Gen Plant Pathol* 83: 240–243.

Jaime-Garcia R, Cotty PJ (2006) Spatial relationships of soil texture and crop rotation to *Aspergillus flavus* community structure in South Texas. *Phytopathology* 96: 599–607.

Johnson DA, Zhang H, Alldredge JR (2006) Spatial pattern of Verticillium wilt in commercial mint fields. *Plant Dis* 90: 789–797.

Jones RAC, Salam MV, Maling TJ, Diggle AJ, Thackray D (2010) Principles of predicting plant virus disease epidemics. *Annu Rev Phytopathol* 48: 179–203.

Jordá C, Lázaro Perez A, Martínez Culebras PV, Lacasa A (2001) First report of *Pepino mosaic virus* on natural hosts. *Plant Dis* 85: 1292.

Jossey S, Babadoost M (2006) First report of *Tobacco ringspot virus* in pumpkin (*Cucurbita pepo*) in Illinois. *Plant Dis* 90: 1361.

Kabbage M, Bockus WW (2002) Effect of placement of inoculum of *Gaeumannomyces graminis* var. *tritici* on severity of take-all in winter wheat. *Plant Dis* 86: 298–303.

Kalpana Sharma, Gossen BD, McDonald MR (2011) Effect of temperature on cortical infection by *Plasmodiophora brassicae* and clubroot severity. *Phytopathology* 101: 1424–1432.

Kapooria RG, Ndunguru J, Clover GRG (2000) First reports of *Soilborne wheat mosaic virus* and *Wheat spindle streak mosaic virus* in Africa. *Plant Dis* 84: 921.

Kashiwazaki S, Ogawa K, Usugi T, Tsuchizaki T (1989) Characterization of several strains of *Barley yellow mosaic virus*. *Ann Phytopathol Soc Jpn* 55: 16–25.

Khan A, Atibalentja N, Eastburn DM (2000) Influence of inoculum density of *Verticillium dahliae* on root discoloration of horseradish. *Plant Dis* 84: 309–315.

Kimber RBE, Shtienberg D, Ramsey MD, Scott ES (2007) The role of seedling infection in epiphytotics of Ascochyta blight in chickpea. *Eur J Plant Pathol* 117: 141–152.

Koch S, Dunker S, Kleinhenz B, Röhrig M, von Tiedeman A (2007) A crop loss-related forecasting model for Sclerotinia stem rot in winter oilseed rape. *Phytopathology* 97: 1186–1194.

Koenig R, Pleij CW, Beier C, Commandeur U (1998) Genome properties of *Beet virus Q*, a new furo-like virus from sugar beet, determined from unpurified virus. *J Gen Virol* 79: 2027–2036.

Konga EB, Sutton JC (1988) Inoculum production and survival of *Gibberella zeae* in maize and wheat residues. *Canad J Plant Pathol* 10: 232–239.

Kora C, McDonald MR, Boland GJ (2005a) Epidemiology of Sclerotinia rot of carrot caused by *Sclerotinia sclerotiorum*. *Canad J Plant Pathol* 27: 245–258.

Kora C, McDonald MR, Boland GJ (2005b) Lateral clipping of canopy influences the microclimate and development of apothecia of *Sclerotinia sclerotiorum* in carrots. *Plant Dis* 89: 549–557.

Lapwood DN, Hering TF (1970) Soil moisture and the infection of young potato tubers by *Streptomyces scabies* (common scab). *Potato Res* 13: 296–304.

Larkin RP (2003) Characterization of soil microbial communities under different potato cropping systems by microbial population dynamics, substrate utilization and fatty acid profiles. *Soil Biol Biochem* 35: 1451–1466.

Larkin RP, Griffin TS, Honeycutt CW (2010) Rotation and cover crop effects on soilborne potato diseases, tuber yield and soil microbial communities. *Plant Dis* 94: 1491–1502.

Larkin RP, Honeycutt CW (2002) Crop rotation effects on Rhizoctonia canker and black scurf of potato in central Maine, 1999 and 2000. *Bio Cult Tests* (online) Report 17: PT0610.1094/BC17

Larkin RP, Honeycutt CW, Olanya OM (2011) Management of Verticillium wilt of potato with disease-suppressive green manures and as affected by previous cropping history. *Plant Dis* 95: 568–576.

Leberton L, Daval S, Guillern-Erckelboudt A-Y, Gracianne C, Gazengel K, Sarniguet (2014) Sensitivity to pH and ability to modify ambient pH of the take-all fungus *Gaeumannomyces graminis* var. *tritici*. *Plant Pathol* 63: 117–128.

Lemanceau P, Alabouvette C (1991) Biological control of *Fusarium* diseases by fluorescent *Pseudomonas* and nonpathogenic *Fusarium*. *Crop Protect* 10: 279–286.

Lenc L, Kwaśna H, Sadowski C (2012) Microbial communities in potato roots and soil in organic and integrated production systems compared by the plate culturing method. *J Phytopathol* 160: 337–345.

Lewis BG (1970) Effects of water potential on the infection of potato tubers by *Streptomyces scabies* in soil. *Ann Appl Biol* 60: 83–88.

Li H, Sivasithamparam K, Barbetti (2007) Soilborne ascospores and pycnidiospores of *Leptosphaeria maculans* can contribute significantly to blackleg disease epidemiology in oilseed rape (*Brassica napus*) in Western Australia. *Austr Plant Pathol* 36: 439–444.

Li X-g, Zhang T-l, Wang X-x, Hua K, Zhao L, Han Z-m (2013) The composition of root exudates from two different resistant peanut cultivars and their effects on the growth of soilborne pathogen. *Internatl J Biol Sci* 9: 164–173 (with English abstract).

Li YP, You MP, Colmer TD, Barbetti MJ (2015) Effect of timing and duration of soil saturation on soilborne Pythium diseases of common bean (*Phaseolus vulgaris*). *Plant Dis* 99: 112–118.

Lô-Pelzer E, Aubertota JN, David O, Jeuffroy MH, Bousset L (2009) Relationship between severity of blackleg (*Leptosphaeria maculans/L. biglobosa* species complex) and subsequent primary inoculum production in oilseed rape stubble. *Plant Pathol* 58: 61–70.

Ma B, Hibbing ME, Kim HS, et al. (2007) Host range and molecular phylogenies of the soft rot enterobacterial genera *Pectobacterium* and *Dickeya*. *Phytopathology* 97: 1150–1163.

Martelli GP, Gallitelli D, Russo M (1968) Tombusviruses. In: Koenig R (ed.), *The Plant Viruses*, Volume 3, Plenum Press, New York, pp. 13–72.

Mayo MA, Pringle CR (1998) Virus taxonomy–1997. *J Gen Virol* 79: 649–657.

Mc Mullen MP, Stack RW, Miller JD, Bromel MC, Youngs VL (1984) *Erwinia rhapontici*, a bacterium causing pink wheat kernels. Proc North Dakota Acad Sci, Grand Forks, ND, USA 38: 78.

Menzies J (1959) Occurrence and transfer of a biological factor in soil that suppresses potato scab. *Phytopathology* 49: 648–652.

Metting FB Jr. (1993) *Soil Microbial Ecology: Applications in Agricultural and Environmental Management*, Marcel Dekker Inc., New York.

Meunier A, Schmit J-F, Stas A, Kutluk N, Bragard C (2003) Multiplex reverse transcription-PCR for simultaneous detection of *Beet necrotic yellow vein virus*, *Beet soilborne virus* and *Beet virus Q* and their vector *Polymyxa betae* Keskin on sugar beet. *Appl Environ Microbiol* 69: 2356–2360.

Mila AL, Carriquiry AL, Zhao J, Yang XB (2003) Impact of management practices on prevalence of soybean Sclerotinia stem rot in the North-Central United States and on farmers' decisions under uncertainty. *Plant Dis* 87: 1048–1058.

Mila AL, Yang XB (2008) Effects of fluctuating soil temperature and water potential on sclerotia germination and apothecial production of *Sclerotinia sclerotiorum*. *Plant Dis* 92: 78–82.

Moh AA, Massart S, Jijakli MH, Lepoivre P (2012) Models to predict the combined effects of temperature and relative humidity on *Pectobacterium atrosepticum* and *Pectobacterium carotovorum* subsp. *carotovorum* population density and soft rot disease development at the surface of wounded potato tubers. *J Plant Pathol* 94: 181–192.

Mojtahedi H, Boydston RA, Thomas PE et al. (2003) Weed hosts of *Paratrichodorus allius* and *Tobacco rattle virus* in the Pacific Northwest. *Amer J Potato Res* 80: 379–385.

Momonoi K, Mori M, Mastuura K, Moriwaki J, Morikawa T (2015) Quantification of *Mirafiori lettuce big-vein virus* and its vector *Olpidium virulentus*, from soil using real-time PCR. *Plant Pathol* 64: 825–830.

Moussart A, Even MN, Lesne A, Tivoli B (2013) Successive legumes tested in a greenhouse crop rotation experiment modify the inoculum potential of soils naturally infested by *Aphanomyces euteiches*. *Plant Pathol* 62: 545–551.

Musoli CP, Pinard F, Charrier A et al. (2008) Spatial and temporal analysis of coffee wilt disease caused by *Fusarium xylarioides* in *Coffea* canephora. *Eur J Plant Pathol* 122: 451–460.

Muzhinji N, Woodhall JW, Truter M, van der Walls JE (2018) Relative contribution of seed tuber- and soilborne inoculum to potato disease development and changes in the population genetic structure of *Rhizoctonia solani* AG3-PT under field condition in South Africa. *Plant Dis* 102: 60–66.

Narayanasamy P (2017) *Microbial Plant Pathogens–Detection and Management in Seeds and Propagules*, Volume 2, Wiley-Blackwell, Chichester, UK.

Naseri B, Davidson JA, Scott ES (2009) Maturation of pseudothecia and discharge of ascospores of *Leptosphaeria maculans* on oilseed rape stubble. *Eur J Plant Pathol* 125: 523.

Naseri B, Shobeiri SS, Tabande L (2016) The intensity of a bean Fusarium root rot epidemic is dependent on planting strategies. *J Phytopathol* 164: 147–154.

Navas-Cortés JA, Landa BB, Méndez-Rodríguez MA, Jimenez-Díaz RM (2007) Quantitative modeling of the effects of temperature and inoculum density of *Fusarium oxysporum* f.sp. *ciceris* races 0 and 5 on development of Fusarium wilt in chickpea cultivars. *Phytopathology* 97: 564–573.

Navas-Cortés JA, Landa BB, Mercado-Blanco J, Trapero-Casas JL, Rodriguez-Jurado D, Jimanéz-Díaz RM (2008) Spatiotemporal analysis of spread of infections by *Verticillium dahliae* pathotypes within a high tree density olive orchard in Southern Spain. *Phytopathology* 98: 167–180.

Nepal A, del Rio Mendoza LE (2012) Effect of sclerotial water content on carpogenic germination of *Sclerotinia sclerotiorum*. *Plant Dis* 96: 1315–1322.

Nischwitz C, Olsen IM, Ramussen S (2004) Effects of irrigation type on inoculum density of *Macrophomina phaseolina* in melon fields in Arizona. *J Phytopathol* 152: 133–137.

Nomura K, Kashiwazaki S, Hibino H et al. (1996) Biological and serological properties of strains of *Barley mild mosaic virus*. *J Phytopathol* 144: 103–107.

Obermeier C, Sears JL, Liu HY, Schlueter KO, Ryder EJ, Duffus JL, Koike ST, Wisler GC (2001) Characterization of distinct tombusviruse that cause diseases of lettuce and tomato in the western United States. *Phytopathology* 91: 797–806.

Ochola D, Ocimati W, Tinzaara W, Blomme G, Karamura EB (2015) Effects of water stress on the development of banana Xanthomonas wilt disease. *Plant Pathol* 64: 552–558.

Parker ML, McDonald MR, Boland GJ (2014) Assessment of spatial distribution of ascospores of *Sclerotinia sclerotiorum* for regional disease forecasting in carrots. *Canad J Plan Pathol* 36: 438–446.

Paula Júnior TH, Rotter C, Hau B 92007) Effects of soil moisture and sowing depth on the development of bean plants grown in sterile soil infested with *Rhizoctonia solani* and *Trichoderma harzianum*. *Eur J Plant Pathol* 119: 193–202.

Penton CR, Gupta VVSR, Tiedje JM et al. (2014) Fungal community structure in disease suppressive soils assessed by 28S LSU gene sequencing. *PLoS ONE* 9 (4): e93893

Pérombelon MCM, Hyman IJ (1988) *Effect of Latent Infection of Erwinia on Yield*, Scottish Crop Res Inst Annu Rep, Dundee, UK.

Pinkerton JN, Kraus J, Martin RR, Schreiner RP (2008) Epidemiology of *Xiphinema americanum* and *Tomato ringspot virus* on red raspberry *Rubus idaeus*. *Plant Dis* 92: 364–371.

Pitman A, Harrow S, Visnovsky S (2010) Genetic characterization of *Pectobacterium wasabiae* causing soft rot disease of potato in New Zealand. *Eur J Plant Pathol* 126: 423–435.

Pivonia S, Cohen R, Kigel J, Katan J (2002) Effect of soil temperature on disease development in melon plants infected by *Monosporascus cannonballus*. *Plant Pathol* 51: 472–479.

Porter LD, Dasgupta N, Johnson DA (2005) Effects of tuber depth and soil moisture on infection of potato tubers in soil by *Phytophthora infestans*. *Plant Dis* 89: 146–152.

Postma J, Scheper RWA, Shilder MT (2010) Effect of successive cauliflower planting and *Rhizoctonia solani* AG2-1 inoculations on disease suppressiveness of a suppressive and conducive soils. *Soil Biol Biochem* 42: 804–812.

Pradhanang PM, Elphinstone JG, Fox RTV (2000) Identification of crop and weed hosts of *Ralstonia solanacearum* biovar 2 in the hills of Nepal. *Plant Pathol* 49: 403–416.

Pringle CR (1999) Virus taxonomy–1999. *Arch Virol* 144: 421–429.

Prior P, Grimault V, Schmit J (1994) Resistance to bacterial wilt (*Pseudomonas solanacearum*) in tomato: present status and prospects. In: Hayward AC, Hartman GL (eds.), *Bacterial Wilt: The disease and its causative agent, Pseudomonas solanacearum*, CAB International, Willingford, UK, p. 209.

Qandah IS, del Rio Mendoza LE (2011) Temporal dispersal patterns of *Sclerotinia sclerotiorum* ascospores during canola flowering. *Canad J Plant Pathol* 33: 159–167.

Qandah IS, del Rio Mendoza LE (2012) Modelling inoculum dispersal and Sclerotinia stem rot gradients in canola fields. *Canad J Plant Pathol* 34: 390–400.

Raaijmakers JM, Paulitz TC, Alabouvette C, Steinberg C, Möenna-Loccoz Y (2008) The rhizosphere, a playground and battle field for soilborne pathogens and beneficial microorganisms. *Plant Soil*10.107/s11104-008-9568-6

Raski DJ, Hewitt WB, Goheen AC, Taylor CE, Taylor RH (1965) Survival of *Xiphinema index* and reservoirs of fan leaf virus in fallowed vineyard soil. *Nematologica* 11: 349–352.

Ratna AS, Rao AS, Reddy AS et al. (1991) Studies on transmission of *Indian peanut clump virus* disease by *Polymyxa graminis*. *Ann Appl Biol* 118: 71–78.

Reddy AS, Hobbs HA, Delfosse P, Murthy AK, Reddy DVR (1998) Seed transmission of *Indian peanut clump virus* (ICPV) in peanut and millets. *Plant Dis* 82: 343–346.

Rekah Y, Shtienberg D, Katan J (2000) Disease development following infection of tomato in basil foliage by airborne conidia of the soilborne pathogens *Fusarium oxysporum* f.sp. *radicis-lycopersici* and *F. oxysporum* f.sp. *basilici*. *Phytopathology* 90: 1322–1329.

Rennie DC, Holtz MD, Turkington TK et al. (2015) Movement of *Plasmodiophora brassicae* resting spores in windblown dust. *Canad J Plant Pathol* 37: 188–196.

Richard B, Bussiére F, Langrume C et al. (2013) Effect of pea canopy architecture on microclimate and consequences on Ascochyta blight infection under field conditions. *Eur J Plant Pathol* 135: 509–524.

Ritchie F, Bain R, Mcquilken M (2013) Survival of sclerotia of *Rhizoctonia solani* AG3PT and effect of soilborne inoculum density on disease development on potato. *J Phytopathol* 161: 180–189.

Robinson DA (2003) *Tobacco Rattle Virus*, AAB Descriptions of Plant Viruses, number 398.

Roca LF, Moral J, Trapero C, Blanco-López MA, Lopez-Escudero FJ (2016) Effect of inoculum density on Verticillium wilt incidence in olive orchards. *J Phytopathol* 164: 61–64.

Roggero P, Lot H, Souche S, Lenzi R, Milne RG (2003) Occurrence of *Mirafiori lettuce big-vein virus* and *Lettuce big-vein virus* in relation to development of big-vein symptoms in lettuce. *Eur J Plant Pathol* 109: 261–267.

Rush CM (2003) Ecology and epidemiology of benyviruses and plasmodiophorid vectors. *Annu Rev Phytopathol* 41: 567–592.

Saadatmand AR, Banihashemi Z, Sepaskhah AR, Maftoun M (2008) Soil salinity and water stress and their effect on susceptibility to Verticillium wilt disease, ion composition and growth of pistachio. *J Phytopathol* 156: 287–292.

Salam MU, Galloway J, Diggle AJ, MacLeod WJ, Maling T (2011) Predicting regional-scale spread of ascospores of *Didymella pinodes* causing Ascochyta blight disease on field pea. *Austr Plant Pathol* 40: 640–647.

Salomone A, Roggero P (2002) Host range, seed transmission and detection by ELISA and lateral flow of an Italian isolate of *Pepino mosaic virus*. *J Plant Pathol* 84: 65–68.

Sanogo S (2007) Asexual production of *Phytophthora capsici* as affected by extracts from agricultural and nonagricultural soils. *Phytopathology* 97: 873–878.

Sapkota R, Olsen MH, Deleuran LC, Nicolaisen M (2016) Effect of *Verticillium dahliae* soil inoculum levels on spinach seed infection. *Plant Dis* 100: 1564–1570.

Satou M, Kubota M, Nishi K (2006) Measurement of horizontal and vertical movement of *Ralstonia solanacearum* in soil. *J Phytopathol* 154: 592–597.

Scarlett K, Tesoriero L, Daniel R, Maffi D, Faoro F, Guest DI (2015) Airborne inoculum of *Fusarium oxysporum* f.sp. *cucumerinum*. *Eur J Plant Pathol* 141: 779–787.

Schmidt CS, Agostini F, Leifert C, Killham K, Mullins CE (2004) Influence of soil temperature and matric potential on sugar beet seedling colonization and suppression of Pythium damping-off by the antagonistic bacteria *Pseudomonas fluorescens* and *Bacillus subtilis*. *Phytopathology* 94: 351–363.

Schoeny A, Jumel S, Rouault F, Le May C, Tivoli B (2009) Assessment of airborne primary inoculum availability and modelling of disease onset of Ascochyta blight in field peas. *Eur J Plant Pathol* 119: 87–97.

Scott JC, Gordon TR, Shaw DV, Koike ST (2010) Effect of temperature on severity of Fusarium wilt of lettuce caused by *Fusarium oxysporum* f.sp. *lactucae*. *Plant Dis* 94: 13–17.

Sharma P, Meena PD, Kumar A, Kumar V, Singh D (2015) Forewarning models for Sclerotinia rot (*Sclerotinia sclerotiorum*) in Indian mustard (*Brassica juncea* L.). *Phytoparasitica* 43: 509–516.

Shipton PJ, Cook RJ, Sutton JW (1973) Occurrence and transfer of a biological factor in soil that suppresses take-all in wheat eastern Washington. *Phytopathology* 63: 511–517.

Short DPG, Gurung S, Koike ST, Klosterman SJ, Subbarao KV (2015a) Frequency of *Verticillium* species in commercial spinach fields and transmission of *V. dahliae* from spinach to subsequent lettuce crops. *Phytopathology* 105: 80–90.

Short DPG, Sandoya G, Vallad GE et al. (2015b) Dynamics of *Verticillium* species microsclerotia in field soils in response to fumigation, cropping patterns and flooding. *Phytopathology* 105: 638–645.

Simon TE, Le Cointe R, Delarue P et al. (2014) Interplay between parasitism and host ontogenic resistance in epidemiology of the soilborne plant pathogen *Rhizoctonia solani*. *PLoS ONE* 9 (8): e105159

Slawiak M, van Beckhoven JRCM, Speksnijder AGCL, Czajkowski R, Grabe G, van der Wolf JM (2009) Biochemical and genetic analyses reveal a new clade of biovar 3 *Dickeya* spp. strains isolated from potato in Europe. *Eur J Plant Pathol* 125: 245–261.

Soesanto L, Termorshuizen AJ (2001) Effect of temperature on the formation of microsclerotia of *Verticillium dahliae*. *J Phytopathol* 149: 685–691.

Sprague SJ, Howlett BJ, Kirkegaard JA (2009) Epidemiology of root rot caused by *Leptosphaeria maculans* in *Brassica napus* crops. *Eur J Plant Pathol* 125: 189–202.

Stanghellini ME, Mohammadi M, Adaskaveg JE (2014) Effect of soil matric water potentials on germination of ascospores of *Monosporascus cannonballus* and zoospores of *Olpidium bornovans*. *Eur J Plant Pathol* 139: 393–398.

Strehlow B, de Mol F, Struck C (2015) Risk potential of clubroot disease on winter oilseed rape. *Plant Dis* 99: 667–675.

Stutz EW, Défago G, Kern H (1986) Naturally occurring fluorescent pseudomonads involved in suppression of black rot of tobacco. *Phytopathology* 76: 181–185.

Suffert F, Montfort F (2007) Demonstration of secondary infection by *Pythium violae* in epidemics of carrot cavity spot using root transplantation as method of soil infestation. *Plant Pathol* 56: 588–594.

Sujkowski LS, Parra GR, Gumpertz ML, Ristaino JB (2000) Temporal dynamics of Phytophthora blight on bell pepper in relation to the mechanisms of dispersal of primary inoculum of *Phytophthora capsici* in soil. *Phytopathology* 90: 148–156.

Taylor CE, Raski DJ (1964) On the transmission of grape fan leaf by *Xiphinema index*. *Nematologica* 10: 489–495.

Tjosvold SA, Chambers DL, Koike ST, Mori SR (2008) Disease on nursery stock as affected by environmental factors and seasonal inoculum levels of *Phytophthora ramorum* in stream water used for irrigation. *Plant Dis* 92: 1566–1573.

Tooley PW, Browning M (2009) Susceptibility to *Phytophthora ramorum* and inoculum production potential of some common eastern forest understory plant species. *Plant Dis* 93: 249–256.

Tooley PW, Browning M (2015) Temperature effects on the onset of sporulation by *Phytophthora ramorum* on *Rhododendron* 'Cunningham's White'. *J Phytopathol* 163: 908–914.

Tooley PW, Browning M, Berner D (2008) Recovery of *Phytophthora ramorum* following exposure to temperature extremes. *Plant Dis* 98: 431–437.

Tooley PW, Browning M, Kyde KL, Berner D (2009) Effect of temperature and moisture period on infection of *Rhododendron* 'Cunningham's White' by *Phytophthora ramorum*. *Phytopathology* 99: 1045–1052.

Tooley PW, Carras MM (2011) Enhanced recovery of *Phytophthora ramorum* from soil following 30 days of storage at 4oC. *J Phytopathol* 159: 641–643.

Toscano-Underwood C, Huang YJ, Fitt BDL, Hall AM (2003) Effects of temperature on maturation of pseudothecia of *Leptosphaeria maculans* and *L. biglobosa* on oilseed rape stem debris. *Plant Pathol* 52: 726–736.

Toth IK, van der Wolf JM, Saddler G et al. (2011) *Dickeya* species: an emerging problem for potato production in Europe. *Plant Pathol* 60: 385–399.

Travadon R, Marquer B, Ribule A et al. (2009) Systemic growth of *Leptosphaeria maculans* from cotyledons to hypocotyls in oilseed rape: Influence of number of infection sites, competitive growth and host polygenic resistance. *Plant Pathol* 58: 461–469.

Tsror (Lakhim) L, Peretz-Alon I (2005) The influence of inoculum source of *Rhizoctonia solani* on development of black scurf on potato. *J Phytopathol* 153: 240–244.

van Bruggen AHC, Semenov A (2000) In search of biological indicators of soil health and disease suppression. *Appl Soil Ecol* 15: 13–24.

van de Graaf L, Wale SJ, Lees AK (2007) Factors affecting the incidence and severity of *Spongospora subterranea* infection and galling in potato roots. *Plant Pathol* 56: 1005–1013.

Van der Plank JE (1963) *Plant Diseases: Epidemics and Control*, Academic Press, New York.

van der Wolf JM, van Beckhoven JRCM (2004) Factors affecting survival of *Clavibacter michiganensis* subsp. *sepedonicus* in water. *J Phytopathol* 152: 161–168.

van der Wolf JM, van Beckhoven JRCM, Hukkanen R, Muller P (2005) Fate of *Clavibacter michiganensis* subsp. *sepedonicus*, the causal organism of bacterial ring rot of potato in weeds and field crops. *J Phytopathol* 153: 358–365.

Vereijssen J, Schneider JHM, Termorshuizen AJ (2005) Root infection of sugar beet by *Cercospora beticola* in a climate chamber and in the field. *Eur J Plant Pathol* 112: 201–210.

Wang A, Lazorvits G (2005) Role of seed tubers in the spread of plant pathogenic *Streptomyces* and initiating potato common scab disease. *Amer J Potato Res* 82: 221–230.

Wang S, Gergerich RC (1998) Immunofluorescent localization of tobacco ringspot nepovirus in the vector nematode *Xiphinema americanum*. *Phytopathology* 88: 885–889.

Ward LI, Delmiglio C, Hill CF, Clover GRG (2009) First report of *Tobacco ringspot virus* on *Sophora microphylla*, a native tree of New Zealand // New Disease Reports–2009 Vol 19, p.28.

Wei F, Fan R, Dong H et al. (2015) Threshold of microsclerotial inoculum for cotton Verticillium wilt determined through wet-sieving and real-time quantitative PCR. *Phytopathology* 105: 220–229.

Weinhold AR, Oswald JW, Bowman T, Bishop J, Wright D (1964) Influence of green manures and crop rotation on common scab of potato. *Amer Potato J* 41: 265–273.

Weller DM, Raaijmakers JM, McSpadden Gardener BB, Thomshow LS (2002) Microbial populations responsible for specific soil suppressiveness to plant pathogens. *Annu Rev Phytopathol* 40: 309–348.

Westphal A, Xiang L (2011) Soil suppressiveness against the disease complex of soybean cyst nematode and sudden death syndrome of soybean. *Phytopathology* 101: 878–886.

Willbur JF, Fall ML, Christopher-Bloomingdale et al. (2018) Weather-based models for assessing the risk of *Sclerotinia sclerotiorum* apothecial presence in soybean (*Glycine max*) fields. *Plant Dis* 102: 73–84.

Willocquet L, Liberton L, Sarniguet A, Lucas P (2008) Quantification of within-season focal spread of wheat take-all in relation to pathogen genotype and host spatial distribution. *Plant Pathol* 57: 906–915.

Wimer AF, Ridcout SL, Freeman JH (2011) Temporal and spatial distribution of tomato bacterial wilt on Virginia's Eastern Shore. *HortTechnology* 21: 198–201.

Wu BM, Subbarao KV (2008) Effects of soil temperature, moisture and burial depths on carpogenic germination of *Sclerotinia sclerotiorum* and *S. minor*. *Phyotpathology* 98: 1144–1152.

Wu BM, Subbarao KV, Liu Y-B (2008) Comparative survival of sclerotia of *Sclerotinia minor* and *S. sclerotiorum*. *Phytopathology* 98: 659–665.

Xing L, Westphal A (2006) Interaction of *Fusarium solani* f.sp. *glycines* and *Heterodera glycines* in sudden death syndrome of soybean. *Phytopathology* 96: 763–770.

Yakabe LE, Parker SR, Kluepfel DA (2014) Incidence of *Agobacterium tumefaciens* biovar 1 in and on 'Paradox' (*Juglans hindsii* x *Juglans regia*) walnut seed collected from commercial nurseries. *Plant Dis* 98: 766–770.

Young CS, Clarkson JP, Smith JA, Watling M, Phelps K, Whipps JM (2004) Environmental conditions influencing *Sclerotinia sclerotiorum* infection and disease development in lettuce. *Plant Pathol* 53: 387–397.

Zadoks JC, Schein RD (1979) *Epidemiology and Plant Disease Management*, Oxford University Press, New York.

# Index